SAFETY, RELIABILITY AND RISK OF STRUCTURES, INFRASTRUCTURES AND
ENGINEERING SYSTEMS

PROCEEDINGS OF THE TENTH INTERNATIONAL CONFERENCE ON STRUCTURAL SAFETY
AND RELIABILITY (ICOSSAR2009), OSAKA, JAPAN, 13–17 SEPTEMBER 2009

Safety, Reliability and Risk of Structures, Infrastructures and Engineering Systems

Editors

Hitoshi Furuta
Kansai University, Osaka, Japan

Dan M. Frangopol
Lehigh University, Bethlehem, PA, USA

Masanobu Shinozuka
University of California, Irvine, CA, USA

With the Assistance of Michiyuki Hirokane, Kansai University

CRC Press
Taylor & Francis Group
Boca Raton London New York Leiden

CRC Press is an imprint of the
Taylor & Francis Group, an **informa** business

A BALKEMA BOOK

Taylor & Francis is an imprint of the Taylor & Francis Group, an informa business

© 2010 Taylor & Francis Group, London, UK

Typeset by Charon Tec Ltd (A Macmillan Company), Chennai, India
Printed and bound in Great Britain by Antony Rowe (A CPI Group Company), Chippenham, Wiltshire

Published by: CRC Press/Balkema
                P.O. Box 447, 2300 AK Leiden, The Netherlands
                e-mail: Pub.NL@taylorandfrancis.com
                www.crcpress.com – www.taylorandfrancis.co.uk – www.balkema.nl

ISBN: 978-0-415-47557-0 (Hbk+CD-ROM)

Safety, Reliability and Risk of Structures, Infrastructures and
Engineering Systems – Furuta, Frangopol & Shinozuka (eds)
© 2010 Taylor & Francis Group, London, ISBN 978-0-415-47557-0

Table of Contents

PLENARY LECTURES

Freudenthal Lecture

Special Lecture

Keynote Lectures

TECHNICAL CONTRIBUTIONS

Monday Morning (MOM) Sessions

Mini-Symposia (MS01) Uncertainty Modeling of Rare/Imprecise Data
[Session Organized on Behalf of IASSAR's CSMSE Umbrella]

Mini-Symposia (MS13) Smart Structural Health Monitoring and Safety Assessments
[Session Organized on Behalf of IASSAR Subcommittee SC4]

Mini-Symposia (MS12) Monte Carlo Simulations Applied to Civil Engineering

Mini-Symposia (MS04) Structural Reliability and Optimization

Mini-Symposia (MS11) Current Status and Future Applications of Probabilistic Seismic Hazard Assessment

General Session (Risk Analysis in Decision Making)

General Session (Bridges and Buildings)

Monday Afternoon (MOA) Sessions

Mini-Symposia (MS01) Uncertainty Modeling of Rare/Imprecise Data
[Session Organized on Behalf of IASSAR's CSMSE Umbrella]

Mini-Symposia (MS13) Smart Structural Health Monitoring and Safety Assessments
[Session Organized on Behalf of IASSAR Subcommittee SC4]

Mini-Symposia (MS12) Monte Carlo Simulations Applied to Civil Engineering

Mini-Symposia (MS04) Structural Reliability and Optimization

Mini-Symposia (MS11) Current Status and Future Applications of Probabilistic Seismic Hazard Assessment

General Session (Bridges and Buildings)

Monday Evening (MOE) Sessions

Mini-Symposia (MS01) Uncertainty Modeling of Rare/Imprecise Data [Session Organized on Behalf of IASSAR's CSMSE Umbrella]

Mini-Symposia (MS13) Smart Structural Health Monitoring and Safety Assessments [Session Organized on Behalf of IASSAR Subcommittee SC4]

Mini-Symposia (MS12) Monte Carlo Simulations Applied to Civil Engineering

Mini-Symposia (MS04) Structural Reliability and Optimization

Organized Session (OS18) Seismic Risk Assessment and Information

General Session (Probabilistic Materials Analysis)

General Session (Monitoring and Maintenance Systems & Structural Health Monitoring)

General Session (Bridges and Buildings)

Tuesday Morning (TUM) Sessions

Mini-Symposia (MS01) Uncertainty Modeling of Rare/Imprecise Data [Session Organized on Behalf of IASSAR's CSMSE Umbrella]

Mini-Symposia (MS13) Smart Structural Health Monitoring and Safety Assessments [Session Organized on Behalf of IASSAR Subcommittee SC4]

Organized Session (OS21) Developing Performance-Based Concrete Design Code

Mini-Symposia (MS04) Structural Reliability and Optimization

Organized Session (OS18) Seismic Risk Assessment and Information

Organized Session (OS05) Vulnerability and Robustness of Structures

Mini-Symposia (MS02) NDT Reliability and Its Use for Maintenance Applications

Organized Session (OS24) Mechanics of Random Media

Organized Session (OS26) Safety Assessment of Offshore and Marine Structures

General Session (Fatigue Reliability)

Tuesday Afternoon (TUA) Sessions

Mini-Symposia (MS12) Monte Carlo Simulations Applied to Civil Engineering

Mini-Symposia (MS13) Smart Structural Health Monitoring and Safety Assessments [Session Organized on Behalf of IASSAR Subcommittee SC4]

Organized Session (OS05) Vulnerability and Robustness of Structures

Mini-Symposia (MS02) NDT Reliability and Its Use for Maintenance Applications

Organized Session (OS14) Non-Destructive Testing of Concrete Structures

General Session (Geotechnical Engineering)

General Session (Offshore and Marine Structures)

General Session (Reliability-Based Design and Regulations & Reliability-Based Optimization and Control & Reliability Theory)

Tuesday Evening (TUE) Sessions

Mini-Symposia (MS06) Uncertainties in Civil Structures & Infrastructure Engineering

Mini-Symposia (MS13) Smart Structural Health Monitoring and Safety Assessments [Session Organized on Behalf of IASSAR Subcommittee SC4]

Organized Session (OS25) Monitoring Performance under Uncertainty

Mini-Symposia (MS10) Structural Safety and Reliability by Means of Structural Health Monitoring

Organized Session (OS27) Hoshiya Memorial

General Session (Design Concepts)

Organized Session (OS12) Inspection and Analysis of Concrete Bridges in Coastal Environments

General Session (Reliability-Based Design and Regulations & Reliability-Based Optimization and Control & Reliability Theory)

Wednesday Morning (WEM) Sessions

Mini-Symposia (MS06) Uncertainties in Civil Structures & Infrastructure Engineering

Mini-Symposia (MS07) Vulnerability Assessment and Risk-Based Life-Cycle Analysis and Design

Mini-Symposia (MS15) System Identification and Structural Health Monitoring [Session Organized on Behalf of IASSAR Subcommittee SC5]

Mini-Symposia (MS16) Risk Assessment of Lifeline Networks and Decision Support

Mini-Symposia (MS17) Earthquake Engineering and Engineering Seismology

Mini-Symposia (MS08) Time-Dependent Reliability Methods and Their Applications

Organized Session (OS07) Performance-Based Design for Steel Structures

General Session (Random Vibration)

Mini-Symposia (MS07) Vulnerability Assessment and Risk-Based Life-Cycle Analysis and Design

Mini-Symposia (MS15) System Identification and Structural Health Monitoring [Session Organized on Behalf of IASSAR Subcommittee SC5]

Mini-Symposia (MS16) Risk Assessment of Lifeline Networks and Decision Support

Mini-Symposia (MS17) Earthquake Engineering and Engineering Seismology

Mini-Symposia (MS08) Time-Dependent Reliability Methods and Their Applications

Organized Session (OS15) Challenging Technology for Condition Screening of Bridge Structures

Organized Session (OS06) Nonlinear Stochastic Dynamics of Structures with Uncertain Mechanical and Geometric Properties [Session Organized on Behalf of IASSAR Subcommittee SC1]

General Session (Stochastic Computational Mechanics & Stochastic Finite Elements)

General Session (Simulation Methods)

Wednesday Evening (WEE) Sessions

Mini-Symposia (MS15) System Identification and Structural Health Monitoring [Session Organized on Behalf of IASSAR Subcommittee SC5]

General Session (Insurance and Management of Risk & Economic Analysis & Probabilistic Risk Analysis & Simulation Methods)

Thursday Morning (THM) Sessions

Organized Session (OS11) Health Monitoring, Structural Monitoring & Control

Mini-Symposia (MS05) Modelling Seismic Action

Organized Session (OS23) New Methods for Non-Gaussian and Pulse Problems in Stochastic Dynamics [Session Organized on Behalf of IASSAR Subcommittee SC2]

Mini-Symposia (MS09) Life-Cycle Reliability and Optimization of Deteriorating Structures

Mini-Symposia (MS14) Computational Methods in Stochastic Mechanics
[Session organized on behalf of IASSAR Subcommittee SC1]

Organized Session (OS04) Risk Evaluation on Geotechnical and
Geo-Environmental Problems

General Session (Earthquake Engineering)

General Session (Structural Systems & System Reliability)

Thursday Afternoon (THA) Sessions

Organized Session (OS11) Health Monitoring, Structural Monitoring & Control

Mini-Symposia (MS05) Modelling Seismic Action

Organized Session (OS23) New Methods for Non-Gaussian and Pulse Problems in Stochastic Dynamics [Session Organized on Behalf of IASSAR Subcommittee SC2]

Mini-Symposia (MS09) Life-Cycle Reliability and Optimization of Deteriorating Structures

Mini-Symposia (MS14) Computational Methods in Stochastic Mechanics [Session organized on behalf of IASSAR Subcommittee SC1]

Organized Session (OS04) Risk Evaluation on Geotechnical and Geo-Environmental Problems

General Session (Earthquake Engineering)

AUTHOR INDEX

Preface

In recent years, safety, reliability, risk and life-cycle management of structures, infrastructures and engineering systems have become emergent and key issues due to the frequent occurrence of natural and man-made disasters (such as earthquakes, hurricanes, tsunamis, terrorist attacks, and airplane, offshore platform, and nuclear power plant accidents), the infrastructure crisis, sustainability issues and global warming. In dealing with these problems uncertainties are unavoidable. The society can no longer afford to ignore congested roads, deficient bridges, aging dams, broken levees and water mains, among others. The importance of costs associated with managing existing structures, infrastructures and engineering systems is increasingly recognized. Decisions regarding requirements for design, continued service, rehabilitation or replacement should be based on multi-objective optimization under uncertainty, in order to balance conflicting requirements such as cost and performance.

In the past four decades, the International Conference on Structural Safety and Reliability (ICOSSAR) organized by the International Association for Structural Safety and Reliability (IASSAR) has provided a valuable opportunity to share the knowledge, experience and information on structural safety and reliability among scientists and engineers. The first ICOSSAR was held in 1969 in Washington, D.C., USA. Since 1977, it has been successfully held every four years at several cities in Europe, USA and Japan, and gathered many engineers in academia, industry, government and private practice. Previous ICOSSARs have been held in Washington, D.C., USA (1969), Munich, Germany (1977), Trondheim, Norway (1981), Kobe, Japan (1985), San Francisco, USA (1989), Innsbruck, Austria (1993), Kyoto, Japan (1997), Newport Beach, USA (2001) and Rome, Italy (2005). Following the past two ICOSSARs in Japan, Kobe in 1985 and Kyoto in 1997, the ICOSSAR2009 is held in Osaka, which is the second biggest city in Japan and the center of commerce and industry.

Safety, Reliability and Risk of Structures, Infrastructures and Engineering Systems consists of a book of abstracts and a CD-ROM containing the full texts of the lectures and papers presented at the Tenth International Conference on Structural Safety and Reliability (ICOSSAR2009) held in Osaka, Japan, September 13–17, 2009. It includes the Freudenthal lecture, the special lecture, the eight keynote lectures and 540 technical papers from 43 countries. This set presents an up-to-date overview and significant novel contributions to the safety, reliability and risk issues of all types of engineering systems. The topics cover major aspects of safety, reliability and risk of structures, infrastructures and engineering systems, with special focus on advanced technologies, analytical and computational methods of risk analysis, probability based design and regulations, smart systems and materials, life cycle cost analysis, damage assessment, social aspects, urban planning, and commercial applications. Emerging concepts as well as state of the art and novel applications of reliability principles in all types of structural systems and mechanical components are included. Civil, marine, mechanical, transportation, nuclear and aerospace applications are discussed.

On behalf of IASSAR and ICOSSAR2009, the Editors would like to thank the authors, organizers of sessions and mini-symposia, and participants for their contributions, the members of the Conference Scientific Committee for their dedicated work, and the members of the International Advisory Committee, IASSAR Award Committee, and Local Organizing Committee for their continuous teamwork. Finally, we would like to thank all the sponsors of ICOSSAR2009, and in particular Osaka Convention & Tourism Bureau, the Maeda Engineering Foundation, the Obayashi Foundation, the Kajima Foundation, and Kansai University

Hitoshi Furuta, Dan M. Frangopol, and Masanobu Shinozuka
July 2009

Safety, Reliability and Risk of Structures, Infrastructures and Engineering Systems – Furuta, Frangopol & Shinozuka (eds)
© 2010 Taylor & Francis Group, London, ISBN 978-0-415-47557-0

Conference Organization

Organizing Association

IASSAR
International Association for Structural Safety and Reliability (http://www.civil.columbia.edu/iassar). The primary objective of IASSAR is to sponsor and oversee the organization of the International Conferences on Structural Safety and Reliability (ICOSSAR) that are held regularly at approximately four year intervals. IASSAR was founded in 1985.

IASSAR (2005–2009)

Hitoshi Furuta	(Japan)	President
Masanobu Shinozuka	(USA)	Executive Vice President
Dan M. Frangopol	(USA)	Chairman of Executive Board
Alfredo H-S. Ang	(USA)	Member of Executive Board
Giuliano Augusti	(Italy)	Member of Executive Board
Robert E. Melchers	(Australia)	Member of Executive Board
Gerhart I. Schuëller	(Austria)	Member of Executive Board
Heki Shibata	(Japan)	Member of Executive Board
Naruhito Shiraishi	(Japan)	Member of Executive Board
Wilson H. Tang	(Hong Kong)	Member of Executive Board

CONFERENCE CHAIR
Hitoshi Furuta, Kansai University, Osaka Japan

INTERNATIONAL SCIENTIFIC COMMITTEE CHAIR
Dan M. Frangopol, Lehigh University, Bethlehem, PA, USA

INTERNATIONAL SCIENTIFIC COMMITTEE
D.M. Frangopol, Lehigh University, Bethlehem, PA, USA (Chair)
H. Adeli, Ohio State University, Columbus, OH, USA
G. Augusti, University of Rome "La Sapienza", Rome, Italy
Z.P. Bažant, Northwestern University, Evanston, IL, USA
J.L. Beck, California Institute of Technology, Pasadena, CA, USA
L.A. Bergman, University of Illinois, Urbana, IL, USA
F. Biondini, Technical University of Milan, Milan, Italy
C. Borri, University of Florence, Florence, Italy
E. Brühwiler, EPFL, Lausanne, Switzerland
C. Bucher, Vienna University of Technology, Vienna, Austria
J.R. Casas, Technical University of Catalonia, Barcelona, Spain
F. Casciati, University of Pavia, Pavia, Italy
K.C. Chang, National Taiwan University, Taipei, Taiwan
H.N. Cho, Hanyang University, Ansan, Korea
M.K. Chryssanthopoulos, University of Surrey, Guilford, UK
M. Ciampoli, University of Rome "La Sapienza", Rome, Italy
J.P. Conte, University of California, San Diego, CA, USA
R.B. Corotis, University of Colorado, Boulder, CO, USA
D. de Leon, Mexico State Autonomous University, Mexico State, Mexico
G. Deodatis, Columbia University, New York, NY, USA
A. Der Kiureghian, University of California, Berkeley, CA, USA
M. di Paola, University of Palermo, Palermo, Italy
S.M.C. Diniz, Federal University of Minas Gerais, Bello Horizonte, Brazil
O. Ditlevsen, Technical University of Denmark, Lyngby, Denmark,

TECHNICAL COMMITTEE
T. Takada (University of Tokyo) Chair
Y. Mori (Nagoya University) Vice-Chair
T. Takahashi (Chiba University)
H. Tanaka (Kyoto University)
S. Katsuki (National Defense Academy)
N. Sato (Chuo University)
Y. Ohtori (CRIEPI)
M. Akiyama (Tohoku University)
H. Idota (Nagoya Institute of Technology)
K. Kawano (Kagoshima University)
Y. Kimura (Kogauin University)
A. Miyamoto (Yamaguchi University)
T. Nagao (NILIM)
S. Sawada (Kyoto University)
T. Sugiyama (Yamanashi University)
M. Suzuki (Shimizu Corporation)
K. Takara (Kyoto University)
S. Unjo (PWRI)
N. Yamamoto (Nippon Kaiji Kyokai)
Y. Zhao (Nagoya Institute of Technology)
Z. Wu (Ibaraki University)
K. Ohdo (Research Institute of Industrial Safety)
N. Nojima (Gifu University)]
H. Nakayasu (Konan University)
Y. Itoh (Nagoya University)
A. Nishitani (Waseda University)
H. Ohtsu (Kyoto University)
H. Ohtsuka (Kyushu University)
T. Hayashikawa (Hokkaido University)
H. Sugimoto (Hokkai Gakuen University)
T. Oshima (Kitami Institute of Technology)
K. Sugiura (Kyoto University)
M. Kishi (Muroran Institute of Technology)
A. Nakajima (Utsunomiya University)
M. Suzuki (Tohoku University)
Y. Chikata (Kanazawa University)
I. Iwaki (Nippon University)
I. Okura (Osaka University)
K. Tateishi (Nagoya University)
T. Mori (Hosei University)
S. Inoue (Tottori University)
H. Yamada (Yokohama National University)
F. Nagao (Tokushima University)
M. Ohtsu (Kumamoto University)

FINANCIAL COMMITTEE
Y. Fujino (University of Tokyo) Chair
M. Yamamoto (Kajima Corporation) Vice-Chair
M. Fujita (Simizu Corporation)
T. Nakayama (Hiroshima Institute of Technology)
A. Sudo (Chizaki Technology)
N. Yasuda (TEPCO)
K. Kamemura (Taisei Corporation)
Y. Shimura (Nippon Steel Corporation)
M. Asano (Nikken Sekkei)
K. Kobayashi (Fujita Corporation)

Safety, Reliability and Risk of Structures, Infrastructures and Engineering Systems – Furuta, Frangopol & Shinozuka (eds)
© 2010 Taylor & Francis Group, London, ISBN 978-0-415-47557-0

List of Organized Sessions and Mini-Symposia

For the papers in each session of Organized Sessions and Mini-Symposia refer to the Table of Contents.

Organized Sessions

OS02 Probabilistic design of wind turbines, organized by J. D. Sorensen.

OS03 Construction risk and safety management, organized by K. Ohdo, T. Hojo, and M. Hirokane

OS04 Risk evaluation on geotechnical and geo-environmental problems, organized by Y. Honjo, A. Haldar, T. Katsumi, and S. Nishimura.

OS05 Vulnerability and robustness of structures, organized by J. Agarwal.

OS06 Nonlinear stochastic dynamics of structures with uncertain mechanical and geometric properties, organized by G. Stefanou, M. Fragiadakis, and M. Papadrakakis.

OS07 Performance-based design for steel structures, organized by S-H. Kim.

OS08 Machine learning in structural reliability and probabilistic mechanics, organized by M. Lemaire, and J-M. Bourinet.

OS11 Health monitoring, structural monitoring & control, organized by A. Nakajima, and T. Obata.

OS12 Inspection and analysis of concrete bridges in coastal environments, organized by M. Suzuki, I. Iwaki, and H. Tsuruta.

OS13 Novel approaches for reliability analysis and statistical structural health monitoring, organized by M. Noori, and Z. Hou

OS14 Non-destructive testing of concrete structures, organized by T. Kamada, and M. Iwanami.

OS15 Challenging technology for condition screening of bridge structures, organized by M. Kawatani, and C. W. Kim

OS16 Safety prediction and evaluation of wind-induced phenomena, organized by H. Yamada, and H. Shirato.

OS18 Seismic risk assessment and information, organized by Y. Mori, and N. Luco.

OS21 Developing Performance-based concrete design code, organized by H. Jeong.

OS23 New methods for non-Gaussian and pulse problems in stochastic dynamics, organized by R. Iwankiewicz.

OS24 Mechanics of random media, organized by L. Graham-Brady, and S. Arwade.

OS25 Monitoring performance under uncertainty, organized by D. M. Frangopol, and A. Strauss.

OS26 Safety assessment of offshore and marine structures, organized by M. Fujikubo, and N. Yamamoto.

OS27 Hoshiya memorial, organized by O. Maruyama.

Mini-Symposia

MS01 Uncertainty modeling of rare/imprecise data, organized by M. Beer, Q.S. Tong and P.K. Kwang

MS02 NDT reliability and its use for maintenance applications, organized by C. Cremona.

MS04 Structural reliability and optimization, organized by A. Der. Kiureghian.

MS05 Modeling seismic action, organized by H. Sandi.

MS06 Uncertainties in civil structures & infrastructure engineering, organized by A.S. Elnashai, N.D. Lagaros and Y. Tsompanakis

MS07 Vulnerability assessment and risk-based life-cycle analysis and design, organized by D.M. Frangopol, and Y. Tsompanakis

MS08 Time-dependent reliability methods and their applications, organized by Y-G. Zhao, X. Liu, and M. Lemaire.

MS09 Life-cycle reliability and optimization of deteriorating structures, organized by F. Biondini, and D. M. Frangopol

MS10 Structural safety and reliability by means of structural health monitoring, organized by N. Catbas, H. Furuta, D.M. Frangopol, J. Casas, and F. Casciati.

MS11 Current status and future applications of probabilistic seismic hazard assessment, organized by T. Takada, and J. Baker.

MS12 Monte Carlo simulations applied to civil engineering, organized by M. Shinozuka.

MS13 Smart structural health monitoring and safety assessments, organized by C-B. Yun.

MS14 Computational methods in stochastic mechanics, organized by R. Ghanem.

MS15 System identification and structural health monitoring, organized by J.L. Beck, and L.S. Katafygiotis.

MS16 Risk assessment of lifeline networks and decision support, organized by J. Song, and L. Duenas-Osorio.

MS17 Earthquake engineering and engineering seismology, organized by H. Morikawa.

Safety, Reliability and Risk of Structures, Infrastructures and
Engineering Systems – Furuta, Frangopol & Shinozuka (eds)
© 2010 Taylor & Francis Group, London, ISBN 978-0-415-47557-0

Conference Sponsors

International Association for Bridge Maintenance and Safety (IABMAS)
International Association for Life-Cycle Civil Engineering (IALCCE)
International Cooperation and Exchange Committee of China Civil Engineering Society (CCES)
Structural Engineering Institute of the American Society of Civil Engineers (SEI-ASCE)
Architectural Institute of Korea (AIK)
Architectural Institute of Japan (AIJ)
Chinese Institute of Civil and Hydraulic Engineering
Computational Structural Engineering Institute of Korea (CSEIK)
Earthquake Engineering Society of Korea (EESK)
Japan Association for Wind Engineering (JAWE)
Japan Concrete Institute (JCI)
Japan Society of Civil Engineers (JSCE)
Japan Welding Society (JWS)
Japanese Society of Steel Construction (JSSC)
The Japan Society of Naval Architects and Ocean Engineers (JASNAOE)
The Japan Society for Aeronautical and Space Sciences (JSASS)
The Japan Society of Mechanical Engineers (JSME)
The Japanese Geotechnical Society (JGS)
The Society of Materials Science, Japan (JSMS)
Korea Concrete Institute (KCI)
Korean Society of Civil Engineers (KSCE)
Korean Society of Steel Construction (KSSC)
Center for Advanced Technology for Large Structural Systems, Lehigh University, PA, USA (ATLSS)
Kansai University, Osaka, Japan (KU)
Center for Creation of Spiral Network for Safety Intelligence, Kansai University, Osaka, Japan (CSNSI)
Osaka Convention & Tourism Bureau (OCTB)
The Maeda Engineering Foundation (MEF)
The Obayashi Foundation (OF)
The Kajima Foundation (KF)

PLENARY LECTURES

Freudenthal Lecture

Safety, Reliability and Risk of Structures, Infrastructures and
Engineering Systems – Furuta, Frangopol & Shinozuka (eds)
© 2010 Taylor & Francis Group, London, ISBN 978-0-415-47557-0

On risk and reliability – Contributions to engineering and future challenges

A.H.-S. Ang
University of California, Irvine, CA, USA

EXTENDED ABSTRACT

The historical development of the reliability approach
to engineering, with emphasis on the applications to
design of civil structures and infrastructures is briefly
reviewed.

The main objective of the reliability approach is
the modeling and analysis of uncertainties – both
the aleatory type and the epistemic type. In light of
unavoidable uncertainties in all aspects of engineering,
the design variables and parameters need to be modeled as random variables and thus require probability
and statistical concepts and methods.

The theoretical basis of this approach was first
introduced by Freudenthal in the 1940s; although conceptually and theoretically sound and more rational
than the traditional safety-factor method for defining
the safety and reliability of an engineering system,
the implementation of the approach by practicing
design professionals was met with considerable skep-
ticism and consternation. Additional developments to
overcome the major obstacles in implementing the
approach were, therefore, required. These obstacles
included the reluctance of the structural designers in
accepting the notion that there might be a finite (albeit
small) probability of failure in the design of a structure.

The additional developments in the last three-four
decades included the introduction of the concept of the
reliability index or *safety index*. The reliability index
is related to the probability of failure; for a specified set of probability distributions, this relationship
is unique. Therefore, for a given value of the reliability index, there is a corresponding probability of
failure; however, its use as the measure of safety or
reliability avoids any explicit reference to a failure
probability, even though such a probability implicitly exists. The introduction of the reliability index
served to spur the adoption of the reliability approach
in the safety assessment and reliability-based design
of structures and infrastructures, including codified
designs of buildings, bridges, ship, marine and offshore structures, applications in geotechnical systems,
as well as applications in seismic and other hazard
analyses.

The central issue underlying the importance of
the reliability approach is the need for a rational
basis for modeling and analysis of uncertainty. In this
regard, uncertainties may be broadly classified into
two types – the *aleatory* and the *epistemic* types.
The aleatory type is data-based which is the natural variability or randomness of nature, whereas the
epistemic type is knowledge-based and arises because
of our inability to perfectly model reality. In either
case, probability and statistics are the proper tools
for its modeling and analysis. The aleatory type can
be assessed largely on the basis of observed data; on
the other hand the epistemic type would need to be
assessed with engineering judgments. By separating
uncertainties into these two distinct types, there is significant practical advantages – namely, that it allows
risk-averseness (on the part of the decision maker)
in the formulation of risk-based decisions. Including,
for example, the specification of a risk-averse target
reliability index for safe design of major structures.

Other major applications of the reliability approach
include specifically the following:

- Reliability against fatigue failure of structural and
  machine components.
- Formulation of risk-based life-cycle cost designs of
  buildings and bridges.
- Formulation of reliability-based maintenance strategies for structures and infrastructures including
  network of bridges.

A number of related and significant developments
are summarized, including innovative computational
techniques such as adaptive and smart Monte Carlo
simulations, random field and stochastic finite element models, and the recently developed probability
density evolution method for numerical evaluation of
dynamic reliability.

Finally, some of the future challenges are identified
and suggested for further developments, including the
following:

- Innovative basis for determining the target reliability for design or retrofit.
- Innovative analytical or numerical (non-Monte
  Carlo) methods assessing the reliability of complex
  structural systems.
- Risk-related problems associated with the impact
  of climate change, or climate cycles, including

increased vulnerability of infrastructures to rising sea level due to global warming.

- Reliability of an integrated critical infrastructure including the interaction and inter-dependence of several systems within an overall system that is essential for maintaining the safety and functionality of a metropolitan area, a geographical region, or an entire country.
- Effective and practical methods for estimating uncertainties, particularly of the epistemic type, will continue to present an important challenge.

The reliability community must continue to persuade, assist and encourage the practicing professionals to adopt the reliability approach in the development of criteria for design and maintenance of engineering structures and systems, and introduce quantitative risk-based information in engineering management and decision-making.

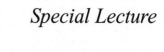

Special Lecture

Safety, Reliability and Risk of Structures, Infrastructures and Engineering Systems – Furuta, Frangopol & Shinozuka (eds)
© 2010 Taylor & Francis Group, London, ISBN 978-0-415-47557-0

Increasingly catastrophic disasters due to global warming

Yoshiaki Kawata

Department of Civil and Environmental Engineering, Kyoto University, Japan

ABSTRACT

What do we need most in a sustainable society under the effect of global warming? I think an appropriate perspective is that we should insure that we do not lose something important to our society. For example, while people are searching for a plan to come out of the current financial crisis that was triggered by the subprime loan issue, there is practically no discussion on "what we should not lose during this crisis." This perspective is pervasive despite the fact that the very foundation of our economy, which is to "produce products and sell them at reasonable prices," is at risk.

The magnitude of human loss due to a large-scale disaster is a major problem. However, another major problem is that a large-scale disaster could completely destroy something important to our society. A large-scale disaster is even more frightening because it could affect us emotionally, rather than simply affecting physical structures and material possessions. A disaster threatens the continuity of society. Despite this fact, people tend not to address or implement measures that reduce or prevent disasters because these measures do not provide a monetary gain. Rather, these measures are only taken seriously immediately after a major damage, and are soon forgotten once the normality of society returns.

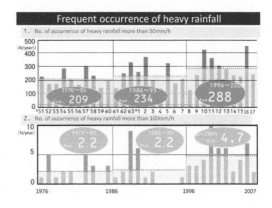

The objectives of this paper to make clear the effect of global warming on storm disasters and to propose how to reduce their damage. With the combination of large scale hazards and vulnerable modern society, the damage will be huge resulting in compound disasters and fused disasters never before experienced in modern urban area. We have defined disasters and in the future our smart (integrated) countermeasures will be focused on disaster reduction through an integrated approach.

Keynote Lectures

Safety, Reliability and Risk of Structures, Infrastructures and Engineering Systems – Furuta, Frangopol & Shinozuka (eds)
© 2010 Taylor & Francis Group, London, ISBN 978-0-415-47557-0

Seismic risk assessment and management of infrastructure systems: Review and new perspectives

A. Der Kiureghian
University of California, Berkeley, CA, USA

ABSTRACT

Civil infrastructure systems, such as transportation, utility, and communication networks, and public facilities for critical services such as health care, security and safety, are backbones of modern societies. Their resilience in face of natural and man-made hazards is vital for the well-being of communities. Earthquakes are a dominant hazard to these systems in many parts of the world. Therefore, the assessment and mitigation of seismic risk for these systems is of vital importance.

The seismic risk of infrastructure systems poses several unique issues, which are not present in the seismic risk assessment of ordinary buildings. These include:

a) Because of their geographically distributed nature, infrastructure systems are potentially subject to a variety of earthquake-related hazards, including ground shaking, fault rupture, liquefaction, and slope instability. Furthermore, the earthquake may lead to other hazards, such as fire and flooding, which may also affect the system components. Hence, a multi-hazard risk assessment approach is necessary for infrastructure systems.

b) The rate of occurrence of earthquakes affecting infrastructures extending over large areas, such as a transportation or power network, can be much larger than the corresponding rate for buildings or systems situated on individual sites. Hence, proper levels of system resilience and redundancy must be provided to assure adequate safety and functionality of the infrastructure.

c) For a given earthquake, the intensities of ground motions affecting the components of a geographically distributed system in general are spatially correlated. Depending on the system configuration, this correlation may have an adverse effect, if the system is highly redundant, or a beneficial effect, if the system lacks redundancy. Therefore, for proper assessment of the seismic risk for infrastructure systems, the ground motion must be considered as a random field with a spatial correlation structure.

d) As alluded to in the previous item, for proper assessment of the seismic risk to an infrastructure, a systems approach to performance assessment must be employed. This requires specification of the system performance in terms of component performances and the system configuration. The system performance may require disciplinary analysis, such as mass balance analysis for a water distribution system, or power flow analysis for a power system.

e) Often the performance of an infrastructure system is dependent on the performance of other infrastructures. For example, the functioning of a water distribution system may depend on the functioning of the power system to empower pumps and other critical components. Since an earthquake may affect both systems, these dependencies must be properly modeled in order to correctly assess the seismic risk.

Depending on the intended application, the requirements for infrastructure seismic risk assessment can be different. For example, in the planning, design, or retrofit of an infrastructure system, the objective might be to determine the required system configuration and component capacities to assure adequate reliability of the system. This may require analysis of the system for scenario earthquakes, or for all earthquakes happening randomly over a specified period of time. For long-term management of the infrastructure system, one may additionally require consideration of various alternatives for inspection, maintenance and repair of the system components so as to minimize the total life-cycle cost. Such an analysis requires cost models for various states of the system and its components, as well as the costs associated with various inspection, maintenance and repair alternatives. Another important application of seismic risk assessment is for post-earthquake decision-making for emergency response and recovery actions. In this application, one is particularly interested in updating the state of the system as information is received about the nature of the earthquake (e.g., the magnitude, location), from sensors (e.g., ground motion recording instruments), or from observation of the states of system components. In the chaotic post-earthquake conditions, this information often is imprecise and evolves in time; therefore, a near-real time probabilistic updating scheme is desirable. The updated information may then be used to make decisions on such matters as dispatching of emergency response crews, closing of facilities, evacuation of people and rapid delivery of recovery actions.

During the past decade, much research has been conducted on seismic risk assessment of infrastructure systems; see the list of references and the references therein. Investigators have used a variety of computational methods to assess the infrastructure seismic risk, including Monte Carlo simulation, linear programming, matrix-based computations, FORM and SORM, GIS, and Bayesian networks. A variety of modeling techniques have been employed, including network and graph theoretic models, multi-scale models, Bayesian network, and various forms of fragility functions and surfaces. Furthermore, a number of measures for system performance involving network connectivity or flow have been formulated. Although much progress has been made, these works have addressed only a portion of the issues (a)-(e) described above. The field is still wide open for creative work and many challenging problems remain to be solved.

This paper will provide a review of some of the key works in the area of infrastructure seismic risk assessment and management. It will provide an assessment of the advantages and disadvantages of different methods and models, and also provide a perspective towards some of the main challenges that remain to be addressed.

REFERENCES

Adachi, T., and B.R. Ellingwood (2008). Serviceability of earthquake-damaged water systems: Effects of electrical power availability and power backup systems on system vulnerability. *Reliability Engineering and System Safety*, 93:78–88.

Basoz, N., M. Williams and A. Kiremidjian (2003). A GIS-based emergency response system for transportation networks. *Proceedings of the Sixth U.S. Conference and Workshop on Lifeline Earthquake Engineering*, August 10–13, 2003, Long Beach, CA; Technical Council on Lifeline Earthquake Engineering, Monograph No.25, pp. 926–935.

Bensi, M., D. Straub, P. Friis-Hansen and A. Der Kiureghian (2009). Modeling infrastructure system performance using BN. Proceedings, 10th ICOSSAR, Osaka, Japan, September 2009.

Der Kiureghian, A., and J. Song (2008). Multi-scale reliability analysis and updating of complex systems by use of linear programming. *Reliability Engineering & System Safety*, 93:288–297.

Dueñas-Osorio, L., J.I. Craig, and B.J. Goodno (2007). Seismic response of critical interdependent networks. *Earthquake Engineering and Structural Dynamics*, 36:285–306.

Kang, W.-H., J. Song, and P. Gardoni (2008). Matrix-based system reliability method and applications to bridge networks. *Reliability Engineering and System Safety*, 93:1584–1593.

Kiremidjian, A., J. Moore, Y-Y. Fan, O. Yazlali, N. Basoz and M. Williams (2007). Seismic Risk Assessment of Transportation Network Systems. *Earthquake Engineering*, 11:371–382.

Lee, R., and A.S. Kiremidjian (2007). Uncertainty and correlation for loss assessment of spatially distributed systems. *Earthquake Spectra*, 23:753–770.

Moghtaderi-Zadeh, M., K. Wood, A. Der Kiureghian and R.E. Barlow (1982). Seismic reliability of lifeline networks. *Journal of Technical Councils*, ASCE, 108:60–78.

Nuti, C., A. Rasulo and I. Vanzi (2007). Seismic safety evaluation of electric power supply at urban level. *Earthquake Engineering and Structural Dynamics*, 36:245–264.

Shinozuka, M., X. Dong, T.C. Chen and X. Jin (2007). Seismic performance of electric transmission network under component failures. *Earthquake Engineering and Structural Dynamics*, 36:227–244.

Werner, S., C. Taylor, S. Cho, J.P. Lavoie, C. Huyck, C. Eitzel, R. Eguchi, and J.E. Moore (2003). New developments in seismic risk analysis of highway systems. MCEER publications http://mceer.buffalo.edu/research/redars/news_ and_publications.asp.

*Safety, Reliability and Risk of Structures, Infrastructures and
Engineering Systems – Furuta, Frangopol & Shinozuka (eds)*
© *2010 Taylor & Francis Group, London, ISBN 978-0-415-47557-0*

Life-cycle performance, management, and optimization of structural systems under uncertainty: Accomplishments and challenges

Dan M. Frangopol

Lehigh University, Bethlehem, PA, USA

ABSTRACT

Our knowledge to model, analyze, design, maintain,
monitor, manage, predict and optimize the life-cycle
performance of structures and infrastructures under
uncertainty is continually growing. However, in many
countries, including the United States, the civil infrastructure is no longer within desired levels of performance and safety. Stable economic growth and social
development of most countries are intimately dependent upon the reliable and durable performance of their
structures and infrastructures. Natural hazards, aging,
and functionality fluctuations can inflict detrimental
effects on the performance of structural systems during their life-cycles. Even the inherently conservative
initial design of structural systems may not protect
a structure from these threats. Natural phenomena
such as earthquakes, hurricanes and floods can create
structural disasters. Aging and/or increased structural performance demand may significantly affect
the vulnerability of constructed facilities. Environmental stressors are the primary factors that drive the
aging process. The effect of structural aging is perhaps
most widely apparent in bridge deterioration, exacer-
bated by increase in traffic over time, but also impacts
other civil infrastructure systems such as buildings
and nuclear power plants (Ellingwood and Mori 1993,
Ellingwood 1998, 2005).

The accurate modeling of structures and the loading
conditions to which they are expected to be exposed
during their life-cycle as well as their possible deterioration mechanisms are major issues of structural
and engineering mechanics, respectively (Schuëller
1998). Uncertainty in the modeling of structures and
randomness in loading phenomena require the use of
probabilistic methods in life-cycle analysis. Explicitly
distinguishing the two types of uncertainty, namely the
aleatory and epistemic, is crucial for the proper handling of a probabilistic analysis approach (Ang and
Tang 2007). Whereas randomness (or aleatory uncertainty) cannot be reduced, improvement in knowledge
or in the accuracy of predictive models will reduce the
epistemic uncertainty (Ang and De Leon 2005).

Ultimately, optimal decisions are to be made
that ensure maintaining or improving reliability of

Figure 1. Schematic representation of life-cycle integrated
management framework.

structural systems under multiple objectives and various constraints. This can only be achieved through
proper integrated risk management planning in a
life-cycle comprehensive framework.

Figure 1 shows a schematic representation of a
life-cycle integrated management framework example. In this framework, tools for structural performance
assessment and prediction, structural health monitoring (SHM), integration of new information (from
SHM and/or inspection), and optimization of strategies (inspection, maintenance, monitoring, repair, and
replacement) are required.

Life-cycle performance assessment is the backbone
of the process which, as shown in Figure 2, requires
current evaluation and future prediction. Uncertainty
is an integral component in all aspects of this or any
life-cycle management framework (Frangopol and Liu
2007).

Decisions regarding civil infrastructure systems
should be supported by an integrated reliability-based
life-cycle multi-objective optimization framework by
considering, among other factors, the likelihood of
successful performance and the total expected cost
accrued over the entire life-cycle with or without SHM
(see Figure 3).

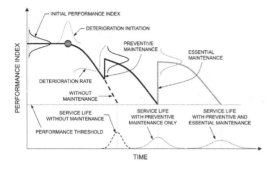

Figure 2. Life-cycle performance profile under uncertainty.

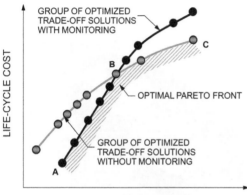

Figure 3. Trade-off solutions between two conflicting objectives with or without SHM, and optimal Pareto front.

The primary objective of this paper is to highlight recent accomplishments in the life-cycle performance assessment, maintenance, monitoring, management and optimization of structural systems under uncertainty. Challenges are also identified.

REFERENCES

Ang, A.H-S. & De Leon, D. 2005. Modeling and analysis of uncertainties for risk-informed decisions in infrastructures engineering. Structure and Infrastructure Engineering, Taylor & Francis, 1(1), pp. 19–21.

Ang, A.H-S. & Tang, W.H. 2007. Probability Concepts in Engineering. 2nd edition, Wiley, New York.

Corotis, R.B. 2009. Societal issues in adopting life-cycle concepts within the political system. Structure and Infrastructure Engineering, Taylor & Francis, 5(1), pp. 59–65.

Ellingwood, B.R. 1998. Issues related to structural aging in probabilistic risk analysis of nuclear power plants. Reliability Engineering and System Safety, 62(3), pp. 171–183.

Ellingwood, B.R. 2005. Risk-informed condition assessment of civil infrastructure: state of practice and research issues. Structure and Infrastructure Engineering, Taylor & Francis, 1(1), pp. 7–18.

Ellingwood, B.R. & Mori, Y. 1993. Probabilistic methods for condition assessment and life prediction of concrete structures in nuclear power plants. Nuclear Engineering and Design, 142, pp. 155–166.

Frangopol, D. M., Kallen, M-J. & van Noortwijk, J. 2004. Probabilistic models for life-cycle performance of deteriorating structures: review and future directions. Progress in Structural Engineering and Materials, John Wiley & Sons, 6(4), pp. 197–212.

Frangopol, D.M., Kong, J.S. & Gharaibeh, E.S. 2001. Reliability-based life-cycle management of highway bridges. Journal of Computing in Civil Engineering, ASCE 15(1), pp. 27–34.

Frangopol, D.M., Lin, K.-Y. & Estes, A.C. 1997. Life-cycle cost design of deteriorating structures. Journal of Structural Engineering, ASCE, 123(10), pp. 1390–1401.

Frangopol, D.M. & Liu, M. 2007. Maintenance and management of civil infrastructure based on condition, safety, optimization, and life-cycle cost. Structure and Infrastructure Engineering, Taylor and Francis, 3(1), pp. 29–41.

Frangopol, D.M. & Messervey, T.B. 2009. Maintenance principles for civil structures. Chapter 89, Encyclopedia of Structural Health Monitoring, C. Boller, F-K. Chang, and Y. Fujino, eds., John Wiley & Sons Ltd, Vol. 4, pp. 1533–1562.

Frangopol, D.M., Strauss, A. & Kim, S. 2008. Bridge reliability assessment based on monitoring. Journal of Bridge Engineering, ASCE, 13(3), pp. 258–270.

Ghosn, M., Moses, F. & Frangopol, D.M. 2010. Redundancy and robustness of highway bridge superstructures and substructures. Structure and Infrastructure Engineering, Taylor and Francis, 6(1-3) (in press).

Kong, J.S. & Frangopol, D.M. 2003. Life-cycle reliability-based maintenance cost optimization of deteriorating structures with emphasis on bridges, Journal of Structural Engineering, ASCE, 129(6), pp. 818–828.

Liu, M. & Frangopol, D.M. 2005b. Multiobjective maintenance planning optimization for deteriorating bridges considering condition, safety, and life-cycle cost, Journal of Structural Engineering, ASCE, 131(5), pp. 833–842.

Messervey, T.B. 2008. Integration of Structural Health Monitoring into the Design, Assessment, and Management of Civil Infrastructure. Ph. D. Thesis, Univ. of Pavia, Italy.

Moan, T. 2005. Reliability-based management of inspection, maintenance and repair of offshore structures. Structure and Infrastructure Engineering, Taylor and Francis, 1(1), pp. 33–62.

Okasha, N.M & Frangopol, D.M. 2010. Time-variant redundancy of structural systems. Structure and Infrastructure Engineering, Taylor and Francis, 6(1-3) (in press).

Schuëller, G.I. 1998. Structural reliability - Recent advances. Proceedings of ICOSSAR'97, Kyoto, Japan, November 1998. A.A. Balkema Publications, pp. 3-35.

Shinozuka, M. 2008. Resilience and sustainability of infrastructure systems. Proceedings of the International Workshop on Frontier Technologies for Infrastructures Engineering, Taipei, Taiwan, pp. 225–244.

van Noortwijk, J.M. & Frangopol, D.M. 2004. Two probabilistic life-cycle maintenance models for deteriorating civil infrastructures. Probabilistic Engineering Mechanics, Elsevier, 19(4), pp. 345–359.

Wen, Y.K. & Kang, Y.J. 2001. Minimum building life-cycle cost design criteria I. Methodology, and II. Applications. Journal of Structural Engineering, ASCE, 127(3), pp. 330–346.

Safety, Reliability and Risk of Structures, Infrastructures and
Engineering Systems – Furuta, Frangopol & Shinozuka (eds)
© *2010 Taylor & Francis Group, London, ISBN 978-0-415-47557-0*

The audacity of change: A transition to the non-stationary and non-linear era

A. Kareem

NatHaz Modelling Laboratory, University of Notre Dame, IN, USA

SYNOPSIS

Historically, typical structures have been designed to resist external loads under the tacit assumption of stationarity and Gaussianity, with the exception of special cases. A lack of an appropriate computational thinking paradigm in this context has precluded the attention these effects deserve, while customary practice prevails, as it draws from well established schemes based on the Fourier transform and the theory of Gaussian processes. Recent recognition of the need to capture the non-stationary and non-Gaussian features, based on the realization that most environmental load effects typically bear these features, has led to many advances in these areas that have matured to the stage that they are ready to be embraced by the design community. Areas where there is a need to account for non-stationary features include: transient wind conditions experienced in thunderstorms by aircrafts aloft and structures near ground exposed to gust fronts, which may lead to an overshoot in pressures, significantly enhancing demand; transient patterns of ground motion in near-field earthquakes imbued in both amplitude and frequency modulations; rogue waves encountered by ships and offshore platforms in open seas and tsunami waves encountered by coastal construction; and a host of other examples. Similarly, in the case of non-linear systems the loads and their effects may significantly depart from Gaussianity, which may result in a response that may not bear any resemblance to the linear/Gaussian response and its extremes may far exceed the Gaussian estimates. Examples, though not exhaustive, include: response of offshore platform systems exposed to nonlinear wave surface profiles that interact nonlinearly introducing splashing and slamming loads and giving birth to new frequency contents, which could potentially excite modes otherwise dormant; aerodynamics of long-span bridges exposed to turbulence that exhibit hysteretic behavior with respect to the wind angle of attack; performance of wind turbines under turbulence.

While the contributions of the Fourier transform to signal processing cannot be denied, the fact that data is often characterized by localized or time-varying features, obscured by the infinite bases of the Fourier transform, prompted a departure from this classical approach towards a time-frequency analysis framework. In the analysis of civil and mechanical systems, the wavelet transform, rooted in a strong mathematical basis, has gained widespread popularity.

In the nonlinear/non-Gaussian domain, a number of recent advances have promoted the use of a number of schemes that range from translational models (static transformation) to rheological models, including? neural networks and Volterra series (akin to Taylor series with memory). The translational processes, in conjunction with Hermite polynomials, have recently gained widespread popularity for the modeling and simulation of non-linear phenomena. Though these models/simulations can faithfully capture the higher-order moments of the process, they lack the ability to reproduce respective higher-order spectral distributions. The application of Volterra series modeling to nonlinear dynamical systems is also becoming more attractive as means of establishing higher order kernels and heir subsequent manipulation is becoming mathematically more tractable as one benefits from its memory features that translational models lack.

The paper highlights recent advances in these areas with applications and introduces the next frontiers to facilitate a much needed change in our way of thinking about and addressing these problems in research, applications and education to ensure the safety and integrity of the built environment.

Safety, Reliability and Risk of Structures, Infrastructures and Engineering Systems – Furuta, Frangopol & Shinozuka (eds)
© 2010 Taylor & Francis Group, London, ISBN 978-0-415-47557-0

Probability density evolution equations: History, development and applications

J. Li

School of Civil Engineering, Tongji University, Shanghai, China

ABSTRACT

The history, development and applications of probability density evolution equations in stochastic dynamics are presented.

It has long been awared that in engineering disciplines such as civil, mechanical and aerospace and aeronautic engineering, to rationally characterize and deal with the randomness involved in the excitations and system properties is of paramount important. For instances, in civil engineering large degree of randomness is involved not only in the extreme dynamic excitations such as the earthquake, strong wind, explosion and the sea wave, but also in the material or geometric properties. Such backgrounds lead to stochastic dynamics, in which one of the most important while challenging problem is how to tackle the systems with coupling of nonlinearity and randomness, which is still open even after around half a century's of endeavors.

Examining the history of the stochastic dynamics since the pioneering work of Eistein and Langevin a century ago will find that there are two parallel traditions: the phenomelogocal tradition along the way of Eistein, Fokker, Planck and Kolmogorov and the physical tradition along the way of Langevin and Itô, although in some aspects these two traditions are not so thoroughly separate and exhibit some mixture. This understanding threw new light on the problem. Actually, the sources and propagations of randomness are embedded in physics and thus the thought of physical stochastic systems should be adopted. In this point of view, the intrinsic relationship between the stochastic systems and the deterministic systems can be revealed upon a unified physical basis (Li, 2006; Li & Chen, 2009).

In view of the new insight, the physical meaning of the principle of preservation of probability is clarified. This principle can be understood from the points of view of the space description and the random event description, respectively (Li & Chen, 2008; 2009), see Figure 1.

In these points of view it is clearly exposed that the essentials of transit of probability density results from the embedded physical mechanism, while the latter are usually described in or can be transformed to state equations. Hence, incorporation of the principle

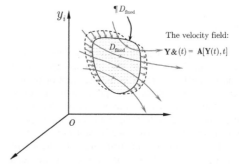

(a) state space description

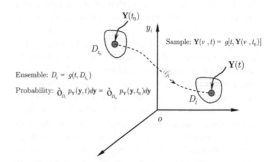

(b) Random event description

Figure 1. The principle of preservatino of probability.

of preservation of probability and the state equation will result in probability density evolution equations. Actually, note that the physical system can be described in a coupled or uncoupled way, then combinations of different descriptions of the principle of preservation of probability and different (coupled or uncoupled) descriptions of the state equation will yield different types of probability density evolution equations. Investigations show that the traditional probability density evolution equations including the Liouville, the Dostupov-Pugachev and the FPK equation can all be successfully re-derived along this line.

Consider an n-DOF nonlinear system

$$\mathbf{M}(\Theta)\ddot{\mathbf{X}}(\Theta) + \mathbf{C}(\Theta)\dot{\mathbf{X}} + \mathbf{f}(\Theta, \mathbf{X}) = -\mathbf{MI}\ddot{X}_g(\Theta, t) \qquad (1)$$

Here Θ are the basic random variables chacterizing the randomness in the initial condition, the excitation and the system properties, which is treated in a unified way.

Note that if the randomness in the system properties is not considered while the excitation $\ddot{X}_g(t)$ is modeled as a white noise, Equation 1 can be transformed to an Itô or Stratonovich's SDE, which is a coupled stochastic state equation. Incorporated it into the principle of preservation of probability from the state space description, a $2n$-dimensional FPK equation can be derived (Chen & Li, 2008)

$$\frac{\partial p_Y}{\partial t} = -\sum_{j=1}^{n} \frac{\partial[p_Y a_j(\mathbf{y},t)]}{\partial y_j} + \sum_{i=1}^{n}\sum_{j=1}^{n} \frac{\partial[p_Y b_{ij}(\mathbf{y},t)]}{\partial y_i\ y_j} = 0 \quad (2)$$

where $p_Y(\mathbf{y},t)$ is shortened by p_Y, $\mathbf{Y} = (\dot{X}, X)^T$ is the state vector.

Likewise, the Liouville and the Dostupov-Pugachev equation can be derived. Because the randomness is treated in different ways from the phenomelogobical point of view, we are led to different probability density evolution equations, of which all are in such a high dimension that the solution is almost impossible except some special simple cases.

What is most important is that, besides the above well-known tratidional probability density evolution equations, when the principle of preservation of probability is incorporated with the uncoupled description of the dynamical system, we result in a new family of generalized density evolution equation (GDEE). Denote the physical quantities of interest as $\mathbf{Z}(t) = (Z_1(t), Z_2(t), \cdots, Z_m(t))^T$, here m is the number of physical quantities, then the GDEE is

$$\frac{\partial p_{Z\Theta}(\mathbf{z},\theta,t)}{\partial t} + \sum_{j=1}^{m} \dot{Z}_j(\theta,t)\frac{\partial p_{Z\Theta}(\mathbf{z},\theta,t)}{\partial z_j} = 0 \quad (3)$$

where $\dot{Z}_j(\theta,t)$'s are the velocities of the physical process for a specified θ.

The GDEE is in obvious contrast to the equations including the Liouville, the D-P and the FPK equation, particularly in the fact that: (1) the GDEE holds for both linear and nonlinear systems with randomness involved separately or simultaneously in the initial conditions, excitations and system property parameters, in other words, the randomness involved is treated in a unified way; and (2) the dimension of GDEE is completely independent to the dimension of the original dynamical system n. Specifically, in the case $m = 1$, which is of most importance from the practical viewpoint, a one-dimensional GDEE is reached. Technically, due to these great advantages the GDEE is applicable to general multi-dimensional nonlinear stochastic dynamical systems without essential difficulties compared to linear or simple systems (Li & Chen, 2004; 2006).

Further, starting with the above thought, new approaches arise for the reliability evaluation (Chen & Li, 2007a). By imposing an absorbing condition on the GDEE or constructing a virtual stochastic process associated with the extreme value, two classes of new approaches based on the GDEE for dynamic reliability can be developed, showing inherent advantages over the existing approaches (Li & Chen, 2005; Chen & Li, 2005; 2007b). In addition, this can also be extended to system reliability evalution without essential difficulties (Li et al, 2007). The above approaches avoid the essential difficulties in traditional theory for dynamical relability and system reliability.

Implementation procedure and algorithm through incorporating the deterministic dynamic analysis and probability density evolution analysis (Li & Chen, 2007; Chen et al, 2007), which is of course the embodiment of the intrinsic relationship between the deterministic systems and stochastic systems as revealed in the methodology, are discussed. The applications to some large-scale complex practical engineering structural systems are illustrated, showing that the proposed methodology is promising. Problems open are outlined.

REFERENCES

Chen, J.B. and Li, J. 2005. Dynamic response and reliability analysis of nonlinear stochastic structures. Probabilistic Engineering Mechanics, 20(1): 33–44.

Chen, J.B. and Li, J. 2007a. Development-process-of-nonlinearity-based reliability evaluation of structures. Probabilistic Engineering Mechanics, 22(3): 267–275.

Chen, J.B. and Li, J. 2007b. The extreme value distribution and dynamic reliability analysis of nonlinear structures with uncertain parameters. Structural Safety, 29: 77–93.

Chen JB, Ghanem R, Li J. 2007. Partition of the probability-assigned space in probability density evolution analysis of nonlinear stochastic structures. Probabilistic Engineering Mechanics, doi:10.1016/j.probengmech.2007.12.017.

Chen, J.B. and Li, J. 2008. A note on the principle of preservation of probability and probability density evolution equation. Probabilistic Engineering Mechanics, doi:10.1016/j.probengmech.2008.01.004.

Li, J. and Chen, J.B. 2004. Probability density evolution method for dynamic response analysis of structures with uncertain parameters. Computational Mechanics, 34: 400–409.

Li, J. and Chen, J.B. 2005. Dynamic response and reliability analysis of structures with uncertain parameters. International Journal of Numerical Methods in Engineering, 62: 289–315.

Li, J. and Chen, J.B. 2006. The probability density evolution method for dynamic response analysis of non-linear stochastic structures. International Journal for Numerical Methods in Engineering 65: 882–903.

Li, J. and Chen, J.B. 2007. The number theoretical method in response analysis of nonlinear stochastic structures. Computational Mechanics, 39(6): 693–708.

Li, J., Chen, J.B. and Fan, W.L. 2007. The equivalent extreme-value event and evaluation of the structural system reliability. Structural Safety, 29: 112–131.

Li, J. and Chen, J.B. 2008. The principle of preservation of probability and the generalized density evolution equation. Structural Safety, 30: 65–77.

Li, J. and Chen, J.B. 2009. Stochastic Dynamics of Structures. John Wiley & Sons.

Safety, Reliability and Risk of Structures, Infrastructures and Engineering Systems – Furuta, Frangopol & Shinozuka (eds)
© 2010 Taylor & Francis Group, London, ISBN 978-0-415-47557-0

Experiments, science and probability theory in building new models for long-term pitting corrosion in marine environments

Robert E. Melchers
Centre for Infrastructure Performance and Reliability, The University of Newcastle, Australia

ABSTRACT

This paper describes some aspects of the processes involved in the development of high-quality probabilistic models for corrosion loss and for maximum pit depth for steels exposed to marine environments. Such models increasingly are required for advanced infrastructure life-cycle management to predict likely future levels of deterioration and for the setting of rational inspection regimes. Most models currently available are empirical. They have high levels of uncertainty. The models described herein are based on appropriate levels of representation of underlying corrosion science fundamentals, including the recognition of the important role of microbiological activity and model calibration using a probabilistic approach to data interpretation. This requires careful interpretation of existing literature data and has required obtaining new data to elucidate particular aspects. A brief overview of this work is given with emphasis on the modeling of maximum pit depth using extreme value distributions. Reinterpretation of existing pit depth data was found to be necessary for consistency with new observations of pit depth development in actual in-situ tests (as distinct from laboratory observations). As a result, the new model provides maximum pit depth as a function of time based on understanding of the underlying pitting process with time. It has shown also that long-term pit depth is most appropriately represented by the Frechet extreme value distribution since this is consistent with the underlying corrosion and bacterial processes. This new interpretation of data provides estimates of pit depth uncertainty as a function of exposure time. It is shown that this can be applied even for existing data sources that were not designed to elucidate probabilistic estimates of uncertainty. The conventional application of the Gumbel extreme value distribution can lead to serious errors in long-term extreme pit depth predictions.

Safety, Reliability and Risk of Structures, Infrastructures and Engineering Systems – Furuta, Frangopol & Shinozuka (eds)
© 2010 Taylor & Francis Group, London, ISBN 978-0-415-47557-0

Reliability-based life cycle assessment of cracks in ocean structures

T. Moan
Centre for Ships and Ocean Structures, Norwegian University of Science and
Technology, Trondheim, Norway

ABSTRACT

Ocean structures intended for exploitation of petroleum and renewable wave and wind energy, for transport as well as for food production are predominantly welded steel structures. Crack-type weld defects are inevitably present in such structures and grow due to cyclic loading and dynamic response amplification, especially in welded joints with high stress concentration in dynamically sensitive structures. Experiences (Moan 2005) show that they may lead to significant operational costs and occasionally to catastrophic accidents. Such adverse consequences are normally limited by adequate design and inspection and repair procedures. The first design codes for offshore structures appeared around 1970 and included fatigue requirements, which were later refined, especially after the fatigue induced total loss of the semi-submersible platform Alexander L. Kielland in 1980. Main inspections are made every 4th–5th year. Yet, more detailed approaches are needed to account for particular features of the different types of ocean structures as well as the uncertainties associated with the predicted behaviour and inspection methods.

Significant developments of structural reliability methodology in general and on Bayesian updating techniques in particular have taken place since the 1980s.

While the initial efforts in structural reliability were related to civil and aeronautical structures, some developments in system reliability and Bayesian updating methods, were directly motivated by needs in the offshore industry

The offshore industry early recognized the need to account for uncertainties in load effects and resistances in design procedures and applied structural reliability analysis to calibrate Ultimate Limit State code requirements for fixed platforms and later for other types of platforms. Practical application of reliability methods was also introduced to deal with the uncertainties in strength degradation and the quality of inspection in establishing design and inspection criteria for fixed platforms, and later for other types of platforms, with due account of the interrelation between design and inspection procedures. Yet, the inspection plans made at the design stage, need to be updated depending upon the findings in connection with condition assessments during fabrication and operation; and especially in connection with service life extension of the structure.

This paper deals with recent developments, validations and applications of reliability-based design, inspection, maintenance and repair procedures of ocean structures, with particular emphasis on design for robustness as well as inspection and repair procedures to limit the occurrence of fatigue failures or fracture. The focus herein is on developments and applications after the review (Moan 1994).

REFERENCES

Moan, T. 1994. Reliability and Risk Analysis for Design and Operations Planning of Offshore Structures. Keynote lecture. *Proc. 6th ICOSSAR, Structural Safety and Reliability.* Rotterdam, ed. Balkema.

Moan, T. 2005. Reliability-based management of inspection, maintenance and repair of offshore structures. *Journal of Structure and Infrastructure engineering* 1(1): 33–62.

Safety, Reliability and Risk of Structures, Infrastructures and
Engineering Systems – Furuta, Frangopol & Shinozuka (eds)
© 2010 Taylor & Francis Group, London, ISBN 978-0-415-47557-0

Performance reliability of port facilities

Masanobu Shinozuka
University of California, Irvine, CA, USA

ABSTRACT

Major ports are important nodes of transportation networks and can play a crucial role in regional, national and international economics. They provide shipping and distribution of goods, and other services related to the transport of cargoes via water. In many countries, trade through ports represent a most dominant mode of transportation compared to other modes such as land and air. For these reasons, the downtime of the port facilities due to natural disasters such as earthquake or typhoon, or manmade disasters as represented by the case of the 2002 labor strike in California (Oakland, Los Angeles, and Long Beach) results in severe economic consequences. However, past experience demonstrated that port facilities tends to be highly susceptible to severe damage due to earthquake. Significant damage was observed not only in the case of a strong earthquake such as the 1995 Hyogoken – Nanbu (Kobe) earthquake, but also under moderate earthquakes as well. This is primarily due to the weak foundation and backfill soils that are common in the waterfront environment. Furthermore, the liquefaction phenomena in the loose, saturated sand beneath port structures present serious causes of damage of the ports facilities. In case of the 1995 Kobe Earthquake, loss due to direct physical damage to the Kobe Port facilities is estimated to be $5.5 billion (1995 value). Also, economic loss exceeded $6 billion within the first 9 months of post-earthquake disruption of port function after the earthquake. More importantly, currently, even though the port facilities have been restored after more than 10 years since the event, the total container volume at the Kobe Port is less than the level at 1994, and it currently ranks #35 in the world ranking down from #5 just before the earthquake event. This indicates that the Kobe Port's operation was not sustainable under an earthquake of the Kobe earthquake intensity from the business competition point of view, as demonstrated by a significant decline of its world ranking after the earthquake. This provides an interesting set of data that can be used to study the nature of sustainability of the port and also other infrastructure systems from a global business point of view.

The main purpose of this presentation is to review a methodology for the evaluation of the seismic effect on the performance of port facilities. The consideration for the performance includes not only robustness and resilience, but also sustainability of the port facility, and hence it involves broader concerns on socioeconomic issues related to the system. The methodology is based on the simulation based engineering and science approach. In this approach, a model of seismic vulnerability of the port facility is developed taking the associated uncertainty of various type into consideration in the form of fragility curves. Typical example of this is the fragility curves for a wharf. These fragility curves can be developed by the seismic response analysis as a function of PGA or any alternative measure of the base- rock ground motion intensity on the consensus based designation of different levels of damage influencing corresponding levels of performance of the wharf allowing the ships to dock. The seismic response analysis mentioned above represents a Monte Calro simulation of the facility response, and the response analysis is repeatedly carried out for each of a large number of probabilistic scenario earthquakes. These scenario earthquakes are developed to be consistent with the well established regional seismic hazard. These fragility curves can then be used to make probabilistic prediction of the damage states and associated direct losses of the port facilities under the seismic hazard in the form of a risk curve.

This presentation uses the Kobe Port as a test bed and carryout the simulation analysis as suggested above. Two highlights are noted. First, the derivation of fragility curves for the typical wharf structure in the Kobe Port is made on the basis of PIANC's damage and wharf performance designation. And second is the use of a random field modeling of the backfill soil property in order to generate by simulation unusually large sea-ward permanent displacement of the Wharf under liquefaction and associated lateral spread of backfill soils.

Safety, Reliability and Risk of Structures, Infrastructures and
Engineering Systems – Furuta, Frangopol & Shinozuka (eds)
© 2010 Taylor & Francis Group, London, ISBN 978-0-415-47557-0

Characterizing geotechnical model uncertainty

W.H. Tang, J. Zhang & L.M. Zhang
Department of Civil and Environmental Engineering, The Hong Kong University of Science
and Technology, Hong Kong SAR, China

1 INTRODUCTION

There are two types of uncertainty that could affect the creditability of a geotechnical analysis, i.e., (1) uncertainty in model input parameters, and (2) uncertainty associated with the geotechnical model itself. Once these uncertainties are characterized, their effects on a design can be quantified through application of reliability theories. Many previous researches have studied how to characterize uncertainties in geotechnical model input parameters. On the other hand, model uncertainty characterization is generally considered as not easy and less work has been done in this respect. As a result, model uncertainty is often not considered in many of present geotechnical reliability analyses due to the lack of knowledge. The focus of this paper is therefore on how to characterize the model uncertainty using observed performance data.

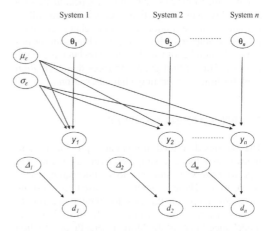

Figure 1. Relationship among variables for model uncertainty characterization.

2 BAYESIAN THEORIES FOR MODEL UNCERTAINTY CHARACTERIZATION

In principle, model uncertainty can be characterized via a systematic comparison between predicted and actual performances of similar geotechnical systems. The difficulty in determining model uncertainty arises from the fact that in calculating the predicted performances, the model input parameters are often not exactly known. Moreover, the actual performances may also be uncertain. Therefore, when model uncertainty characterization is attempted it is often found that model uncertainty is mixed with uncertainty in model input parameters and uncertainty in the actual performances.

A Bayesian formulation to characterize geotechnical model uncertainty when model input parameters are uncertain and the observed system response is subjected to error is described in this paper. Fig. 1 shows the relationship among uncertain variables considered in this Bayesian formulation. In this figure, θ_i, y_i, Δ_i, and d_i denote the uncertain model input parameters, actual performance, observational uncertainty, and observed performance of the ith geotechnical system, respectively; μ_ε and σ_ε denote the mean and standard deviation of the model correction factor ε, which is a variable used to represent the effect of model uncertainty; and n denotes the number of geotechnical systems. It is clear from this figure that the effects of model uncertainty, uncertainty in model input parameters, and observational uncertainty on the disparity between model predictions and observed performances are different. While the uncertain model input parameters and the observational uncertainty associated with a system only affect the disparity between model prediction and observed performance for that system, the model uncertainty is the common but usually unknown factor affecting the disparity between model predictions and observed performances of all systems.

Based on the relationship among variables shown in Fig. 1, model uncertainty parameters μ_ε and σ_ε can be updated simultaneously with uncertain model input parameters θ_i ($i = 1, 2, \ldots, n$) as well as observational uncertainty variables Δ_i ($i = 1, 2, \ldots, n$) using the observed performance data d_i ($i = 1, 2, \ldots, n$) from multiple systems. The joint posterior distribution of these uncertain variables summarizes the knowledge about all uncertain variables, including μ_ε and σ_ε. Markov chain Monte Carlo (MCMC) simulation (e.g.,

Gelman et al. 2004) can be employed to calculate the posterior distributions of the uncertain variables. The Bayesian formulation described above can consider multiple ways to model the model uncertainty utilizing various types of observed data.

Although model uncertainty characterization as described above is versatile, it may be computationally intensive when the number of variables involved is large. To facilitate professional application, a simplified formulation for model uncertainty characterization has also been sought. This is achieved by analytically eliminating uncertain model input parameters and observational uncertainty variables from the original Bayesian formulation in an approximate way. Although uncertain model input parameters and observational uncertainty variables do not appear in the simplified Bayesian formulation, their effects can still be considered in the model uncertainty characterization process. Compared with the original Bayesian formulation, the number of variables to be updated is greatly reduced. This in turn greatly reduces the calculation work for model uncertainty characterization.

3 APPLICATION EXAMPLES

The theories described in this paper provide a quantitative basis for assessing the effects of quality and type of observed data on model uncertainty characterization. Three examples are used to illustrate the above Bayesian theories for model uncertainty characterization, i.e., (1) an example on studying the model uncertainty of the Morgenstern-Price's method (Morgenstern-Price 1965) for slope stability analysis using centrifuge data, (2) an example about determining the model uncertainty of a pile capacity model with field pile load test data, and (3) an example for characterizing the model uncertainty of a liquefaction potential evaluation model based on field performance data.

In the slope example the effect of test uncertainty on model uncertainty characterization is quantified. Model uncertainty characterization under different levels of test uncertainty reveals that that there is a tolerable test uncertainty within which the characterized model uncertainty is not sensitive to test uncertainty. However, beyond that tolerable test uncertainty, further increase in test uncertainty would impair the ability of the centrifuge tests to identify model uncertainty. The slope example also illustrates the dependence of characterized model uncertainty on the quality of the information available for model uncertainty characterization.

In determining the model uncertainty of pile capacity models, it is common that some piles used for calibration were not loaded to failure. For a pile not loaded to failure, its actual capacity is only known to be larger than the maximum load applied in the load test. Traditionally, it is often assumed that the capacity of a pile not loaded to failure is the maximum load applied in the load test. This is called the assumption of ignoring data censoring in this paper. The information of whether the pile has failed during pile load test can be incorporated in model uncertainty characterization process using the theories described in this paper. It is found that if data censoring is ignored, both the mean and the standard deviation of the model correction factor are underestimated. The mean is underestimated because ignoring data censoring underestimates the capacity of a pile not loaded to failure. The standard deviation is underestimated because ignoring data censoring assumes explicit knowledge of pile capacity, thus implying more accurate information than deserved. Hence, it would be more effective in characterizing model uncertainty. Nevertheless, since the assumption of ignoring data censoring does not reflect the real state of the piles that did not fail, the results from this convenient assumption are only apparent and are not realistic.

In the liquefaction model example, the model proposed in Juang et al. (2003) for calculating the factor of safety of soil against liquefaction during seismic loading based on cone penetration test (CPT) data is investigated. This example illustrates the conceptual difference between whether or not the model uncertainty is considered in a probabilistic analysis. When model uncertainty is not considered, the limit state boundary separating the safe domain from the failure domain is a single deterministic surface. When model uncertainty is considered, the location of the boundary is uncertain, and a spectrum of boundaries is drawn instead to separate the safe domain from the failure domain. The boundary spectrum is characterized by the probability density function of the model correction factor. If model uncertainty is neglected, the level of uncertainty in the reliability analysis will be underestimated, making the prediction apparently more informative.

REFERENCES

Gelman, B. A., Carlin, B.P., Stem, H.S. & Rubin, D.B. 2004. *Bayesian data analysis*. 2nd Ed., London: Chapman & Hall.

Juang, C.H, Yuan, H., Lee, D.H. & Lin, P.S. 2003. Simplified cone penetration test-based method for evaluating liquefaction resistance of soils. *ASCE Journal of Geotechnical and Geoenvironmental Engineering* 129(1): 66–80.

Morgenstern, N.R. & Price, V.E. 1965. The analysis of the stability of general slip surfaces. *Geotechnique* 15(1): 79–93.

TECHNICAL CONTRIBUTIONS

Monday Morning (MOM) Sessions

Mini-Symposia (MS01) Uncertainty Modeling of
Rare/Imprecise Data
[Session Organized on Behalf of IASSAR's CSMSE Umbrella]

Safety, Reliability and Risk of Structures, Infrastructures and
Engineering Systems – Furuta, Frangopol & Shinozuka (eds)
© 2010 Taylor & Francis Group, London, ISBN 978-0-415-47557-0

A fuzzy-based approach to comprehensive modeling and analysis of systems with epistemic uncertainties

M. Hanss & S. Turrin
*Institut für Angewandte und Experimentelle Mechanik, Universität Stuttgart,
Stuttgart, Germany*

ABSTRACT

A common problem in the numerical simulation of real-world systems is the fact that exact values for the parameters of the models can exhibit a high level of uncertainty. This non-determinism in numerical models may arise as a consequence of different sources, motivating some categorization of uncertainties. Although other classifications are possible in almost the same manner (e.g. (Möller 2004)), the following categorization (Oberkampf 2007) proves to be well-suited in this context: aleatory uncertainties, such as natural variability or scatter, on the one side, and on the other side, epistemic uncertainties, which arise from an absence of information, rare data, vagueness in parameter definition, subjectivity in numerical implementation, or simplification and idealization processes employed in the modeling procedure.

While aleatory uncertainties have successfully been taken into account by the use of probability theory, the modeling of epistemic uncertainties still remains a challenging topic. As a practical approach to solve this limitation, an interdisciplinary methodology to comprehensive modeling and analysis of systems is presented which allows for the inclusion of uncertainties – in particular of those of epistemic type – from the very beginning of the modeling procedure. This approach is based on fuzzy arithmetic, a special field of fuzzy set theory, which has gained practical relevance after the introduction of the Transformation Method (Hanss 2002). By this technique, the uncertainties can be assessed in a well-defined way and included in the mathematical model, and one can demonstrate how the uncertainties are propagated through the calculation procedure. The Transformation Method avoids the possibly serious drawbacks of conventional fuzzy arithmetic (Hanss 2002) and allows the determination of the degrees of influence of each fuzzy parameter, quantifying the proportion to which the uncertainty of each model parameter contributes to the overall uncertainty of the model output.

As practical and illustrative examples of two different areas of the engineering sciences, the fuzzy-based approach to comprehensive modeling and analysis of uncertain systems is applied to the simulation of automotive crash in structural dynamics as well as to the simulation of landslide failure in geotechnical science and engineering (Musso 2002). In both applications, epistemic uncertainties are considered which arise from some lack of knowledge, from simplification in modeling as well as from subjectivity in implementation.

The main achievement of comprehensive modeling can be seen in the fact that the intentional inclusion of uncertainties in the modeling procedure from the very beginning leads to a somehow more sincere and honest numerical simulation which reflects both the benefits and the limitations of the available information about the system model. This also offers a new perspective on the comparison of numerical simulation and experimental results and evokes the authors' proposal to reconsider the current modus operandi in modeling of dynamical systems for the future by redirecting the increasing computer power to the inclusion of uncertainties, rather than to a further refinement of the often already very complex models.

REFERENCES

Möller, B. & Beer, M. 2004. *Fuzzy Randomness âŁ" Uncertainty in Civil Engineering and Computational Mechanics*, Springer: Berlin.

Oberkampf, W. L. 2007. Model Validation under Both Aleatory and Epistemic Uncertainty. In *Proc. of NATO AVT-147 Symposium on Computational Uncertainty in Military Vehicle Design*, Athens, Greece.

Hanss, M. 2005. *Applied Fuzzy Arithmetic – An Introduction with Engineering Applications*, Springer: Berlin.

Hanss, M. 2002. The transformation method for the simulation and analysis of systems with uncertain parameters, *Fuzzy Sets and Systems* 130, No. 3, 277–289.

Musso, A., Provenzano, P. & Selvadurai, A. P. S. 2002. Stability of motion of detrital reservoir banks In *Proc. of the 2nd International Structural Engineering Conference – ISEC02*, Rome, Italy.

Safety, Reliability and Risk of Structures, Infrastructures and
Engineering Systems – Furuta, Frangopol & Shinozuka (eds)
© 2010 Taylor & Francis Group, London, ISBN 978-0-415-47557-0

Dynamic reliability of structures using adapted process

S.K. Au

Department of Building and Construction, City University of Hong Kong, HKSAR, China

ABSTRACT

Importance sampling is a variance reduction technique that can be used for estimating failure probabilities of engineering systems. Its successful implementation relies critically on a good choice of an importance sampling density that is intended to generate failure samples more frequently. For linear structures subjected to stochastic excitations, very efficient important sampling densities have been constructed using design point excitations (Au & Beck 2001; Yuen & Katafygiotis 2005; Jensen & Valdebenito 2007). The picture is quite different for nonlinear hysteretic structures, however. Importance sampling has not been successful with hysteretic systems, even in the SDOF case. The determination of design point excitations was thought to be the major hurdle for hysteretic systems. An efficient algorithm has been recently developed for determining the design point excitations of SDOF elasto-plastic systems (Au 2006a; Au 2006b). A follow-up numerical study (Au et al 2007), however, showed that simply using an importance sampling density as a weighted sum of Gaussian PDFs among the design points only leads to a limited gain in efficiency. Essentially, the random occurrence of plastic excursions de-synchronizes the critical excitation from the response and undermining its effectiveness in producing failure samples. This motivates research into developing a change of distribution with a mean shift that can 'adapt' to the response so that its future action can be more effective. This idea may appear subtle or even illegitimate at first glance, because the mean shift in the importance sampling density is by definition fixed. However, in the context of stochastic simulation problems where there is an embedded temporal structure in the random variables, it is legitimate when the mean shift as a stochastic process has the property that its next future value depends on information only up to the present. In the theory of stochastic process this is called an 'adapted process'.

This paper presents an importance sampling method using adapted process for estimating the first passage probability of both linear and hysteretic structures subjected to stochastic excitations. The importance sampling distribution corresponds to shifting the mean of the excitation to an 'adapted' stochastic process whose future is determined based on information only up to the present. Previous work ignored the effect of white noise in the design of the adapted process. In this paper we adopt a stochastic control approach that explicitly address the issue. The optimal control law is determined by a control potential, which satisfies the Bellman's equation, a nonlinear partial differential equation on the response state-space. Numerical results for a single-degree-of freedom elasto-plastic structure shows that the proposed method leads to significant improvement in variance reduction over importance sampling using design points reported recently.

REFERENCES

Au SK. Critical excitation of SDOF elasto-plastic systems. Journal of Sound and Vibration 2006a; 296(4-5):714–733.
Au SK. Sub-critical excitations of SDOF elasto-plastic systems. Journal of Non-linear Mechanics 2006b; 41(9):1095–1108.
Au SK, Beck JL. First excursion probability for linear systems by very efficient importance sampling. Probabilistic Engineering Mechanics 2001; 16(3):193–207.
Au SK, Lam HF, Ng CT. Reliability analysis of SDOF elasto-plastic systems: Part I: critical excitation. Journal of Engineering Mechanics 2007; 133(10):1072–1080.
Jensen, HA, Valdebenito, MA. Reliability analysis of linear dynamical systems using approximate representations of performance functions. Structural Safety 2007; 29(3), 222–237.
Yuen KV, Katafygiotis LS, An efficient simulation method for reliability analysis of linear dynamical systems using simple additive rules of probability. Probabilistic Engineering Mechanics 2005; 20:109–114.

Safety, Reliability and Risk of Structures, Infrastructures and Engineering Systems – Furuta, Frangopol & Shinozuka (eds)
© 2010 Taylor & Francis Group, London, ISBN 978-0-415-47557-0

Comparison of uncertainty models in the analysis of offshore structures

M.Q. Zhang, M. Beer, S.T. Quek & Y.S. Choo
Department of Civil Engineering, Center for Offshore Research & Engineering,
National University of Singapore, Singapore

ABSTRACT

In this paper selected uncertainty models are compared with one another in view of their capabilities in dealing with rare and imprecise information. The effects of the different models on the results of a reliability assessment are investigated with focus on corrosion of offshore structures.

Variations of structural parameters due to uncertainty can be modeled and processed appropriately with the aid of probabilistic methods; these are well-established and widely applicable in engineering practice (Schenk & Schuëller 2005, Spanos & Deodatis 2007). A problematic issue, however, can be the probabilistic modeling of the available information if this appears in the form of rare and imprecise data. Examples are uncertain quantities for which mere bounds or linguistic expressions are known. A traditional probabilistic modeling then requires subjective assumptions, which may introduce unwarranted information if they cannot be justified sufficiently. This may lead to biased computational results and to wrong decisions with potential for serious consequences.

It is, thus, advisable to take account of the entire uncertainty according to its nature to obtain realistic results (Helton & Oberkampf 2004). A variety of non-traditional mathematical models and approaches have been formulated for this purpose; see Fellin et al. (2005) and Klir (2006). The usefulness and capabilities of these models and approaches have already been demonstrated in the solution of practical problems in civil and mechanical engineering (Helton & Oberkampf 2004, Möller & Beer 2008).

In the present study, the probabilistic model for marine corrosion of steel provided in Melchers (2003) is investigated in the context of limited information. This model contains several uncertain parameters. Due to rare and imprecise information, the bias factor in the model cannot be specified precisely and is merely known in the form of bounds. The solutions obtained with respect to pure probabilistics, interval modeling, fuzzy modeling, and imprecise probabilities are comparatively scrutinized.

The various uncertain forms of the corrosion model are applied to a simple numerical example of a steel plate according to Melchers (2003). The upper bound of the failure probability is computed with the different uncertainty models via sampling methods. All probabilistic quantities are processed with direct Monte Carlo simulation. For calculations with interval and fuzzy quantities the algorithm of α-level optimization according to Möller & Beer (2004) is adopted. Computations with imprecise probabilities are realized with a combination of both methods. The effects of the different uncertainty models are investigated by comparing the information contents in the results and the numerical efficiency. Practical applicability is demonstrated. A reliability analysis is performed for a jacket-type offshore platform with a limit state defined by a push-over analysis.

REFERENCES

Fellin, W., Lessmann, H., Oberguggenberger, M. and Vieider, R. (eds.) 2005, *Analyzing uncertainty in civil engineering.* Springer, Berlin Heidelberg New York.

Helton, J.C. and Oberkampf, W.L. (eds.) 2004, Special Issue on Alternative Representations of Epistemic Uncertainty. *Reliab Eng Syst Safety*, Vol. 85(1–3), 1–369.

Klir, G.J. 2006, *Uncertainty and information: foundations of generalized information theory.* Wiley, Hoboken.

Melchers, R.E. 2003, Probabilistic Model for Marine Corrosion of Steel for Structural Reliability Assessment. *Journal of Structural Engineering* Vol. 129(11), pp. 1484–1493.

Möller, B. and Beer, M. 2004. *Fuzzy Randomness – Uncertainty in Civil Engineering and Computational Mechanics.* Springer, Berlin.

Möller, B. and Beer, M. 2008. Engineering Computation Under Uncertainty – Capabilities of Non-Traditional Models. *Computers & Structures*, Vol. 86(10), pp. 1024–1041.

Schenk, C.A. and Schuëller, G.I. 2005. *Uncertainty Assessment of Large Finite Element Systems.* Springer, Berlin Heidelberg New York.

Spanos, P.D. and Deodatis, G. (eds.) 2007. *Computational Stochastic Mechanics.* Millpress, Rotterdam.

*Safety, Reliability and Risk of Structures, Infrastructures and
Engineering Systems – Furuta, Frangopol & Shinozuka (eds)
© 2010 Taylor & Francis Group, London, ISBN 978-0-415-47557-0*

Copulas and T-norms: Mathematical tools for modelling propagation of errors and interactions

W. Näther

TU Bergakademie Freiberg, Germany

ABSTRACT

In this presentation, we will demonstrate that such
similar tools like T-norms and copulas are useful for
different kinds of error propagation as well as for
interaction models.

Traditionally there are at least three approaches
for propagation of errors. Firstly there is the classical error propagation using calculus and a linear
Taylor series approximation. Secondly we know propagation of absolute errors by interval mathematics.
Thirdly the Gaussian error propagation is known which
presents approximately the standard deviation of the
output if the standard deviations of the inputs and the
input-output transfer function are given.

Firstly, we point out that the well known extensional
principle from Fuzzy Set Theory is nothing more than a
soft variant of error propagation by interval mathematics. This typical worst case scenario of error propagation can be dampened by use of modified extensional
principles where the strong $\min -T-$norm is changed
by a weaker $T-$norm. Some examples are discussed
in the presentation. Of special interest in this context
are parametric families of T-norms. Some of which
are discussed in this paper.

Error propagation is deeply influenced by dependencies or interactions between the input variables.
But only the general Gaussian error propagation takes
into account these dependencies by considering correlations between the variables. As a simple example
consider two random (input-) variables X and Y with
$VarX = 1$ and $VarY = 1$, correlation $\rho(X, Y)$ and the
sum $X + Y$ as output. Obviously, since $Var(X + Y) =
2 + 2\rho(X, Y)$, the variance of the sum can taken any
value between zero and four depending on the value
of $\rho(X, Y)$.

Only linear dependencies can be identified by
correlations. A more general tool for describing dependencies between random variables are copulas, well
known since 1960 (Schweizer and Sklar) but applied
just in the last years, for example in finance mathematics. In this presentation, we will introduce copulas
and discuss some special cases. Consider, for example,
two random variables X and Y where $Y = g(X)$ with
a monotonically increasing g. This dependency leads
to the strongest possible copula, nevertheless $\rho(X, Y)$
may be nearly zero.

The classical approach for identifying interactions
uses a parametric stochastic model for input and output variables. Some efforts are done in the last years to
integrate interactions directly in the measure of evaluation. Obviously, if the measure of evaluation is a
probability measure for disjoint events A and B we
always have $P(A \cup B) = P(A) + P(B)$, i.e. interaction between A and B does not change this relation.
However, non-additive set functions Q (so called fuzzy
measures) with $Q(A \cup B) < Q(A) + Q(B)$ (negative
synergy) or $Q(A \cup B) > Q(A) + Q(B)$ (positive synergy) are suitable for modelling interactions. For some
kind of fuzzy measures the deviation from additivity
can be described by $T-$conorms. We point out that
modelling interactions by fuzzy measures (especially
by ordered weighted averages) needs a smaller number
of parameters than traditional modelling.

REFERENCES

Alsina, C., Frank, M.J. and Schweizer, B., 2006. Associa-
tive Functions – Triangular Norms and Copulas. World
Scientific, Singapore.

Safety, Reliability and Risk of Structures, Infrastructures and
Engineering Systems – Furuta, Frangopol & Shinozuka (eds)
© 2010 Taylor & Francis Group, London, ISBN 978-0-415-47557-0

Multivariate models for random sets generated by non-parametric methods

Th. Fetz & M. Oberguggenberger
Unit of Engineering Mathematics, University of Innsbruck, Austria

ABSTRACT

Let $g(x_1, \ldots, x_n)$ be a limit state function of a structure depending on n variables $x_1, \ldots, x_n$ which are assumed to be uncertain. The uncertainty for each variable is modelled from measurement data by a *non-parametric* method using *Tchebycheff's inequality*

$$P\left(|X - \mu| > \frac{\sigma}{\sqrt{\alpha}}\right) \le \alpha. \quad (1)$$

The inequality shows that $I_\alpha = [\mu - \frac{\sigma}{\sqrt{\alpha}}, \mu + \frac{\sigma}{\sqrt{\alpha}}]$ is a confidence interval of level $1 - \alpha$. Denoting by I_α^c its complement, we get

$$P(I_\alpha^c) \le \alpha = \bar{P}(I_\alpha^c) \quad (2)$$

where $\bar{P}(I_\alpha^c)$ is the upper probability of the values outside the confidence interval, cf. Oberguggenberger et al. (2007); Oberguggenberger et al. (2008); Oberguggenberger & Fellin (2008). Since confidence intervals I_α parameterized by $\alpha \in (0, 1]$ are nested, a fuzzy set or a random set can be constructed from these intervals. The goal is to compute upper bounds for the probability $\bar{P}(g(x_1, \ldots, x_n) \le 0)$ of failure. For this we have to model the *joint uncertainty* of all variables involved. There are two generic ways to do that depending on which uncertainty model is used:

Fuzzy sets: The advantage is the low computational effort, but from a probabilistic point of view there is no non-parametric way of taking independence or interactions into account.

Random sets: Here we would have the possibility to model different types of independence and unknown interaction Fetz (2004); Fetz & Oberguggenberger (2004), but the main disadvantage is the high computational effort.

Our new approach here is somewhere between the above two approaches. We only combine intervals of the same confidence level α, which also reduces the computational effort, but then we use random sets in the following way: For a variable x_k a *local random set* based on level α consists of two focal sets $I_{k,\alpha}, I_{k,\alpha}^c$ and weights $m(I_{k,\alpha}^c) = \bar{P}(I_{k,\alpha}^c) = \alpha$ and $m(I_{k,\alpha}) = 1 - \alpha$. To obtain the joint uncertainty of several variables we construct the joint random set assuming random

set independence or using *Fréchet's bounds* for the weights. The result for a joint confidence set $I_{1,\alpha} \times I_{2,\alpha}$ for two variables is then an inequality of the same style as the above inequality (2), but with different right hand sides $f(\alpha)$ dependening on the method used to combine the random sets:

$$P((I_{1,\alpha} \times I_{2,\alpha})^c) \le f(\alpha) \quad \text{with} \quad (3)$$

$$f(\alpha) = \begin{cases} \alpha & \text{lower bound,} \\ 2\alpha - \alpha^2 & \text{random set independence,} \\ \min(2\alpha, 1) & \text{upper bound.} \end{cases}$$

The sets $I_{1,\alpha} \times I_{2,\alpha}$ are nested again, but the level is now $f(\alpha)$. That means the joint confidence set is written as $J_{f(\alpha)} = I_{1,\alpha} \times I_{2,\alpha}$ which leads to the following formula for the upper probability of failure:

$$\bar{P}(g(x_1, x_2) \le 0) \le \inf\{f(\alpha) : g(J_{f(\alpha)}) \cap (-\infty, 0] = \emptyset\}.$$

An example similar to Melchers (1999) is presented where we model the uncertainty of the variables by Tchebycheff's inequality and compute the probability of failure for the different approaches.

REFERENCES

Fetz, T. 2004. Multi-parameter models: rules and computational methods for combining uncertainties. In W. Fellin, H. Lessman, R. Vieider, & M. Oberguggenberger (eds) *Analyzing Uncertainty in Civil Engineering.* Berlin: Springer.

Fetz, T. & Oberguggenberger, M. 2004. Propagation of uncertainty throughmultivariate functions in the framework of sets of probability measures. *Reliability Engineering and System Safety,* 85(1–3): 73–87.

Melchers, R. E. 1999. *Structural Reliability Analysis and Prediction.* Chichester: Wiley.

Oberguggenberger, M. & Fellin, W. 2008. Reliability bounds through random sets: nonparametricmethods and geotechnical applications. *Computers & Structures,* 86: 1093–1101.

Oberguggenberger, M., King, J. & Schmelzer, B. 2007. Imprecise probability methods for sensitivity analysis in engineering. In G. de Cooman, J. Vejnarova, & M. Zaffalon (eds) *ISIPTA'07, Proceedings of the Fifth International Symposium on Imprecise Probability: Theories and Applications,* (pp. 317–326). Prague: Action M Agency, SIPTA.

Safety, Reliability and Risk of Structures, Infrastructures and Engineering Systems – Furuta, Frangopol & Shinozuka (eds)
© 2010 Taylor & Francis Group, London, ISBN 978-0-415-47557-0

Uncertainty in building and the effect of human intervention

F. Knoll
Nicolet, Chartrand, Knoll, Montreal, Quebec, Canada

ABSTRACT

The paper attempts to find a relationship between the theoretical concept of uncertainty with respect to the parameters of the building process and the practice which recognizes the main source of significant deviations from he correct value of those parameters to be human intervention. Human action is the main contaminating agent in the building process but at the same time also the filtering agent which usually succeeds in achieving acceptable performance. This fact must therefore be reflected prominently in any assessment of uncertainty and if that uncertainty shall be minimized, this is where we must begin.

Attempts have been made in the past to find modeling parameters which would permit a mathematical treatment of human intervention in the building process. To this author's knowledge, none of these endeavours was pursued to an useful end, in other words, a model permitting research on the basis of forensic data from real life situations.

The enormous complexity and variety of reality may be cited as the main reason for this as it surpasses what a single researcher or even a small team seems to be able to handle. In this paper the author tries to lay out the rudiments of a new attempt to "come to grips" with the problem, identifying some characteristics of error generation as well as their correction, both of which are rooted entirely in the individual behaviour of humans.

Experience shows that real life data often comes incomplete, contradictory and very often contentious since in many cases attempts to find the truth are stopped short – judges tend to favour settlements out of court rather than drawn-out battles clogging the courts' agenda.

The uncertainty coming of the fact that most data is less than unequivocal and clear must not necessarily prevent the gathering of insight through concerted research which ought to include cases where errors were caught in time and corrected, albeit at extra costs

The paper focusses on the filtering which occurs simultaneously with the generation of errors during the construction process. It is effective to the point of creating a socially acceptable general degree of reliability in building construction. Better knowledge about how, when and where it works, or the circumstances which hinder it will eventually permit to develop rationally founded strategies to improve the filtering. At this time these strategies are entirely based on the personal experience and judgment of the participants which may at the same time answer to incentives or interests foreign to the task which is to reduce the effect of Murphy's law: "If something can go wrong it will".

Mini-Symposia (MS13) Smart Structural Health Monitoring and Safety Assessments
[Session Organized on Behalf of IASSAR Subcommittee SC4]

Mini-Symposia (MS13) Smart Structural Health Monitoring and Safety Assessment
[Session Organized on Behalf of IASSAR Subcommittee SC4]

Safety, Reliability and Risk of Structures, Infrastructures and
Engineering Systems – Furuta, Frangopol & Shinozuka (eds)
© 2010 Taylor & Francis Group, London, ISBN 978-0-415-47557-0

Wireless impedance sensor nodes for structural health monitoring

S. Park
Sungkyunkwan University, Suwon Gyonggi, Korea

H. Shin, H.-J. Kim & C.B. Yun
KAIST, Daejeon, Korea

ABSTRACT

Cost-effective and reliable structural health monitoring (SHM) of critical members is very essential for safe operation of civil infrastructures such as bridges, buildings, power plants, and pipeline systems. In recent years, a large amount of research has been focused on utilizing the impedance-based SHM techniques (Linag et al., 1996; Park et al., 2003). Particularly, with a current trend of SHM heading towards ubiquitous self-contained sensor environments, a wireless impedance sensor node was developed (Mascarenas et al., 2007). The wireless impedance sensor node incorporates a low-cost miniaturized impedance circuit chip that can measure and record the electrical impedance of a piezoelectric sensor, an on-board microcontroller that performs local computing for signal processing, and a radio frequency (RF) telemetry module that transmits the structural health diagnostic results to a base station. In this context, the main goal of this study is to develop an improved wireless impedance sensor node shown in Figure 1 with a multiplexer and an external temperature sensor for simultaneous achievement of both functions of structural damage detection and sensor-self diagnosis (Park et al., 2008). Three main objectives are as follows: (1) to develop reliable and adjustable wireless impedance sensor nodes in terms of both hardware and software, (2) to verify the feasibility of the wireless impedance sensor node for both functions of structural damage identification and sensor self-diagnosis, and (3) to propose a temperature effects-free sensor self-diagnosis algorithm. The block diagram of the sensor node is displayed in Figure 2.

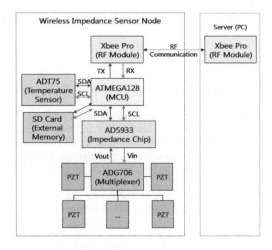

Figure 2. Block diagram of the impedance sensor node.

The performances of the new wireless impedance sensor node have been demonstrated in an experimental study to detect cut damage in an aluminum plate and to diagnosis PZT sensor and/or bonding defects at the sensor node by itself. Finally, this paper summarizes several issues and future works that deal with energy harvesting and power delivery/management issues of the sensor nodes.

REFERENCES

Liang, L., Sun, F.P., and Rogers, C.A. (1996) 'Electromechanical impedance modeling of active material systems,' Smart Materials and Structures, 5: 171–186.

Park, G., Sohn, H., Farrar, C.R. and Inman, D.J. (2003) 'Overview of impedance-based health monitoring and path forward,' The Shock and Vib. Digest, 35(6), 451–463.

Mascarenas, D.L., Todd, M.D., Park, G., and Farrar, C.R. (2007) 'Development of an impedance-based wireless sensor node for structural health monitoring,' Smart Materials and Structures, 16: 2137–2145.

Park, S., Park, G., Yun, C.B., and Farrar, C.R. (2008) 'Sensor self-diagnosis using a modified impedance model for active sensing-based SHM,' *Structural Health Monitoring*, in press.

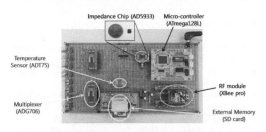

Figure 1. New prototype of wireless impedance node.

Safety, Reliability and Risk of Structures, Infrastructures and Engineering Systems – Furuta, Frangopol & Shinozuka (eds)
© *2010 Taylor & Francis Group, London, ISBN 978-0-415-47557-0*

Development of a wireless piezoelectric transducer excitation technique using laser and optoelectronic devices

H.J. Park, H. Sohn & C.B. Yun
Korea Advanced Institute of Science and Technology, Daejeon, Korea

J. Chung
CyTroniq Co., Ltd, Asan, Korea

I.B. Kwon
Korea Research Institute of Standards and Science, Deajeon, Korea

ABSTRACT

In recent years, nondestructive testing (NDT) has seen increasingly wider use for structural health monitoring and damage detection applications. Among the NDT methods, guided wave based NDT techniques have attracted much attention from re-searchers, because they are not only sensitive to small defects, but also can propagate over a long dis-tance in a plate-like structure.

These guided waves in a structure can be gener-ated and sensed by a variety of techniques (Su et al. 2006). Our study proposes a new scheme for PZT excitation and sensing based on laser and optoelec-tronic technologies where power as well as data can be transmitted via laser. This paper mainly focuses on the excitation aspect of the ultimate goal (Fig. 1). First, a laser is used as the power source and an arbitrary waveform such as a toneburst signal is generated using an electro-optic modulator (EOM) (Smith et al. 1995, Wilson et al. 1998). This modulated laser is ampli-fied by an erbium-doped fiber amplifier (EDFA) and emitted to a photodiode (Kasap et al. 2001). The pho-todiode converts the light into an electrical power and a transformer increase the voltage level of the con-verted electrical power. Finally, this electrical power can excite a PZT transducer and the guided waves are generated in a structure. In particular, the feasibility of the power transmission aspect of the proposed wire-less scheme has been experimentally demonstrated in a laboratory setup (Fig. 2).

The results confirm that the laser-based guided wave generation technique exhibits the reasonable performance by comparing the conventional wiring guided wave generation method. Utilizing the pro-posed technology, a PZT transducer can be perma-nently attached to a structure with no needs of power supply and thus a self-sufficient PZT transducer unit with little additional electronic components can be envisioned. Although the technology to sense the generated guided waves is not be described in this paper, some potential techniques for sensing can be

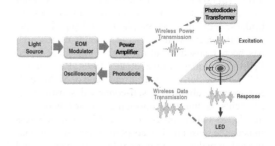

Figure 1. Overall Schematics of the optic-based wireless power and data transmission system.

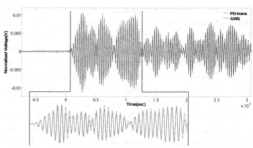

Figure 2. The guided waves generated by an arbitrary waveform generator and laser.

developed by utilizing other optoelectronic devices such as light-emitting diodes (LEDs) or laser vibrom-eters. These are also being investigated by the authors.

REFERENCES

Kasap, S.O. 2001. *Optoelectronics and Photonics: Principles and Practices*. New Jersey: Prentice Hall.
Smith, S.D. 1995. *Optoelectronic Devices*. New Jersey: Prentice Hall.
Su, Z. & Ye, L. & Lu, Y. 2006. Guided Lamb waves for iden-tification of damage in composite structures: A review. *Journal of Sound and Vibration*. 295: 753–780.
Wilson, J. & Hawkes, J. 1998. *Optoelectronics: An introduc-tion*. New Jersey: Prentice Hall.

Safety, Reliability and Risk of Structures, Infrastructures and
Engineering Systems – Furuta, Frangopol & Shinozuka (eds)
© 2010 Taylor & Francis Group, London, ISBN 978-0-415-47557-0

Landslide detection using microwave

H.C. Rhim, S.W. Kim & J.Y. Song
Yonsei University, Seoul, Korea

ABSTRACT

In this paper, the measurements of the dielectric properties of soil are made using both a dielectric probe kit
and ground penetrating radar. Soil samples are oven-
dried until there is no change in weight, we make soil
samples with moisture content varying from 5 to 35
%. The dielectric constants are first measured using
a dielectric probe kit including an open ended coaxial
probe and network analyzer from 600 MHz to 2.0 GHz
as shown in Figure 1.

And two different antennas with 900 MHz and
1.0 GHz are connected to the GPR system, we obtained
GPR data from soil samples (Fig. 2).

Figure 1. Network analyzer and dielectric probe kit.

Figure 2. GPR system for moisture measurements.

Data obtained from the measurements can be used
to establish relationship between the dielectric constants and moisture content of soil. This information
can be further utilized in predicting the possibility of
landslide by relating the variation of water movement
inside soil to the initiation of soil movement prior
to the landslide. The measurement setup is described
and sample results are provided.and has important
implications for upland water quality.

REFERENCES

Weihermüller, L. 2007. Mapping the spatial variation of soil
water content at the field scale with different ground
penetrating radar techniques. *Journal of Hydrology* 340:
205–216.
Singh, K.K.K. 2003. Ground penetrating radar study for
hydrogeological conditions related with mining activity. Environmental Geology 44: 20–27.
Turesson. 2006. Water content and porosity estimated
from ground-penetrating radar and resistivity. Journal of
Applied Geophysics 58: 99–111.
Lunt, I.A. & Hubbard, S.S. & Rubin, Y. 2005. Soil moisture content estimation using ground-penetrating radar
reflection data. Journal of Hydrology 307: 254–269.
Galagedare, L.W. & Parkin, G.W. & Redman, J.D. 2005.
Measuring and Modeling of Direct Ground Wave Depth
Penetration Under Transient Soil Moisture Conditions.
Subsurface Sensing Technologies and Applications 6(2).
Stoffregen, H. & Zenker, T. & Wessolek, G. 2002. Accuracy
of soil water content measurements using ground pene-
trating radar: comparison of ground penetrating radar and
lysimeter data. Journal of Hydrology 267: 201–206.
Fang Ji & Kawasaki Akiyuki & sadohara satoru. 2005. Devel-
opment of a Slope Failure Management System Using
ArcGIS Server. ESRI International User Conference.
Cheng & Field, D. K. & Wave. 1989. Electromagnetics
Addison Wesley Pub. Co. Massachusetts 703.

*Safety, Reliability and Risk of Structures, Infrastructures and
Engineering Systems – Furuta, Frangopol & Shinozuka (eds)*
© 2010 Taylor & Francis Group, London, ISBN 978-0-415-47557-0

FRP-concrete debonding detection using distributed brillouin fiber optic strain sensor

M. Imai & M. Feng

Department of Civil and Environmental Engineering, University of California, Irvine, CA, USA

ABSTRACT

For retrofitting concrete beams or slabs with fiber
reinforced polymer (FRP) composites, debonding of
FRP sheets is an issue of great concern. One of
the most preferable methods to inspect debonding is
to measure strain, because strain on FRP is almost
constant at debonded areas. Unlike discrete sensors,
Brillouin-based fiber optic sensor is promising for
debonding detection because of its capability of measuring strain distribution along the entire length of
the sensing fiber. Using two encountered lightwaves,
recent progress has been made to increase the spatial
resolution of Brillouin-based senor. If the lightwave
interaction length is sufficient, only the steady state of
the induced acoustic wave needs to be considered. If
the length is shorter than the phonon lifetime, the influence of the transient state cannot be ignored. Thus, an
analysis is necessary to be developed to extract meaningful information from measured Brillouin spectrum
in the case of the narrow pulse lightwave which enables
high spatial resolution.

This paper proposes a model that takes into consideration of both the steady and the transient Brillouin
interaction state. Assuming that the transient term has
an analogous effect to the steady state term, new two
parameters are introduced. A simple test is conducted
for phenomenological determination of the coefficients. The proposed model indicates that the signal
intensity distribution at the specific frequency, which
corresponds to strain at debonded region, is sensitive to
occurrence of debonding. For evaluation of the model,
an experiment is carried out on a reinforced concrete beam specimen retrofitted with glass FRP sheets
which sensing fibers are mounted on. The specimen is
subjected to bending, and strain on the FRP sheets is
determined by stimulated Brillouin-based fiber optic
sensor. Simultaneously, measured Brillouin spectrum
is analyzed by the proposed model and the debonding
location and length is estimated. The results agree well
with test observation, thus its analytical and experimental study demonstrates the effectiveness of the
proposed model that incorporates not only the steady

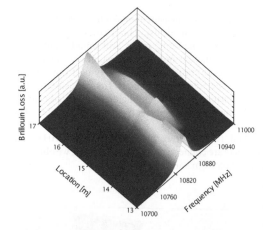

Figure 1. Analytical result of distributed Brillouin loss
spectrum.

Brillouin interaction state. Using the model, the study
provides a simple method for detecting FRP-concrete
debonding without baseline.

REFERENCES

Bao, X., Brown, A., DeMerchant, M. & Smith, J. 1999. Characterization of the Brillouin-loss spectrum of single-mode
fibers by use of very short (<10-ns) pulses, *Optics Letters*,
24(8), 510–512.
Bayvel, P. & Radmore, P. M. 1990. Solutions of the SBS
equations in single mode optical fibres and implications
for fibre transmission systems, *Electronics Letters*, 26(7),
434–436.
Horiguchi, T. & Tateda, M. 1989. BOTDA-Nondestructive
measurement of single-mode fiber attenuation characteristics using Brillouin interaction: theory, *Journal of
Light-wave Technology*, 7(8), 1170–1176.
Sutherland, R. L. 2003. *Handbook of Nonlinear Optics*, Marcel Dekker, Inc., Basel, Switzland.
Niu, H. & Wu, Z. 2001. Peeling-off criterion for FRP-
strengthened R/C flexural members, *Proceedings of International Conference on FRP Composite in Civil Engineering*, Hong Kong, 571–578.

Safety, Reliability and Risk of Structures, Infrastructures and Engineering Systems – Furuta, Frangopol & Shinozuka (eds)
© *2010 Taylor & Francis Group, London, ISBN 978-0-415-47557-0*

Measurement and diagnosis of concrete cracks using image processing and fuzzy theory

B.Y. Lee, J.K. Kim & H. Myung
Department of Civil and Environmental Engineering, KAIST, Daejeon, Korea

ABSTRACT

A structure is required to be sustained during the target service life without any severe problems in safety, serviceability and durability. Most of reinforced concrete structures, however, degraded serviceability and durability due to a very low tensile strength of concrete material even though it has quite high strength on compression. Especially, concrete structures with cracks due to tensile stresses show very fast degradation on serviceability and durability. Therefore, measuring and diagnosing cracks are essential to decrease direct or indirect effect of cracks on the safety, durability, and serviceability of concrete structures. In this regard, it is necessary to accurately measure the width, length, and direction of the cracks and deduce the causes of cracking based on the measuring data. However, because inspectors generally measure cracks manually, a great deal of time and energy is required to take measurements and to compile relevant data.

This paper presents the crack measurement technique for detection and analysis of cracks in digital image of concrete surface. This technique composes of an automatic measurement process of crack characteristics such as width, length, and orientation based on image processing and inference and recognition process of crack patterns, which includes horizontal, vertical, diagonal(−45°), diagonal(+45°), and random cracks, based on artificial neural network. Diagnosis technique based on fuzzy theory also is presented.

To verify the applicability of proposed technique, measurement and diagnosis program have been developed and a series of experimental investigations was carried out. A comparison of the original crack images with the results obtained by the proposed technique shows that the proposed technique can accurately detect and measure crack images. Presented crack pattern classifier based on ANNs can effectively classify cracks into 5 categories. Presented crack diagnosis program based on fuzzy theory presented same results with those by human and the possibility of each crack causes which is quantitative result.

REFERENCES

Adlassnig, K.P., Kolarz, G., 1982. CADIAG-2: Computer-Assisted Medical Diagnosis using Fuzzy Subsets (Edited by Gupta, M.M. and Sanchez, E., Approximate Reasoning in Decision Analysis), North Holland, pp.219–247.

Ammouche, A., Breysse, D., Hornain, H., Didry, O., and Marchand, J., 2000. A New Image Analysis Technique for The Quantitative Assessment of Microcracks in Cement-Based Materials, *Cement and Concrete Research*, 30(1), 25–35.

JCI, 1980. Guideline for crack investigation, repair/ strengthening method.

Foresee, F.D. and Hagan, M.T., 1997. Gauss-Newton approximation to Bayesian Regularization, *Proceedings of the 1997 International Joint Conference on Neural Networks*, pp.1930–1935.

Haralick, R.M, and Linda, G.S., 1992. *Computer and Robot Vision*, Volume I, Addison-Wesley, pp. 28–48.

Mackay, D.J.C., 1992. Bayesian interpolation, *Neural Computation*, 4(3), 415–447.

Otsu, N.A., 1979, Threshold Selection Method from Gray Level Histogram, *IEEE Transactions on Systems*, SMC-9(1), 62–66.

Richard, O.D., Peter, E.H., and David. G.S., 2001. *Pattern Classification*, 2nd ed., John Wiley & Sons, Inc.

Seul, M., O'Gorman, L., and Sammon, MJ., 2000. *Practical Algorithms for Image Analysis*, Cambridge University Press.

Stroeven, P., 1973. Some Aspects of the Micromechanics of Concrete, Ph. D. Thesis, Stevin Laboratory, Technological University of DELFT.

Zadeh, L. A., 1965. *Fuzzy sets*, Information and Control. 8, 338–353.

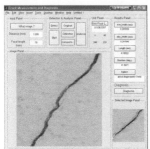

Figure 1.   Developed measurement and diagnosis program.

Mini-Symposia (MS12) Monte Carlo Simulations Applied to Civil Engineering

Safety, Reliability and Risk of Structures, Infrastructures and
Engineering Systems – Furuta, Frangopol & Shinozuka (eds)
© *2010 Taylor & Francis Group, London, ISBN 978-0-415-47557-0*

Hot spot stress and stress distribution at web-to-flange welded joints of corrugated web girder

Z.Y. Wang

Department of Civil Engineering, Sichuan University, Chengdu, Sichuan, P.R. China
Department of Civil Engineering, University of Nottingham, Nottingham, UK

Y.Y. Chen

Shenzhen Municipal Design & Research Institute Co., Ltd, Shenzhen, P.R. China

Q.Y. Wang

Department of Civil Engineering, Sichuan University, Chengdu, Sichuan, P.R. China

ABSTRACT

Girders with corrugated steel web have been of increasing interest in bridge design and construction since its web is supposed to own adequate out-of-plane stiffness and negligible bending capacity. Also, corrugated steel web plates have been suggested as a substitute to traditional concrete or plane steel plate webs with stiffeners and a better way of weight saving to meet requirements for bridge girders. Currently, increasing concerns regarding the fatigue behavior of the joint between corrugated web and flange plate have been reported in several experimental studies. However, due to rather difficulty in weld profile modeling and complexity of the stress field in the vicinity of the welds, the web-to-flange welded joints have not been sufficiently considered by the majority of recent studies regarding fatigue behavior of corrugated web girder.

This study investigated the stress distribution along weld throat plane and at the hot spot region for web-to-flange welded joints of corrugated web girder. As the stress distribution along the weld is very important for the prediction of the fatigue life of welded structural components, the analysis with sufficient consideration of weld profile can be of great benefit to further study the fatigue behavior of corrugated web girder. A three dimensional finite element model has been developed. Based on the numerical results, the effect of the geometrical parameters on the stress distribution on weld throat plane along external & internal ridge were analyzed and compared. In this study, the hot spot stresses were defined at the intersection points of parallel and inclined weld. Three different stress extrapolation techniques for predicting hot-spot stresses at the weld toes were investigated and compared with Dong's method. Thereafter, a parametric study has been carried out to study the effect of four popularly used geometrical parameters on stress gradient in the vicinity of the intersection points of parallel and inclined weld.

It is found that the weld joints at intersection points of parallel and inclined fold may become plastic much earlier than that in other parts of the web-to-flange welded joints working in low stress state, which suggest much attention need to be paid to the stress distribution at these intersection points in fatigue process. Besides, the significant influence of corrugated angle not only on stress ratio of the end and middle of the inclined weld along external weld ridge, but on hot spot stress of welded joints has been revealed.

REFERENCES

Abbas HH, Sause R, Driver RG. Behavior of corrugated web I-girders under in-plane loads. *J Eng Mech* 2006; 132(8):806–14.

Anami. K and Miki. C, Fatigue strength of welded joint made of high-strength steels. *Structure Engineering Mater*, 86–94, 2001.

Dong P. A structural stress definition and numerical implementation for fatigue analysis of welded joints. *International Journal of Fatigue* 2001; 23:865–76.

Japan Society of Steel Construction, Fatigue design recommendation for steel structure, 1993.4. (In Japanese)

Miki. C, Tateishi .K, Yamamoto. Y and Miyauchi. M, A study on the local stress approach to the fatigue assessment, *Journal of Structural Engineering*, JSCE (Japanese Society of Civil Engineering), Vol. 38A: 989–997, 1992 (in Japanese).

Niemi. E. Stress determination for fatigue analysis of welded components, Technical report IIS/IIW-1221-93, International Institute of Welding, Cambridge, United Kingdom, 1995.

Sherif A. Ibrahim, Wael W. El-Dakhakhni, Mohamed Elgaaly. Behavior of bridge girders with corrugated webs under monotonic and cyclic loading. Engineering Structures. 2006(28):1941–1955

Takesita A, Yoda T, Sato K, Sakurada M. Fatigue tests of a composite girder with corrugated web. *Proceedings of annual conference of the Japan society of civil engineering*, vol. 52, 1997, p. 122–3, (in Japanese).

Safety, Reliability and Risk of Structures, Infrastructures and
Engineering Systems – Furuta, Frangopol & Shinozuka (eds)
© 2010 Taylor & Francis Group, London, ISBN 978-0-415-47557-0

FE analysis of the influence of column axial load on the cyclic behavior of bolted endplate joint

Z.Y. Wang
Department of Civil Engineering, University of Nottingham, Nottingham, UK
Department of Civil Engineering, Sichuan University, Chengdu, Sichuan, P.R. China

W. Tizani
Department of Civil Engineering, University of Nottingham, Nottingham, UK

Q.Y. Wang
Department of Civil Engineering, Sichuan University, Chengdu, Sichuan, P.R. China

ABSTRACT

Recent design methods of steel-framed structure assume beam-to-column connection as one of the most important structural elements since the reliability and efficiency of the joint have great influences not only on the structural components but also on the behavior of the whole structure. Since the end of 20th century, bolted endplate joint has been paid increasing attention as a type of improved beam-to-column connection in seismic area by both industry and academia. As an important point to study the ductility the joint possesses, many experimental and numerical studies have been carried out to study cyclic behavior of iso-lated bolted endplate joint configuration. However, few researches were reported on the cyclic behavior of bolted endplate joint with sufficient consideration of the simultaneous presence of column axial load.

The aim of the present study is to obtain insights into the influence of column axial load on the behavior of bolted endplate joint under cyclic loading. The application of finite element modeling will explore related variables and potential failure modes, which could complement the experimental studies. A series of ANSYS elastic-plastic large deflection finite element analyses (FEA) have been carried out in the present paper. The monotonic behavior of the column flange plate analytical model subjected to combined bending and axial loading was examined at first to provide related evidence for following numerical analysis model of joint, and the findings were compared with the relevant equations. Then the analytical model of bolted endplate joint was investigated and tests were carried out to study the cyclic behavior of this component of bolted endplate joints subjected to column axial loading. The verification of the numerical results with experiment results is also reported.

Based on the numerical analysis results, this paper demonstrates how the column axial load affect the moment–rotation hysteresis loops, energy dissipation capacity and failure modes of the joint subjected to cyclic loading. It shows that the cyclic behavior of bolted endplate joint seems to be significantly influ-enced especially under high column axial load levels. An explanation for this phenomenon is also made with the comparison of the corresponding theory and experiment.

REFERENCES

Ahmed. B & Nethercot. D. A. Effect of column axial load on composite connection behavior. *Engineering Structures*, Vol. 20, Nos 1–2, pp. 113–128, 1998

Bahaari, M.R. & Sherbourne,A.N. 1997. 3D simulation of bolted connections to unstiffened columns-II: extend end-plate connections. *Journal of Structural Engineering*, Vol. 40, No. 3, pp. 189–223, 1997

CEN, Eurocode 3, prEN-1993-1-8, 2003, Part 1: General rules. Eurocode 3: design of steel structures. Stage 49 draft, CEN, European Committee for Standardization, Brussels, Belgium

CEN, Eurocode 3, prEN-1993-1-8, 2003, Part 3: Design of joint. Eurocode 3: design of steel structures. Stage 49 draft, CEN, European Committee for Standardization, Brussels, Belgium

De Lima, L.R.O; Simoes da Silva, L.; Vellasco, P.C.G. da S.; de Andrade, S.A.L. Experimental evaluation of extended endplate beam-to-column joints subjected to bending and axial force. *Eng Struct.* 2004; 26: 1333–47.

European Convention for Constructional Steelwork, Recom-mended Testing Procedures for Assessing the Behaviour of Structural Elements under Cylic Loads. Technical Com-mittee 1. TWG 1.3-Seismic Design, Publ. No. 45, 1986, p. 12.

Grecea. D, Stratan .A, Ciutina. A, Dubina. D, Rotation capac-ity of MR beam-to-column joints under cyclic loading. In: *Connections in Steel Structures V – Amsterdam – June 3–4, 2004*

Mazzolani. F.M: *Moment resistant connections of steel frames in seismic areas*, E&FN SPON Press. 2000. pp. 102–105

Zepeda. J. A, Itani. A.M., Sahai. R. Cyclic behavior of steel moment frame connections under varying axial load and lateral displacements. *Journal of Construction Steel.* 59:(2003); 1–25

Safety, Reliability and Risk of Structures, Infrastructures and Engineering Systems – Furuta, Frangopol & Shinozuka (eds)
© 2010 Taylor & Francis Group, London, ISBN 978-0-415-47557-0

Estimation of failure probabilities by Markov chain simulation-based approximate method

Hongzhe Dai & Wei Wang
Harbin Institute of Technology, Harbin, China

ABSTRACT

In order to predict the failure probability of a complicated structure, the structural responses usually need to be estimated by a numerical procedure, such as finite element analysis. When the performance functions are implicit, such calculations require additional efforts and might be time-consuming. To reduce the computational effort required for reliability analysis, the support vector machine (SVM) method has been developed recently to cope with the issues with implicit performance functions. The main role of the SVM is simply a classifier though it has been applied in structural reliability community successfully, on the other hand, the training samples for the SVM, as published so far, are mostly generated randomly around the basic random variables. This is not appropriate for many cases, especially for the problems of small failure probability. Because only with the enough large number of samples, can the results achieve the satisfied accuracy. Therefore, the method is computationally expensive.

To overcome the above issues in the present methods, this paper develops an approximate method combing the Markov chain simulation (Metropolis 1953, Au & Beck 1999) and the support vector regression algorithm (Hurtado 2004). In the proposed method, the training samples are firstly simulated as the states of a Markov chain, an approximate limit state function (LSF) is then re-constructed from these samples via support vector regression machine, the Monte Carlo simulation method is finally adopted to compute the failure probability for the above approximate LSF. The purpose of the Markov chain is to generate the training samples for SVM which can adaptively distributed these sample points in the region of higher probability density in the failure domain, while the SVM is to produce a surrogate of the finite element solver that provides the value of the LSF. The computational cost of the proposed method comes mainly from the simulation of the Markov chain samples to re-construct the approximate LSF and this is proportional to the number of function calls of the indicator function $I_f(\mathbf{x})$. In order to decrease the Markov chain samples needed to construct the LSF and thus increase the computational efficiency of the proposed method, this paper introduces an adaptive Metropolis algorithm (Haario et al 2001) to generate the Markov chain. This adaptive Markov chain simulation method is based on the concept that the adaptive Metropolis algorithm can effectively increase the rate of convergence of the classical Metropolis-Hasting algorithm and thus can greatly decrease the number of function calls of the indicator function $I_f(\mathbf{x})$.

The numerical examples show that the re-constructed LSF using the above Markov chain samples can well approximate the true LSF around the design point and thereby can be used as an alternative as the real one. In addition, the proposed method can achieve comparable accuracy when compared with traditional method with much fewer samples, indicating that it can provide accurate and computationally efficient estimates of failure probability. Therefore, the proposed Markov chain-based approximate method qualifies as a comprehensive tool in structural reliability analysis.

REFERENCES

Au, S.K. & Beck, J.L. 1999. A new adaptive importance sampling scheme for reliability calculations. *Structural Safety* 21: 135–158.

Haario, H. & Saksman, E. & Tamminen, J. 2001. An adaptive Metropolis algorithm. *Bernoulli* 7: 223–242.

Hurtado, J.E. 2004. *Structural reliability: statistical learning perspectives*. Heidelberg: Springer.

Metropolis, N. & Rosenbluth, A.W. et al. 1953. Equations of state calculations by fast computing machines. *Journal of Chemical Physics* 21: 1087–1092.

Safety, Reliability and Risk of Structures, Infrastructures and Engineering Systems – Furuta, Frangopol & Shinozuka (eds)
© 2010 Taylor & Francis Group, London, ISBN 978-0-415-47557-0

A study on design traffic live load for a cable-stayed bridge based on the vehicle axle load survey

Tsutomu Nishioka, Yoshihei Horie & Masatsugu Shinohara
Hanshin Expressway Company Limited, Osaka, Japan

Osamu Aketa
Sogo Engineering Incorporated, Osaka, Japan

ABSTRACT

A new design traffic live load for a cable-stayed bridge is proposed in this study. A probabilistic model of the traffic live load is developed by the vehicle axle load survey. Monte Carlo simulation is conducted to get the expected values of the annual peak response forces and their traffic live load. It turns out that the range of the safety indices among structural members by the proposed load is narrower than that by the conventional one, that means the proposed load is more reasonable than the conventional one from the viewpoint of the reliability design.

ACKNOWLEDGEMENT

This study is carried out in commission of Hanshin National Road Works Office, Kinki Regional Development Bureau, Ministry of Land, Infrastructure and Transport. We deeply express our appreciation to them.

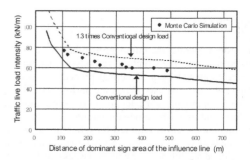

Figure 2. Relationship between the traffic live load intensity and the distance of dominant sign area of the influence line.

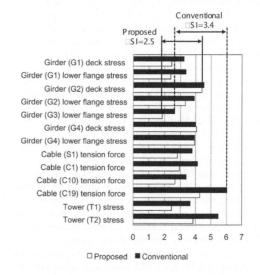

Figure 3. Comparison of the safety indices between the conventional and the proposed design load.

REFERENCES

Hanshin Expressway Public Corporation (HEPC), 1986. The research report of the design load of Hanshin Expressway. 36–86. Osaka.

Honshu-Shikoku Bridge Authority (HSBA), 1995. The specifications of bridge superstructures. Kobe.

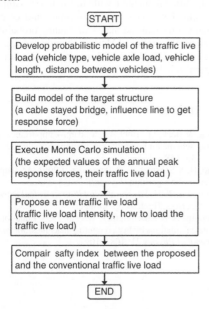

Figure 1. The flow chart of this approach.

Safety, Reliability and Risk of Structures, Infrastructures and Engineering Systems – Furuta, Frangopol & Shinozuka (eds)
© 2010 Taylor & Francis Group, London, ISBN 978-0-415-47557-0

An efficient Bayesian method for taking into account the modeling uncertainties in the evaluation of structural reliability

F. Jalayer, L. Elefante, I. Iervolino & G. Manfredi
Department of Structural Engineering, University of Naples Federico II, Italy

ABSTRACT

A significant portion of the total seismic risk in Italy, evaluated in economic terms, comes from the damages endured by the existing buildings. As a result, more recent Italian seismic guidelines (e.g., OPCM, NTC) pay particular attention to seismic assessment of existing structures, which is distinguished from that of the new construction by lack of information about both the original features and the current state of building. These modeling uncertainties can be classified into two groups; the uncertainty in the mechanical properties of the construction materials and the uncertainty in the structural detailing (a.k.a. the defects). This work employs the Bayesian probability framework in order to make robust estimation of the structural reliability and its dispersion (mean and standard deviation) based on a limited number of structural analyses. The resulting probabilistic assessments of the structural performance are compared to the corresponding assessments based on simulation-based reliability methods and extensive structural analyses. In order to evaluate the overall effect of modeling uncertainties on reliability assessment of existing buildings, the structural modeling uncertainties are taken into account together with the uncertainties in the ground motion representation.

REFERENCES

Ministero delle infrastrutture DM 14 gennaio 2008. Norme Tecniche per le Costruzioni (NTC), Gazzetta Ufficiale della Repubblica Italiana, 29, 14 gennaio, 2008 (in Italian).

Ordinanza del Presidente del Consiglio dei Ministri (OPCM) n. 3274, Norme tecniche per il progetto,la valutazione e l'adeguamento sismico degli edifici, Gazzetta Ufficiale della Repubblica Italiana n. 105 del 8-5-2003 (Suppl. Ordinario n. 72), 2003 (in Italian).

Safety, Reliability and Risk of Structures, Infrastructures and
Engineering Systems – Furuta, Frangopol & Shinozuka (eds)
© 2010 Taylor & Francis Group, London, ISBN 978-0-415-47557-0

An alternative to Monte Carlo simulations for dynamic problems

J. Huh
Chonnam National University, Yeosu, Korea

A. Haldar
University of Arizona, Tucson, AZ, USA

ABSTRACT

A time domain nonlinear reliability analysis technique
is presented to estimate reliability of dynamic systems
as an alternative to the classical Monte Carlo simulations (MCS) technique. Variance-reduction techniques
(VRTs) are generally used to improve efficiency in
MCS by altering the sampling methods, correlation
methods, and special methods. By utilizing a special
input scheme or sampling method, a new simulation-
based reliability evaluation method is proposed. It is
ideally suitable for the reliability analysis of nonlinear
dynamic systems represented by finite elements. In
this class of problems, the limit state or performance
functions are implicit in nature and also function of
time. To approximately generate the implicit performance function in the failure region, the response
surface method (RSM), the stochastic finite element
method (SFEM) proposed by the authors, and the
first-order reliability method (FORM) are integrated.

The damage to steel structures caused by the
Northridge earthquake of 1994 is a major motivation behind this study. To mitigate the problem, in
post-Northridge steel connections, flexibility is being
introduced in many different forms. It introduces a
major source of energy dissipation but it is also a major
source of nonlinearity. To improve efficiency in the
deterministic finite element method to consider flexibility of connections, the stress-based finite element is
used in the mathematical formulation. In the assumed
stress-based finite element method, the tangent stiffness can be expressed in explicit form, the stresses of
an element can be obtained directly, fewer elements
are required to describe a large deformation configuration, and integration is not required to obtain the
tangent stiffness.

Moment rotation (M-θ) curves are generally used
to consider flexibility in the connections. In this study,
the Richard four-parameter moment-rotation model is

used to represent loading behavior. The unloading and
reloading parts of the M-θ curves are theoretically
constructed using the Masing rule.

Two types of second order polynomial are used
for the response surface. Saturated design and central
composite design (CCD) are used for the experimental sampling points. To generate the response surface
in the failure region, it is linked with FORM. The initial center point is assumed to be the mean values of
the random variables for the first iteration. By conducting nonlinear FEM, the responses are calculated
using saturated design for a second order polynomial without cross terms; and a limit state function is
generated. Using the explicit expression for the limit
state function and FORM, the reliability index β, the
corresponding coordinates of the checking point are
obtained for each random variable. The coordinate of
the new center point is obtained by applying the linear
interpolation scheme. The updating of the center point
continues until it converges to a predetermined tolerance level. In the final iteration, the information on the
most recent center point is used to formulate the final
response surface using CCD with a full second order
polynomial. The FORM method is then used to calculate β and the corresponding most probable failure
point (MPFP).

To demonstrate the desirability and application
potential of the method, the reliability is estimated for
both serviceability and strength limit states. All the
design parameters including the four parameters of the
Richard model are considered to be random variables.
Two steel frames with different flexibilities of connections are then excited by actual recorded earthquake
time histories and the corresponding reliabilities are
estimated. The accuracy of the method was estimated
by calculating the corresponding risk using MCS of
100,000 cycles. It was shown that the proposed method
can estimate reliability with similar accuracy with only
about 100 to 200 cycles of simulation.

Mini-Symposia (MS04) Structural Reliability and Optimization

Safety, Reliability and Risk of Structures, Infrastructures and
Engineering Systems – Furuta, Frangopol & Shinozuka (eds)
© 2010 Taylor & Francis Group, London, ISBN 978-0-415-47557-0

Integration of reliability methods into a commercial finite element software package

I. Papaioannou, A. Düster & E. Rank
Chair for Computation in Engineering, TU München, Munich, Germany

H. Heidkamp & C. Katz
SOFiSTiK AG, Oberschleißheim, Germany

ABSTRACT

In order to motivate the incorporation of the reliability concept in the structural design and analysis procedure, the engineering community is in need of finite element (FE) software with capabilities of including the stochastic nature of input parameters.

In this paper, a reliability tool that is being developed as part of the SOFiSTiK FE software package is presented. SOFiSTiK is a FE program with particular emphasis on civil and structural engineering (SOFiSTiK AG 2008). The program is made up of a modular structure and provides communication by a very efficient database. The reliability tool is programmed in a stand-alone fashion. The tool controls the deterministic analyses by running the corresponding SOFiSTiK modules, while the data-exchange is achieved by interaction with the database (Fig. 1).

The reliability analysis may be assessed either by the first order reliability method (FORM) or by a variety of simulation techniques. In the FORM, an optimization problem is solved in order to find the minimum distance of the limit state function to the origin of the equivalent standard normal space (Liu & Der Kiureghian 1991). For the solution of this program, a series of optimization algorithms have been implemented. These include the standard HLRF method and its improved version, the sequential quadratic programming method and an improved version of the gradient projection method.

The implemented simulation methods may be divided into two categories. In the first, the sampling is done directly, i.e. without previous information. Such methods are the direct Monte Carlo method with Latin hypercube sampling (LHS) for a better variance reduction rate (Olsson et al. 2003), and the directional simulation (DS) (Bjerager 1990). The DS is further combined with smart techniques that select an optimum directional sample distribution (Nie & Ellingwood 2000). In the second category, initial calculations are used in order to reduce the variance of the produced samples. These include importance sampling (IS) based on the design point (Bjerager 1990), such as standard IS, axis orthogonal IS, radial IS and directional IS, and two adaptive techniques.

The reliability methods may also be combined with an approximation of the response surface, utilizing a polynomial meta-model (Faravelli 1989).

The paper gives a pair of geotechnical FE reliability analysis examples, to demonstrate the applicability of the tool to industrial problems.

REFERENCES

Bjerager, P. 1990. On computational methods for structural reliability analysis. *Structural Safety* 9: 79–96.

Faravelli, L. 1989. Response surface approach for reliability analysis. *Journal of Engineering Mechanics, ASCE* 115(12): 2763-2781.

Liu, P.-L. & Der Kiureghian, A. 1991. Optimization algorithms for structural reliability. *Structural Safety* 9: 161–177.

Nie, J. & Ellingwood, B. R. 2000. Directional methods for structural reliability analysis. *Structural Safety* 22: 233–249.

Olsson, A., Sandberg, G. & Dahlblom, O. 2003. On Latin hypercube sampling for structural reliability analysis. *Structural Safety* 25: 47–68.

SOFiSTiK AG 2008. *SOFiSTiK analysis programs version 23.0*. Oberschleißheim: SOFiSTiK AG.

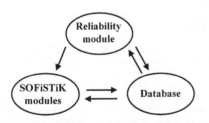

Figure 1. Communication between the reliability module and the SOFiSTiK program.

Safety, Reliability and Risk of Structures, Infrastructures and
Engineering Systems – Furuta, Frangopol & Shinozuka (eds)
© 2010 Taylor & Francis Group, London, ISBN 978-0-415-47557-0

Reliability-based topology optimization of frame structure using SLSV approach

Yutaka Hirano & Nozomu Kogiso
Osaka Prefecture University, Sakai, Osaka, Japan

Shinji Nishiwaki, Kazuhiro Izui & Masataka Yoshimura
Kyoto University, Kyoto, Japan

ABSTRACT

This research applies SLSV (Single-Loop-Single-Variable) method (Chen and Neill 1997) to reliability-based topology optimization (RBTO) problem for frame structure to improve computational efficiency. The design problem is formulated to minimize the structural volume of frame structure in terms of cross-sectional area of each frame element under the two reliability constraints. The two mode reliability criteria consist of the mean compliance and mean eigenfrequency under variations on applied loads and nonstructural mass.

The SLSV method is known as an efficient technique for decoupling the nested iteration loops consisting of the reliability analysis loop inside of the optimization loop by onverting the reliability constraint into equivalent deterministic one.

Through numerical example, the computational efficiency and the accuracy of the reliability approximation by SLSV method are demonstrated in comparison with those by the conventional double-loop method (Mogami and Kogiso 2006). The ground structure is shown in Figure 1, where the applied load and the nonstructural mass with normal distributed variations are applied at the center of the right-hand side. The reliability-based optimum configurations obtained by SLSV method under target reliability of $\beta_t = 3.0$ for the mean compliance criterion β_l and the mean eigenrequency criterion β_Λ is illustrated in Figure 2, that has almost the same volume with identified configuration as one obtained by the conventional double-loop method. The computational performance of the SLSV method is compared with the conventional double-loop method in Table 1. It is indicated that SLSV method has sufficient reliability approximation accuracy, regardless that the method evaluate the reliability only by approximation. Note that the reliability indices for the optimum design in Table 1 are evaluated by th first-order reliability method. It is also found that SLSV method significantly reduces the number of limit state function evaluations (NFE) and CPU time in comparison with the conventional double-loop method.

This result indicates that SLSV method has sufficient ability for practical RBTO problems.

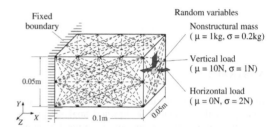

Figure 1. Ground structure of 3D frame.

Figure 2. Reliability-based optimum configuration ($\beta_t = 3.0$)

Table 1. Comparison of computational performance.

	SLSV	Double-loop
β_l	3.024	3.000
β_Λ	2.988	2.997
Volume (mm^3)	63581	63791
NFE (g_1)	616	6358
NFE (g_2)	616	10075
CPU time (sec.)*	1262	8424

* Pentium 4, 3.4 GHz

REFERENCES

Chen, X., Hasselman, T. K., and Neill, D. J. (1997). Reliability-Based Structural Design Optimization for Practical Applications. In *Proc. 38th AIAA SDM Conference*. AIAA-97-1403.

Mogami, K., Nishiwaki, S., Izui, K., Yoshimura, M., and Kogiso, N. (2006). Reliability-Based Topology Optimization of Frame Structures for Multiple Failure Criteria Using Topology Optimization Techniques. *Structural and Multidisciplinary Optimization 32*, 299–327.

Safety, Reliability and Risk of Structures, Infrastructures and Engineering Systems – Furuta, Frangopol & Shinozuka (eds)
© 2010 Taylor & Francis Group, London, ISBN 978-0-415-47557-0

Updating fragility estimates for reinforced concrete bridges using nondestructive testing

Q. Huang, P. Gardoni & S. Hurlebaus
Texas A&M University, College Station, TX, USA

ABSTRACT

Existing bridges in the US are suffering from aging and deterioration due to harsh environmental exposure conditions and/or are damaged by natural (earthquakes, hydrologic forces, etc.) or man-made (collisions, fire, road salt, etc.) hazards. With limited funds available for the maintenance of aging and degrading bridges, it is critical to evaluate the performance of existing bridges in the field such that the bridge owners can optimize the allocation of the resources for maintenance, repair, and/or rehabilitation of bridge systems. Accurately estimating the up-to-date in-place structural properties is a critical part to evaluate the bridge performance.

Compared with the widely used visual inspection, which gives a subjective assessment and is unable to detect hidden damages in a structure, vibration-based nondestructive testing (NDT) can be used to identify damage at an earlier stage in an objective manner. In this paper, we propose a methodology to identify the structural properties of a reinforced concrete (RC) bridge with modal data obtained from vibration responses using NDT. The structural properties are then used to update the shear and deformation fragility estimates of the bridge. Modal data can be easily extracted from a vibration testing, making the proposed approach of practical value.

Finite element model (FEM) updating is a straightforward method to determine the changes in the physical properties of a structure by calibrating or updating the structural parameters using vibration measurements. However, due to the computational cost and the inverse ill-conditioning problem, FEM updating is not practical for identifying damages at the element level. On the other hand, investigating the changes of structural vibration characteristics before and after damage can provide the needed information about the changes in the physical properties of a structure at the element level. In this case, a reliable baseline is critical for the successful performance.

In this study, a Bayesian model updating approach is used to develop the baseline structure using eigenfrequencies, and the damage index method (DIM), which is based on mode shape curvature changes, is used to identify the local damages. The proposed framework makes use of the advantages of these two techniques and compensates for their drawbacks. Additionally, two types of uncertainties can be easily accounted for: modeling error due to the inexactness in the computer-simulated model and the uncertainties in the structural properties of the real structure, and measurement error reflecting the uncertainties in the response measurement.

The proposed framework is illustrated using a simulated FEM of the Lavic Road Overcrossing Bridge, a typical box-girder RC highway bridge in California. Through this case study, we found that a Bayesian approach can effectively be used to update a preliminary FEM based on the frequencies of the first four bending modes of the bridge, providing a more accurate baseline for the DIM. With this baseline, the DIM successfully detects the change in the local stiffness. Finally, the shear and deformation fragility of the Lavic Road Overcrossing Bridge is computed using the updated structural properties.

Safety, Reliability and Risk of Structures, Infrastructures and Engineering Systems – Furuta, Frangopol & Shinozuka (eds)
© 2010 Taylor & Francis Group, London, ISBN 978-0-415-47557-0

Reliability-based vulnerability requirements in the design of ship panels for blast loads

S.J. Pahos & P.K. Das
Department of Naval Architecture and Marine Engineering, Universities of Glasgow and Strathclyde, Glasgow, UK

ABSTRACT

A computational methodology for defining Reliability-Based Vulnerability Requirements (RBVRs) using probabilistic Finite Element (FE) analysis is described in this paper. Previous reliability-based work on the response of typical stiffened panels from blast loads of a 4 kg explosive charge highlighted the most influential design parameters against the arisen stresses. The reliability index was found to be negative as an indication that typical ship panels found in merchant vessels are vulnerable against blast loads. The next step forward is to prudently increase these contributive strength variables, pointed out by the sensitivity analysis, in order to raise the reliability index, and subsequently reduce vulnerability. The challenge lies in finding the golden means of this increase so that the proposed methodology is cost-effective and operationally realisable. The geometric variation of the most contributive design strength variables is based on a rational increase from the inventory of modern shipyards. Manufacturing-induced distortions are also accounted in the probabilistic design. All the above are incorporated in a parametrically-defined macro file subjected to the probabilistic characteristics of strength and load variables. ANSYS probabilistic design analysis (PDA) is used as a basis for determining the response surface of the accounted designs. The blast effect from the detonation of 4 kg PETN Improvised Explosive Device was realised in LS-DYNA solver, while the PDA was realised with ANSYS implicit. Reliability assessment is performed with CALREL software and more precisely First Order Reliability Method. The reliability index of the stiffened panel, loaded by the impulse of 4 kg PETN charge at a selective range of standoff distances, was found to gradually increase as the panel is reinforced. The resulting response surface of the reliability index as a function of the most contributive variables was found in this work. The generated response surface allows for quick, yet reliable, estimates during the design stage regarding the vulnerability of stiffened panels against blast loads. It is aspired that the proposed methodology is to have a significant impact on establishing RBVRs for ship panels at an early stage of the design. This work is putting forward a methodology that can be used for developing future rules and requirements for asymmetric threats in the marine environment. Ultimately, this work gives a better appraisal of the vulnerability of typical ship panels as a step forward in enhancing safety at sea and fighting the global war on terror.

Safety, Reliability and Risk of Structures, Infrastructures and
Engineering Systems – Furuta, Frangopol & Shinozuka (eds)
© 2010 Taylor & Francis Group, London, ISBN 978-0-415-47557-0

Reliability-based optimization of flexible pavements accounting for deterioration and the effects of rehabilitation actions

Vighnesh P. Deshpande, Ivan D. Damnjanovic & Paolo Gardoni
Zachry Department of Civil Engineering, Texas A&M University, College Station, TX, USA

ABSTRACT

Flexible pavements represent the major part of the U.S. transportation infrastructure. It is estimated that flexible pavements comprise approximately 60 percent of the total paved public roads in the U.S., or approximately 500,000 miles (Highway Statistics 2005). With utilization and aging, the condition of pavements deteriorates. As a result, periodic rehabilitation and preventive maintenance actions are needed to sustain pavement functionality and safety.

In pavement engineering, preventive maintenance is a planned strategy of treatments. Examples of treatments are fog sealing, microsurfacing, crack sealing and other treatments designed to slow down the deterioration process and improve the functionality without increasing the pavement structural capacity. On the other hand, rehabilitation actions such as pavement overlays represent activities that increase pavement structural capacity.

One of the most important considerations in pavement design is the life-cycle cost. Life-cycle cost includes all the costs incurred during pavement life such as: initial construction cost, preventive maintenance costs, rehabilitation costs, as well as users' costs (e.g., time related, vehicle operating, safety, and environmental costs) (Abaza 2002).

Timely planned application of rehabilitation and maintenance actions is essentially a cost effective means for obtaining desired performance and service life for the pavement structures. Delaying the application of these actions beyond serviceability level requires more extensive, time consuming and costly actions, and makes the pavement section unusable for an extended time, thus causing inconvenience to public by disrupting the traffic. On the other hand, very early

application of maintenance and rehabilitation actions adds little or no benefits.

The performance of pavement structures is generally associated with large uncertainties that arise from two different sources: uncertainty in the pavement utilization and uncertainty in the pavement response. Hence, it is important to explicitly account for these uncertainties while developing optimal design strategies.

The concept of reliability has been used in pavement engineering applications to account for the probabilistic nature of the pavement performance. However, current models do not explicitly consider the effect of rehabilitation actions on the pavement reliability, which is an important shortcoming for their effective implementation in determining optimal rehabilitation strategies.

This paper develops a reliability-based optimization model that can be used for determining optimal rehabilitation strategies (time and design) for flexible pavements. Pavement fragility curves, defined as the conditional probability of failure given a specified level of demand, are used in the reliability analysis. To facilitate the use of the fragility curves in the optimization problem, a closed-from fragility model is developed to expresses the fragility estimates in terms of decision variables.

In this paper, a multi-objective reliability-based optimization model is developed. Since the developed model is nonlinear and non-convex, a multi-objective genetic algorithm (MOGA) is used for the optimization.

The results from a numerical example show that the developed optimization model can be efficiently used in determining the optimal rehabilitation strategies and cost-reliability tradeoffs.

Safety, Reliability and Risk of Structures, Infrastructures and Engineering Systems – Furuta, Frangopol & Shinozuka (eds)
© 2010 Taylor & Francis Group, London, ISBN 978-0-415-47557-0

Reliability-based optimized rating equation of deteriorating bridges using loading and corrosion data

Baidurya Bhattacharya
Department of Civil Engineering, IIT Kharagpur, India

Degang Li
Greenhorne & O'Mara, Inc., Laurel, MD, USA

Michael J. Chajes
Department of Civil Engineering, University of Delaware, Newark, DE, USA

ABSTRACT

When bridges are designed, the future service conditions and loads on it can only be estimated. Once put into service, a bridge is subjected to loading and deterioration processes, some of which may be out of the envelope considered in the design process. An accurate estimate for the remaining life and load carrying capacity of an inservice bridge is needed so that public safety is not compromised, and, at the same time, scarce resources are not unnecessarily spent in bridge repair.

Once built, the best model of the bridge is the bridge itself and inservice data should be used as much as possible. This paper presents a probability-based methodology for load rating bridges that can accommodate detailed site-specific in-service structural deterioration and response data in an LRFR (load and resistance factor rating) format.

A new model of gross section loss, C, due to corrosion occurring with an exponentiated Ornstein-Uhlenbeck type stochastic noise (η) is considered:

$$\frac{dC}{dt} = \begin{cases} 0 & , t \leq T_I \\ \beta(t-T_I)^\gamma \, e^{\eta(t)} & , t > T_I \end{cases} \tag{1}$$

where β, γ are random parameters of the model, T_I is the random corrosion initiation time, and the evolution of the Ornstein-Uhlenbeck process is given by:

$$\frac{d\eta(t)}{dt} = -k\eta + \sqrt{D}\,\zeta(t) \tag{2}$$

ζ is the Gaussian white noise, and k, D are drift and diffusion coefficients.

The proposed methodology also allows a stationary, dependent live load-effect sequence; the dependence is of a weakened mixing type that asymptotically decreases to zero with increasing separation. In addition, uncertainties in field measurement, modeling uncertainties and Bayesian updating of the empirical distribution function are considered to obtain an extreme value distribution of the time-dependent maximum live load.

An illustrative example utilizes in-service peak strain data from ambient traffic collected on a high-volume steel girder bridge. The limit state considered is plastic collapse of the critical girder cross-section during a specified service life. In-service load and aging resistance factors, optimized for an inventory of bridges, are developed for different service lives. The rating factors thus obtained can aid bridge owners to schedule the time and extent of future repairs more efficiently than allowed by the current standard practices.

REFERENCES

Bhattacharya, B. 2008. The Extremal Index and the Maximum of a Dependent Stationary Pulse Load Process Observed above a High Threshold. *Structural Safety* **30**(1): 34–48.

Bhattacharya, B., D. Li & M. J. Chajes 2008. Bridge rating using in-service data in the presence of strength deterioration and correlation in load process. *Structure and Infrastructure Engineering, Taylor and Francis*, **4**(3): 237–249.

NCHRP (1999). Manual for condition evaluation and load and resistance factor rating of highway bridges. Washington, D.C., Transportation Research Board, National Research Council.

Mini-Symposia (MS11) Current Status and Future Applications of Probabilistic Seismic Hazard Assessment

Safety, Reliability and Risk of Structures, Infrastructures and
Engineering Systems – Furuta, Frangopol & Shinozuka (eds)
© 2010 Taylor & Francis Group, London, ISBN 978-0-415-47557-0

Probabilistic seismic risk analysis for building portfolio

M. Wang
The University of Waterloo, Waterloo, Ontario, Canada

K. Tanaka & T. Takada
The University of Tokyo, Tokyo, Japan

ABSTRACT

Seismic risk management becomes more and more
important in the disaster prevention and mitigation for
local government. There are two kinds of risk, individual and aggregate. The former is basically applied to a
critical facility at a specific site such as nuclear power
plant, while the latter could be aggregate risk of multiple buildings either located at a single site or widely
distributed at the different sites. This study focuses on
the aggregate risk assessment for the building portfolio
at a single location instead of for spatially distributed
portfolio.

The site-specific risk analysis cannot directly be
applied to the aggregate risk analysis for certain technical problems such as the correlation of the ground
motion. The estimation of the probability of simultaneous failure, P_f, is of much concern in the building portfolio. The analytical expression of P_f for two building
is formulated. Suppose that the limit state function,
G, for a single building can simply be defined as
$G = \ln R - \ln S_A$ where R is the resistance of the building
and S_A is the response spectral acceleration. Assuming
both R and S_A are independent log-normal variables,
G follows normal distribution with parameter μ_G and
σ_G^2. As far as two buildings are concerned, the probability of simultaneous failure is equal to probability
that both G_1 and G_2 are smaller than 0, that is

$$P_f = \int_{-\infty}^{0} \frac{|\Sigma|^{\frac{1}{2}}}{2\pi} \exp\left[-\frac{1}{2}(\mathbf{g}-\mathbf{\mu_G})^{\mathrm{T}} \Sigma^{-1} (\mathbf{g}-\mathbf{\mu_G})\right] d\mathbf{G} \quad (1)$$

where $|\cdot|$ denotes the determinant,

$$g = \begin{bmatrix} g_1 \\ g_2 \end{bmatrix}, \mathbf{\mu_G} = \begin{bmatrix} \mu_{G1} \\ \mu_{G2} \end{bmatrix}, \Sigma = \begin{bmatrix} \sigma_{G1}^2 & \rho_G \sigma_{G1}\sigma_{G2} \\ \rho_G \sigma_{G1}\sigma_{G2} & \sigma_{G2}^2 \end{bmatrix},$$

and $\rho_G = \dfrac{\rho_{\ln R}\sigma_{\ln R1}\sigma_{\ln R2} + \rho_{\ln SA}\sigma_{\ln SA1}\sigma_{\ln SA2}}{\sigma_{G1}\sigma_{G2}}$. Equation

Equation 1 can easily be extended to more than two
buildings. Given the loss definition, the total risk can
be obtained in a straightforward manner.

As can be seen, the accuracy in the estimate of P_f
depends on the appropriate estimate of the ground
motion ($\mu_{\ln SA}$ and $\sigma_{\ln SA}$), correlation of the ground
motion ($\rho_{\ln SA}$) between the two different buildings, and
the correlation of the resistance (ρ_R) between the two

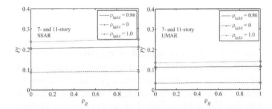

Figure 1. Comparison of P_f for 7-story and 11-story buildings with the ground motions predicted by SSAR and
UMAR.

buildings. In this paper, the site-specific attenuation
relationship (SSAR) for S_A is developed within the
Bayesian framework in light of the new observations
at the specific site. Based on the high-density observations from recent earthquakes, the correlation model of
S_A between the different periods is developed as well.

A numerical example of a building portfolio is illustrated for three steel buildings under an earthquake
scenario by incorporating the ground motion predicted
by SSAR, the correlation of ground motion ($\rho_{\ln SA}$) and
the correlation of resistance (ρ_R) Figure 1 shows the
P_f estimated for 7-story and 11-story buildings. The
fragility curves for moderate damage are adopted from
Nakamura & Nakamura (2001).

The following observations are notable: 1)
the P_f using SSAR ground motion is larger
than that using ground motion obtained from
the Uchiyama-Midorikawa attenuation relationship
(UMAR, Uchiyama & Midorikawa 2006); 2) the
P_f with consideration of the correlation of ground
motion lies between those using perfectly correlated
and independent ground motion; 3) the P_f increases
with increasing ρ_R, however, the effect of ρ_R is
slight because the uncertainty is predominated by the
uncertainty of the ground motion.

REFERENCES

Nakamura, T. and Nakamura, T. 2000. Study on a characteristic of earthquake motion and logarithmic standard deviation of seismic fragility curve (SFC) based
on dynamic response analysis. *Summaries of Technical
Papers of Annual meeting, AIJ* B-1: 39–40.

Uchiyama, Y. & Midorikawa, S. 2006. Attenuation relationship for response spectra on engineering bedrock considering effects of focal depth. *Journal of Structural and
Construction Engineering, AIJ* 606: 81–88 (in Japanese
with English abstract).

Safety, Reliability and Risk of Structures, Infrastructures and
Engineering Systems – Furuta, Frangopol & Shinozuka (eds)
© 2010 Taylor & Francis Group, London, ISBN 978-0-415-47557-0

Is PSHA an option for earthquake early warning?

I. Iervolino

Dipartimento di Ingegneria Strutturale, Università degli Studi di Napoli Federico II, Naples, Italy

ABSTRACT

Due to a large development of regional networks in
recent years worldwide, and because of the current
advances of real-time seismology, the question of
using earthquake early warning systems (EEWSs) for
site-specific applications (Figure 1) is rising. Hybrid
EEWS' are of current interest as cost-effective solutions for seismic risk mitigation, although efficiency
evaluation and feasibility analysis for earthquake engineering applications is still debated.

Seismologists have recently developed several
methods to estimate the magnitude (M) of an event
given limited information of the P-waves for real-
time applications. Similarly, the source-to-site distance (R) may be rapidly determined by analyzing the time and order of the seismic stations
detect the developing earthquake. Consequently, given
a vector of measures informative for the mag-
nitude, $\{\tau_1, \tau_2, \ldots, \tau_n\}$, and the sequence of stations triggered by the event, $\{s_1, s_2, \ldots, s_n\}$, the
probability density functions (PDFs) of M and
R, $f(m|\tau_1, \tau_2, \ldots, \tau_n)$ and $f(r|s_1, s_2, \ldots, s_n)$ respectively,
may be available. Thus, it is possible to compute in real-
time the probabilistic distribution (or hazard curve) of
a ground motion intensity measure at a site of interest
(Iervolino et al., 2006) as in Eq. (1), which also requires
an attenuation relationship, $f(im|m, r)$, available for
the chosen IM.

Nevertheless, such prediction involves significant
uncertainty and therefore effectiveness of EEW for
engineering applications requires proper assessment.

This paper presents a review of the work of the
author and co-workers about the real-time adaption of
probabilistic seismic hazard analysis or PSHA (Cornell, 1968) and of the performance-based earthquake
engineering framework.

The procedure for early warning predictions of engineering ground motion parameters is presented first,
then some issues related to the involved uncertainty
and its influence on the alarm issuance decision (e.g.,
false/missed alarm occurrences) are discussed. Secondly, a prototypal terminal based on real-time PSHA
and able to issue the alarm for the EEWS under development in the Campania region (southern Italy) is
shown. Finally, it is discussed how to set EEW alarm
thresholds, for structures/infrastructures, based on the
expected loss conditional to the information provided
in real-time by the seismic network.

REFERENCES

Cornell CA. (1968). Engineering seismic risk analysis. *Bull Seism Soc Am*, 58:1583–606.
Iervolino I., Convertito V., Giorgio M., Manfredi G., Zollo A. (2006). Real-time risk analysis for hybrid earthquake early warning systems. *Journal of Earthquake Engineering*, 10:867–885.

$$f_n(im) = \iint_{m \, r} f(im|m,r) \; f(m|\tau_1, \tau_2, \ldots, \tau_n) \times$$
$$\times f(r|s_1, s_2, \ldots, s_n) \, dr \, dm \tag{1}$$

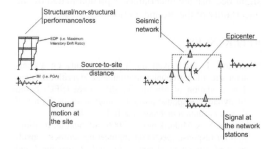

Figure 1.   Hybrid EEWS sketch (Iervolino et al., 2006).

Safety, Reliability and Risk of Structures, Infrastructures and
Engineering Systems – Furuta, Frangopol & Shinozuka (eds)
© 2010 Taylor & Francis Group, London, ISBN 978-0-415-47557-0

Prediction of inelastic structural response using an average of spectral accelerations

M. Bianchini & P.P. Diotallevi

*Department of Structural, Transport, Hydraulic, Survey & Territory Engineering, University of Bologna,
Bologna, Italy*

J.W. Baker

Department of Civil & Environmental Engineering, Stanford University, Stanford, CA, USA

ABSTRACT

Performance-Based Earthquake Engineering (Cornell
& Krawinkler 2000) aims to quantify the seismic reliability of a structure at a site. For that purpose, Probabilistic Seismic Demand Analysis (PSDA, Shome et
al. 1998) quantifies the mean annual frequency of
exceeding a specified value of the significant seismic
response of a building, known as engineering demand
parameter, EDP. Usually, PSDA is decoupled in the
ground motion hazard and nonlinear dynamic analyses
(such as incremental) through an intermediate variable
known as the ground motion intensity measure, IM,
which quantifies the characteristics of a record that are
important to predict the inelastic structural response.
Conventional scalar IMs are the peak ground acceleration, PGA, and the pseudo-spectral acceleration at the
first-mode period, $S_a(T^{(1)})$, or simply S_a.

Some desirable properties are needed to define
an optimal IM in PSDA of inelastic systems, such
as hazard computability, efficiency, sufficiency and
scaling robustness. Since IM defines also the seismic hazard curve at a given site, the availability of
ground motion prediction model and the simplicity in
performing the analysis for a given IM leads to the hazard computability property. However, one of the most
important desirable properties consists in the efficiency of IM, i.e., in reducing the standard deviation
of the $\ln(IM_{cap}|EDP)$, which, in turn, reduces the num-
ber of nonlinear dynamic analyses preserving the same
accuracy in seismic performance estimation (Fig. 1).

Other desirable properties of an optimal IM are
the sufficiency and the scaling robustness. The first
property requires that the probability distribution of
the EDP given an IM is conditionally independent of
the other ground motion parameters (e.g., magnitude,
distance, ε, etc.), whereas the latter implies there is
no statistically strong relationship between the EDPs
and the scale factors used in scaling the amplitude of
records.

Since structural response of inelastic multi-degree-
of-freedom systems is sensitive to multiple period T_i,
then an IM which averages elastic spectral acceleration

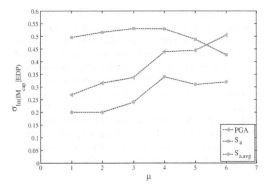

Figure 1. Comparison between the standard deviation
of $\ln(IM_{cap}|EDP)$ for several median ductility capaci-
ties assuming PGA, S_a and $S_{a,avg}$ as IM. The inelastic
multi-degree-of-freedom system has an elastic fundamental
period equal to 1.8 s and a number of storeys of 18.

values over a certain range of periods might be a useful and convenient predictor of structural response of
inelastic systems. The present work shows the desirable IM properties of $S_{a,avg}(T_1, \ldots, T_n)$, or briefly
$S_{a,avg}$, to be used in PSDA (Bianchini et al., in prep.).
$S_{a,avg}$ is defined as the geometric mean of a certain
number of $S_a(T_i)$ and it is applied to demand assessment of inelastic multi-degree-of-freedom systems. It
was found that $S_{a,avg}$ can be used as a useful and practical predictor of structural response compared with
other scalar-based IMs.

REFERENCES

Bianchini, M., Cornell, C.A. & Baker, J.W. A ground motion
intensity measure based on an average of spectral accelerations. *Earthquake Engng. Struct. Dyn.* (in prep.).
Cornell, C.A. & Krawinkler, H. 2000. *Progress and challenges in seismic performance assessment.* Berkeley:
PEER Center News.
Shome, N., Cornell, C.A., Bazzurro, P. & Carballo, J.E.
1998. Earthquakes, records, and nonlinear responses.
Earthquake Spectra 14(3): 469–500.

Safety, Reliability and Risk of Structures, Infrastructures and Engineering Systems – Furuta, Frangopol & Shinozuka (eds)
© 2010 Taylor & Francis Group, London, ISBN 978-0-415-47557-0

Efficient approach to vector-valued probabilistic seismic hazard analysis of multiple correlated ground motion parameters

P. Bazzurro, P. Tothong & J. Park

AIR Worldwide Corporation, San Francisco, CA, USA

ABSTRACT

In recent years, the site-specific seismic hazard analysis has become a common tool in the engineering community to determine the likelihood that ground motions of different amplitude may be observed at a designated site. Conventionally this computation is only done for a scalar ground motion intensity measure–such as peak ground acceleration or spectral acceleration–although the methodology for computing the joint (i.e., vector-valued) hazard has been proposed almost 10 years ago. The knowledge of joint hazard at a site is valuable in many applications where more than one ground motion parameters is needed for predicting a structure's response. However, the joint hazard analysis has not yet been adopted in the engineering seismology community mainly due to the lack of a computational tool to perform the vector-valued seismic hazard analysis (VPSHA). This article presents a VPSHA methodology that requires only the manipulation of results from the scalar PSHA as opposed to performing a direct integration of the jointly normal distribution, as originally proposed. This alternative VPSHA approach is not only appealing because bypasses the need of writing a VPSHA code but also because it can be used to compute the hazard of a pool of ground motion parameters larger than that allowed by the original method.

Safety, Reliability and Risk of Structures, Infrastructures and
Engineering Systems – Furuta, Frangopol & Shinozuka (eds)
© 2010 Taylor & Francis Group, London, ISBN 978-0-415-47557-0

Construction of live management system for highway networks in a seismic disaster emergency

A. Kawamoto
Aratani Construct Consultant Inc., Tottori, Japan

W. Shiraki
Kagawa University, Kagawa, Japan

K. Yasuda
Newjec Inc., Tokyo, Japan

N. Ito
CAE Inc., Tottori, Japan

M. Dogaki
Kansai University, Osaka, Japan

ABSTRACT

In Japan, with the sudden occurrence of the natural disaster such as the earthquake, heavy rain fall, sediment disaster, etc., social infrastructure facilities frequently catastrophic damaged, so that much of human life and huge fortune were lost. It is fear that the disastrous earthquake named Toukai, Tounankai, and Nankai will occur more than 50% of high probabilities within 30 years in the future. Therefore, each local government that will be damaged already proceeded to consider the emergency response plan mainly in the urban area which is a densely populated region.

In general, the emergency response plan is composed of the three steps such as emergency response, emergency compatible and the restoration and revival, and then the contents in each step are organized in time series. Among these, the emergency response which makes a hit within 1 day from the earthquake occurring is the most important phase which does the activity of the reduction and the prevention of the lifesaving, the after damage. Immediately after earthquake occurring, an epicenter, the depth of the epicenter, the scale of the earthquake and earthquake intensity distribution at each place are released from the Meteorological Agency. If the reliability assessment of the highway network in the target area is quickly done at this point, the response at the initial stage can be done for reduction of the damage. In such plans, however, it is not treated various damage scenario patterns with considering the lost of performance of the social infrastructure facilities such as bridges and roads.

In early studies, based on the live design concept, we proposed a new strategic maintenance approach for protection against disasters management of existing social infrastructure stock as well as for long-range management at normal service time.

In this paper, based on this live management[1]–[3], the safety evaluation of road-networks for feasible routes from a fire station to two hospitals and one harbor in Akashi City which is a typical satellite town near Kobe City is examined immediately after earthquake using a simulation method. As the result, three safety road-network routes are selected as important routes at an emergency, and the effective emergency countermeasures are considered. The effectiveness of the live management approach is demonstrated by the example.

REFERENCES

A. Kawamoto, W. Shiraki, K. Yasuda, N. Ito and M. Dogaki: Concept of live design for maintenance of infrastructures, Proc. of Safety Problems, Japan Society of Civil Engineers, Vol.1, pp.73–78, 2006.11 (Japanese).

A. Kawamoto, W. Shiraki, K. Yasuda, N. Ito and M. Dogaki: A proposal of constructing live design database (LDDB) for infrastructural maintenance, the domestic symposium (JCOSSAR2007) collected papers, Vol.6, pp. about the safety and the reliability at the sixth building197–202, 2007.6(Japanese).

A. Kawamoto, W. Shiraki, K. Yasuda, N. Ito and M. Dogaki: Construction of live management system for highway networks in a seismic disaster emergency, the structural engineering collected papers, Japan Society of Civil Engineers, Vol.54A, pp.152–161, 2008.3(Japanese).

Safety, Reliability and Risk of Structures, Infrastructures and Engineering Systems – Furuta, Frangopol & Shinozuka (eds)
© *2010 Taylor & Francis Group, London, ISBN 978-0-415-47557-0*

Evaluation of seismic performance rate of each structural element of highway bridge system using Multi state system approach

H. Morisaki
Pacific Consultants Co., Ltd., Osaka, Japan

S. Okawa
Kansai University, Osaka, Japan

W. Shiraki
Kagawa University, Kagawa, Japan

N. Ito
Kansai University, Osaka, Japan CAE Inc., Tottori, Japan

M. Dogaki
Kansai University, Osaka, Japan

ABSTRACT

Authors are trying to apply MSS[1] to seismic design of the bridge, and to evaluate the influence degree that the state of performance demonstrating of an individual structural element that composes the bridge and the performance demonstrating of an individual structural element cause in the state of the performance demonstrating of the entire bridge quantitatively. In this study, as a second stage of study of weight putting of the damage extent between structural elements, the structure analysis that changes the performance of structural element (bridge shoe) of the bridge system designed with present design code[2] is executed. And, the influence level that the performance of bridge shoe causes for performance of entire system is considered. Moreover, it proposed the method of setting the earthquake-proof performance demonstrating rate to the entire bridge of each structural element.

The entire analysis model is shown in Fig. 1. Superstructure and substructure were modeled by beam element. A nonlinear characteristic was evaluated with plasticity spring given to base of pier.

Analysis condition was assumed to be 4 cases that changed the rigidity of shoe. (Case 0~Case 3)

Here, performance rate of individual structural element was defined as regularized numerical value of individual structural element displacement by ultimate displacement of individual structural element "This performance rate shows a relative performance to the entire bridge system of an individual structural element." (Refer to Table1)

Here, displacement coefficient is value that ultimate displacement of each structural element is regularized by

Table 1. Performance rate table.

Case	Pier	Shoe on abutment	Falling prevention device
Case 0	0.81	0.43	0.01
Case 1	0.85	0.59	0.01
Case 2	0.86	0.71	0.07
Case 3	0.87	0.73	0.08

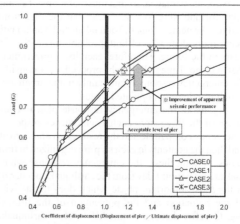

Figure 2. Load – Coefficient of displacement curve (case.0~case.3).

the ultimate displacement of pier, and the value indirectly expresses performance rate of entire bridge.

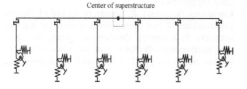

Figure 1. Analytical model chart.

REFERENCES

A. Lisnianski, and G. Levitin, "Multi-state System Reliability - Assessment, Optimization and Applications - " (2003) World Scientific, Publish-Ing Co., Ltd.

Japan Road Association, "Specifications for highway bridge part?; Seismic design", (2002).

Organized Session (OS02) Probabilistic Design of Wind Turbines

Safety, Reliability and Risk of Structures, Infrastructures and Engineering Systems – Furuta, Frangopol & Shinozuka (eds)
© 2010 Taylor & Francis Group, London, ISBN 978-0-415-47557-0

Extrapolation of extreme response for wind turbines based on field measurements

H.S. Toft
Aalborg University, Denmark

J.D. Sørensen
Aalborg University, Denmark
Risø-DTU, Denmark

ABSTRACT

The characteristic loads on wind turbines during operation are among others dependent on the mean wind speed, the turbulence intensity and the type and settings of the control system. These parameters must be taken into account in the assessment of the characteristic load.

The characteristic load is normally determined by statistical extrapolation of the simulated response during operation according to (IEC 61400-1 2005). However, this method assumes that the individual 10 min. time series are independent and that peaks extracted by the Peak Over Threshold method are independent. In the present paper two new methods for loads extrapolation are presented.

The first method (method 1) is based on the same assumptions about independence of the individual extremes and the 10 min. time series as the existing method. However, for this method the statistical extrapolation is only performed for a limited number of mean wind speeds where the extreme load is likely to occur. The wind speed for which the response can start to become critical is denoted the storm wind speed and determined from the nominal wind speed for the wind turbine and the turbulence intensity.

The second method (method 2) for load extrapolation divides the 10 min. mean wind speeds into storms which are assumed independent. The assumption about independence is secured by adding a time separation between the storms and combined two storms into one if the mean wind speed between the storms is not below 80% of the storm wind speed. The characteristic load is determined by statistical extrapolation of the extreme load in each storm.

RESULTS

In order to compare the existing method in (IEC 61400-1 2005) with the two new methods proposed in this paper the characteristic load is calculated for the same wind turbine using the three different methods. The calculated characteristic loads are given in table 1 where the loads are normalized with the characteristic load calculated for IEC 61400-1 without statistical uncertainty.

For the first method (method 1) the calculated characteristic load without statistical uncertainty is approximate 3% higher than using (IEC 61400-1 2005). For the second method (method 2) the characteristic load without statistical uncertainty is approximate 11% higher. The significant increase in the characteristic load by using this method can be due to that the assumptions of independence of the extremes and the 10 min. time series in the (IEC 61400-1 2005) method are not satisfied.

Based on the present study it is recommended that load extrapolation for wind turbines during operation is performed by the second method where the characteristic load is determined based on the extreme loads in each storm. The advantage of this method is that the load extrapolation is performed based on independent extremes leading to a statistically more accurate determination of the characteristic load. The drawback of the method is the large amount of measurement or simulations required in order to perform the load extrapolation.

Table 1. Characteristic loads for the three methods.

Method	Characteristic load without stat. unc.	with stat. unc.
IEC 61400-1	1.000	1.037
Method 1	1.025	1.162
Method 2	1.106	1.348

Safety, Reliability and Risk of Structures, Infrastructures and
Engineering Systems – Furuta, Frangopol & Shinozuka (eds)
© 2010 Taylor & Francis Group, London, ISBN 978-0-415-47557-0

Generation of synthetic turbulence in arbitrary domains

L. Gilling & S.R.K. Nielsen
Department of Civil Engineering, Aalborg University, Aalborg, Denmark

N.N. Sørensen
*Wind Energy Division, National Laboratory for Sustainable Energy, Risø-DTU,
Technical University of Denmark, Roskilde, Denmark*

ABSTRACT

A new method for generating synthetic turbulence is
presented. The method is intended for generating a
turbulent velocity field with a fine spatial resolution
in a domain covering a small moving part of the rotor
area of a wind turbine. An example of such a domain
is shown in Figure 1.

To generate the synthetic turbulence field shown in
the figure by the methods of Mann (1994, 1998) or
Veers (1984) the domain should cover the entire rotor
area. To obtain a fine resolution in the area of interest a
fine resolution should be used in the entire rotor plane.
This would require an unfeasible number of points in
the resolved domain.

In the method of this paper the synthetic turbulence
is generated in a number of points that are allowed to
move in time. Thereby, the domain in Figure 1 only
consists of 16 points moving in time. This approach
gives a large saving in required computer memory,
which allows generating turbulence fields with fine
resolution.

The proposed method can generate synthetic turbulence with the correct correlation in space and the
correct auto-spectrum. Like the Sandia method by
Veers it is based on one-dimensional Fourier transform of correlated auto-spectra for each point in the
resolved domain. The main difference from the Sandia method is that here the points in the domain are
allowed to move in time.

The three-dimensional cross-correlation tensor of
the frozen turbulence field is determined from an
analytical or empirical expression. By the Wiener-
Khinchin relation the cross- and auto-spectra are
determined for all points in the resolved domain. The
spectral information is factored and multiplied by
Gaussian complex random numbers before it is Fourier
transformed to give the time dependent velocity field.

The presented method reduces to the Sandia method
with an alternative coherence function for applications

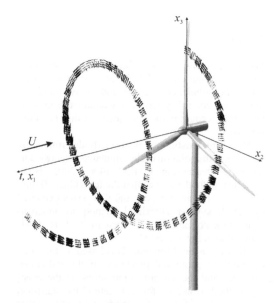

Figure 1. Inflow turbulence for a large eddy simulation of
a section of a rotating blade. Each of the cones represents a
velocity vector in the frozen turbulence field. The resolution
here is only 4×4 in the $x_2 - x_3$-plane.

where the full turbulence field covering the entire rotor
area is needed.

REFERENCES

Mann, J. 1994. The spatial structure of neutral atmospheric
surface-layer turbulence. *J. fluid mech.* 273. 141–168.
Mann, J. 1998. Wind field simulation, *Prob. engng. mech.*
13(4) 269–282.
Veers, P.S. 1984. *Modeling stochastic wind loads on vertical axis wind turbines.* SAND83-1909, Sandia National
Laboratories, Albuquerque.

Safety, Reliability and Risk of Structures, Infrastructures and
Engineering Systems – Furuta, Frangopol & Shinozuka (eds)
© *2010 Taylor & Francis Group, London, ISBN 978-0-415-47557-0*

Risk-based inspection and maintenance planning optimization of offshore wind turbines

J.G. Rangel-Ramírez & J.D. Sørensen
Aalborg University, Aalborg, Denmark

ABSTRACT

Wind power installations have become the second largest contributor to installation of electricity capacity in the European Union during the last decade. With this increase in production capability and size, technical and economical efforts should be intended for an optimal and suitable life-cycle planning of their components, assuring an acceptable risk state. The deterioration processes such as fatigue and corrosion are typically affecting offshore structural systems. The damage decreases the system performance, thus not fulfilling the established safety criteria. To control this deterioration, the inspection/maintenance activities are developed, representing the most relevant and effective means of control. The Risk-based inspection planning (RBI) methodology, based on Bayesian decision theory, represent an important tool to identify the suitable strategy to inspect and control the deterioration in structures such as wind turbines. During the last decades (Madsen et al. 1987 and Thoft-Christensen and Sørensen 1987) RBI approach has been applied to the oil and gas industry, giving a theoretical background that can also be applied for offshore wind turbines (OWT). Unlike other structures, OWTs represent low risk of human injury allowing allocation of a lower reliability level.

In water depths of about 20 m to 50 m, the use of jacket and tripod structures represents a feasible option that improves technical aspects concerning structural redundancy, damage distribution and dynamical behavior. An important OWT part is the transition node between the jacket or tripod and the tubular tower. The transition node is a critical design element, needing special careful design concerning the fatigue performance. Offshore Wind farms need additional technical efforts. Wake effects, coming from the decrease of wind velocity behind OWT, increase the turbulence resulting in decrease in OWT fatigue life. The turbulence intensity represents an important aspect to consider because of its effect on OWT's fatigue life. In this work, a code-based model (Frandsen, S. 2005 and Sørensen et al. 2007) is used for the efficient standard deviation of wake turbulence. A framework for optimal inspection and maintenance planning is applied for OWT, addressing the analysis of fatigue prone details in cast iron and welded steel at the jacket or tripod steel support structures. In wind park location and single/alone OWT are taken into account by using a turbulence model. For RBI planning the fracture mechanics model is usually calibrated to result in the same reliability level as the code-based SN model and inspection plan obtained with a maximum acceptable annual probability of failure. The results show earlier inspections times coming out in-wind farm sites due to the increase of fatigue coming from wake turbulence. It is noted that in all cases the design parameters are determined by deterministic design such that the design criteria is exactly satisfied. This inspection optimization approach represents a viable method to outline inspection plans aimed at OWT, regarding its application to large structural systems.

REFERENCES

Benjamin J.R., Cornell C.A., 1970. Probability, statistics and decision for civil engineering. McGrawHill, New York.

Frandsen, S., 2005. Turbulence and turbulence-generated structural loading in wind turbine clusters, Risø National Laboratory, Denmark, Report R1188.

Madsen, H.O., Skjong R., Kirkemo F., 1987. Probabilistic Fatigue Analysis of Offshore Structures – Reliability Updating Through Inspection Results. In Integrity of Offshore Structures 3, edited by Faulkner D, Cowling MJ, Incecik A, Elsevier Applied Science.

Raiffa H., Schlaifer R., 1961. Applied Statistical Decision Theory. Cambrige University Press, Cambridge, Mass.

Sørensen, J.D., Frandsen, S. & Tarp-Johansen, N.J., 2007. Fatigue reliability and effective turbulence models in wind farms. Applications of statistics and Probability in Civil Engineering – Kanda, Takada & Furuta (Eds).

Thoft-Christensen P., Sørensen J.D., 1987. Optimal strategy for inspection and repair of structural systems. Civil Engineering Systems, 4, 94–100.

Safety, Reliability and Risk of Structures, Infrastructures and Engineering Systems – Furuta, Frangopol & Shinozuka (eds)
© 2010 Taylor & Francis Group, London, ISBN 978-0-415-47557-0

Setting the frame for up-scaled off-shore wind turbines

J.D. Sørensen
Aalborg University & Risø DTU, Denmark

T. Chaviaropoulos
CRES, Greece

P. Jamieson
Garrad Hassan, UK

B.H. Bulder
ECN, The Netherlands

S. Frandsen
Risø DTU, Denmark

ABSTRACT

Wind turbines with a rated power of 5-6 MW are now being designed and produced. Within the EU supported UpWind research project (UpWind 2008) a cost model is being developed for up-scaling of wind turbines up to 20 MW. These wind turbines are expected to have a rotor diameter of approx. 250 m and a hub height of approx. 150 m. The optimal design of wind turbines can be obtained using a life-cycle approach where all relevant benefits and costs are included. Further, a constraint on a maximum acceptable probability of failure is in general added. Offshore wind turbines are characterized by a low risk of human injury in case of failure when compared to onshore wind turbines, and to civil engineering structures in general. It is therefore relevant to assess the optimal design on the basis of minimizing the total life-cycle costs of the turbine (farm) without a reliability constraint. Especially for offshore wind turbines costs related to operation and maintenance can be significant and have to be included. One reason is that maintenance can only be performed under certain weather conditions, which affects the availability of the system strongly.

A theoretical framework for risk-based optimal design of large wind turbines is developed. Three levels of formulations are considered: 1) a risk/reliability-based formulation, 2) a deterministic, code-based formulation and 3) a crude deterministic formulation. These formulations are described in the paper.

In the crude, deterministic formulation generic models for the costs are formulated directly as function of the design parameters and using basic up-scaling laws adjusted for technology improvement effects. The optimal design is obtained as the design which minimises the total expected costs per MWh (levelised production costs). The main design parameters are selected as the rotor diameter, the hub height, the tip speed and the wind turbine separation (in wind farms). In a more detailed modelling the following parameters could be added to the list of design parameters: cross-sectional dimensions defining geometry of blades, tower, . . .), operation & maintenance (O&M) strategy, etc. Further, the following design conditions are fixed for a given site: wind farm size (in terms of MW and / or geographic area of wind farm), wind climate and terrain: mean wind speed & turbulence, wave and current climate (offshore), water depth, soil conditions and distance from land and nearest harbour.

The costs model is formulated on the basis of a life-cycle approach including capitalised costs to planning, fabrication, installation, operations & maintenance, inspection & repair and demolition. The main up-scaling parameter is typically the rotor diameter. The cost models are basically formulated as function of this design parameter using an up-scaling factor with an up-scaling exponent (3. for example) and a technology improvement factor.

An important overall design decision is to find the optimal balance between a relatively expensive initial design with low failure rates of the main components (blades, gearbox etc.) and therefore low O&M costs, and a relatively inexpensive initial design with higher failure rates and thus high O&M costs. Further, there is also a decision to find the optimal ratio of corrective and preventive maintenance.

The paper presents the overall framework for cost-optimal up-scaling of offshore wind turbines, and illustrative examples are given.

Safety, Reliability and Risk of Structures, Infrastructures and
Engineering Systems – Furuta, Frangopol & Shinozuka (eds)
© 2010 Taylor & Francis Group, London, ISBN 978-0-415-47557-0

Numerical estimation of fatigue life of wind turbines due to shadow effect

P. Thoft-Christensen, R.R. Pedersen & S.R.K. Nielsen
Department of Civil Engineering, Aalborg University, Aalborg, Denmark

ABSTRACT

Due to the stagnation of the wind field upwind of
tower, the blades will meet a smaller lift during the
tower passage, which gives rise to periodic stress
variations. Along with the turbulent and aeroelastic
oscillations this is a major reason for fatigue of the
blades. The width of the stagnation zone depends on
the diameter of the tower. Hence, if the diameter can
be reduced the accumulated damage (crack growth,
debonding, delamination, etc.) will decrease accordingly. In this paper the impact on damage accumulation
is sought reduced with an alternative tower design, see
Figure 1.

The tower is designed as a tripod structure with several tubes of smaller radii $R_1 = R_2 = R_3$ and a mutual
distance $2h = 8\,\text{m}$, compared to the mono tower with
a radius R_1. The mean damage evolution of a blade is
determined and compared to the case, where a standard
tubular mono tower is used. The expected damage at
the blade hub is estimated, where a lifetime of 20 years
is assumed. Figure 2 shows the geometry of a single
rotating blade of a horizontal axis up-wind turbine with
a rotational velocity $\Omega = 1.26\,\text{rad/s}$. The structural
model of blade and tower is based on low-dimensional
modal analysis, where the blade characteristics are
taken from a 5 MW reference wind turbine, where
$L = 65\,\text{m}$ and $H = 88\,\text{m}$. The applied turbulence model
is based on a frozen turbulence model translated into
the rotor area by the mean wind.

The expected damage value for the blade is lower for
the tripod system compared to the same analysis using
a mono-tower. In this analysis, the structural design of
the tripod tower is not considered. Furthermore, only
one wind direction is included. However, the analysis suggests that the alternative tower design shows
promise for improving the fatigue life of the blade.

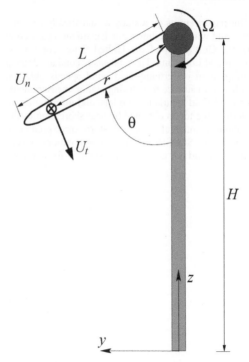

Figure 2. Tower and rotating blade with the wind velocity
components U_t and U_n.

Table 1. Expected damage values.

Tower system	Expected Damage E(D)
Mono tower	0.251
Tripod tower	0.014

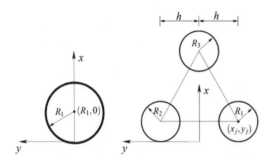

Figure 1. Principal sketch of the two tower configurations.

Safety, Reliability and Risk of Structures, Infrastructures and
Engineering Systems – Furuta, Frangopol & Shinozuka (eds)
© 2010 Taylor & Francis Group, London, ISBN 978-0-415-47557-0

Modeling the joint wind-surge hazard due to hurricanes in the Gulf of Mexico

Y. Wang & D.V. Rosowsky
Zachry Department of Civil Engineering, Texas A&M University, College Station, TX, USA

W. Pang
Department of Civil Engineering, Clemson University, Clemson, SC, USA

ABSTRACT

This paper reports on a study to statistically describe the joint wind-surge hurricane hazard along a section of the Gulf Coast of the United States comprising Texas and Louisiana. The analysis is conducted using information on ground-level wind speeds obtained either from surface monitoring station measurements or wind field simulation models, and information on storm surge levels obtained by near shore monitoring stations. Numerical models for storm surge prediction (such as the SLOSH model) are computationally very intensive and are therefore not suitable for real-time emergency management purposes. With the increasing availability of accurate surge measurements, however, a joint wind-surge model can be defined (e.g. piecewise along the coastline) using regression analysis. The joint wind-surge hazard is considered herein as it varies from the point of landfall of the hurricane and by location along the section of Gulf Coast considered. Good comparison is observed between the storm surge predictions based on the SLOSH model and the wind-surge model developed herein. Finally, possible applications of joint hazard model for infrastructure design and assessment are discussed.

Organized Session (OS03) Construction Risk and Safety Management

Safety, Reliability and Risk of Structures, Infrastructures and
Engineering Systems – Furuta, Frangopol & Shinozuka (eds)
© 2010 Taylor & Francis Group, London, ISBN 978-0-415-47557-0

Research on safety management system in construction work

T. Hojo
Monotsukuri Institute of Technologists, Saitama, Japan

K. Ohdo
National Institute of Occucpational Safety and Health, Tokyo, Japan

ABSTRACT

This paper analyzes an example of a construction accident, examines the background which results in an accident, and adds consideration to a safety management system. It is the Westgate Bridge collapse accident in Australia which was taken up as an example of a construction accident in this paper. At the time of 1970 bridge construction, a box girder carried out compression buckling, and then collapsed, and many casualties came out.

The accident investigating committee was established in the state government immediately after the accident, and investigation about the burden sharing of cause investigation of a technical side and each organization which result in the accident was conducted. Firstly, the outline of an accident was introduced, the technical factor was analyzed from the contents of the investigation report. Subsequently, the measure circumstances to construction of each organization were considered.

As a result, it is being argued that it is the leading cause of an accident that the shortage of examination of the structural design of the consultant who took charge of bridge designing, and the directions to the construction contractor about the construction method was not enough. When these causes are analyzed, it should be considered a factor with more major indirect factors, such as problem of a system and each organization, which generated the technical problem.

There are many points which should be learned from this accident investigation or subsequent correspondence. The direct cause of this accident is buckling collapse by the shortage of girder rigidity when installing, but lack of cooperation between an authority and a designer, and a constructor is considered as the potential factor actually. As a result of analyzing the factor, it was shown rather than the technical factor which is a direct cause of an accident in case of this accident that the indirect human factor by the contract, an organization, governing structure, etc. has had serious influence. Even if the indirect human factor is hidden under the direct human factor which is actually in sight, it is rather big and often important

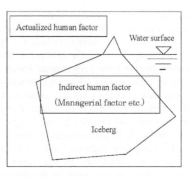

Figure 1. Concept of Human Factor.

problem. This relation is shown in Fig. 4. In case of this accident, it can be said that an indirect human factor becomes clearer than a direct human factor.

Finally, it is shown the collapsed girders are saved and exhibited within the campus in the Monash University, and it is considering as an aid of the educational activities to the student and engineer who study civil engineering. About such an accident example, analysis is performed from a viewpoint of not only a technical side but management, and it is thought that it is a useful method to try the device efficiently employed in future education in order to improve safety in the engineering field, and it will be needed increasingly from now on. It is considered to be an effective method to analyze the past accident and disaster example, and to use for safety education as the example.

REFERENCES

Hojo, T., Ohdo, K. and Maekawa, Y. 2007. Research on Changes in Safety Management Techniques, *Proceedings of '07 panel discussion on Safety Problems, Japan Society of Civil Engineers*: CD-ROM.

Hojo, T. 2005. Safety Management in Construction Engineering Education, *Proceedings of the XVIIth World Congress on Safety and Health at Work*: CD-ROM.

Parliament of Victoria. 2000. Report of Royal Commission into the Failure of West Gate Bridge.

Safety, Reliability and Risk of Structures, Infrastructures and
Engineering Systems – Furuta, Frangopol & Shinozuka (eds)
© 2010 Taylor & Francis Group, London, ISBN 978-0-415-47557-0

Target reliability index determination in risk share consensus making

R. Katade & S. Katsuki

Department of Civil and Environmental Engineering, National Defense Academy, Yokosuka, Japan

ABSTRACT

Recently, the structural design codes in Japan are changing drastically to the performance based design system. There are some interpretations for "performance based design" among the civil engineers. Most of engineers interpret the performance based design as the design harmonizing some requirement performance, e.g. safety, initial cost, life cycle cost, environmental performance and so on. Some of engineers consider that the performance based design system states clearly the hold performance of the structure supplied by the public organizations from the view point of their accountability. In this sense, one of the typical backgrounds is given by the performance matrix proposed by SEAOC Vision2000. The others interpret the performance based design as an application of design and/or supply system of the globally standardized structural design concept represented by ISO2394, i.e. the reliability based design system.

In the reliability based design especially level I or II, the target reliability index plays an important part of design and/or design manual. There are some methods to determine the target reliability index. Most popular and easy method to apply for developing new design code is the code calibration method in which the target reliability index should be adjusted the equivalent safety level of existent structures designed by conventional design codes. The fact that the safety level of new design manual is equivalent to the conventional one may be strongly accountable to the citizen, but not necessarily convenient for giving the citizen a safety or reliability level selection opportunity corresponding to the significance of the structure as shown in ISO2394 or Vision2000.

This paper presents a new idea to determine the target reliability index from the risk share concept between user, e.g. tax payer, and the engineer, e.g. in-house engineer of national or local government as a code writer. The total risk of structural failure can

be divided into two parts. First one, which relates the fact that structure will fail against the particular load exceeding the contracted or defined level of design process, is shared by user. Second one, which relates the fact that the structure will fail against smaller load which is defined in design code, should be responsible to engineer. The proposed method to determine the reliability index is strongly corresponding to the second risk.

Based on this concept, the following inverse function to determine the target reliability index from the shared risk and optimal shared risk corresponding to total cost are presented in this paper.

$$\beta_T = g_1\left(P_{sf}, P_{fA}\right) - \frac{g_3\left(P_{sf}, P_{fA}, \theta\right)}{g_2\left(P_{sf}, P_{fA}, \theta\right)}$$

$$g_1\left(P_{sf}, P_{fA}\right) = 1.57\widetilde{P}_{sf} + 3.93 P_{fA}$$

$$g_2\left(P_{sf}, P_{fA}, \theta\right) = 1.06\widetilde{P}_{sf}{}^2 + 6.03 P_{fA}{}^2 + 2.91\theta^2 + 24.8\widetilde{P}_{sf} P_{fA}$$
$$- 21.4 P_{fA}\theta - 3.1_2\widetilde{P}_{sf}\theta + 1.87\widetilde{P}_{sf} + 14.1 P_{fA} + 1.6\theta - 0.44$$

$$g_3\left(P_{sf}, P_{fA}, \theta\right) = -0.25\widetilde{P}_{sf}{}^2 - 4.47 P_{fA}{}^2 + 1.5\theta^2 + 1.85\widetilde{P}_{sf} P_{fA}$$
$$- 1.74 P_{fA}\theta + 2.75\widetilde{P}_{sf}\theta + 1.5\widetilde{P}_{sf} + 4.38 P_{fA} - 6.51\theta + 2.92$$

where, β_T: target reliability index, P_{sf}: warranty liability occurrence probability, P_{fA}: contractual probability, θ: angle of limit state line, $\widetilde{P}_{sf} = \sqrt{-\ln P_{sf}}$.

The decision making process of the target reliability index of a check dam is discussed by using proposed method.

REFERENCES

SEAOC Vision 2000. 1995. *Performance based seismic engineering of buildings.*
Takeshi Nagao. 2000. Reliability based design way for caisson type breakwaters, *Proc. of JSCE*, No.689/I-57: 173–182.

Safety, Reliability and Risk of Structures, Infrastructures and Engineering Systems – Furuta, Frangopol & Shinozuka (eds)
© 2010 Taylor & Francis Group, London, ISBN 978-0-415-47557-0

Difference between the stated purpose and actual practice of safety education at construction sites

Michiyuki Hirokane
Kansai University, Osaka, Japan

Katutoshi Ohdo
National Institute of Occupational Safety and Health, Tokyo, Japan

Shigeo Hanayasu
Yokohama National University, Kanagawa, Japan

Yasuhiro Kamada
Graduate School of Kansai University, Osaka, Japan

ABSTRACT

Recently, safety education has become more important. Though the attitude towards, and measures for, safety education depend on the individual company, many workers feel that their safety education is superficial at best. We analyzed the results of a questionnaire survey, "Safety Education at Construction Sites," that asked workers who are in charge of safety at their company about their safety education. The answer for Question 6, a question that asks workers about the degree of difference between the practice and stated purpose of safety education at their company, let one applicable item choose from 4 items, "Yes", "Yes, relatively", "No, I don't feel the difference so much", "No". Ten questions in Section 2 of this survey ask workers about their practice of safety education. We extracted the relationship between the degree of difference they feel and the practice.

This survey contains a multi-response type question that let workers select multiple answers items from six items. To deal with the background information of multi-response data with fewer variables in C4.5, we used a Boolean approach and simplified all the 38 combinations of answers into 7 terms.

We applied C4.5 in two ways. One used six variables standing for whether each item in the multi-response typed question was selected or not, and seven variables, stand for whether each term in the simplified expression applied to the answer pattern or not.

In the former method, 13 rules are extracted, and three rules resulted in "Yes" and 8 rules resulted in "Yes, relatively." For example, an extracted rule shows the reflection of qualifications to their salary and allowance while they must pay their expanse to participate third party seminar. The rule results in "Yes". Another rule also has reflection of qualifications to their salary and allowance, but resulted in "Yes, relatively." This rule shows that their company has their own safety educational program and pays their expense to participate third party training. Therefore, it is possible for companies to make the perceived difference smaller by providing proper support to their workers.

In the latter method, 10 rules are extracted, but most rules resulted in "No." For example, a rule that resulted in "No." shows that the company includes acquirement of qualification in their educational training, but do not promote it as a company and do not order their workers participate third party seminar. In short, there is a tendency that workers do not feel a difference if they have high motivation to acquire third party training even if the company was not supportive.

Overall, t here is a tendency that promoting safety educations makes workers feel difference between their stated purpose and actual practice. On the other hand, there are some cases wherein the degree of difference workers feel would change if answer patterns included the same factor into themselves. It is important to select their contents of safety education that suit for their construction sites.

It appears that there are also some cases that make workers feel a difference when their practice is not solid enough; there are also some cases where workers do not feel a difference if they have high motivation. These cases suggest that it is important for safety education to keep worker motivation high and support them as well as enrichment of their action.

REFERENCES

Misumi, K. 2002. The Logic of Multi-responses in Social Research: Exploration by Boolean Approach. Bulletin of the Graduate School of Social and Cultural Studies, Kyushu University Vol. 8: 57–65.

Japan Society of Civil Engineers. 2005. The Report of the Survey about Safety Education in Construction Sites.

Quinlan, J.R. 1993. C4.5: Programs for Machine Learning. Morgan Kaufmann Publishers, Inc.

Safety, Reliability and Risk of Structures, Infrastructures and Engineering Systems – Furuta, Frangopol & Shinozuka (eds)
© 2010 Taylor & Francis Group, London, ISBN 978-0-415-47557-0

Study on mitigation of fall risk from scaffolds in construction industry

K. Ohdo, Y. Toyosawa, S. Takanashi, Y. Hino & H. Takahashi
National Institute of Occupational Safety and Health, Kiyose, Japan

ABSTRACT

Fall accidents are a serious problem in the construction industry in Japan, and approximately 40% of fatal accidents during construction are caused by workers' falls. Therefore, the Japanese Ministry of Health, Labour and Welfare established a committee in our institute for considering the countermeasures.

The committee's work experimentally confirmed the effectiveness of using plastic sheets as a covering around scaffolds to protect against falls (a method widely used in Japan) by the human dummy. In all of the experimental cases, the human dummy did not fall from the scaffolds, and it was found that the plastic sheets were effective for fall protection, given a perfectly installed sheet.

However, in some cases, the dummy almost fell from the scaffolds. Then, the space between the plastic sheet and the work platform of the scaffolds was spread widely, and the dummy had a possibility of falling from the space.

The ideal sheet installation method is difficult to apply at all construction sites. The falling space in the case of typical installations is larger than that in the case of an ideal installation. Therefore, the risk of falling into the space is not small, when the typical installation method of the sheet is used.

The plastic sheets are usually bound to the pipes of the scaffolds by fiber ropes. However, the ropes often loosen, creating a space that tends to be spread open by the worker's weight. Alternatively, the plastic sheets are occasionally bound to the scaffold pipes by a special coupler. This method appears to make it possible easily to reduce the opening of the space, but the effect has not yet been fully explored.

Therefore, the effectiveness of the special coupler was confirmed experimentally in preventing the space from spreading open easily. In the experiment, the strengthening effect of the plastic sheets was also confirmed, and a small piece of sheeting was piled with the plastic sheets to add strength. Table 1 shows the experimental cases in this study.

Figure 1 shows the results of the experiment. The results were compared with the statistical data for the hip depth of Japanese males from the ages of 18 to 59 from Japanese body size data 1992-1994 (Research

Table 1. Experimental cases.

Experimental cases	Condition
Case 1	The plastic sheets bound to the scaffolds by fiber ropes.
Case 2	The plastic sheets bound to the scaffolds by special couplers.
Case 3	The plastic sheets bound to the scaffolds by special couplers with strengthening from a small piece of sheeting.

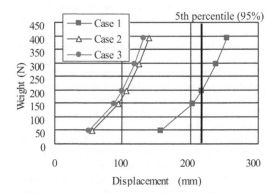

Figure 1. Experimental results.

Institute ... 1997)). From Figure 1, more than 95% of persons might be protected from falling using these installation methods by Case 2 and 3. Therefore, it was found that these installation methods are effective in preventing the spread of the space between the work platform and the plastic sheets, and it is possible to easily reduce the risk of falling from the space with this method.

REFERENCES

Research Institute of Human Engineering for quality Life 1997. Japanese body size data 1992–1994. Tokyo: Research Institute of Human Engineering for quality Life.

Safety, Reliability and Risk of Structures, Infrastructures and Engineering Systems – Furuta, Frangopol & Shinozuka (eds)
© 2010 Taylor & Francis Group, London, ISBN 978-0-415-47557-0

Risk assessment for scaffolding work in strong winds

S. Charuvisit
Civil Engineering Department, Kasetsart University, Bangkok, Thailand

K. Ohdo, Y. Hino & S. Takanashi
National Institute of Occupational Safety and Health, Kiyose, Japan

ABSTRACT

Construction accidents caused by wind are one type of serious accident that has a high ratio of accidents, more than 10%, compared to other kinds of serious accident, both in Japan and overseas. For accidents caused by wind, accidents related to scaffolding-based work are important and need to be seriously concerned as they include many types of accidents such as scaffold collapse, worker falls, etc. Based on the data of "the disaster investigation report", "the construction industry safety health and sanitation year book" and the previous study of Phongkumsing et al. 2002, many workers are injured or die every year from accidents related to strong winds acting on scaffolds.

To prevent these accidents, this study aims to evaluate the risk of scaffolding-based work under strong wind conditions by reviewing the experimental results of wind acting on the scaffolds both from wind tunnel and field measurement, and conducting a reliability analysis to examine the transferred loads on the scaffold support ties to a building. Wind tunnel experiments were conducted on the scaffold scale models to study the wind pressures acting on the scaffolds, and on the actual scaffolds to study the fall risk of work conducted on the scaffolds. Field measurement was conducted on actual scaffolds to measure wind forces on the scaffolds surface and transferred loads to the support ties.

The experimental results of the scale models show that the pressure acting on the interior surface of the scaffolds (the side of the scaffolds that faces the building) caused either by wind entering the gap between the scaffolds and building, or passing through wall openings, is a significant factor in scaffolds stability. Moreover, the wind direction that caused the most severe case of wind pressures acting on the scaffolds is found to be other direction that is not normal to the scaffolds surface due to shape, size and openings of the scaffolds and the building.

From the field measurement, the transferred loads at the scaffolds support ties to the building were compared with the computational results from the FEM analysis, and the system probability of failure was analyzed for each wind direction. It is found that the surface loads acting on the scaffolds by wind were not properly and entirely transferred to the scaffold support ties due to some scaffold assembly errors, and these errors reduce the system reliability by increasing the probability of failure more than 10 times. Thus, the wall opening and the construction errors should be considered in the design code for wind loads and overall safety analysis of scaffolds.

Finally, the wind tunnel experiments to study the fall risk of assembling and dismantling works on the actual scaffolds show the wind speed limits of such works exposed to uniform and gust flow. Those wind speed limits were evaluated by grading experienced Japanese workers' perception of risks. From the results of wind speed limits, they were compared to the results of previous study from Murakami et al. (1982) on the limits of other works exposed to uniform flow. As the wind speed limits obtained from this study are lower as about 8 m/s, it can be implied that the scaffolding works have a higher risk than other works such as walking and carrying load, changing of posture.

REFERENCES

Engineering News-Record 1988-2001 (weekly). New York: McGraw-Hill.
Japan Construction Safety and Health Association 1989-2002 (annual). *Construction industry safety and health yearbook.* Tokyo: Japan Construction Safety and Health Association (in Japanese)
Murakami, S., Deguchi, K. and Takahashi T., 1982. Acceptable Wind Speed for Windy Working Environment High Above Ground, Institute of Industrial Science, 34(3): 21–24
Phongkumsing, S., Ohdo, K. and Hino Y., 2002. Investigation on Labor Accidents caused by Wind, Proceedings of the 32nd National Symposium on Engineering Safety, No. Sh-3

General Session (Damage Analysis and Assessment)

Safety, Reliability and Risk of Structures, Infrastructures and Engineering Systems – Furuta, Frangopol & Shinozuka (eds)
© 2010 Taylor & Francis Group, London, ISBN 978-0-415-47557-0

Effects of climate change on cyclone damage estimation

Yue Li

Department of Civil and Environmental Engineering, Michigan Technological University, Houghton, MI, USA

Mark G. Stewart

Centre for Infrastructure Performance and Reliability, School of Engineering, University of Newcastle, Callaghan, NSW, Australia

ABSTRACT

Numerous global warming and climate change studies warn that the built environment will be affected by climate change through rising sea levels and altered patterns of natural hazards. This has led to a growing interest in the potential impact of climate change on building damage from predicted changes in climatology of tropical cyclones over the next 100 years. The Intergovernmental Panel on Climate Change (IPCC 2007) predicts a warming of the climate with associated increases in the severity and or frequency of wind hazards, namely, tropical cyclones (hurricanes). Recent studies also suggest that tropical cyclone intensities are likely to increase in the next 50-100 years as a result of warmer ocean temperatures. Cyclone intensity can be described by the maximum wind speed closes to the surface, which can be directly related to design wind speed and damage estimation.

The paper proposes a risk-based framework for assessment of economic damage caused by tropical cyclones due to the increases in wind speeds resulting from climate change. In addition to the direct losses, cyclones cause tremendous social disruption for extended periods of time. The framework contains the following key ingredients – probabilistic modeling of the occurrence and intensity of cyclone, time-dependent increase in annual maximum wind speed from global warming, and vulnerability function to represent the economic risks and losses conditioned in wind speed. Regional dynamics including increasing house numbers built with different building codes is integrated into the framework.

Residential construction in North Queensland in Australia is chosen to demonstrate the potential impact of climate change on wind damage estimation from a regional perspective. The worst-case scenario of a 25% increase in cyclone intensity by 2050 will result in almost triple the total insured damage for North Queensland which will comprise approximately 250,000 houses by 2050. In comparison, an increase

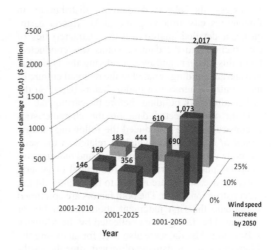

Figure 1. Regional cumulative damage losses in North Queensland with different time references.

in wind speed of 10% will cause 56% addition damage losses for the same region (Figure 1)

The proposed framework provides an effective tool for the insurance industry, policy makers and emergency planners for risk-based decision making on strategies need for climate change adaptation. The decisions include whether to adopt high wind standard building codes or mandatory retrofit during house transactions to minimize the potential impacts of climate change. This framework can form the basis for improvements in construction practices and building codes to adapt to potential impacts of climate change, appropriate underwriting by the insurance industry, and sociopolitical response of the community to wind hazards.

REFERENCES

IPCC 2007. Intergovernmental Panel on Climate Change.

Safety, Reliability and Risk of Structures, Infrastructures and Engineering Systems – Furuta, Frangopol & Shinozuka (eds)
© *2010 Taylor & Francis Group, London, ISBN 978-0-415-47557-0*

Evaluation of buildings quality and soil condition in Boumerdes city using damage data following the 2003 Algeria earthquake

A. Meslem, F. Yamazaki & Y. Maruyama

Department of Urban Environment Systems, Chiba University, Chiba, Japan

ABSTRACT

To assess the observed damage distribution in Boumerdes city following the 2003 Algeria earthquake, a detailed analysis on the characteristics of construction and the damage status was conducted. From this analysis and by considering the typicality of Algerian buildings related to the regional culture of the country, a number of major factors based on the real conditions of buildings before the earthquake.

Using the collected data on the characteristics of construction we introduced the following factors: *period of construction*, which is related to the seismic code version that has been used; and *building use category*: public or private for residential or general activities use, which are related to the quality during the construction phase. The both factors are combined to evaluate the first judgment about the quality of building. The damage data collected by the Ministry of Housing, Algeria, were also used for analyzing the different causes of damage observed after the earthquake. The observed damage categories of buildings have been taken into account as a factor to correlate the first judgment on the quality of building, based on expert opinion. The influence of described factors has been observed from the experience of the past earthquakes. The building quality was defined in the similar manner as the damage grade. This study has been conducted for a total of 2794 buildings existed in the city of Boumerdes.

Five classes were used to determine the building quality as follows: Good, Acceptable, Medium, Poor, and Very Poor). The distribution of building quality in the city of Boumerdes is shown in Figure 1. The results from this analysis show that the buildings associated with poor and very poor qualities are mostly private buildings. The constructions with good and acceptable qualities are mostly for public buildings. The examination on the relationship between the building damage and building quality (Figure 2) showed some variation at several locations, which is considered due to soil condition. Microtremor observation

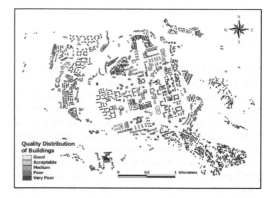

Figure 1. Distribution of buildings with respect to the quality classification in Boumerdes city.

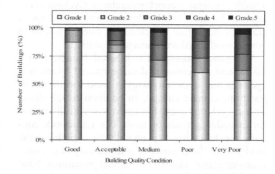

Figure 2. Relationship between of building quality and observed damage grade.

was conducted at several locations in the city. The measured H/V ratios showed flat shape at most of the locations with small amplitude, having small peaks corresponding to period ranging from 0.4 to 0.9 s. In the area with heavy damage, the peak is around 0.7 to 0.9 s. Detailed geotechnical information is needed to better analysis of the soil response characteristics and explain the damage distribution.

Safety, Reliability and Risk of Structures, Infrastructures and Engineering Systems – Furuta, Frangopol & Shinozuka (eds)
© *2010 Taylor & Francis Group, London, ISBN 978-0-415-47557-0*

Low-cycle fatigue effects on steel moment frames

James C. Anderson, Navid Nastar & Fabian Rojas
Department of Civil & Environmental Engineering, University of Southern California, CA, USA

ABSTRACT

Following the Northridge Earthquake (1994), the SAC steel project was initiated to investigate the causes of widespread damage to the connections of steel moment frame buildings that were discovered following the earthquake. The published results of these studies were concentrated on local connection defects that could potentially have initiated the observed cracks. It was also considered that much of the observed connection damage was the result of yielding of the material in the connection region and the formation of plastic hinges. However, damage to some buildings could not be reconciled by the use of these failure mechanisms. This has led to a renewed interest in the effects of low-cycle fatigue on the members and connections in steel buildings that are either elastic or have been driven into the weakly nonlinear range.

Several factors have been identified that can contribute to potential low-cycle fatigue failure in steel buildings. Fatigue is primarily a function of stress level and number of stress cycles. Conditions that could cause a stress level that is at or just below yield include moderate to strong earthquake ground motions, localized stress concentrations and low structural damping. Factors increasing the number of stress cycles experienced in the building include the frequency of occurrence of moderate to strong earthquakes and the contribution of higher modes of vibration which increase the number of stress cycles. The contribution of higher modes can also be increased by structural framing that incorporates vertical setbacks.

The current study uses a comprehensive analysis procedure for low-cycle fatigue that is based on the Palmgren-Miner method. Although fatigue curves relating cyclic stress to number of cycles to failure (S-N) are readily available for conditions of high-cycle fatigue there is only a limited amount of experimental data available for low-cycle fatigue. Using this limited experimental data, the authors have extended the S-N curves for the high-cycle fatigue range into the low-cycle range for use in this study.

Several instrumented high rise buildings in Los Angeles have recorded dynamic response to earthquakes beginning with the San Fernando earthquake (1971) and extending through the Northridge earthquake (1994). Also included in this time frame are smaller earthquakes such as the Whittier Narrows earthquake (1987). A linear time history analyses is conducted on a 16 story building that was instrumented during this time period and has a lateral force system consisting of steel moment frames in both directions. The current paper presents the procedure used for the low cycle fatigue evaluation. The method of rainflow cycle counting, developed by Prof. T. Endo, is used to evaluate the highly irregular variations of load with time and permit the use of the Palmgren-Miner Method for estimating the fatigue life that has been expended during the seismic event. Finally, the pattern of cumulative fatigue at critical locations in the building will be evaluated.

Safety, Reliability and Risk of Structures, Infrastructures and Engineering Systems – Furuta, Frangopol & Shinozuka (eds)
© 2010 Taylor & Francis Group, London, ISBN 978-0-415-47557-0

A comparative study on equivalent linear models to assess occurrence of building damage

Y. Niikawa & M. Kohiyama

Graduate School of Science and Technology, Keio University, Yokohama, Japan

ABSTRACT

In estimating probability of seismic damage of a building, the maximum value of building response to input earthquake motion is often evaluated to assess existence of damage. In order to calculate the maximum response value, the nonlinear time history analysis or the modal analysis is employed. However the former method has a problem that it requires longer calculation time. On the other hand, the latter analysis employs an equivalent linear model and it can reduce calculation time. Most of equivalent linear models in past studies were proposed to obtain the reliable maximum response value. But this study investigates equivalent linear models which can assess occurrence of building damages with high accuracy for several threshold values.

In this study, 3- and 10-story buildings were analyzed. The stiffness of each story was given by the A_i distribution of the Building Standard Law of Japan. Examined threshold values are 1/200, 1/150 and 1/100 rad for story drift angle. The number of input earthquake motions is 900. The accuracy of each model is compared based on receiver operating characteristics (ROC) (Van Trees, 1968).

This study considers a response of a model with bilinear force-displacement as a true response, and accuracy of three equivalent linear models are examined. These models use the equivalent stiffness and the equivalent damping factor. The nonlinear time history analysis was conducted to assess which story is damaged. As a result, the response exceeds the threshold value most frequently in the first story. Hence an equivalent model with the equivalent stiffness at the first story and the initial stiffness at the second and above is examined.

The equivalent stiffness k' is calculated in two ways. The first method for calculating k' assumes that the maximum story drift depends on earthquake story shear force and k' is given to have equal story shear force at threshold story drift D to that of the nonlinear model as shown in Figure 1. The second method assumes that the maximum story drift depends on earthquake input energy and k' is given to have equal earthquake input energy at D as shown in Figure 2.

Regarding examined three equivalent linear models, Model A uses equivalent stiffness of Figure 1 for all stories, Model B uses the stiffness of Figure 1 at the first story and Model C uses the stiffness of Figure 2 at the first story.

In this study, the equivalent damping factor of the first story is obtained by expanding the energy balance method (Minami & Midorikawa, 1992) to a multi-degree-of-freedom system. Modal analysis is conducted by using the evaluated equivalent damping. The findings are described below:

1. Model B, a linear model with equivalent stiffness based on a story shear force at the first story had the highest accuracy.

2. The accuracy decreased when a threshold value became larger because the nonlinear behavior of the second story and above was not considered in the equivalent linear model.

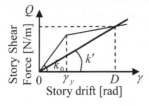

Figure 1. Equivalent stiffness based on the story shear force.

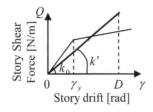

Figure 2. Equivalent stiffness based on the earthquake input energy.

REFERENCES

Van Trees, H.L. 1968. *Detection, Estimation, and Modulation Theory*: 23-46. New York: Wiley.

Minami, T. & Midorikawa, M. 1992. Procedure for response spectrum analysis. *Seismic Loading – strong motion prediction and building response*: 129-140. Tokyo: Architectural Institute of Japan (in Japanese).

*Safety, Reliability and Risk of Structures, Infrastructures and
Engineering Systems – Furuta, Frangopol & Shinozuka (eds)
© 2010 Taylor & Francis Group, London, ISBN 978-0-415-47557-0*

Impact dynamic behavior on CFTA girder and reliability assessment

Trinh Thai Trung & Min Chul Jeong
*School of Civil, Environmental, and Architectural Engineering, College of Engineering, Korea University,
Seoul, Korea*

S.A. Yi
Structure Research Department, Korea Institute of Construction Technology, Korea

Jung Sik Kong
*School of Civil, Environmental, and Architectural Engineering, College of Engineering, Korea University,
Seoul, Korea*

ABSTRACT

One of the principle aims of engineering design is
the assurance of the system performance within the
constraint of economy. Indeed, the assurance of performance, including safety, is the primarily the responsibility of engineers. In this study, the impact dynamic
behavior of an innovative girder named Concrete-
Filled and Tied steel Tubular Arch or CFTA girder was
studied. The girder consists of steel plate frame, arch
concrete and outside tendons. CFTA girder has several advantages compare to the conventional types of
girders such as bucking prevention by concrete filling, increase the stiffness and durability of concrete
due to confinement effect and aesthetics and economical matter due to arch concrete. In this study, impact
dynamic simulation and analysis were performed to
investigate the dynamic responses of the girder. Furthermore, the reliability assessment developed for
CFTA girder will facilitate the intuitive sense in the
design, and the evaluation of the bridges' performance

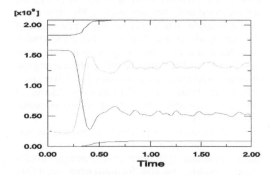

Figure 2.   Strain- Time relationship of 04 tendons.

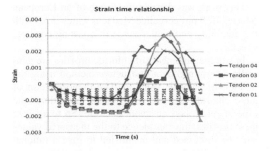

Figure 3.   Strain- Time relationship of 04 tendons.

REFERENCES

A. Shinghal, A.S Kiremidjian. Method for Probabilistic evaluation of Seismic structural damage. *Journal of Structual
Engineering.* (1996)
Ang AH-S, Cornell CA. Reliability bases of structure safety
and design. *Journal of Structural Engineering*, ASCE
1974; 100(9):1755–69.

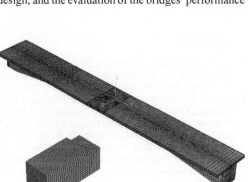

Figure 1.   Finite Element Model of CFTA and truck.

Safety, Reliability and Risk of Structures, Infrastructures and
Engineering Systems – Furuta, Frangopol & Shinozuka (eds)
© 2010 Taylor & Francis Group, London, ISBN 978-0-415-47557-0

Effect of error measurement of chloride profiles on reliability assessment

S. Bonnet & F. Schoefs
Research Institute in Civil Engineering and Mechanics, GeM, UMR CNRS 6183, University of Nantes, France

J. Ricardo & M. Salta
Laboratório Nacional de Engenharia Civil, Lisbon, Portugal

ABSTRACT

Under the marine environment, the main cause of
degradation of reinforced concrete structure is the corrosion of steel bars. It is due to the presence of chloride
ions in the concrete porosity. When the chloride ions
concentration reaches a critical value, steel bars are
not protected and corrosion begins. The determination of chloride content and corrosion rate is one of
the main goals for the structure service life prediction. Non destructive monitoring techniques are under
development to detect and quantify corrosion rate and
chloride content with good accuracy [1]. Silver/silver
chloride electrodes are used to monitor chloride in
situ but these sensors are affected by environmental
exposure [2, 3].

The deterministic or probabilistic models for service life predictions of structures used total chloride
profiles determining by destructive techniques to calibrate the models. This study focused on the uncertainty
in measured chloride ingress profiles by destructive
techniques.

This work was done in the frame of an European
Interreg III B project called Medachs which deals with
the durability and maintenance of coasting concrete
structures.

In situ measurements of chloride profiles are generally determined by a destructive method using the
titration of chloride solution. The determination of
these chloride profiles is a manual procedure: the
experimenter plays an important rule. So to analyze
these profiles, the confidence to the measured values
must be discussed and quantified to give a rational
decision aid-tool for repair or inspection planning.

As the bound chloride are participating to the corrosion process this study focused only on the uncertainty
in measuring total chloride content.

Repetitive tests have to be done on laboratory specimens. These tests were done on concrete casted in lab
and stored in the same chloride solution. Four different
researchers from two different laboratories have established the total chloride profiles which were analyzed
by a statistical approach. A probabilistic modeling of
errors (human and protocol) were suggested. After
the description of the material and the experimental

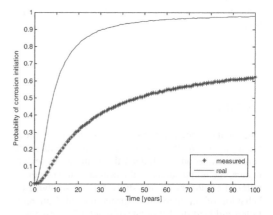

Figure 1. Evolution of Preal and Pmeas with time
($x = 3$ cm).

procedure, the study provides the statistical analysis,
the probabilistic modeling and the discussion on the
effect of error in measurements for reliability and life-
cycle assessment. Two applications are then selected.
The first concerns the estimation of the probability of
detection of corrosion initiation, from on site destructive tests. The second compares the evolution of the
probability of corrosion initiation with (P_{meas}) and
without taking into account the error of measurement
(P_{real}) (figure 1).

REFERENCES

Ahmad S. Reinforcement corrosion in concrete structures,
its monitoring and service life prediction – a review. Cem
Conc Comp 2003 ; 25 : 459–71.

Montemor MF, Alves JH, Simoes AM, Fernandes JCS,
Lourenço, Costa AJS, Appleton AJ, Ferreira MGS. Multi-
probe chloride sensor for in situ monitoring of reinforced
concrete structures. Cem Conc Comp 2006 ; 28: 233–36.

Atkings C.P., Carter M.A., Scantlebury J.D. Sources of
error in using silver/silver chloride electrodes to monitor chloride activity in concrete. Cem Conc Res 2001; 31:
1207–11.

Rosquoët F., Bonnet S., Schoefs F., Khelidj A. Chloride
propagation in concrete harbour structure, International
RILEM conference on "Advances in Concrete through
Science and Engineering", Québec (Canada), 2006.

General Session (Risk Analysis in Decision Making)

*Safety, Reliability and Risk of Structures, Infrastructures and
Engineering Systems – Furuta, Frangopol & Shinozuka (eds)
© 2010 Taylor & Francis Group, London, ISBN 978-0-415-47557-0*

Seismic risk curve in infinity time period

Y. Kawakami & T. Nakamura
Shinozuka Research Institute, Tokyo, Japan

M. Hoshiya
Advanced Research Laboratories, Musashi Institute of Technology, Tokyo, Japan

ABSTRACT

Seismic risk information has been employed for
assessment of seismic performance, decision-making
of seismic reinforcement, securitization of real estates,
etc. Seismic risk information is, for instance, probable
maximum loss, seismic life cycle cost, etc. Expected
value of seismic loss, which is frequently utilized for
seismic life cycle cost, and probable maximum loss
indicate only one point in a probability distribution of
seismic loss. Therefore, it is difficult to explain whole
aspect of seismic risk from only these indices.

Hence, we have focused on seismic risk curve
defined by exceedance probability function of seismic
loss, and proposed a method to extend the evaluation
period of seismic risk curve to multiple-year. However,
evaluation period is generally unspecified, uncertain
and there is no clear rule to determine. Thus, to set the
evaluation period on the basis of well-founded reason
is difficult.

The present work is intended to propose an evaluation method of seismic risk curve in infinity time
period. The procedure is as follows: 1) Probability
density function of earthquake occurrence is set on
the basis of Poisson process and renewal process; 2)
Seismic risk curve in multiple-year is formulated; 3)
Discount rate is applied to the variable of probability density function of seismic loss, and discounted
seismic risk curve in multiple-year is formulated; 4)
Seismic risk curve in infinity time period is finally
formulated. After examining characteristics of seismic
risk curve in infinity time period, cost-effectiveness of
seismic retrofit for a building located in Tokyo, Japan,
is discussed as an examination of applicability.

The major results of this paper are summarized
as follows: 1) Seismic risk curve in infinity time
period is not so different from that in service life
assumed by engineers. Against the evaluation period
without clear rule to determine, infinity time period
taking discount rate into consideration may provide
one criterion which is relatively consistent with service life assumed by engineers; 2) Seismic loss greatly
decreases when discount rate is taken into consideration. Therefore, decision-making is greatly affected
by whether discount rate is considered or not; 3) Seismic risk curve of current state intersects with that of
retrofitted state at one point. The seismic retrofit is
unexplainable in the range of exceedance probability higher than the intersection; 4) Seismic retrofit
is relatively difficult to explain in case of large discount rate compared with small one; 5) Seismic risk
curve in infinity time period enables decision-making
based on economic rationality and decision-maker's
risk tolerance level.

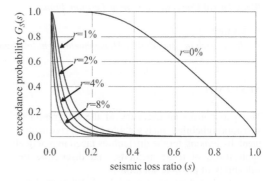

Figure 1. Seismic risk curve in infinity time period with
each discount rate r.

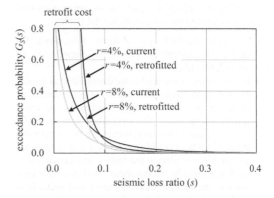

Figure 2. Comparison of seismic risk curve between current
state and retrofitted state with each discount rate r.

Safety, Reliability and Risk of Structures, Infrastructures and Engineering Systems – Furuta, Frangopol & Shinozuka (eds)
© *2010 Taylor & Francis Group, London, ISBN 978-0-415-47557-0*

"SRA into SRA" – Structural reliability analyses into system risk assessment: Activities of an ESReDA project group

E. Ardillon
EDF-R&D, Industrial Risk Management Department, Chatou, France

D. Proske
University of Natural Resources and Applied Life Sciences, Vienna, Austria

A. Chateauneuf
Blaise Pascal University, Clermont-Ferrand, France

A. Lannoy
IMdR – French Institute of Risk Management, Gentilly, France

ABSTRACT

This paper gives an overview of the integration of structural reliability analyses into system risk assessment, by looking for the difficulties faced in practical applications of these methods, and discussing the valuability of possible solutions.

In the last decades structural reliability analyses (SRA) have been increasingly applied in many industrial branches. Although they constitute a helpful tool for safety and reliability assessments, some questions arise when using them in an industrial context:

- Are they the only probabilistic framework for degradation modeling? Can they be applied to all degradation phenomena?
- In most cases industrial companies have to ensure the safety of both their facilities (considered as systems) and the components constituting these facilities, especially the structures; how to make sure that these two requirements can be met consistently? That target reliability levels for structures are acceptable considering the failure consequences on the facility and its environment? And that reliability assessment for complex structures and systems is performed adequately?

These two distinct issues have been investigated in the framework of the ESReDA Project Group: "SRA into SRA" (i.e. "Structural Reliability Analyses into System Risk Assessment"). ESReDA (European Safety & Reliability Data Association) is a European excellence network, whose objective is to promote and harmonize European research, application and training in the fields of dependability of industrial facilities. In particular, ESReDA has an everlasting activity in Structural Reliability through some of its Working Groups.

Regarding the first issue, alternative solutions are available for the modeling of degradation or

degradation kinetics, if only a poor physical model exists. In this case, stochastic processes like the (generalized) gamma process possibly including covariates may provide a possible solution to model the degradation evolution, although difficulties may arise when calibrating the process parameters. For kinetics modeling, the use of the Cox statistical model may be relevant and has been performed in industrial applications. In any case, it is necessary to get a sufficient amount of data.

Regarding the second issue, the links between structural reliability assessments, reliability target values, risk assessments of passive components and of the industrial systems in which they are integrated, social acceptance of risks, have been investigated. In particular, the definition of risk measures by quality of life measures is proposed and may have advantages for engineers, but the existence of an optimal safety (an implicit assumption of these approaches) is questionable since it does not account sufficiently for subjective risk perception. Finally, it appears that risk assessments are not sufficient to manage risks: the global management of risks includes risk assessments, but also relies on mitigation measures, disaster management and post disaster management.

Finally, the three practical methods applied to specify reliability target values for structures are presented: : implicit method, expert judgment and cost-benefit balancing. The first one is based on either the comparison with existing codes of practice or the analysis of existing *acceptable* structures. It is globally accepted by all the decision-makers. These three approaches are complementary as the obtained reliability levels reflect a certain confidence in the way to deal with system safety (qualitative information, imprecise quantitative data, etc.). It could be proposed to combine these three approaches, and to introduce some Bayesian tools for setting an expert system for decision making relative to safety targets for engineering systems.

*Safety, Reliability and Risk of Structures, Infrastructures and
Engineering Systems – Furuta, Frangopol & Shinozuka (eds)
© 2010 Taylor & Francis Group, London, ISBN 978-0-415-47557-0*

Risk informed management of road infrastructure

D. Zonta, R. Zandonini & F. Bortot

Department of Mechanical and Structural Engineering, University of Trento, Italy

ABSTRACT

This contribution introduces a risk-based approach to
bridge management as implemented by the Department of Transportation of the Autonomous Province of
Trento (APT). The APT Bridge Management System
(BMS) has been operative since 2004, and inspections
are currently being carried out on the whole bridge
stock. The trained inspectors have allocated the collected data on two different levels. The first-level data
describe the bridge location and geometry, and give a
general characterization of the structure; second-level
data point out in detail the Condition State CS of all
the elements of the bridge. CS is evaluated on the basis
of a procedure that acknowledges the general rules of
the AASHTO Commonly Recognized (CoRe) Standard Element System (AASHTO 1997), in order to
conserve compatibility with PONTIS (Thompson et
al. 1998) evaluation and deterioration models

All data are gathered in a Data Base where they
are automatically analyzed. The probabilistic models
adopted aim to calculate reliability and risk associated with each bridge; to assess the evolution of the
condition state in time; to simulate long-term scenarios for facilities development by critically comparing
alternative Maintenance, Repair and Reconstruction
(MR&R) strategies; to prioritize MR&R actions for
the whole stock.

The ranking criterion adopted is based on a principle highlighting those plans which, given a certain
budget, will minimize in the future the risk of facing
unacceptable events within the whole stock during a
given time span. Risk is defined as a function of the
damage caused by the given unacceptable event (Adey
& Bruhwiler 2003). The definition of unacceptable
event and damage is an owner-involving act related to
the management policy, which includes technical, economical, social and ethical issues. The effectiveness of
an intervention scenario is ranked by a prioritization
index, which takes into account both future social risk
and economic demand for MR&R. Several unacceptable events are alleged to affect the overall risk level of
a bridge: failure of a principal element; failure of a secondary element; pile collapse due to scour; accidents
due to sub-standard guardrails; earthquake.

Figure 1. Distribution of risk associated with structural
failure.

The analysis of the statistical relationships between
bridge age and condition allows us to articulate prediction models and to identify the causes leading more
frequently to abnormal degradation and their aftermath. The results of the analyses are part of the Data
Base, which becomes the main information source
used by APT to plan the future development of the
regional road network. The system is accessed by a web
interface, and the user can interact with it, for example to visualize inventory data and inspection results,
to request an analysis as well as to visualize its results.

By referring to some practical cases, this paper aims
to illustrate both how the APT's BMS is currently used,
and how it is applied to budget programming.

REFERENCES

AASHTO 1997. *Guide for Commonly Recognized Structural Elements*. Washington, DC: American Association
of State Highway and Transportation Officials, Inc.
Adey, B. & Bruhwiler, E. 2003. Risk-based approach to
the determination of optimal interventions for bridges
affected by multiple hazards. *Engineering Structures* 25:
903–912.
Thompson, P.D., Small, E.P., Johnson, M. & Marshall, A.R.
1998. The Pontis bridge management system. *Structural
Engineering International* 8(4): 303–308.

Safety, Reliability and Risk of Structures, Infrastructures and
Engineering Systems – Furuta, Frangopol & Shinozuka (eds)
© 2010 Taylor & Francis Group, London, ISBN 978-0-415-47557-0

Study on risk-informed in-service inspection for nuclear power plant piping

T. Koriyama, M. Yamashita, S. Miura & H. Fujimoto
Japan Nuclear Energy Safety Organization (JNES), Tokyo, Japan

ABSTRACT: Risk-informed in-service inspection for piping in consideration of both internal and seismic events was studied for a typical BWR5 plant in Japan.

1 INTRODUCTION

This article identifies significant piping segments (segments mean portions of piping, each of which has the same effect on safety when a rupture occurs) in terms of risk by utilizing Level 1 PSA for internal and seismic events during power operation for a typical BWR5 plant in Japan. It also considers the classifications of piping segments in terms of risk and inspection methods corresponding to those classifications.

2 EVALUATION OF PIPING RUPTURE PROBABILITY

Piping segment rupture probabilities were determined from the database of the OECD-NEA Piping Failure Data Exchange project for internal events. Since seismic initiating event frequencies are provided as functions of peak ground acceleration (PGA) levels, earthquake-caused piping segment rupture probabilities are given at PGA levels by fragility analyses. The seismic piping segment rupture probabilities are added to the existing random rupture probabilities induced by the internal events.

3 EVALUATION OF EFFECTS ON ACCIDENT SEQUENCES

In a similar way to the methodology for internal events, the level 1 PSA model adopted at JNES for seismic events during power operation was utilized as a basis for evaluating the effects of ruptures in piping segments on accident sequences, based on the Westinghouse Owners Group (WOG) methodology. Core Damage Frequency (CDF) induced by each segment rupture was determined by the use of a surrogate component which had the equivalent level of effect on accident sequences to that of the rupture in the segment.

4 RISK SIGNIFICANCE INDICATORS, RUPTURE IMPORTANCE, AND CORRESPONDED INSPECTION METHODS

Referring the WOG methodology, the risk significance indicator given by CDF and the rupture importance that depends on a degradation mechanism were used for the classifications of piping segments. It is because the risk significance indicator can be used to review the priority levels of inspection activities, and the rupture importance can determine the effectiveness and necessity of inspections. Inspection methods corresponding to the classification were investigated.

Non-destructive examinations were added to leak examinations for segments of the resultant high safety significance determined by the categorization criteria composed with both the risk significance indicator and rupture importance. The performed engineering evaluation showed that the changes from current examinations gave around 19% reduction of segments subject to both leak and non-destructive examinations within the total segments. Deterministic insights and engineering judgments on top of risk significance should be applied to get the final decision of inspection methods. In that process, an expert panel, which has yet to be established in this study, will take the important roles.

Aimed at the contribution to enhancing the scientific rationality of piping inspections, technical knowledge for utilizing risk information in piping inspection was accumulated.

Safety, Reliability and Risk of Structures, Infrastructures and Engineering Systems – Furuta, Frangopol & Shinozuka (eds)
© *2010 Taylor & Francis Group, London, ISBN 978-0-415-47557-0*

Study on seismic performance of wooden houses in snowy region by seismic response analysis – In 2-stories wooden houses built in Sapporo

Takahiro Chiba, Makihiko Munakata & Tsukasa Tomabechi
Hokkaido Institute of Technology, Sapporo, Japan

Toru Takahashi
Graduate School of Engineering, Chiba University, Chiba, Japan

Takeyoshi Uematsu
Hokkaido Northern Regional Building Research Institute, Asahikawa, Japan

ABSTRACT

To examine relationship between snow load and seismic performance of wooden houses, the authors carried out seismic response analysis. 62 wooden houses built in Sapporo were analyzed. Snow depth of 0.0m, 1.0m 1.5m, 2.0m used for seismic design were set in the analysis. Based on the past study, the hysteresis characteristics that put the bilinear element on the slip element were used by the analysis (Araki et al. 2005). In the analysis, to examine snow load dependence of relationship between scale of strong motion and the seismic performance of the houses, maximum response displacement of the houses were calculated.

Figure 1 shows the relationship between the maximum velocity of ground motion and the ratio of object houses' response exceeds safety limit displacement. In addition, the fragility curve (total collapse) of the past study was installed in the figure (Murao et al. 2002). In most cases, calculated ratio was smaller than the Murao's fragility curve. When snow depth was thicker than 1.0m, the influence of snow depth was small. The correlation between the maximum velocity and the ratio exceeds safety limit displacement was high. In 2F on the wooden houses, the ratio exceeds safety limit displacement increased with snow depth increase. However, the ratio more than the safety limit displacement was smaller than in each snow depth.

As stated above, 2F floor area of many wooden houses in the analysis was smaller than 1F, and the wooden houses were analyzed that collapsed in 1F due to snow load on 1F roof. The damage increment of the wooden houses with a snow depth increase was clarified. The influence of snow load contribution to ratio that exceeds damage limit displacement was large in 2F. Even if the seismic reinforcement would be done only in 1F, it might be collapsed in 2F. Therefore, in the seismic reinforcement of wooden houses in snowy region, it is necessary to consider not only 1F but also 2F.

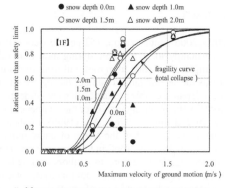

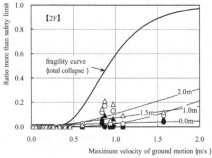

Figure 1. Relationship between the maximum velocity and the ratio that exceeds safety limit.

REFERENCES

Osamu Murao, et al. 2002. Building Fragility Curves for the 1995 Hyogoken-Nanbu Earthquake Based on CPIJ & AIJ'S Survey Results with Detailed Inventory. Journal of structural and construction engineering. Transactions of AIJ. pp.185–192.

Yasuhiro Araki, et al. 2005. A Study on Hysteresis Model for Earthquake Response Analysis of Timber Structures. Jour-nal of structural and construction engineering. Transactions of AIJ. pp.79–85.

Safety, Reliability and Risk of Structures, Infrastructures and
Engineering Systems – Furuta, Frangopol & Shinozuka (eds)
© 2010 Taylor & Francis Group, London, ISBN 978-0-415-47557-0

Evaluation on accumulation shape and variability of snow-slide from roof for human safety in snowy country

T. Takahashi & M. Suzuki
Chiba University, Chiba, Japan

T. Chiba & T. Tomabechi
Hokkaido Institute of Technology, Sapporo, Japan

ABSTRACT

In 2005–2006 winter season, 152 people were killed by snow mainly due to accidents during snow removal, for example, snow-slide from roof, fall from roof, etc. That was the second largest number during recent 60 years, in Japan. Reliable evaluation of snow-slide from the roof is very important for reduce the number of loss of human. Nakamura (1978a), Abe (1996), Takita and Watanabe (1998) studied the shape of snow mound by snow-slide from the roof, but the estimation error was not negligible. Therefore, the authors assumed not only friction between snow and roof material, but also air resistance and ground to roof snow density ratio caused by impact of snow block on the ground snow.

To measure the air resistance for snow-slide, the authors performed experiments of snow-slide on February 18 and February 20, 2008 at Hokkaido Institute of Technology. Totally 37 pairs of snow density and snow weight were chosen and slid 3 times each. Therefore, totally 111 times of slide were performed. Each slide was recorded by two digital video cameras, those located on the roof and on the ground. Sometimes the snow block was divided into some fragments. Therefore, developed muzzle velocity and falling velocity of totally 196 fragments were calculated based on video frame numbers. Probabilistic distributions of air resistance coefficient were calculated in 3 categories (around 3 kg fragments, around 1 kg fragments, and smaller than 1 kg fragments). The authors assumed friction between snow and roof material based on Takakura et al. (1998). The difference of snow density before and after the slide was evaluated based on the survey of Nakamura (1978b). The authors examined the assurance of these coefficients using 11 cases of field surveys performed by Abe (1996) and Takakura (2005). The value of relative error for the shape of snow accumulation bank was around 11 to 21 percent. That is around half of the errors by Nakamura, Takita or Abe. Therefore, the importance of air resistance for

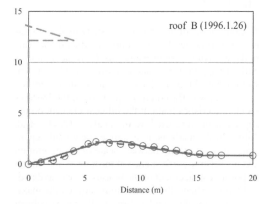

Figure 1. Estimated snow mound (red line) and observed data (circle). Relative error: 20.2%.

evaluation of the shape of snow accumulation bank by snow slide was clarified in this paper.

REFERENCES

Abe, O. (1996) Shape of Snow Banks Formed under the Very Large Sloped Roofs – Measurement and estimation, *Proc. of JSSE*, Vol.13, pp.159–164 (in Japanese).

Nakamura, H. (1978a) Shape of Snow Banks after Sliding, *Seppyo* Vol.40, No.1, pp.37–41.

Nakamura H. (1978b) Difference of Snow Density after Sliding, *Report of the National Research Center for Disaster Prevention* No.19, pp.239–242.

Takakura, M. et al. (1998) Effect of Temporal Changes in the Quality of Snow on Friction Between Snow and Steel Sheets on the Sloped Roof, *Journal of Structural and Construction Engineering, AIJ*, No.510, pp.45–50.

Takakura, M. et al. (2005) The Range within Snow Sliding Down from Sloped-roof in Detached Houses, *Journal of Architecture and Building Science*, No.21, pp.57–60.

Takita, M. and Watanabe, M. (1998) Chart and Table for Evaluating Roof Snow Accumulation after Sliding, *Journal of JSSE*, Vol.14, No.2, pp.126–132 (in Japanese).

General Session (Bridges and Buildings)

General Section (Bridges and Buildings)

*Safety, Reliability and Risk of Structures, Infrastructures and
Engineering Systems – Furuta, Frangopol & Shinozuka (eds)*
© 2010 Taylor & Francis Group, London, ISBN 978-0-415-47557-0

Model tests and simulation analysis on steel-concrete joint part of hybrid cable-stayed bridge

J. He, R. Liu, J.M. Wu, X. Kong, Y.Q. Liu & A.R. Chen
Department of Bridge Engineering, Tongji University, Shanghai, China

M.Y. Hu
Hubei E dong Changjiang Expressway Bridge Co., L.T.D, Hubei, China

T. Yoda
Department of Civil and Environmental Engineering, Waseda University, Tokyo, Japan

ABSTRACT

The innovative concept of using steel girders as main spans and concrete girders as side spans has been applied and developed world-widely in continuous girder bridges, cable-stayed bridges, and suspension bridges. Jointing two different kinds of girders at their appropriate places in a reasonable way will greatly improve the performance of bridges and will easily realize long-span bridges. The structural types and features of the joint part were introduced. In view of the existing steel-concrete joint parts of hybrid cable-stayed bridges, there are two kinds of structural form: with or without steel cells. As to the positions of bearing plate, three types of steel-concrete transition section within steel cells, which are front bearing plates, back bearing plates, both front and back bearing plates; while there are two types without steel cells, such as back bearing plates and back bearing plates combined with bottom plates. The shear connectors such as studs and PBL on the bearing plates connect the concrete and steel in the transition section.

Finite element analysis and model tests on key parts of the bridge are essential for design and construction. The model test with the scale of 1:2 for the joint part of the hybrid girder of the cable-stayed bridge is carried out on site. In the experiments, distributions of normal stresses, crack resistance ability, slip between concrete and steel, and deflection were carefully measured to investigate the mechanical performance, force transmission, stiffness matching and failure mechanism of the joint part of the hybrid girder. A 3D finite element model with material and geometrical nonlinear property was built to analyze the joint part. The test data were investigated comparing with the FEA results to evaluate the load carrying capacity and reliability of the joint part. From the comparisons of the test and FEA, it is found that the concentrated stress at the loading section can be reduced and transfer uniformly to the concrete through the transition region and joint part, the distribution and values of the FEA basically agree well with the test results. In addition, the bearing plate carries about half of the axial load, and the other half is transferred to the concrete girder through studs and PBL connectors. This leads to the conclusion that the ratio is rational and the structure shows good performance and reliability under the design load. Relatively small slip between concrete and steel denotes the connection soundness between two different materials. The findings from the tests and FEA were considered to be of special significance to design the bridge.

REFERENCES

Gimsing NJ. 1983. *Cable-supported bridges*. Chichester: John Wiley.
LIU Yu-qing. 2005. *Composite Bridge*, Beijing: China Communications Press.
LIU Yu-qing, LIU Rong et al. 2008. *Research on Safety, Reliability and Durability of E dong Bridge*, Shanghai, China.
Ohlsson, S. 1986. *Modal Testing of the Tjorn Bridge*, Proceedings of the 4th International Modal Analysis Conference, Kissimee, FL, 1: 599–605.
R. Saul, T.G. Lovett and S. Hopf, 1995. *Design and construction of the Kap Shui Mun Bridge at Hong Kong*. Bridges into the 21st century: Proceedings of the International Conference, The Hong Kong Institution of Engineers, Hong Kong.
T. Eiichi, S. Daisaku and K. Masaaki et al. 2002. *Kiso River Bridge Design and Erection*, Kawada Technical Report, 21: 42–47.
Virlogeux M, Vu B, Abel-Michel H, et al. 1994, *Design of the Normandie Bridge*. AFPC conference: Cable-Stayed and Suspension bridges. Deauville: 606–630.
Willam, K.J., Warnke, E.P. 1975. *Constitutive model for the tri-axial behavior of concrete*, Proceedings of International Association for Bridge and Structural Engineering. Bergamo, Italy: ISMES.

Safety, Reliability and Risk of Structures, Infrastructures and Engineering Systems – Furuta, Frangopol & Shinozuka (eds)
© 2010 Taylor & Francis Group, London, ISBN 978-0-415-47557-0

Influence of shear rigidity in vertical and horizontal frames on strength of prefabricated scaffolds

H. Takahashi, K. Ohdo & S. Takanashi
National Institute of Occupational Safety and Health, Tokyo, Japan

The vertical load on prefabricated scaffolds holds 'live weight', i.e. the people and materials on them. Expressly, standard prefabricated scaffolds might be used as the concrete support. Therefore, excessive vertical load is also likely to act on prefabricated scaffolds. The buckling modes of scaffolds are illustrated in Figure 1. They include member buckling, when each story of the scaffold curves, and total buckling, when the entire side of the scaffold curves (Mori, Y., Mae I. and Kunimori M. 1962). The buckling load for total buckling is smaller than that for member buckling because the buckling length in total buckling is greater than that in member buckling. Therefore, the member buckling is stronger than the total buckling.

Recent studies (Mori, Y., Mae I. and Kunimori M. 1962) confirm that buckling comes about mainly when the stiffening member in the vertical frame is shorting. Therefore, total buckling happens when the shear rigidity in the vertical frame is deficient.

On the other hand, the horizontal frame spans the vertical frame. Shear rigidity of the vertical frame is the influencing factor when scaffolds are buckling. Therefore, it is thought that the shear rigidity of the horizontal frame also influences the strength of scaffolds. However, a design method based on this new knowledge was not examined.

In this study a buckling analysis of standard prefabricated scaffolds was executed, using the shear rigidity of the vertical and the horizontal frames as parameters, to investigate the strength performance of prefabricated scaffolds.

The results of this study can be summarized as follows:

1. When the junction between the scaffold and the ground is pinned, as for a highest risk situation and the scaffold is between 2 and 10 stories, we conclude that that the strength of the scaffold is decided by the shear rigidity of the vertical and horizontal frames of the lowest story of the scaffold, regardless of the number of stories.

2. When the shear rigidity of the vertical frame k_s is greater than k_{cr} of the following equation, i.e. $k_{cr} < k_s$, the prefabricated scaffolds will experience member buckling in spite of the shear rigidity of the horizontal frame k_h.

$$k_{cr} = \frac{\pi^2 E I_e}{h_0^3} = \frac{2\pi^2 E\left(I_0 + I_s \dfrac{h_s}{h_0}\right)}{h_0^3} \quad (1)$$

Member buckling will also occur in the case of $k_{cr} > k_s$, when the value of k_h is $k_{cr} - k_s (= k_{crh})$ or more.

REFERENCES

Mori, Y., Mae I. and Kunimori M. 1962. On the Load-carrying Capabilities of the Steel Tubular Vertical Frames which are Used for Supporting the Concrete Bridge Mold, *Research Report of the Research Institute of Industrial Safety*, 3: 1–8, in Japanese.

*Safety, Reliability and Risk of Structures, Infrastructures and
Engineering Systems – Furuta, Frangopol & Shinozuka (eds)
© 2010 Taylor & Francis Group, London, ISBN 978-0-415-47557-0*

A proposal of hybrid structure with carbon steel and non-magnetic steel for MAGLEV

M. Hirohata & Y.-C. Kim
Joining and Welding Research Institute, Osaka University, Osaka, Japan

Recently, research and development of magnetically levitated vehicle (MAGLEV) as a new high-speed transportation system have been advanced. The MAGLEV runs by repulsion between superconductive magnets on the vehicle and coils on the railway. However, when ordinary steels are used for the girder as the guide way, it is known that loss of energy occurs by magnetically resistance force. Although the non-magnetic steels need to be used, they are more expensive than ordinary steels. Therefore, it was proposed that the non-magnetic steels were used only in the area where magnetism was generated and the ordinary steels were used in the other area where magnetism was not generated.

Assuming the hybrid structure assembled by the non-magnetic steel (High manganese non-magnetic steel: Hi-Mn steel) and the ordinary steel (SM490Y steel: SM steel), a series of research has been carried out. Young's modulus of Hi-Mn steel is about 80% of that of ordinary steel. Yield stress and tensile strength of Hi-Mn steel are higher than those of SM steel. At present, the mechanical behavior of the hybrid steel structure assembled by dissimilar steels with different Young's moduli is not elucidated. Therefore, it is necessary that the safety and reliability of the hybrid steel structure is evaluated.

In this paper, in order to propose the hybrid structures assembled by Hi-Mn and SM steels for the guide way of MAGLEV, the structural characteristics and mechanical behavior of hybrid steel I-girder assembled by Hi-Mn and SM steels were elucidated based on the results of the elastic-plastic large deformation analysis. Moreover, the method for ensuring the structural performance of the hybrid girder the same as that of the girder assembled by SM steel was investigated from the viewpoint of the safety and reliability of the girder.

The analysis was carried out on the following three models, the girder with only SM steel (SM model), the girder with only Hi-Mn steel (Hi-Mn model) and the hybrid girder with Hi-Mn and SM steels (Hybrid model). Although the shapes of these girders were the same, in the case of Hybrid model, the lower flange and the lower side of web were SM steel, and the upper flange, upper side of web and stiffeners were Hi-Mn steel.

From the results of analysis, the yielding load and ultimate load of Hybrid model were larger than those of SM model because the yield stress and the tensile strength of Hi-Mn steel were larger than those of SM steel. It could be said that the high strength of Hi-Mn steel was exploited in Hybrid model. However, the bending stiffness of Hybrid model was lower than that of SM model because Young's modulus of Hi-Mn steel was about 80% of that of SM steel.

In designing a hybrid structure assembled by dissimilar steels, whose Young's moduli and yield stress were different from each other, to make the most of merits of each material and to compensate their demerits with each other were natural demands. In order to increase the bending stiffness of Hybrid model the same as that of SM model, enlarging the length of part with SM steel in the web of Hybrid model was proposed. It would be more economical than enlarging the part with Hi-Mn steel because SM steel was cheaper than Hi-Mn steel. And it would be effective in increasing the stiffness of Hybrid model because Young's modulus of SM steel was larger than that of Hi-Mn steel.

By enlarging 6% of the height of girder to the length of part with SM steel in web, the bending stiffness of Hybrid model was almost the same as that of SM model. Moreover, the yielding load of Hybrid model was 10% increased and the ultimate load of Hybrid model was 21 % increased compared with those of Hi-Mn model.

Safety, Reliability and Risk of Structures, Infrastructures and Engineering Systems – Furuta, Frangopol & Shinozuka (eds)
© *2010 Taylor & Francis Group, London, ISBN 978-0-415-47557-0*

Reduction of residual displacements of concrete columns by high strength rebars

Ali Reza Khaloo, Mostafa Tazarv & Yousef Javid

Department of Civil Engineering, Sharif University of Technology, Tehran, Iran

1 ABSTRACT

Displacements and drifts are major concerns in the seismic-performance design methodology. Moreover, residual displacements are gradually getting higher attention in performance based design. There are several examples of damaged bridges that have lost their serviceability and safety due to large residual displacements of their piers (columns) after severe earthquake. After 1995 Hyogo-ken Nanbu earthquake (Japan), more than 100 reinforced concrete bridge columns experienced a tilt angle of more than 1 degree (1.75% drift). These columns had to be removed and new columns built because of the difficulty of setting the superstructure back to the original alignments and levels (Kawashima, 2000). So, controlling the residual displacement of bridges would keep fundamental infrastructures safe and functional. Even though seismic codes ensure life safety for bridges, large residual displacement after severe earthquake may make the bridge useless. Reduction of residual displacement is also necessary for minimizing the rehabilitation cost and also maximizing the serviceability of bridges. In this research, a new method will be proposed in order to reduce the residual displacements of concrete columns as well as bridge's piers. Several new structures have been developed to reduce residual displacements in which post-tensioning technique as well as unbonded strands and rebars were essence of these methods. Last innovative method for reduction of residual displacements of concrete columns has been developed by Mahin & Sakai (2004a, 2004b and 2006). They replaced some mild rebars of concrete columns with a longitudinal post-tensioning tendon at the center of columns section. In this paper, a new method will be proposed to reduce residual displacements by the use of High Strength Rebars (HSR). This method is different from the most researched method for reduction of residual displacement, i.e., post-tensioning technique. Some HSR will be replaced or added to the column's section as closed as possible to the center of columns. The optimized position and percentage of high strength rebar will also be determined to minimize residual displacements. Result of analyses revealed that this method can effectively reduce residual displacements. High strength rebars distributed near the core of column section will give the column a self-centering tendency. Simultaneously consideration of cyclic and dynamic analyses indicates that the best position for HSR is the center of the column section. But, due to construction difficulties, radial distribution of HSR in the radius of 25% radius of column section was selected. Adding 40% area of mild reinforcements as HSR to the concrete column will dissipate adequate energy to restrict residual displacements in acceptable range. Dynamic analyses indicated that proposed method can reduce residual displacements of RC column more than 50% that is comparable to the previous innovative method in this field. Experimental study on this method may enhance the gained results and undoubtedly will show the real efficiency of this method for reduction of residual displacements.

REFERENCES

Kawashima, K. 2000, Seismic design and retrofit of bridges, *Proc. of 12th World Conference on Earthquake Engineering*, CD-ROM No. 2828, New Zealand Society for Earthquake Engineering, Auckland, New Zealand.

Mahin, S. A. and Sakai, J. 2006, Use of Partially Prestressed Reinforced Concrete Columns to Reduce Post-Earthquake Residual Displacements of Bridges, *Fifth National Seismic Conference on Bridges & Highways*, San Francisco, CA, September 18–20

Sakai, J. and Mahin, S. A. 2004a, Analytical investigations of new methods for reducing residual displacements of reinforced concrete bridge columns, *PEER-2004/02*, Pacific Earthq. Engrg. Res. Center, Univ. of California at Berkeley, California.

Sakai, J. and Mahin, S. A. 2004b, Mitigation of residual displacements of reinforced concrete bridge columns, *Proc. of 20th US-Japan Bridge Engineering Workshop*, pp. 87–102, Washington D.C., USA.

Safety, Reliability and Risk of Structures, Infrastructures and Engineering Systems – Furuta, Frangopol & Shinozuka (eds)
© 2010 Taylor & Francis Group, London, ISBN 978-0-415-47557-0

Random fluctuation of internal forces in rough beams under moving oscillators

G. Muscolino

Dipartimento di Ingegneria Civile & CIDiS (Centro Interuniversitario di Dinamica Strutturale), Università di Messina, Italy

A. Palmeri

School of Engineering, Design and Technology, University of Bradford, UK

A. Sofi

Dipartimento di Arte, Scienza e Tecnica del Costruire, Università "Mediterranea" di Reggio Calabria, Italy

1 ABSTRACT

Surface roughness plays a fundamental role in the analysis of bridge-vehicle dynamic interaction since it represents one of the most important vibration sources during the passage of vehicles over a bridge. In the literature, the bridge structure is often treated as an Euler-Bernoulli beam, while vehicles are modelled as SDoF oscillators, moving with constant or variable speed along the beam. Moreover, the roughness is effectively represented as a zero-mean stationary Gaussian random field, fully characterized by the Power Spectral Density (PSD) function. As a result, the dynamic response of the supporting beam in terms of transverse deflections and internal forces turns out to be described by time-dependent random fields, which have to be characterized in a probabilistic sense. For this purpose, Monte Carlo simulations are generally resorted to, while just in a handful of papers alternative stochastic approaches have been investigated (see e.g. Schenk & Bergman 2003; Muscolino et al. 2007).

The dynamic response of the oscillator-beam system is usually obtained by applying the classical modal analysis, based on a Conventional Series Expansion (CSE) of the solution in terms of the eigenfunctions of the undamped and unloaded supporting structure (see e.g. Muscolino et al. 2008). The rate of convergence of this technique is quite good as far as the beam's transverse displacements are of interest. On the contrary, when the evaluation of bending moment and shear force is performed, the CSE converges poorly, since higher order derivatives of the eigenfunctions are required (Frýba 1999). To overcome this drawback, correction methods have been recently proposed in a deterministic setting (e.g. Pesterev & Bergman 2000; Biondi & Muscolino 2005; Bilello et al. 2008).

In this paper, the problem of evaluating the second-order statistics of internal forces in a simply supported beam with random surface roughness crossed by a linear oscillator is tackled. In the context of the CSE, the set of first-order ordinary differential equations, with time-dependent coefficients, governing the

second-order statistics of the oscillator-beam response is derived and solved with an efficient step-by-step procedure. In a second stage, a Quasi-Static (QS) correction term, consisting of a time-dependent random field, is proposed in order to improve the convergence of the CSE. This novel contribution is formulated by extending the Mode Acceleration Method (MAM), originally introduced by Williams (1945) for deterministic problems. It is shown that the QS correction term so obtained can be probabilistically characterized starting from the second-order statistics of the CSE, which makes the proposed technique computationally very efficient. The accuracy is proved by the numerical results presented and discussed in the paper.

REFERENCES

Bilello, C., Di Paola, M., Salamone, S., 2008. A correction method for the analysis of continuous linear one-dimensional systems under moving loads. *J SOUND VIB,* 315: 226–238

Biondi, B., Muscolino, G. 2005. New improved series expansion for solving the moving oscillator problem. *J SOUND VIB,* 281: 99–117

Frýba, L., 1999. *Vibrations of Solids and Structures under Moving Loads.* London: Thomas Telford

Muscolino, G., Palmeri, A., Sofi, A., 2007. Random vibration of linear oscillators moving on rough beams. *Proc. 10th Int. Conf. on Application of Statistic and Probability in Civil Engineering (ICASP10)*

Muscolino, G., Palmeri, A., Sofi, A., 2008. Absolute versus relative formulations of the moving oscillator problem. *INT J SOLIDS STRUCT,* doi:10.1016/j.ijsolstr.2008.10.019 (In press)

Pesterev, A.V., Bergman, L.A., 2000. An improved series expansion of the solution to the moving oscillator problem. *J VIB ACOUST,* 122: 54–61

Schenk, C. A. and Bergman, L. A., 2003. Response of continuous system with stochastically varying surface roughness to moving load. *J ENG MECH-ASCE,* 129: 759–768

Williams, D., 1945. Dynamic loads in aeroplanes under given impulsive loads with particular reference to landing and gust loads in a large flying boat. *Great Britain Aircraft Establishment Report SME 3309 and 3316*

Safety, Reliability and Risk of Structures, Infrastructures and Engineering Systems – Furuta, Frangopol & Shinozuka (eds)
© 2010 Taylor & Francis Group, London, ISBN 978-0-415-47557-0

Estimating extreme highway bridge traffic load effects

C.C. Caprani

Department of Civil & Structural Engineering, Dublin Institute of Technology, Ireland

E.J. OBrien

School of Architecture, Landscape & Civil Engineering, University College Dublin, Ireland

ABSTRACT

In the reliability analysis of a bridge structure, traffic induced load effects are some of the most variable parameters. Because of this, more accurate estimation of the distributions of traffic load effect can result in improvement in the accuracy of calculated safety levels. Traditionally, an extreme value analysis is performed using either the block maxima approach, through the Generalized Extreme Value (GEV) distribution, or the peaks-over threshold approach, using the Generalized Pareto Distribution (GPD). A recently proposed model, the Box-Cox-GEV distribution (Bali 2003), which nests both the GEV and GPD models, is used in this study. In this way, the data itself determines its domain of attraction, be it GEV or GPD. In addition, the framework of composite distribution statistics (CDS), recently presented by Caprani et al. (2008) is used in conjunction with the Box-Cox-GEV model. This framework allows for the different distributions that result from different loading event types. This combination of novel approaches is applied to real traffic data for a range of bridge lengths and load effects.

Weigh-in-motion data obtained from the A6 motorway, France, is used as the basis of a Monte Carlo simulation of 1000 days of truck traffic. Daily maximum load effect data for different bridge lengths, load effects, and loading event types is calculated. Based on this data set, a conventional approach (synthesized from the literature), a GEV-CDS model, and a Box-Cox-GEV-CDS model are used to estimate characteristic traffic load effect.

Application of the Box-Cox-GEV model shows that the data lies strongly in the domain of attraction of the GEV model for a large range of the thresholds considered. An optimum threshold and model parameter set are identified. It is also shown that the Box-Cox-GEV model provides stable estimates of characteristic load effect for thresholds up to the sample mean. Through likelihood ratio testing, it is shown that the GEV and GPD models are strongly rejected in favor of the Box-Cox-GEV model for almost all cases considered.

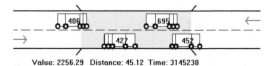

Value: 2256.29   Distance: 45.12   Time: 3145230

Figure 1. An example 4-truck loading event on a 2-span 40 m long bridge.

Comparison of the lifetime load effect predictions from each of the stated methods is performed. It is shown that, on average, the GEV-CDS model estimates slightly lower load effect (-2.3%) than the conventional method. Conversely, the Box-Cox-GEV model, within the CDS framework, on average, estimates higher load effect (+4.9%) than the conventional method.

An interesting critical combination of bridge length and influence line shape was found for one considered load effect. It was found that the lifetime value for this load effect is sensitive to the tail behavior of the distribution of load effect for 4-truck loading events; an example is shown in Figure 1.

In summary, the combination of two recently proposed methods, the Box-Cox-GEV and composite distribution statistics approach, is shown to provide further statistical information about bridge traffic load effect. In doing so, the influence that subjective decision-making can have on the results is minimized. The output from this proposed approach is readily applicable to a full reliability analysis.

REFERENCES

Bali, T.G. 2003. The generalized extreme value distribution. *Economics Letters* 79: 423–427.

Caprani, C.C., OBrien, E.J. & McLachlan, G.J. 2008. Characteristic traffic load effects from a mixture of loading events on short to medium span bridges. *Structural Safety* 30(5): 394–404.

Mini-Symposia (MS01) Uncertainty Modeling of Rare/Imprecise data
[Session Organized on Behalf of IASSAR's CSMSE umbrella]

Safety, Reliability and Risk of Structures, Infrastructures and Engineering Systems – Furuta, Frangopol & Shinozuka (eds)
© 2010 Taylor & Francis Group, London, ISBN 978-0-415-47557-0

On new cautious structural reliability models in the framework of imprecise probabilities

L.V. Utkin

Saint-Petersburg State Forest Technical Academy, St.Petersburg, Russia

I. Kozine

Technical University of Denmark, Kgs. Lyngby, Denmark

ABSTRACT

Uncertainty of parameters in engineering design has been modeled in different frameworks such as interval analysis, fuzzy set and possibility theories, random set theory and imprecise probability theory. The authors of this paper for many years have been developing new imprecise reliability models and generalizing conventional ones to imprecise probabilities. The theoretical setup employed for this purpose is imprecise statistical reasoning (Walley 1991), whose general framework is provided by upper and lower previsions (expectations). The appeal of this theory is its ability to capture both aleatory (stochastic) and epistemic uncertainty and the flexibility with which information can be represented.

The previous research of the authors related to generalizing structural reliability models to imprecise statistical measures is summarized in Utkin & Kozine (2002) and Utkin (2004). The presupposed input for the imprecise structural reliability models was some probabilistic measures (precise or imprecise) of strength and stress. While the accepted premises are meaningful and practical in some applications, they do not cover many other cases the reliability analyst faces in practice. Often the above mentioned inputs do not exist and the analyst has only some judgments or measurements (observations) of values of stress and strength. How to utilize this available information for computing the structural reliability and what to do if the number of judgments or measurements is very small? Developing models enabling to answer these two questions has been in the focus of the new research the results of which are described in the paper.

In this paper we describe new models for computing structural reliability based on measurements of values of stress and strength and taking account of the fact that the number of observations may be rather small. The approach to developing the models is based on using the *imprecise Bayesian inference models* (Walley 1996). These models provide a rich supply of coherent imprecise inferences that are expressed in terms of posterior upper and lower probabilities. The probabilities are initially vacuous, reflecting prior ignorance, become more precise as the number of observations increase.

The new imprecise structural reliability models are based on imprecise Bayesian inference and are imprecise Dirichlet, imprecise negative binomial, gamma-exponential and normal models. The models are applied to computing cautious structural reliability measures when the number of events of interest or observations is very small. The main feature of the models is that prior ignorance is not modeled by a fixed single prior distribution, but by a class of priors which is defined by upper and lower probabilities that can converge as statistical data accumulate. Numerical examples illustrate some features of the proposed approach.

REFERENCES

Utkin, L.V. 2004. An uncertainty model of the stress-strength reliability with imprecise parameters of probability distributions, *Zeitschrift fuer Angewandte Mathematik und Mechanik (Applied Mathematics and Mechanics)* Vol.84, 688–699.

Utkin, L.V. & Kozine, I.O. 2002. Stress-strength reliability models under incomplete information, *International Journal of General Systems* Vol.31, 549–568.

Walley, P. 1991. *Statistical Reasoning with Imprecise Probabilities*. London: Chapman and Hall.

Walley, P. 1996. Inferences from multinomial data: Learning about a bag of marbles, *Journal of the Royal Statistical Society, Series B* Vol.58, 3–57. with discussion.

*Safety, Reliability and Risk of Structures, Infrastructures and
Engineering Systems – Furuta, Frangopol & Shinozuka (eds)*
© *2010 Taylor & Francis Group, London, ISBN 978-0-415-47557-0*

Fuzzy extensions in state-space filtering – Some applications in geodesy

H. Kutterer & I. Neumann

Geodätisches Institut, Leibniz Universität Hannover, Germany

ABSTRACT

State-space filtering is an important task in many
engineering disciplines. A typical example of a state-
space filter is the recursive Kalman filter (KF). As it
is widely used in the applications, it will be mainly
considered in this paper. The KF combines in an optimal way intrinsic information on a system which is
described by a set of linear differential equations and
additional information based on measurements of the
system state. It allows to predict future or unobservable system states and to estimate system parameters
which are not or only weakly determined (adaptive
filtering). Conventionally, the related uncertainty is
assessed in a stochastic framework: measurement and
system errors are modeled using random variables
or stochastic processes, respectively. However, as this
approach neglects all components of the uncertainty
budget which are due to non-stochastic effects the
reported uncertainty measures are too optimistic. For
a more adequate representation and quantification a
more general formulation is required.

In this study, the KF is extended with respect to
imprecision which is due to insufficient knowledge
about model and measurements. It is based on a previous study where imprecise quantities are modeled
as fuzzy intervals with a random mean point which
reflect the epistemic type of uncertainty. The respective fuzzy extension of the KF is rigorously based
on Zadeh's extension principle. This previous work is
now extended with respect to mainly two items: (i)
the extension to adaptive filtering in order to estimate
unobservable system parameters within the filtering
process, and (ii) studies on the improvement of the
efficiency of the numerical computations. Item (i)
addresses in particular the uncertainty of the estimated system parameters in the frame of an extended
uncertainty budget taking the uncertainty of the prior
information on such parameters into account. Item (ii)
is concerned with the fact that recursive operations on
fuzzy intervals are tending to significantly overestimate the spread of the actual fuzzy results. In order to
prevent this effect, the recursive formulation of the KF

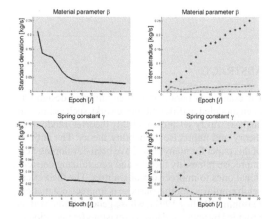

Figure 1. Standard deviations (left) and imprecision (interval radii) of the parameters determined by adaptive filtering.

has been resolved in the previous work; this yields a
time-consuming algorithm.

Two relevant applications in geodesy are presented
and discussed with respect to the extended methodology for the first time. In both cases the respective modeling of uncertainty and the use of the KF technique
are shown. The first application is concerned with
the geodetic monitoring of control processes based on
positioning devices. The focus is put on the numerical
efficiency of the extended filter algorithms. Structural
monitoring with geodetic techniques is considered in
a simplified scenario as a second application. In this
case material constants are introduced as uncertain
system parameters which are estimated adaptively (see
Figure 1).

REFERENCES

Neumann, I. & Kutterer, H. 2007. A Kalman Filter exten-
    sion for the analysis of imprecise time series. *Proceed-
    ings of the 15th European Signal Processing Confer-
    ence (EUSIPCO)*, 3–7 September 2007, Poznan, Poland,
    (CD-Proceedings).

Safety, Reliability and Risk of Structures, Infrastructures and Engineering Systems – Furuta, Frangopol & Shinozuka (eds)
© 2010 Taylor & Francis Group, London, ISBN 978-0-415-47557-0

A comparison of distribution-free entropy quantile functions using probability weighted moments from complete or censored samples

Jian Deng
Department of Civil Engineering, The University of Waterloo, Waterloo, Ontario, Canada
School of Resources and Safety Engineering, Central South University, Changsha, China

M.D. Pandey
Department of Civil Engineering, The University of Waterloo, Waterloo, Ontario, Canada

ABSTRACT

This paper presents a distribution-free procedure for estimating the quantile function (QF) of a non-negative random variable using the principle of entropy subject to constraints specified in terms of integer or fractional probability weighted moments estimated from complete or censored sample data. When the prior distribution is uniform, principle of minimum cross-entropy (CrossEnt) reduces to the principle of maximum entropy (MaxEnt).

The estimation of extreme quantiles corresponding to small probabilities of exceedance (POE) is commonly encountered in the risk analysis of engineering systems. Although the entropy approach leads to the most unbiased distribution, the main problem remains the accurate estimation of moments from limited data. The estimates of higher order moments (order > 2) from small samples (size less than 50) tend to be highly biased. The entropy distribution derived from poor moment estimates would lead to inaccurate quantile values. This difficulty can be circumvented by using the probability weightd moments (PWM) in place of ordinary moments.

We would compare the performances of the quantile function estimation procedure based on fractional PWMs (FPWM) with those based on classical integer PWMs (IPWM). Meanwhile, we compare the performances of the principle of maximum entropy and the principle of minimum cross entropy.

Two example distributions are considered to evaluate the performance of the proposed method: Generalized Pareto distribution and Weibull distribution. A practical example was presented to estimate a quantile corresponding to 99.9% probability. The results show that PPWM based CrossEnt quantile function results considerable accurate estimates under different censoring thresholds.

Examples are presented for illustration. Conclusions can be made from Figure 1 that the accuracy of the FPWMs based quantile function is superior to that estimated from the use of conventional IPWMs.

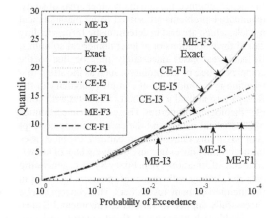

Figure 1. Approximation of Pareto distribution tail region by IPWM/FPWM based MaxEnt/CrossEnt QFs.
[Note: Symbols CE-Ik (or CE-Fk) ($k = 1, 2, \ldots$) represent the kth order IPWM (or FPWM) solution of CrossEnt QF. ME-Ik (or CE-Fk) means IPWM/FPWM based MaxEnt QF. eg, CE-I3 is the tail function using the first three IPWMs.]

A properly-selected prior distribution in the entropy principle would play a positive role in estimation of quantile function.

REFERENCES

Deng, J. and Pandey,M.D., 2009. Estimation of minimum cross entropy quantile function using fractional probability weighted moments. *Probabilistic Engineering Mechanics* 24(1):43–50.

Deng, J., Pandey, M.P., Gu, D., 2009. Extreme quantile estimation from censored sample using partial cross entropy and fractional partial probability weighted moments. *Structural Safety* 31(1):43–54.

Pandey, M.D., 2000. Direct estimation of quantile functions using the maximum entropy principle. *Structural Safety* 22(1), 61–79.

Pandey, M.D., 2001. Extreme quantile estimation using order statistics with minimum cross-entropy Principle. *Probabilistic Engineering Mechanics* 16(1): 31–42.

Safety, Reliability and Risk of Structures, Infrastructures and Engineering Systems – Furuta, Frangopol & Shinozuka (eds)
© 2010 Taylor & Francis Group, London, ISBN 978-0-415-47557-0

Hierarchical Fuzzy Inference System (HFIS) modeling for environmental issues

M. Elektorowicz & A. Amid

Department of Building, Civil, and Environmental Engineering, Concordia University, Montreal, Canada

ABSTRACT

In science, engineering and many other areas of study, quantitative problems are solved using mathematical models, which are cast in deterministic form, yielding closed form solutions or at least numerical solutions. In many real cases uncertainty may be due to the complex nature of the problem, such as the stress distribution in a component of complex geometry or the flow pattern around a bluff body in a fluid with complex rheological properties. These are not uncertainties of a statistical nature and the corresponding concept is of possibility rather than probability. In some cases the uncertainty raises because of having lack of knowledge and information. Then, Fuzzy System, generating data and dealing with vague information, has been in the attention of many researchers. This concept can be successfully applied to various environmental issues, e.g. health risk assessment, flood control, soil erosion, water resource management, aquifer pollution, wetland design, soil pollution, biogas generation and transport in landfill.

This paper focus on uncertainty problems observed during soil remediation processes, particularly when soil is contaminated with heavy metals having an important impact on health of inhabitants. Assessing the mobility (fate) of heavy metals within contaminated sites is a key constituent of prioritizing contaminant sites for remediation (or other purposes). For this reason, characteristics of the soil, specifically those related to sorption of heavy metals ought to be well studied; therefore, every individual function involved in the fate of heavy metals in soil has to be selected. The effect of all of these parameters can be aggregated together. The major objective of this research was introducing a new methodology for designing models suitable for decision-making in environmental management. Subsequently, the hierarchical fuzzy model approach was applied in this study. For the assessment and identification of soils vulnerable to heavy metals, fate of metals, attenuation of metals, or mobility of metals a hierarchical fuzzy inference system (HFIS) was developed.

In this tool, soil parameters involved in natural attenuation (sorption) of heavy metals were selected (pH, CEC, SSA, Clay content, OM, and Moisture). The selected parameters were introduced to a hierarchical fuzzy inference system in order to rank the vulnerable area in the topsoil. An investigation for finding a solution technique to predict the sorption potential in the soil was performed, and subsequently, a fuzzy inference engine HFIS was designed. The developed model was validated by studying different case studies

The designed model is suitable for decision-making in environmental management, particularly where the development of brownfields presents a common challenge for municipal and provincial (states) authorities.

Safety, Reliability and Risk of Structures, Infrastructures and Engineering Systems – Furuta, Frangopol & Shinozuka (eds)
© *2010 Taylor & Francis Group, London, ISBN 978-0-415-47557-0*

Robustness assessment of structures incorporating generalized uncertainty models

J.-U. Sickert, S. Pannier, W. Graf & C. Jenkel
Institute for Structural Analysis, TU Dresden, Germany

ABSTRACT

Actually, structural engineers are increasingly confronted with the task to design structures, which ensure beside high reliability and safety levels during their life-cycle also a robust performance. This evolves due to the occurrence of unforeseen impacts, e.g., hazardous environmental conditions or terrorism. While the terms reliability and safety level are well-defined, an overall mathematical definition of the linguistic well-known term of "robust structures" is missing so far. The proposed approach based on (Beer & Liebscher 2008) contributes to the discussion how to quantify robustness in order to compare different structural designs. Thereby, robustness is assessed according to the definition offered by (Taguchi et al. 2004) which is enhanced to consider generalized uncertainty models and different uncertain structural processes.

The proposed robustness assessment approach bases on a time-dependent reliability analysis of structures under consideration of stochastic and non-stochastic uncertainty. In order to incorporate both stochastic and non-stochastic uncertainty the generalized uncertainty model fuzzy randomness is adopted. Consequently, the reliability analysis results in a time-dependent fuzzy failure probability, see (Sickert et al. 2005). This fuzzy function may be discretized in a set of fuzzy quantities associated with crisp time points. For selected time points the robustness is computed. Therefore a robustness measure is introduced, which considers different approaches of robustness as a unity. Thus, the robustness measure evaluates consequences of uncertainty as well as variations in the load and alteration process on the structural responses. Thereby, the robustness assessment requires measuring the uncertainty of quantities. Measures for uncertainty of random and fuzzy quantities are known, e.g. in (Wu & Mendel 2007). They are introduced for the application in robustness assessment and evaluated by way of examples. Additionally, advanced measures to specify fuzzy random quantities are proposed and also introduced in robustness assessment.

Beside measuring robustness by itself here the interest is focused implicit on designing robust structures under consideration of the generalized uncertainty model fuzzy randomness. The aim is to reduce the uncertainty of result quantities and extend the safety margins in order to become independent to the uncertainty of input quantities to a large extent as well as to capture extreme events. Thereby, reliability analysis is repeatedly performed considering selected stress processes in order to determine a robust structure in the sense that the structure will resist a high number of different – also extreme – load and alteration processes. The introduced measure is applied by way of examples. Thereby the focus is set on engineering applications such as RC folded plate structures under consideration of the governing nonlinearities and uncertainties. The robustness measure may be also utilized to compare different designs in the framework of a robustness-based optimization.

REFERENCES

Beer, M. & Liebscher, M. (2008). Designing robust structures – a nonlinear simulation based approach. *Computers & Structures 86*, 1102–1122.

Sickert, J.-U., Graf, W. & Reuter, U. (2005). Application of fuzzy randomness to time-dependent reliability. In G. Augusti, G. Schuëller & M. Ciampoli (Eds.), *Safety and Reliability of Engineering Systems and Structures Proceedings of the 9th Int. Conference on Structural Safety and Reliability, ICOSSAR'05*, Rome, Rotterdam, pp. 1709–1716. Millpress. CD-ROM, Doc. MS0702.

Taguchi, G., Chowdhury, S. & Wu, Y. (2004). *Taguchi's Quality Engineering Handbook*. New York: Wiley.

Wu, D. & Mendel, J. M. (2007). Uncertainty measures for interval type-2 fuzzy sets. *Information Science 177*, 5378–5393.

Safety, Reliability and Risk of Structures, Infrastructures and Engineering Systems – Furuta, Frangopol & Shinozuka (eds)
© 2010 Taylor & Francis Group, London, ISBN 978-0-415-47557-0

Uncertainty modelling and sensitivity analysis of tunnel face stability

W. Fellin & A. Kirsch
University of Innsbruck, Unit of Geotechnical and Tunnel Engineering, Innsbruck, Austria

J. King
Austrian Academy of Sciences, Breath Research Unit, Dornbirn, Austria

M. Oberguggenberger
University of Innsbruck, Unit of Engineering Mathematics, Innsbruck, Austria

An important issue in the construction of shallow tunnels, especially in weak ground conditions, is the tunnel face stability. It has been a topic of research until the present day to predict the necessary support pressures for shield tunnelling. A variety of theoretical and numerical models have been proposed. In this paper, we use laboratory experiments performed at the University of Innsbruck combined with modern methods of uncertainty and sensitivity analysis for assessing adequacy, predictive power and robustness of the models.

The laboratory experiments were performed with various types of sand in a box with a piston moving in a horizontal cylinder. The experimental record consisted in a sample of force-displacement curves, the onset of a nearly horizontal plateau defining the necessary support pressure.

The principal aim of this research was to validate and compare the predictions of theoretical and numerical models with the experimental results, among them variations of the Horn model, the Ruse-Vermeer model, the Kolymbas model, and a Finite Element model based on the Mohr-Coulomb law.

Robustness and adequacy of the models can be best understood by means of sensitivity analysis. When combining experimental data and theoretical models, sampling based sensitivity analysis – with its recently developed powerful statistical indicators – suggests itself as a suitable approach.

In the models under scrutiny, the output parameter was the necessary support pressure p_f. The input (soil) parameters possessing the largest degree of random variability were identified as the actual void ratio e and the loose and dense state void ratios e_l and e_d, respectively. All these parameters were estimated in small-scale laboratory experiments. Another important model parameter is the friction angle φ of the soil. This parameter is estimated by means of a linear model $\varphi \approx \beta_0 + \beta_1 I_d$, with the relative density I_d (which in turn is a function of e, e_l and e_d). In order

to assess the influence of the regression coefficients β_0, β_1 on the output p_f, we needed to determine their statistical distribution. We achieved this by means of the so-called resampling technique, producing a large bootstrap sample of the experimental data and thereby simulating the joint distribution of β_0 and β_1. We believe that this is a novel method for obtaining joint distributions – including correlations – of geotechnical data.

As an application of the statistical data model, we could assess the ranges of the output parameter p_f by means of the First-Order-Second-Moment-Method and compare them with the experimental results. The model with the best fit was then scrutinized further: we calculated the sensitivities of the output p_f with respect to the five input parameters described above. Here we used Monte Carlo simulation based on the input distributions obtained before. Going beyond the rather crude picture obtained by scatterplots, we computed stronger statistical measures of sensitivity, such as partial correlation coefficients. These indicators are designed so as to remove hidden influences of co-variates. This method lends itself to a further application of resampling, allowing to determine the statistical significance of the resulting sensitivities.

Further, we demonstrate that this approach is applicable even in computationally expensive Finite Element calculations.

The basis for this presentation is formed by the investigations of (Kirsch 2008), and the methods developed in (Oberguggenberger et al. 2009).

REFERENCES

Kirsch, A. 2008. *On the face stability of shallow tunnels.* Ph.D. thesis, University of Innsbruck.

Oberguggenberger, M., King, J. & Schmelzer, B. 2009. Classical and imprecise probability methods for sensitivity analysis in engineering: A case study. *International Journal of Approximate Reasoning. 50*: 680–693.

Mini-Symposia (MS13) Smart Structural Health Monitoring and Safety Assessments
[Session Organized on Behalf of IASSAR Subcommittee SC4]

Safety, Reliability and Risk of Structures, Infrastructures and Engineering Systems – Furuta, Frangopol & Shinozuka (eds)
© 2010 Taylor & Francis Group, London, ISBN 978-0-415-47557-0

Hybrid health monitoring of PSC girder bridges using MEMs-based smart sensors

J.H. Park, D.S. Hong & J.T. Kim
Pukyong National University, Busan, Korea

C.B. Yun
Korea Advanced Institute of Science and Technology, Daejeon, Korea

ABSTRACT

In PSC girder bridges, the compressive prestress makes more economical use of the concrete by allowing all of the concrete section to play some part in supporting external loadings. For the PSC girder bridge, the prestress force of tendon and the flexural stiffness of concrete girder are two important parameters that should be secured for its serviceability and safety against external loadings and environmental conditions.

However, although a proper monitoring system is installed in a PSC bridge, the following problems still remain to be solved: (1) the costs associated with installation and upkeep of monitoring systems, which employ wires for the transfer of measurements to a centralized data server, can be high and (2) data repositories with high capacity is needed for future engineering analysis. In order to overcome these problems, many wireless monitoring systems have been developed.

In this study, a MEMs-based smart sensor is developed for autonomous SHM of PSC girder bridges. Figure 2 shows the schematic of the serial operation scheme for the hybrid SHM in structures. The basic idea of the serial operating scheme is that the global SHM methods using acceleration-based smart sensor node (Acc-SSN) are used for monitoring damage occurrence in global structure level and, as a subsequent operation, the local SHM methods using impedance-based smart sensor node (Imp-SSN) are used for classifying damage types in local member level. Finally, the performance of the developed system and commercial systems is evaluated using a PSC girder model as shown in Figure 1 and Figure 3.

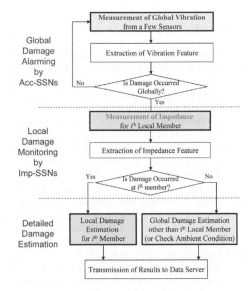

Figure 2. Serial operation scheme for hybrid SHM.

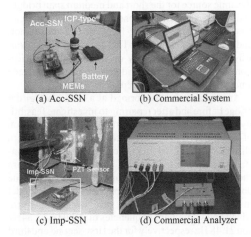

(a) Acc-SSN   (b) Commercial System

(c) Imp-SSN   (d) Commercial Analyzer

Figure 3. Developed smart sensors and commercial measurement systems.

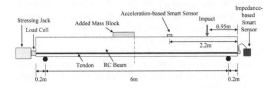

Figure 1. Experimental setup for PSC girder.

Safety, Reliability and Risk of Structures, Infrastructures and
Engineering Systems – Furuta, Frangopol & Shinozuka (eds)
© 2010 Taylor & Francis Group, London, ISBN 978-0-415-47557-0

Modal identification of short and medium span bridges under moving loads

C.-W. Kim, M. Kawatani & R. Ozaki
Department of Civil Engineering, Kobe University, Kobe, Japan

J. Hao
Construction Project Consultants Inc., Tokyo, Japan

ABSTRACT

This study intends to investigate the feasibility of modal identification of a short span bridge by means of a multivariate autoregressive model that provides information relating to natural frequencies, damping constants.

Modal identification of bridge structures based on the system output such as vibration data of bridges is one of the important methods for model updating and bridge health monitoring (Doebling et al, 1996). Ambient vibrations induced by traffic and wind thus have been adopted as dynamic data for long span bridges. For short and medium span bridges, which are insensitive (or sometimes impassive) to the wind-induced vibration, normal traffic excitations are important dynamic sources (Kim and Kawatani, 2009). However, the traffic-induced vibration is a kind of nonstationary process (Kim and Kawatani, 2008) that becomes stronger with decreasing span length. Field experiments report that the dominant frequency of the wheel load of heavy vehicle, which is a major dynamic source for the short and medium span bridges and affects dynamic responses of the bridges, scatters near the first bending frequency of those short and medium span bridges. This is one of the reasons why limit amount of study on the modal identification of short and medium bridges using traffic-induced ambient vibration is reported.

An experiment is performed at a steel plate girder bridge before and after repairing to investigate feasibility of this method in case of traffic-induced vibration of the bridge which is repaired. Examples of pattern change of identified frequencies and damping constants according to repairing by the AR model are shown in Figure 1. Observations demonstrate that the dominant frequency is well identified by the AR model in comparing with those from Fast Fourier Transform which gives dominant frequencies of 4.09Hz, 12.54Hz and 21.48 Hz respectively for the first, second and third modes. On the other hand, estimated damping constants have large variance. After repairing, dominant

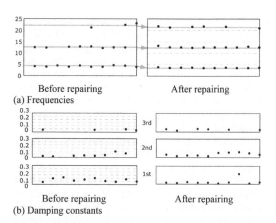

(a) Frequencies

(b) Damping constants

Figure 1. Estimated frequencies and damping constants by the AR model.

frequencies are dropped in comparison with before repairing. Comparing the identification results, for example, between before repairing and after repairing such as pattern change of identified parameters encourages the use of the method for long term health monitoring even for short span bridges.

REFERENCES

Doebling, S.W., Farrar, C.R., Prime, M.B. and Shevitz, D.W. 1996. Damage identification and health monitoring of structural and mechanical systems from changes in their vibration characteristic. A literature review, Los Alamos National Laboratory Report; LA-3070-MS.

Kim, C.W. and Kawatani, M. 2008. Pseudo-static approach for damage identification of bridges based on coupling vibration with a moving vehicle, *Structure and Infrastructure Engineering*. 4(5): pp.371–379.

Kim, C.W. and Kawatani, M. 2009. Challenge for a drive-by bridge inspection, *Proc. of the 10th Int. conf. on Structural Safety and Reliability (ICOSSAR2009)*, 13–17 Sept, Osaka, Japan.

Safety, Reliability and Risk of Structures, Infrastructures and
Engineering Systems – Furuta, Frangopol & Shinozuka (eds)
© 2010 Taylor & Francis Group, London, ISBN 978-0-415-47557-0

Modal analysis of the Yeondae Bridge using a reconfigurable wireless monitoring system

J. Kim & J.P. Lynch
University of Michigan, Ann Arbor, MI, USA

D. Zonta
University of Trento, Trento, Italy

C.B. Yun
Korea Advanced Institute of Science & Technology, Daejeon, Korea

J.J. Lee
Sejong University, Seoul, Korea

ABSTRACT

Wireless sensors have evolved into a viable alternative to traditional tethered sensors. The advantages offered by wireless sensors are their low costs and easy installation. Especially for large structures, such as long-span bridges, elimination of wiring represents a significant cost-savings. Another advantage wireless sensors enjoy over tethered sensors is their ability to be easily reconfigured. In particular, system reconfiguration is attractive for short-term dynamic testing where a small set of sensors can be repeatedly employed in different installation locations.

In this study, a rapid-to-deploy wireless monitoring system is utilized to monitor the vibration response of the 180 meter long Yeondae Bridge. The bridge consists of a steel box girder section that continuously spans across five supports. Using 20 wireless sensors with MEMS accelerometers interfaced, the wireless monitoring system is installed to measure the bridge response to truck traffic. The elimination of wiring allows the wireless monitoring system to be reconfigured at will. The monitoring system sensors are installed using three different spatial configurations so that the acceleration response of the bridge can be measured at 50 sensor locations evenly distributed across the bridge length. Using traditional off-line output-only modal identification techniques, *i.e.* frequency domain decomposition (FDD), mode shapes were successfully obtained by combining acceleration data collected from the three system topologies. Mode shapes are compared to those predicted by a finite element model of the Yeondae Bridge with excellent agreement found.

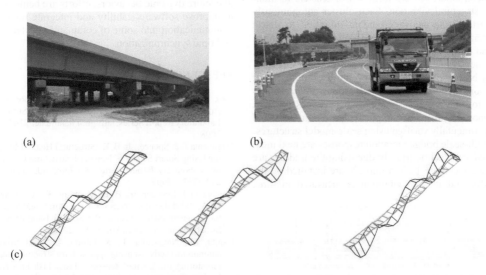

(a) (b)

(c)

Figure 1. Yeondae Bridge: (a) perspective view, (b) forced vibration testing, and (c) extracted modes at 2.246 Hz, 2.637 Hz, and 3.337 Hz.

Safety, Reliability and Risk of Structures, Infrastructures and Engineering Systems – Furuta, Frangopol & Shinozuka (eds)
© *2010 Taylor & Francis Group, London, ISBN 978-0-415-47557-0*

Structural health monitoring system development and full-scale bridge vibration measurement using smart sensors

T. Nagayama, M. Ushita, H.M. Dinh & Y. Fujino
University of Tokyo, Tokyo, Japan

B.F. Spencer, Jr., J.A. Rice, S.A. Jang, K.A. Mechitov & G.A. Agha
University of Illinois at Urbana-Champaign, IL, USA

ABSTRACT

Densely distributed smart sensors can provide rich information for structural health monitoring using their computational and wireless communication capabilities. Though smart sensor technology has seen substantial advances during recent years, fundamental functionalities for structural monitoring such as synchronized sensing and reliable communication have just begun to be provided as middleware services (Nagayama et al 2009).

Utilizing these middleware services, two structural monitoring systems are proposed. Both systems employ distributed processing of synchronously obtained vibration data. The size of data collected in dense smart sensor networks is substantial and therefore central data processing systems are not scalable to a large number of sensors. The systems are designed to handle structural vibration data in a decentralized manner. Information from sensors in the neighborhood is utilized in data processing while sensors far away from each other do not need to communicate. In the first system, smart sensors form sensor clusters to monitor structural vibration. Each cluster autonomously measures and analyzes random vibration to detect damage in a distributed and collaborative manner. (Gao & Spencer 2008; Nagayama & Spencer 2007). In the second system, smart sensors measure railway bridge vibrations based on train schedule. Smart sensors then extract vibration responses corresponding to train passages and apply data processing. The proposed systems are implemented on Imote2s and experimentally verified using scale-model structures.

Full-scale bridge vibration responses are next measured at a historic truss bridge using the middleware services (Figure 1). Ten Imote2s are installed on the bridge and its free vibration is measured in three

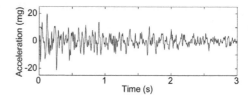

Figure 2. Vertical acceleration response.

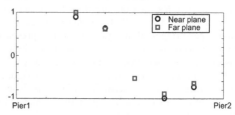

Figure 3. 2nd vertical modeshape.

directions (Figure 2). Offline analyses of the collected data reveal modal properties of the vibration including modeshapes (Figure 3). To advance the capability to capture dynamic behaviors, efforts are being made to improve software stability and integrate multihop communication into some of command transfer tasks and time synchronization.

REFERENCES

Gao, Y. and Spencer Jr., B.F.. Structural health monitoring strategies for smart sensor networks. Newmark Structural Laboratory Report Series, http://hdl.handle.net/2142/8802, 2008.

Nagayama T. & Spencer, Jr. B. F. "Structural Health Monitoring Using Smart Sensors," Newmark Structural Engineering Laboratory Report Series 001 http://hdl.handle.net/2142/3521, 2007.

Nagayama T. Spencer, Jr., B. F., Mechitov, K. A., & Agha, G. A. "Middleware services for structural health monitoring using smart sensors" the Special Issue of Smart Structures and Systems, Vol. 5, No. 2, 2009.

Ushita, M., Nagayama, T., & Fujino, Y. "A distributed autonomous active-sensing approach for structural health monitoring using smart sensors." Proc. 11th East Asia-Pacific Conference on Structural Engineering & Construction (EASEC-11), Taipei, Taiwan, 2008.

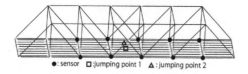

●: sensor   □:jumping point 1   △ : jumping point 2

Figure 1. The truss bridge and sensor locations easurement.

Safety, Reliability and Risk of Structures, Infrastructures and
Engineering Systems – Furuta, Frangopol & Shinozuka (eds)
© *2010 Taylor & Francis Group, London, ISBN 978-0-415-47557-0*

Fiber optic smart monitoring of railway structures

K.S. Kim

Department of Materials Science and Engineering, Hongik University, Chungnam, Korea

Sung Gyu Cho & Myeong Se Kim

ICES, Business Incubation Center, Hongik University, Chungnam, Korea

ABSTRACT

For monitoring of railway structures, optical fiber sensors are very convenient. The fiber sensors are very small and do not disturb the structural properties. They also have several merits such as electro-magnetic immunity, long signal transmission, good accuracy and multiplicity of one sensor line. Strain measurement technologies with fiber optic sensors have been investigated as a part of smart structure. In this paper, we investigated the possibilities of fiber optic sensor application to the monitoring of railway structures. Fiber optic sensors showed good durability and long term stability for continuous monitoring of the railway structures as well as good response to the structural behaviors during construction.

In Won-hyo Tunnel KTX 13-4 construction site, the automatic measurement was enforced with the FBG sensors at total 13 sections. Figure 1 showed the installation areas of FBG sensors on the ending point of the tunnel. Basically, it was designed to be installed on each 30 m distance. However, 15 m distance on the entrance area which could be weak of safety, and observed the changes of diameter of the tunnel.

During two months after the installation of sensors, the displacement change relatively showed a lot; however, after that during about six month, the change was almost not shown. After 14 February 2007, it showed a sharp displacement change. Therefore, we can conjecture that after section 378 km 310 lower part digging

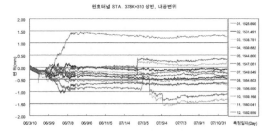

Figure 2. Section 378 km 310 Upper Part Displacement Change Measurement Data.

was started at 14 February 2007, the displacement change was occurred on the upper part of the tunnel.

The point of sharp displacement change occur time shows different between the locations of the measurement, such as Figure 2. This could be explained as the result of the construction, which was processed separately as two pieces from the middle point of the tunnel. After the digging of the lower part, and after the stabilization of the tunnel, sharp displacement change was no longer shown.

This measurement process was continued as a very difficult situation which was performed during construction; however, many sensors are showing good data from the beginning to the end. Compare with the measured data of originally installed electronic sensors in the tunnel or subway construction, this measured data from fiber optic sensors have a remarkable reliability.

REFERENCES

A. D. Kersey, K. P. Koo and M. A. Davis, "Fiber Optic Bragg Grating Laser Sensors", SPIE, Vol. 2292 Fiber Optic and Laser Sensors XII, pp. 102–112, (1994)

W. W. Morey, J.R. Dunphy, and G. Meltz, "Mul-tiplexing Fiber Bragg Grating Sensor", SPIE, Vol. 1586, Paper #22, Boston, pp. 216–224. (1994)

Kim, K. S., L. Kollar and G. S. Springer, "A Model of Embedded Fiber Optic Fabry-Perot Temperature and Strain Sensors", J. of Composite Materials Vol. 27, pp. 1618–1662, (1993)

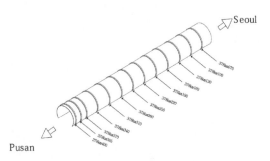

Figure 1. Installation Area of FBG Sensor at KTX 13-4 Construction Site.

Mini-Symposia (MS12) Monte Carlo Simulations Applied to Civil Engineering

Safety, Reliability and Risk of Structures, Infrastructures and Engineering Systems – Furuta, Frangopol & Shinozuka (eds)
© 2010 Taylor & Francis Group, London, ISBN 978-0-415-47557-0

Stochastic simulation of local elasto-plastic properties of composite materials

L. Graham-Brady & K. Acton
Johns Hopkins University, MD, USA

ABSTRACT

Homogenization methods are often a useful tool for addressing random composite materials, as they provide a simplified yet physically relevant representation of the material microstructure. Homogenization methods applied at the macro-scale provide average constitutive properties, thereby obscuring effects that appear locally in the materials, including stress concentrations. These local effects may be especially important when analyzing nonlinear material behavior such as plasticity, in which local yielding may provide the mechanism for initiation of material failure. Meso-scale homogenization provides an intermediate solution by allowing for local variations in constitutive properties. Using this technique, some local effects are retained, while taking advantage of the simplifications provided by homogenization. This work examines plastic behavior in composites using meso-scale modeling, specifically, the moving window Generalized Method of Cells (GMC).

Former research by the co-authors has investigated the optimal level of smoothing (i.e., the optimal window size) to be applied to pixelized two-phase microstructures with a range of contrast ratios of elastic constants and yield stresses. This model was based on a comparison with an analytical model of a single fiber in an infinite matrix under hydrostatic loading. Having chosen the optimal level of smoothing for a given material, the model is expanded here from a single inclusion model to a 2D plane strain model which includes a random scattering of circular inclusions; this can be considered to model a transverse section of a fibrous composite. In the current work, a comparison is made between a two-phase finite element model of the material with an irregular mesh, and a windowed finite element model of the material with a regular, pixelized mesh.

The elastoplastic properties obtained using the moving-window approach described above provides samples that serve as the basis of stochastic simulation. By calculating the statistics that describe the randomly varying elastoplastic properties, it is then straightforward to apply stochastic simulation techniques to generate further samples of these properties. The simulated properties are used as input to finite element analyses to measure variability in such quantities as the global stress that leads to initiation of plastic strains and the area fraction of yielded material at given global stress levels. These quantities can be viewed as some measure of material reliability, since they are a reflection of the damage levels that vary between different samples of a nominally identical material.

Safety, Reliability and Risk of Structures, Infrastructures and
Engineering Systems – Furuta, Frangopol & Shinozuka (eds)
© 2010 Taylor & Francis Group, London, ISBN 978-0-415-47557-0

Analysis of the value of information in the design of resilient water distribution networks

M. Comboul & R. Ghanem
University of Southern California, Los Angeles, CA, USA

ABSTRACT

Water distribution networks are centralized around urban areas, which are major consumption sites. Unnoticed, the pollution of a municipal drinking system with harmful chemicals could have a catastrophic impact on public health. It is crucial to anticipate the potential contamination of a water system whether in the form of an accidental infiltration of pollution due to an aging/damaged system or an intentional release of chemical or biological agent. Water networks should be monitored accordingly to prevent a contaminant from spreading undetected. However, in a large water distribution network, the system scale and budget constraints usually prohibit the monitoring of the entire structure so where should it be observed to guarantee rapid threat detection? The optimal sensor placement problem in water distribution networks constrained on the budget or the number of sensors has been abundantly undertaken in the past. Primarily, the difficulty comes from the absence of information about the potential contamination events. The nature of the pollution we want to apprehend can take many forms in terms of source of injection, type of contaminant, and instance of occurrence, the impact of which is closely correlated with the flow pattern and the water demands. In fact, two identical chemicals injected at the same source but at different times or dates will likely generate distinct impacts given that the flow patterns driving the pollution spread and the water demands will change. The water demands, which drive the flow pattern, fluctuate in time depending on the time of the day, the day of the week and the season; they reflect the spatial distribution of the population. Small perturbations in the nodal demand values can extensively modify the flow pattern, thus it may be unfeasible to identify all possible flow patterns.

Yet, current literature on the subject for the most part ignores those effects and assumes deterministic demands.

The objective of this study is to find an optimal sensor layout that is robust to unforeseen fluctuations in the nodal demands as well as the possible sensor failures. The EPANET object library allowed us to build an efficient Monte Carlo simulation tool for the study of water network responses to uncertainty in the demand parameters and contamination scenarios. For each demand realization, we perform a hydraulic simulation and for each contamination scenario we perform a water quality simulation. Finally, the expected damage reduction over all demand patterns and all scenarios is maximized over all possible sensor placements. For our optimization formulation, we built on the paper by (1) and introduced uncertainty in the demands and an imperfect sensor probability of detection that is proportional to the contaminant concentration at the time it reaches the sensor. The fast greedy algorithm developed in (1) allowed us to find a near-optimal solution in a reasonable amount of time. The comparison of the different outcomes obtained using deterministic/stochastic demands and perfect/imperfect sensors demonstrate the high affectibility of the resulting sensor placement in relationship to those initial assumptions.

REFERENCE

[1] Leskovec J., Krause A., Guestrin C., Faloutsos C., Van-Briesen J., Glance N., 2008. Cost-effective outbreak detection in networks. In 13th ACM SIGKDD International Conference on Knowledge Discovery and Data Mining.

Safety, Reliability and Risk of Structures, Infrastructures and Engineering Systems – Furuta, Frangopol & Shinozuka (eds)
© 2010 Taylor & Francis Group, London, ISBN 978-0-415-47557-0

Integrated approach to assess the impact of tsunami disaster

Shunichi Koshimura & Shintaro Kayaba
Disaster Control Research Center, Graduate School of Engineering, Tohoku University, Japan

Masashi Matsuoka
National Institute of Advanced Industrial Science and Technology, Japan

ABSTRACT

A research project is underway by the support of Industrial Technology Research Grant Program of New Energy and Industrial Technology Development Organization (NEDO), to assess the impact of major tsunami disaster in the aftermath of the event. The authors propose a research framework in developing a method to search and detect the impact of tsunami disaster by integrating numerical modeling, remote sensing and GIS technologies, which consist of four damage mapping efforts, i) Regional hazard/damage mapping effort to search the potential impacted region, ii) Damage estimation effort using the numerical modeling of tsunami inundation and fragility function for structural damage, iii) Regional damage detection effort using SAR imagery, iv) Local damage mapping effort using the analysis of high-resolution optical satellite imagery to detect the extent of tsunami inundation zone and the structural damage.

The method is implemented to the recent tsunami event, the 2007 Solomon Islands earthquake tsunami, to identify the structural damage probabilities within the inundation zone, combined with the post-tsunami survey data.

The development of the present method is now underway in developing automatic damage detection algorithms, the real-time tsunami inundation modeling, building house/structure inventory database, and high-resolution merged bathymetry-topography data. Especially, the global deployment of developing fragility functions including damage mapping with satellite imagery is one of the most significant issues. Also, another significant barrier to overcome is developing the database of high-resolution bathymetry and topography grid in world-wide scale.

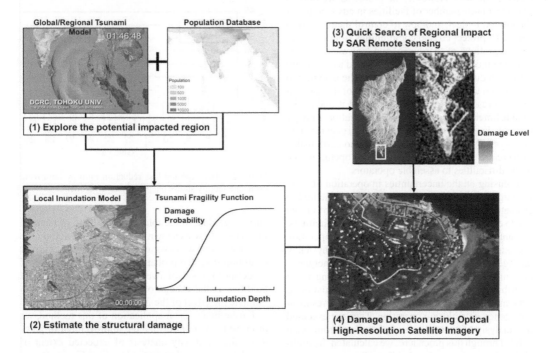

Figure 1. Structure of the present study.

Safety, Reliability and Risk of Structures, Infrastructures and Engineering Systems – Furuta, Frangopol & Shinozuka (eds)
© 2010 Taylor & Francis Group, London, ISBN 978-0-415-47557-0

An evaluation method of tsunami vulnerability in metropolitan area considering the functionality of coastal facilities

S. Suzuki & Y. Kawata
Disaster Prevention Research Institute, Kyoto University, Kyoto, Japan

S. Koshimura
Disaster Control Research Center, Graduate School of Engineering, Tohoku University, Sendai, Japan

ABSTRACT

Pacific coast of southwest Japan is under the threat of extensive disaster due to the great earthquake and tsunami that are supposed to occur along the Nankai Trough within next 30 years. As tsunami disaster preparedness and mitigation effort, local governments are maintaining coastal facilities, such as seawalls and land locks. And they are also preparing emergency response systems to operate and utilize these facilities. Especially in Osaka, more than 400 gates in 60 km-length of seawall are in operation and are supposed to be closed prior to the tsunami arrival (within 90 minutes). The maintenance and operation of these facilities are important requisites to reduce damage and loss. However, the problems are how to manage a large number of facilities in advance under the limited budget, and how to control them at emergency. To protect coastal communities from tsunami, the integrations of 'hardware' and 'software' aspects of tsunami countermeasures are required as a tsunami disaster reduction system. Because of the unexpected damage on the facilities, the feasibility of these countermeasures has great uncertainty. And it makes the establishment of tsunami disaster mitigation strategy difficult. Failure of countermeasures will occur if there are physical damages in facilities due to earthquake, or there are malfunction of emergency operation due to the difficulties to assemble operators.

Focusing on the uncertainties in operation of closing large number of gates, we developed a method to evaluate the vulnerability of the waterfront of Osaka and to evaluate the effect of improvement of facilities. The uncertainty in the failure of gate-closing operation consists of the probabilities if the gates are physically damaged by strong ground motion and if emergency personnel are able to operate the gate. Using these probabilities, we carry out Monte Carlo simulation to generate possible status of all the gates as a result of emergency gate closing operation. Then we executed the analysis of tsunami inundation flow from open gates. Through this procedure, we calculated the probability distribution and probability of non-exceedance

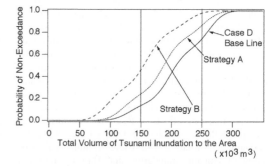

Figure 1. Example of the non-exceedance probability curves that shows the effect of mitigation strategies.

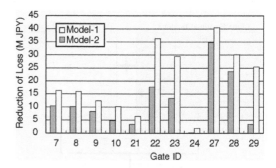

Figure 2. The expected loss reduction ratio by improving each gate facility shows the priority of investment.

curve regarding to the total water volume of tsunami inundation flow as shown in Figure 1. By setting the acceptable total water volume, we can know the current condition from the probability of non-exceedance. If we define the result of investment in improving coastal facilities as the change of the uncertainties, we can evaluate the effect of the improvement.

Furthermore, as an application of this method, we discussed how to prioritize the improvement of gates using the sensitivity analysis of expected extent of inundation area and expected loss (e.g. Figure 2).

Safety, Reliability and Risk of Structures, Infrastructures and Engineering Systems – Furuta, Frangopol & Shinozuka (eds)
© 2010 Taylor & Francis Group, London, ISBN 978-0-415-47557-0

Seismic performance simulation of water systems considering uncertainties in ground motions and pipeline damages

G.-Y. Liu & H.-Y. Hung

National Center for Research on Earthquake Engineering, Taipei, Taiwan

ABSTRACT

Water systems are one of the most essential infrastructures in modern societies. The disruption of water supply following earthquakes may cause serious inconvenience to the daily life of people in the disastrous areas. Medical caring, sanitation, fire-fighting and so forth may be seriously affected, too. In this study, a technology for the assessment of seismic performance of water pipe networks has been developed based on earthquake scenario simulation and pipe flow hydraulic analysis. Figure 1 depicts the flow chart of the proposed assessment procedure. The pipe damage model summarized by Shi and Wang (Shi 2006, Wang 2006) was adopted. This model includes pipe break and various types of pipe leaks, and the probability that each will occur in various pipe materials, and also the hydraulic models and parameters for each type of pipe damage (Figure 2). The software EPANET (Rossman 2000) was employed the water network hydraulic analysis. The proposed assessment technology is capable of dealing with uncertainties raised from either the strong motion attenuation or the pipe repair rate.

The water system of Taipei metropolitan area (Figure 3) has been employed as a test bed for case study.

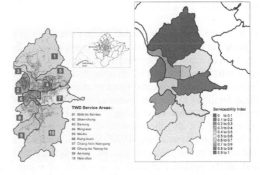

Figure 3. Taipei water system.   Figure 4. Assessment results.

The system serviceability following earthquakes has been quantified (Figure 4, of an M7.5 event related to the Hsincheng fault). It seems that the network characteristics rather than other factors have a predominant effect on a system's seismic performance. The influence of uncertainties in strong motion attenuation and pipe repair rate has been investigated. They were preliminarily found of little or no influence to the assessment results. Also, there is no clear tendency that these uncertainties will affect the speed of convergence. Further investigations to bridge the behavior of system performance and network complexity for water and other lifeline systems are highly recommended.

REFERENCES

Rossman, L. A. 2000. EPANET 2 User's Manual, EPA, Cincinnati, OH.

Shi, P. 2006. Seismic Response Modeling of Water Supply Systems, Ph.D. Dissertation, Cornell University, Ithaca, NY.

Wang, Y. 2006. Seismic Performance Evaluation of Water Supply Systems, Ph.D. Dissertation, Cornell University, Ithaca, NY.

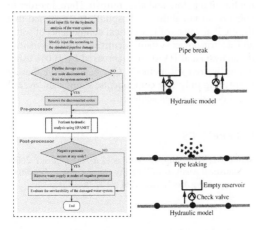

Figure 1. Assessment procedure.   Figure 2. Pipe damage models.

Safety, Reliability and Risk of Structures, Infrastructures and
Engineering Systems – Furuta, Frangopol & Shinozuka (eds)
© 2010 Taylor & Francis Group, London, ISBN 978-0-415-47557-0

Enhancement of sustainability of nonstructural systems in critical facilities

S. Ray Chaudhuri & M. Shinozuka

Department of Civil and Environmental Engineering, University of California, Irvine, CA, USA

ABSTRACT

In a severe earthquake event, critical facilities are
expected to remain operational. This is because facilities such as acute care hospital must assist the seismically injured and otherwise traumatized people with
immediate medical attention and necessary care. For
these facilities to remain operational, not only their
building structures must remain safe for continued
occupancy, but also their non-structural components
including elevators, stairs, HVAC systems, sprinklers
and other fire suppression systems, communications
and utility systems, as well as a variety of equipment, must remain functional. Significant studies have
been made to seismically upgrade and retrofit the
non-structural components in these critical facilities.
However, there are no definitive codes, regulations
and guidelines developed as yet that take into consideration their design, manufacturing, qualification
test, installment, operation and maintenance from the
system performance point of view. A number of studies separately dealt with some of these issues without
significantly recognizing the systems aspect.

The present study explores the system aspect and
demonstrates systems analysis for the ultimate purpose of the development of a framework for design
codes and guidelines for non-structural components
in order to enhance sustainability of nonstructural
systems in critical facilities. Without such recognition, codes and guidelines may be developed for each
type of non-structural components, as often observed
so far, implicitly allowing them to be designed at
a specific risk level of seismically induced failure
or malfunction. The probabilistic systems analysis
involves development of component fragility (developed based on observed and experimental data or
analytical simulations coupled with engineering judgments) and in principle utilizes event and fault tree
procedures to evaluate the system fragility defined
by the probability that the system will not perform

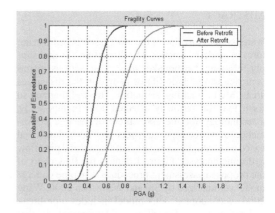

Figure 1. Fragility of hospital before and after retrofit.

its intended purpose (Shinozuka et al. 2000). Finally,
enhancement of sustainability by improving performance of nonstructural components is demonstrated
by using numerical model of a representative hospital
building, a suit of ground motions of different hazard
levels, and utilizing the concept of average sensitivity index. Figure 1 shows the over fragility curve for
the hospital before and after retrofit. Here, in order to
enhance sustainability of hospital, it is assumed that
the median values of few critical components of water
and electric power systems identified through sensitivity analysis and HVAC are increased by 100% using
innovative retrofit methods. One can observe that by
retrofitting only few components, significant overall
fragility reduction is achieved.

REFERENCES

Shinozuka, M. & Feng, M. Q. & Lee, J. & Naganuma, T. 2000.
Statistical Analysis of Fragility Curves. *ASCE Journal of
Engineering Mechanics*. 126(12) 1224–1231.

Mini-Symposia (MS04) Structural Reliability and Optimization

Safety, Reliability and Risk of Structures, Infrastructures and
Engineering Systems – Furuta, Frangopol & Shinozuka (eds)
© 2010 Taylor & Francis Group, London, ISBN 978-0-415-47557-0

Discrete-continuous variable optimization of stochastic dynamical systems

H.A. Jensen & D.S. Kusanovic

Department of Civil Engineering, Santa Maria University, Valparaiso, Chile

For many structural optimization problems the design variables must be selected from a list of discrete values. For example, cross-sectional areas for truss members, in practice, have to be chosen from a list of commercially available member sizes. In addition, the growing use of composite materials also underlines the importance of being able to treat structural optimization problems where some or even all of the design variables are discrete.

In this work attention is directed to the structural synthesis of stochastic dynamical systems. In particular, the reliability-based optimization of linear and non-linear systems involving discrete sizing type of design variables and subject to stochastic excitation is considered. The reliability-based optimization problem is formulated as the minimization of an objective function for a specified reliability. The probability that design conditions are satisfied within a given time interval is used as measure of system reliability. The basic mathematical programming statement of the structural optimization problem is converted into a sequence of explicit approximate primal problems of separable form. For this purpose, the objective function and the reliability constraints are approximated by using a hybrid form of linear and reciprocal approximations. Specifically, a conservative approximation is considered in the present formulation. The approximations are combined with an efficient simulation technique to generate explicit expressions of the reliability constraints in terms of the discrete design variables.

The explicit approximate primal problems are solved by constructing continuous explicit dual functions, which are maximized subject to simple non-negativity constraints on the dual variables. The evaluation of the dual function is direct since it requires the minimization of a series of one-dimensional minimization problems. A gradient projection type of algorithm is used to find the solution of each dual problem. The efficiency of the method is demonstrated by presenting a numerical example of a non-linear system subject to stochastic ground acceleration. The number of reliability estimations required during the optimization process is very small. In fact, that number represents a small fraction of the total number of reliability estimations to be performed in discrete variable algorithms that treat the problem directly in the primal space (i.e. branch and bound techniques).

The proposed method represents a structural synthesis capability for discrete sizing problems which exhibits a computational efficiency approaching those achievable in pure continuous variable problems. Thus, this technique is expected to be useful in discrete variable structural synthesis of stochastic dynamical systems

Safety, Reliability and Risk of Structures, Infrastructures and
Engineering Systems – Furuta, Frangopol & Shinozuka (eds)
© 2010 Taylor & Francis Group, London, ISBN 978-0-415-47557-0

Bayesian network as a framework for structural reliability analysis in infrastructure systems

Daniel Straub
Technical University Munich, Germany

Armen Der Kiureghian
University of California, Berkeley, CA, USA

ABSTRACT

The authors have investigated the combination of
Bayesian networks (BN) with structural reliability methods (SRM) elsewhere (Straub and Der
Kiureghian, Proc. IFIP WG 7.5 conference, Toluca,
Mexico, 2008). Structural reliability methods allow
computation of failure probabilities for structural components or systems defined by continuous random
variables. The BN methodology facilitates updating of
the model with evidence, such as results of inspections
or observations of hazards and system performances.
The resulting combination of the SRM and BN is
termed enhanced Bayesian network (eBN). The eBN
facilitates the integration of the detailed modeling of
a structure through SRM into the wider context of risk
and decision analysis in complex civil systems, in particular for systems with evolving information, where
the capabilities of the BN for Bayesian updating can
be exploited. For this reason, the eBN concept is well
suited for multi-scale probabilistic analysis of large
infrastructure systems and, ultimately, for decision
optimization, at the planning stage and in near-real-
time during the operational phase. This paper outlines
a generic framework for infrastructure risk analysis
based on the eBN framework and demonstrates its
application.

Figures 1 and 2 illustrate the eBN modeling of an
exemplarily infrastructure system. The model includes
both the temporal dimension and the spatial dimension, and is able to update the full probabilistic model
with any information in an integral manner. Such
information includes measurements of material properties, observations of structural performances and
monitoring of environmental loads. The eBN, like any
other model, has its computational limitations that
are related to the complexity of the network structure, which can be analyzed through a novel concept,
the Markov envelopes. Overall, it is concluded that
the eBN provides a powerful tool and framework for
decision support in large infrastructure systems in
near-real-time.

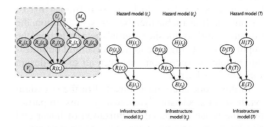

Figure 1. eBN model of the *j*th structure within the full
model of the infrastructure system (temporal dimension).
$R_j(t)$ is the capacity at time t, M_{j4} is the measurements of
structural element 4, H_j is the hazard at location j, D_j is the
deterioration of structure j and E_j is the performance of the
structure.

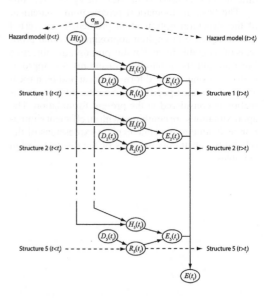

Figure 2. eBN model of the infrastructure at an instance
of time (spatial dimension). The model includes dependence
among the hazard at different locations (through H) and
between different hazard events (through σ_{Hi}).

Safety, Reliability and Risk of Structures, Infrastructures and
Engineering Systems – Furuta, Frangopol & Shinozuka (eds)
© 2010 Taylor & Francis Group, London, ISBN 978-0-415-47557-0

Application of advanced simulation methods for the reliability and redundancy analysis of bridge structures

Feng Miao & Michel Ghosn

Department of Civil Engineering, The City College of New York/CUNY, New York, NY, USA

ABSTRACT

Following several recent failures and catastrophic collapses due to man-made and natural hazards such as the collapse of the World Trade Center towers in 2001 and the recent I-35W Mississippi River bridge in Minnesota, there has been increased interest in upgrading the bridge design philosophy from a member-oriented approach to a system-oriented approach in order to ensure that bridge structures possess adequate levels of structural safety, redundancy and robustness. To account for the high levels of uncertainty associated with the quantification of the levels of redundancy of complex structures while remaining consistent with the reliability basis of recent bridge design code developments, system reliability methods must be applied.

Several recent studies have focused on the progressive collapse mechanisms of existing buildings and on developing guidelines for designing buildings with high levels of redundancy (Marjanishvili 2004). Existing criteria to reduce the risk of progressive collapse have been developed for buildings using traditional deterministic methods. More recently, Ellingwood (2006) proposed an outline describing how to account for member and load uncertainties in developing progressive collapse guidelines. These probability-based guidelines are similar to those proposed earlier by Ghosn & Moses for evaluating the redundancy and system safety of bridge systems.

The reliability procedures followed by Ghosn & Moses (1998) used simplified models because of the difficulty of obtaining accurate estimates of the reliability of complex structural systems using existing techniques. However, recent research has led to the development of efficient reliability analysis methods capable of obtaining accurate estimates of the reliability of complex systems with large numbers of random variables and low probabilities of failures.

In this paper, an advanced Markov-Chain-based simulation method proposed by Au & Beck (2001) and known as "Subset Simulation" is used to evaluate the reliability of bridge systems under overloads. System reliability is evaluated for different limit states including overloading and the risk of collapse of bridges that sustained local failure following the occurrence of fatigue or collision type hazards. The application of system factors following the approach proposed by Ghosn & Moses (1998) would serve to improve overall system safety to meet pre-set target reliability levels.

The analysis of a prestressed concrete bridge and a truss bridge example illustrate the reliability analysis process. The results of these reliability analyses can then be used to penalize non-redundant bridges by requiring that their members be more conservatively designed than equivalent redundant systems. Alternatively, members of bridges having adequate levels of redundancy and overall system safety would be allowed to use lower safety factors.

REFERENCES

Marjanishvili 2004. Progressive Analysis Procedure for Progressive Collapse. Journal of Performance of Constructed Facilities, ASCE, May: 79–85.

Ellingwood, Bruce R. 2006. Mitigating Risk from Abnormal Loads and Progressive Collapse, Journal of Performance of Constructed Facilities, ASCE, November: 315–323.

Au, S.K. & Beck, J. 2001. Estimation of small failure probabilities in high dimensions by subset simulation. Probabilistic Engineering Mechanics 16:263–277.

Ghosn, M. & Moses, F. 1998. NCHRP Report 406, Redundancy in Highway Bridge Superstructures. Transportation Research Board–National Research Council. National Academy Press, Washington D C.

Safety, Reliability and Risk of Structures, Infrastructures and Engineering Systems – Furuta, Frangopol & Shinozuka (eds)
© 2010 Taylor & Francis Group, London, ISBN 978-0-415-47557-0

Reliability analysis of fracture in piezoelectric components with a random microstructure

C.V. Verhoosel

Faculty of Aerospace Engineering, Delft University of Technology, Delft, The Netherlands

M.A. Gutiérrez

Faculty of Mechanical, Maritime and Materials Engineering, Delft University of Technology, Delft, The Netherlands

ABSTRACT

Piezoelectric ceramics are materials that exhibit relatively large deformations through application of an electric field and vice versa. This electromechanical coupling makes piezoelectrics suitable for many applications, such as MEMS (micro electromechanical systems). The pronounced piezoelectric effect on small scales is one of the driving forces of miniaturisation of piezoelectric components. Since the control of the production processes of piezoelectric components on this small scale is more difficult than on larger scales, the structural properties of piezoelectric MEMS are subject to relatively large uncertainties. These uncertainties can have a significant influence on both the performance and reliability of the components. Numerical tools for assesing these issues are important for the further development of piezoelectric microsystems.

In this contribution the fracture process of a thin PZT (Lead Zirconate Titanate) specimen is studied. For the considered specimen, a significant spatial variation in properties is experimentally observed. The idea of the proposed model is to simulate crack nucleation and propagation in a homogenised specimen with random fields for the bulk and cohesive properties. These random fields are generated by simulation of the random microstructure.

From SEM images it is observed that voids are present in the granular microstructure. The microstructural geometry is described using random fields for the porosity and average grain size. In order to derive the homogenised properties of the microstructure, the granular structure needs to be simulated. The grains are modelled using a Poisson Voronoi diagram (Okabe et al. 1992), with periodic boundary conditions. The voids are modelled by assuming that some of the Voronoi cells represent voids instead of grains. The parameters used for the simulation of the microstructure are determined on the basis of real statistical data obtained from SEM images. Three-dimensional finite element simulations are used to obtain the homogenised properties of the microstructure at specific points on the specimen (Verhoosel and Gutiérrez 2008). Random fields for these properties are created on the basis of the homogenisation results.

Fracture of piezoelectric ceramics such as PZT has been studied both experimentally and numerically. In this contribution, crack nucleation and growth is modelled by means of the partition of unity method (Babuska and Melenk 1997) in combination with cohesive zone models. The probability of occurrence of failure below a specified critical load is determined using reliability analysis (Gutiérrez and Krenk 2004). The design point is computed using semi-analytical expressions for the sensitivities. The proposed model is demonstrated by means of numerical simulations.

REFERENCES

Babuska, I. and J. M. Melenk (1997). The partition of unity method. *International Journal for Numerical Methods in Engineering 40*(4), 727–758.

Gutiérrez, M. A. and S. Krenk (2004). *Encyclopedia of computational mechanics*, Chapter Stochastic finite element methods, pp. 657–681. Wiley.

Okabe, A., B. N. Boots, and K. Sugihara (1992). *Spatial tessellations: concepts and applications of Voronoi diagrams*. Chichester, England; New York: Wiley.

Verhoosel, C. V. and M. A. Gutiérrez (2009). Modelling inter- and transgranular fracture in piezoelectric polycrystals. *Engineering Fracture Mechanics 76*(6), 742–760.

Safety, Reliability and Risk of Structures, Infrastructures and Engineering Systems – Furuta, Frangopol & Shinozuka (eds)
© 2010 Taylor & Francis Group, London, ISBN 978-0-415-47557-0

Approximate optimization of general systems with high dimensional uncertainties and multiple reliability constraints

Jianye Ching

National Taiwan University, Taipei, Taiwan.

Wei-Chi Hsu

National Taiwan University of Science and Technology, Taipei, Taiwan

ABSTRACT

A novel approach is proposed to solve reliability-based optimization (RBO) problems where the uncertainty dimension can be large and where there may be many reliability constraints. The basic idea is to transform all reliability constraints in the target RBO problem into non-probabilistic ordinary ones by a theorem of equivalence between safety factor (η^*) and reliability ($1-P_F^*$).

Three numerical examples are investigated to verify the proposed novel approach, including a retaining wall subjected to earthquake excitation and system uncertainties. Eight limit states such as lateral sliding, rotation, bearing capacity failure, etc. are considered as the reliability constraints. The design parameters considered in this example include the height of the retaining wall θ_1 and the weight of the wall θ_2. The objective of this example is to minimize the volume of the concrete used to construct the wall subject to

the eight reliability constraints that the probabilities of failure of the first four limit states should not exceed 10^{-3} and that the probabilities of failure of the last four limit states should not exceed 10^{-2}.

With a single run of subset simulation, the estimated $\eta^* - P_F^*$ relationships are shown in Figure 1. In Figure 2, the resulting safety-factor feasible sets are shown, where the thick lines are the feasibility boundaries. The reliability feasible sets obtained by the brute-force analysis for the eight limit states are shown in Figure 2, where the regions with label O indicate the feasible region, while the label × regions are infeasible. The comparison shows that the safety-factor feasible sets are close to the actual reliability feasible sets. These results show that the proposed approach is effective in converting reliability constraints into safety-factor constraints for this example.

Figure 1. The estimated $\eta^* - P_F^*$ relationships (solid line) for the eight limit states.

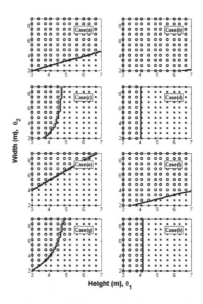

Figure 2. The safety-factor feasible sets versus the reliability feasible sets, where the thick lines are the feasibility boundaries for the safety factors, and its comparison with the actual reliability feasible regions (the region with label O).

Mini-Symposia (MS11) Current Status and Future Applications of Probabilistic Seismic Hazard Assessment

Safety, Reliability and Risk of Structures, Infrastructures and
Engineering Systems – Furuta, Frangopol & Shinozuka (eds)
© 2010 Taylor & Francis Group, London, ISBN 978-0-415-47557-0

Development of conditional probabilistic seismic hazard map

S. Fukushima
Tokyo Electric Power Services Co., Ltd., Tokyo, Japan

T. Hayashi
Tokio Marine & Nichido Risk Consulting Co., Ltd., Tokyo, Japan

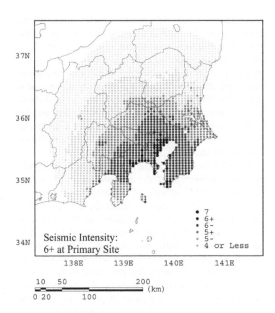

Figure 1.   Conditional seismic hazard map for Kanto region.

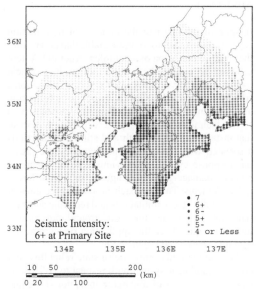

Figure 2.   Conditional seismic hazard map for Kansai region.

ABSTRACT

Probabilistic seismic hazard technique developed by Cornell (1968) has been sophisticated in step with the advances in elemental technology such as empirical attenuation relations. Recently, the probabilistic hazard maps estimated by this technique have been widely used for the setting of earthquake loads for structures or for the calculation of earthquake insurance rate. It must be noted that, in drawing up these hazard maps, the each points on the maps are calculated independently, since the dominant earthquake for each points may be different so that it is not adequate for the study of facilities distributed in wide area. For example, the estimation of seismic risk of the infrastructure such as electricity, gas and water supply needs to consider the events that multiple facilities are damaged simultaneously. On the other hand, the hazard map of scenario earthquake is also employed for such purposes. However, the damage by unidentified faults in recent years emphasizes the need for probabilistic hazard analysis for wide areas.

This paper proposes the conditional seismic hazard map which is the expectation of the ground motion intensity at the secondary site on the condition that the given ground motion intensity at the primary site occurs. This hazard map enables to consider probabilistically the correlation of ground motion for multiple sites. As application, the two areas in Kanto and Kansai regions of Japan are employed, followed by the conditional seismic hazard maps corresponding to some ground motion intensities at the primary site as shown in Figures 1 and 2.

REFERENCES

Cornell, C. A. 1968. Engineering Seismic Risk Analysis, *Bulletin of the Seismological Society of America: Vol.58, No.5,* pp.1583–1606.

Safety, Reliability and Risk of Structures, Infrastructures and
Engineering Systems – Furuta, Frangopol & Shinozuka (eds)
© 2010 Taylor & Francis Group, London, ISBN 978-0-415-47557-0

About efficient estimation of the seismic reliability of complex structural systems

Orlando Díaz-López & Luis Esteva
Institute of Engineering, National University of Mexico

ABSTRACT

In spite of the merit of the concept of deformation capacity as an indicator of the variable that determines the collapse condition of a complex structural system subjected to earthquake ground motion, its practical use in quantitative terms shows severe limitations, which arise from the sensitivity of the deformation at the instant of impending collapse to variables such as the lateral deformation pattern of the system at that instant and the strength and stiffness reductions associated with the damage produced by the previous response cycles, by previous seismic events or by other damage-generating agents, such as differential settlements. The approach proposed here aims at both, avoiding these limitations and reducing substantially the sizes of the samples of ground motion time histories required for the application of incremental dynamic analysis (IDA) methods.

According to this study, the seismic reliability of a complex nonlinear system with respect to collapse is expressed in terms of a damage indicator (D) equal to the reduction of the secant lateral stiffness of the system, where the latter is measured by the ratio of the base shear to the peak roof displacement relative to the ground at the instant when that displacement is reached:

$$D = (K_0 - K)/K_0$$

Here, K_0 is the tangent initial lateral stiffness, as determined with the aid of a pushover analysis and K is the secant value of that stiffness at the instant when the peak value of the roof lateral displacement is reached. Collapse occurs when $K = 0$, which corresponds to a stiffness-reduction index equal to unity. For a given intensity, the safety level of the system can be expressed in terms of Cornell's safety index $\beta = \bar{Z}/\sigma_Z$, where $Z = -\ln D$ is the safety margin.

In practical applications it may be convenient to measure the seismic capacity of the system of interest by Y_C, the minimum value of the seismic intensity required to produce collapse; the safety margin is then expressed as the natural logarithm of the ratio of the minimum intensity causing collapse to that acting on the system. In order to estimate the value of β for a given intensity it is necessary to estimate the mean value and the dispersion of a new safety margin, now defined as $Z = \ln(Y_C/y)$, where Y_C was defined above and y is the acting intensity. The mean value and the dispersion of Z have to be estimated from a sample of pairs of values of the acting intensity and the stiffness-reduction index (D).

In order to obtain accurate estimates of the probabilistic parameters of the collapse intensity, with a reasonably small sample of dynamic response time histories, it is necessary to select the interval of the ground motion intensities included in the sample in such a manner that a large proportion of the resulting values of the stiffness-reduction index D will be slightly smaller than unity. This is achieved with the aid of the Montecarlo algorithm presented in this study. According to it, a first estimate of the median value and the dispersion of the intensity producing collapse is obtained with a small sample of earthquake excitations. A second sample is then generated, based on these statistical parameters. The efficiency of the approach is measured by two main indicators: a) the ratio of the number of cases where collapse does not occur to the total number of dynamic response simulations, and b) the ratio of the number of cases where the value of the secant-stiffness reduction index lies within a specified interval of values slightly smaller than unity to the total number of dynamic response simulations. For the cases studied here, maximum values of those indicators were respectively equal to 0.7 and 0.48; the latter value corresponds to an acceptance interval $0.8 \leq d < 1.0$. They were obtained with sample sizes equal to 20 and 60, for the initial estimate and the final Montecarlo simulation process, respectively.

Safety, Reliability and Risk of Structures, Infrastructures and Engineering Systems – Furuta, Frangopol & Shinozuka (eds)
© *2010 Taylor & Francis Group, London, ISBN 978-0-415-47557-0*

Sensitivity of probabilistic seismic hazard results to background seismic activity models

N. Yilmaz
Earthquake Research Department, General Directorate of Disaster Affairs, The Ministry of Public Works and Settlement, Ankara, Turkey

M.S. Yucemen
Civil Engineering Department, Middle East Technical University, Ankara, Turkey

ABSTRACT

In the probabilistic seismic hazard procedures, the past earthquake records that can not be associated with any one of the specific faults are treated as background seismic activity. Contribution of background seismic activity to seismic hazard is generally calculated by using two different models, namely: background area source with uniform seismicity and spatially smoothed seismicity model.

The spatially smoothed seismicity model developed by Frankel (1995) assumes that future earthquakes will occur in the vicinity of past earthquakes. In this model, earthquakes that are not assigned to major seismic sources are assumed to be potential seismic sources and are spatially distributed to cells of a grid. In the alternative model, background area sources are delineated and over these background area sources seismic characteristics are assumed to be spatially homogeneous.

In this study, sensitivity of seismic hazard results to the models used to describe background seismic activity is investigated. For this purpose, two case studies are carried out for a large (a country) and a small region (a province). These two case studies involve Jordan and the Bursa province, which is located in Turkey. Seismic databases are compiled for these two regions. In compiling these databases the raw data in the earthquake catalogs are processed as follows: Magnitudes reported in different scales are converted to a common magnitude scale. Earthquake clusters are identified and dependent events (fore and after shocks) are eliminated by defining appropriate space and time windows. So an alternative database containing only main shocks is obtained to ensure statistical independence.

Using the resulting seismic databases, seismicity parameters are determined for background seismic activity and seismic hazard analyses are carried out by using both spatially smoothed seismicity model and background area source with uniform seismicity.

Spatial variation of the differences between the seismic hazard values obtained from these two models is examined. For this purpose maps displaying the spatial variation of the differences in PGA values obtained from the spatially smoothed seismicity model and the background area source with uniform seismicity are plotted for return periods of 475 and 2475 years. In order to construct these maps differences between the PGA values obtained from these two models are calculated at each grid point covering the whole region under consideration. In the case of Bursa province the influence of the assumptions with respect to catalog completeness and dependence are also reflected to the difference maps.

Based on the results of these two case studies, it is observed that for background seismic activity, the use of spatially smoothed seismicity model or the alternative background area source with uniform seismicity affects the results. Spatially smoothed seismicity model gives higher seismic hazard values at the regions where the epicenters of earthquakes cluster. On the other hand, nearly a spatially uniform seismic hazard distribution is obtained in the case of background area source with uniform seismicity. Therefore, background area source model with uniform seismicity is expected to give higher seismic hazard values compared to the spatially smoothed seismicity model for sites located far away from clustering regions of past earthquake epicenters; i.e. where the epicenters of earthquakes are scarce or no earthquakes have occurred in the past.

REFERENCES

Frankel, A., 1995. Mapping seismic hazard in the Central and Eastern United States, Seismological Research Letters, 66(4): 8–21.

Safety, Reliability and Risk of Structures, Infrastructures and Engineering Systems – Furuta, Frangopol & Shinozuka (eds)
© *2010 Taylor & Francis Group, London, ISBN 978-0-415-47557-0*

Comparison of vulnerability of new high-rise concrete moment frame structures using HAZUS and nonlinear dynamic analysis

Nilesh Shome & Paolo Bazzurro

AIR Worldwide, San Francisco, CA, USA

ABSTRACT

The HAZUS Loss Model (FEMA 2003) has been developed to estimate losses for earthquake risk mitigation of federal, state, and local governments in the United States. The HAZUS methodology uses the Capacity Spectrum Method which is a nonlinear static analysis procedure to calculate the peak building response for damage prediction. The structures in the HAZUS methodology are represented by nonlinear SDOF systems and the building response characteristics are defined by the building capacity curves. The capacity curve provides the pushover displacement of structures for laterally applied earthquake loads. The capacity curves of the buildings in HAZUS are based on a considerable degree of engineering judgment and expert opinion. Recently researchers and practitioners have started using the results of nonlinear time-history analysis to estimate losses from earthquakes since nonlinear time-history analysis provides the most accurate results of seismic demand of structures. In this paper we are using the nonlinear time-history analysis of realistic 2-D representations of structures to estimate the damage. The structure considered in this paper is a 12-story concrete moment-resisting-frame (MRF) structure designed according to the ASCE 7-02 (ASCE 2002) design requirements. This building has been designed as part of the Applied Technology Council ATC-63 Project (ATC 2007). The details of the model and design of the structure can be found in Haselton and Deierlein (2007). The structure was designed for general high seismic sites in California for NEHRP-D soil class ($V_s \approx 400\,\mathrm{m/s}$). The nonlinear time history analysis of the structure has been carried out using the OpenSEES open source software package (OpenSEES 2008). The story-drift and peak-floor acceleration results from nonlinear time-history analysis for a suite of ground-motion records have been used to estimate the vulnerability of the structure. The methodology that we have developed in this paper is similar to the methodology proposed by Ellingwood et al. [2007]. Although Ellingwood et al. have considered the maximum inter-story drift (which is the maximum drift among all the stories in the analysis), we have considered the inter-story drift distribution along the height. This approach helps to take into account the non-uniform damage distribution along the height, especially for tall buildings. The collapse (complete damage) of structure in the proposed approach, unlike HAZUS, is estimated from the non-converging results or presumed collapse at 10% maximum inter-story drifts. The fragility curves as well as the mean damage functions developed in this study are compared with the HAZUS high-code concrete frame structures to demonstrate the improvement of the results from the proposed approach.

REFERENCES

American Society of Civil Engineers (ASCE7-02) 2002. *Minimum Design Loads for Buildings and Other Structures*. Reston, VA.

Federal Emergency Management Agency 2003. *Multihazard loss estimation methodology, Earthquake model*, HAZUS-MH MR3-Technical Manual, Washington, DC.

Ellingwood , B.R., Celik, O.C., & Kinali, K. 2007. Fragility assessment of building structural systems in Mid-America. *Earthquake Engineering & Structural Dynamics*, 36 (13): 1935–1952.

ATC-63/FEMA P695 2007. *Quantification of Building Seismic Performance Factors*. Applied Technology Council, CA/ Federal Emergency Management Agency, Washington, DC.

Haselton, C. & Deierlein, G. 2007. Assessing Seismic Collapse Safety of Modern Reinforced Concrete Moment Frame Buildings, Report No. 156, *The John A. Blume Earthquake Engineering Center*, Stanford University, Stanford, CA.

Open System for Earthquake Engineering Simulation (Opensees) (2008). *Pacific Earthquake Engineering Research Center*, University of California, Berkeley, http://opensees.berkeley.edu/ (last accessed August 1, 2008).

Safety, Reliability and Risk of Structures, Infrastructures and Engineering Systems – Furuta, Frangopol & Shinozuka (eds)
© *2010 Taylor & Francis Group, London, ISBN 978-0-415-47557-0*

Statistical analysis of corrosion process along French coasts

F. Schoefs
GeM, UMR CNRS 6183, Nantes Atlantic University, Nantes, France

J. Boéro & B. Capra
OXAND S.A., Avon, France

R. Melchers
Center of Infrastructure Performance and Reliability, The University of Newcastle, Australia

ABSTRACT

Steel structures in sea or estuary area are subjected to corrosion process. The phenomenon is very complex due to the nature of the environment and material, and the type of the structure. This phenomenon needs to be characterized and modeled for structural analysis which accounts for the loss of thickness. Moreover, due to the randomness of the corrosion process, probabilistic models are needed for structural reliability. Ultrasonic residual measurements allow determining profiles of loss of thickness, identifying the areas which are the most affected by corrosion and stating the assumptions for modeling the random corrosion process. Several inspection campaigns have been performed during the three last decades in some French harbours. Thus a great number of ultrasonic residual measurements are now available for structures in various environments. The protocol is recommended by the CETMEF (French Center for Maritime and Fluvial Technical Studies :Engineering centre of the French Ministry of Public Works): at a given level, this protocol suggest to perform three geometrical measurements at a given location to assess the loss of thickness as the average of these three readings. Generally, the amount of data is more than 1000 per structure and allows performing a statistical analysis of the spatial distribution of the loss of thickness according to the different areas (mainly tidal and immersion areas). Moreover environmental parameters (temperature, PH, oxygen, salinity, conductivity, nutrients) have been measured since a decade and are available.

This study began within the MEDACHS framework (Marine Environment Damage to Atlantic Coast Historical and transport works or Structures-Interreg IIIB Project funded by EC 2005-2007) and is performed now in the GEROM framework (Risk management of French harbor structures: stakes, current practices and needs – Experience feedback of owners). The steel structures are sheet-piles, on-pile and on-sheet-pile wharves. This paper focuses on sheet-piles seawalls and coffer-dams. More than 30 000 data are available: they correspond to measurements in 4 harbours and

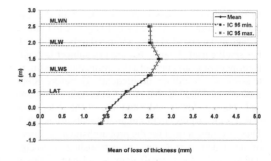

Figure 1. Mean of loss of thickness and Confidence Interval according to the depth – Wharf BO1.

concern more than 20 quays. This paper focuses on the spatial dependence of the corrosion process and we present only results of three of them where the distance between data is small (0.20 m horizontally and 0.10 m vertically) – see Figure 1).

In the first part, the paper presents the data available to study the spatio(-temporal) fields of corrosion for steel sheet-piles.

The second part presents the geo-statistical modeling of the spatial variability of the steel sheet piles corrosion; and more particularly the study of the stationarity of the process of corrosion according to the length and the depth along three marine structures. Finally, a first level of stochastic modeling is suggested.

REFERENCES

Clément, A., Boéro, J., Schoefs, F., Memet, J.B. 2008. Overview of corrosion impact for marine structures: analysis of French structures in several harbours, *1st International Conference MEDACHS, Lisboa, Portugal, 28–30 January 2008.*

Melchers, R.E. 2003. A new model for marine immersion corrosion in structural reliability, *Applications of Statistics and Probability in Civil Engineering*, Ed: A Der Kiureghian, S Madanat and JM Pestana, Millpress, Rotterdam, 599–604.

Safety, Reliability and Risk of Structures, Infrastructures and
Engineering Systems – Furuta, Frangopol & Shinozuka (eds)
© 2010 Taylor & Francis Group, London, ISBN 978-0-415-47557-0

Probabilistic assessment of small earthfill dams subject to adverse climate effects

M.-C. Preziosi & T. Micic
City University London, UK

ABSTRACT

In recent years, it has been recognized in the UK
that dam failure events are more likely due to climate
change and associated risks will need to be evaluated.
For small earthfill dams that are not subject to regular inspections carrying out a deterministic assessment
to assess safety will not be adequate. Therefore, more
sophisticated models that reflect uncertain conditions
and obtain a clearer understanding of the risks associated with various environmental threats are needed.
Thus, a simple model for the probabilistic assessment
of small dam performance is established, a precipitation model for a small dam is defined, notional
reliability for a small dam that is subject to diverse
precipitation scenarios is derived.

The physical model is based on a generic, small,
homogenous earthfill embankment, as illustrated in
Figure 1.

The behaviour of the embankment is predicted in
terms of its failure modes. Here, it is assumed that
embankment failure is due to the overall saturation of
the embankment and the instability of the embankment fill as a result of low shear strength. The stability
of a given slope is evaluated using limit equilibrium
method, namely, the sliding block method or wedge
method. Thus, the associated linear limit state functions are established as functions of the total active
pressures acting on the upstream/downstream sections, pore water pressure from the reservoir acting on
the upstream section, passive earth pressure and the
shear strength. A long established FOSM (First Order
Second Moment) is used to evaluate the likelihood of
failure. Height of the embankment, crest width, headwater height, internal friction for the soil, cohesion and
the moisture content of the soil are treated as uncertain.

To ascertain how climate change affects different
failure modes associated with small dams the effect of
precipitation on the slope stability of the embankment
is evaluated. To take the rainfall into account, it was
assumed that it had been raining continuously for 3
hrs at a specific rate. Table 1 shows the sample reliability indices obtained for the upstream slope failure
for a particular London clay embankment. Effects of
the rainfall infiltration through embankment fill are
evident.

Table 2. Reliability indices for London clay
embankment after 3hr rainfall for 1:3 upstream and
1:3 downstream slope gradients, when the soil degree
of saturation is varied

β	Upstream slope failure		
Sr	25%	50%	90%
Initially	6.0783	5.3153	4.7502
3hr rainfall	4.3784	3.5188	2.5987

Sensitivity factors associated with the London clay
embankment show that, with climate effects represented as the 3hr rainfall, soil properties remain the
most important variables. The methodology can be
expanded for consideration of alternative profile of
the dam, soil properties for the embankment fill and
rainfall rates. Results are a useful tool as they can
provide site-specific information for decision-making
regarding necessary remedial or maintenance work
required to counteract climate change effects and, ultimately, will enable better management and improved
risk assessments in respect to climate effects for these
structures.

REFERENCES

Bromhead, E. N., 1998, *Stability of Slope*. Taylor & Francis
(eds.).
Halder, A. & Mahadevan, S. 2000, Probability, Reliability
and Statistical methods in Engineering Design. NY: Wiley
(ed.)

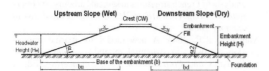

Figure 1. Cross section of the embankment.

General Session (Materials & Probabilistic Materials Analysis)

Safety, Reliability and Risk of Structures, Infrastructures and
Engineering Systems – Furuta, Frangopol & Shinozuka (eds)
© 2010 Taylor & Francis Group, London, ISBN 978-0-415-47557-0

Fire-induced damage in prestressing steels after fire

J.M. Atienza & M. Elices

Department of Materials Science. Universidad Politécnica de Madrid (UPM), Madrid, Spain

ABSTRACT

Following a fire, if no collapse happens, there is a possibility of fire-induced damage. Then, the question which unavoidably arises is if the structure is still safe; what the load bearing capacity of the structure is and how it has been affected by the fire. Appropriate knowledge of the behavior of construction materials after a fire is of major importance for answering these questions; For this end, the behavior of prestressing steel wires after a fire is the subject of this work.

Measurements on loaded steel wires used for pre-stressed concrete during and after several simulated fires have been performed. A detailed study of the pre-stressing steel behavior was performed, examining the two principal aspects that could affect the performance of the structure after fire:

- The residual mechanical properties of the wires after fire: Extensive literature is available on the mechanical properties of structural steel at high temperatures. However no attention has been paid to the mechanical residual properties after fire. In this work, wires were subjected to a complete cycle of heating and cooling while loaded. Figure 1 shows the variation of tensile strength at room temperature after fire, for the different maximum temperatures reached during the fire simulations.
- The stress relaxation losses produced during fire: Temperature has a great influence on the stress relaxation; an increment of temperature produces a large increase in stress relaxation losses and, consequently, the prestressed compressive load provided to concrete will be significantly reduced after fire. In this work, stress relaxation tests at different temperatures and at different initial loads were carried out. In Figure 2, an example of the time-stress relaxation losses curves obtained in this work is depicted as a function of temperature for a fire of 4 hours.

Fire safety should consider not only the performance of the structure during the fire but also the behavior of the structure after cooling. Even if a fire does not give rise to apparent damage in the prestressed structure, mechanical properties of materials as well as load distribution can be affected.

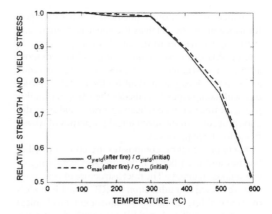

Figure 1. Relative residual strength and yield stress at room temperature, after heating at different temperatures.

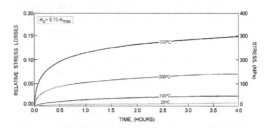

Figure 2. Stress-relaxation losses (lost load/maximum load) as a function of temperature, for an initial stress of 70% of the wire strength.

After fire, a verification of residual load-bearing capacity is necessary to determine if the structure can be maintained in use. This paper seeks to give a simplified estimation of the non-visible fire-induced damages produced in steel wires, with the hope that this information can be useful for engineers in the structural evaluation after a fire.

Safety, Reliability and Risk of Structures, Infrastructures and Engineering Systems – Furuta, Frangopol & Shinozuka (eds)
© 2010 Taylor & Francis Group, London, ISBN 978-0-415-47557-0

Effect of micro structure on mechanical properties of woven fabric composite

Y. Fujita, T. Kurashiki & M. Zako

Osaka University, Osaka, Japan

ABSTRACT

Thought a fiber reinforced plastic (FRP) has high performance properties such as the specific strength and stiffness, a variation in the mechanical properties is unavoidable due to some uncertainties like a random array of mono-filament, heat, pressure and curing time and so on. It is very important for the design of structure with composites to consider the variation. In this paper, the uncertainty has been described and an evaluation procedure of the mechanical properties of woven fabric has been proposed.

Many of the fabric composites are consist of fiber bundles and the fiber bundles are also consisting of continuous filaments. The microstructure of fiber bundles are treated as unidirectional FRP. In the FE models of woven fabric composites, equivalent mechanical properties of unidirectional FRP are applied in the element in the fiber bundles to prevent huge number of elements and a lot of computational time. The mechanical properties of unidirectional FRP are calculated by theoretical method but they can't evaluate the dispersion of the mechanical properties. Filaments in the fiber bundles array randomly and the micro structure becomes uncertain. The mechanical behavior should vary because of it. To characterize the dispersion of mechanical properties of woven fabric composites, it is important to consider the micro scale behaviors. Based on these backgrounds, the purpose of this study is to characterize the mechanical properties of unidirectional FRP in woven fabric composites considering random fiber arrangement.

In this study, FEM has been used to estimate the effect of random fiber arrangement. As a proposal method, firstly several FE models with randomly arrayed fibers have been generated and carried out damage development analysis for each models. The mechanical behaviors for these models have been simulated and the effects of the array on the stiffness, the strength and Poisson's ratio have been evaluated. In addition, the residual thermal stress of manufacturing process has also been considered to estimate strength correctly. As the temperature difference, −50 degree C and −80 degree C are provided to consider strain release and investigate the effect on mechanical behaviors. FE model of hexagonal model has also been generated to compare with random model. Figure 1 shows typical example of FE models.

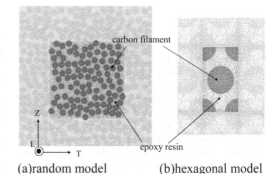

(a)random model   (b)hexagonal model

Figure 1.   Example of FE models.

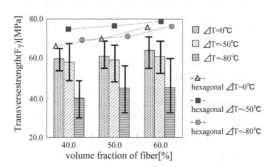

Figure 2.   Numerical result of transverse strength.

As the result, it is recognized that the stiffness for longitudinal direction increases along with the volume fraction of fibers (V_f) and the dispersions of the stiffness and Poisson's ratio for longitudinal direction are small. On the other hand, the transversal strength has been dispersed. Figure 2 shows the numerical result of transverse strength. From this figure, it is recognized that the fiber arrangement and residual thermal stresses have a great influence on the transversal strength. Therefore, it is important for estimation of the dispersion of the strength on woven fabric composites to consider the fiber arrangement and residual thermal stresses.

However, corresponding ΔT to each volume fraction of fiber should be provided because transverse strength decreases generally when volume fraction of fiber increases. When volume fraction of fiber is large, ΔT will be large.

Safety, Reliability and Risk of Structures, Infrastructures and Engineering Systems – Furuta, Frangopol & Shinozuka (eds)
© 2010 Taylor & Francis Group, London, ISBN 978-0-415-47557-0

Macroscopic mechanical behavior of metal foam influenced by mesoscopic randomness

D. Schwarzer & C. Proppe

Institut für Technische Mechanik, Universität Karlsruhe (TH), Germany

1 ABSTRACT

Due to their useful properties in lightweight construction and due to their excellent behavior in energy absorption for example in crash mechanics, metal foams became an interesting, often utilized and investigated material. For the determination of the mechanical properties of foams without the help of expensive experiments, a way for computing these properties and their distributions is searched. With the help of these distributions, computations can be done in order to predict, for example with the help of stochastic FEM, the failure probabilities of mechanical components made of foam. This work concentrates on the description of the elastic properties by distribution functions.

In order to get information on the macroscopic behavior of a mechanical structure consisting of metal foam, mesoscopic three dimensional volume elements of a foam are randomly sampled including the effects of inhomogenities like varying thickness along a ligament, pre-deformed ligaments, imperfections, partially closed cell faces or non-planar cell faces. These volume elements are analyzed via the finite element method. Because of their irregular structure, the size of the volume elements is smaller than the so called representative volume elements (RVE) and therefore only bounds for the mechanical properties can be computed by introducing kinematic uniform (KUBC) and static uniform boundary conditions (SUBC), respectively (Kanit et al. 2003, Ostoja-Starzewski 2006). The elastic properties of these volume elements and the influences of the introduced inhomogenities are estimated and evaluated. The mesoscopic model is validated by comparing with results of the computations to results found in literature and in experiments.

In order to perform the step from the mesoscale to the macroscale, the distributions of the mechanical properties of the volume element are computed and compared to different analytic distribution functions, which were fit to the calculated distribution by an optimization. Also the Kolmogorov-Smirnov test is done. The result is that the Young's modulus is normal distributed. In a next step the autocorrelation function is also determined by cutting out cubes of a macroscopic sample moving through the complete volume of the sample. For each cube the mechanical properties are computed and therefore the autocorrelation functions can be approximated by an analytical function of the following type:

$$\widetilde{\psi}_{0EE}(x) = e^{-c|x|}(u\cos(w|x|) + v\sin(w|x|)). \tag{1}$$

With the help of this function, the Karhunen-Loeve expansion can be calculated in order to precede to the next step in describing the scatter of the material parameters (Ghanem 1991). This expansion can be used in methods like the stochastic finite element method, which will be done in future work.

In this work the step to the macroscale will be done via numerical simulations with the help of these results and the statistical quantities of the mechanical properties. A local varying stiffness can be computed and inserted into the macromechanical model. In this way the propagation of uncertainties from the mesoscale to the macroscale can be assessed. This procedure is used to predict the scatter of the eigenfrequencies for aluminum foam beams.

REFERENCES

Ghanem, R.D., Spanos, P.D. 1991: Stochastic Finite Elements: A spectral approach. Springer-Verlag.

Kanit, T., Forest, S., Galliet, I., Mounoury, V. & Jeulin, D. 2003. Determination of the size of the representative volume element for random composites: statistical and numerical approach. *International Journal of Solids and Structures* 40: 3647–3679.

Ostoja-Starzewski, M. 2006. Microstructural Randomness and Scaling in Mechanics of Materials, Boca Raton: Chapman&Hall/CRC.

Safety, Reliability and Risk of Structures, Infrastructures and
Engineering Systems – Furuta, Frangopol & Shinozuka (eds)
© 2010 Taylor & Francis Group, London, ISBN 978-0-415-47557-0

Spatial variability modelling of GFRP panels

S. Sriramula & M.K. Chryssanthopoulos

Faculty of Engineering and Physical Sciences, University of Surrey, Guildford, Surrey, UK

Keywords: Uncertainly; Composite materials; Stochastic fields; Correlation length; Experimentation

ABSTRACT

In the light of both experimental and analytical results, it is being suggested that considering the spatial variability of mechanical properties in FRP structures could improve performance estimates, as compared to a traditional random variable based probabilistic approach. However, the analytical and computational complexity associated with such models, coupled with the lack of experimentally based random field descriptions, has hitherto restricted the scope of such studies. This paper is a contribution towards filling the gap in experimentally based characterization of the spatial variability of FRP composites; it describes the experimental approach, and the associated modeling technique, adopted for probabilistic spatial characterisation of a set of GFRP panels. These panels are obtained from Mondial House, a state-of-the-art 45m building constructed at London in 1974 that was demolished after 32 years of service for property development reasons. Coupons from these panels are obtained in a pattern suitable for spatial variability modelling as per two cutting plans attempting to capture spatial correlations over a range of coupon spacings. A spatial plot of compression strength (X_C) for a typical panel is shown in Figure 1, where it can be seen that the variability over a $\sim$1 m^2 area is substantial.

Statistical analysis of the two sets of data support the hypothesis that these two sets can be considered as one group. The characteristics of random fields associated with the tensile and compressive properties and thickness of the composite panels are considered in the paper. The random fields are assumed to be spatially statistically homogeneous so that the primary and higher-order moments are independent of coupon location. Typical results of the auto and cross correlation patterns for a number of mechanical properties in tension and compression are presented and discussed. Alternative auto correlation functional forms are evaluated and correlation lengths are estimated, e.g. a typical auto correlation plot for compression strength is shown in Figure 2. The correlation lengths for the experimentally obtained variables suggest that the higher the variability, the smaller is the correlation length. These aged and weathered panels are expected to provide an indication of the upper bound on the variability of FRP composites. Further studies are currently undertaken on alternative factory made composite material systems in a pristine condition. It is hoped that such studies can aid in formulating appropriate SFEA and multi-scale methods for composite structures.

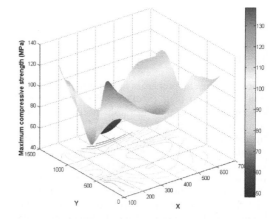

Figure 1.   Spatial variation of X_C for a typical panel.

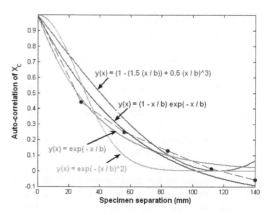

Figure 2.   Auto correlation estimates of X_C.

Safety, Reliability and Risk of Structures, Infrastructures and Engineering Systems – Furuta, Frangopol & Shinozuka (eds)
© *2010 Taylor & Francis Group, London, ISBN 978-0-415-47557-0*

Prediction of service life for concrete structures with crack: Probabilistic approach for carbonation

H.-W. Song
Yonsi University, Seoul, Korea

U.-J. Na
Ministry of Land, Transport, and Maritime Affairs, Busan, Korea

S.-J. Kwon
University of California Irvine, Irvine, CA, USA

ABSTRACT

Deterioration of reinforced concrete (RC) structures is attributed to the adverse environmental conditions that the structures undergo during their lifetimes. In general, carbonation and chloride penetration are considered as two major deterioration processes. Concrete structures in urban areas or in underground sites such as subways are generally exposed to carbonation due to acid rain or high concentration of carbon dioxide (CO_2). Once corrosion of reinforcement is initiated in a concrete structure, it progresses almost at a steady rate. This corrosion of reinforcement shortens the service life of the structure by causing surface cracking and subsequent spalling of the cover concrete. Since the rate of corrosion directly affects the service life of RC structures, service life prediction and evaluation has been considered as one of very important issues related to durability design for both new and existing structures.

A lot of researches have been conducted to develop reliable prediction models for carbonation. For the evaluation of service life, the so-called limit state design concept for durability has been introduced in the concrete specifications of many countries. Related to probabilistic analysis, various approaches have been attempted to consider the uncertainties of input variables in the analysis is. For the carbonation behavior, various parameters such as carbon dioxide diffusion coefficient, cover depth, and skin effect of cover concrete have been considered as random variables through statistical distributions. However, these applications using probabilistic approaches have been generally limited to the sound concrete without considering crack effect. This is due to the difficulty to evaluate the crack effect on the carbon dioxide diffusion characteristics. However, early-aged cracks may occur in concrete structures so that the crack effect on carbon dioxide diffusion behavior should be considered for accurate prediction of service life in RC structures.

In this study, the early-aged crack is defined as cracking due to material characteristics such as hydration heat and drying shrinkage. This crack is neither progressive nor critical to the structural safety but generally becomes a main route of deteriorating agent, which may lead to the more rapid steel corrosion. To consider the crack effects on carbonation behavior, field investigations for twenty seven RC columns, which had been exposed to carbonation in urban city area for 18 years, were conducted. Through the field investigation, diffusion coefficients of concrete with different crack width are obtained. Using the relationship between diffusion coefficients of cracked concrete and crack widths, the service life is predicted and discussed. For the probabilistic analysis, carbon dioxide diffusion coefficient, cover depth, binding capacity and concentration of carbon dioxide at the surface of cover concrete are considered as statistical variables. To decide the criterion of durability failure, it is assumed that corrosion process begins when the carbonation front reaches reinforcement steel bar. Based on the results of the field investigations, the service life of cracked concrete is predicted through the probabilistic methodology framework using Monte Carlo Simulation.

REFERENCES

Song, H.-W., Kwon, S.-J., Byun, K.J. & Park, C.-K. 2006. Predicting carbonation in early-aged cracked concrete, *Cement and Concrete Research*, 36(5): 979–989.

CEB-FIP. 2006. *Model Code for Service Life Design*, International Federation for Structural Concrete (fib), Task Group 5.6.

Song, H.-W., Cho, H.-J., Park, S.-S., Byun, K.-J. & Maekawa, K. 2001. Early-age cracking resistance evaluation of concrete structure, *Concrete Science Engineering*, 3: 62–72.

Safety, Reliability and Risk of Structures, Infrastructures and
Engineering Systems – Furuta, Frangopol & Shinozuka (eds)
© 2010 Taylor & Francis Group, London, ISBN 978-0-415-47557-0

Estimation of fatigue life of textile composites based on damage propagation analysis and fatigue test of UD composites

S. Hanaki, M. Yamashita & H. Uchida
Graduate School of Engineering, University of Hyogo, Hyogo, Japan

M. Zako
Graduate School of Engineering, Osaka University, Osaka, Japan

S.V. Lomov & I. Verpoest
Department of Materials Engineering, Katholieke Universiteit Leuven, Leuven, Belgium

ABSTRACT

A conventional method to estimate the fatigue life of textile composite based on damage propagation analysis has been proposed. In the proposed method, the stress distribution under cyclic loading is evaluated by FEM. On the other hand, each element of fiber bundle is considered as unidirectional material and its material properties are estimated by fatigue test results for unidirectional composites.

The stress distribution in the textile structure is estimated by FEM and SACOM software (Zako, M.et al.1995) is used in this study. SACOM has a function for damage development analysis. The damage mode, which affects strongly the mechanical behavior, is classified into mode L, $T\&LT$, $Z\&ZL$ and TZ. The mode L represents the fibre breaking; the others represent the transverse and shear cracking. The cumulative damage in each element is evaluated by Minor's damage rule. At each stress cycle, the cumulative damage is calculated according to the results of stress analysis and $S - N$ property of unidirectional material.

The proposed method is applied to the fatigue test for plain woven fabric. To validate obtained results, the fatigue test results obtained by Nishikawa et al (Nishikawa, Y. et al. 2006) is selected. In this calculation, meso-model of textile structure is prepared. For the geometric modeling of internal textile structure, the Wise Tex software package (Lomov, SV. et al. 2001) is applied. Fig.1 shows estimated $S-N$ curve for plain woven CFRP. Experimental results obtained by Nishikawa et al (Nishikawa, Y. et al. 2006) are also plotted in this graph. It is confirmed that the estimated $S - N$ curve shows good agreement in high cycle fatigue region. To consider the effect of scatter of input data on estimated fatigue life of textile composite, Monte Carlo simulation is carried out. In this analysis, the stress level of fatigue limit is treated as random valuable. Fig.2 shows log-normal plots of obtained fatigue life. It is also confirmed that fatigue life distribution is well approximated by log normal distribution.

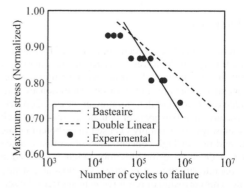

Figure 1. Obtained $S - N$ curve for plain woven CFRP.

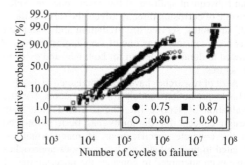

Figure 2. Log-normal plots of fatigue life distribution.

REFERENCES

Zako M, Takano N, Tsujikami T, Uetsuji Y. 1995. A proposition for fracture behavior analysis of composites, *Proceedings of the 38th Japan congress on materials research*, 163–169.
Nishikawa Y, Okubo K, Fujii T, Kawabe K. 2006. Fatigue crack constraint in plain-woven CFRP using newly-developed spread tows, *International Journal of Fatigue*, 28, 1248–1253.
Lomov SV, Huysmans G, Luo Y, Parnas R, Prodromou A. 2001. Verpoest I, et al. Textile composites: modelling strategies, *Composites Part A*, 32, 1379–1394.

General Session (Deterioration Modeling)

Safety, Reliability and Risk of Structures, Infrastructures and Engineering Systems – Furuta, Frangopol & Shinozuka (eds)
© 2010 Taylor & Francis Group, London, ISBN 978-0-415-47557-0

Solar radiation effect on long-term performance of rubber bearings

Paramashanti, Y. Itoh & Y. Kitane
Department of Civil Engineering, Nagoya University, Japan

ABSTRACT

Previous studies by Itoh (2006) have revealed that heat oxidation is one of the major deterioration factors for rubber and that temperature has a significant influence on the deterioration process of rubber. The equivalent shear stiffness of a rubber bridge bearing increases over time due to the deterioration of rubber, and the higher the temperature, the faster it increases. A yearly average ambient temperature is currently used to predict aging of rubber bearings. It is assumed in the aging estimation that rubber in the bearing and at the surface has the same temperature as its surrounding ambient temperature because rubber bearings are thought to be usually in the shadow as they are installed between superstructure and substructure. However, in the actual environment, there are many bearings that are exposed to solar radiation that causes bearing temperature to increase. To evaluate the effect of solar radiation on the bearing temperature, bearing surface temperatures were measured for a bearing installed in an elevated highway in Nagoya, Japan. To understand a temperature variation in a bearing, a bearing model was manufactured, and a temperature measurement was carried out.

The long-term shear stiffness changes based on the yearly average ambient temperature and bearing's internal temperature are calculated, and the difference is discussed.

The measurement results show that bearing surface temperatures are significantly higher than the ambient temperature when the surface is exposed to solar radiation. The maximum difference can be more than 20°C.

From the measurement of bearing's internal temperature, it is resulted that bearing internal temperature is always greater than the ambient temperature when bearing is exposed to solar radiation.

Solar radiation results in the increase of bearing internal temperature. By using the bearing internal temperature variation in the long-term performance prediction, it will result in a greater aging effect than the case using the yearly average ambient temperature, implying that the long-term performance prediction of a bearing exposed to solar radiation is not accurate when the yearly average ambient temperature is used to predict the aging effect.

REFERENCES

Itoh, Y., Gu, H. S., Satoh, K. and Kutsuna, Y. 2006a. Experimental investigation on aging behaviors of rubbers used for bridge bearings. *Journal of Structural Mechanics and Earthquake Engineering, JSCE*, 808/I-74: 17–32.
Itoh, Y., Gu, H. S., Satoh, K., and Yamamoto, Y. 2006b. Long-term deterioration of high damping rubber bridge bearing, *Journal of Structural Mechanics and Earthquake Engineering, JSCE*, 62(3): 595–607.
Itoh, Y., Satoh, K., Gu, H. S. and Yamamoto, Y. 2006c. *Study on the deterioration characteristics of natural rubber bearings, Journal of Structural Mechanics and Earthquake Engineering, JSCE*, 62(2): 255–266.

Safety, Reliability and Risk of Structures, Infrastructures and
Engineering Systems – Furuta, Frangopol & Shinozuka (eds)
© 2010 Taylor & Francis Group, London, ISBN 978-0-415-47557-0

A comparison of statistical models for visual inspection data

M.J. Kallen
HKV Consultants, Lelystad, The Netherlands

ABSTRACT

Large groups of structures like bridges, pavements
and sewer systems, are often inspected visually and
their condition is quantified based on a discrete scale.
Markov chains have traditionally been used to model
the uncertain rate at which these structures progress
through a condition scale. In order to determine optimal strategies for inspections and maintenance activities, these Markov chains must be fitted to the data
obtained in the field. For this purpose, quite a few
models and methods have been proposed in the past.
These are reviewed here and references to applications
in the field of civil engineering are given. Such models are typically applied to inspection data of bridges,
pavements, sewer systems and water distribution networks. Three model categories are distinguished in the
paper:

1. regression-based models which are represented by
   some objective function which provides a measure
   for the difference between the data and the model;
   the model parameters are estimated by minimizing this objective function subject to the relevant
   constraints.
2. models based on the principle of maximum likelihood estimation which are represented by a likelihood function which provides the probability of
   the parameters given the data; the model parameters are estimated by maximizing the likelihood
   function subject to the relevant constraints.
3. other models, like non-parametric models and
   Bayesian models; these are only briefly mentioned.

Models based on the principle of maximum likelihood estimation are generally more flexible in terms
of model building when compared to regression-based
models. However, well-defined regression models are
no less suitable for the purpose of parameter estimation
in Markov chains. A choice for either type will therefore largely be a matter of personal preference. On the
other hand, the choice for a specific model within one
of the categories above mostly depends on the type of
data available to the modeller. Three types of data may
be distinguished:

Type I: observations of the state itself and represented
by realizations $x(t)$ of the process $X(t)$,
Type II: aggregated data in the form of relative frac-
tions of proportions represented by $y(t)$, and
Type III: count data in the form of the number of
transitions represented by realizations $n(t)$ of some
counting process $N(t)$.

Each of these models will perform differently under
different situations. However, some models will perform poorly under all conditions. Most notably, these
are the models which ignore the dependence between
successively observed states and the (ordered) probit
or logit models. The latter models are well suited for
categorical data, but not the type of categorical data
which arrises from a Markov chain.

Note that this paper does not include a review of
all models which are theoretically possible. Rather,
it provides an overview of those models which have
been applied in the field of civil engineering. The
intended audience are decision makers looking for
models to analyse data which they obtained through
visual inspections.

*Safety, Reliability and Risk of Structures, Infrastructures and
Engineering Systems – Furuta, Frangopol & Shinozuka (eds)
© 2010 Taylor & Francis Group, London, ISBN 978-0-415-47557-0*

Reliability analysis for pipe failure due to corrosion

Hongan Lin
ReliaSoft Asia, Singapore

ABSTRACT

Corrosion in Pipes is inevitable especially in process
industries. The cost of pipe failure can be expensive,
and knowing when to perform preventive maintenance is desirable to the company. In this presentation, a method to perform reliability analysis using
degradation data obtained during inspection will be
presented.

Degradation analysis involves the measurement and
extrapolation of degradation or performance data that
can be directly related to the presumed failure of the
product in question. Many failure mechanisms can be
directly linked to the degradation of part of the product,
and degradation analysis allows the user to extrapolate
to an assumed failure time based on the measurements
of degradation or performance over time.

First of all, it is necessary to be able to define a
level of degradation or performance at which a failure is said to have occurred. With this failure level
of performance defined, it is a relatively simple matter to use basic mathematical models to extrapolate
the performance measurements over time to the point
where the failure is said to occur. In addition, multi-
level degradation can be defined and estimated. For
example, engineers can set design limit and destructive
limit to assess different level of risk.

Secondly, after the level of failure (or the degradation level that would constitute a failure) is defined,
the degradation for multiple units over time needs to
be measured. Once this information has been recorded,
the next task is to extrapolate the performance measurements to the defined failure level in order to
estimate the failure time.

With this TTF data set, this presentation will attempt
to answer the following questions:

1. The time by which 1% of the pipe corrosion will
   degrade beyond critical point;
2. Given the cost of corrective maintenance and preventive maintenance, when is the optimum time to
   replace the pipe.

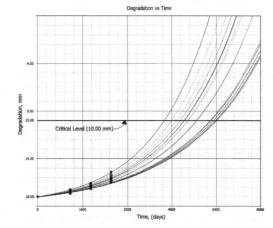

Figure 1. Extrapolating degradation data to expected
Time-to-Failure (TTF).

This method can also be used in other situations
where the equipment/component failure is due to a
characteristic that degrade with usage time.

REFERENCES

2007 Life Data Analysis Reference, ReliaSoft Publishing.
www.Weibull.com
Software Analysis: Weibull++ 7

*Safety, Reliability and Risk of Structures, Infrastructures and
Engineering Systems – Furuta, Frangopol & Shinozuka (eds)
© 2010 Taylor & Francis Group, London, ISBN 978-0-415-47557-0*

Stochastic assessment of chloride ingress into concrete matrix

E. Bastidas-Arteaga & M. Sánchez-Silva
Department of Civil and Environmental Engineering, Universidad de los Andes, Bogotá, Colombia

A. Chateauneuf & P. Bressolette
LaMI – Politech'Clermont-Ferrand, Aubière, France

F. Schoefs
GeM – University of Nantes, Nantes, France

ABSTRACT

Although durability is one of the main properties of
reinforced concrete structures, experience has shown
that in chloride-contaminated environments, corrosion
induced by chloride ingress affects significantly their
operational life. Chloride ingress is treated as a diffusion problem governed by Fick's second law where the
chloride concentration inside the concrete is estimated
as a function of an error function complement (Tuuti,
1982). This classical approach evaluates the apparent
diffusion coefficient as constant in time and space,
and assumes that the chloride concentration in the surrounding environment remains constant and that the
concrete is saturated.

This paper presents a stochastic approach to assess
chloride ingress accounting mainly for: (1) the chloride binding capacity; (2) the random nature of temperature, humidity and chloride concentration in the
surrounding environment; (3) concrete aging; and (4)
convection. Chloride-ingress is described by a system
of three partial differential equations (PDEs) representing: chloride transport, moisture transport, and
heat transfer (Saetta et al., 1993). The flow of chlorides into concrete is estimated by solving the system
of PDEs simultaneously. The solution methodology
combines a finite element formulation with finite
difference and considers Robin boundary conditions.

The time to corrosion initiation is defined as the
time at which the chloride concentration at the cover
thickness reaches a threshold value and corrosion
begins. The probability of corrosion initiation is determined based on this definition. Given the complexity
of the procedure used to estimate the chloride ingress,
the proposed model uses Monte Carlo simulation with
Latin hypercube sampling. The proposed probabilistic
approach distinguishes between two types of random
variables "time-invariant" and "time-variant". The latter are treated as stochastic processes which are modeled by Karhunen-Loève expansion and log-normal
noises.

The proposed approach is illustrated by a numerical example where the factors controlling the chloride

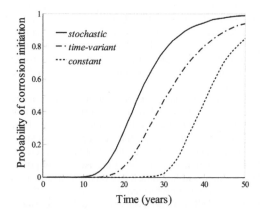

Figure 1.   Impact of type of weather model.

ingress were studied. For instance, Figure 1 illustrates
about the effect of the type of weather model on the
probability of corrosion initiation. Overall behavior
indicates that the probability of corrosion initiation
increases when the randomness and the seasonal variations of humidity and temperature and convection
are considered; and decreases when chloride binding
is taken into account. These results stress the importance of including the effects and the random nature of
temperature, humidity, surface chloride concentration,
chloride binding and convection for a comprehensive
lifetime assessment.

REFERENCES

Saetta, A.V. Scotta, R.V. & Vitaliani, R.V. 1993. Analysis of
chloride diffusion into partially saturated concrete. *ACI
Materials Journal* 90:441–451.
Tuutti, K. 1982. Corrosion of steel in concrete. *Swedish
Cement and Concrete Institute.*

Safety, Reliability and Risk of Structures, Infrastructures and Engineering Systems – Furuta, Frangopol & Shinozuka (eds)
© 2010 Taylor & Francis Group, London, ISBN 978-0-415-47557-0

An efficient computational framework for probabilistic deterioration modeling and reliability updating

Daniel Straub

Technical University Munich, Germany

ABSTRACT

The modeling of deterioration is subject to significant uncertainty, which arises from a simplistic representation of the actual physical processes (typically through empirical or semi-empirical models) and from limited information of material, environmental and loading characteristics. This uncertainty has been addressed in stochastic models of the deterioration processes. Through Bayesian updating these models can also account for the effect of additional information on the uncertainty, with information arising, e.g., from inspection and monitoring. In the past, stochastic deterioration models have been applied in the context of life-cycle optimization and inspection planning, but the application in practice has been relatively limited. A main reason for this is that the evaluation of the stochastic deterioration and inspection models through structural reliability methods or simulation techniques either requires expert knowledge or is too computationally demanding (when applying robust simulation methods).

Based on recent work of the author (Straub, in press), this paper presents a computational framework for evaluating stochastic deterioration models, which can accommodate most common stochastic deterioration models. The framework is based on the dynamic Bayesian network (DBN) methodology that can be interpreted as a generalization of the Markov chain. However, unlike the Markov chain model, the proposed framework allows to efficiently include non-ergodic random variables, which are the dominant source of uncertainty in most engineering models of deterioration.

The intention of this paper is to summarize the framework and demonstrate its application to reliability analysis and updating for pipelines and process equipment subject to localized corrosion. It is demonstrated that the framework enables the evaluation of time-variant reliability problems conditional on any type of information. Exemplarily, Figure 1 shows the reliability of a pipe conditional on inspection and monitoring results. In addition, the framework allows updating any of the uncertain model parameters. This

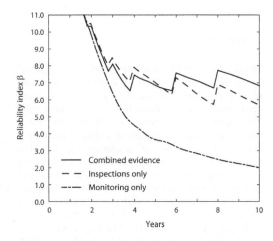

Figure 1. The reliability index of a pipe subject to CO2 corrosion, conditional on monitoring of the process conditions and inspections of the pipe.

facilitates learning about the model parameters from information obtained in-service. Finally, the DBN model can be readily expanded to model jointly the performance of a large number of components, by introduction of common parameters as parent nodes in the DBN.

The presented application demonstrates that the DBN framework allows including inspection results and monitoring data through Bayesian updating in a robust manner, i.e., the computations can be performed automatically. Such robust computations have a strong potential for applications in the asset integrity management of engineering systems.

REFERENCE

Straub, D. (in press). "Stochastic Modeling of Deterioration Processes through Dynamic Bayesian Networks." *Journal of Engineering Mechanics, Trans. ASCE*, in press.

Safety, Reliability and Risk of Structures, Infrastructures and
Engineering Systems – Furuta, Frangopol & Shinozuka (eds)
© 2010 Taylor & Francis Group, London, ISBN 978-0-415-47557-0

Climate change, time-dependent corrosion damage and safety of deteriorating RC structures

J. Peng

Institute of Bridge Engineering, School of Civil Engineering, Hunan University, Changsha, Hunan, China

M.G. Stewart

Centre for Infrastructure Performance and Reliability, School of Engineering, The University of Newcastle, Callaghan, NSW, Australia

ABSTRACT

The durability of RC structures is adversely affected by environmental stressors. A common and serious stressor is carbon dioxide (CO_2) which can cause depassivation of the protective film of steel reinforcement (known as carbonation). Carbon dioxide is always present in the atmosphere and its concentration is higher in the vicinity of its sources – in industrial and densely populated regions which tend to have the highest proportion of built infrastructure. The present concentration of atmospheric CO_2 is about 380 ppm. However, climate change and global warming studies predict that the level of CO_2 may increase to over 1000 ppm by the year 2100. The paper will assess how increases in atmospheric CO_2 levels will affect carbonation-induced corrosion damage and safety loss to reinforced concrete structures. The 2007 Intergovernmental Panel on Climate Change (IPCC) reported future CO_2 concentrations for six emission scenarios. The two emission scenarios considered herein are A1FI and B1 emission scenarios. Both these scenarios assume that there are no controls/regulations to mitigate CO_2 emissions. The annual CO_2 concentration growth-rate is 1.9 ppm per year since 2000, and so the best case scenario after 2010 would be that the CO_2 concentration is kept stable at 2010 levels (386 ppm) due to reduction and stabilisation of CO_2 emissions. This scenario is called the 'best mitigation' scenario. The carbonation depth (x_c in cm) is predicted from Yoon et al. (2007), but corrected to allow for modelling uncertainties. The mean corrosion current density (i_{corr}) is assumed equal to 0.25 $\mu A/cm^2$ with a COV of 1.0 as typical for moderate deterioration induced by carbonation. Probabilistic methods are used as there is significant uncertainty and variability of atmospheric CO_2 levels, deterioration mechanisms, material properties, dimensions, strength and loading. The time-dependent structural reliability analysis will predict the probability of corrosion initiation, mean proportion of corrosion (cover) damage and the probability of failure (collapse) of typical reinforced concrete beams over the next 100

years when the CO_2 concentration increases with time over the next 100 years for a typical RC beam. The nominal resistance in RC beam is obtained from design condition $\phi R_{nom} = 1.2 G_n + 1.6 Q_n$, in which $\phi = 0.9$. It follows that nominal flexural capacity is $M_{nom} = (1.2 G_n + 1.6 Q_n)/0.9$. The nominal capacity depends on the live-to-dead load ratio $\rho = Q_n/G_n$ where G_n and Q_n are design dead and live loads respectively. The ultimate to nominal flexural resistance (M_u/M_{nom}) for a RC beam is modeled as normal distribution with mean of 1.135 and COV of 0.085. These statistics include the random variability of ME, f'_c, d, f_y, and A_{st}. The reduction in bond between concrete and steel is ignored. Statistic parameters for a stochastic office floor load form these uniformly distributed loads produce a bending moment (S). The time period is taken as 100 years, with extraordinary live load effects updated annually and sustained live load effects updated every 8 years resulting in $k = 100$ load events. The bottom main longitudinal and transverse reinforcement is 16-mm or 27-mm diameter bars. The specified concrete compressive strength is 30 MPa and water-cement ratio is 0.5. Concrete design cover is 30 mm. The live-to-dead load ratio ($\rho = 0.5 - 1.5$) did not affect the overall trend of the results presented herein, hence results to follow are for $\rho = 1.0$. The likelihood and extent of corrosion damage and strength failure of RC beam will be evaluated at annual time increments for a design life of 100 years. A reliability analysis of a RC beam including various emission and mitigation scenarios found that for the worst case scenario the probability of corrosion initiation is 720% higher than a scenario based on maximum mitigation of CO2 emissions. It is found that for the worst case scenario the mean proportion of corrosion damage is 540% higher than the best mitigation scenario. It was also found that the worst emission scenario increased the probability of strength failure by only 6% when compared to the structural reliability for the best mitigation scenario. If the worst emissions scenario is viewed as the most likely scenario then increasing design cover by approximately 3–18 mm may be needed to ameliorate corrosion damage.

Organized Session (OS13) Novel Approaches for Reliability Analysis and Statistical Structural Health Monitoring

Safety, Reliability and Risk of Structures, Infrastructures and Engineering Systems – Furuta, Frangopol & Shinozuka (eds)
© 2010 Taylor & Francis Group, London, ISBN 978-0-415-47557-0

Application of support vector machine for reliability assessment and structural health monitoring

M.N. Noori
California Polytechnic State University, San Luis Obispo, CA, USA

Y. Cao
North Carolina State University, Raleigh, NC, USA

Z. Hou & S. Sharma
Worcester Polytechnic Institute, Worcester, MA, USA

ABSTRACT

The paper presents two applications of the support vector machine (SVM): one for structural reliability assessment and the other for structural health monitoring. Applying the SVM technique to a structural system for which training data are available, the data can be classified into two classes and a hyper plane that separates the two classes can be determined. If needed, a nonlinear boundary surface can be constructed by a kernel mapping.

In reliability analysis, the training data are generated by the Monte Carlo simulation (MCS) based on the given probabilistic distribution of uncertainties and the hyper plane found can be viewed as an approximate limit state function that defines the safe and failure regions. The trained hyper plane is tested by the test data also generated by MCS. The classification error rate, defined as a ratio of the unsuccessful testing cases to the total size of the test data, is used to evaluate the performance of the SVM classifier. The approximate limit state function can be used by some established approaches for reliability analysis. In this study, MCS is used to generate samples and each sample represents a point in the space of random variables that may fall in either the safe region or the failure region. For a large number of such points, the ratio of number of the points in the safe region to the total number of points gives the estimated structural reliability. Two cases studies are provided: one for a linear state function and the other for a nonlinear state function. The results are compared with the exact solution and those by other approaches such as FORM and SORM. Effects of the kernel function on the classification error rate and the accuracy of the failure probability are investigated.

In structural health monitoring, the SVM classifier is used to classify the structure into two classes: damaged and undamaged. The training data are the vibration measurements for the undamaged class supplemented by data for the damaged class artificially created by shifting the first two natural frequencies of the healthy class. A cepstrum-like technique is utilized to improve performance of the SVM classifier in the presence of noises. The trained SVM hyper plane can later be used for structural health monitoring when new vibration data become available. The SVM trained hyper plane can be adapted by including the new measurement data in the training set and the evolving hyper plane may indicate the current status of the structural health condition. The methodology is illustrated for health monitoring of a linear four-DOF building model. Effects of the percentage shift of the natural frequencies in generation of data in the damaged class, the noise level, and the kernel function on the classification error rate are studied.

Good results are obtained for all the test examples. The results demonstrate great promises of SVM for reliability analysis and structural health monitoring.

REFERENCES

Madsen, H. O., Krenk, S., & Lind, N. C. 1986. Methods of Structural Safety. *Prentice Hall Inc., Englewood Cliffs, New Jersey.*

Mita, A. & Hagiwara, H. 2003. Qualitative Damage Diagnosis of Shear Structures Using Support Vector Machines. *KSCE Journal of Civil Engineering*, 7(6), pp. 683–689

Widodo, A. & Yang, B.-S. 2007. Support Vector Machines in machine condition monitoring and fault diagnosis. *Mechanical Systems and Signal Processing*, 21, pp. 2560–2574.

Worden, K. & Lane, A. J. 2001. Damage Identification Using Support Vector Machines. *Smart Materials and Structures*, 10, pp. 540–547

Safety, Reliability and Risk of Structures, Infrastructures and
Engineering Systems – Furuta, Frangopol & Shinozuka (eds)
© 2010 Taylor & Francis Group, London, ISBN 978-0-415-47557-0

Vulnerability curve for a timber house considering the maximum bearing force variation

K. Yamada

Toyota National College of Technology, Toyota, Japan

A vulnerability curve is made from a given seismic hazard. This vulnerability curve means the macroscopic damage function. Nowadays, we need the vulnerability curve for a given building to estimate its read-estate value. This means that we have to consider the structural strength variation of a given building and the variation of estimated ground motions. There is no attempt to author's knowledge. In this paper, I propose the generation method of the vulnerability curve for a Japanese traditional timber house in consideration of the maximum shear force variation.

The seismic damage of a house occurs in the balance of the earthquake motion at the house's site and the seismic performance of the house. The earthquake motion is affected by an earthquake fault, and the surface ground vibration property at the site. Both the seismic performance of the house and maximum and the transient behavior of the earthquake motion have an effect on the maximum displacement response of the house. Nowadays, we know the variation of seismic performance and the variation of the surface ground vibration property. So this report deals with these variations to calculate the vulnerability curve. The damage probability of a house is calculated by the next 4 steps.

First, the earthquake motion at the engineering bedrock with its event probability is defined by a theoretical method.

Second, a number of surface earthquake motions are calculated by SHAKE in consideration of the variation of the surface ground vibration property.

Third, the response analyses of a house are executed against each surface earthquake motion in consideration of the variation of seismic performance. These calculations are executed by Monte Carlo simulation.

Forth, the distribution of the maximum displacement responses gives the damage probability. In this report, I discuss the minimum total number of simulation instances for one house against one surface ground motion.

A Japanese traditional wooden house has bearing walls and non-structural walls that cause 1/3 of its horizontal resistance force. Analysis model consists of both wall elements and rigid floor elements. Aseismic elements of this analysis model are bearing walls and non-structural walls. Their analysis model is supposed as equivalent spring, which is proportional to both their wall length. The variation coefficient of the maximum resisting force is supposed as 0.3. Non-structural wall is fixed on the both side of inner wall, and fixed on the inner side of outside wall. A height of storey is 2,700 mm. Mass matrix consists of consistent mass matrix.

Five surface ground motions are adopted to check the influence of the transient behavior of earthquake motions. Their peak ground velocities of earthquakes are modified into 50 cm/s. The mid-acceleration method is applied with 0.005 s time subtraction.

The maximum displacement response is defined as the largest relative storey displacement between 1st floor and 2nd floor, which is almost approximately larger than the relative storey displacement between 2nd floor and roof.

The results may be summarized as follows: the minimum number of instances for one house is 20 for the average and standard variation of a vulnerability curve. A simplified vulnerability curve generation method with one instance is also proposed. The 50 percentile of this simplified method is 10 cm/s greater than the 50 percentile of the proposed method with 20 instances.

Safety, Reliability and Risk of Structures, Infrastructures and
Engineering Systems – Furuta, Frangopol & Shinozuka (eds)
© 2010 Taylor & Francis Group, London, ISBN 978-0-415-47557-0

Earthquake response analysis for traditional wooden buildings using semi-rigid frame model

S. Matsumoto
Hiroshima University, Higashi-Hiroshima, Japan

Y. Suzuki
Ritsumeikan University, Kusatsu, Japan

ABSTRACT

It has not been clarified that the structure-mechanical characteristics for the traditional wooden buildings because of the horizontal plane of structure such as roof or floor construction, the foundation stone of column base which does not hold the waist tie beam, Orthogonal vertical plane of structure, and complicated joints such as the connected beam to column and so on. The vibration test was carried out at the E-defense (Hyogo Earthquake Engineering Research Center) to clear the effect of seismic response characteristic and earthquake resisting performance for traditional wooden buildings. These traditional wooden buildings have soil walls for earthquake resisting element. In this study, we remarked the particularity for tradition construction method on the column base and floor plane of structure.

In this paper, we illustrate the modeling for soil walls and the beam-column joints using on the semi-rigid frame analysis. The soil wall element is formulated by the composite elements consist of one horizontal nonlinear spring element and three rigid beams pined other elements (See Fig.1). The nonlinear characteristics of walls and beam-column joints are evaluated by slip-bilinear relationship model obtained from previous experimental studies. We compare the present earthquake response analysis with full-scale vibration test on E-defense. And we investigate the validity of this modeling and characteristics of the solution on earthquake response for Japanese traditional wooden buildings. The numerical example is shown in Figure 2. The nonlinear characteristic of column-beam joint is formulated by the semi-rigid beam element and the relationships between bending moment and relative rotation angle of the column-beam joint are evaluated by slip bilinear restoring force characteristics based on previous experimental tests. The story deformation angle of X13 plane of structure inputted BCJ-L2 wave 300 gal is shown in Figures 3.

As the result, it was confirmed that the effectiveness of modeling for the wooden building by the rotation spring element at the element ends and shear spring element for soil wall. These results can be applied

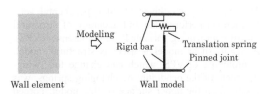

Figure 1. The wall model (rocking rigid body model).

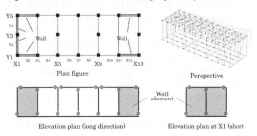

Figure 2. Numerical example.

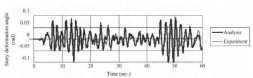

Figure 3. The story deformation angle of X13 plane of structure.

to the research of the structure health monitoring in future.

REFERENCES

Suzuki, Y. Gotou, M. Saito, Y. Kamada, T. Shimizu, H. and Nakamura, I., 2007. Vibration tests of traditional wooden buildings by E-Defense shaking table Part 1 Objectives and outline of vibration tests, Summaries of Technical Papers of Annual Meeting Architectural Institute of Japan, C-2, pp.515–516, (in Japanese)

Matsumoto, S. and Fujitani, Y., 2007. Study on analytical method for wooden structures with sheathed shear walls, International Conference on Computational Methods, International Conference Center Hiroshima, Japan, CD-ROM, p36

Safety, Reliability and Risk of Structures, Infrastructures and Engineering Systems – Furuta, Frangopol & Shinozuka (eds)
© 2010 Taylor & Francis Group, London, ISBN 978-0-415-47557-0

Structural identification of existing building using variable pendulum sensor

Y. Kobayashi & T. Furukawa
Nagoya University, Nagoya, Aichi, Japan

ABSTRACT

This paper reports the result of experimental verification on performance and accuracy of structural identification using Variable Pendulum Sensor (VPS). A SDOF shear model is used as structural model. A prototype of VPS, which can change the swinging cycle of the pendulum by changing springs, was installed on the upper surface of the model. Forced vibration test was carried out, and responses of both the structure and the VPS were measured before and after changing the springs of the VPS. As the physical dynamical parameters, the mass, the shear stiffness, and the damping coefficient of the model are directly identified from the observed response data, by prediction error method and using reduplicated state-space model derived from the equations of motion before and after the shifting of the swinging cycle of the pendulum.

The experimental structural model consists of the VPS, the structural model, viscous dampers, and an excitation device. To estimate the dynamical characteristics of the structural model, system identification are carried out using the observed data. Six cases of identification are carried out based on the selection of the column of the structural model and the selection of the spring of VPS. 300 times of different identification trials are carried out using the same experimental data with 300 random sets of initial value factors. Accuracy of the identification is evaluated stochastically.

Result of the structural system identification using the response data of the structural model with VPS shows that the proposed VPS can successfully estimate the physical dynamical parameters, e.g. mass, stiffness, and damping coefficient of single degree of

Table 1. The case classification for the system identification.

	Model's columns	VPS springs	
Case 1	Slender (1.95Hz)	Soft (2.92Hz)	Middle (4.90Hz)
Case 2	Slender (1.95Hz)	Soft (2.92Hz)	Hard (6.68Hz)
Case 3	Slender (1.95Hz)	Middle (4.90Hz)	Hard (6.68Hz)
Case 4	Thick (3.54Hz)	Soft (2.92Hz)	Middle (4.90Hz)
Case 5	Thick (3.54Hz)	Soft (2.92Hz)	Hard (6.68Hz)
Case 6	Thick (3.54Hz)	Middle (4.90Hz)	Hard (6.68Hz)

Table 2. Identification results about median and convergence probability.

	Median / true value				Convergence probability	
	%				%	
	m_1	k_1	c_1	Error	$+-10\%$	$+-15\%$
Case 1	94.3	86.2	94.3	8.4	0.0	99.7
Case 2	94.8	86.5	104.6	7.8	0.3	86.0
Case 3	94.7	86.1	90.8	9.5	0.0	94.7
Case 4	92.5	92.2	96.5	6.3	100	100
Case 5	92.5	89.8	88.2	9.8	21.7	97.0
Case 6	92.4	91.2	94.2	7.4	99.3	99.3

freedom structural model to an accuracy of 15 % of relative error. The accuracy of the estimation depends on the selection of the springs of the VPS. It indicate that to improve the accuracy of the estimation, further study about the methodology of tuning the sifting swinging cycle of the pendulum should be needed.

REFERENCES

Furukawa, T., Ito, M., Izawa, K. & Noori, M. N. 2005. System identification of base-isolated building using seismic response data. *Journal of engineering mechanics*. Vol.131 No.3, ASCE: 268-275.

Figure 1. Variable Pendulum Sensor and Experimental structural model.

Safety, Reliability and Risk of Structures, Infrastructures and
Engineering Systems – Furuta, Frangopol & Shinozuka (eds)
© 2010 Taylor & Francis Group, London, ISBN 978-0-415-47557-0

A risk-based maintenance optimization methodology for pipelines with corrosion defects

Y.-T.J. Wu
Applied Research Associates, Inc., Raleigh, NC, USA

C.P. Hsiao
Chevron Energy Technology Company, Richmond, CA, USA

A. van Roodselaar
Chevron Energy Technology Company, Houston, TX, USA

ABSTRACT

Pipelines, especially oil transmission pipelines, are subject to environment-induced corrosion mechanisms. The growing of corrosion defects can potentially lead to burst and leak failures if without proper monitoring and maintenance programs. The consequences of the failures range from simple local repairs to major replacements that cause interruptions of oil production. In addition, certain failures may create significant safety and environmental issues. To manage the risks, in-line inspection (ILI) tools such as the magnetic flux leakage device are increasingly being used to detect the locations of the metal-loss with defect size estimates to provide information for pipeline operators to make decisions on maintenance. This paper presents a risk-based maintenance optimization (RBMO) methodology based on ILI data. A RBMO framework is summarized in Figure 1. The objective of RBMO is to optimize the life-cycle risk by including ILI inspections, repair/replacement time and decision thresholds, and other mitigation strategies.

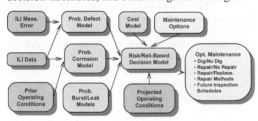

Figure 1. A RBMO Methodology Framework.

The sampling-based simulation approach has been implemented in a PipeRisk software with the following features: (1) Capable of handling thousands of defects each with 10 random variables. (2) Builds on a system PRA framework that integrates data, failure predictive models, and cost model for maintenance optimization. (3) Builds on a flexible segment-based analysis framework that analyzes multiple segments with different pipe properties, allowable risks, failure consequences, and corrosion rates. (4) Based on computationally robust random simulations to model complex maintenance scenarios. (5) Implements several importance-sampling methods including a two-stage RBMO approach, a failure sample generator using Markov-Chain Monte Carlo, and an Adaptive Stratified Importance Sampling (ASIS) method for accurately and efficiently computing probability-of-failure. Figs. 2–3 shows sample results.

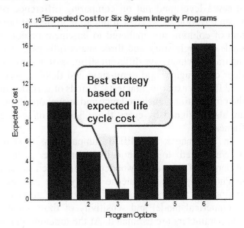

Figure 2. Expected Costs for Six Strategies.

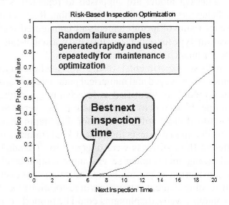

Figure 3. ILI Time Optimization.

Safety, Reliability and Risk of Structures, Infrastructures and
Engineering Systems – Furuta, Frangopol & Shinozuka (eds)
© 2010 Taylor & Francis Group, London, ISBN 978-0-415-47557-0

Numerical evaluation for shaking-table test results of full-scale wooden frames supported by unanchored foundation

Y. Mukai
Nara Wemen's University, Nara, Japan

Y. Suzuki
Ritsumeikan University, Kusatsu, Japan

ABSTRACT

To investigate for non-linear seismic behaviors of wooden frame structures constructed by traditional jointing and built-up way, shaking-table tests were carried out at "E-defence" in Hyogo, Japan in 2007. Emphases are put on evaluating interactions of multi-spans frames which are connected to each neighboring other with comparatively flexible floor diaphragm at eaves level, and put on comparing difference of seismic responses whether foundation girders at bottom of columns are anchored to basement or not. 3 kinds of single-story and three spans full-scale testing frames which have different stiffness of horizontal floor diaphragm; rigid, semi-rigid and flexible, were used for this experimental tests. 2 kinds of variation at their foundations, anchored type and unanchored type, are also designed by every those 3 kinds of testing frames.

In this paper, simplified lumped-mass system is considered to numerically reproduce seismic responses of examining wooden frames (such as seen in Photograph 1). Parametric surveys are executed to find out adequate value of structural parameters of numerical model and sensitivity of those identified parameters are estimated. At the touching plane between column and foundation, virtual small mass in numerical model are supposed to represent sliding behaviors of unanchored foundation. Restoration characteristics of spring elements are configured as tri-linear slip type for upper structure and normal bi-linear type for unanchored foundation.

At first, experimental results for anchored-type frames are surveyed and numerical parameters of analytical model are identified as to accurately simulate non-liner behaviors of upper structures by every variation of different stiffness of floor diaphragm. Each set of shaking table tests for those 3-kind frames with different floor stiffness were operated for 3 kinds of wall-arrangements under several input motion levels. By considering extent of plastic deformation at every anti-seismic element, parameters of restoration models of columns, walls, diaphragms could be properly estimated. After those calibrating procedures, friction

Photograph 1.   Testing frame (Unanchored type).

and slipping behaviors between bottom of wooden columns and flat supporting plates made of stone are numerically investigated for the unanchored cases.

Thorough those studies, following features are assured. 1) To trace accurately out time-history of non-linear behaviors of testing frames, it is significant to identify not only parameters beyond elastic-limit of spring elements, but also elastic-domain stiffness. 2) While sliding behaviors of the unanchored model can be fairly simulated by using simply-modeled restoration element, slipping gap between column and basement is quite sensitively affected by assumption of sliding (yielding) point of restoration model. 3) Equivalent friction force ratio is estimated to comparatively small value, and sliding behavior of the unanchored model is considered as depending on interaction related to upper structures stiffness.

REFERENCES

Suzuki, Y. et al. 2007. Simulations of the Vibration Tests of E-Defense of Traditional Wooden Houses (Part 1 to 4). *Summaries of Technical Papers of Annual Meeting (2007-Kyushu) Architectural Institute of Japan (AIJ), C-1 (Structures III)*, pp.531–538 (in Japanese).
Suzuki, Y. (ed.) 2007. *Technical report: Investigation for anti-seismicity of the wooden structures built by traditional Japanese constructing techniques.* Kyoto: Researching Group to shaking-table experiment for traditional wooden constructions. (in Japanese).

General Session (Bridges and Buildings)

Safety, Reliability and Risk of Structures, Infrastructures and
Engineering Systems – Furuta, Frangopol & Shinozuka (eds)
© 2010 Taylor & Francis Group, London, ISBN 978-0-415-47557-0

Assessment and characteristic analysis of safety for shear failure on deteriorated RC bridges due to chloride attack based on field testing

H. Kano
Japan Bridge Engineering Center, Osaka, Japan

H. Morikawa
Kobe University, Kobe, Japan

ABSTRACT

The safety evaluation of RC bridge is executed for the bending moment mainly. However, there is concern that the shear failure may become predominant on considering the uncertainties of rebar corrosion, execution management, material strength and so on. Therefore, in this paper, the deterioration prediction on three target bridges was carried out newly by the previous method (Yuasa & Morikawa, 2006) in consideration of the uncertainties included in the deterioration factors, and the shear safety based on various indices was evaluated. In addition, various safety evaluation indices were also examined by comparing with the flexural safety evaluation result.

The principal findings obtained in this paper are as follows.

(1) The comparison of time dependent degradation in safety by three indices which are factor of safety FS, safety margin M, and safety index β, was carried out in the three target bridges. For "Bridge S", the safety index β fell below the limit value of target reliability index 2.08 in 75 years old because of the progress of degradation, though the other indices still stayed over the threshold in 100 years old. It is the case where the importance of the safety evaluation in consideration of uncertainty of deterioration factor is suggested on evaluating safety (Figure 1).

(2) The conversion of the load carrying capacity R and stress resultant S into the safety value where the reliability of 95% was secured was attempted. For "Bridge S", the converted FS in consideration of uncertainties fell below the limit value in almost the same age as the safety index β (Figure 1).

(3) When the concrete strength is high and the corrosion of rebar progresses early at the bottom of girder like "Bridge I", there are a lot of possibilities that the flexural failure becomes predominant like a normal girder. However, there are also possibilities that the shear failure becomes predominant when it is expected that the concrete strength is low level, or that the uncertainties of

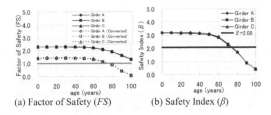

(a) Factor of Safety (*FS*) (b) Safety Index (*β*)

Figure 1. Safety Evaluation Result on "Bridge S".

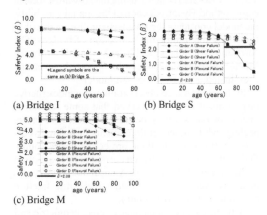

(a) Bridge I (b) Bridge S

(c) Bridge M

Figure 2. Comparison of *β* on Target Bridges.

concrete strength is large and the stirrup on the side of girder corrodes earlier than the bottom rebar like "Bridge S" or "Bridge M". In this case, it is considered that the shear safety evaluation is available (Figure 2).

REFERENCES

Yuasa, K. & Morikawa, H., 2006, Shear Resisting Performance and Safety of RC Bridges with Chloride Induced Deterioration Based on Field Testing, *Proceeding of the Concrete Structure Scenarios, JSMS*, Vol.6, 29–36.

Safety, Reliability and Risk of Structures, Infrastructures and
Engineering Systems – Furuta, Frangopol & Shinozuka (eds)
© 2010 Taylor & Francis Group, London, ISBN 978-0-415-47557-0

System reliability analysis of existing RC slab bridges incorporating spatial variation of damages

X.M. Wang, X.F. Shi, X. Ruan, Z.S. Tian & S.B. Geng
Department of Bridge Engineering, Tongji University, Shanghai, China

ABSTRACT

In accordance with notion of system identification, the paper divides bridge assessment method into two categories: Inverse-problem class Method and Forward-problem class Method. A bridge probability assessment model considering the brittle fracture of the corner corroded reinforcements is proposed to solve the phenomenon that the diseases are mainly concentrated in the outer corner of the exterior slabs in assembly RC Slab Bridges, using the Forward-problem class Method, By introduction of the spatial variability, the model considers the inconsistent disease distribution along the longitudinal direction of bridges. The correlation formula is established, which is fit for the Monte Carlo sampling. A simple supported slab bridge is evaluated using the proposed model.

By the appearance observation and nondestructive inspection, it is often observed that some reinforcements corrode severely because of protective layers peeling, while other reinforcements have lower corrosion ratio owing to their protective layers sound. The reinforcements with high corrosion ratio are often in the corner of the section. A specialized resistance model is established to consider the brittle fracture caused by high corrosion ratio.

With the changes of damage degree, geometric and mechanical parameters of slabs along bridge span, structural resistance changes also. So the whole structure can't only be assessed by the mid-span section since the spatial variability in resistance and effects. In order to get the flexural failure probability of the structure, the paper establishes a series model of multi-sections.

Monte-Carlo simulation is used herein to calculate the failure probability of the slab. PSAS1£¨Probability Structural Analysis Software Version 1£©is employed to calculate the reliability index and the failure probability of each sections.

By using the adaptive importance sampling method, the program PSAS1 can not only solve the failure probability of small probability event which can not be solved by the direct Monte-Carlos method, but also take the correlation of the multi limit state into account accurately.

Based on the proposed assessment process and results presented the following conclusions are made:

1. It maybe seriously underestimate the failure probability if the assessment of simply supported slab is based on mid-span section, especially for the structure in the presence of disease. The system assessment method consisting of several sections in mid-span region can solve the problem.

2. Bridge assessment method belonging to Forward-Problem Class can obtain the evaluation parameters directly and identify credible system model based on inspection data. Consequently, it would improve accuracy of the assessment and predication effectively. With development of the detection technology, this kind of method would solve more practical problems.

3. The model proposed in the paper is suitable for the RC slab or beam. Its further application for PRC structure needs the development of the detection technology for some key parameters such as the degree of prestress loss resulted form pipeline friction and long-term effects.

ACKNOWLEDGEMENT

This research is supported by National High-tech Research & Development Program (863 Program) from the Ministry of Science and Technology of China (Grant No. 2007AA11Z107) and Western region science & technology development projects from the Ministry of Communication (Grant No. 2007 318 822 31).

REFERENCES

Almusallam AA. 2001.Effect of degree of corrosion on the properties of reinforcing steel bars. Construct Build Mater 15:361–8.

Safety, Reliability and Risk of Structures, Infrastructures and Engineering Systems – Furuta, Frangopol & Shinozuka (eds)
© 2010 Taylor & Francis Group, London, ISBN 978-0-415-47557-0

Safety assessment of RC bridges strengthened by using CFRP sheet related to chloride-induced deterioration

F. Peng & H. Morikawa
Kobe University, Kobe, Japan

ABSTRACT

Due to the large volume of existing RC structures, in which such damage occurred by various cause, durability concerns have grasped considerable attention worldwide. And now, it is obviously the issue of how to maintain in service existing concrete infrastructure has been important for society. In recent years, the use of Fiber Reinforced Polymers (FRP) sheet has been established as one of the most promising methodologies in the field of structural rehabilitation of concrete structures.

In this study, the effect of deteriorated reinforced concrete bridge strengthened with Carbon Fiber Reinforced Polymer (CFRP) sheet was evaluated by reliability analysis. And this paper addressed a way of managing bridge before and after strengthened with CFRP sheet by safety assessment.

The actual existing RC Bridge F located in Awaji Island, Japan, damaged by chloride-induced deterioration, was assessed in this study. It is composed of three main girders (from ocean-side, A-girder, B-girder, C-girder). In this paper, firstly the systematic and detailed inspection was carried out by both the Non-Destructive Inspection and the truck loading test at the age of 46. From the results of the material physical and chemical tests (were conducted for the core test specimen sampled from girders), it was clarified that the ingress of chloride ions into concrete plays a crucial role in reinforcing bar corrosion and, hence, for the durability of A-girder, B-girder. And then the deterioration prediction of girders was evaluated based on the methodology proposed by the authors [1].

When the Monte Carlo simulations for evaluation of the probability model of the stress resultant and the probability model of the load carrying capacity, were proceeded, it is necessary to evaluate and to model the peeling property of CFRP sheet bonded on girders, and finally to verify the applicability of the proposed method to smeared crack type FEM analysis. This paper shows that the sheet bond constitutive law has been studied by two ways. One is getting it from RC beam test (different dimension from the girder) [2]. And the other is from the database which

was constructed by collecting experimental results of many research institutions. This database stores large amounts of information of RC members in consideration of strengthening by FRP sheet, and were studied by multiple linear regression analysis.

Finally, the safety index "β" was calculated based on the probability model of safety margin evaluated from the stress resultant and the load carrying capacity. When the "β" of the analysis result after strengthening with CFRP sheet was compared with the before, the following remarks were derived:

1. The method, which was proposed in this study as managing bridge before and after strengthening with CFRP sheet by reliability analysis of safety assessment, was demonstrated to be effective.
2. In spite of different sheet bond constitutive law, the results of analysis presents that the load carrying capacity and safety index of each girder increased with strengthening.
3. By the reliability analysis, the difference of two sheet bond constitutive laws was clarified. The evaluation based on "RC beam test" is relative conservative in comparison with the evaluation from database.

REFERENCES

H. Morikawa, Y. Morita and D. Kojima, "Evaluation of Deterioration and Safety on RC Bridges with Chloride Induced Deterioration Considering Uncertainties Included in The Evaluation Process," *Journal of Materials, Concrete structures and Pavements*, JSCE, No.809/V-70, pp.117–130, Feb. 2006.

H. Morikawa, T. Kamotani and H. Kajita, "Evaluation of Delamination Characteristic on RC Beams Bonded with CFRP Sheet and Application to Smeared Crack Type FEM Analysis," *Journal of Materials, Concrete structures and Pavements*, JSCE, No.802/V-69, pp.15–31, Nov., 2005.

Safety, Reliability and Risk of Structures, Infrastructures and Engineering Systems – Furuta, Frangopol & Shinozuka (eds)
© 2010 Taylor & Francis Group, London, ISBN 978-0-415-47557-0

Comparative seismic performance of the steel building systems

H.Y. Chang, H.Y. Yeh & J.Y. Chen
Department of Civil and Environmental Engineering, National University of Kaohsiung, Kaohsiung, Taiwan

C.C.J. Lin
National Center for Research on Earthquake Engineering, Taipei, Taiwan

ABSTRACT

The text in this paper is for visual purpose only. No rights can be taken from this.

The seismic performance of steel buildings can vary significantly depending on the adopted lateral load resisting systems (e.g. Wei 2006; Mayes et al 2005). This paper presents the result of a reliability-based seismic performance assessment for steel building systems. The analyzed 6- and 20-story steel office buildings have three kinds of different bracing systems, including moment-resisting frames (MRFs), eccentrically braced frames (EBFs) and buck-ling-restrained braced frames (BRBFs) (Wei 2006). The buildings had the same floor plans (5 × 3 bays) with 9-m bay spacing while the height was 4.5 m at the first story and 4.0 m at the remaining floors. The buildings were designed for the peak ground acceleration (PGA) of 0.428 g (value from seismic hazard curve that has a 10% chance of exceedance in 50 years) in accordance with Taiwan's building codes. All the beam-column connections were moment-resisting. Lateral loads were also resisted by a total of four braced frames, two in each direction.

The performance of the buildings was assessed using three-dimension nonlinear time history analyses with a suite of earthquake ground motions scaled to increasingly higher intensity (Curadelli and Rieba 2004; Lin and Tsai 2004) Following that, the fragilities (i.e. failure probabilities, P_f) for immediate occupancy (IO) and life safety (LS) were evaluated using the drift limits of 0.7% and 2.5%, respectively (FEMA2000). It was found that for the IO limit states, the lateral load resisting systems can only make a small difference to the failure probability of the 6-story buildings. For the LS limit states, the difference in the probability of failure using MRFs, EBFs and BRBFs becomes significantly large. Compared to the 6-story building, the 20-story buildings are not easily affected by the lateral load resisting systems. For the IO limit states, the lateral load resisting systems cannot make any marked distinction to the failure probability of the 20-story buildings using MRFs, EBFs and BRBFs. At more severe limit states (i.e. the LS limit states), the distinction in the probability of failure reduces to a small extent. The added EBFs and BRBFs have shown different effects in enhancing the seismic performance of low-rise and high-rise steel buildings.

Table 1. Limit-state PGAs for 6- and 20-story buildings.

(a)

6-story building		MRF	EBF	BRBF
$\theta \geq 0.7\%$	PGA (g)	0.171	0.300	0.318
$P_f = 10\%$	Normalization	1.000	1.754	1.860
$\theta \geq 2.5\%$	PGA (g)	0.625	1.191	1.390
$P_f = 50\%$	Normalization	1.000	1.906	2.224

(b)

20-story building		MRF	EBF	BRBF
$\theta \geq 0.7\%$	PGA (g)	0.310	0.365	0.410
$P_f = 10\%$	Normalization	1.000	1.177	1.322
$\theta \geq 2.5\%$	PGA (g)	1.343	1.700	1.610
$P_f = 50\%$	Normalization	1.000	1.265	1.191

REFERENCES

Wei, C.Y. (2006). "A study of local-buckling BRB and cost performance of BRBF." Thesis for the M.S. degree at National Taiwan University.

Curadelli, R. O. and Rieba J. D. (2004). "Reliability based assessment of the effectiveness of metallic dampers in buildings under seismic excitations." Engineering Structures, 26, 1931–1938.

Mayes, R. L., Goings, C. B., Naguib, W. I. and Harris, S. K. (2005) "Comparative Seismic Performance of Four Structural Systems" Proceedings of 2005 SEAOC Convention, San Diego, California.

Lin, B. Z. and Tsai, K. C. (2006). Platform of inelastic structural analysis for 3D systems - PISA3D R2.0.2 users' manual. National Center for Research on Earthquake Engineering.

Federal Emergency Management Agency (2000). FEMA-356: Prestandard and commen-tary for the seismic rehabilitation of buildings. Washington, D. C.

Safety, Reliability and Risk of Structures, Infrastructures and Engineering Systems – Furuta, Frangopol & Shinozuka (eds)
© 2010 Taylor & Francis Group, London, ISBN 978-0-415-47557-0

Simulation of the loads induced by running

C. Sahnaci & M. Kasperski

Department of Civil and Environmental Engineering Sciences, Research Team EKIB,
Ruhr-University Bochum, Bochum, Germany

ABSTRACT

The modern architecture tends to design lightweight and slender structures which is leading to an increased susceptibility to vibrations. Accordingly, the evaluation of serviceability against vibrations due to human excitation has gained more and more importance. More then ever it is important to determine and describe action effects on a structure to avoid vibration levels affecting the serviceability or even the safety. Men-induced dynamic loads can be subdivided into two major groups of activities addressing basically two groups of structures. Rhythmic activities on the spot like foot stamping, bobbing and jumping have to be considered for the design of grandstands or assembly halls. The basic locomotion forms walking and running have to be considered for the design of pedestrian bridges. The basic definition of the locomotion 'walking' is that at least one foot has contact to the ground; if between two contact phases a flight phase occurs, the corresponding locomotion form is called 'running'. While in recent years the loads induced by walking have been intensively studied sound and reliable data for the loads induced by a running person are scarce. A major reason for the absence of good data probably is that tests for running require a large experimental effort. Recent studies make use of data acquired from treadmill device tests. However, treadmill devices are an inappropriate test arrangement leading to an underestimation of the load amplitudes.

As optimum method for monitoring the loads for consecutive steps for natural running a longer walkway is required which is mounted on force-sensors. Such an experimental setup has been realized worldwide for the first time at the Ruhr-University Bochum by the Research Team EKIB. A 16 m measuring platform has been constructed which is completely mounted on force-sensors. To allow an almost natural running a 4 m lead-in section and a 8 m lead-out section was added at both ends of the platform leading to a total walkway length of 28 m. The width of the walkway has been chosen with 1.5 m. The basic aim of the experiments at the Ruhr-University Bochum is to identify the randomness and imperfections in the load process running. Two different subsets or scenarios have to be distinguished. The first scenario comprises 'short distance running with a specific purpose', e.g. catching a train. To study this subset of running, almost any volunteer can be analysed. The second scenario summarizes all running forms as they occur in sport and leisure, e.g. jogging, sprinting or even running a marathon. For the respective experiments, only practicing runners should be investigated. The paper presents the extended tests performed at the Ruhr-University Bochum and discusses in detail the randomness in the running parameters and their interdependencies. The randomness in the load amplitudes in regard to the basic frequency content is analysed. The paper finally gives some recommendations for a load model which can be implemented in a simulation.

Safety, Reliability and Risk of Structures, Infrastructures and
Engineering Systems – Furuta, Frangopol & Shinozuka (eds)
© 2010 Taylor & Francis Group, London, ISBN 978-0-415-47557-0

Dynamic response analysis of Shinkansen train-bridge interaction system under moderate earthquakes

X. He
Graduate School of Engineering, Hokkaido University, Sapporo, Japan

M. Kawatani
Graduate School of Engineering, Kobe University, Kobe, Japan

T. Hayashikawa
Graduate School of Engineering, Hokkaido University, Sapporo, Japan

S. Nishiyama
Technical Division, Nikken Sekkei Civil Engineering, Ltd., Tokyo, Japan

ABSTRACT

Japan is located in an earthquake-prone region, and the earthquake-proof capacity of the Shinkansen bridge-train interaction system is always a concern considering the extremely high-speed of the bullet trains. However, the dynamic response of the bridge-train interaction system under earthquakes is an extremely complicated phenomenon and is not adequately elucidated.

In this study, assuming that the structural behavior remains elastic during a moderate earthquake, an analytical approach to simulate the seismic response of the Shinkansen bridge-train interaction system is established. A bullet train car model idealized as three-dimensional sprung-mass dynamic system, which can express the motions of the car in both vertical and horizontal directions, is developed. The depiction of the car model is shown in Figure 1. A typical high-speed railway reinforced concrete viaduct in the form of rigid portal frame is adopted for the seismic response analysis. The viaduct with track structures is modeled with three-dimensional finite elements as shown in Figure 2. The dynamic differential equations of the bridge are derived using modal analysis technique. To discuss the seismic response of the bridge-train interaction system, three actual recorded moderate ground motions with different frequency components downloaded from Kyoshin Network of National Research Institute for Earth Science and Disaster Prevention are adopted as the seismic loads. Newmark's β step-by-stet numerical integration method is employed to solve the coupled differential equations of the bridge-train interaction system subjected to seismic loads.

Employing the seismic analytical procedure of the bridge-train interaction system established in this study, the seismic responses of the bridge and the train are simulated and evaluated. The dynamic effect of the train on the seismic response of the bridge is examined. The influence of the input ground motion on the response of the train is investigated.

The main conclusions can be summarized as follows. The seismic response of the bridge-train interaction system is rather complicated and dependent on both the

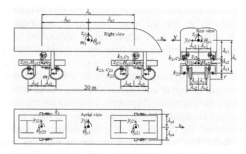

Figure 1.   15-DOF bullet train car model.

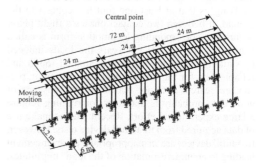

Figure 2.   Finite element bridge model.

characteristics of the bridge and the ground motions. The phenomenon that the train can act as a damper to the bridge during the earthquake is confirmed by the analytical results. According to the discussion of the seismic responses of the bridge-train interaction system, the fact can be recognized that it is not completely rational to consider the train merely as additional mass to the bridge structure, which is the case in the seismic design code in Japan, because in some cases the bridge responses may even be underestimated. In the design practice based on performance-based design, effort should be paid to achieve safer but more economic bridge structures by considering the train load as appropriate dynamic system.

Mini-Symposia (MS01) Uncertainty Modeling of Rare/Imprecise Data
[Session Organized on Behalf of IASSAR's CSMSE Umbrella]

Safety, Reliability and Risk of Structures, Infrastructures and Engineering Systems – Furuta, Frangopol & Shinozuka (eds)
© 2010 Taylor & Francis Group, London, ISBN 978-0-415-47557-0

On the impact of robust statistics on imprecise probability models: A review

T. Augustin
Department of Statistics, Ludwig-Maximilians University, Munich, Germany

R. Hable
Department of Mathematics, University of Bayreuth, Bayreuth, Germany

ABSTRACT

Especially in the engineering sciences, credibility of statistical statements is crucial because wrong conclusions can be very expensive or even life-threatening, and, therefore, careful and cautious modeling of all kinds of uncertainty is inevitable (e.g. Möller & Beer (2008), Oberguggenberger et al (2008)). The raise of robust statistics originates in the discovery that even small deviations of the real world from the assumed statistical model (called ideal model) can make the optimal statistical method break down. Therefore, robust statistics is concerned with statistical methods which may be less perfect in the ideal model but still lead to reliable conclusions if the ideal model is only approximately true. More recently, the theory of imprecise probabilities was developed as a general method to cope with the multi-dimensional nature of uncertainty and to handle also non-stochastic uncertainty (ambiguity) in an adequate manner. Mathematically, imprecise probability can be interpreted as providing upper and lower bounds on the involved probabilities, also providing a powerful framework for sensitivity analysis.

The strong connections between robust statistics and the theory of imprecise probabilities go far beyond their similar purposes – a fact which is frequently neglected. In robust statistics, small deviations of the ideal model are modeled by certain neighborhoods around the ideal model. As a matter of fact, nearly all commonly used neighborhoods are imprecise probabilities. That is, a considerably large part of robust statistics can be seen as a special case of imprecise probabilities. Therefore, it seems quite promising to address problems in the theory of imprecise probabilities by trying to generalize results of robust statistics. In this review paper, we present some cases where this has already been done successfully and highlight some cases where the connections between robust statistics and imprecise probabilities are most striking. This is done for the two main parts of frequentist statistics – namely hypothesis testing and estimation.

One of the main difficulties in the application of imprecise probabilities is that many problems are computationally expensive. However, quite often, the hypothesis testing problem can be solved by considering a simple testing problem between certain least favorable distributions which drastically reduces the computational effort (Huber-Strassen theory). This proceeding can also be extended to more general decision problems.

While ideas originating from robust statistics had a strong impact on hypothesis testing under imprecise probabilities, the theory of robust estimation has hardly been investigated in that context. Though only very few estimation problems have been analytically solved for finite sample size in the frequentist theory of robust estimation, these results are particularly interesting for the theory of imprecise probability as it seems to be promising to retransmit these results into the theory of imprecise probabilities in order to extend them within this new setup. Since many estimation problems have been solved in robust asymptotic statistics, it seems to be promising to resort to asymptotics also in the theory of imprecise probabilities whenever exact finite sample optimality results are not attainable. This also provides a strong motivation for developing an asymptotic theory of imprecise probability since such a theory is still at an early stage.

REFERENCES

Oberguggenberger, M., King, J. & Schmelzer, B. 2009. Imprecise probability methods for sensitivity analysis in engineering. International Journal of Approximate Reasoning, in press.

Moeller, B. & Beer, M. 2008. Engineering computation underuncertainty - Capabilities of non-traditional models. Computer & Structures, 86(10): 1024–1041

Safety, Reliability and Risk of Structures, Infrastructures and Engineering Systems – Furuta, Frangopol & Shinozuka (eds)
© 2010 Taylor & Francis Group, London, ISBN 978-0-415-47557-0

Random set analysis for tunnel excavation – Application to real project

H.F. Schweiger & A. Nasekhian
Graz University of Technology, Graz, Austria

R. Pöttler
ILF Consulting Engineers, Rum/Innsbruck, Austria

ABSTRACT

In order to carry out a full probabilistic design in practical geotechnics subjective assumptions about probability density function of parameters are often made while in many cases the results of geotechnical investigations are set valued rather than being precise and point valued. Peschl (2004) has extended Random Set Theory to be combined with the finite element method, called Random Set Finite Element Method (RS-FEM) to deal with this kind of uncertainty in geotechnics doing semi probabilistic analysis (e.g. reliability analysis). They illustrated the applicability of the developed framework to practical geotechnical problems and showed that RS-FEM is an efficient tool conducting reliability analysis in geotechnics in early design phases being highly complementary to the so-called observational method.

To demonstrate the applicability and efficiency of RS-FEM in geotechnical practice, a real case study has been chosen, namely a tunnel excavation located in the south of Germany. Application of a reliability analysis based on the Random-Set-Finite-Element-Method (RS-FEM) is presented for a real project. The novelty in this paper lies in the fact that a) the analysis is done before the start of the project, i.e. no updated information on material properties is available (the data provided by the geotechnical report for different sections of the tunnel are the only source of information) and b) that the proposed method is compared to a conventional design analysis.

The basic variables were obtained from two sources, first, from the geotechnical report which is based on a number of laboratory and in situ tests and second, from expert knowledge that is derived from previous experiences.

The resulting range of results is shown in Figure 1. They clearly indicate that conventional design analysis based on particular parameters sets defined at the design process cannot reflect the behaviour in situ, which can be expected. One of the important features of this kind of analysis is that once in situ

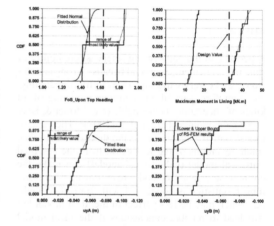

Figure 1. Some of the Results of RS-FEM analysis in terms of cumulative distribution functions.

measurements are available the quality of the numerical model can be judged. If measurements fall outside the range of calculated displacements than either the model itself was not appropriate or the range of parameters did not reflect the in situ behaviour. This result may have important consequences in cases where e.g. the ground investigation programme was insufficient and therefore design and construction has been based on incorrect assumptions. Unfortunately the start of the project has been delayed so that comparison with in situ measurements cannot be presented at this stage but this will be done once measurements become available.

REFERENCE

Peschl, G.M. 2004. Reliability Analysis in Geotechnics with the Random Set Element Method. Dissertation. Graz University of Technology.

Safety, Reliability and Risk of Structures, Infrastructures and Engineering Systems – Furuta, Frangopol & Shinozuka (eds)
© 2010 Taylor & Francis Group, London, ISBN 978-0-415-47557-0

Estimation of design impact forces of debris flows

D. Proske, J, Suda & J. Hübl

University of Natural Resources and Applied Life Sciences, Vienna, Austria

ABSTRACT

Mountain regions are exposed to a number of natural hazards such as avalanches, debris flows, rock falls and rock avalanches, flash floods and landslides. Mitigation measures are often employed to reduce the risks of hazards to humans and human settlements to an acceptable level. The design of structural mitigation measures are often not regulated and chosen arbitrary. Despite this situation, the design should at least comply with current rules for the design of structures. This measure has not yet been put in place. Currently in Austria a new code of practice for the design of structural mitigation (concrete) measures against debris flow, is under development. This code deals with the design of debris flow barriers in terms of load cases, such as reinforcement details, static and dynamic loads. One of the major tasks to establish this new code is the preparation of flow impact forces for the design process. In this background document all known techniques for the estimation of such debris flow impacts are investigated in terms of prediction quality. Furthermore, also included are theoretical works, miniaturized testing (including tests conducted by the authors) and known real world measurements. The formulas are further compared (based on sensitivity) against unknown input variables. This investigation has been extended to include weighting factors according to the First Order Reliability Method.

Despite the Institute of Mountain Risk Engineering having one of the greatest data files on natural hazard events (starting around 500 A.C.), the knowledge of debris flow in certain regions, very often lacks a sufficient amount of data for statistical analysis. Furthermore data sets are heavily corrupted due to climate change, changing geomorphologic conditions and changing flora. Besides this the reporting quality of early events is extremely low. Therefore further techniques have been used, such as Fuzzy sets, Rough sets and Grey numbers. Awareness of the uncertainty and indeterminism of the data heavily influences the choice of the design impact force and can not be neglected. Furthermore partial safety factors for this event have also been chosen.

REFERENCES

Scotton, P. (1996): Dynamic Impact of Debris Flows: Experimental Study. Dipartimento di Ingegneria Civile ed Ambientale. Universita Di Trento.

Tiberghien, D., Laigle, D., Naaim, M., Thibert, E. and Ousset, F. (2007): Experimental investigation of interaction between mudflow and obstacle. Debris-flow Hazards Mitigation: Mechanics, Prediction and Assessment. 4th International Conference on Debris-flow Hazards Mitigation, Chengdu, China.

Ishikawa, N., Inoue, R., Hayashi, K., Hasegawa, Y. & Mizuyama, T. (2008): Experimental Approach on measurement of impulsive fluid force using debris flow model. Interprevent 2008

Hübl, J. & Holzinger, G. (2003): Entwicklung von Grundlagen zur Dimensionierung kronenoffener Bauwerke für die Geschiebebewirtschaftung in Wildbächen: Kleinmaßstäbliche Modellversuche zur Wirkung von Murbrechern. WLS Report 50 Band 3, Im Auftrag des BMLFUW VC 7a

Armanini, A. (1997): On the dynamic impact of debris flows. Recent developments on debris flows, Lecture notes in Earth Sciences 64, Springer Verlag, Berlin

Lichtenhahn, C. (1973): Die Berechnung von Sperren in Beton und Eisenbeton. Kolloquium über Wildbachsperren. Mitteilungen der Forstlichen Bundesanstalt Wien, Heft 102, 1973, pp 91–127

Watanabe, M. & Ikeya, H. (1981): Investigation and analysis of volcanic mud flows on Mount Sakurajima, Japan. Erosion sediment transport measurement. International Association on Hydrol. Florence, Science Publication, 133, 1981, pp 245–256

Zhang, S. (1993): A comprehensive approach to the observation and prevention of debris flow in China. Natural Hazards, 7, pp 1–23

Totschnig R & Hübl J (2008) Historische Ereignisdokumentation. University of Natural Resources and Applied Life Sciences, Institute of Mountain Risk Engineering, Vienna

Safety, Reliability and Risk of Structures, Infrastructures and Engineering Systems – Furuta, Frangopol & Shinozuka (eds)
© *2010 Taylor & Francis Group, London, ISBN 978-0-415-47557-0*

Generalized reliability theory of structures simultaneously considering random, fuzzy and incomplete information

D.G. Lu & G.Y. Wang
School of Civil Engineering, Harbin Institute of Technology, Harbin, China

Y.B. Liu
School of Building Engineering, Dalian Nationalities University, Dalian, China

P. Zhang
School of Building Engineering, Southwest Petroleum University, Chengdu, China

ABSTRACT

We classify uncertain information into three categories: randomness of future events, fuzziness of objective cognition, and incompleteness of subjective cognition. It has been found by us that incompleteness is a kind of weak uncertainty in nature compared with randomness and fuzziness, since it can be incorporated in the random or fuzzy variables in the form of parameters by using "subjective probability" or "subjective membership degree", respectively.

As we all know, in classical structural reliability theory, the uncertainty can be generally broken down into another three types, that is, physical uncertainty, statistical uncertainty and modeling uncertainty. The former is aleatory, while the latter two are epistemic. In this paper, the statistical and modeling uncertainties are viewed as two specific kinds of incompleteness, since they can be described by means of subjective probability or Bayesian statistics.

When incomplete parameters in random variables are treated as subjective probability, we call this phenomenon as "dual randomness". Likewise, when incomplete parameters in fuzzy variables are treated as subjective membership degree, we call this phenomenon as "dual fuzziness". Meanwhile, random information and fuzzy information are often coupled or mixed in many fields, which result in "random fuzziness" or "fuzzy randomness" situations.

The classical reliability theory of structures only deals with the physical uncertainty in nature. In this paper, we introduce four kinds of generalized reliability theories of structures. The first is named as "dual random reliability theory", which considers both randomness and incompleteness. This kind of reliability problems has been well solved by many researchers using the methodologies of Bayesian statistics. The second is called as "dual fuzzy reliability theory". To solve this kind of reliability problems, a new definition of cumulative distribution function for fuzzy variables is introduced, and then, the incompleteness

in the satisfaction degree is dealt with by using fuzzy expectation operation. The third kind of generalized reliability problems is named "random-fuzzy reliability theory", in which random factors are presented in the membership function as parameters. We have utilized fuzzy probability theory to arrive at the reliability probability. The fourth kind of generalized reliability problems is called "fuzzy-random reliability theory", in which there still exist two cases, one case is that fuzzy factors are presented in the distribution functions of random variables as parameters; the other is that fuzzy-random information lies in the fuzzy membership functions. For these two different cases, we introduce two different kinds of fuzzy-random variables to derive at the fuzzy reliability index respectively.

REFERENCES

Der Kiureghian, A. 1989. Measures of structural safety under imperfect states of knowledge. *Journal of Structural Engineering* 115(5): 1119–1140.

Liu Y.B., Qiao Z., Wang G.Y. Fuzzy random reliability of structures based on fuzzy random variables. Fuzzy Sets and Systems, 1997; 86: 345–355.

Wang, G.Y. 1992. *Theory of Soft Design in Engineering.* Beijing: Science Press.

Wang, G.Y. & Wang, W.Q. 1986. Fuzzy reliability analysis of aseismic structures. *Acta Mechanica Sinica* 2(4): 322–332.

Wang, G.Y., Wang, W.Q., and Duan, M.Z. 1987. Fuzzy random reliability analysis of aseismic structures. *Proceedings of the Fifth Canadian Conference on Earthquake Engineering.* Ottawa, Canada.

Wang, G.Y., Wang, W.Q., and Ou, J.P. 1987. Generalized reliability of engineering systems. *Proceedings of the Second International Fuzzy Systems Association Congress.* Tokyo, Japan.

Wang, G.Y. & Zhang, Y. 1992. The theory of fuzzy stochastic process. *Fuzzy Sets and Systems* 51: 161–178.

Zhang, Y. & Wang, G.Y. 1993. *Theory of Fuzzy Stochastic Dynamic Systems.* Beijing: Science Press.

Safety, Reliability and Risk of Structures, Infrastructures and Engineering Systems – Furuta, Frangopol & Shinozuka (eds)
© 2010 Taylor & Francis Group, London, ISBN 978-0-415-47557-0

Correlations between uncertainty theories and their applications in uncertainty propagation

Kejiang Zhang & Gopal Achari
Department of Civil Engineering, University of Calgary, Calgary, AB, Canada

ABSTRACT

Different types and multisource uncertain information occur in characterization of contaminated sites. Three kinds of uncertainty are identified, i.e. random uncertainty which is inherently random in nature, informal uncertainty which is caused by a lack of information such as when the number of observations is not sufficient, and lexical uncertainty which occurs when linguistic variables such as "high" "medium", "low", "near" and "far" are used. Traditionally, stochastic methods were widely used to represent all types of uncertainties. With the advent of new uncertainty theories such as fuzzy set theory, possibility theory, evidence theory, and random sets, the proper representations of different kinds of uncertainties are revisited. Informal and lexical uncertainties can be better represented using fuzzy set theory and its extension, possibility theory, evidence theory, and random sets.

When one theory best represents one parameter whereas another theory may be more suitable for another, the hybrid propagation of these uncertainties will occur. In this paper, the correlations amongst different uncertainty theories are investigated. A probability density function (pdf) can be discretized into a set of random sets by dividing an approximate support of a pdf into n intervals and a basic probability assignment is defined as the integral of the pdf in the corresponding interval. A fuzzy number can also be divided into a set of consonant intervals by using level-cuts. A possibility distribution (the membership function of a fuzzy set can be used as a possibility distribution) can be transformed into a family of probability distributions with the upper and lower bounds. A possibility measure and its dual, necessity measure can be used to construct a family of probability measures. A unimodal probability density function with bounded support can be transformed into a less specific possibility distribution which has the same support and a possibility distribution with upper semi-continuous, unimodal, support bounded can also be transformed to a pdf. Basic probability assignment function (m), Belief function (Bel), and plausibility function (Pl)

are three fundamental definitions of evidence theory. When the sets are nested, the belief measure and plausibility measures become necessity measure and possibility measure, respectively. Random sets are sets defined on some probability space where values are represented as sets rather than points. The distribution function of a random finite set is a belief function and the distribution functional of a random closed set is a plausibility function.

Three methods were proposed to propagate hybrid uncertainty (i.e., the random and non-random uncertainty): (1) use a probability-possibility transformation to convert the probability distributions to the corresponding possibility distributions, (2) transform the possibility distributions into probability distributions, and (3) hybrid method. The results obtained from (1) and (3) are similar though (1) provides the outer bound of the results of (3), and the results of method (2) are contained in the results of method (1).

The effects of dependencies amongst uncertain parameters on the simulation results were also investigated and quantified by evaluating the belief and plausibility measures and then using them to approximate the probability measure based on evidence theory. A simple case study was used to explain this methodology. The correlations amongst uncertainty theories facilitate the propagation of hybrid uncertainty through different models.

*Safety, Reliability and Risk of Structures, Infrastructures and
Engineering Systems – Furuta, Frangopol & Shinozuka (eds)*
© 2010 Taylor & Francis Group, London, ISBN 978-0-415-47557-0

Dynamic analysis of a cable-stayed bridge with uncertain structural parameters

M.V. Rama Rao & S. Vandewalle
Department of Computer Science, Katholieke Universiteit Leuven, Belgium

M. De Munck
Department of Mechanical Engineering, PMA, Katholieke Universiteit Leuven, Belgium

D. Moens
Department of Applied Engineering, De Nayer, Lessius Hogeschool, Associatie K.U. Leuven, Leuven, Belgium

ABSTRACT

This paper focuses on the transient dynamic analysis of a cable-stayed bridge with uncertain structural parameters, subjected to an impact load. The uncertainty is located in the Young's modulus and mass density of concrete and steel. These parametric uncertainties are quantified by triangular membership functions. The bridge is analyzed using the α-sublevel technique. In order to analyze the time-domain response of the bridge under impact loading, an interval extension of Wilson's theta method is developed and applied for the direct evaluation of the uncertain dynamic response of the structure. The interval problem is based on the global optimization strategy. A Kriging-based response surface methodology adapted to multiple output analysis of FE models is introduced for this purpose. This method is applied and compared to a classical direct optimization approach. The work demonstrates the effectiveness of the Kriging-approach based fuzzy finite element method in evaluating the dynamic response of the cable-stayed bridge with multiple uncertainties.

The multiple-output response surface based methodology is as follows. In the first step of the procedure, a small space filling design (for example a Latin hypercube design) is generated and all objective functions are calculated at these response points by the FE solver. Using this information, initial response surfaces are created. Additional points are best selected by the adaptive procedure in step two instead of being randomly selected in this step. In the second step, a large space filling design is calculated. These points are not yet response points; only the few most promising points from this set will become real response points for which a finite element analysis will be performed at the end of this step. For each of these points, the function value and the expected error on the function value are estimated using the calculated response surfaces.

For each of these candidate response points, the average maximum improvement or AMI is calculated. The sum of squares selects the average maximum improvement over all output quantities. The candidate response points with the highest AMI are then selected and added to the response point set. Only in these points, an FE analysis is performed. Finally, all response surfaces are recalculated or updated with the new information.

In the numerical example, Wilson's theta method is applied to perform a transient structural analysis of a cable-stayed bridge subjected to dynamic loading. Uncertainty is introduced into the mass and stiffness properties of the structural system. The analysis is performed by optimizing the displacement response at each time step. It is observed that compared to the global and local optimisation routines, the Kriging response surface approach yields equally accurate results at only a small fraction of the computational cost.

REFERENCES

Moens, D. and Vandepitte D. 2005. A survey of non-probabilistic uncertainty treatment in finite element analysis. *Computer Methods in Applied Mechanics and Engineering*, 194:1527–1555.

Rama Rao. M.V. and Ramesh Reddy R. 2007. Analysis of a cable-stayed bridge with multiple uncertainties – a fuzzy finite element approach. *Structural Engineering and Mechanics*, 27(3):263–276.

Wilson E.L., Farhoomand I., and Bathe K.J. 1973. Nonlinear analysis of complex structures. International Journal of Earthquake Engineering and Structural Dynamics, Vol.1:241–252.

Mini-Symposia (MS13) Smart Structural Health Monitoring and Safety Assessments
[Session Organized on Behalf of IASSAR Subcommittee SC4]

Safety, Reliability and Risk of Structures, Infrastructures and
Engineering Systems – Furuta, Frangopol & Shinozuka (eds)
© 2010 *Taylor & Francis Group, London, ISBN 978-0-415-47557-0*

Experimental verification of the adaptive extended Kalman filter with unknown inputs for damage identification of structures

Hongwei Huang
Department of Bridge Engineering, Tongji University, Shanghai, China

Jann N. Yang
Department of Civil & Environmental Engineering, University of California, Irvine, CA, USA

Li Zhou
College of Aerospace Engineering, Nanjing University of Aeronautics and Astronautics, Nanjing, China

ABSTRACT

Time domain analysis methodologies based on measured vibration data, such as the least square estimation and the extended Kalman filter, have been studied and shown to be useful for the on-line tracking of structural damages. The traditional extended Kalman filter (EKF) method requires that all the external excitation data (input data) be measured or available, which may not be the case for many structures. Recently, an adaptive extended Kalman filter approach with unknown inputs (excitations), referred to as AEKF-UI, has been proposed to identify the structural parameters as well as the unmeasured excitations [Yang et al (2007b)].

In the AEKF-UI approach, the recursive solutions for the estimations of the extended state vector $\mathbf{Z}$ and the unknown input $\mathbf{f}^*$, namely $\hat{\mathbf{Z}}_{k+1|k+1}$ and $\hat{\mathbf{f}}^*_{k+1|k+1}$ (for $k = 1, 2, \ldots$), are given in the following.

$$\hat{\mathbf{Z}}_{k+1|k+1} = \hat{\mathbf{Z}}_{k+1|k} + \mathbf{K}_{\mathbf{Z},k+1}$$
$$\bullet [\mathbf{y}_{k+1} - \hat{\mathbf{h}}_{k+1|k} - \mathbf{D}^*_{k+1|k}(\hat{\mathbf{f}}^*_{k+1|k+1} - \hat{\mathbf{f}}^*_{k|k})] \tag{1}$$

$$\hat{\mathbf{f}}^*_{k+1|k+1} = \mathbf{S}_{k+1} \mathbf{D}^{*T}_{k+1|k} \mathbf{R}^{-1}_{k+1} (\mathbf{I} - \mathbf{H}_{k+1|k} \mathbf{K}_{\mathbf{Z},k+1})$$
$$\bullet (\mathbf{y}_{k+1} - \hat{\mathbf{h}}_{k+1|k} + \mathbf{D}^*_{k+1|k} \hat{\mathbf{f}}^*_{k|k}) \tag{2}$$

$$\hat{\mathbf{Z}}_{k+1|k} = \hat{\mathbf{Z}}_{k|k} + \int_{k\Delta t}^{(k+1)\Delta t} \mathbf{g}(\hat{\mathbf{Z}}_{t|k}, \mathbf{f}, \hat{\mathbf{f}}^*_{t|k}, t)\, dt \tag{3}$$

in which $\hat{\mathbf{h}}_{k+1|k}$, $\mathbf{K}_{\mathbf{Z},k+1}$, $\mathbf{S}_{k+1}$, $\mathbf{H}_{k+1|k}$ and $\mathbf{D}^*_{k+1|k}$ are the intermediate matrices to be computed recursively.

In this paper, the AEKF-UI approach will be verified using the experimental data obtained in [Wu et al (2006), Zhou et al (2008)]. A series of experimental tests using a scaled 3-story building model have been conducted recently. In the experimental tests, white noise excitations were applied to the top floor of the model, and different damage scenarios were simulated and tested. These experimental data will be used to verify the capability of the AEKF-UI approach in tracking the structural damage. The tracking results for the stiffness of all stories, based on the AEKF-UI approach, are compared with the stiffness predicted by the finite-element method. Experimental results demonstrate that the AEKF-UI approach is capable of tracking the structural damage and unknown excitations with reasonable accuracy.

ACKNOWLEDGEMENT

This research is partially supported by the U.S. National Science Foundation Grant No. CMS-0554814, and the National Natural Science Foundation of China Grant Nos. 50478037 and 50808138.

REFERENCES

Wu, S., Zhou, L., and Yang, J. N. 2006. Experimental study of an adaptive extended Kalman filter for structural damage identification. *Proc. 4th International Conference on Earthquake Engineering*, Paper No. 161, CD Rom, Taipei, Taiwan, Oct. 12–13, 2006.

Yang, J. N., Pan, S. W., and Huang, H. W. 2007b. An adaptive extended Kalman filter for structural damage identification II: unknown inputs. *Structural Control and Health Monitoring, Vol. 14, No.3, pp. 497–521.*

Zhou, L., Wu, S., and Yang, J. N. 2008. Experimental studies of an adaptive extended Kalman filter for structural damage identification. ASCE, *J. Infrastr. Systems*, 4(1), 42–51.

Safety, Reliability and Risk of Structures, Infrastructures and
Engineering Systems – Furuta, Frangopol & Shinozuka (eds)
© 2010 Taylor & Francis Group, London, ISBN 978-0-415-47557-0

Reliability analysis of an existing RC structure updated by inspection data

I. Yoshida
Musashi Institute of Technology, Tokyo, Japan

M. Akiyama
Tohoku University, Sendai, Japan

S. Suzuki
Tokyo Electric Services Co'Ltd, Tokyo, Japan

ABSTRACT

The uncertainties of model parameters for deterioration should be updated depending on the quantity and quality of the inspection data. Consequently more credible information can be obtained for decision-making of reasonable maintenance. Reliability estimation of existing structures is however more difficult than that of newly constructed structures because the uncertainties related to limit states should be updated by inspection or measurement data. Probability density function (PDF) of the updated model parameters are not expressed in simple form when a nonlinear equation or non-Gaussian variable is involved. MC based methods for non-linear filtering or Bayesian updating technique have been developed since 1990's (Gordon et al., 1993, Kitagawa, 1996, Ristic et al., 2004). This line of methods are referred by many terms, such as MC filter, Bootstrap filter, recursive MCS, sequential MCS, the Sampling Importance Resampling (SIR) method or the Sequential Importance Sampling with Resampling (SISR). In this paper, the term Sequential MCS (SMCS) is used to refer the MCS based updating method.

The reliability estimation of existing structure requires updating process. When performing a probabilistic assessment of the reliability of deteriorating structures, we often need to integrate the results of different inspections in time and/or methods. A new framework is described based on the above mentioned SMCS in this study. Figure 1 shows the outline of the method. The proposed method is demonstrated through a numerical example of reliability analysis as to deteriorating RC structures. Corrosion of rebar due to chloride attack is a major mechanism for the deterioration of RC structures. Several models are proposed to estimate the limit states for the chloride attack. Limit state exceeding probabilities of a RC structure are calculated using one of the deterioration model (Akiyama et al., 2006), in which many uncertainties such as wind speed, diffusion coefficient, corrosion speed and so on, are involved. Several scenarios of inspection and observed data are assumed,

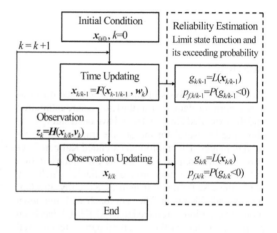

Figure 1. Flowchart of the proposed method.

including the visual inspection and detailed inspection. The visual inspection gives rank of deterioration level, from 1 to 4, whereas detailed inspection gives profiling of chloride concentrations through depth. The limit state exceeding probabilities are estimated for each scenario case.

REFERENCES

Akiyama, M., Suzuki, M., Matsuzaki, H., Dang, T. H. and Li, Y. F., 2006., Application of Reliability Theory to the Durability-Design of Concrete Structures Subjected to Corrosion, Proceedings of the 2nd fib congress, CD-ROM(ID: 15–6)

Gordon, N., Salmond, D. and Smith, A., 1993., A novel approach to nonlinear/non-Gaussian Bayesian state estimation, Proceedings of IEEE on Radar and Signal Processing, 140, pp. 107–113

Kitagawa, G., 1996., Monte Carlo Filter and Smoother for Non-Gaussian State Space Models, Journal of Computational and Graphical Statistics, Vol. 5, No. 1, pp. 1–25

Ristic, B., Arulampalam, S . and Gordon, N., 2004., Beyond the Kalman Filter, Particle Filters for Tracking Applications, Artech House

Safety, Reliability and Risk of Structures, Infrastructures and Engineering Systems – Furuta, Frangopol & Shinozuka (eds)
© *2010 Taylor & Francis Group, London, ISBN 978-0-415-47557-0*

Adaptive Monte Carlo filter for identification of abrupt changes in structural parameters

M. Chung & C.B. Yun
Smart Infra-Structure Technology Center, Department of Civil and Environmental Engineering, KAIST, Republic of Korea

T. Sato
Seismic Disaster Institute, Kobe Gakuin University, Japan

ABSTRACT

Structural health monitoring (SHM) of the existing civil infra-structures becomes an important subject, because bridges and buildings are often subjected to gradual damage or sudden collapse caused by fatigue loads or unexpected large loads such as earthquakes. To prevent the severe damage on structures due to these unexpected events, health monitoring has been facilitated to large civil structures using new technologies developed in the field of sensor, measurement, signal processing, and system identification. System identification techniques have been used to estimate the system characteristics of large structures based on the measured data. In time domain identification, the stochastic filter methods have been used to identify the time-varying and invariant parameters a dynamic structural system. Commonly, the unknown parameters of a system are recursively predicted and estimated by the measured information with noise. And the uncertainties of initial values and system modeling as well as observation noise were evaluated based on Gaussian probability function.

The Kalman filter (KF) (Kalman 1960) has been most widely used for the identification of time-varying systems in various engineering field. Since then, to overcome the limitations of the KF, many techniques have been developed such as the extended Kalman filer (Yun & Shinozuka 1980, Hoshiya & Saito 1984) and Sequential Monte Carlo (or Particle filter) (Gordon et al. 1993). Recently, Monte Carlo filter (MCF), one of particle filter, developed by Kitagawa (1996) is used to identify the significant parameters for a nonlinear non-Gaussian system. This method has a great potential and versatile filtering approach to estimate the system parameters.

In this study, a new algorithm for adaptive MCF is proposed by introducing the variable forgetting factors on the process and observation noises, because the conventional MCF does not have a sufficient time tracking ability for abrupt changes in structural parameters. The proposed method improves the likelihood of the state vector by increasing the uncertainties of the concurrent noise distributions using the weight gains which are the first order function of the forgetting factor less than the one. The gains are expressed as the ratios of the covariance matrices of the concurrent noise distributions to their initially assumed distributions and each forgetting factor is applied to the noise distributions at each time step when the changes of unknown parameters are larger than a threshold value. Thus better estimates can be obtained for the abrupt changing parameters that do not depend on the past data. In the adaptive MCF, three cases are considered. In the first case, the forgetting factor is applied only to the process noise, while it is applied only to the observation noise in the second case. In the last case, however, the forgetting factors are applied to both noises. Validations of the proposed method have been performed on the identification of abrupt changes in the stiffness and damping parameters of a shear building model under a strong earthquake excitation. The results indicate that the identified parameters show more rapid tracking ability for the abrupt changes and better convergence to the true values than those by the ordinary MCF.

REFERENCES

Gordon, N., Salmond, D. & Smith, A. 1993. Novel approach to nonlinear/non-gaussian Bayesian state estimation. *IEE Proceedings-F* 140: 107–113.

Hoshiya, M. & Saito, E. 1984. Structural Identification by Extended Kalman Filter. *Journal of Engineering Mechanics*, ASCE 110(12): 1757–1770.

Kalman, R.E. 1960. A New Approach to Linear Filtering and Prediction Problems. *Journal of Basic Engineering, Transaction of the ASME*, March, 82D(1): 35–45.

Kitagawa, G. 1996. Monte Carlo Filter and Smoother for Non-Gaussian Nonlinear State Space Model. *Journal of Computational and Graphical Statistics* 5(1): 1–25.

Yun, C.B. and Shinozuka, M. 1980. Identification of nonlinear structural dynamics systems. *Journal of Engineering Mechanics* 8(2): 187–203.

Safety, Reliability and Risk of Structures, Infrastructures and
Engineering Systems – Furuta, Frangopol & Shinozuka (eds)
© 2010 Taylor & Francis Group, London, ISBN 978-0-415-47557-0

Identification of nonlinear behavior of RC bridge piers using extended Kalman filter

K.-J. Lee

Daelim Industrial Company Limited, Seoul, Korea

C.B. Yun

Korea Advanced Institute of Science and Technology(KAIST), Daejeon, Korea

ABSTRACT

For the health monitoring of civil infrastructures, it is important to identify the nonlinear behavior related to structural damage. The extended Kalman filter (EKF) has been widely used for the parameter estimation of a nonlinear structural dynamic system. Nonetheless, when the EKF is applied to a complex system or a practically large structural system, a few implementation and numerical problems may arise. To overcome the numerical difficulties in obtaining the state transition matrix and improve the divergence of conventional EKF the sequential modified extended Kalman filter (SMEKF) is proposed in which the EKF with finite difference scheme is used for the state estimation and the nonlinear sequential prediction error method is for the parameter identification.

The forces induced on a bridge structure with reinforced concrete (RC) piers during major earthquakes may exceed the yield capacity of some piers and cause large inelastic deformations and damages in the piers. It is, important to recognize that a highly versatile model is required to closely reproduce the hysteretic behavior, in which several important aspects of hysteretic loops can be included, viz., stiffness degradation, strength deterioration, pinching behavior, and variability of hysteretic areas at different deformation levels under repeated load reversals. However, the model should also be simple to implement.

In this study, identification of the nonlinear hysteric behavior of a RC bridge pier subjected to earthquake loads is carried out. Only the acceleration measurements of the input earthquake motion and bridge responses are utilized, which are the easiest quantities in dynamic measurements, particularly for bridges with long-spans. The modified Takeda model is used to describe the hysteretic behavior of the RC member with a small number of parameters, in which nonlinear behavior is described by various rules of loading and reloading rules rather than an analytical expression. Then a two-step approach so called the sequential modified extended Kalman filter (SMEKF) algorithm

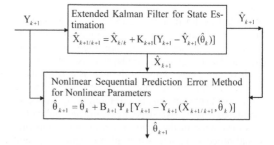

Figure 1. Sequential Modified Extended Kalman Filter (SMEKF) Algorithm.

is proposed to identify the unknown parameters and the state vector separately in two steps, so that the size of the problem for each identification procedure may be reduced and possible numerical problems may be avoided. Mode superposition with a modal sorting technique is also proposed to reduce the size of the identification problem for the nonlinear dynamic system. Example analyses are carried out for a continuous bridge model with a RC pier subjected to earthquake loads in the longitudinal and transverse directions.

REFERENCES

Lee, K.J. and Yun, C.B. 2008. Parameter identification for nonlinear behavior of RC bridge piers using sequential modified extended Kalman filter, *Smart Structures and Systems*, 4(3): 319–342.

Roufaiel, M.S.L. and Meyer, C. 1987. Analytical modeling of hysteretic behavior of R.C. Frames, *J. Struct. Engrg.*, ASCE, 113(3): 429–443.

Schei, T.S. 1997. A finite-difference method for linearization in nonlinear estimation algorithm, *Automatica*, 33(11): 2053–2058.

Mini-Symposia (MS12) Monte Carlo Simulations Applied to Civil Engineering

Safety, Reliability and Risk of Structures, Infrastructures and Engineering Systems – Furuta, Frangopol & Shinozuka (eds)
© *2010 Taylor & Francis Group, London, ISBN 978-0-415-47557-0*

Cost-benefit analysis of bridge seismic retrofit in a highway transportation network

S. Banerjee & M. Shinozuka
University of California, Irvine, CA, USA

M. Sgaravato
University of Padua, Italy

ABSTRACT

This research studied socio-economic effect of the seismic retrofit implemented on bridges in the Los Angeles area highway transportation network. The study is performed in two parts; in the first part, an integrated analytical framework is developed to evaluate the network performance under 47 scenario earth-quakes which represent the regional seismic hazard consistent with USGS seismic hazard maps. This frame-work used realistic models to represent the post-earthquake change in origin-destination (OD) characteristics of network traffic flow and bridge repair/restoration process. Monte Carlo simulations are performed to predict the post-event damage of highway bridges, and eventually, of the entire net-work. In this predictive simulation, two sets of fragility curves of bridges with and without seismic retrofit are used. Analysis evaluated post-earthquake network per-formance in the form of social cost arising from driver delay and loss of opportunity in the degraded network. This social cost is the expected loss in the system due to future seismic events. Result showed that the expected loss is much less if all bridges in the network are retrofitted. The second part of the paper focused on the evaluation of benefit from seismic retrofitting which is expressed as the summa-tion of loss avoidance from social cost and bridge repair/restoration cost. Thus estimated benefit is compared with bridge retrofit cost to investigate the benefit-cost ratio. The study showed that taking Caltrans as the stakeholder, bridge seis-mic retrofit is cost-effective when loss avoidance from social cost is considered.

REFERENCES

Chang, E.S., Shinozuka, M. & Moore, J. 2000. Probabilistic earthquake scenarios: extending risk analysis methodolo-gies to spatially distributed systems. *Earthquake Spectra* 16(3): 557–572.

Federal Emergency Management Agency (FEMA). 1999. Earthquake loss estimation methodology: HAZUS 99 (SR2). Washington DC: FEMA. Technical manual.

Shinozuka, M., Feng, M. Q., Kim, H., Uzawa, T., & Ueda, T. 2003a. Statistical analysis of fragility curves. Report MCEER-03-0002, Multidisciplinary Center for Earth-quake Engineering Research (MCEER), The State Uni-versity of New York at Buffalo, NY.

Shinozuka, M., Murachi, Y., Dong, X., Zhou, Y., & Orlikowski, M.J. 2003B. Effect of seismic retrofit of bridges on transportation networks. *Earthquake Engineer-ing and Engineering Vibration* 2(2): 169–180.

Southern California Association of Governments (SCAG). 1997. 1997 Model validation and summary. CA: SCAG.

Zhou, Y. & Shinozuka, M. 2006. Development of proba-bilistic scenario earthquakes for seismic risk analysis of spatially distributed systems. *8th National Conference of Earthquake Engineering*, San Francisco.

Zhou, Y. 2006. Probabilistic seismic risk assessment of high-way transportation network. PhD thesis, University of California, Irvine.

Zhou, Y., Banerjee, S., & Shinozuka, M. 2009. Socio-economic effect of seismic retrofit of bridges for high-way transportation networks: a pilot study. *Structure and Infrastructure Engineering*, Published online.

Safety, Reliability and Risk of Structures, Infrastructures and
Engineering Systems – Furuta, Frangopol & Shinozuka (eds)
© 2010 Taylor & Francis Group, London, ISBN 978-0-415-47557-0

Probabilistic evaluation of seismic performance for Vincent Thomas Bridge

Debasis Karmakar, Samit Ray Chaudhuri, Ho Lee & Masanobu Shinozuka
Department of Civil and Environmental Engineering, UC Irvine, Irvine, CA, USA

ABSTRACT

Throughout the history of suspension bridges, their tendency to vibrate under different dynamic loadings such as wind, earthquake, and traffic loads has been a matter of concern. In particular, after the disastrous failure of Tacoma Narrows Bridge in 1940, wind-induced vibrations of suspension bridges have mostly been studied. However, due to damage of this type of bridge in recent years including that occurred to the under-construction Akashi Kaikyo Bridge during the 1995 Kobe (Hyogo-ken Nanbu) earthquake, concerns regarding seismic safety of suspension bridges have gained significant attention.

The Vincent Thomas Bridge which connects Terminal Island with San Pedro and thus serves both Los Angeles and Long Beach ports, two of the busiest ports in the west coast of USA. The seismic performance of the Vincent Thomas Bridge is under scrutiny in recent years following the finding that the main span of the bridge crosses directly over the Palos Verdes Fault. It is believed that this fault has the capacity to produce an earthquake of magnitude (Mw) of 7.25 when a return period of 1000 years is considered. Considering the importance of this bridge in terms of carrying cargo from Long beach and Los Angeles ports, in spring 2000, the bridge underwent a major retrofit using visco-elastic dampers. Before the retrofit, an evaluation of the anticipated improvement of seismic performance of the bridge was conducted with the goal that the retrofit will eliminate the risk of collapse of the structure while allowing the bridge to be closed for repair when subjected to postulated 'safety evaluation earthquake' (i.e. an earthquake with a return period of 950 years). This study highlighted the collapse of tower-shaft as one of the major concerns under many other possible effects (Ingham et al. 1997).

Previous studies dealing with the Vincent Thomas Bridge have so far shown that the after retrofit vibrational characteristics obtained analytically using commercial software ADINA are more than 10% off the frequencies identified on the basis of the measurement data obtained by on-site instrumentation under ambient loading conditions (e.g. He et al. 2008). Therefore, there is a need for further investigation to accurately predict the vibrational characteristics of this bridge. In addition, estimating seismic vulnerability of this retrofitted bridge under expected scenario earthquakes is also very important.

In this paper, a member-based detailed three dimensional Finite Element (FE) as well as panel-based simplified models of the bridge are developed. In order to show the appropriateness of these models, eigenproperties of the bridge models are evaluated and compared with the system identification results obtained using ambient vibration. In addition, model validation is also performed by simulating the dynamic response during the 1994 Northridge earthquake and comparing with the measured response. Finally, considering a set of strong ground motions in the Los Angeles area, nonlinear time history analyses are performed and the ductility demands of critical sections are presented in terms of fragility curves. The study shows that a ground motion with PGA of 0.9g or greater will result in plastic hinge formation at one or more tower locations with a probability of exceedance of 50%. Also, it is found that the effect of damper is minimal for low to moderate earthquakes and high for strong earthquakes.

REFERENCES

He, X., Moaveni, B., Conte, J.P., and Elgamal, A., 2008. Modal identification study of Vincent Thomas Bridge using simulated wind-induced ambient vibration data. *Computer-Aided Civil and Infrastructure Engineering*, 23: 373–388.

Ingham, T.J., Rodriguez, S., amd Nader, M. 1997. Nonlinear Analysis of the Vincent Thomas Bridge for Seismic Retrofit. *Computers and Structures* 64(5/6): 1221–1238.

Safety, Reliability and Risk of Structures, Infrastructures and
Engineering Systems – Furuta, Frangopol & Shinozuka (eds)
© 2010 Taylor & Francis Group, London, ISBN 978-0-415-47557-0

Service life prediction for cracked RC structures using probability approach: Chloride attack through field investigations

S.-J. Kwon
University of California Irvine, Irvine, CA, USA

U.-J. Na
Ministry of Land, Transport, and Maritime Affairs, Busan, Korea

H.-W. Song
Yonsei University, Seoul, Korea

M.Q. Feng
University of California Irvine, Irvine, CA, USA

ABSTRACT

Reinforced Concrete (RC) structures are generally exposed to various environmental conditions during their lifetimes so that the service life of them is greatly reduced by the deteriorations. Chloride attack has been identified as one of the major deteriorations of RC structures. Since chloride ion rapidly penetrates into the concrete and breaks the passive film of reinforcement, it directly causes steel corrosion which results in cracking, delamination, and spalling of concrete cover. For the chloride behavior in sound concrete, various models and analysis techniques have been proposed. Advanced techniques for modeling chloride behavior were also introduced using the characteristics in early-aged concrete like porosity and saturation. The increased diffusion and permeation in chloride behavior due to occurred crack are considered in these researches. Even though these previous models can estimate the behavior of chloride diffusion reasonably, these techniques still require further studies to examine their limitations due to the uncertainties of variables.

A probabilistic durability design method was attempted to consider the uncertainties of input variables such as material properties and geometry. Related with the chloride attack, several parameters such as surface chloride content, cover depth, and diffusion coefficient were considered as random variables through statistical distributions applying Monte Carlo Simulation. Recently, probabilistic methods using spatial variability are introduced to the evaluation of service life considering steel corrosion and the limit state of probability for durability safety are considered in Concrete Specification. However, these applications using probabilistic approaches have been generally limited to the sound concrete structures without considering crack effect which occurred at early age. This is because it is difficult to evaluate the crack effect on the chloride diffusion characteristics.

Cracks due to hydration and shrinkage occur unavoidably in concrete structures. These cracks can be the main routes for steel corrosion. Through the cracks, deteriorating agents like chloride ion may penetrate into the concrete and also can be one of the major reasons of service life reduction in RC structures exposed to chloride attack.

In this paper, the service life in cracked concrete exposed to the marine environment conditions is predicted considering the crack effect on chloride diffusion. Carbonation behavior in cracked RC columns is also studied from another field investigation. For chloride attack, apparent diffusion coefficients in sound and cracked concrete with $0.1 \sim 0.3$ mm crack width are obtained in port structures through field investigation. The diffusion coefficients are greatly changed with occurred crack width. The relationship between crack width and diffusion coefficient is obtained through regression analysis. Design parameters such as cover depth, surface chloride content, and exposed period are also considered for probability of durability failure-P(durability). Utilizing intended P(durability) for design and Monte Carlo Simulation, the service life is calculated and the results are compared with those from deterministic method.

REFERENCES

CEB-FIP, 2006, Model Code for Service Life Design, the International Federation for Structural Concrete (fib), Task Group 5.6.
Tang, L. & Joost G. 2007. On the mathematics of time-dependent apparent chloride diffusion coefficient in concrete, Cement and Concrete Research, 37:589–595.
Song, H.-W., Pack, S.W., Lee, C.H., & Kwon, S.-J. 2006. Service life prediction of concrete structures under marine environment considering coupled deterioration, J. of Restoration of Buildings and Monuments, 12: 265–84.
Ishida, T. & Maekawa, K. 2003. Modeling of durability performance of cementitious materials and structures based on thermo-hygro physics. RILEM PRO 29, Life Prediction and Aging Management of Concrete Structures, 1:39–49.

Safety, Reliability and Risk of Structures, Infrastructures and
Engineering Systems – Furuta, Frangopol & Shinozuka (eds)
© 2010 Taylor & Francis Group, London, ISBN 978-0-415-47557-0

Accounting for system complexities in reliability estimates of light frame wood roofs using Monte Carlo simulation

A.V. Gleason, B.G. Nielson & B. Shanmugam
Clemson University, Clemson, SC, USA

Y. Li
Michigan Tech University, Houghton, MI, USA

ABSTRACT

The growing concern of the structural capability of light frame wood residential structures exposed to high wind events has increased interest in evaluating their reliability as a step towards performance based engineering. Special aspects of a roof system that need proper attention in such studies include the distribution of the load applied to the sheathing panels, redistribution of the load after connection failure and inclusion of the spatial variation of wind pressures – i.e. stochastic field. Performing the reliability analysis, considering the complexity of the system and load distribution, becomes a very difficult – if not impossible – task using classical reliability approaches.

The present study illustrates the role of Monte Carlo simulation in dealing with such complexities. This method involves running thousands of tests with key parameter values treated as probabilistic values. Because this method addresses the random nature of load and structural behavior instead of simply taking the extreme or average values, more reliable data is provided for decision making relative to wind exposed structures.

In this study, a simulation based finite element analysis of a simple 12-truss wood roof system is given. The model roof system utilizes an analytical model for the roof-to-wall connection developed in a past study (Shanmugam et al. 2009) to capture the behavior of the critical link in the uplift load path. This model includes rigid support in the downward direction but nonlinear behavior in the uplift direction.

Wind loads are assumed to follow a lognormal distribution (Rosowsky & Cheng 1999a). To account for spatial variation, a correlation matrix is developed from database wind pressures. This matrix is based on a cubic equation relating the correlation in pressures at two given points relative to the distance between points. In conjunction with the statistical distribution, this matrix was used to generate pressures over discrete regions of the roof.

Simulations for wind speeds of 161 km/h (100 mph) and 210 km/h (130 mph) are run as a demonstration of this method. Failure estimates recognizing the spatial variation of both wind pressures and roof-to-wall connections have been achieved using a crude Monte Carlo simulation. These simulation techniques used in conjunction with finite element modeling packages may be effectively used to capture system behaviors and complexities.

REFERENCES

Rosowsky, D.V. & Cheng, N. 1999a. Reliability of light-frame roofs in high-wind regions. I: Wind loads. *Journal of Structural Engineering* 125(7): 725–733.
Shanmugam, B., Nielson, B.G. & Prevatt, D.O. 2008b. Probabilistic descriptions of in-situ roof to top plate connections in light frame wood structures. *2008 ASCE Structures Congress, Vancouver, BC, Canada.*
Shanmugam, B., Nielson, B.G. & Prevatt, D.O. 2009. Probabilistic and analytical models for roof components in existing light-framed wood structures. *Engineering Structures.* (In Review)

Safety, Reliability and Risk of Structures, Infrastructures and Engineering Systems – Furuta, Frangopol & Shinozuka (eds)
© *2010 Taylor & Francis Group, London, ISBN 978-0-415-47557-0*

Tracking uncertainties from component level to system level fragility analyses through simulation

B.G. Nielson
Clemson University, Clemson, SC, USA

K.R. Mackie
University of Central Florida, Orlando, FL, USA

1 INTRODUCTION

Seismic fragility curves of highway bridges can be an effective tool for communicating the seismic vulnerability that a bridge or bridge portfolio may possess. As such, many approaches for generating these fragility curves have been developed; however, the analytical approaches have gained significant popularity because of their notable versatility. One overwhelming need is shared by all analytical methods - the need to track, and if possible, minimize uncertainties throughout the process. As the structural systems and failure domains considered become more complex and computational power has become more accessible, simulation tools have proven useful. The objective of this paper is to illustrate the versatility and power of a simulation-based approach in quantifying the varying sources of uncertainty in seismic bridge fragility modeling.

The bridge model considered in this paper is based on a commonly occurring configuration of multi-span simply-supported (MSSS) bridges in the central and southeastern parts of the United States. Analytical damage fragility curves are developed using probabilistic seismic demand models coupled with discrete damage states through the use of simulation techniques. Two general classes of uncertainty exist, namely aleatory (inherent randomness – non-reducible) and epistemic (knowledge based – reducible). Using a number of simulation studies, this paper looks to quantify uncertainties arising from ground motion randomness and epistemic uncertainty from as-built conditions (i.e., bridge component modeling parameters). The dispersion of the seismic demand conditioned on the ground motion intensity measure (IM) is analyzed to determine the relative contributions from randomness in the ground motion suite and bridge modeling uncertainties. Five simulation cases are performed to achieve this end.

The first case evaluates the subject bridge by varying the ground motions while including uncertainties in the bridge modeling parameters. The second case assumes median values for all modeling parameters while varying the ground motions only. This directly assesses the aleatory uncertainty in the seismic demand. The third case incorporates modeling uncertainty while considering only a single ground motion. This analysis is used to directly quantify the epistemic uncertainty due to modeling parameters. This treatment is actually carried out using four different ground motions having varying characteristics – each ground motion is used for the full set of simulations. Cases four and five repeat cases one and two, respectively, only the incidence angle for all ground motions is taken to be constant. For the sake of completing a fragility analysis without introducing additional sources of uncertainty, an assumed value of dispersion in capacity is maintained as constant throughout.

A total of 11 modeling parameters are identified that are seen as having an appreciable impact on the seismic response of this type of bridge. Each bridge parameter has been assigned an appropriate probability distribution to model the epistemic uncertainty. The simulation procedure adopted for generating probabilistic seismic demand models incorporates the combination of the Latin hypercube sampling and the cloud method of analysis. For case one, 80 bridge realizations are generated using the Latin hypercube technique. Each bridge realization is then paired with one of the 80 California ground motions in the suite. The ground motion is then applied to the bridge at some randomly selected incident angle. The fragility curve for the bridge system is found by integrating the component joint demand models over all possible failure domains using crude Monte Carlo simulation may be performed for this task.

Results indicate that the inclusion of aleatory randomness from the larger portfolio of motions dominates. Incident angle of ground motions does not play as significant of a role as originally believed.

Mini-Symposia (MS04) Structural Reliability and Optimization

Safety, Reliability and Risk of Structures, Infrastructures and
Engineering Systems – Furuta, Frangopol & Shinozuka (eds)
© 2010 Taylor & Francis Group, London, ISBN 978-0-415-47557-0

Single-loop system Reliability Based Design Optimization (RBDO) using Matrix-based System Reliability (MSR) method

T.H. Nguyen, J. Song & G.H. Paulino
University of Illinois, Urbana, IL, USA

ABSTRACT

This paper proposes a single-loop system reliability based design optimization (SRBDO) approach using the recently developed matrix-based system reliability (MSR) method. A single-loop method (Liang et al. 2007) was employed to eliminate the inner loop of SRBDO that evaluates probabilistic constraints. The MSR method (Song & Kang 2008; 2009) computes the system failure probability and its parameter sensitivities efficiently and accurately through efficient matrix calculations. The SRBDO/MSR approach proposed in this paper is uniformly applicable to general systems including series, parallel, cut-set and link-set system events. The sensitivity of the system failure probability with respect to design variables or constraints provided by the MSR method facilitates the use of gradient-based optimization algorithms. Two numerical examples demonstrate the proposed approach.

In the first example, the cross-sectional areas of the six members of a statistically indeterminate truss structure are determined for minimum total weight with a constraint on the system failure probability satisfied (MacDonald & Mahadevan 2008). The system fails if two members fail; therefore, the system event is described by 15 minimal cut sets each of which consist of two component events. The results are compared with those in the previous study in order to demonstrate the merits of the SRBDO/MSR approach. In particular, it is observed that the accurate system reliability estimates during the SRBDO/MSR reflect the symmetric conditions between diagonal members and between non-diagonal members in the optimal design (i.e. area) and the component failure probability (i.e. reliability index)

The second example demonstrates the application of the proposed approach to topology optimization (TO). Topology optimization seeks for optimal distribution of material in the predefined domain (Wang et al. 2007). In this study, the uncertainties of loads are considered in component based RBTO (CRBTO), and system reliability based TO (SRBTO). The influences

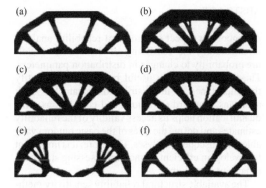

Figure 1. Optimal topologies by (a) deterministic TO; (b-c) CRBTO; and (d-f) SRBTO/MSR.

of the statistical correlation and the types of constraints, i.e. deterministic, probabilistic (component) and probabilistic (system) on the optimal topology are investigated as well as shown in Figure 1.

REFERENCES

Liang, J., Mourelatos, Z., & Nikolaidis E. 2007. A single-loop approach for system reliability-based design optimization, *Journal of Mechanical Design*, 129: 1215–1224.

MacDonald M, & Mahadevan S., 2008. Design Optimization with system-level reliability. *Journal of Mechanical Design*, 130.

Song, J., & Kang, W.-H. 2009. System reliability and sensitivity under statistical dependence by matrix-based system reliability method. *Structural Safety*, 31(2): 148–156.

Song, J., & Kang, W.-H. 2008. Estimation of structural robustness against progressive collapse based on system reliability, *Proc. Inaugural International Conference of the Engineering Mechanics Institute*, May 18~21, Minneapolis, MN, USA.

Wang, S., Sturler, E., & Paulino, G.H. 2007. Large-scale topology optimization using preconditioned Krylov subspace methods with recycling, *International Journal for Numerical Methods in Engineering*, 69: 2441–2468

Safety, Reliability and Risk of Structures, Infrastructures and Engineering Systems – Furuta, Frangopol & Shinozuka (eds)
© 2010 Taylor & Francis Group, London, ISBN 978-0-415-47557-0

Quasi Monte Carlo simulation for structural reliability sensitivity estimation

Wei Wang & Hongzhe Dai
Harbin Institute of Technology, Harbin, China

ABSTRACT

An important aspect of structural reliability analysis is calculating the sensitivity of the estimated failure probability to changes in distribution parameters. These sensitivities are useful because they quantify the importance of the distribution parameters such as means, standard deviations and correlations. This quantification helps us access validity of the reliability estimates and define the roles of the basic random variables in subsequent analyses. Therefore, it is important to develop an efficient method for assessing structural reliability sensitivity.

The available structural reliability sensitivity methods are mostly based on the corresponding structural reliability methods, including the first and second order reliability methods, the finite-difference method and the simulation method. Although there are many reasons for choosing one method over another, the most widely employed method is the Monte Carlo (MC) simulation method for complex problems. But the MC method is far more computationally demanding because the estimation of reliability sensitivity requires a considerable amount of model simulations. The challenge is to minimize the computational cost for achieving the required accuracy of reliability sensitivity, and many efforts have been made in this direction recently. Wu (1994) proposed a reliability sensitivity method based on the CDF of the structural response variable, the normalized reliability sensitivity coefficient is expressed as an expectation of the partial derivative of the PDF, wherein the sampling-based method can be used to compute the reliability sensitivity. Au (2005) presented a method of reliability-based design sensitivity analysis by efficient simulation, and this simulation is in fact a subset simulation on the basis of Markov Chain Monte Carlo. Lu *et al* (2008) provided a reliability sensitivity method that is based on the line sampling which is a new technique for evaluating high-dimensional reliability problems.

Although the above methods require much fewer samples than crude MC, in some cases their quantity can still be considered as large, especially for problems of small failure probability or problems that require a large number of costly finite element (FE) analyses in each sampling cycle.

To overcome the above issues, this paper introduced QMC method as an alternative and efficient simulation procedure for structural reliability sensitivity analysis. The main reason behind this situation is that QMC method can often achieve a given accuracy with far fewer samples and effectively decrease the total computational cost when compared with the traditional MC method. Three kinds of low-discrepancy sequences, the GLP, H-W and Halton, have been investigated and verified to have high fidelity in representing the uniform distribution on the unit hypercube. Thus, a new procedure, in which the low-discrepancy sequences are combined with the IS technique, is developed and used successfully to assess the reliability sensitivity. Numerical examples show that the proposed QMC procedure gives noticeably higher accuracy than MC method with the same number of samples, in addition, it has a faster rate of convergence than existing MC method. Therefore, the proposed QMC method qualifies as a comprehensive tool in structural reliability sensitivity analysis.

REFERENCES

Au, S.K. 2005. Reliability-based design sensitivity by efficient simulation. *Computers and Structures* 83: 1048–1061.

Lu, Z.Z. & Song, S.F. & Yue, Z.F. & Wang, J. 2008. Reliability sensitivity method by line sampling. *Structural Safety* 30: 517–532.

Wu, Y.T. 1994. Computational methods for efficient structural reliability and reliability sensitivity analysis. *AIAA Journal* 32: 1717–1723.

Safety, Reliability and Risk of Structures, Infrastructures and
Engineering Systems – Furuta, Frangopol & Shinozuka (eds)
© 2010 Taylor & Francis Group, London, ISBN 978-0-415-47557-0

Life cycle cost analysis for retrofit of critical infrastructure subject to multiple hazards

F. Jalayer, D. Asprone, A. Prota & G. Manfredi
Department of Structural Engineering, University of Naples Federico II, Naples, Italy

ABSTRACT

Design and assessment of critical civil infrastructure can be considered as a decision making problem in which the desired performance objectives, defined in terms of a set of design parameters, are optimized subject to a number of constraints. In the context of performance-based design, several performance objectives (e.g., minimize initial cost of construction, ensure life-safety in case of extreme and rare events) can be considered for a set of [discrete] limit states. In order to implement the performance objectives in a decision making framework, it is desirable to quantify and measure these objectives in terms of a common benchmark variable. The life-cycle cost has been proposed by many (Wen 2001, Faber and Rackwitz 2004) as a suitable benchmark performance variable. Lifecycle cost, which is historically identified as an economic term expressed in monetary units, accounts for initial costs of construction of facility, the regular costs of its maintenance and functionality over time, loss of revenue in case of damage, repair/replacement costs, social losses including eventual loss of life and end-of-life recycling costs. The evaluation of life-cycle cost is subject to several sources of uncertainty, such as the occurrence and the intensity of critical hazards, the resistance of the infrastructure and the service life itself. Therefore, the lifecycle cost is generally evaluated in terms of its expected value over the lifetime of the infrastructure.

The present study aims to apply the life cycle cost criteria to retrofit design of an existing critical infrastructure located in a seismic zone. Given the importance of the infrastructure, an unexpected strong explosion is considered to be plausible through its lifetime. Hence, the performance-based retrofit design of the infrastructure needs to be conducted on a multiple-hazard basis (i.e., earthquake and blast in this case). The retrofit design involves decision making between a set of viable options which can be evaluated and compared in terms of their corresponding life cycle cost and subject to reliability constraints. In particular, for each retrofit option, the corresponding life cycle cost is evaluated by calculating in monetary terms, the direct cost of the installation of the retrofit solution, the maintenance cost of the retrofitted system, the repair/replacement costs in case of damage, and the social costs including eventual loss of life and end-of-life recycling costs. After the low-cost option is identified among the set of options, the system reliability for the corresponding retrofitted infrastructure needs to be verified against an acceptable threshold. In this work, the system reliability is calculated taking into account both blast and earthquake hazards (Asprone et al. 2009).

REFERENCES

Wen YK. Reliability and performance-based design. Structure Safety (23): 407–428, 2001.
Faber MH, Rackwitz R. Sustainable decision making in Civil Engineering. Structural Engineering International (3): 237–242, 2004.
Asprone D., Jalayer F., Prota A. and Manfredi G. Proposal of a probabilistic model for multi-hazard risk assessment of structures in seismic zones subjected to blast for the limit state of collapse. In press, Structural Safety (2009).
Ellingwood, B. R., Mitigating Risk from Abnormal Loads and Progressive Collapse, Journal of Performance of Constructed Facilities, (20): 315–323, 2006.

Safety, Reliability and Risk of Structures, Infrastructures and
Engineering Systems – Furuta, Frangopol & Shinozuka (eds)
© 2010 Taylor & Francis Group, London, ISBN 978-0-415-47557-0

FE-based reliability analysis of the buckling of shells
with random shape, material and thickness imperfections

V. Dubourg, C. Noirfalise, J.-M. Bourinet & M. Fogli

IFMA & Université Blaise Pascal, Laboratoire de Mécanique et Ingénieries, Aubière Cedex, France

ABSTRACT

Shell structures are widely used in many industries due to their load carrying capacity *vs.* weight optimal ratio. By their very nature, the mechanical behavior of these lightweight structures is rather sensitive to slight variations in geometry, material properties, boundary conditions or loads. In an effort to address the effects of these variations which are most often of random nature (*e.g.* uncertain shape or thickness, due to the manufacturing process, spatial variations of material properties), the present work aims at quantifying the level of safety of such structures with respect to buckling which is a major concern for structural engineers involved in shell design. It is proposed here to have recourse to a highly efficient and robust Finite Element (FE) method in order to capture the intricate mechanical behavior of shells and to make use of an efficient simulation technique for assessing small probabilities of failure of interest.

In the proposed approach, we use an original method (Cochelin 1994) known as the Asymptotic-Numerical Method (ANM). It was proved to be less computationally demanding than more conventional incremental-iterative methods offered in usual general purpose FE codes. The ANM takes into account material and geometric nonlinearities, follower forces and boundary conditions *w.r.t.* local axes. Another major advantage comes from its ability to accurately detect bifurcation and limit point loads in a fully automatic and robust manner, which makes it very suitable for stochastic analyses. The present work is based on the 7-parameter formulation Büchter and Ramm shell element, which avoids thickness locking through the EAS concept. The stochastic model involves various random fields over the shell. We assume here some randomness in the shape of the structure which differs from the perfect one and we also take into account the spatial random variation of material properties and thickness over the shell.

In order to assess the failure probability of such a challenging structural reliability problem involving random fields and a computationally demanding FE buckling problem, it is proposed to make use of subset simulation (Au & Beck 2001). This method outperforms crude Monte Carlo in estimating small probabilities of failure, as it converts the initial problem to another more tractable one which basically consists in finding a few larger conditional probabilities of intermediate failure events. For completeness, a FORM approach is also used to provide significant local sensitivity results.

The proposed methodology is applied for illustration purpose to a structure derived from the Scordelis-Lo shell roof, taken from reference (Scordelis & Lo 1961). The shape imperfection is assumed to be a linear combination of a few most critical elastic buckling modes, whose coefficients are considered as random variables. Random variations of thickness and material properties (Young's modulus and yield strength) are taken into account by means of lognormal isotropic and homogeneous stochastic fields. These random fields are represented in Karhunen-Loève expansions, based on an exponential covariance function. It is worth pointing out that solving such FE-based structural reliability problems is rather computationally demanding and it is made possible here by combining efficient methods at both mechanical and stochastic levels (ANM and subset simulation respectively), on a multiprocessor computer platform allowing distributed computing.

REFERENCES

Au, S.-K. & Beck, J.L. 2001. Estimation of small failure probabilities in high dimension by subsets simulations. Probabilistic Engineering Mechanics 16 (4): 263–277.

Cochelin, B. 1994. A path-following technique via an asymptotic-numerical method. Computers and Structures 53: 1181–1192.

Scordelis, A.C. & Lo, K.S., 1961, Computer analysis of cylindrical shells, American Concrete Institute Journal 61: 539–561.

Safety, Reliability and Risk of Structures, Infrastructures and Engineering Systems – Furuta, Frangopol & Shinozuka (eds)
© 2010 Taylor & Francis Group, London, ISBN 978-0-415-47557-0

A review of recent features and improvements added to FERUM software

J.-M. Bourinet, C. Mattrand & V. Dubourg

IFMA & Université Blaise Pascal, Laboratoire de Mécanique et Ingénieries, Aubière, France

ABSTRACT

The development of FERUM (**F**inite **E**lement **R**eliability **U**sing **M**atlab) software as an open-source Matlab® toolbox was initiated in 1999 under Prof. A. Der Kiureghian's leadership at the University of California at Berkeley (Der Kiureghian *et al.* 2006). This general purpose structural reliability code was developed and maintained by T. Haukaas, with the contributions of many researchers. Since 2003, this code is no longer officially maintained.

The present paper aims at presenting the main features and capabilities of a new version of this open-source code based on the main contributions of the first author and the help of a few Ph.D. students at IFMA. The main concepts of FERUM are preserved and this paper lists the major changes operated in this code. FERUM should still be viewed as a development platform for testing new methods and applying them to various applications.

Regarding FERUM framework, the main structure of input data is preserved. Besides, major changes have been brought to version 3.1. The *gfun.m* function now allows sending multiple simultaneous calls to the *g* function. This change makes it possible to carry out simulations based on analytically defined limit-state functions vectorized in the Matlab sense, in a very efficient manner. Another potential use is distributed computing on multiprocessor computer platform. Other modifications brought to FERUM were besides driven by this important change, in order to adapt all existing reliability algorithm so that they take advantage of multiple simultaneous calls to *g*.

Reliability methods such as FORM, SORM, Monte Carlo Simulation (MCS) and Importance Sampling (IS) are still available in this new version. Directional Simulation (DS) based on deterministic directions is implemented and allows structural reliability analyses for low-dimensional random spaces. Subset Simulation recently introduced by S.-K. Au and J. Beck (Au & Beck 2001) is also available and represents an interesting alternative to MCS in estimating small probabilities of failure.

All these methods are still based on the so-called Nataf model. The underlying $\mathbf{R}_0$ correlation matrix is now obtained by numerical Gauss integration. An effort has also been made in the new version to calculate accurate sensitivities to distribution parameters by numerical integration too, including sensitivities to correlation. Other features regarding sensitivity and importance measures have also been added. FERUM now offers extra capabilities regarding global sensitivity measures, based on Sobol' indices (Sobol 1993). These indices are evaluated through various sampling techniques such as crude MCS and Quasi-Monte Carlo (QMC) simulations or by means of a limit-state function surrogate using Support Vector Machines.

FERUM now also offers capabilities regarding RBDO analysis on the basis of J.O. Royset's works (Royset & Polak 2004, 2007). The implemented method consists in a brute-force outer optimization loop over the reliability evaluation. The outer optimization loop makes use of the Polak-He optimization algorithm. The RBDO analysis can be carried out using FORM, MCS or DS.

REFERENCES

Au, S.-K. & Beck, J.L. 2001. Estimation of small failure probabilities in high dimension by subsets simulations. *Probabilistic Engineering Mechanics* 16(4): 263–277.

Der Kiureghian, A. & Haukaas, T. & Fujimura, K. 2006. Structural reliability software at the University of California, Berkeley. *Structural Safety* 28(1-2): 44–67.

Royset, J.O. & Polak, E. 2004. Reliability-based optimal design using sample average approximations. *Probabilistic Engineering Mechanics* 19(4): 331–343.

Royset, J.O. & Polak, E. 2007. Extensions of stochastic optimization results to problems with system failure probability functions, *Journal of Optimization Theory and its Application* 132(2): 1–18.

Sobol', I.M. 1993. Sensitivity estimates for nonlinear mathematical models. Mathematical *Modelling and Computational Experiments* 1: 407–414.

Safety, Reliability and Risk of Structures, Infrastructures and Engineering Systems – Furuta, Frangopol & Shinozuka (eds)
© 2010 Taylor & Francis Group, London, ISBN 978-0-415-47557-0

Reliability-based integrity assessment of structural systems subject to heterogeneous deterioration and large-scale inspection uncertainty

M.A. Maes, M.R. Dann & K.W. Breitung
University of Calgary, Calgary, Canada

ABSTRACT

Deterioration affects the reliability and the safety of virtually all types of structures and infrastructure. The present paper focuses on the posterior lifetime estimation of structural systems affected by corrosion and wear, but in the specific case where (1) only imprecise inspection data are available to estimate the parameters of the deterioration model; and (2) considerable heterogeneity is observed in the degradation paths of distinct corrosion features within different structural units. This situation occurs quite often in a variety of structural contexts. A typical example is the assessment of future onshore/offshore oil/gas pipeline integrity, for which the two single largest sources of degradation are internal pitting corrosion and micro-bacterial corrosion: both types of degradation can be well described by a stochastic process such as a specific transformation of a Brownian motion with drift, or a time-transformed gamma process with independent increments. But the true challenges are that internal inspections (e.g. using an intelligent pig) can only be performed infrequently and suffer from considerable uncertainty due to sizing errors, detectability, and reportability. Individual corrosion pits/features are also subject to considerable heterogeneity between degradation paths beyond what can be accounted for by conditioning on explanatory variables. In other words, spatial heterogeneity can partly be explained by cause-effect models capturing exposure, sensitivity, material variability, etc., but we must also account for additional local aleatory effects. Both scenarios suggest the use of Bayesian models and the use of hierarchical models accounting for surrogate random effects within the stochastic deterioration processes. The present paper focuses on both the framework and the mathematical techniques that are most convenient in the processing of large sets of imprecise inspection data using heterogeneous deterioration models with the objective of making informed lifetime integrity prognoses. An important aspect of the framework is that the inspection data usually suffer from sizing uncertainty (e.g. regarding corrosion pit depth) that is of the same order of magnitude as the uncertainty associated with random growth increments between inspections, as well as important detectability issues. The model variables that need to be estimated or updated in a Bayesian context include the basic parameters of the underlying stochastic process, the clock-time transformation parameters needed to match the basic corrosion process (e.g. initiation time, power law or sigmoid transformations, plateau modeling) and the random effects causing heterogeneity between degradation paths within structural units or systems. Numerical examples illustrate how the different components of the hierarchical protocol affect lifetime and integrity assessment.

Organized Session (OS18) Seismic Risk Assessment and Information

Safety, Reliability and Risk of Structures, Infrastructures and Engineering Systems – Furuta, Frangopol & Shinozuka (eds)
© 2010 Taylor & Francis Group, London, ISBN 978-0-415-47557-0

Virtual building inspection for open risk assessment software

J. Mitrani-Reiser and N.P. Jones
Johns Hopkins University, Baltimore, MD, USA

H. Ryu
Stanford University, Stanford, CA, USA

N. Luco
U.S. Geological Survey, Golden, CO, USA

ABSTRACT

Extensive damage to woodframe housing during the Loma Prieta and Northridge earthquakes emphasized the vulnerability of this type of structure and strengthened the need to develop methodologies that quantify the risk of woodframe structures being uninhabitable or "red-tagged" after a strong seismic event. There is a current initiative within the USGS to make seismic risk assessment openly available to owners of woodframe houses, which comprise the largest population of buildings exposed to seismic hazard. This initiative is working towards an interactive web-based tool, known as ResRisk-WH (Residential Risk – Woodframe Houses), that queries homeowners for basic descriptions of their homes (e.g., location, size, age, etc.) for use in a subsequent automated risk analysis. An essential component of this analysis process is to quantify the seismic performance of woodframe housing in terms of building safety (i.e., probabilities of red tagging).

This paper demonstrates how the probability of building safety can be estimated through a "virtual inspector" using the probabilities of damage of the building's components and applying current building inspection guidelines. An analytical model, otherwise known as the *virtual inspector* (Mitrani-Reiser 2007; Mitrani-Reiser and Beck 2007), was developed to estimate the probabilities that the building will be tagged at various levels of ground shaking with the well-known red, yellow, and green safety placards, based on current U.S. guidelines. This model may also be used to locate the areas of higher probability of damage in the structure. The *virtual inspector* probabilistically estimates building safety by matching up the damage descriptions from the fragility functions of the structural and nonstructural components with the damage descriptions from inspection guidelines.

The virtual inspection methodology is divided into two modules: (1) the rapid evaluation phase, which evaluates the structural integrity of a building and the probability of a red-, yellow- or a green-tag being posted, based on what would be a speedy inspection of the exterior structural components; and (2) the detailed evaluation phase, which includes a more thorough inspection of the structural components in the exterior as well as the interior of the building. This paper will describe the established close-form method for estimating safety tagging of structures using these safety evaluation guidelines. This paper will present the extension of this application into the ResRisk-WH framework, using Monte-Carlo simulations to quantify the probability of safety tagging of woodframe structures. The methodology presented here will provide a framework for automated virtual tagging of structures to estimate the forcible closure of woodframe housing as well as a means to estimate the probability of person displacement conditioned on seismic hazard intensity.

REFERENCES

Mitrani-Reiser, J., 2007. An ounce of prevention: probabilistic loss estimation for performance-based earthquake engineering, PhD Thesis in Applied Mechanics. California Institute of Technology, Pasadena, California.

Mitrani-Reiser, J. and Beck, J.L., 2007. "Incorporating Losses Due to Repair Costs, Downtime and Fatalities in Probabilistic-Based Earthquake Engineering," *Proc., Computational Methods in Structural Dynamics and Earthquake Engineering*, Crete, Greece, June 2007.

Safety, Reliability and Risk of Structures, Infrastructures and
Engineering Systems – Furuta, Frangopol & Shinozuka (eds)
© 2010 Taylor & Francis Group, London, ISBN 978-0-415-47557-0

Seismic risk communication with owners in structural design

K. Hirata & T. Ishikawa
Japan Women's University, Tokyo, Japan

ABSTRACT

Structural performance decision-making for a building is to determine the risk of failure probability concerning human life and property. However, private building owners have poor experience and knowledge about risk or decision-making for structural performance. As a result, structural engineers, with support of design codes, decide all performance and risk without consulting or explaining to owners. As in other professions, structural engineers need to begin to communicate with owners on matters of risk.

This paper describes citizen's attitudes using questionnaire results as a basic survey to establish a framework of risk-communication methods with building owners. This survey specifies contents and requirement for risk communication for seismic safety focusing on the owner's trust of engineers and decision-making for performance. The goal of communication is a reasonable selection of seismic safety performance level.

The first citizen's attitudes survey was carried out on the web in 2006 and obtained 535 people's opinions. The respondents were men and women more than 30 years old, who have lived in a relatively high seismic hazard area (Tokyo, Hokkaido, and Hyogo). Their answers didn't show difference between genders. There was a little difference of living area or experience of big earthquake. This survey defines the developing process for citizens with three steps in risk communication. The first step is recognition that risk and information disclosure is needed. The second step is owner's involvement in decision-making and deeper understanding of risks. The last step is reasonable decision-making of owners supported by structural designers. The results of this first survey indicate that Japanese citizens have reached the second step with understanding risks based on relatively correct knowledge, but they cannot yet stand on the third step. For establishing a process of risk communication based on trust, we professionals should understand the present requirement of owners and users and then we have to support improved decision making by owners.

A Second survey was carried out on the web in 2008 with 580 citizens. The respondents live in all prefectures and are from 30 years to 60 years old. This survey showed that respondents feel that there is no or little information about soil, piles, or material strength when they buy a brand new home. They also feel unsafe, and it means distrust to professionals. Over the half of respondents think self-check on trust of the design result. Many respondents were undecided on numerous parts of the survey, indicating that much needs to be done to improve risk communication. They are so interested in risk communication, and they need not a commentary but explanation of engineer. The explanation of performances before the agreement is especially needed.

Risk communication is essential to owners but it also has difficulty under present circumstances. Contents of communication which designers should explain before the agreement and method of decision-making are also considered in this paper.

REFERENCES

Hirata, K., Ishikawa, T., Sexsmith, R. G., and Haukaas, T., 2005. Evaluation of Target Seismic Safety Level and Safety Consciousness from User Surveys, Safety and Reliability of Engineering Systems and Structures, 9th International Conference on Structural Safety and Reliability, Rome, Italy, 3361–3368.

Safety, Reliability and Risk of Structures, Infrastructures and Engineering Systems – Furuta, Frangopol & Shinozuka (eds)
© 2010 Taylor & Francis Group, London, ISBN 978-0-415-47557-0

Reduction of seismic risk of overflow of refugees by upgrading existing non-conforming school buildings and wooden houses

Yasuhiro Mori
Department of Environmental Engineering and Architecture, Nagoya University, Nagoya, Japan

Takeshi Yamaguchi
City of Nagoya, Nagoya, Japan

Hideki Idota
Department of Architecture and Civil Engineering, Nagoya Institute of Technology, Nagoya, Japan

Kota Ibuki
Department of Environmental Engineering and Architecture, Nagoya University, Nagoya, Japan

ABSTRACT

More than 6400 people lost their lives during Kobe Earthquake in 1995; about 84% of them became victims of structural collapse of old wooden houses and wooden apartment buildings which did not meet the current seismic requirement enacted in 1981. There are about 11 million of such existing non-conforming wooden houses today in Japan, and upgrading of their seismic performance is essential for disaster mitigation. Many of the local governments offer financial support for the upgrading provided that the upgraded performance meets the current design requirement. However, even with the financial support, the owners still have to spend about one million yens on average. Such large expenditure makes the owners reluctant to upgrade their houses and accordingly, more than 99% of existing non-conforming wooden houses remain untouched.

The authors have proposed to upgrade a large number of such houses to the level even lower than the one required by the current design requirement rather than upgrading small number of houses satisfying current design requirement (Mori, et al., 2007). Such a strategy is much more cost effective from the viewpoint of life safety for the whole society and also for a house owner. However, since houses could be heavily damaged, the feasibility of such a strategy should be investigated from various point of views.

Many people would have to evacuate from their houses at an event of strong ground motion and take refuge to public buildings such as school buildings and gymnasiums. If the number of refugees is too large, there could be an overflow of refugees as seen after Kobe earthquake. Also as many of existing school buildings and gymnasiums does not meet current design requirement, their availability as refuge is in question. Many of local government are trying to upgrade such non-conforming buildings.

This paper investigates the effectiveness of upgrading existing non-forming school buildings and gymnasiums as well as wooden houses in Nagoya, Japan from the viewpoint of the reduction of risk of refugees to overflow. Buildings and gymnasiums in public elementary schools and junior high schools are considered. The risk is estimated for each elementary school district taking into account the seismic performance level of the school buildings and gymnasiums based on the seismic diagnosis. Although people would take refuge with various reasons such as loss of energy supply, only those people who are forced to take refuge because of the severe damage on their houses are considered.

It is shown in this paper that upgrading of school buildings can reduce the refugees, but the effect is limited. Upgrading of wooden houses is essential for reducing the refugees to overflow intensively; target seismic grade lower than 1.0 still has a large effect on the reduction of the risk. Such a strategy is in accordance with the one for saving lives (Mori, et al., 2007).

Risk maps of refugees to overflow in each elementary school district of the City of Nagoya are also presented. Such information is useful for a local government to give priority to upgrading non-conforming existing school buildings and the wooden houses in a district with higher risk than the others. Such risk information is also useful for individuals such as house owners to make decision whether they upgrade their houses and/or where they would take refuge.

REFERENCES

Mori, Y., T. Yamaguchi, and H. Idota. (2007) *Optimal Strategy for Upgrading Existing Non-Conforming Wooden Houses*, Application of Probability and Statistics in Civil Engineering, Kanda, Takada, and Furuta (eds), 7pp. (CD-Rom)

Safety, Reliability and Risk of Structures, Infrastructures and
Engineering Systems – Furuta, Frangopol & Shinozuka (eds)
© 2010 Taylor & Francis Group, London, ISBN 978-0-415-47557-0

Decision making based on mitigation effectiveness of deterioration risk of RC buildings damaged by chloride ingress

Chien-Kuo Chiu

Department of Construction Engineering, National Taiwan University of Science and Technology, Taipei, Taiwan

Takafumi Noguchi

Graduate School of Engineering, Department of Architecture, The University of Tokyo, Tokyo, Japan

ABSTRACT

Generally, structural members, beams, or columns whose reinforcing steel components are corroded by carbonation or chloride ingress are appropriately repaired/retrofitted after investigating the cause of corrosion. However, besides minimizing the life-cycle cost (LCC) (Mori et al. 1994, Frangopol et al. 1997, Val et al. 2005, Chiu et al. 2007, Kanemastu et al. 2001), including the deterioration risk few studies discuss how an optimal maintenance plan can be envisaged beforehand to increase the mitigation effectiveness of the deterioration risk. Therefore, there is a need for a system that can be used to determine a maintenance plan based on the mitigation effectiveness of the deterioration risk in terms of structural capacity and serviceability. In this paper, we describe a system that can estimate the LCC, including the deterioration risk (induced by chloride ingress), of a building resulting from failure and severe spalling/cracking during earthquakes (Figure 1). In addition, five repair/retrofit techniques or strategies considered and build probabilistic models in this paper to reflect the effect of each strategy. Finally, based on the mitigation effectiveness of deterioration risk, including the reduction ratio of LCC and the benefit-cost ratio of maintenance, the optimal maintenance strategy and period can be decided and case studies are used to discuss the application of the system.

According to the results of the case study, maintenance technologies with steel supplementing, e.g.

Table 1. Statistics for deterioration parameters.

Uncertainty	Distribution	C.O.V
Surface chloride content (C_o)	Lognormal	10%
Depth of concrete cover (x)	Lognormal	20%
Apparent diffusion coefficient (D_p)	Lognormal	30%
Critical threshold chloride concentration (C_{limit})	Uniform	1.0-1.2 kg/m³
Corrosion rate (V_{corr})	Lognormal	50%

TYPE-IV and TYPE-V of this research, are effective to reduce the LCC of an RC building located in the regions with chloride ingress and high seismic hazard, e.g. Japan and Taiwan.

Although recurrence periods of earthquakes have been assigned in this paper, users can change them to fit in with seismic design codes considered. In future, instead of specified recurrence periods of earthquakes, seismic hazard analysis will be included in this system and develop more useful estimation models.

REFERENCES

Mori, Y., and Ellingwood, B. R. 1994. Maintaining reliability of concrete structures. II: Optimum inspection/repair, *Journal of Structure Engineering (ASCE), 120(3)*, 846–862.

Frangopol, D. M., Lin, K. Y., and Estes, A. C. 1997. Life-cycle cost design of deteriorating structures, *Journal of Structure Engineering (ASCE), 123(10)*, 286–297.

Val, D. V. 2005. Effect of different limit states on life-cycle cost of RC structures in corrosive environment, *Journal of Infrastructure Systems (ASCE), 11(4)*, 231–240.

Chiu, C. K., Kanematsu, M., Noguchi, T., and Nagai, H. 2007. Optimization maintenance plan by minimizing life-cycle cost including deterioration risk due to chloride attack, *Journal of Structural and Construction Engineering (AIJ), 616, 41–47. (in Japanese)*

Kanematsu, M. and Noguchi, T. 2001. Study on optimization method of maintenance and repair scheme by applying a genetic algorithm, *Proceedings of JCI Symposium on Evaluation and Maintenance Planning for Combined Deterioration of Concrete Structures, 51–54. (in Japanese)*

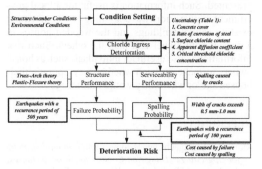

Figure 1. Flowchart depicting how deterioration risk induced by chloride ingress is evaluated.

General Session (Probabilistic Materials Analysis)

Safety, Reliability and Risk of Structures, Infrastructures and Engineering Systems – Furuta, Frangopol & Shinozuka (eds)
© *2010 Taylor & Francis Group, London, ISBN 978-0-415-47557-0*

Evaluation method of *S-N* data of composite materials based on probabilistic fatigue model of UD composites

K. Mukoyama, S. Hanaki & H. Uchida
Graduate School of Engineering, University of Hyogo, Hyogo, Japan

M. Zako
Graduate School of Engineering, Osaka University, Osaka, Japan

S.V. Lomov & I. Verpoest
Department of Materials Engineering, Katholieke Univesiteit Leuven, Leuven, Belgium

ABSTRACT

Probabilistic fatigue properties of unidirectional composite are evaluated using the data sets collected from database and published papers. For mathematical expressions of $S - N$ curve, 8 types of $S - N$ models in the Standard Evaluation Method of Fatigue Reliability for Metallic Materials (The Japanese Society of Material Science. 2004) are prepared. From numerical result, it is confirmed that single or double logarithmic linear model is optimum for UD-CFRP. On the other hand, double logarithmic curve model is the optimum one for UD-GFRP.

To evaluate scatter of the fatigue test data quantitatively, each data set is normalized by tensile strength. As a result, each data can be gathered into narrow region. However, some data plots still exist out of common region and statistical test for differentiating between data sets based on analysis of variance is carried out.

Using the merged data set, fatigue life and strength distributions are evaluated. Fig.1 shows fatigue life and strength distribution of UD-GFRP. Fatigue life distribution is well approximated by weibull distribution. However, the adaptability decreases at high or law stress level as shown in Fig.1(a). On the other hand, fatigue strength distribution is well approximated by normal distribution. From these results, it is confirmed that fatigue life distribution shows complex distribution pattern and varies by changing the stress level for evaluation. On the other hand, fatigue strength distribution shows relatively simple distribution pattern and are approximately identical at different number of cycles.

The developed model is expanded to apply the data sets which indicate plural failure modes. Fatigue test results for plain woven GFRP (Tanimoto,T. et al. 1979) is selected as an example and a data separation algorithm based on failure mode is proposed. Fatigue life distribution is approximated by mixed log-normal distribution and the occurrence probability is used to

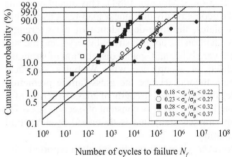

(a) Fatigue life distribution on Weibull probability paper.

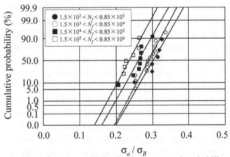

(b) Fatigue strength distribution on normal probability paper.

Figure 1. Fatigue life and strength distribution for UD-CFRP.

determine the number of cycles which separate the data sets. As a result, it is confirmed that reasonable $S - N$ curves were obtained.

REFERENCE

The Society of Material Science. 2004. *Standard Evaluation Method of Fatigue Reliability for Metallic Materials-Standard Regression Method of S-N Curves-*, JSMS-SD-6-04 (in Japanese).

Safety, Reliability and Risk of Structures, Infrastructures and
Engineering Systems – Furuta, Frangopol & Shinozuka (eds)
© 2010 Taylor & Francis Group, London, ISBN 978-0-415-47557-0

Stochastic modeling of architectural glass strengths

S.G. Reid
University of Sydney, Australia

ABSTRACT

Current developments in the theoretical modeling of
glass strength are based mainly on a glass failure
prediction model developed at Texas Tech University (Beason & Morgan, 1984). This model accounts
for the geometrically non-linear load-response of thin
glass plates, the time-dependence of glass strength
(dependent on the applied stress-history), and the statistical variability and spatial distribution of strengths
(modeled in accordance with Weibull's statistical
strength theory). Hence the glass failure prediction
model describes the strength of glass in terms of a
notional failure stress (a constant 60-second stress for
a unit stressed area) that is described by a Weibull
distribution.

However, glass test results are often reported with
regard to the peak loads resisted by glass plates, the
peak stresses developed in the glass plates, or the
critical stresses at the fracture origins (which are normally different from the peak stresses). The associated
strengths are commonly described by fitting Weibull
distribution functions, but the strength statistics and
fitted distribution functions are not directly proportional to those for the equivalent 60-second stresses
described by the glass failure prediction model.

The paper examines two aspects of glass strength
modeling that have not been dealt with before. Firstly,
the paper examines the relationship between the glass
failure prediction model and the distribution of the
critical stresses at the fracture origins for non-uniform
stress fields. The paper shows how the glass failure prediction model can be used to model the distribution of
the critical stresses at fracture origins, with particular
reference to the non-uniform stress fields developed
in ring-on ring tests on glass plates.

Secondly, the paper examines the influence of a
crack propagation threshold in relation to the Weibull
distribution of strengths described by the glass failure prediction model. The paper describes a stochastic
crack growth model of strength for glass with flaw
sizes that are Pareto distributed (corresponding to a
Weibull distribution of strength) and examines the
influence of a crack propagation threshold based on
a threshold value of the stress intensity factor.

Results are presented concerning the distribution
of critical stresses at fracture origins in a non-uniform
stress field, compared with the strength distribution
described by the glass failure prediction model (based
on the strength of the total stressed area). It is shown
that although the critical stresses at fracture origins
are generally lower and more variable than the maximum applied stresses, the shape of the respective
distribution functions is similar. Therefore, in accordance with the glass failure prediction model, the
critical stresses at fracture origins are approximately
Weibull-distributed.

Results are also presented concerning the influence
of a crack propagation threshold, considering alternative crack growth models to account for a threshold
value of the stress intensity factor K_I. It is shown that
if the threshold simply prevents crack growth when K_I
is below the threshold, then it has a negligible effect on
the strength distribution for a ramp load. However, if
crack growth is a function of the amount by which K_I
exceeds the threshold, then the strength distribution
for a ramp load is strongly influenced by the crack
propagation threshold. Nevertheless, the distribution
of failure stresses does not exhibit a stress threshold.

REFERENCES

Beason, W.L. & Morgan, J.R. 1984. Glass failure prediction
model. *Journal of Structural Engineering*, ASCE, 110(2):
197–212.

Safety, Reliability and Risk of Structures, Infrastructures and
Engineering Systems – Furuta, Frangopol & Shinozuka (eds)
© 2010 Taylor & Francis Group, London, ISBN 978-0-415-47557-0

Statistical analysis towards the identification of accurate probability distribution models for the compressive strength of concrete

G. Gasparini, S. Silvestri, T. Trombetti & C. Ceccoli
Department DISTART, University of Bologna, Bologna, Italy

ABSTRACT

Structural design deeply involves material strengths, which are obviously affected by uncertainty (Ang and Tang 2007, Melchers 1999, Conte 2001). As far as the material "concrete" is concerned, building codes (Italian codes, Eurocodes, ACI regulations) typically state that it should be identified and classified by means of its conventional characteristic uniaxial compressive cubic strength at 28 days. Although the fundamental importance of the problem of the evaluation of the characteristic compressive strength of concrete in structural design, the statistical analysis geared to identify an accurate probabilistic model of the concrete strength has not been a central issue of the research works in the field for many years.

This paper describes the results of an investigation performed to obtain the compressive strength statistic characteristics of a production of about half a million cubic meters of concrete. This amount of concrete was produced over a five-year period.

With the aim of obtaining the more adequate representation of the actual probability distribution of the concrete compressive cubic strength, the three following distributions are selected and considered in the statistical analysis reported in this research work: the Normal distribution; the Shifted Lognormal distribution; the Gumbel distribution.

The techniques for estimating the parameters values from available observational data (measurements) of a random variable and for deriving the appropriate probabilistic model for a random variable are embodied in the methods of statistical inference, in which information obtained from sampled data is used to represent the corresponding information regarding the population from which the samples are derived. Inferential methods of statistics, therefore, provide a link between the real world and the idealized probability models assumed in a probabilistic analysis.

The results have shown that the Shifted Lognormal distribution (Fig. 1) is capable of capturing the characteristics of the experimental data much better than all the other distributions examined. The Gumbel distribution has also shown good results. On the other hand, the Normal distribution, which is so often used for

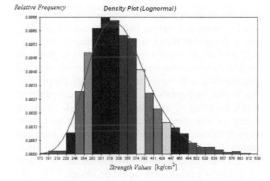

Figure 1. Shifted Lognormal distribution: theoretical PDF and experimental frequency histogram.

statistical representations, has displayed worse results in the capacity of capturing the characteristics of the experimental data.

On the other hand, the Italian code and, to some extent, the Eurocode substantially base the evaluation of concrete properties upon a Normal distribution. It is therefore advisable that design codes will encompass the possibility for the engineer to evaluate the concrete properties based upon these more refined statistical models.

REFERENCES

Ang, A.H.S. and Tang, W.H. (2007), Probability Concepts in Engineering. Emphasis on Applications to Civil and Environmental Engineering, 2nd Edition, John Wiley and Sons, Inc.

Conte, J.P., (2001). Statistics, Probability and Reliability. Course at the University of California at San Diego, fall 2001, 4.68-4.74.

Melchers, R.E. (1999), Structural Reliability Analysis and Prediction, 2nd Edition, John Wiley and Sons, Inc.

Safety, Reliability and Risk of Structures, Infrastructures and Engineering Systems – Furuta, Frangopol & Shinozuka (eds)
© 2010 Taylor & Francis Group, London, ISBN 978-0-415-47557-0

Probabilistic assessment of masonry compressive strength

N. Mojsilović & M.H. Faber
Institute of Structural Engineering, ETH Zurich, Switzerland

ABSTRACT

Structural masonry is a construction material composed of units and mortar, two components i.e. materials with quite different mechanical properties. However, masonry is usually treated as a unique, more or less homogeneous material. The key masonry material characteristic is the compressive strength perpendicular to the bed joints, f_x. It would be desirable to obtain information in regard to f_x through intensive testing of masonry specimens according to appropriate test standards. In this way a proper statistical treatment of the test data would be possible. In particular, appropriate probability distribution functions and corresponding parameters would be straightforwardly assessable.

On the other hand, the masonry compressive strength perpendicular to the bed joints can be put in relation to the corresponding unit and mortar compressive strengths. The components' strengths are easy to obtain and numerous tests are usually performed during the production; those results are thus statistically well defined and appropriate probabilistic models may be established.

The first important issue to be considered in establishing a probabilistic model is thus whether to base the modeling on test results from large masonry panels, or test results from the masonry components, units and mortar. Based on the experience of the authors, probabilistic models developed on the basis of tests on large masonry elements are generally more reliable, i.e. associated with less model uncertainty. However, the obvious drawback is that such tests are more expensive and labor intensive. The two approaches can, however supplement each other such that the performance of masonry structures established on the basis of tests on the components individually can be updated based on a few large scale tests on masonry using Bayesian updating.

Firstly, the masonry compressive strength taken from the code will be treated as random variable and the procedure of the updating will be presented. The prior probabilistic model (before updating) for f_x is assumed normal distributed with known standard deviation σ_{fx} and an uncertain mean value μ_{fx}. The mean value is assumed normal distributed with known mean value μ' and standard deviation σ'. Here, these two values are introduced based on the engineering experience which is characteristic of the Bayesian way of thinking. Assuming that n tests for determining f_x are performed on large specimens and test results are represented by the vector $\hat{f}_x$ having mean value $\bar{x}$, the prior probabilistic model for the mean value of f_x can be updated using natural conjugate distribution. Based on the new observations i.e. test results the posterior parameters μ'' and σ'' can be calculated and the posterior distribution function can be established. More general, both of the prior distribution parameters μ_{fx} and σ_{fx} can be assumed as normally distributed random variables. After obtaining the test results, the posteriori distribution can be readily derived. In this case the natural conjugate distribution is Normal Inverse Gamma-2 distribution.

Secondly, for illustrative purposes an example is calculated. Masonry compressive strength obtained using the European structural masonry code empiric formula (which combines the compressive strengths of masonry components – units and mortar) is updated, using the results from additional tests on masonry specimens, according to the procedure outlined in the foregoing. It could be shown that the Bayesian updating (even with only one additional test) can considerably update the prior distribution of the masonry compressive strength: in this example the mean value has lessen (according to the results of the additional tests) and the standard deviation has decreased.

Finally, recommendations for practical applications and for future work in this area are provided.

Safety, Reliability and Risk of Structures, Infrastructures and Engineering Systems – Furuta, Frangopol & Shinozuka (eds)
© 2010 Taylor & Francis Group, London, ISBN 978-0-415-47557-0

Probabilistic fracture mechanics guidelines and templates

D.S. Riha, R.C. McClung & J.M. McFarland
Southwest Research Institute, San Antonio, TX, USA

S.S. Pai
NASA Glenn Research Center, Cleveland, OH, USA

ABSTRACT

Fracture control is a standard practice for critical structural components in NASA applications. Standard guidelines have been developed to define when fracture mechanics (also known as damage tolerance) calculations are required, and templates have also evolved to identify how the fracture mechanics (FM) calculations are to be performed.

Those guidelines and templates address a deterministic calculation of the safe life from a FM standpoint. Inputs to the calculation are generally either standard, defined quantities (e.g., standard initial crack sizes based on the applicable nondestructive evaluation (NDE) method) or nominal values (e.g., average material fatigue crack growth (FCG) rate and fracture toughness data). In some special cases, conservative bounding values (e.g., upper bound FCG rate data or lower bound toughness) are required. The so called "safe life" limit for NASA applications is generally defined by performing the FCG life computation based on these nominal inputs and then dividing the life by a factor of four to account for material variability.

However, this well-established paradigm for deterministic fracture control may be inadequate in certain situations. In some cases, the calculation may produce an overly conservative result that cannot be sustained due to program economic or schedule factors. In other cases, the conservatism (and hence the safety) of the result cannot be guaranteed due to various uncertainties in the analysis. Furthermore, the existing deterministic fracture control paradigm does not produce any quantitative measure of the risk of fracture, and so there is no understanding of how the propensity for fracture contributes to the total risk of system failure. Therefore, the current deterministic fracture control calculation does not (and cannot) support the quantitative computation of total system reliability, even when failure by fracture is one of the most significant possible causes of system failure.

A probabilistic fracture mechanics (PFM) analysis can address these shortcomings. An initial guidance document has been developed that describes the conditions when a PFM analysis should be performed and qualitative directions on how to perform this analysis (SwRI, 2007).

These PFM guidelines and templates describe a comprehensive approach to reliability assessment of complex fracture problems. This degree of complexity and level of effort is not feasible, however, for PFM evaluations at the design stage. At the design stage, a quick answer is needed to support preliminary reliability assessments (including system risk), and the available information is insufficient to support a comprehensive analysis. Instead, a simple "Level 1" assessment is needed that generates a preliminary measure of reliability for comparison purposes. It appears that a simple PFM analysis based on three key random variables—initial crack size (based on generic production NDE POD curves), life scatter (based on historical materials data), and loads variability (based on engineering estimates of average and worst-case load environments)—may be useful for this purpose.

This paper provides an overview of the PFM guidance document. In addition, demonstration of the initial "Level 1" approach is provided utilizing the NESSUS® (SwRI, 2006) probabilistic analysis software and the NASGRO® (SwRI, 2008) fracture mechanics software.

REFERENCES

Southwest Research Institute (SwRI) and NASA Johnson Space Center. 2008. NASGRO® Fracture Mechanics and Fatigue Crack Growth Analysis Software, Ver. 5.2.

Southwest Research Institute. 2007. Probabilistic Fracture Mechanics Guidelines and Templates. Report for NASA Contract NAS3-02142.

Thacker, B.H., Riha, D.S, Fitch, S.K. Huyse, L.J., and Pleming, J.B. 2006. Probabilistic Engineering Analysis Using the NESSUS Software. Structural Safety, Vol. 28, No. 1-2, pp. 83–107.

Safety, Reliability and Risk of Structures, Infrastructures and Engineering Systems – Furuta, Frangopol & Shinozuka (eds)
© 2010 Taylor & Francis Group, London, ISBN 978-0-415-47557-0

Decision support simulator for post-earthquake emergency management based on bayesian updating theory

N. Nojima

Department of Civil Engineering, Gifu University, Gifu, Japan

ABSTRACT

In post-earthquake emergency response, collection of actual damage information is time-consuming. Therefore, many kinds of rapid damage estimation systems have been developed on the basis of real-time processing of seismic intensity information, building inventory, and fragility relations. However, since initial damage estimation is highly uncertain, decision makers are often involved in a trade-off between rapidness and preciseness of important actions.

This problem stems from the fact that conventional damage estimation systems are not capable of bridging the gap between the initial damage estimates and the actual damage situations. Considering the balance of rapidness and preciseness in emergency response, decision support systems should have functions to incorporate actual damage information, update damage estimates, and eliminate uncertainty in the initial damage estimates.

In this view of the problem, a decision support simulator was developed for post-earthquake emergency response. Two kinds of damage information are integrated on the basis of Bayesian updating theory: one is rapid damage assessment, and the other is actual damage information.

Assuming multi-rank damage, the prior PDF of damage probability is modeled by Dirichlet distribution subject to sequential update incorporating the real damage information. The PMF of the number of damage is obtained as Dirichlet-multinomial distribution which is also to be updated. Focusing on the highest rank of damage, corresponding marginal distributions are obtained as beta PDF and beta-binomial PMF. Sequential probability ratio test is then applied for engineering judgment of activation of emergency response.

Through numerical examples, major finding are derived as follows.

1) Recognizing and quantifying the uncertainty inherent in the initial damage estimates is essential for subsequent updating of damage estimates in Bayesian manner (Figure 1).
2) Sequentially updated estimates converge to the actual situation with increasing information on the

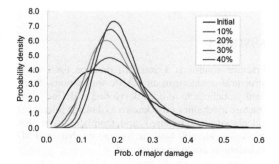

Figure 1. Bayesian updating process of probability distribution of the damage probability (beta PDF).

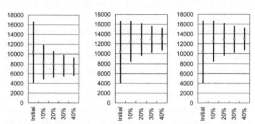

(a) Initial estimates appropriate (b) Initial estimates overestimate (c) Initial estimates underestimate

Figure 2. Updating process of the number of building damage (average ± standard deviation).

real damage (Figure 2). Even if the initial estimates are erroneous, flexible response can be performed in accordance with the actual situations.

3) By incorporating real information, the variance of updated damage estimates can be reduced. Only 10% progress of damage investigation significantly improves the accuracy (Figure 2).
4) According to prescribed parameters, the decision support simulator suggests suitable decisions. The proposed method is capable of supporting emergency responders by providing judgment criteria for emergency actions within restrictions of time and resources.

General Session (Monitoring and Maintenance Systems & Structural Health Monitoring)

Safety, Reliability and Risk of Structures, Infrastructures and Engineering Systems – Furuta, Frangopol & Shinozuka (eds)
© 2010 Taylor & Francis Group, London, ISBN 978-0-415-47557-0

Research on introduction of mahalanobis distance into bridge asset management as index of total integrity

Katsushi Aso
Member of JSCE, Nihonkai co.Ltd., Kanazawa, Japan

Yasuo Chikata
Member of JSCE, Graduate School of Natural Science & Technology, Kanazawa University, Kanazawa, Japan

ABSTRACT

In the Asset management of the bridge, it is necessary to use total integrity of each structure for the targeted deterioration forecast of the structure and the priority level decision etc. of the repair and reinforcement as an index. Up to now, the way of setting the weight coefficient has been important and difficult prob-lem, because the corrections corresponding to the bridge form such as long-span bridges, the earthquake environment, the natural condition and the traffic environment are needed. Therefore the establishment of an appropriate evaluation method would be needed immediately.

The trial method for applying to the bridge total integrity evaluation in this paper is the one that is called MT system one of the techniques of Taguchi-method. The MT system uses the distance of Mahalanobis as the standard by the method of judging how a certain individual shifts from the center of a normal group. When a normal space is on the edge of the population, it is easy to identify the one of abnormality because the Mahalanobis distance from a normal space grow. The MT system is composed of three techniques of MT method (Mahalanobis-Taguchi), MTA method (Mahalanobis-Taguchi adjoint), and TS method (Taguchi-Schmidt) now. In this paper, the MTA method was selected for the bridge total integrity evaluation from among these three techniques. There is a quantitative index as a total integrity of each bridge, what we call Brigde Health Index (BHI) calculated by the method of weighted mean by using the weight coefficient between materials. In this paper, firstly the verification of the practicality of this method was done by the comparison with BHI and examined whether this method being able to substitute BHI. By using this method a person who is not the specialist could easily evaluate the total integrity of bridge based on appropriate inspection results, so it is effective. This method can simultaneously consider a lot of items

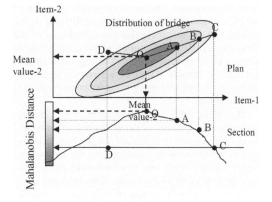

Figure 1. Concept of abnormal judgment.

and evaluate total integrity without the weight coefficient of each item separately(containing the weighting factor inside). If the importance degree is considered as an item using this characteristic, the Mahalanobis distance is expressed as priority in which the total integrity and the importance degree evaluation are unified. In this paper, the method of calculating the index that shows the priority level using the Mahalanobis distance is shown. The Mahalanobis distance of the bridge with the inspection data showing deterioration is large from a normal space. If there is a history of periodic inspection, the deterioration curve that expresses the secular distortion by the Mahalanobis distance can be made. In this paper, the deterioration curves by the Mahalanobis distance that used the real inspection data are shown.

REFERENCES

Aso, K., Chikata, Y.: Appliction of extended Mahalanobis distance to the evaluation of total bridge integrity based on inspection results, Journal of structure Eng., Vol.52A, pp.151–162, 2006, 3. (In Japanese)

Safety, Reliability and Risk of Structures, Infrastructures and Engineering Systems – Furuta, Frangopol & Shinozuka (eds)
© 2010 Taylor & Francis Group, London, ISBN 978-0-415-47557-0

A method to detect bolt loosening for a simple structure by Hilbert spectrum

H. Morikawa
Department of Built Environment, Tokyo Institute of Technology, Yokohama, Japan

T. Murakami
JFE R&D Corporation, Kawasakai, Japan

ABSTRACT

To identify damage of structure, various techniques has been proposed. The hammering test is an one of simple and quick methods and frequently used. A few trained experts can recognize the differences of the tones generated by the hammering and detect possible damage of a structure. Since this is very difficult skill for untrained people, it must be very useful to develop alternative techniques using some instruments instead of human ears.

The applicability of methods using free vibration data to detect damage of a cantilevered column steel model is discussed in this study. Firstly, time histories of free vibration were measured by the hammering tests under the conditions in which some bolts at the base plate are artificially loosen. The damaged levels are controlled by the torque of bolt: that is, normal, 75%, 50%, 25%, and no torque. When the torque of bolt is very small (less than 50%), it is easy to recognize the bolt loosing by the tone of sound during the vibration. This means that any mathematical techniques are not necessary. Thus, the object of this research is concentrated to detect slight damage such as 75% torque of normal condition.

For this purpose, Fourier and time-frequency analysis are applied to the time histories obtained by the hammering test. Since the differences between predominant period of normal and 75% torque are very small, it is very difficult for Fourier analysis to detect slight damage. On the other hand, Hilbert-Huang Transform (HHT) (Huang et al., 1998), which is a one of time-frequency analysis, can recognize such the slight damage by considering the time properties of predominant period. Figure shows the results of the Hilbert spectrum for cases of normal condition and slight damage (75% torque). From this figure, it is observed that the duration time of second mode, whose frequency is about 70 Hz, is significantly different: the duration time is shortened by the damage.

The results suggest that the damping ratio is affected by damage and it can be detected by Hilbert spectrum readily. From this, we can conclude that the Hilbert spectrum is useful tool to detect damage for

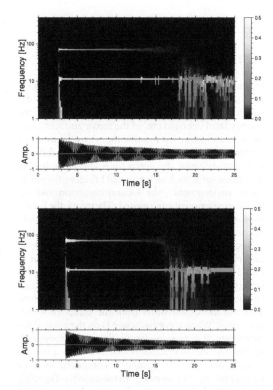

Figure 1. Hilbert spectra for the data of hammering test.

a kind of simple structure. The obtained results, however, are limited for a special sample with a few experiments. As future developments, the physical background and statistical stability will be discussed through the numerical and experimental approaches.

REFERENCES

Huang, N.E., Shen, Z., Long, S.R., Wu, M.C., Shin, H.H., Zheng, Q., Yen, N.-C., Tung, C.C. & Liu., H.H. 1998. The empirical mode decomposition and the Hilbert spectrum for nonlinear and non-stationary time series analysis. *Proc. of the Royal Society of London.* Vol. 454: 903–995.

*Safety, Reliability and Risk of Structures, Infrastructures and
Engineering Systems – Furuta, Frangopol & Shinozuka (eds)
© 2010 Taylor & Francis Group, London, ISBN 978-0-415-47557-0*

Evaluation of surface roughness of a corroded steel angle exposed in oceanic environment for 19.5 years

N. Watanabe & Y. Itoh

Nagoya University, Nagoya, Japan

ABSTRACT

This study evaluated surface roughness of a corroded steel angle member exposed to oceanic environments for 19.5 years. This member was cut into pieces, and their shape ware measured by the laser distancemeter by a 0.3mm pitch. The measured surface geometry was decomposed into components of waviness and roughness by using a multi resolution analysis, and they were assessed by 3-D surface characterization parameters. In 3 groups of parameters i.e., amplitude hybrid, and functional parameters, were dealt with. Applicability of proposed simulation of corroded surface was examined. As a simulation method, the normal probability field characterized by variogram parameters of sill and range was used.

Roughness characteristics of corroded member assessed by 3-D surface characterization parameters were followings. The value of skewness (S_{sk}) and kurtosis (S_{ku}) characterized the shape of pitting. At three regions of over Tokyo peil (T.P.), near the high hater level (H.W.L.), and near the low water level (L.W.L.), the big values of skewness and kurtosis were observed such that the dominant scale was 4.8~9.6 mm at the near of H.W.L., while it was 9.6~19.2 mm over T.P. and at the near of L.W.L.. Hybrid parameters (R.m.s. slope (S_q), Arithmetic mean summit (S_{sc}), and developed surface area ratio (S_{dr})) had the relation with the value of sill divided by the value of range in normal probability field model. The value of hybrid parameters with simulated shape was higher than that of actual shape.

A series of examines showed that a multi resolution analysis and 3-D surface parameter can characterize the shape of corrosion such that R.m.s. deviation characterizes the large scale of form, skewness and kurtosis characterizes the scale and distribution of pitting, and hybrid parameters (R.m.s. slope, Arithmetic mean summit, developed surface area ratio) characterizes the small scale of form. It was shown that surface roughness of actual member is characterized corresponding to various corrosion environments including atmospheric, splash, tidal, and submerged zones. From comparison between actual shape and simulations, it

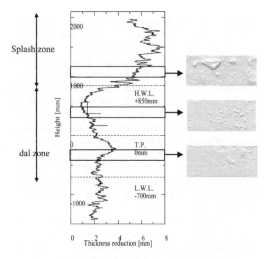

Figure 1. Distribution of residual thickness and shape of corrosion on back side of plate.

was observed that hybrid parameters had relationship to variogram parameters. It shows that the shape information measured roughly enable to estimate the value of hybrid parameters of actual shape with using the model of normal probability field to uniform corrosion shape.

REFERENCES

Jiang, X. Q.et al. 1999, Three-dimensional surface characterization for orthopaedic joint prostheses, Proc Instn Mech Engrs Vol. 213 Part H,: 49–68

Kainuma. S., et al. 2005, Time dependent corrosion behavior of structural steel members in boundary with concrete; Proceedings of JSCE No.780, 97–114

Safety, Reliability and Risk of Structures, Infrastructures and Engineering Systems – Furuta, Frangopol & Shinozuka (eds)
© 2010 Taylor & Francis Group, London, ISBN 978-0-415-47557-0

On the use of wavelet coefficient energy for structural damage diagnosis

H. Noh & A. S. Kiremidjian

Department of Civil and Environmental Engineering, Stanford University, CA, USA

ABSTRACT

Efficient and reliable estimation of damage immediately after a large earthquake is of great importance for reducing potential injuries and business interruption. Previous work on damage detection has focused on the use of ambient vibrations obtained before the occurrence of an earthquake and immediately after such an event. In this paper, the response motion obtained during an earthquake is used to determine if a change has occurred due to the strong shaking. Since earthquake motion is nonstationary, previously used autoregressive models for damage detection do no apply. However, use of wavelet coefficient energies of the output vibration signals is a particularly suitable approach for detecting damage in the structure from earthquake strong motion.

The model that will be described in the paper uses the Morlet wavelet to characterize the earthquake motion. The energy of the wavelet coefficients at a particular scale is defined as the square of the Euclidean norm of the wavelet coefficients vector as defined by Nair & Kiremidjian (2007). Similarly the energy of the wavelet coefficients at a time is defined. It is found that the energies of the fifth, sixth and seventh dyadic scales of the Morlet wavelet are the largest among all scales, and changes in the structure appear to be manifested as changes in these energies. In the model, it is hypothesized that an initial recording from a strong but a non-damaging earthquake has occurred at some point prior to a damaging event providing baseline energies for that structure. The model is tested with acceleration data for a series of tests performed on a reinforced concrete bridge column subjected to successively increasing earthquake ground motion applied by the University of Nevada, Reno (UNR) shake table (Choi, 2007).

Wavelet coefficients of response signals from different inputs clearly show different patterns as represented in Figure 1. As the intensity of the input motion increases, the peaks of wavelet coefficients shift both in time and in scale. These patterns are extracted out as *DSF*'s using the wavelet coefficient energy. With larger damage, a migration of these energies can be observed, and changes in them are found to be indicative of the onset of damage.

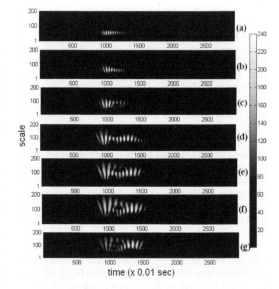

Figure 1. Wavelet coefficients of the acceleration response at the top of the column for different *DP*'s: (a) *DP* 1; (b) *DP* 3; (c) *DP* 5; (d) *DP* 7; (e) *DP* 9; (f) *DP* 11; (g) *DP* 13

REFERENCES

Choi. H., Saiidi, M. & Somerville, P. 2007. Effects of Near-fault Ground Motion and Fault-rupture on the Seismic Response of Reinforced Concrete Bridges, Report No. CCEER-07-06, Report submitted to the California Department of Transportaion, Department of Civil and Environmental Engineering, University of Nevada, Reno, Nevada.

Nair, K. K. & Kiremidjian, A. S. 2007. A Wavelet Based Damage Detection Algorithm for Structural Health Monitoring, submitted to the *ASCE Journal of Engineering Mechanics*.

Safety, Reliability and Risk of Structures, Infrastructures and Engineering Systems – Furuta, Frangopol & Shinozuka (eds)
© *2010 Taylor & Francis Group, London, ISBN 978-0-415-47557-0*

Health monitoring of water distribution systems

J. Liang
Tianjin University, Tianjin, China

D. Xiao
Nankai University, Tianjin, China

X. Zhao & H. Zhang
Tianjin University, Tianjin, China

ABSTRACT

Water distribution system is one of the most critical urban lifeline systems. Under severely cold weather, or due to corrosion of pipes, deformation of ground, etc., breaks often occur to urban water distribution systems and the breaks cannot be easily located, especially immediately after the events.

In recent years, supervisory control and data acquisition (SCADA) technology has been applied to water distribution systems. In recent years, supervisory control and data acquisition (SCADA) technology has been applied to water distribution systems, in which water pressure and/or flow rate at some selected positions such as water resources, pumps and pipes are monitored online by remote terminal units (RTUs) and transmitted to main terminal unit (MTU) by radio or internet. It is expected that monitoring information for water pressure and/or flow rate at resources, pumps and pipes is collected online for more efficient operation and control of the water distribution systems. However, in reality the number of monitoring stations is limited, and the monitoring stations for water pressure or/and flow rate are usually decided by experience or simply evenly distributed, and more importantly, there are few effective methodologies available, therefore, it is difficult to locate a break precisely and timely in practice.

This paper discusses two key issues in health monitoring of water distribution systems, including the methodology for optimal monitoring of water pressure or flow rate in a water distribution system for the purpose of more efficient monitoring, and the methodology to locate a break in a water distribution system by monitoring water pressure online at three nodes in the water distribution system. It is shown that the methodologies provide an effective and practical way for health monitoring of water distribution systems.

REFERENCES

Liang, J. 2003. Online monitoring of seismic damage in a water delivery system, *Proceedings of 6th U.S.*

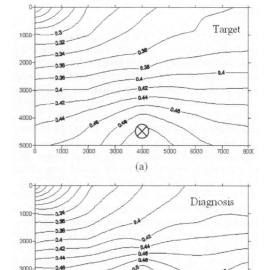

Figure 1. Break location of one pipe link for $A_d/A_0 = 0.05$.

Conference and Workshop on Lifeline Earthquake Engineering, ASCE, Long Beach, 464–473.

Liang, J. 2004. Online detection and location of seismic damage in a water distribution system, *Proceedings of 13th World Conference on Earthquake Engineering*, Vancouver, Canada, Paper No.1177, 12 pages.

Schulte, A.M. & Malm, A.P. 1993. Integrating hydraulic modeling and SCADA systems for system planning and control, *Journal of AWWA* (7): 62–66.

Takakuwa, T. 1978. *Pipeline Network Analysis and Design*, Tokyo: Morikita Publishing Company. (in Japanese)

Safety, Reliability and Risk of Structures, Infrastructures and
Engineering Systems – Furuta, Frangopol & Shinozuka (eds)
© 2010 Taylor & Francis Group, London, ISBN 978-0-415-47557-0

Damage assessment of reinforced concrete bridge decks using fuzzy ensemble system

Hitoshi Furuta & Koichiro Nakatsu
Department of Informatics, Kansai University, Takatsuki, Japan

Hiroshi Hattori & Wataru Adachi
Graduate School of Informatics, Kansai University, Takatsuki, Japan

Yasutoshi Nomura
Organization of Advanced Science and Technology, Kobe University, Kobe, Japan

ABSTRACT

In order to establish a rational management program for bridge structures, it is necessary to evaluate the structural damage of existing bridges in a quantitative manner. However, it is difficult to avoid the subjectivity of inspectors when visual data are used for the evaluation of damage or deterioration. In this research, an attempt is made to develop an optimal bridge maintenance system by using a health monitoring technique. The damage of Reinforced Concrete (RC) bridge decks is evaluated with the aid of digital photos and pattern recognition. So far, neural network has been applied to judge the damage state of RC bridge decks. However, there are still such problems that learning data are not enough, the best feature selection is difficult and recognition accuracy is not satisfactory. In order to solve these problems, the fuzzy ensemble classifier system is applied here, which uses fuzzy rule-based systems assigning a weight value to each of the suggested classes from the classifier systems.

The pattern recognition is performed based on the features obtained from the object. The feature extraction is very important to control the recognition performance. However, the appropriate selection of the feature is difficult, because there are various features. Therefore, an attempt is made to develop a system that can select appropriate features with the classifier systems consisting of various feature vectors. Moreover, the classifier systems are constructed with two algorithms that are fuzzy logic and neural network. Because, the recognition performance is different according to the combination of the algorithm and the feature.

The proposed classifier ensemble system consists of fuzzy classifiers and neural network classifiers and fuzzy ensemble (Figure 1). Fuzzy classifiers and neural network classifiers are constructed according to each feature. Those are tried to construct classifiers based on various characters. The fuzzy ensemble estimates weight value to each classifier result. The best classifier result is selected from classifiers results using this weight value.

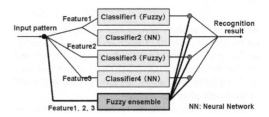

Figure 1. Proposed classifier ensemble system.

In this study, the damage of RC bridge decks is evaluated with the aid of digital photos and pattern recognition. In general, the procedure for extracting the characteristics of cracks showing up on concrete decks through digital images and the classification of the damage level based on the characteristics are used in the typical pattern recognition system. Several classifiers were constructed according to the features that extracted from digital photos of RC bridge decks and the algorithms that are fuzzy logic and neural network. From the examination, it was confirmed that the recognition results of these classifiers are quite different. In the proposed classifier ensemble system, it selects the classifier effectively from these classifiers and the recognition performance is improved. In future, to include more features and algorithms to the proposed classifier ensemble system expects more improvement of recognition performance.

REFERENCES

Duda, R. O., Hart, P. E., and Stork, D. G. 2001. Pattern Recognition. A Wiley-Interscience Publication

Li Zhang, Gang Sun, Jun Guo. 2004. Feature selection for pattern classification problems. IEEE Computer and Information Technology. pp.233–237.

T. Nakashima, G. Nakai and H. Ishibuchi. 2003. Constructing fuzzy ensembles for pattern classification problems. Systems, Man and Cybernetics, IEEE International Conference on vol.4, pp. 3200–3205.

General Session (Bridges and Buildings)

Safety, Reliability and Risk of Structures, Infrastructures and
Engineering Systems – Furuta, Frangopol & Shinozuka (eds)
© 2010 Taylor & Francis Group, London, ISBN 978-0-415-47557-0

Highway bridge assessment for dynamic interaction with critical vehicles

D. Cantero, E.J. OBrien & A. González
University College Dublin, Dublin, Ireland

B. Enright
Dublin Institute of Technology, Dublin, Ireland

C. Rowley
Buro Happold Ltd, London, UK

ABSTRACT

Freight transport has increased by 45% in Europe over
the past decade (Eurostat 2008) and this trend seems
likely to continue into the medium term future. A possible solution to the resulting capacity problem that
this creates would be to increase the permitted gross
vehicle weight to 60t and the number of axles to 8
(OBrien et al. 2008). The effect on highway bridges
of this potentially significant change in traffic configuration is currently under evaluation, although Weigh
In Motion (WIM) records have already observed high
frequencies of extremely heavy vehicles, with weights
well in excess of the normal legal maximum in some
heavily trafficked highways. These extreme vehicles,
with gross weight in excess of 100 tonnes, tend to be
either mobile cranes with very closely spaced axles or
low loaders with much longer wheelbases. Such vehicles would be expected to have special permits and
escort vehicles, but were recorded mixed with normal traffic and travelling close to the speed limit of
80km/h. Whether or not the legal limit for trucks without permits was changed, it is reasonable to expect
that cranes and crane-type vehicles will govern the
design/assessment of some bridges in some circumstances. Therefore, it is needed to assess their dynamic
interaction with bridges and the allowance that needs
to be made for dynamics.

So, this paper reviews the dynamic effects of large
cranes on short to medium span bridges and compares
them to common 5-axle articulated trucks, focussing
on the mid-span bending moment load effect. To
account for the variability in vehicle characteristics,
more than 40,000 vehicle-bridge interaction events are
computed using Monte Carlo simulation.

WIM measurements were taken in 2005 at a heavily
trafficked site in the Netherlands. A significant feature
of the gathered data is the high population of extremely
heavy vehicles with daily occurrences of vehicles over
100t. Figure 1 shows an example of a typical crane and
a 5-axle truck of the type used for comparison.

The vehicle-bridge interaction model used for the
Monte Carlo simulation consists of an articulated

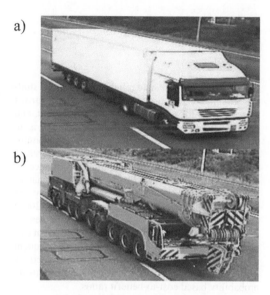

Figure 1.   Photo examples of recorded WIM trucks, a) 5-axle
truck, b) crane.

3-dimensional truck model and a finite element plate
model representing a simply supported orthotropic
bridge. The solution of the vehicle moving at constant
speed over the bridge with an uneven road profile is
obtained by an iterative procedure.

The results obtained show that Dynamic Amplification Factor (DAF) values for cranes are significantly
reduced when compared to 5-axle articulated trucks.

REFERENCES

Eurostat 2008. *EU energy and transport in figures*. Luxem-
bourg: Office for official publications of the European
communities. Full document at: http://ec.europa.eu.
OBrien, E., Enright, B. & Caprani, C. 2008. Implications
of Future Heavier Trucks for Europe's Bridges. *Transport
research arena Europe*.

Safety, Reliability and Risk of Structures, Infrastructures and Engineering Systems – Furuta, Frangopol & Shinozuka (eds)
© 2010 Taylor & Francis Group, London, ISBN 978-0-415-47557-0

Expansion joints and bridge structural reliability

P. Bradford
Watson Bowman Acme/BASF, NY, USA

J.E. Padgett
Rice University, TX, USA

R. DesRoches
Georgia Institute of Technology, GA, USA

ABSTRACT

Bridge structure limit state reliability (fragility) studies have many end uses, including design, retrofit prioritization, life cycle costing, risk assessment, and code development. A structure's fragility curve is an expression of the behaviors of many different components synthesized into a simple metric that addresses the basic question of "will structure performance be adequate" for a given level of loading?" By characterizing bridge structure types, the concept can be extended to entire classes of structures. Nielson (2005) classified 88% of all highway bridges in the central and southeastern United States into eight different categories appropriate for fragility studies. Subsequent studies (Padgett and DesRoches, 2008) of structures in these categories ordered retrofit measures using probability based cost-to-benefit ratios.

Structural fragility investigations require accurate subsystem reliability characterizations. This study establishes reliability models for expansion joints during seismic events, with subsequent integration into bridge system fragility models. Uncertainty in the expansion joint model is addressed via a series combination of two reliability blocks, the first random variation in nature and the second epistemic. Design capacity-demand reliabilities and material variability are used to formulate the random variation block, while validation is used to characterize the second. A two parameter beta function (Harr, 1987) is used as the probability density function for the validation block, with the primary parameters being knowledge and predisposition. These, in turn, are affected by analysis, test/field experience, simplicity, and similarity.

The impact of various levels of damage to the bridge or its components on the probable allowable traffic carrying capacity can be assessed using relationships that model the expected time-dependent restoration following an event (Padgett, 2007). If severe, damage of the joint system and joint neighborhood can inhibit passage of post-event emergency response vehicles. Fragility analyses including expansion joint reliability models coupled with damage-functionality relationships help determine if a critical bridge structure can meet code requirements for post-event traffic flow needs.

Safety, Reliability and Risk of Structures, Infrastructures and
Engineering Systems – Furuta, Frangopol & Shinozuka (eds)
© 2010 Taylor & Francis Group, London, ISBN 978-0-415-47557-0

Analytical and experimental studies on seismic behavior of buildings with mid-story isolation

K.C. Chang, B.H. Lee, M.H. Lin & C.C. Chiang
Department of Civil Engineering, National Taiwan University, Taipei, Taiwan

J.S. Hwang
Department of Construction Engineering, National Taiwan University of Science and Technology, Taipei, Taiwan

S.J. Wang
National Center for Research on Earthquake Engineering, Taipei, Taiwan

ABSTRACT

As an alternative to base-isolated buildings, mid-story
isolated buildings in which the isolation system is
typically designated at the top of the first or other
lower stories of the buildings can practically satisfy
the architectural concerns, potentially facilitate the
construction in site and effectively utilize the limited
available space. This is particularly meaningful for
the design of seismically isolated buildings located
at highly populated areas where to put the isolation
system beneath the basement is extremely difficult.
Since 2000, Koh and Kobayashi have studied vibration
characteristics, seismic responses and possible adverse
effects due to the existence of the structure below the
isolation system for mid-story isolated buildings.

This paper aims at dynamic characteristics and seismic responses of the buildings with the mid-story
isolation system composed of the seismic bearings
revealing a bi-linear hysteretic property based on a
simplified three-lumped-mass structural model. The
equivalent lateral force procedure is investigated in
order to extensively include the mid-story isolation
design in the design guidelines. The composite damping ratio considering the effects of structures below
and above the isolation system is derived in this study.
In addition, an equivalent lateral force design philosophy considering the fundamental and higher mode
responses is provided to rationally and effectively
design the structure below the isolation system. It can
facilitate engineers to prevent the improper design of
the substructure and superstructure in the preliminary
design process. It is worthy of noting that the undesired
dynamic response attributed to the modal coupling
of the higher modes (i.e., the modal natural frequencies of the higher modes are almost identical) should
be paid more attention. The modal coupling of the
higher modes occurs within a certain frequency ratio
bandwidth, which can be prevented using the proposed
prediction formula. As a result, the mid-story isolation
system is effective in controlling the seismic responses
as desired if the design for structures below and above
the isolation system is appropriate.

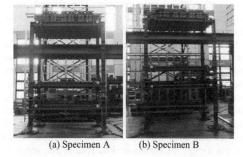

(a) Specimen A   (b) Specimen B

Figure 1. Installation of test structures on the shaking table.

The experimental and numerical validations were
performed to compare the 1/4 scaled down test structural models with and without the modal coupling
effect, as shown in Figures 1(a) and 1(b), respectively. For the test structure with the modal coupling
effect (i.e., Specimen B), the acceleration transmitted to the substructure is enlarged significantly and
even more than twice as large as PGA, and the
acceleration response at the isolation layer is slightly
larger than PGA. It is evident that the modal coupling effect results in the undesired seismic response
and should be avoided. Consequently, both structures
above and below the isolation system should be taken
into account in the preliminary design process for
mid-story isolated buildings.

REFERENCES

Koh, T. & Kobayashi, M. 2000. Vibratory Characteristics and
Earthquake Response of Mid-Story Isolated Buildings.
*Memoirs of the Institute of Sciences and Technology, Meiji
University* 39(12): 97–114.

Koh T. & Kobayashi, M. 2004. Analytical Study of Modal
Coupling Effect on Mid-story Isolation System by
Eigen Value Analysis and Random Vibration Analysis.
*Summaries of Technical Papers of Annual Meeting Archi-
tectural Institute of Japan (Hokkaido)*: 21167–21168.

Kelly, J. M. 1990. Base Isolation: Linear Theory and Design.
Earthquake Spectra 6(2): 223–244.

Kelly, J. M. 1996. *Earthquake-Resistant Design with Rubber.*
2nd ed. London: Springer Verlag.

Safety, Reliability and Risk of Structures, Infrastructures and Engineering Systems – Furuta, Frangopol & Shinozuka (eds)
© 2010 Taylor & Francis Group, London, ISBN 978-0-415-47557-0

Live load for long span

A.S. Nowak & M. Lutomirska
University of Nebraska, Lincoln, NE, USA

ABSTRACT

The available live load models were developed for short and medium span bridges. The present paper deals with the development of live load for long span structures. Loads are considered as random variables, with regard to magnitude and frequency occurrence. In contrast to short and medium spans, long span live load must be considered as a multiple presence of many trucks. The available data base includes weigh-in-motion (WIM) truck surveys and videos of the traffic taken on selected bridges. Using the actual truck and traffic data, the statistical parameters are determined such the bias factor (ratio of the mean to nominal value) and coefficient of variation. The results can serve as a basis for the reliability analysis and code calibration (calculation of load and resistance factors).

For short span bridges, the maximum live load is caused by the heaviest trucks. For two lane bridges, the extreme load is usually the result of two side-by-side trucks. On the other hand, for long span bridges, the roadway is not necessarily entirely covered by the heaviest trucks. As the load length increases, the load per unit length decreases, because cars and other light vehicles are injected into the traffic stream. Traffic mix (truck and cars) and headway distance determine live loads. For longer spans, load distribution is close to uniform. For longer spans, two scenarios are considered, a random traffic and a traffic jam situation.

Random traffic is a mixture of cars and trucks moving with highway speed, and the percentage trucks in the mix is usually site-specific. The extreme load effect can be determined by simulations, including truck configurations, weights, headway distances and speed. For longer spans, occurrence of overloaded but single trucks does not affect the total live load effect and it asymptotically approaches a line of average trucks, separated by average headways and cars. The load effect can be increased by the dynamic load. However, the field observations indicate that dynamic load factor (DLF) decreases for multiple trucks.

In traffic jam situation, the traffic is forced to a crawling speed. It is likely to have trucks traveling in one lane and cars using the faster lanes. The headway distance can be small and it can be considered equal to bumper-to-bumper. For two or more lanes, even in a traffic jam situation, the other lanes can have a mixture of cars (majority) and occasional trucks. In consideration of the maximum loading condition, since the traffic is stationary, no allowance is made for impact (dynamic load). However, if the traffic starts to move, the distance between the vehicles increases and the load intensity gets reduced, allowing for impact.

Multiple presence for long span bridges changes in comparison to short span bridges. First of all, simultaneous occurrence of trucks in one lane is more likely for longer spans. However, in practice, the span length has no influence on the frequency of side-by-side truck occurrence. Furthermore, long span bridges have usually more than two lanes and, therefore, site-specific traffic patterns must be considered.

The obtained parameters of live load are compared with the design live loads for bridges specified in Northern American and European Codes. Equivalent uniformly distributed loads are calculated for a wide spectrum of span length.

REFERENCES

ASCE Committee on Loads and Forces on Bridges. 1981. Recommended Design Loads for Bridges. *ASCE Journal of Structural Engineering* 107 (7): 1161–1213.
Buckland, P.G. at al. 1980. Proposed Vehicle Loading of Long-Span Bridges. *ASCE Journal of Structural Division* 106: 915–932.
Nowak, A.S. at al. 1994. Effect of Truck Loading on Bridges. *Research Report UMCE 94–22*

Safety, Reliability and Risk of Structures, Infrastructures and
Engineering Systems – Furuta, Frangopol & Shinozuka (eds)
© 2010 Taylor & Francis Group, London, ISBN 978-0-415-47557-0

Reliability models for bridges with light-weight SCC

P. Paczkowski
University of Nebraska, Lincoln, NE, USA

M. Kaszynska
Szczecin University of Technology, Szczecin, Poland

A.S. Nowak
University of Nebraska, Lincoln, NE, USA

ABSTRACT

The objective of the paper is to develop the statistical models and perform the reliability analysis for bridge girders designed using the light-weight SCC. The resulting reliability indices are compared with made of ordinary concrete.

Light-weight high-performance concrete (LWHPC), with unit weight of 1,850-2,000 kg/m^3, can have a compressive strength up to 120 MPa. LWHPC is made using only light-weight aggregates or combinations of light-weight aggregates with ordinary aggregates. The remaining components are the same as for ordinary high-performance concrete, i.e. plasticizers, superplasticizers, and mineral additives (silica fume and fly-ash). The light-weight concrete has a reduced unit weight by about 20–30% compared to the ordinary high-performance concrete. Therefore, the structural components are much lighter. This is important for longer span bridges with dominating dead load because it can help to reduce the total load effects. Weight reduction is also important when planning the transportation of precast members. The differences in mechanical properties of the light-weight concrete compared to ordinary concrete include: lower modulus of elasticity, more linear stress-strain relationship, lower ductility in compression, lower tensile strength, and more brittle structure. An important feature of light-weight concrete is a good bond between aggregate and cement paste, mostly due to a rough surface texture of the lightweight aggregate and improved hydration at the interface as a result of internal curing water being released from the pores of the aggregate.

In this study, the considered load combination includes three components of dead load, static live load and dynamic load. The capacity of a bridge depends of on the resistance of its components. The component resistance, R, is determined by the material strength and dimensions of the element. R is considered as random variable due to uncertainty that can be put into the following categories: Material Factor M: strength of material, modulus of elasticity cracking stress, Fabrication Factor F: variability of geometry, and Professional Factor P: reflecting the accuracy of the analytical model. The statistical parameters of material factor are determined from the test data provided by the industry. The reliability analysis procedure used in this calibration includes the following steps:

(1) Prepare input data:
(2) Calculate load parameters
(3) Determine the statistical parameters of R
(4) Calculate the reliability index, β,

The reliability indices are calculated for bridge girders made with ordinary concrete and light weight SCC. Two limit states are considered, moment carrying capacity and shear capacity. The calculated percentages of dead load and live load vs. span length indicate that for moments the trend in relationship is non-linear and for shears it is linear.

The reliability indices for girders with the light-weight SCC are slightly higher than for those with ordinary concrete. Reliability indices for moments are higher than those for shear.

The reliability analysis is performed for several possible values of the resistance factor. The selection criterion for the resistance factor is closeness to the target reliability index. The results indicate that the same resistance factor can be applied for girders with light-weight SCC and ordinary concrete.

Safety, Reliability and Risk of Structures, Infrastructures and
Engineering Systems – Furuta, Frangopol & Shinozuka (eds)
© 2010 Taylor & Francis Group, London, ISBN 978-0-415-47557-0

Probabilistic condition assessment of PC box girder bridge based on field test data

X. Ruan & X.F. Shi
Tongji University, Shanghai, China

D.M. Frangopol
Lehigh University, Bethlehem, PA, USA

ABSTRACT

Field test technologies can provide additional information for highway bridges. This information helps
engineers reach a more accurate bridge condition
assessment (Frangopol et al. 2008 a, b). In this paper,
a three span post-tensioned prestressed concrete (PC)
continuous box girder bridge is taken as an example
to illustrate this process by model updating. In addition, a probabilistic condition assessment process is
provided.

The bridge is located on the highway system
in China. It is a post-tensioned prestressed concrete continuous box girder bridge with three spans:
65m+100m+65m. The bridge was built by free balance
cantilevering method in 1996. The main girder has a
single-cell box cross section, with the height varying
from 5.6m (over pier section) to 2.2m (mid span section). The girder holds two lines of traffic in same
direction.

The bridge alignment was found abnormal in early
1999. A periodically appearance and alignment survey
started from that time, and some results are indicated in
this paper. Calculation using model updating based on
filed test research are carried out, including post dead
load model based on pavement thickness measurement
and rigidity parameters based on load testing.

A safety assessment is carried out based on design
specification in China (Ministry of Communication
of China 2004), both by theoretical and updating
calculation model.

Compared with design code based assessment,
probability-based assessment describes the safety condition of the bridge in a more realistic way. A
probability-based assessment is carried out to reach
a more comprehensive understanding of the condition
of the bridge. Moment reliability indices are in the
interval 5.23 -12.67, and shear reliability indices are
in the interval 4.90 -12.15 (see Figure1).

Comparisons of moment and shear reliability
indices obtained based on theoretical and updating
models are also given in the paper. It is found that

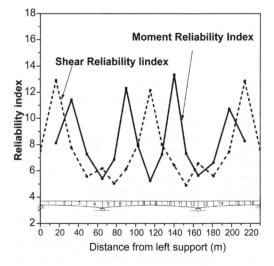

Figure 1. Moment and shear reliability indices for all
seventeen cross-sections (based on updating model).

results based on the updating model produce lower
reliability indices than those associated with the theoretical model.

REFERENCES

Frangopol, D.M., Strauss, A., & Kim, S.Y. 2008a. Bridge
Reliability Assessment Based on Monitoring. *Journal of
Bridge Engineering* 13(3): 258–270.
Frangopol, D.M., Strauss, A., & Kim, S.Y. 2008b. Use of
Monitoring Extreme Data for the Performance Prediction
of Structures: General Approach. *Engineering Structures*
30:3644–3653
Ministry of Communication of China 2004. *Reinforced Con-
crete and Pre-stressed Concrete Bridges and Culverts
(JTG D62- 2004)*. Beijing: China Communication Press.

Tuesday Morning (TUM) Sessions

Mini-Symposia (MS01) Uncertainty Modeling of Rare/Imprecise Data
[Session Organized on Behalf of IASSAR's CSMSE Umbrella]

Safety, Reliability and Risk of Structures, Infrastructures and
Engineering Systems – Furuta, Frangopol & Shinozuka (eds)
© 2010 Taylor & Francis Group, London, ISBN 978-0-415-47557-0

Interval estimates of structural reliability

H. Zhang

School of Civil Engineering, University of Sydney, NSW, Australia

R.L. Muhanna

School of Civil & Environmental Engineering, Georgia Institute of Technology, Atlanta, GA, USA

ABSTRACT

Reliability analysis is often complicated by the existence of epistemic uncertainty, which is due to lack of knowledge or incomplete information. For example the probability distributions of random variables are often estimated from a limited sample of data, thus the distribution parameters are subject to some uncertainty. This additional source of epistemic uncertainty can play an important role in reliability analysis.

The unknown parameters can be considered by two approaches, namely the Bayesian approach and the confidence interval approach. In the Bayesian approach the unknown distribution parameters are treated as (Bayesian) random variables, and the total (aleatory and epistemic) uncertainties are characterized by the expected value of the failure probability (e.g. Kiureghian, 2008). In this paper the confidence interval approach is employed to describe the unknown distribution parameters. The confidence interval approach is conceptually simpler, and requires less subjective judgment than the Bayesian approach. The reliability analysis needs to consider a family of distributions whose parameters are within the confidence intervals. Consequently, the failure probability is uncertain and itself varies in an interval. The width of this interval reflects the uncertainty in the reliability due to the uncertainties in the distribution parameters. A method for computing the interval estimate of

failure probability is introduced. Monte Carlo simulation is combined with the interval analysis to propagate the aleatory and epistemic uncertainties through reliability analysis. In each simulation, for each basic variable an interval is randomly generated to include all possible random numbers generated from a family of distributions. The performance function is then checked over the entire width of the simulated interval. A key step herein is to compute the range of the structural response when the system inputs vary in an interval vector. This can be efficiently achieved by using the interval finite element analysis (Muhanna et al., 2007).

This paper demonstrates the feasibility of computing an interval estimate of reliability when the distribution parameters are uncertain. An example is given to compare the proposed approach with the Bayesian approach.

REFERENCES

Kiureghian, A. D. (2008). "Analysis of structural reliability under parameter uncertainties." *Probabilistic Engineering Mechanics*, 23, 351–358.

Muhanna, R. L., Zhang, H., and Mullen, R. L. (2007). "Combined axial and bending stiffness in interval finite-element methods." *J. Struct. Engrg., ASCE*, 133(12), 1700–1709.

Safety, Reliability and Risk of Structures, Infrastructures and
Engineering Systems – Furuta, Frangopol & Shinozuka (eds)
© 2010 Taylor & Francis Group, London, ISBN 978-0-415-47557-0

Calibrating resistance factors of single bored piles based on incomplete test results

Jianye Ching
National Taiwan University, Taipei, Taiwan

Horn-Da Lin & Ming-Tso Yen
National Taiwan University of Science and Technology, Taipei, Taiwan

ABSTRACT

More recently, reliability-based design approaches
have emerged as a new design paradigm for pile
ultimate capacities because reliability is a consistent
measure of uncertainties. For reliability-based designs,
one designs a pile so that its failure probability is in an
acceptable range. Moreover, a single factor, called the
resistance factor, is often used to quantify resistance
uncertainties to facilitate reliability-based designs.

In the case that ultimate capacities of the test piles
in a region are known, the resistance factors of that
region can be calibrated conveniently based on the
load test database. However, in practice many load
tests are not conducted to ultimate capacity failures
but only to two times of the design loads depending on
the local code provisions. Taking the bored piles of the
Taipei City in Taiwan as an example, there are about
57 load tests in the local database, but among them,
only 8 of them were conducted to failures. Figure 1
illustrates the load-deformation curves of two piles in
the database, one was loaded to failure while the other
was not. By Davisson's criteria, the ultimate capacity
of the first pile can be identified, while that of the second pile is not identifiable because of the incomplete
load-deformation curve.

This paper addresses an issue often encountered in
calibrating resistance factors for pile ultimate capacities based on load test database. In practice, many pile
load tests are not conducted to failures but only to a
multiple (e.g.: 2) of the design load. This leads a difficult situation of incomplete information: for these test
results, the ultimate bearing capacities of the test piles
are unknown. How can these test results still be used to
calibrate resistance factors of piles? A full probabilistic framework is proposed in this research to resolve
this issue.

A local pile test database of Taipei (Taiwan) is presented for demonstration. The analysis results show
that the inclusion of the incomplete pile load test data
helps in calibrating the resistance factors. Moreover,
it is found that the calibrated resistance factors are
consistent to the safety factors that are adopted in the
current Taiwan design code.

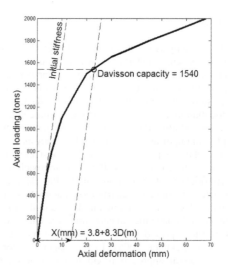

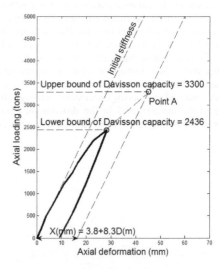

Figure 1. Complete (Top) versus incomplete (Bottom)
load-deformation curves. For the incomplete curve, the lower
bound of the Davisson capacity is simply taken to be the maximum load, while the upper bound is taken to be the load at
point A.

Safety, Reliability and Risk of Structures, Infrastructures and
Engineering Systems – Furuta, Frangopol & Shinozuka (eds)
© 2010 Taylor & Francis Group, London, ISBN 978-0-415-47557-0

Fuzzy finite element analysis based on reanalysis technique

L. Farkas, D. Moens, D. Vandepitte & W. Desmet
Department of Mechanical Engineering, K.U. Leuven, Belgium

ABSTRACT

Non-deterministic approaches are gaining momentum
in the field of numerical modelling techniques. The
ability to include non-deterministic properties is of
great value for a design engineer. It enables realistic
reliability assessment that incorporates the uncertain
aspects of the design. Furthermore, the design can be
optimized for robust behaviour under varying external influences. In this context, interval and fuzzy
approaches are becoming increasingly popular for the
analysis of numerical models that incorporate uncertainty in their description. In the interval approach,
parameter uncertainties are considered to be contained
within a predefined range. The fuzzy approach extends
this methodology by introducing a level of membership representing to what extent a numerical value is
member of the set of possible values for the considered
uncertain parameter.

Based on the α-sublevel technique, the fuzzy analysis requires the consecutive solution of a number of
related interval problems. For that reason, much attention in recent literature goes to the actual solution
and implementation of interval finite element analysis. The solution of the interval FE problems at the
core of the fuzzy analysis is often based on black-box
approaches such as vertex sampling or optimisation
techniques. Recent developments by the authors in
this context focus on the reduction of the computational burden when performing black-box algorithms.
In the Reduced Optimisation (RO) technique, the cost
of the global optimisation is decreased by excluding
the uncertain parameters with monotonic effect on the
output [1]. Another way to accelerate the optimisation is by using a surrogate model, which replaces
the real response of the analysis based on only a few
computed values [2]. In order to further decrease the
computational cost of these black-box approaches, this
paper introduces a novel reanalysis-based FE method
(ReFEM). This technique is making use of the high
degree of similarity between the consecutive FE analyses when applying search or sampling strategies for
solving the interval FE problem. The method focuses
on the system regeneration as well as the solution
part of the FE analysis. On the one hand, the element stiffness matrices are reformulated as explicit
functions of the uncertain parameters. This formulation increases efficiency by avoiding the numerical
integration at each reassembly of the element stiffness
matrices. Secondly, a fast solution of the modified FE
system is obtained by using the results from existing
FE solutions based on the Conjugate Gradient (pCG)
method [3]. This iterative method is proved to be an
efficient solver for linear positive definite symmetric
systems arising in static structural FE analysis. This
paper shows the general applicability of this approach
in the context of Fuzzy FE analysis. The method is
demonstrated on a mid-sized test case, yielding reduction of up to two orders of magnitude in computational
time.

REFERENCES

[1] Farkas L., Moens D., Vandepitte D., Desmet W., *Application of fuzzy numerical techniques for product performance analysis in the conceptual and preliminary
design stage*, Computers & Structures, Vol. 86, No. 10,
pp 1061–1079, 2008
[2] De Munck M., Moens D., Desmet W. , Vandepitte D.,
*A response surface based optimisation algorithm for
the calculation of fuzzy envelope FRFs of models with
uncertain properties*, Computers & Structures, Vol. 86
No. 10, pp 1080–1092, 2008
[3] Saad Y., *Iterative methods for sparse linear systems*,
PWS, 1996

Safety, Reliability and Risk of Structures, Infrastructures and Engineering Systems – Furuta, Frangopol & Shinozuka (eds)
© *2010 Taylor & Francis Group, London, ISBN 978-0-415-47557-0*

Probabilistic study of the strength of steel scaffold systems

H. Zhang, T. Chandrangs & K.J.R. Rasmussen

School of Civil Engineering, University of Sydney, NSW, Australia

ABSTRACT

Steel scaffold systems provide temporary support for the construction of bridges, office and residential buildings, car parks and similar structures. They are used as platforms while building the formwork for reinforced concrete, and are usually heavily loaded under the weight of fresh concrete, formworks, stacked material and workers. The consequences of failure of scaffold systems involve high risk to the safety of workers and the public as well as legal and financial risks. It is crucial to ensure that scaffold systems are safe and reliable.

This paper studies the ultimate strength of multi-story steel support scaffold systems through a rational statistical framework and advanced nonlinear finite element analysis. Steel scaffold systems are characterized by the large variations of geometric and mechanical parameters, notably joint stiffness, initial geometric imperfections (member out-of-straightness and frame out-of-plumb), and load eccentricity. The statistical data for these "physical uncertainties" were acquired through field measurements and laboratory tests. The structural behavior of steel scaffold frame is predicted by advanced finite element analysis which accounts for the material and geometric nonlinear ($P - \delta$ and $P - \Delta$) effects. The column-to-beam joints are modeled as semi-rigid joints. The relation between the moment and the rotation of the semi-rigid joints is described by a tri-linear curve. The "model error" of the advanced analysis model is estimated by comparing the predicted values to fifteen full scale scaffold tests conduced at the University of Sydney (CASE, 2006). The "physical uncertainties" and the model error are then propagated through Monte Carlo simulation to estimate the statistical distribution of the strength of scaffold systems. A 3-story 1 bay × 1 bay steel scaffold system was considered as an example.

The work presented in the paper is part of an ongoing research effort to develop a system reliability based design method with advanced analysis for steel scaffold systems. With known statistics for the strength of scaffold systems, the reliability of scaffold systems can be readily estimated by taking into account the load uncertainty (Ellingwood et al., 1980).

REFERENCES

CASE (2006). "Tests of formwork subassemblies and components." *Investigation report No. S1499*, Centre for Advanced Structural Engineering, School of Civil Engineering, University of Sydney.

Ellingwood, B., Galambos, T. V., MacGregor, J. G., and Cornell, C. A. (1980). "Development of a probability based load criterion for American National Standard A58." NBS special publication 577, National Bureau of Standards, Washington D.C.

*Safety, Reliability and Risk of Structures, Infrastructures and
Engineering Systems – Furuta, Frangopol & Shinozuka (eds)*
© 2010 Taylor & Francis Group, London, ISBN 978-0-415-47557-0

Study on modeling error in strong nonlinear area of RC shear walls in fragility evaluation of nuclear power plant buildings

N. Nakamura & T. Suzuki
Takenaka Corporation, Inzai, Japan

M. Fushimi & S. Miyazaki
The Kansai Electric Power Co. Inc,, Osaka, Japan

Y. Hino
Obayashi Corporation, Tokyo, Japan

Y. Fujita
Shimizu Corporation, Tokyo, Japan

H. Sugita
Kajima Corporation, Tokyo, Japan

K. Moriyama
Taisei Corporation, Tokyo, Japan

ABSTRACT

The Regulatory Guide for Reviewing Seismic Design of Nuclear Power Reactor Facilities was revised in September 2006. These revisions noted the existence of risk (residual risk) due to seismic motion magnitude exceeding that of the seismic design standards. Evaluation methods based on the latest techniques, such as probabilistic safety analysis (PSA), have therefore been incorporated in order to quantitatively evaluate this risk, and aggressive approaches are sought to establish these methods. In fragility analysis, which is a step of PSA, the uncertainty is divided into two types: uncertainty due to randomness that is intrinsic to the data and phenomena (aleatory uncertainty: βr), and uncertainty related to the knowledge and awareness of the analysis and modeling methods (epistemic uncertainty: βu). This study investigates a primary factor of βu, which is the error due to modeling in the strongly nonlinear area of the building for seismic analyses (βu_N).

A lumped mass bending-shear model is a representative seismic analysis model of current nuclear power reactor facilities. For seismic PSA, the model is supposed as the building response analysis model also. The method in reference is used to evaluate the restoring force characteristics of reinforced concrete (RC) shear walls, which are the primary structural elements of the building. Furthermore, attention was focused on RC shear wall shear failure as the dominant failure mode of reactor core damage.

This analysis model has been widely used in facility design. In investigations during the design phase, shear walls may reach the weakly nonlinear area under the seismic motion standards. The response values of the shear walls at these levels have been verified as being virtually validated.

However, in seismic PSA, investigation is performed on large input levels that exceeded the design-phase input levels, and in this situation the shear walls may reach the strongly nonlinear area. Up until now, the appropriate values for the variation in modeling error in this type of strongly nonlinear area (hereinafter, the limit of which is represented by βu_N) were not known. This study conducted an investigation with the goal of estimating βu_N.

Although it is desirable to evaluate βu_N by comparison with experimental results, the dynamic experiments of shear walls in nuclear reactor facilities that have been conducted are few and not sufficient. A three-dimensional FEM model that is thought to accurately model the nonlinear behavior of RC shear walls is therefore used.

Therefore in this paper, βu_N was estimated using a 3D FEM model which is supposed to be able to accurately estimate the nonlinear behavior of the RC shear walls. First, it was confirmed in the comparison between a 3D FEM model and the dynamic test results that the accuracy of this model is favorable. Then, in comparison of this model and the lumped mass model, a value of about 0.1 was obtained for βu_N.

REFERENCES

Nuclear Safety Commission of Japan: The Regulatory Guide for Reviewing Seismic Design of Nuclear Power Reactor Facilities, 2006

Atomic Energy Society of Japan: A Standard for Procedure of Seismic Probabilistic Assessment for Nuclear Power Plants 2007, AESJ-SC-P006: 2007, 2007 (in Japanese)

Japan Electric Association: Technical Guidelines for Aseismic Design of Nuclear Power Plants Supplement, JEAG4601-1991, 1991 (in Japanese)

*Mini-Symposia (MS13) Smart Structural Health Monitoring
and Safety Assessments
[Session Organized on Behalf of IASSAR Subcommittee SC4]*

Safety, Reliability and Risk of Structures, Infrastructures and
Engineering Systems – Furuta, Frangopol & Shinozuka (eds)
© *2010 Taylor & Francis Group, London, ISBN 978-0-415-47557-0*

US-Korea collaborative research for bridge monitoring testbeds

C.B. Yun, H. Sohn & K.Y. Koo
Department of Civil & Environmental Engineering, KAIST, Daejeon, South Korea

M.L. Wang
Department of Civil & Material Engineering, Northeastern University, Boston, MA, USA

Y.F. Zhang
Department of Civil & Environmental Engineering, University of Maryland, College Park, MD, USA

J.P. Lynch
Department of Civil & Environmental Engineering, University of Michigan, Ann Arbor, MI, USA

ABSTRACT

This paper presents the current status of a US-Korea collaborative research on bridge health monitoring testbeds. The objective of the project is to integrate and validate cutting-edge sensors and structural health monitoring methods under development for monitoring the long-term performance and structural integrity of highway bridges. Emerging sensors and monitoring technologies are investigated such as: (1) reference-free local damage detection using mode conversion of polarized piezoelectric ceramic sensors, (2) piezoelectric paint sensors offering a blending of the high piezoelectric activity of the organiz synthetic polymer, (3) wireless active sensor nodes for local damage detection based on electro-mechanical impedance measurement, (4) elasto-magnetic (EM) sensors to estimate the tensile stress state of cables, tendons, or reinforcement bar, (5) wireless sensor networks for economical health monitoring without labor-intensive tethering works, (6) robust vibration-based global monitoring using deflections obtained from modal flexibility, (7) vision-based bridge deflection measurement using a commercial low-cost camcorder with high fidelity, and (8) bridge rating based on system identification technique using ambient vibration measurement without blocking the passing traffic.

As to the testbeds, three bridges on the test road by the Korea Expressway Corporation (KEC) are under investigation as in Figure 1. They are of different structural types as a PC box-girder bridge, a steel box-girder bridge, and a steel plate-girder bridge as in Figure 2 (Lee *et al.*, 2004). The traffic on the test road section may be controlled, so that tests on the bridges may be carried out with the ordinary traffic or the calibrated traffic loads. Currently, basic facilities

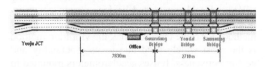

Figure 1. KEC test-road with three bridges.

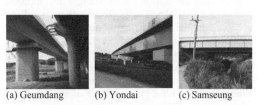

(a) Geumdang   (b) Yondai   (c) Samseung

Figure 2. Three test-road bridges.

including internet lines have been constructed on the testbeds, and the participants carried out the preliminary tests by connecting their own measurement and monitoring systems in the local area network environment at the bridge site. The participants can access and control their measurement systems by using Remote Desktop in Windows XP through internet. In addition to these three highway bridges, a cable-supported bridge is going to be selected as the fourth testbed structure for this study in the near future. Researchers interested in this testbed are encouraged to join in the collaborative research.

REFERENCES

Lee, C.G., Yi, J. H., Cho, S. & Yun, C.B. 2004. *Development of Integrated System for Smart Evaluation of Load Carrying Capacity of Bridges*. Technical Report, Korea Highway Corp., Korea.

Safety, Reliability and Risk of Structures, Infrastructures and
Engineering Systems – Furuta, Frangopol & Shinozuka (eds)
© 2010 Taylor & Francis Group, London, ISBN 978-0-415-47557-0

Development and application of a low-cost automated wireless tension estimation system for cable structures

S. Cho

Department of Civil and Environmental Engineering, KAIST, Daejeon, South Korea

J.P. Lynch

Department of Civil and Environmental Engineering, University of Michigan, Ann Arbor, MI, USA

C.B. Yun

Department of Civil and Environmental Engineering, KAIST, Daejeon, South Korea

ABSTRACT

Cable tension force is one of the most important structural parameters for structural health monitoring (SHM) of cable-supported bridges during construction and operation. Among the available measurement methods for cable tension force, the vibration method is the most widely used due to its practicality and cost-effectiveness. It uses accelerometers mounted to the surface of the cable to capture the natural frequencies and the tension forces. Though the vibration method is practical and cost-effective, there is still an economic hurdle in realizing monitoring systems for cables on many bridges, such as the high cost of the system components and the intense labor and time for the installation. Alternatively, wireless sensors are now considered as a viable substitute for tethered sensors. When dealing with a large number of sensors for a civil structure's SHM system, wireless communication between sensors and data repositories is very attractive in terms of the system cost. Besides the economical benefit, wireless sensors can accommodate a computational capability which can infer some important information from measured data.

The work presented in this study builds upon *prior* research to create a low-cost automated wireless tension force estimation system (WTFES) for cable structures. The low-cost hardware is composed of a wireless sensing unit, a MEMS accelerometer, and a signal conditioning circuit. The wireless sensing unit was developed by Wang *et al.* (2005) and employs a 16-bit multi-channel analog digital converter (ADC), an 8-bit microcontroller, 128 kB random access memory (RAM), and a 900 MHz wireless transceiver with a whip antenna. The MEMS accelerometer is interfaced with a signal conditioning circuit to mean-shift, amplify, and anti-alias filter the sensor output (acceleration) prior to interfacing to the wireless sensor. For the automated software, a modern vibration method (Zui *et al.*, 1996) which considers the sag condition of a cable is embedded into the microcontroller to estimate tension force using the natural frequencies of the cable. To extract the natural frequencies without human intervention, a peak picking algorithm is proposed considering typical properties of the Fourier spectrum of a vibrating cable.

To validate the proposed WTFES, a laboratory experiment which tests a scaled down version of a cable on a cable-stayed bridge was conducted. A series of forced vibration tests are executed to investigate the feasibility of the system for various sensing locations, cable tension forces, and sags. From the laboratory tests, it was found out that the proposed system can accurately calculate the 3 lower natural frequencies of the cable using the proposed automated peak-picking method on the Fourier spectra. Using the three natural frequencies, the present method estimated the tension forces of the cable with minimal error.

The further validation of the proposed system are going to be carried out on a cable-stayed bridge in Korea through "a US-Korea collaboration project on bridge health monitoring test bed" supported by the National Science Foundation (NSF) and Korean Science and Engineering Foundation (KOSEF).

REFERENCES

Kim, B.H. and Park, T., 2007, "Estimation of Cable Tension Force Using Frequency-based System Identification Technique," Journal of Sound and Vibration, 304, pp. 660–676.

Wang, Y., Lynch, J.P. and Law, K.H., 2005, "Validation of an Integrated Network System for Real-time Wireless Monitoring of Civil Structures," Proceedings of the 5th International Workshop on Structural Health Monitoring, Stanford, CA, September 12–14.

Lynch, J.P., Wang, Y., Loh, K., Yi, J.H., and Yun, C.B., 2006, "Performance monitoring of the Geumdang Bridge using a dense network of high-resolution wireless sensors," Smart Materials and Structures, 15, pp. 1561–1575.

Zui, H., Shinke, T. and Namita, Y., 1996, "Practical Formulas for Estimation of Cable Tension by Vibration Method," Journal of Structural Engineering, ASCE, 122(6), pp. 651–656.

Safety, Reliability and Risk of Structures, Infrastructures and Engineering Systems – Furuta, Frangopol & Shinozuka (eds)
© 2010 Taylor & Francis Group, London, ISBN 978-0-415-47557-0

Application of reference-free crack detection techniques to in-situ bridge monitoring

C.G. Lee, S. Park & H. Sohn
Department of Civil and Environmental Engineering, KAIST, Daejeon, Republic of Korea

S.B. Kim
Department of Civil and Environmental Engineering, Carnegie Mellon University, PA, USA

ABSTRACT

A new Lamb wave based NDT technique, which does not rely on previously stored baseline data, has been developed for crack monitoring in plate-like structures. The reference-free NDT technique takes advantage of two collocated lead zirconate ti-tanate transducers (PZTs) pairs placed on the top and bottom surfaces of a plate. In a thin uniform elastic medium, the change in thickness causes propagating Lamb modes to be transformed to other modes. In the previous study, it has been shown that converted Lamb wave modes due to crack for-mation can be extracted from instantaneously meas-ured signals.

In this study, the proposed approach is further advanced so that it can be also applied to in-service bridges as shown in Figure 1. In this paper, it was mainly assumed that the thickness variation due to a stiffener was considered as damage instead of crack because real defects could not be formatted in real structures

In this study, the concept of mode conversion due to the structural defect was used for making a decision. In order to identify damage, first, two mode conversions were extracted by decomposing individual Lamb modes. Next, total 4 time signals, M1 and M2 which contain two mode conversions and M3 and M4 which are composed of one mode conversion respectively, were composed by using decoupled mode conversions and these time signals were transformed to frequency domain signals. Finally, it was decided whether the structure was intact or damaged by calculating the energy ratio of the transfer impedances and then the following damage classification statement can be established: "If the minimum of out-of-class energy differences defined as Diff 1 is larger than the maximum of within-class energy differences which is Diff 2, a crack exists. Otherwise, there is no crack". In Figure 2, the results of decision criteria are illustrated.

Consequently, since the damage-sensitive features were extracted through the transfer imped-ances, the decision making could be determined by

(a) Buffalo Creek Bridge in USA	(b) Samseung Bridge in Korea

Figure 1. The overview of tested bridges in this study.

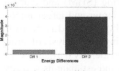 

(a) Pristine condition.	(b) Damaged condition.

Figure 2. Decision making using the energy ratio of the transfer impedances.

instantly establishing the decision boundaries which were decided by using the energies of four transfer impedances without pre-determined thresholds.

REFERENCES

Balageas, D., Fritzen, C. & Güemes, A. 2006. *Structural Health Monitoring*. London: ISTE Ltd.

Cho, Y. 2000. Estimation of Ultrasonic Guided Wave Mode Conversion in a Plate with Thickness Variation. *IEEE transactions on ultrasonics, ferroelectrics, and frequency control* 47(3): 591–603.

Kim, S. B & Sohn, H, 2007. Instantaneous reference-free crack detection based on polarization characteristics of piezoelec-tric materials. *Smart Materials and Structures* 16(6): 2375–2387

Yamanaka K., Nagata Y. & Koda T. 1991. Selective excitation of single-mode acoustic waves by phase velocity scanning of a laser beam. Applied Physics Letter 58: 1591–1593

Wilcox P. D., Lowe M. J. S. & Cawley P. 2001 Mode and transducer selection for long range Lamb wave inspection Journal of Intelligent Material Systems and Structures 12: 553–565

Safety, Reliability and Risk of Structures, Infrastructures and
Engineering Systems – Furuta, Frangopol & Shinozuka (eds)
© 2010 Taylor & Francis Group, London, ISBN 978-0-415-47557-0

The strategic planning of safety management network for infrastructure

Y.S. Kim, S.S. An, S.R. Ahn & D.S. Yang
KISTEC, Goyang, Republic of Korea

ABSTRACT

For the last four decades, Korea economy has been developed rapidly and significantly. As a consequence, the size of Korea economy (in GDP) became one of the top ten countries in 2007. Like other countries, the civil infrastructure has played an important role in the development, by underpinning different parts of the society. Among infrastructure must put safety as their foremost consideration, and, together with precise design, construction, safety management is a must to secure their safe usability. However, as civil structures are subject to loads and other environmental effects, the condition of many civil infrastructure systems is deteriorating.

Structural health monitoring (SHM) and damage identification are assuming larger and larger importance in civil engineering, SHM is defined as the use of in-situ, non-destructive sensing and analysis of structural characteristics in order to identify if a damage has occurred, define its location and estimate its severity, evaluate its consequences on the residual life of the structure.

This paper studied the safety management network system of infrastructure which constructed smart sensors, closed- circuit television (CCTV) and monitoring system. This safety management of infrastructure applied to bridge, cut slop and tunnel, embankment and playground etc. The system applied to technologies of standardization guidelines, data acquirement technologies, data analysis and judgment technologies, system integration setup technology, and IT technologies. It was constructed safety management network system of various infrastructure to improve efficient management and operation for many infrastructure.

Integrated safety management network system of infrastructure consisted of the real-time structural health monitoring system of each infrastructure, integrated control center, measured data transmission using internet web-based, collecting data using server, early alarm system which the dangerous event of infrastructure occurred.

Integrated control center consisted of conference room, control room to manage and analysis the data, server room to present the measured data and to collect the raw data. Early alarm system proposed realization of warning and response within 5 minute or less through development of sensor-based progress report and propagation automation system using the media such as MMS, VMS, EMS, FMS, SMS and web services of report and propagation.

Based on this, the most effective u-Infrastructure Safety Management System is expected to be stably established at a less cost, thus making people's life more comfortable. Information obtained from such systems could be useful for maintenance or structural safety evaluation of existing structures, rapid evaluation of conditions of damaged structures after an earthquake, estimation of residual life of structures, repair and retrofitting of structures, maintenance, management or rehabilitation of historical structures

REFERENCES

Aktan, A.E., Tsikos, C.T., Catbas, F.N., Crimmelsman, K. and Barrish, R. 1999. Challenges and Opportunities in Bridge Health Monitoring, *proceedings of the 2nd International Workshop on Structural Health Monitoring*, Stanford, CA, USA.

De Stefano, A. 2007. Structural identification and health monitoring on the historical architectural heritage, *Key Engineering Materials* Vol. 347, pp. 37–54.

Kraemer, P. and Fritzen, C-P. 2007. Sensor Fault identification using autoregressive models and the Mutual Information concept , *Key Engineering Materials* Vol. 347, pp. 79–116

Mufti, A. 2001. *Guidelines for Structural Health Monitoring*, University of Manitoba, ISIS, Canada.

Omenzetter, P., Brownjohn, J.M.W. and Moyo, P. 2004. Identification of unusual events in multi-channel bridge monitoring data, *Mechanical Systems and Signal Processing*, Vol. 18, pp. 409–430.

Silkorsky, C. 1999. Development of a Health Monitoring System for Civil Structures using a Level IV Non-Destructive Damage Evaluation Method, *proceedings of the 2nd International Workshop on Structural Health Monitoring*, Stanford, CA, USA.

Tiiljsten, B., Hejll, A. and James, G. 2007. Carbon Fiber-Reinforced Polymer Strengthening and monitoring of the Grandal Bridge in Sweden, *ASCE Journal of Composites for Construction*, Vol. 11, pp. 227–235.

Safety, Reliability and Risk of Structures, Infrastructures and Engineering Systems – Furuta, Frangopol & Shinozuka (eds)
© 2010 Taylor & Francis Group, London, ISBN 978-0-415-47557-0

Bridge displacement monitoring techniques using strain measurement

K.T. Park, B.C. Joo, C. Lee & Y.K. Hwang
Korea Institute of Construction Technology, South Korea

ABSTRACT

For the safety of the bridges, periodic monitoring, maintenance, and repairs are required. Maintenance and structural safety of the bridges may be guaranteed by the application of a health monitoring system. It provides valuable information conducive to detailed inspection, repair and rehabilitation of the bridges. For these reasons, long-term monitoring systems are largely integrated and various kinds of sensors are used in the bridge monitoring system. Especially, the strain sensors are greatly applied due to the importance and usefulness of the strain values in evaluating the safety level of a bridge.

The monitoring displacement of a bridge allows for an estimation of input dynamic excitation characteristics and has an important role in the evaluation of structure status determination. In many bridges, the vertical displacements are the most relevant parameters to be monitored in both short and long term period. The LVDT, tiltmeter, or the laser deflection meters are the general methods used for displacement measurements. However, these methods can not be applied effectively if the bridge height is so high that the installation of a sensor is difficult. In such cases, the greater part of the reliability and accuracy of measured results are usually uncertain.

In this paper, to overcome such problems, two techniques for calculating displacement from measured strain are proposed; the iteration of data averaging and lowpass filtering of measured strain. To verify these techniques, laboratory and field tests are carried out. Dynamic tests on steel pipes in laboratory were performed using dial gauge for displacement measurements and strain gauge for strain measurements. The dynamic vibration of the specimen was achieved by means of a rubber hammer at optional position to simulate real-time in the case of vehicle crossing the bridge, and the measured times were 7 minutes in 4 trials. To verify the filtering method in the field, the measured data from the steel box girder bridge, which is located in the national highway in South Korea, were applied to calculate the displacement by filtering. The measured spans are the 3 continuous steel box girders, which are located in the start point of bridge.

The results of this paper are as follows. (1) Data averaging and lowpass filtering techniques are performed to calculate real-time displacement from measured strain. To verify these methods, the measured and calculated displacement of a bridge were compared. (2) Data averaging method is time-consuming, and the lowpass filtering method is simpler than data averaging. If FFT analyzing and filtering on measured strain can be performed simultaneously in the field, it is possible to calculate displacement using measured strain by lowpass filtering method without dial gauge or LVDT. (3) Proposed techniques would be more effective in calculating displacement of steel bridges with sound conditions. But, this method has limitations in the case of the bridge having varied sections or damages at the measured point or support, so additional researches are needed in this field.

ACKNOWLEDGEMENTS

This research was supported by a grant(Code No: 06B05) from Construction Core Technology Program funded by Ministry of Land, Transport and Marine Affairs of Korean Government.

REFERENCES

Inaudi D. *et al* (1998) *Vertical deflection of a pre-stressed concrete bridge obtained using deformation sensors and inclinometer measurements.* ACI Structural Journal, Vol. 95, No. 5, pp. 518–526.

KICT *et al* (2000) *Development of bridge inspection techniques using fiber optic sensors*, 1 Year Report (in Korean), R&D/99I-09, MOCT of Korea.

KICT *et al* (2002) *Development of bridge inspection techniques using fiber optic sensors, Final Report (in Korean)*, R&D/99I-09, MOCT of Korea.

Organized Session (OS21) Developing Performance-Based Concrete Design Code

*Safety, Reliability and Risk of Structures, Infrastructures and
Engineering Systems – Furuta, Frangopol & Shinozuka (eds)*
© *2010 Taylor & Francis Group, London, ISBN 978-0-415-47557-0*

Probabilistic models for mechanical properties of concrete in Korea

J.S. Kim & J.H. Shin
Department of Civil Engineering, Seokyeong University, Seoul, Korea

J.H. Moon
Korea Institute of Construction Materials, Seoul, Korea

K.M. Lee
Department of Civil and Environmental Engineering, Sungkyunkwan University, Suwon, Korea

ABSTRACT

In Korea, the design of reinforced concrete members is based on the ultimate strength design method, which adopts load factors and strength reduction factors. The factors are introduced to take into account the uncertainties in both loads and resistances and determined with the use of reliability concepts. In order to perform reliability based code calibration, the probabilistic models for loads and resistances should be obtained from realistic local data. As in many other countries, there is no enough and reliable statistical data in Korea. To overcome this drawback, the authors started to develop the original probabilistic models for concrete and steel produced in Korea.

As a first step, the probabilistic models of mechanical properties of concrete, compressive and tensile strength, are developed based on a number of published and unpublished sources in Korea and additional laboratory tests done by the authors.

For the ready mixed concrete company data of 24 MPa and 30 MPa of compressive strengths, the coefficients of variation are 6.65% and 4.96%, which shows that there is less variation with increasing compressive strength. For the additional laboratory data including higher strength concrete, the coefficients of variation of 24 MPa, 40 MPa, 50 MPa and 60 MPa are 14.27%, 3.43%, 2.01% and 2.02% respectively. It shows the similar trends those of ready-mixed concrete company data, which means that the magnitudes of variance do not increase as the compressive strength increase with the same proportion. The visual comparisons of histograms and proposed Normal distribution curves make it proper to use of Normal distribution for compressive strength of concrete for both ready-mixed concrete and laboratory data.

This paper also gives new relationships between compressive and splitting tensile strength and elastic modulus of concrete used in Korea.

In Korean design code, the elastic modulus of normal weight concrete may be determined with the specified design strength by Eq. (1).

$$E_c = 8,500\sqrt[3]{f_{ck} + 8\,MPa} \tag{1}$$

The regression analysis was performed with 245 data with the exponent of 1/3 and derived equation is as follows;

$$E_c = 8,680\sqrt[3]{f_{ck} + 8\,MPa} \tag{2}$$

The coefficient in Eq.(2) is little greater than that Eq.(1), which means that Korean design code underestimates the elastic modulus. In the regression analysis performed to obtain the equation, the correlation coefficient was very low. The regression with the n value of 0.3 as in Eurocode also shows low value of correlation coefficient. It seems that the large scatter of experimental data results in low values of correlation coefficients and further studies are required for the form of equations.

REFERENCES

Ang, A.H-S and Tang, W.H. 2007, *Probability Concepts in Engineering Planning and Design, Vol. 1, 2nd Ed.*, Hoboken, John Wiley & Sons, Inc.
BS EN 1992-1-1 2004, *Eurocode 2: Design of concrete structures-part 1-1: General rules and rules for buildings*, The British Standards Institutio.
Comite Euro-International du Beton, 1993, *CEB-FIP Model Code 1990*, Thomas Telford.
Eibl, J. (ed.) 1994, *Concrete Structure Euro-Design-Handbook*, Karlsruhe, Ernest & Sohn.
KCI 2007, *Korean Standard Design Code for Concrete Structures (in Korean)*, Seoul, Kimoondang.
Minitab 2005, *Academic Minitab (in Korean)*, Seoul, Minitab Inc.

Safety, Reliability and Risk of Structures, Infrastructures and
Engineering Systems – Furuta, Frangopol & Shinozuka (eds)
© 2010 Taylor & Francis Group, London, ISBN 978-0-415-47557-0

Moment redistribution for earthquake design of reinforced concrete moment-resisting frames

Tae-Sung Eom
Catholic University of Daegu, Gyeongsan, Korea

Hong-Gun Park & Sung-Gul Hong
Seoul National University, Seoul, Korea

Jae-Yo Kim
Kwanwoon University, Seoul, Korea

ABSTRACT

Moment-redistribution is developed by the changes in the member stiffness caused by inelastic deformations. In addition, moment-redistribution is affected by the plastic mechanism of the structure, member stiffness loading condition, and the allowable story drift. Therefore, the allowable magnitude of redistributed moment for earthquake design should be determined on the basis of the deformation capacity of the member considering these design parameters.

As shown in Figure 1, the overall load-transfer mechanism of the moment-resisting frame can be divided into three loading stages according to the development of the plastic hinges. At elastic stage (Figure 1(a)), all members behave elastic. At inelastic stage I (Singe-hinged behavior, Figure 1(b)), plastic hinge develops at one end of a beam. At inelastic stage II (Double-hinged behavior, Figure 1(c)), plastic hinges develops at both ends of a beam. Overall moments and rotations of the members are calculated by summing the increments developed at the three loading stages. In the present study, the relationships between the redistributed moment and the plastic rotation of a member caused by the single- and double-hinged behaviors were formulated. Based on the relationships, an earthquake design method for moment-resisting frames was developed to address moment-redistribution, using the results from elastic analysis. The proposed design method is performed following the procedure of 1) elastic analysis, 2) the determination of redistributed moments, and 3) the evaluation of member deformations and safety.

The proposed method is convenient because moment-redistribution can be performed based on the results from elastic analysis, considering the plastic mechanism intended by engineer. In addition, the inelastic deformations required for moment-redistribution can be evaluated and therefore, the safety of each member can be evaluated.

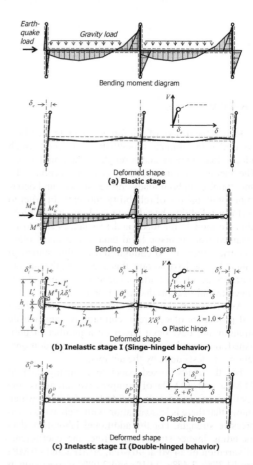

Figure 1. Load-transfer mechanism of moment-resisting sub-frame subjected to combined gravity and earthquake loads.

REFERENCES

American Concrete Institute 2008. *Building code requirements for structural concrete*, ACI 318-08 and ACI 318R-08. Farmington Hills: Michigan.

European Committee for Standardization 2004. *Eurocode 2: Design of concrete structures – part 1-1: General rules and rules for buildings*. British Standards Institute, BS EN 1992-1-1:2004, EC2. London.

Architectural Institute of Korea 2005. *Korean building code – structural*, KBC 2005. Seoul.

Paulay, T. & Priestley, M.J.N. 1992. *Seismic design of reinforced concrete and masonry buildings*, John Wiley and Sons Inc. New York.

Safety, Reliability and Risk of Structures, Infrastructures and
Engineering Systems – Furuta, Frangopol & Shinozuka (eds)
© 2010 Taylor & Francis Group, London, ISBN 978-0-415-47557-0

Implementation of performance in concrete design code

S. Shin
Inha University, Incheon, Korea

I. Paik
Kyongwon University, Songnam, Korea

Y.-S. Chung
Chung-Ang University, Ansung, Korea

ABSTRACT

Performance-based design is an upcoming issue in the design codes for civil structures. Many international design codes introduced in recent years acknowledge the performance concept and open their doors to it. Even if the idea of performance seems to be easily understood by civil engineers, however, it is not easy to figure out how such an idea can be melted down in the codes practically. The paper presents a conceptual way of implementing the performance-based design into a design code in a systematic manner. The paper limits its focus on the safety and serviceability for a design. Basically, the reliability index combined with the probability of failure can be utilized as the reference value in defining the level or quality of a design. The design working life depending on the importance of a structure is also inter-related as a guideline in leveling the required quality of a civil structure.

To determine the performance level, the use of partial factors may be practical especially for the safety evaluation. Since the specific selection of partial factors may be dependent on the code, however, a general procedure of implementing the performance is proposed in Figure 1. In Figure 1, the second stage of the procedure is proposed in two different ways but both cases use the same probability of failure in 1 yr as the reference.

A sample study is presented for the determination of safety factors for the target reliability indexes as defined by each performance levels. PSC-composite

sections are used for the case study for bridge girders with spans of 20 to 40 meters. Figure 2 presents a sample result for the calibration of the resistance factors in different resistance factor formats. Figure 3 shows the load factors calibrated for the target reliability indexes.

REFERENCES

Paik, I, Hwang, E-S. & Shin, S. 2008. Reliability analysis of concrete girder bridges using measured statistical data. In C-K. Choi (ed.), *Proc. 4th intern. conf. on advances in structural engineering and mechanics, Jeju, 26–28 May 2008*. Daejeon: Techno Press.

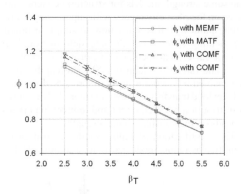

Figure 2. Variation of resistance factors with respect to β_T.

Determine reference design working life depending on structural importance level

↓

Determine 1 yr probability of failure or corresponding reliability index depending on performance level

or

Determine target reliability index corresponding to required design working life

↓

Determine partial factors according to the reliability index for performance level

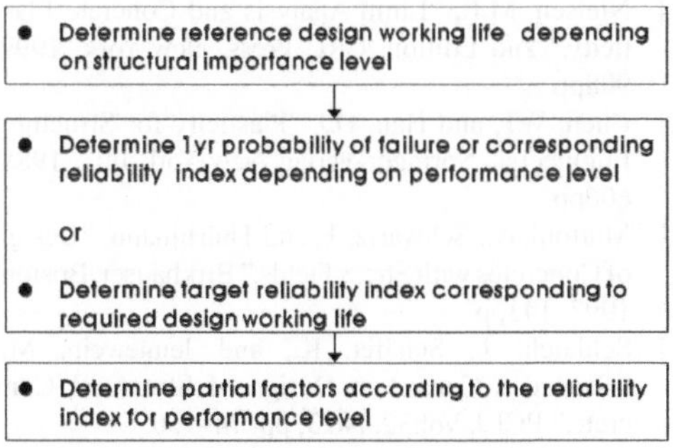

Figure 1.   Procedure of implementing performance.

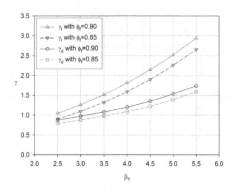

Figure 3.   Variation of load factors with respect to β_T.

Safety, Reliability and Risk of Structures, Infrastructures and Engineering Systems – Furuta, Frangopol & Shinozuka (eds)
© *2010 Taylor & Francis Group, London, ISBN 978-0-415-47557-0*

Performance-based shear design for structural concrete

Sung-Gul Hong
Seoul National University, Seoul, Korea

Sang-Gi Jang
Building Structural Design Team, Sam-Sung Construction Co., Seoul, Korea

Woo-Young Lim
Department of Architecture, Seoul National University, Seoul, Korea

ABSTRACT

Performance-based design for structural concrete requires how to rationally estimate deformation of members and/or systems at various load levels with deformation limits before or after yielding. To estimate deformation with corresponding resistances several physical models based on cracked concrete have been developed. However, such models require tedious numerical calculation procedures. As for strength current some strength models for structural concrete are based on the plasticity theory with suppressing brittle failure modes. In this study how to simply estimate the yielding points and ultimate deformation limits of shear critical members are proposed by stress fields and failure mechanisms. Shear strength model is explained by decomposition in terms of direct and indirect load transfers as shown in figure 1. The degree of distribution between direct and indirect load paths depends on their stiffness and strength.

The classification of failure modes is also easily identified by failure mode matrix for a combined STM

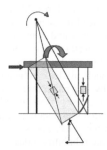

Shear-flexural Compression

Figure 2. Flexural-shear compression failure mechanism for compatibility-based strength interaction.

which is resulted by superposing elementary STM relying on direct and indirect load paths.

Shear strength degradation with increase of deformation and deformation limit at ultimate are discussed by failure mechanisms compatible with stress fields. Also compatibility-based strength interaction curves associated with shear are proposed and thereby deformation limits dependent on failure mechanisms are estimated.

REFERENCES

[1] Nielsen, M.P., "Limit Analysis and Concrete Plasticity," 2nd Edition, CRC Press, New York, 1999, 908pp.
[2] Chen, W.F, and Han, D.J. "Plasticity for Structural Engineers," Springer-Verlag New York Inc., 1988, 606pp.
[3] Muttoni, A., Schwartz, J., and Thürlimann, "Design of Concretes with Stress Fields," Birkhäuser, Boston, 1997, 143pp.
[4] Schlaich, J., Schäfer, K., and Jennewein, M., "Toward a Consistent Design of Structural Concrete," PCI J. Vol.32, No.3, pp. 74–150.
[5] Marti, P., "Basic Tools of Reinforced Concrete Beam Design," ACI Journal, Jan.-Feb., 1985, pp. 46–56.

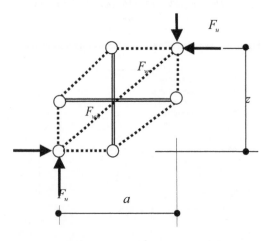

Figure 1. Shear strength model in terms of direct and indirect load transfers.

Safety, Reliability and Risk of Structures, Infrastructures and Engineering Systems – Furuta, Frangopol & Shinozuka (eds)
© 2010 Taylor & Francis Group, London, ISBN 978-0-415-47557-0

Service life assessment of RC structures exposed to carbonation by determining the safety factors

H.-W. Song, S.-W. Pack, K.Y. Ann & C.-H. Lee
Yonsei University, Seoul, Korea

ABSTRACT

In the present study, reinforced concrete (RC) column structures exposed to carbonation were surveyed, and data on the carbonation depth (X_{CO2}), considering different states of cover concrete: sound, cracked, and joint concrete and cover depth (X_C) were obtained. The distribution of parametric values for the RC columns in carbonation environments was examined to calculate the safety factors. 56, 25, and 32 replications of carbonation depth X_{CO2} for sound, cracked, and joint concrete, respectively, were obtained, and 47 replications of concrete cover depth X_C were investigated. Their average values of X_{CO2} accounted for 11.62, 24.66, and 17.43mm for sound, cracked, and joint concrete respectively, and the average of X_C was observed as 57.87mm. To find an optimum probability distribution model for the variables of X_{CO2} and X_C, the goodness-of-fit test was conducted. The histograms of X_{CO2}, irrespective of different states of concrete cover, are closer to a Weibull distribution, while the X_C value is log-normally distributed.

The safety factors used to the service life analysis of concrete structures exposed to carbonation were calculated by reliability-based theory, and then service life of reinforced concrete column structures was assessed. When 1.28 of reliability index β, corresponding to 10% of the probability of corrosion initiation, was taken as the threshold ratio to determine the safety factors, the safety factor (SF_{CO2}) accounted for 1.49, 1.44, and 1.45 for sound, cracked, and joint concrete, respectively, as listed in Table 1. The values of SF_{CO2} were verified by the Monte Carlo Simulation technique. Therefore, the safety factors obtained in the present study can be used to a reasonable partial factor method of service life design and assessment of concrete structures exposed to carbonation. By using the SF_{CO2}, time to carbonation-induced corrosion of steel accounted for 202, 48 and 95 years for sound, cracked and joint concrete, respectively.

Table 1. Safety factors for carbonation depending on different states of cover concrete.

States of cover concrete	Durability resistance factor (ϕ_c)	Environmental load factor (γ_c)	Safety factor (SF_{CO2})
Sound concrete	0.776	1.157	1.49
Cracked concrete	0.758	1.095	1.44
Joint concrete	0.759	1.102	1.45

Finally, the safety factors depending on different casting methods, such as precast and cast-in-place concrete were discussed. SF_{CO2} calculated for precast concrete accounted for 1.31, 1.25, and 1.26 for sound, cracked, and joint concrete, respectively. It was found that differences of the safety factors between precast and cast-in-place concrete structures accounted for 0.19. An increase in the SF_{CO2} implies a significance decrease in the corrosion free life of concrete structures exposed to carbonation. Thus, corrosion free life of RC structures exposed to carbonation can be sensitive depending on different casting methods even under same carbonation environment.

REFERENCES

Ang, A.H.-S. & Tang, W.H. 1990. *Probability Concepts in Engineering Planning and Design*, John Wiley & Sons, Vol. II.

Song, H.-W., Kwon, S.-J., Byun, K.J. & Park, C.-K. 2006a. Predicting carbonation in early-aged concrete, *Cement and Concrete Research*, 36: 979–989.

Song, H.-W., Pack, S.-W., Lee, C.-H. & Kwon, S.-J. 2006b. Service life prediction of concrete structures under marine environment considering coupled deterioration, *Restoration of Buildings and Monuments*, 12(4): 265–284.

Mini-Symposia (MS04) Structural Reliability and Optimization

Safety, Reliability and Risk of Structures, Infrastructures and
Engineering Systems – Furuta, Frangopol & Shinozuka (eds)
© 2010 Taylor & Francis Group, London, ISBN 978-0-415-47557-0

Optimal design of corroded reinforced concrete structures by using time-variant reliability analysis

Y. Aoues, E. Bastidas-Arteaga & A. Chateauneuf
LaMI, Blaise Pascal University, Polytech'Clermont-Fd, Campus des Cézeaux, Aubière, France

ABSTRACT

The design of reinforced concrete structures involves several types of uncertainties related to concrete and steel properties, structural dimensions, depth of the concrete cover and load fluctuations. All these uncertainties contribute to make the performance of RC structures different from the expected one. In addition, when a RC structure is located in a corrosive environment, the corrosion can significantly influence long-term performance. The deterioration of RC structures is a major problem for many companies and countries, where high costs are involved to maintain, repair or replace degraded structures. For example, CC Technologies Laboratories Inc. (2001) show that the annual direct cost of corrosion in USA highway bridges is estimated to be 8.3 billions of dollars.

Structural optimization is frequently applied for effective cost reduction of engineering systems. As the rational approach consists in finding the best compromise between cost reduction and safety assurance, the structural reliability theory provides an appropriate approach to take account for uncertainties (Ditlevsen & Madsen 1996). The Reliability-Based Design Optimization (RBDO) allows us to reach effectively balanced cost-safety con?gurations (Frangopol & Moses 1994). However, for designing RC structures located in corrosive environment the RBDO problem should use the time-variant instead of the time-invariant reliability analysis, where load and environmental fluctuations are modeled by stochastic processes.

Several works are focused on developing efficient and robust approaches for RBDO problem (Chateauneuf, & Aoues 2008). However, few studies are interested by Time-Variant Reliability-Based Design Optimization (TV-RBDO). The classical formulation of the TV-RBDO consists in minimizing an objective function defined by the initial and expected failure costs under reliability constraints (Kuschel & Rackwitz 2000). This formulation is not suitable for real engineering structures because considerable time consumption is involved and convergence can hardly be achieved.

In this work, a new methodology of TV-RBDO of RC structures is developed on the concept of

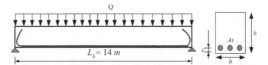

Figure 1. Rectangular RC bridge girder.

Table 1. Comparison of the optimal costs of the RC girder and numerical performances.

	Classical. TV-RBDO	Proposed TV-RBDO
Expected total cost C_T (k€)	3.69	3.70
Initial cost C_I (k€)	3.67	3.68
Expected failure cost C_F (€)	16.57	16.82
CPU (s)	603	21

decoupling time-variant reliability analysis and the optimization procedures. The proposed approach is based on transforming the TV-RBDO problem into a sequence of equivalent deterministic design optimization sub-problems. This transformation is defined by the mean of optimally calibrated safety factors, linking the reliability requirement to the equivalent deterministic optimization. The numerical example on RC bridge girder shows the efficiency and the good-standing of the proposed method, through comparison with classical TV-RBDO approach.

REFERENCES

Chateauneuf, A & Aoues, Y. 2008. Advances in solution methods for Reliability-Based Design Optimization. In Y. Tsompanakis & N.D. Lagaros & M. Papadrakakis (eds), *Structural Design Optimization Considering Uncertainties*. Taylor & Francis.

Ditlevsen, O & Madsen, H.O. 1996. *Structural reliability method*. New York: John Wiley and Sons.

Kuschel, N & Rackwitz, R. 2000. Optimal design under time-variant reliability constraints. *Structural Safety* 22(2): 113–27.

Safety, Reliability and Risk of Structures, Infrastructures and
Engineering Systems – Furuta, Frangopol & Shinozuka (eds)
© 2010 Taylor & Francis Group, London, ISBN 978-0-415-47557-0

Optimal evacuation and shut-down decisions in the face of emerging natural hazards

K. Nishijima, M. Graf & M.H. Faber
Institute of Structural Engineering, ETH Zurich, Zurich, Switzerland

ABSTRACT

Building structures, infrastructure systems and industrial facilities (hereafter jointly referred to as engineered facilities) are often built and operated on locations where natural hazards may take place; implying significant risks. As part of the overall strategy of risk management for such engineered facilities the optional decision to shut down operation and to evacuate people and assets in the face of an emerging hazard event plays an important role. Important examples where such strategies are presently utilized include refineries and fixed offshore platforms subject to tropical storms, storm surges and tsunamis, but also urban habitats and public infrastructure subject to events such as storms, floods, landslides, avalanches and volcanic eruptions.

The important characteristics of the decision situation described above include that: various indicators that contain information on the severity of the hazard continuously become available prior to the impact of the hazard; decision makers have options for reducing risks which may be commenced at any time supported by the observed indicators. Here the typical problem arises that waiting will imply more information but might also reduce available time for evacuation and other loss reduction activities; decision makers can/should make decisions in near-real time in response to the indicators available.

The scope of the present paper is to present a decision framework for the decision problem outline in the foregoing taking basis in the Bayesian pre-posterior decision analysis (Raiffa & Schlaifer (1961)) and then to show the use and the advantages of the presented framework with an illustrative example as well as to discuss the computational difficulties and possible approaches to avoid the difficulties.

The example considered in the present paper is the decision situation where a decision maker is faced to decide whether or not the operation of an offshore platform should be shut down in the emergence of a typhoon event. The decision situation considered in

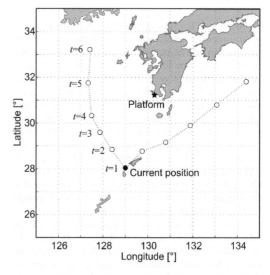

Figure 1. Illustration of the decision situation considered in the example. (Two possible transitions of the typhoon are shown with dashed lines. The location of the offshore platform for which the decision maker is responsible and the current position of the typhoon are also shown. The discretized times shown in the figure correspond to the time interval of six hours.)

the example is illustrated in Figure 1. In this example the probabilistic typhoon model presently under development at the group of Risk and Safety at ETH Zurich (Nishijima & Faber (2007)) is employed.

REFERENCES

Nishijima, K., and Faber, M. H. 2007. A Bayesian framework for typhoon risk management; *12th International Conference on Wind Engineering, 12ICWE*, Cairns, Australia.
Raiffa, H., and Schlaifer, R. 1961. *Applied Statistical Decision Theory*. Cambridge: Cambridge University Press.

Safety, Reliability and Risk of Structures, Infrastructures and Engineering Systems – Furuta, Frangopol & Shinozuka (eds)
© 2010 Taylor & Francis Group, London, ISBN 978-0-415-47557-0

Fragility of conventional woodframe structures built in regions of low-to-moderate seismicity

David V. Rosowsky
Zachry Department of Civil Engineering, Texas A&M University, College Station, TX, USA.

WeiChiang Pang
Department of Civil Engineering, Clemson University, Clemson, SC, USA.

Yue Wang
Zachry Department of Civil Engineering, Texas A&M University, College Station, TX, USA.

Bruce R. Ellingwood
School of Civil Engineering, Georgia Institute of Technology, Atlanta, GA, USA.

ABSTRACT

This paper presents seismic fragility analyses of conventional (non-engineered) residential woodframe buildings located in the Central United States, a region of low-to-moderate seismicity. These conventional buildings were not engineered to withstand strong seismic loads and hence are potentially vulnerable to future earthquakes. Since the buildings considered in this study generally do not have any seismic anchorage devices (e.g., hold-downs), a fragility assessment considering only lateral drift is not realistic. More likely failure mechanisms of these conventional buildings were expected to be wall uplift and sill plate splitting. The results of fragility analysis of a set of representative buildings on two different foundations, slab-on-grade and cripple wall, confirmed that uplift and plate splitting often were the controlling limit states. The effectiveness of two mitigation strategies, (1) increasing the number of sheathing-to-framing nails and (2) adding anchor bolts, were examined through fragility analysis. The fragility assessment and mitigation concepts are illustrated using selected one and two-story residential structures taken to be representative of conventional woodframe construction, typical of the Central United States.

Safety, Reliability and Risk of Structures, Infrastructures and
Engineering Systems – Furuta, Frangopol & Shinozuka (eds)
© *2010 Taylor & Francis Group, London, ISBN 978-0-415-47557-0*

Formulation of stress-strain relations of structural aluminum alloys

I. Okura

Department of Civil Engineering, Osaka University, Osaka, Japan

T. Ishikawa

Department of Environmental Engineering and Architecture, Nagoya University, Nagoya, Japan

T. Nagao

Research & Development Center, Nippon Light Metal Company, Ltd., Shizuoka, Japan

ABSTRACT

In Japan, aluminum structures are designed according to the Japanese Specifications for Design and Fabrication of Aluminum Civil Structures (JSDFACS) (Japanese Aluminum Association 1998). In the JSD-FACS, the width-to-thickness ratio of plate elements and girders are based on the Japanese Specifications for Highway Bridges (JSHB) (Japan Road Association 2002), where the Young's modulus of 200 GPa and the yield stress of steels are replaced by 70 GPa and the 0.2% proof stress of aluminum alloys, respectively. The relation between stress and strain of steels is straight until the yield stress, and it is horizontal after that. However, the relation of aluminum alloys shows a non-linearity around the 0.2% proof stress. Furthermore, the degree of curvature near the 0.2% proof stress is considerably different between 5000- and 6000-series aluminum alloys (Mazzolani 1995). Accordingly the provisions in the JSDFACS may not reflect the actual behavior of plate elements and girders made of structural aluminum alloys.

Under the above background, the objective of this research is to formulate the relations between stress and strain of structural aluminum alloys.

The Ramberg-Osgood form is usually used for the stress-strain relations of structural aluminum alloys, which is given by

$$\varepsilon = \frac{\sigma}{E} + 0.002 \left(\frac{\sigma}{\sigma_{0.2}} \right)^n \qquad (1)$$

where ε and σ are the strain and stress, respectively, E is Young's modulus, $\sigma_{0.2}$ is the 0.2% proof stress, and n is the strain-hardening parameter, which characterizes the strain hardening behavior in the non-elastic region.

Tensile tests are carried out for structural aluminum alloys 6061-T6, 6005C-T5 and 5083-O. Then the probability functions of E and n are determined by application of a probability-statistical approach to the tensile test results, and the following conclusions are provided:

1) The Young's modulus is expressed by a normal distribution. The Young's modulus of 70 GPa used in ordinary structural design is lower than the value corresponding to the 5% probability of not being exceeded on the unfavorable side.
2) The strain-hardening parameter is expressed by a lognormal distribution. The values of the strain-hardening parameter corresponding to the 5 % probability of not being exceeded on the unfavorable side are 29.1 and 5.3 for the 6061-T6 and 6005C-T5 alloys and the 5083-O alloy, respectively.
3) The relations between stress and strain are formulated by the following equations:

$$\varepsilon = \frac{\sigma}{70000} + 0.002 \left(\frac{\sigma}{\sigma_{0.2}} \right)^n \qquad (\sigma \leq \sigma_{0.2}) \qquad (2)$$

$$\sigma = \sigma_{0.2} \qquad (\sigma > \sigma_{0.2}) \qquad (3)$$

where n is 29.1 and 5.3 for the 6061-T6 and 6005C-T5 alloys and the 5083-O alloy, respectively, and σ and $\sigma_{0.2}$ are in MPa.

REFERENCES

Japanese Aluminum Association 1998. Japanese Specifications for Design and Fabrication of Aluminum Civil Structures. (in Japanese)
Japan Road Association 2002. Japanese Specifications for Highway Bridges, Par 2 Steel Bridges. (in Japanese)
Mazzolani, F.M. 1995. Aluminium Alloy Structures, Second Edition, E & FN Spon, 59–79.

Safety, Reliability and Risk of Structures, Infrastructures and Engineering Systems – Furuta, Frangopol & Shinozuka (eds)
© 2010 Taylor & Francis Group, London, ISBN 978-0-415-47557-0

Reliability analysis method based on sample weight for slope stability

Zhi-gang Yang
College of Water Conservancy and Hydropower Engineering, Hohai University, Nanjing, China
School of Architectural Engineering, Nanchang University, Nanchang, China

Tong-chun Li, Miao-lin Dai & Cheng Li
College of Water Conservancy and Hydropower Engineering, Hohai University, Nanjing, China

Keywords: reliability analysis of slope stability; sample's weighting coefficient; *t*-distribution; BAYES formula

At present the main safety evaluation method for slope stability is based on the single safety coefficient according to *Power Industry Standard Design Specification for Slope of hydropower and Water Conservancy Project DL/T5353-2006 of the People's Republic of China*. The control criteria of slop stability analysis by a single safety factor, whether evaluated by rigid limit equilibrium method or assisted by finite element method, are difficult to reflect the actual situation of complicated high slope comprehensively and reliably. This paper presents a reliability method based on sample weight analysis. The method can improve accuracy and reduce the workload of numerical analysis as well. At first, with the distribution characteristics and digital features of random variables, the minimal sample size of every random variable is extracted according to small sample t-distribution under a certain expected value. Secondly, take the weight coefficient of the sample as the contribution to the random variables in the extracted sample, and figure out the weight coefficient of sample combination using the BAYES formula. And then, extract the minimal sample combination according to the value of sample weight coefficient under a certain expected level. Take these sample combination as the input value for systematic analysis of slope, and the random sample output of stability safety factor can be obtained. At last, according to the one-to-one mapping relationship between sample combination and weight coefficient, the reliability index of slope stability and safety can be obtained by multiplication principle. The slope stability analysis on left bank of BaiHeTan is taken as an example, and the analyses results are also given in this paper. It shows that the method of reliability in complicated high slop is reasonable and practicable.

Organized Session (OS18) Seismic Risk Assessment and Information

Safety, Reliability and Risk of Structures, Infrastructures and Engineering Systems – Furuta, Frangopol & Shinozuka (eds)
© *2010 Taylor & Francis Group, London, ISBN 978-0-415-47557-0*

An interactive web tool for quantitative seismic risk assessment of woodframe houses ("ResRisk"): A progress report

N. Luco, E.M. Martinez & E.H. Field
United States Geological Survey, Golden, Colorado & Pasadena, CA, USA

K.A. Porter
University of Colorado at Boulder, Boulder, CO, USA

H. Ryu
Stanford University, Stanford, CA, USA

J. Mitrani-Reiser
Johns Hopkins University, Baltimore, MD, USA

ABSTRACT

The largest and arguably most important portion of the exposure to earthquakes in the U.S.A. is comprised of woodframe houses. According to HAZUS-MH (2003a) woodframe houses constitute about 87% of the buildings in California, for example. At present, quantitative seismic risk assessments of individual woodframe houses are seldom performed, at least in part because (i) most homeowners, and indeed many engineers, are not yet aware that such assessments are possible, and (ii) calling on an engineer to conduct the desired analyses is prohibitively expensive for the average homeowner, or else not profitable for the engineer. While qualitative assessments of seismic risk are currently available (e.g., from www.HomeRisk.com for California), there remains a need for a quantitative risk assessment tool that is readily and freely available to the homeowner. Such a tool can be used to evaluate alternative risk-mitigation strategies, whether through structural retrofit, insurance, enhanced building codes, emergency response, or other means. The United States Geological Survey (USGS) is working towards meeting this need, via an extension of the USGS National Seismic Hazard Maps. The ultimate goal is to develop a freely-available interactive web tool, called ResRisk, that automatically computes the seismic risk – e.g., probabilities of mandatory evacuation and of repair cost exceedances – for a homeowner-described woodframe house.

In contrast to other types of structures, the relatively standardized nature of woodframe construction in the U.S.A. makes it possible to quantify, with reasonable rigor, the seismic risk for a house using attributes provided by a homeowner or, more generally, a non-engineer – e.g., house age, square footage, floor plan, number of stories, foundation type, wall finishes, whether the house has been seismically retrofitted, and if the house is on a slope. Furthermore, the same engineering and statistical analyses applied to perform seismic risk assessments for other types of structures (e.g., see the Pacific Earthquake Engineering Research Center Testbeds – http://www.peertestbeds.net/) can be automated for woodframes houses. More specifically, ResRisk combines (i) an earthquake ground shaking hazard curve for the location of the house with (ii) probabilistic fragility and vulnerability models for the house relating ground motion amplitude to likelihood of, respectively, mandatory evacuation or earthquake-damage repair costs, as examples. Item (i) can be obtained from the USGS National Seismic Hazard Maps (most recently Petersen et al., 2008), and item (ii) can be computed "on the fly" via non-linear dynamic structural analysis simulations. The computed seismic risk must then be communicated in terms that are useful to the homeowner, e.g., chances in some time window of incurring earthquake damage leading to mandatory evacuation or to repair costs larger than the deductible on an insurance policy, and how these odds are changed by mitigation measures such as retrofit. ResRisk will also compare these chances to the seismic risk for other houses (retrofitted or not), other geographic locations, and perhaps even other non-seismic risks (e.g., along the lines of Mori et al., 2007).

REFERENCES

FEMA, 2003. Multi-Hazard Loss Estimation Methodology, Earthquake Model, HAZUS-MH. Washington, DC.
Petersen MD, Frankel AD, Harmsen SC, Mueller CS, Haller KM, Wheeler RL, Wesson RL, Zeng Y, Boyd OS, Perkins DM, Luco N, Field EH, Wills CJ, and Rukstales KS, 2008. Documentation for the 2008 Update of the National Seismic Hazard Maps. USGS Open-File Report 2008–1128.
Mori Y, Yamaguchi T, and Idota H, 2007. Optimal strategy for upgrading existing non-conforming wooden houses. Proceedings of the 10th International Conference on Applications of Statistics and Probability in Civil Engineering, Tokyo, Japan.

Safety, Reliability and Risk of Structures, Infrastructures and
Engineering Systems – Furuta, Frangopol & Shinozuka (eds)
© 2010 Taylor & Francis Group, London, ISBN 978-0-415-47557-0

Seismic performance evaluation for groups of historic wooden houses

Y. Hayashi
Graduate School of Engineering, Kyoto University, Kyoto, Japan

S. Higa
Kishirou Architectural Design Office, Kyoto, Japan

Y. Hasebe
Graduate School of Engineering, Kyoto University, Kyoto, Japan

M. Matsuda
WASS, Toyo University, Tokyo, Japan

M. Koshihara
Institute of Industrial Science, The University of Tokyo, Tokyo, Japan

ABSTRACT

In Japan, there is a growing importance in preservation of historic districts and streetscapes of wooden houses like the Important Preservation District for Groups of Historic Buildings. However, most of fatalities come from the collapse of old wooden houses according to the recent earthquake damage statistics. Therefore, the seismic retrofit of old town houses to mitigate earthquake damage is very important. This paper proposes simple methods to evaluate of seismic performance for all the traditional wooden houses in a district. In this paper, historic wooden houses in Narai located in Nagano prefecture in Japan are studied. The houses were originally constructed in the 17th Century on and have various features particular to post town houses. The old town houses have narrow entries, are long in depth. The neighboring houses touch mutually. Therefore, the movement of townhouses is mutually affected. We call this effects house-house interaction (HHI) effects.

First, in this study, to identify the vibration characteristics, ambient vibration tests for almost all the houses, about 230 houses in total, along the old highways are conducted. Three to five houses are measured simultaneously to investigate the HHI effects. In the measurements, ten wireless and GPS-synchronized handheld vibration recorders with three acceleration sensors inside are used. Considering the vibration characteristics of houses and HHI effects, houses are classified into neighborhood groups. Next, rigorous seismic performance evaluation is carried out for twenty-four selected houses individually and for nine grouped houses. Finally, based on the results of ambient vibration measurement and rigorous seismic performance evaluation, we have proposed two types of approximate evaluation methods of yield base shear coefficient for the neighborhood groups. One method is evaluation from natural frequency (Fig. 1(a)), and another is from the rough floor plan of houses (Fig. 1(b)). By applying these two methods,

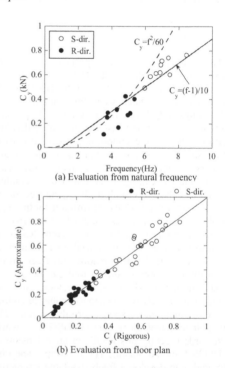

Figure 1. Evaluation of the yield base shear coefficient.

houses with poor seismic performance can be easily specified from all the houses in a district.

In summary, the following conclusions are drawn.

(a) The investigated old town houses show clear and complex HHI phenomenon.
(b) The seismic performance evaluation and seismic retrofitting for such old town houses should be performed considering the HHI effects of a neighborhood group.
(c) We have proposed two types of approximate evaluation methods of yield base shear coefficient for the neighborhood groups.

Safety, Reliability and Risk of Structures, Infrastructures and
Engineering Systems – Furuta, Frangopol & Shinozuka (eds)
© 2010 Taylor & Francis Group, London, ISBN 978-0-415-47557-0

Probabilistic assessment of human casualties due to earthquake-induced building damage in Japan

P. Towashiraporn, T. Lai & J. Guin
AIR Worldwide Corporation, Boston, MA, USA

E. Ikuta
Graduate School of Human Life Science, Osaka City University, Osaka, Japan

K. Fujimura
AIR Worldwide (Japan) LLC, Tokyo, Japan

ABSTRACT

In recent decades, earthquakes have caused a significant number of deaths and injuries worldwide. Unlike many other natural disasters, earthquakes are difficult to predict and could strike in populated areas without much warning, potentially resulting in substantial human losses. Costs related to deaths and medical treatment for injuries from earthquakes sometimes become major burdens to individuals, casualty insurance companies, as well as local governments. Estimates of the potential consequences of future earthquakes on human lives provides a useful tool for risk managers to develop proper action plans and better resource allocation in the event of a major earthquake. This paper presents a probabilistic model for assessing casualty risk as a direct result of building damage from earthquakes in Japan. First, a module for analytically quantifying structural damage from local ground-shaking parameters and structural-vulnerability relationships is developed. A database of buildings and their occupants is constructed based on census data. The dependence of casualty rates on building damage states is then derived from a comprehensive casualty database obtained from the 1995 Hyogoken-Nanbu (Kobe) earthquake. This casualty database includes data from the Hyogo Prefecture coroners' reports, medical records of patients treated at more than 90 hospitals in Kobe and its neighboring cities after the earthquake, as well as information on the damage states of the buildings where each death or injury occurred. Uncertainties surrounding both the building damage and the casualty rates are taken into consideration in this model. Results from the model simulation are validated against actual casualty counts reported following the 1995 earthquake. Finally, the model is used to estimate casualty risk as a result of a hypothetical worst-case-scenario earthquake. The estimated casualty counts from this earthquake warn of the worst that could happen in Japan, while casualty risk maps (Figure A1), which show the spatial

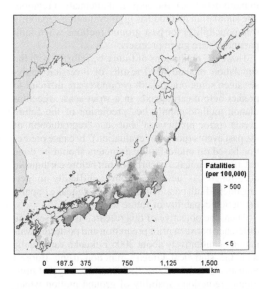

Figure A1. Estimated Fatalities (per 100,000 people) with 0.1% Annual Probability of Exceedance.

distribution of the estimated fatalities with a certain annual exceedance probability, could pave the way for more appropriate mitigation and emergency plans for future earthquakes.

SELECTED REFERENCES

Ikuta, E., Miyano, M., Nagashima, F., Nishimura, A., Tanaka, H., Nakamori, Y., Kajiwara, K., & Kumagai, Y. 2004. Measurement of the Human Body Damage caused by Collapsed Building. *Proceedings of the 13th World Conference on Earthquake Engineering, Vancouver, Canada, August 1–6, 2004.* Paper 628, CD-ROM, Mira Digital Publishing.

National Institute of Building Sciences. 2003. Multi-hazard Loss Estimation Methodology HAZUS-MH MR3. *Technical Manual.*

Safety, Reliability and Risk of Structures, Infrastructures and Engineering Systems – Furuta, Frangopol & Shinozuka (eds)
© *2010 Taylor & Francis Group, London, ISBN 978-0-415-47557-0*

Minimum prediction capability of ground motion

M. Ohbuchi & T. Takada
University of Tokyo, Tokyo, Japan

ABSTRACT

In recent years, seismic design and assessment of civil engineering structures based on simulated strong ground motion are prevailing in Japan. If macro- and micro-fault parameters are given, an Irikura's method (Irikura 1986) can simulate ground motion in more detail than the empirical attenuation relation. The simulation method has high evaluation capability (reproducibility) for past ground motions when fault parameters are given precisely.

However, predictions of future waveforms with the simulation method need results of inversion analyses since some micro-fault parameter are difficult to predict before earthquake. In a strict sense, the simulation method cannot be "prediction of the future events (prior prediction)" but just "reproduction of the past events (posterior evaluation)" because prediction based on micro-fault parameters should be done not after earthquake occurrence but before earthquake occurrence. As described above, difficulty in getting micro-fault parameters before earthquake impairs prediction capability of future waveforms.

Then, the objective of this research is to assess the difference between prior prediction and posterior evaluation quantitatively about 2005 Fukuoka earthquake with the statistical Green's function using Boore's Statistical Model (Boore, 1983), and to suggest minimum prediction capability of ground motion which is required in engineering application. Here, "Prediction" means simulation method undertaken without any information acquired only after earthquake occurrence. Meanwhile, "Evaluation" means simulation method with both prior and posterior information.

Figure 1 compares the result of RMS. The index is root mean square error of between simulation and observed data (true value) defined by Equation (1). In the figure, the diagonal line means that prediction and evaluation are completely same, an area under the line means evaluation is better than prediction. As points in the figure are plotted around the diagonal line, the difference between prediction and evaluation is considerably small.

$$RMS = \sqrt{\frac{1}{n}\sum_{k=1}^{n}\left(\log X - \log x_{ob}\right)^2}$$

$$= \sqrt{E\left[\varepsilon^2\left(\mu_X,\sigma_X\right)\right]} = \sqrt{\mu_X^2 + \sigma_X^2} \qquad (1)$$

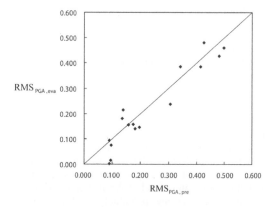

Figure 1. Several examples of asperity.

The past estimation of strong ground motion is improved by consideration of detailed information such as micro-fault-parameters. The detailed information of fault is given only after earthquake occurrence so that the past method isn't priori prediction. However, the result of this study gives suggestion of possibility that the simulation without detailed information can be appropriate to engineering application.

REFERENCES

Boore, D.M. 1983. Stochastic simulation of high-frequency ground motions based on seismological models of the radiated spectra; *Bulletin of the Seismilogical Society of America*, 73: 1865–1894.
Irikura, K. 1986. Prediction of strong acceleration motions using empirical Green's function. *Proc. 7th Japan Earthquake Engineering symposium*: 127–132 (in Japanese)

*Safety, Reliability and Risk of Structures, Infrastructures and
Engineering Systems – Furuta, Frangopol & Shinozuka (eds)
© 2010 Taylor & Francis Group, London, ISBN 978-0-415-47557-0*

Earthquake countermeasures for wood houses to mitigate seismic loss and total cost

Y. Hasebe
Graduate School of Engineering, Kyoto University, Kyoto, Japan

A. Saratani
Daiwa House Industry, Nara, Japan

T. Morii & Y. Hayashi
Graduate School of Engineering, Kyoto University, Kyoto, Japan

ABSTRACT

It is said that plate subduction-zone earthquakes and
inland earthquakes may occur in several decades in
Japan. However, many residents do not try to perform seismic retrofit, even if their wooden houses have
poor seismic performance and will suffer extensive
damage if a severe earthquake occurs. In order to motivate residents to implement an effective earthquake
countermeasure for their houses, this paper proposes
evaluation indexes to select the most cost effective
earthquake countermeasure for their own houses and
to make up the long-term implementation planning.
In our study, maintenance, repair and rebuilding are
included in earthquake countermeasures as well as
seismic retrofitting.

First, we present a seismic risk evaluation procedure for wooden houses, which can consider aged
deterioration affected by biodeterioration and maintenance status as well as their seismic performance
before and/or after the seismic retrofitting and local
seismic hazard. Then, this paper proposes two indexes
to evaluate and assess the earthquake countermeasures. The one is a Loss Reduction Index (LRI), and
this index means an expected reduction in seismic
loss by the earthquake countermeasure. The other is
a Total Cost Reduction Index (TCRI), which is an
expected reduction in the total cost of initial cost, the
cost of earthquake countermeasures and seismic loss
over elapsed years from specified time.

In order to demonstrate the validity of the proposed indexes and discuss how to prepare for the
coming big earthquakes, we have performed some
case studies changing location, earthquake countermeasure techniques and that timing. The locations of
houses are Osaka city, Kyoto city, Nagoya city, Tsu city
and Kochi city in Japan to compare influence of the
local seismic hazard shown in Figure 1. The feature
of houses is changed by seismic performance, age and
durable years. Earthquake countermeasure techniques
are seismic retrofitting to strengthening or improving

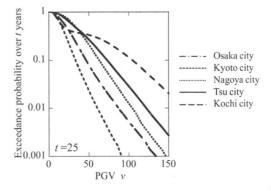

Figure 1.   Seismic hazard curves over t years.

deformation performance, durability enhancement to
increase durable years, and rebuilding.

Finally, by the sensitive analysis, it confirms that
these proposal indexes can judge cost-effective timing
and technique of earthquake countermeasure depending on location.

REFERENCES

Hayashi, Y. 2007. Seismic risk evaluation of wood house
considering aged deterioration and long-term occurrence
probability of earthquakes. Saratani, A & Morii, T (eds),
2nd international conference on urban disaster reduction:
27–29.
Engineering Affairs Division of Construction Minister's Sec-
retariat/Japan Institute of Construction Engineering1986.
Durability enhancement technology of wood house,
Gihodo shuppan, Japan (in Japanese).
Saratani, A. 2006. Fragility function with peak ground
velocity of wood houses considering aged deterioration.
Morii, T & Hayashi, Y (eds), *Proc. Japan Earthquake
Engineering Symposium*, Japan, No.0011. (in Japanese).
National Research Institute for Earth Science and Disaster
Prevention. 2006. Seismic Hazard Map. http://www.j-
shis.bosai.go.jp/doc/SeismicHazardMap.pdf (in Japanese).

Organized Session (OS05) Vulnerability and Robustness of Structures

*Safety, Reliability and Risk of Structures, Infrastructures and
Engineering Systems – Furuta, Frangopol & Shinozuka (eds)
© 2010 Taylor & Francis Group, London, ISBN 978-0-415-47557-0*

Risk-informed approaches for achieving robustness and for mitigating disproportionate collapse of buildings

Bruce R. Ellingwood

School of Civil and Environmental Engineering, Georgia Institute of Technology, Atlanta, GA, USA

ABSTRACT

A progressive (or *disproportionate*) collapse of a structure is one that initiates from *local* damage and, rather than being arrested in the structure by its capability to bridge around the damaged area, propagates to a final damage state that involves a major portion of the structure. Although there is a natural tendency due to recent events to associate such collapses with malevolent attack, other events that are outside the scope of customary structural engineering practice – gas explosions, vehicular collisions, severe fires, and extreme values of environmental loads, and design/construction errors – also can stress the building system to well beyond its normal design envelope. All buildings are susceptible to progressive collapse in varying degrees. A *robust* structural system is inherently capable of absorbing minor damage resulting from design or construction errors or from extreme environmental or man-made events. Although engineers generally recognize the need to engineer robust structural systems, specific design strategies and criteria for progressive collapse resistance have proved difficult to develop and implement in professional practice. At the same time, the public has become increasingly aware of building safety issues as a result of well-publicized collapses due to natural and man-made disasters.

In recent years, changes in building and construction practices and sociopolitical challenges have highlighted the importance of hazards that historically either have not been viewed as significant (explosions or detonations) or have been dealt with through deemed-to-satisfy clauses rather than through formal structural calculations (severe fires). At the same time, the public has become increasingly aware of building safety issues as a result of well-publicized collapses due to natural and man-made disasters. Proposals for achieving robustness and for minimizing the likelihood of progressive collapse now are being introduced into building codes, standards and regulations.

The level of risk arising from the uncertain nature of low-probability hazards and their consequences remains undetermined in recent code proposals aimed at minimizing the likelihood of progressive collapse, making problematic their cost-effectiveness. Concepts of structural reliability analysis and probabilistic risk assessment supported by advanced computational simulation tools can be used to identify specific threat scenarios, to assess the capability of a building to withstand damage without the development of a general structural collapse, and to test the effectiveness of various proposed risk mitigation strategies. This paper summarizes methods for assessing structural robustness and the risk of disproportionate collapse found in national codes, standards and regulatory documents, and concludes with an identification of analytical and experimental research needed to advance risk-informed strategies for mitigating progressive collapse of buildings.

REFERENCES

Ellingwood, B.R. and D.O. Dusenberry (2005). "Building design for abnormal loads and progressive collapse," *Computer-aided Civil and Infrastruct. Engrg.* 20(5): 194–205.

Ellingwood, B.R. (2006). "Mitigating risk from abnormal loads and progressive collapse," J. *Perf. of Constructed Facilities, ASCE* 20(11):315–323.

NIST (2007). Best practices for reducing the potential for progressive collapse in buildings, *NISTIR 7396*, National Institute of Standards and Technology, Gaithersburg, MD (194 pp).

Safety, Reliability and Risk of Structures, Infrastructures and
Engineering Systems – Furuta, Frangopol & Shinozuka (eds)
© 2010 Taylor & Francis Group, London, ISBN 978-0-415-47557-0

Robustness of structures as a design criterion

M. Beer
Department of Civil Engineering, National University of Singapore, Singapore

M. Liebscher
DYNAmore GmbH, Stuttgart-Vaihingen, Germany

ABSTRACT

In this paper a new numerical approach is proposed
for designing robust structures. This can be coupled
with any arbitrary nonlinear computational model for
structural analysis.

The proposed approach is focused on incorporating the issue of structural robustness in the design
process, which is of increasing interest in industry.
The engineer has to find a design which ensures a
proper behavior of the structure despite uncertainty
of structural, environmental and design parameters.
As no general definition exists for structural robustness, a respective measure is developed based on
the understanding of robustness according to (The
American Supplier Institute 2005), which was introduced by GENICHI TAGUCHI and is well-established
in consumer goods industry. A structure is considered as robust if it can withstand a whole spectrum of
occasional or frequent fluctuations of environmental
conditions without noticeable effects on its serviceability. This concerns the system performance under
normally fluctuating conditions – in contrast to the
occasional understanding of robustness as appropriate
performance of a system under exceptional conditions. Moreover, the robustness measure is defined in
conjunction with a non-stochastic uncertainty model
according to fuzzy set theory. This enables the consideration of merely vaguely predictable fluctuations
of structural parameters, which is particularly suitable in early design stages when clear probabilistic
specifications are problematic.

On this basis the numerical approach for designing robust structures is formulated. Design parameters
are specified as uncertain over a value range reasonable for design. Further uncertain parameters are
quantified according to the available information. All
uncertainty is processed through a structural analysis by simulation, such as a Monte Carlo simulation.
This leads to a set of points in the space of the
design parameters and associated structural responses.
By means of design constraints permissible design
parameter combinations are identified and lumped
together via cluster analysis. From the clusters continuous alternative design domains are constructed. These
are modeled using fuzzy sets and processed through
the uncertain structural analysis, again, to verify their
permissibility. For each design variant, the robustness
is computed as the ratio between the uncertainty of
the fuzzy input quantities and the uncertainty of the
associated fuzzy structural responses.

The optimum design variant is determined with a
three-criteria optimization problem. Criterion one is
the traditional optimization criterion, such as a minimum mass or minimum cost. Criterion two is the
maximum robustness of the structure. This aims at a
structural behavior that is only marginally affected by
uncertainty and by changes in the design parameters
during the production process and service life of the
structure. Criterion three concerns a maximum size of
the suitable domains in the space of design parameters. This provides the engineer with comfortable
decision margins during construction or production,
respectively.

The proposed design method is demonstrated by
way of an example.

REFERENCES

The American Supplier Institute 2005. *Robust design*. Technical Report, The American Supplier Institute.

Safety, Reliability and Risk of Structures, Infrastructures and Engineering Systems – Furuta, Frangopol & Shinozuka (eds)
© 2010 Taylor & Francis Group, London, ISBN 978-0-415-47557-0

Performance of earthquake resistant buildings against progressive collapse

L. Lin & N. Naumoski
University of Ottawa, Ottawa, Ontario, Canada

S. Foo
Public Works and Government Services, Gatineau, Quebec, Canada

M. Saatcioglu
University of Ottawa, Ottawa, Ontario, Canada

ABSTRACT

Progressive collapse analyses were conducted on reinforced concrete frame buildings designed according to the seismic provisions of the National Building Code of Canada. The purpose of the analyses was to estimate the vulnerability of the buildings to progressive collapse. Three moderately ductile buildings designed for Ottawa, which is in a moderate seismic zone, and three ductile buildings designed for Vancouver, which is a high seismic zone, were used in the analyses. The buildings considered are 5-storeys, 10-storeys and 15-storeys high. The plans of both the Ottawa and the Vancouver buildings have the same dimensions, i.e., 30 m × 18 m as shown in Figure 1. The spans in the transverse direction of the Ottawa and the Vancouver buildings are also the same, i.e., three spans of 6 m for both buildings. However, the number and the dimensions of the spans in the longitudinal directions of the buildings are different. The Ottawa buildings have four spans of 7.5 m each, and the Vancouver buildings have five spans of 6 m each. The floor heights are 4.0 m for all buildings.

The performance of the buildings against progressive collapse is evaluated following the widely used guidelines for the design and assessment of buildings against progressive collapse, namely "Progressive Collapse Analysis and Design Guidelines" by the U.S. General Services Administration. Different scenarios were considered by removing first storey columns of the buildings in order to simulate the loss of columns due to bomb blasts. The following scenarios of column removals were considered: (i) column on the perimeter, approximately at the middle of the long side of the building, (ii) column on the perimeter, approximately at the middle of the short side of the building, (iii) column at the corner of the building, and (iv) interior column. Elastic static analysis was conducted for each of these scenarios using three-dimensional models. Demand/capacity ratios obtained for primary structural members were used in the assessment of the vulnerability to progressive collapse.

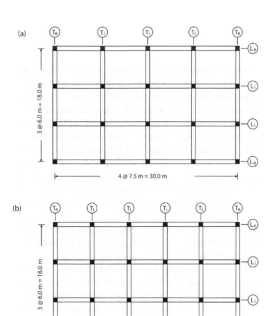

Figure 1. Floor plans of buildings; (a) buildings in Ottawa, and (b) buildings in Vancouver.

The study leads to two general observations: buildings in Vancouver are more robust than buildings in Ottawa in terms of progressive collapse resistance, and that taller buildings are more at risk to progressive collapse than shorter buildings. Another interesting finding is that the vulnerability of seismically designed buildings to progressive collapse depends greatly on the differences between the spans of the longitudinal and the transverse frames, i.e., larger differences between the spans lead to higher vulnerability.

Safety, Reliability and Risk of Structures, Infrastructures and
Engineering Systems – Furuta, Frangopol & Shinozuka (eds)
© 2010 Taylor & Francis Group, London, ISBN 978-0-415-47557-0

A novel optimization method under limited uncertainty

S. Bhattacharjya & S. Chakraborty
Bengal Engineering and Science University, Shibpur, India

A. Haldar
University of Arizona, Tucson, AZ, USA

A. Reyes-Salazar
Facultad de Ingeniería, Universidad Autónoma de Sinaloa, Culiacán, Sinaloa, México

ABSTRACT

A novel robust design optimization (RDO) formulation is proposed here which is specifically appropriate when complete probabilistic information about the design variables (DVs) and design parameters (DPs) are not available. The currently available RDO methods put equal importance to each individual gradient of the objective function and the constraints. But, it is well known that all the gradients are not of equal importance. This has been successfully explored in reducing the number of random variables in reliability analysis. The concept is utilized here in RDO. It is expected to be more appropriate and efficient to use the weighted value of the gradients in accordance with their importance instead of equal weight to define the robustness measure of the objective function.

The methodology proposed here is a heuristic approach which allows the use of importance factor obtained from the respective sensitivity information. To improve the robustness, a new measure of robustness is proposed by redefining the performance dispersion using associated importance of the gradients of the performance function with respect to uncertain variables. The basic idea to improve the robustness of the performance using new index is also extended to the constraints. Essentially, the proposed RDO formulation becomes two-criterion equivalent deterministic optimization problem where the weighted sum of the mean and weight factor-based dispersion index is optimized.

The RDO formulation is elucidated by considering the cost optimization of a steel tubular column hinged at both ends subjected to a compressive load. The uncertainty information available on the DVs and DPs is incomplete, i.e., only the ranges of variations are prescribed but not their distributions. For comparison, the results obtained by the conventional two criteria RDO procedure are also obtained. The Pareto fronts as obtained by the proposed and conventional RDO methods are shown in Figure 1. It can be clearly

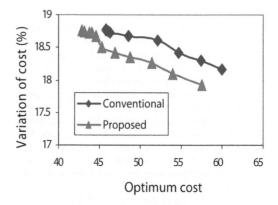

Figure 1. Comparison of Pareto-front as obtained by proposed and conventional RDO method.

observed from the figure that for a specific level of variation if the desired objective is to minimize the cost of the column, the proposed method provides lesser cost compared to the conventional RDO approach. Furthermore, for a prescribed cost of the column, the dispersion of the cost due to uncertainty as obtained by the present approach is less than that of obtained by the conventional RDO approach.

Further results are provided in the full paper. The results of numerical study show that the trend and variations of the proposed RDO results are in conformity with the well established two criteria RDO problems indicating its potential over the conventional approach. Since all the variables are not equally important in capturing the presence of uncertainty, the proposed approach yields more efficient Pareto front. However, how much robustness can be achieved over the conventional method is problem dependent. It may be noted that the approach being generic in nature, can be applied to more complex structures and extended to RDO of general multi objective optimization problems.

Safety, Reliability and Risk of Structures, Infrastructures and
Engineering Systems – Furuta, Frangopol & Shinozuka (eds)
© 2010 Taylor & Francis Group, London, ISBN 978-0-415-47557-0

Robustness of structures, EU COST action TU0601

T. Vrouwenvelder
TU-Delft, The Netherlands

J.D. Sørensen
Aalborg University, Denmark

ABSTRACT

Robustness of structures has been recognized as a desirable property because of a several high system failures, such as the Ronan Point Building in 1968, where the consequences were deemed unacceptable relative to the initiating damage. After the collapse of the World Trade Centre, the robustness has obtained a renewed interest, primarily because of the serious consequences related to failure of the advanced types of structures. In order to minimize the likelihood of such failures many modern building codes consider the need for robustness in structures and provide strategies and methods to obtain robustness. In fact, in all modern building codes, one can find a statement (in this or a slightly different form): "total damage resulting from an action should not be disproportional to the initial damage caused by this action".

During the last decades there have been significant efforts to quantify aspects of robustness. When modelling robustness, system effects are very important. However, the primary criteria in building code are related to design and verification of sufficient reliability of components. It should also be noted that redundancy in systems is closely related to robustness. In principle redundant system are believed to be more robust than non-redundant systems – but this is not always the case as illustrated by the failures of the Ballerup Super Arena and the Bad Reichenhall icehall.

Given this interest in the topic a special EU sponsored COST action TU0601 (http://www.cost-tu0601.ethz.ch/) has been initiated by the Joint Committee on Structural Safety (JCSS) in cooperation with IABSE. Within this European research project close cooperation is maintained with experts from North America, in particular NIST, where a significant research into robustness of structures has taken place for understandable reasons.

The main objective of the COST Action TU0601 is to provide the basic framework, methods and strategies necessary to ensure that the level of robustness of structural systems is adequate and sufficient in relation

to their function and exposure over their life time and in balance with societal preferences in regard to safety of personnel and safeguarding of environment and economy. The purpose of the project is to assess measures and requirements for robustness in a risk based perspective.

Traditionally robustness is considered to be an inherent property of a structure defined through the design of the structure. However, a risk based approach considers robustness from a life cycle perspective to be a product of design together with strategies for operation, maintenance and control as well as appropriate emergency response measures. Assessment of robustness necessitates the consideration of the possible scenarios leading to collapse, their probability of occurrence as well as the corresponding consequences.

Whereas structural design codes at the present time are focusing on ensuring structures with appropriate safety in regard to component failures and thereby are implicitly addressing mainly vulnerability aspects of the structural performance, the aim of the present COST Action is to establish a better understanding on the aspects related to the robustness, i.e. focusing on how structures may be designed, operated and maintained such that potential damages are sustained with an appropriately high level of safety.

Given this main objective the project focuses on a number of main issues:

- Theoretical Framework
- Models for Exposures, Structural behaviour and Consequences of failures.
- Design to improve robustness
- Examples, Dissemination of knowledge and guidelines

The end users of the results of the project will comprise pre-codification and codification committees on structural design as well as actors in the building and construction industry including design, architectural and consulting companies as well as contractors and owners of structures at both a public and private level.

Safety, Reliability and Risk of Structures, Infrastructures and
Engineering Systems – Furuta, Frangopol & Shinozuka (eds)
© 2010 Taylor & Francis Group, London, ISBN 978-0-415-47557-0

Probabilistic failure model for progressive collapse analysis

S.M. Marjanishvili & L. Leininger
Hinman Consulting Engineers, Inc., San Francisco, CA, USA

ABSTRACT

This paper proposes a methodology to describe various
structural systems based on their level of robustness to
resist a disproportionate progressive collapse following an initial structural perturbation. Procedures are
applied for estimating the probability of failure and
robustness level with regards to the progressive collapse potential of steel structures. Uncertainties are
treated in a probabilistic framework with probability distributions assigned to the parameters. Structural
resistance and load demands are treated as variables
with probability distribution functions. The results
are summarized in the form of robustness diagram
derived from progressive collapse diagrams developed in previous study (Marjanishvili & Buscemi,
2005). Progressive collapse diagrams, which relate
expected ductility ratio (**m**) and Demand Capacity
Ratio (*DCR*) were originally derived for deterministic progressive collapse assessments. In this study,
progressive collapse response charts are extended to
include estimated variability for nonlinear systems as
determined through computer simulations employing
fast-running algorithms. These algorithms depend on
the resisting mechanisms of structural systems, such
as flexure, shear and stability loss or catenary action.
The approach adopted in this study is based on the
principle that the structural robustness is the quality of structure, which is evaluated not only by the
mean value (i.e., determination of progressive collapse
potential) but also by the sensitivity of the structural
performance (i.e. structures ability to gracefully withstand unforeseen and extreme loading and damage
conditions.

REFERENCES

Agarwal, J., England, J., August 2008, "Recent developments
in robustness and relation with risk", Proceedings of the
Institution of Civil Engineers, Vol. 161, Issue SB4.
Alimoradi, A., Pezeshk, S., Foley, C.M., June 2007, "Prob-
abilistic Performance-Based Optimal Design of Steel
Moment-Resisting Frames. II- Applications", Journal of
Structural Engineering, ASCE, pp. 767

Ang, A.H.S., Tang W.H., 2007, "Probability Concepts in
Engineering –Emphasis on Applications to Civil and
Environmental Engineering", John Wiley & Sons, Inc.
Ang, A.H.S., Tang W.H., 1975, "Probability Concepts in Engi-
neering Planning and Design-Vol.2, Decision, Risk and
Reliability", John Wiley & Sons, Inc.
Baker, J.W., Schubert, M., Faber M, 2008, "On the
Assessment of Robustness", Structural Safety, Vol. 30,
pp 253–267.
Benjamin, J.R, Cornell, C.A., 1970, "Probability, Statistics
and Decision for Civil Engineers" McGraw-Hill, Inc.,
New York.
Clough, R.W., and Penzien, J., 1975, Dynamics of Structures,
McGraw-Hill, New York, (2nd edition 1993)
Department of Defense (DoD), 2005, Unified facilities cri-
teria (UFC), design of buildings to resist progressive
collapse UFC 4-023-03, DoD.
Ellingwood, B.R., 2006, "Load and Resistance Factor Criteria
for Progressive Collapse Design", REC 06 Wksp.
General Services Administration (GSA), 2003, Progressive
collapse analysis and design guidelines for new federal
office buildings and major modernization projects. GSA.
Kang, Z., 2005, Robust Design Optimization of Structures
under Uncertainties, Dr. –Ing. Thesis, Intitut fur Statik
und Dynamic der Luft- und Raumfahrkonstruktionen,
Universtat Stuttgart.
Marjanishvili, S., 2004, " Progressive Procedures for Progres-
sive Collapse", Journal of Performance of Constructed
Facilities, Vol. 18, No. 2
Marjanishvili, S., Buscemi, N., 2005, "SDOF Model for Pro-
gressive Collapse Analysis", Structures Congress 2005,
ASCE, SEI
Marjanishvili, S., Agnew, E., 2006, "Comparison of Various
Procedures for Progressive Collapse Analysis", Journal of
Performance of Constructed Facilities, Vol. 20, No. 4
Pretlove, A.J., Ramsen, M, Atkins, A.G., 1991, "Dynamic
effects in progressive failure of structures" Int. J. Impact
Eng., 11(4), 539–546
Rjanitsin, A.R., 1978. Theory of reliability of structural
design. Stroiisdat, Moscow. (in Russian)
Schmidt, B.J., Bartlett, F.M., 2002, "Review of Resistance
Factor for Steel: Data Collection", Canadian Journal of
Civil Engineering, Vol. 29, pp 98–108
TSWG, 23 March 2004, "Vehicle Borne Improvised Explo-
sive Devices in Worldwide Terrorism, A Contemporary
Open Source Analysis".

Mini-Symposia (MS02) NDT Reliability and Its Use for Maintenance Applications

Safety, Reliability and Risk of Structures, Infrastructures and Engineering Systems – Furuta, Frangopol & Shinozuka (eds)
© 2010 Taylor & Francis Group, London, ISBN 978-0-415-47557-0

Uncertainty analysis of the local flexibility method for damage identification

Edwin Reynders & Guido De Roeck

Department of Civil Engineering, K.U. Leuven, Leuven, Belgium

ABSTRACT

The Local Flexibility method [Reynders & De Roeck 2008] is a technique which can be used for static as well as dynamic (i.e., vibration based) damage identification. The method uses measured receptance or flexibility. Combination of (the quasi-static limit of) flexibility with a virtual load that causes stresses in the vicinity of one measurement location only yields a virtual static deformation, that is inversely proportional to the local stiffness. The ratio of static deformation, obtained before and after occurrence of a change in stiffness, is equal to the inverse local stiffness ratio. The method can be adapted so that, in the dynamic case, unscaled mode shapes and therefore output-only methods can be used to construct the flexibility matrix, assuming that the mass change is negligible.

Here, an uncertainty analysis of the method is presented. Starting from the covariance matrix of the measured data (in the dynamic case: the covariance matrix of the measured eigenfrequencies, damping ratios, and mode shapes), the variances of the local stiffness ratios, estimated from the method, are obtained using a sensitivity analysis. Since the relative uncertainty on the measured modal parameters is generally low, a first-order approximation is sufficient for most applications. Under the assumption that the measured modal parameters obey a multivariate normal distribution (which is, for most modal parameter estimation algorithms, asymptotically valid), it follows from the first-order perturbation analysis that the estimated local stiffness ratios are normally distributed as well, and consequently the procedure followed enables to estimate confidence intervals on them.

As a simulation example, a simply supported IPE100 beam of 3 m length, where the vertical displacements are measured at 7 equidistant points, is considered. Only the first unscaled mode, that would result from an output-only modal analysis, is retained for the construction of the flexibility matrix. The damage scenario consists of a stiffness reduction of 20% between measurement points 2 and 4.

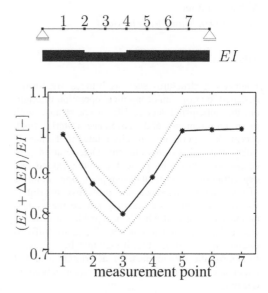

The relative bending stiffness $(EI + \Delta EI)/EI$, together with its 95% confidence interval, calculated using the method presented here, is shown below. The standard deviation of the eigenfrequency is assumed to be 1%, that of the damping ratio 10%, and that of each of the mode shape components 1%. These values are realistic in practical situations [Reynders et al 2008].

REFERENCES

E. Reynders and G. De Roeck. A local flexibility method for vibration-based damage identification in hyperstatic beam structures. In *Proceedings of the 7th European Conference on Structural Dynamics: Eurodyn 2008*, Southampton, UK, July 2008.

E. Reynders, R. Pintelon, and G. De Roeck. Uncertainty bounds on modal parameters obtained from Stochastic Subspace Identification. *Mechanical Systems and Signal Processing*, 22(4):948–969, 2008.

Safety, Reliability and Risk of Structures, Infrastructures and
Engineering Systems – Furuta, Frangopol & Shinozuka (eds)
© 2010 Taylor & Francis Group, London, ISBN 978-0-415-47557-0

Determination of POD curves for T-joints

C. Cremona
Commissariat Général au Développement Durable, Paris, France

M. Grasset
Laboratoire Central des Ponts et Chaussées, Nantes, France

R. Leconte
CETE Lyon, Bron, France

A. Nussbaumer
EPFL, ICOM, Lausanne, Switzerland

ABSTRACT

Engineering structures such as bridges are ideally
designed to ensure an economical operation throughout the anticipated service life in compliance with
given requirements and acceptance criteria. Such
acceptance criteria are typically related to the people
safety and risk to environment. Deterioration processes will always be present to some extend and
depending on the adapted design philosophy in terms
of degradation allowance and protective measures,
the deterioration processes may reduce the system
performance beyond what is acceptable. In order to
ensure that the given acceptance criteria are fulfilled
throughout the service life, it may thus be necessary
to control the development of deterioration and if
required execute protecting and/or repair actions in
due time.

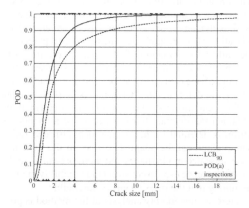

Figure 1. Example of POD (probability of detection) and
LCB (lower confidence bound) curves for the UT technique
(team 1) and hit/miss results (cross symbols).

To ensure safety by inspection implies the identification of what to inspect, how to inspect, where to
inspect and how often to inspect. Within this context, the reliability of a Nondestructive Evaluation
(NDE) technique is particularly important; for welded
joints damaged by fatigue, it is defined as the quantitative measure of its effectiveness in detecting flaws
of specific type and size. NDE procedure capability is characterized in terms of the probability of
detection as a function of the crack size. This function is the probability that a specific flaw is detected
in particular inspection conditions. Within the french
national research project MIKTI, several research
institutes has been involved in various research tasks
for assessing usual and enhanced reliability of several
techniques (ultrasonic, magnetic particles). This paper
will present the principal results obtained by several
NDT techniques on more than 40 joints, the analysis
of the detection and false alarms and the determination
of *POD* curves for the various techniques.

Most of the inspection results are recorded only in
terms of whether or not a flaw was found. Data of
this nature are called hit/miss data, and an analysis
method for this data type evolved from the original

binomial characterization. This is commonly used,
for example, for data from penetrant liquid inspection, visual inspection tests but also ultrasonic or
magnetic particles testing. The basic idea behind this
method is to estimate de probability on detection $P\hat{O}D$
for any given crack size from the hit/miss data by
applying regression analysis or a maximum likelihood
procedure.

The analysis of hit/miss data from the MIKTI
research project allows qualifying the reliability of
magnetic particles and ultrasonic tests through the concepts of probability of detection and probability of
false alarm. The model used is a log-odds function and
the estimation of its parameters has been performed by
the maximum likelihood function.

From the study of the different *POD* curves it comes
that the UT technique, for the two teams concerned by
the control, provides the best results compared to the
MT technique. The obtained information is nevertheless very valuable for crack detection in steel bridges
and represents the first attempt to provide crack detection probabilities for typical T-joint tested to fatigue in
parallel.

Safety, Reliability and Risk of Structures, Infrastructures and
Engineering Systems – Furuta, Frangopol & Shinozuka (eds)
© 2010 Taylor & Francis Group, London, ISBN 978-0-415-47557-0

Inspection intervals for bridge welded joints: Guidelines for the Eurocode 3 applications

J. Duchereau
Ecole des Mines de Paris, Fontainebleau, France

C. Cremona
Commissariat Général au Développement Durable, Paris, France

ABSTRACT

Engineering structures such as bridges are ideally designed to ensure an economical operation throughout the anticipated service life in compliance with given requirements and acceptance criteria. Such acceptance criteria are typically related to the people safety and risk to environment. Deterioration processes will always be present to some extend and depending on the adapted design philosophy in terms of degradation allowance and protective measures, the deterioration processes may reduce the system performance beyond what is acceptable. In order to ensure that the given acceptance criteria are fulfilled throughout the service life, it may thus be necessary to control the development of deterioration and if required execute protecting and/or repair actions in due time. Such considerations have led to shift the design philosophy for new structures (for instance as ships, offshore structures, aircrafts… and in civil engineering) from safe-life or fail-safe to *damage tolerance safety by inspection*. A structure is said damage tolerant if it has a reasonable damage growth life, such that the damage can be detected during one of the affordable scheduled inspection before it can precipitate a failure. To ensure safety by inspection implies the identification of what to inspect, how to inspect, where to inspect and how often to inspect. Even though inspections may be used as an effective means for controlling degradations and thus imply a potential benefit they may also have considerable impact on the operation of the structure and other economical consequences themselves. Eurocode 3 states that it is possible to move from a fail-safe to a damage tolerance approach. For this purpose, reduced partial safety factors (from 1.35 to 1.00) are proposed to bridge designers and owners but nothing is said about the quality of the inspection and the number of inspection times for applying such an approach. Two different analyses are presented in this paper to provide target objectives for inspection quality and intervals with regards to Eurocode 3. The first one consists in performing a inspection/repair (IR) programme such that the reliability index for the damage-tolerant curve is larger than the fail-safe reliability index at 100 years. The IR programme implies that repair is systematically performed as soon as a crack is detected. Consequently, it is intuitive to say that smaller will

be the detection threshold (or the NDE technique efficiency) larger will be the number of times that the joint will be repaired.

For the second approach, a welded joint is inspected by a NDE technique. If no crack is detected, it does not mean that no crack is present but that the inspection technique is not able to provide a positive detection outcome. This is eventually due to the fact that no crack exists but also to the possibility that the existing crack is slightly smaller than the smallest detectable crack size by the technique. The conservative assumption is therefore to assume that a crack exists and the flaw size is the detection threshold. With this hypothesis it is then possible to calculate the crack growth until a critical crack size is reached. As the analysis is probabilistic this calculation has to be turned out into the computation of the time interval between the detection time and the time for which the 3.8 target reliability index is reached. This analysis has been per-formed on several composite bridges. The two approaches provide different results. This is not surprising since the first approach is base on an inspection/repair programme while the second one is relaying on the minimum time interval between the detectable crack size and the allowable reliability level. This latter approach is more conservative. If an extensive 20 yrs inspection programme is scheduled (in addition to the detailed and routine inspections programmes), then it comes that the NDE reliability can only be achieved with a mean detection level (for a 30% coefficient of variation) smaller or equal than 0.3 mm. The first analysis provides wider inspection intervals for the same detection levels (around 30 years). Using less sensitive inspection techniques implies to reduce the inspection interval between 5 to 15 years (10% fractile). The first approach provides similar results (around 20 years). Inspections are performed more intensively but they are compatible with usual detailed inspection programmes.

In conclusion, this paper is making a first attempt to provide sense to the Eurocode 3 proposal to use reduced fatigue partial factors by transferring the decrease in design safety to an increase in inspection safety. If the inspection/repair programme leads to a 20 years average inspection intervals with 0.3–0.6 mm average detection level, the time-to allowable reliability level insists on narrow intervals (10 years).

Safety, Reliability and Risk of Structures, Infrastructures and Engineering Systems – Furuta, Frangopol & Shinozuka (eds)
© *2010 Taylor & Francis Group, London, ISBN 978-0-415-47557-0*

Structural reassessment of prestressed concrete beams by Bayesian updating

C. Cremona
Commissariat Général au Développement Durable, Paris, France

G. Byrne
Ecole Polytechnique, Palaiseau, France

C. Marcotte
CETE Nord-Picardie, Lille, France

ABSTRACT

The VIPP are simple span viaducts made of precast concrete girders prestressed by post-tension. This type of construction was largely developed after the second world war, between the years 1955 to 1970, thanks to its girders launching system which allowed the crossing of non classical obstacles with reasonable height (10 to 25 m above ground-level, range between 30 to 60 m). This technique is not longer used, largely competed by other construction techniques making it possible to carry out more economic and safer redundant structures with respect to rupture. The enthusiasm which reigned at the construction time of the first VIPP generation, moreover resulted in many design and execution mistakes, and by the absence of corrective maintenance actions to restore the defective waterproofing of these bridges.

The detailed description of the problems encountered in this family of bridges shows how much the diagnosis of these structures is essential to conclude a relevant revaluation of their performance. It is in this context that several studies were performed by the authors' organisms in order to develop diagnosis techniques and methods. As questions concerning life-cycle and serviceability of these existing structures appeared, a probabilistic approach is introduced in this paper; its capacity to take into account design, construction, material properties, and on site investigations uncertainties is in particular highlighted. Those are decisive points to re-evaluate the performance of such existing structures.

Inspection information can provide valuable information concerning the structural performance; it is therefore essential to take them into account in an efficient manner. Bayesian updating provides a general framework which can be used in the analysis of the structural performance for VIPP prestressed girders. This paper presents such an application on two real VIPPs, the Merlebach (Fig.1) and the Vauban bridges. The reassessment analysis was performed on one of the five bridge girders, based on prestress measurements by the crossbow method when the bridge was still operated. This measurement technique enables to analyse the cross-section reliability in terms of stresses rather than strains. It provides immediate access to the applied forces and moments located at the instrumented cross-section. This information concerning the prestress force can be used for updating the reliability of different cross-sections at the Serviceability Limit State (S.L.S.). This paper presents such applications showing in particular the difference between two VIPP designs. Reliability index and partial factors updating (Fig.2) are the major scope of this paper showing how the structural performance can be efficiently appraised by Bayesian updating.

Figure 1. Merlebach bridge.

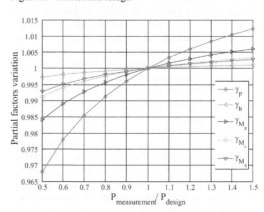

Figure 2. Partial factor variations based on prestress measurement average value (CoV = 10%).

Organized Session (OS24) Mechanics of Random Media

Safety, Reliability and Risk of Structures, Infrastructures and Engineering Systems – Furuta, Frangopol & Shinozuka (eds)
© 2010 Taylor & Francis Group, London, ISBN 978-0-415-47557-0

Measurement and modeling of spatially varying strength in Parallel Strand Lumber

Sanjay Arwade
Civil & Env. Engg., University of Massachusetts, Amherst, MA, USA

Russell Winans
Civil & Env. Engg., Cornell University, Ithaca, NY, USA

Peggi L. Clouston
University of Massachusetts, Amherst, MA, USA

ABSTRACT

Structural composite lumber is widely used in the United States and worldwide in increasingly demanding structural applications. This paper presents measurements of the spatial variability of the strength of a specific type of structural composite lumber, Parallel Strand Lumber (PSL), and describes a computational model for prediction of the strength of PSL. The characterizations of strength variability described in this paper should prove useful in preliminary reliability analysis of PSL members and structures, and is the type of data that may eventually allow for the re-evaluation of resistance factors used in the design of structures using PSL. The computational model is expected to find use in the evaluation and development of new designs for PSL, including the inclusion of new species and the directed mixture of species in the composite.

There has been substantial investigation of the spatial variability of elastic and strength properties in solid lumber (Lam and Varoglu 1991), but this work represents, to the author's knowledge, the first such investigation for PSL. Results show that PSL has some very favorable characteristics such as reduced variability of strength compared to similarly solid lumber. The computational model builds upon previous work that developed tools for the prediction of the strength of composite wood cross sections (Clouston 2006; Clouston and Lam 2001). The current model adds the significant feature of accounting for variation in strength along the length of the structural member.

The measurements are conducted by compression tests on specimens machined from a larger original PSL member. Results show that the variance of PSL strength is substantially lower than for solid lumber of equivalent size and that the strength random process has a correlation length on the order of several hundred millimeters. The computational model accounts for uncertainty in the strand length, strand grain orientation, and strand inelastic properties. The model uses an orthotropic yield surface in the strength calculation. The computational model validates well with respect to the mean strength, but underpredicts the variance of the strength. Investigation of the sources of this disagreement are ongoing.

REFERENCES

Clouston, P. (2006). Characterization and strength modeling of parallel strand lumber. *Holzforschung 61*, 392–399.

Clouston, P. and F. Lam (2001). Computational modeling of strand-based wood composites. *Journal of Engineering Mechanics 127*, 844–851.

Lam, F. and E. Varoglu (1991). Variation of tensile strength along the length of lumber part 1: Experimental. *Wood Science and Technology 25*, 351–359.

Safety, Reliability and Risk of Structures, Infrastructures and
Engineering Systems – Furuta, Frangopol & Shinozuka (eds)
© 2010 Taylor & Francis Group, London, ISBN 978-0-415-47557-0

Multiscale modeling of random heterogeneous materials

X.F. Xu

Stevens Institute of Technology, Hoboken, NJ, USA

ABSTRACT

In mechanics and materials, homogenization and self-consistent methods were the major tools until the breakthrough of multiscale research in 1990s. A variety of multiscale methods have been developed to upscale fine-scale equations into effective coarse ones. Determinism however has remained a universal strategy in the multiscale research, by assuming fine-scale information is completely known. The drastic advances made in the past decade are therefore focused on deterministic systems.

A crucial characteristic of complex materials is uncertainties of the constituent elements and interactions among them. For complex systems characterized with high-dimensions and various uncertainties, a strategy of Multiscale Stochastic Modeling (Xu 2007, Xu et al, 2008, Xu et al 2009) is proposed to follow a path stemming from renormalization group theory as shown in Fig. 1. The new path represents a stochastic version of the existing path developed from homogenization theory to deterministic multiscale modeling.

Asymptotic theories of classical micromechanics are built on a fundamental assumption of large separation of scales. For random heterogeneous materials the scale-decoupling assumption however is inapplicable in many circumstances ranging from conventional failure problems to small-scale engineering systems. Development of new theories and formulations for scale-coupling mechanics is to have significant impacts on diverse disciplines.

Based on the recent work of (Xu 2007, Xu 2009, Xu and Chen 2009a, Xu and Chen 2009b) an overview of multiscale modeling on random heterogeneous materials is presented in this paper. In Section 2 the generalized variational principles for boundary value problems of random heterogeneous materials are stated. Based on the generalized variational principles, formulations and methods are introduced in Section 3 and 4 for the scale-decoupling and scale-coupling problems, respectively.

Random heterogeneous materials are characterized with multiscale, high dimensions, and uncertainty,

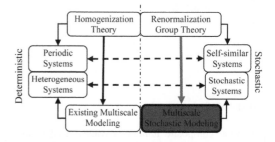

Figure 1. Proposed multiscale stochastic modeling methodology (Xu and Chen, 2009b).

which demands the use of both multiscale and stochastic modeling techniques. Integration of stochastic methods into a multiscale framework, or development of multiscale modeling in a stochastic setting, is becoming an emerging research frontier. It is expected development of so-called multiscale stochastic modeling methodology will not only provide innovative engineering and computational tools such as multiscale stochastic finite element method (MSFEM), but contribute to multidisciplinary research on diverse complex systems.

REFERENCES

Xu, X.F. 2007. A multiscale stochastic finite element method on elliptic problems involving uncertainties. *Comput. Methods Appl. Mech. Engrg.* 196 (25–28): 2723–2736.

Xu, X.F. 2009. Generalized variational principles for uncertainty quantification of boundary value problems of random heterogeneous materials. *ASCE Journal of Engineering Mechanics*. In Press

Xu, X.F. & Chen X. 2009a. Stochastic Homogenization of Random Multi-phase Composites and Size Quantification of Representative Volume Element. *Mechanics of Materials*. 41 (2): 174–186.

Xu, X.F. & Chen X. 2009b. A Multiscale formulation for uncertainty quantification of finite body random heterogeneous materials. Submitted.

Safety, Reliability and Risk of Structures, Infrastructures and Engineering Systems – Furuta, Frangopol & Shinozuka (eds)
© 2010 Taylor & Francis Group, London, ISBN 978-0-415-47557-0

A multi-resolution Bayesian framework for the identification of spatial variability of material properties

P.S. Koutsourelakis

School of Civil and Environmental Engineering, Cornell University, NY, USA

The prodigious advances in computational modeling of deformation processes and the development of highly non-linear, multiscale models poses several challenges in parameter identification. The predictive ability of these models depends strongly on assigning the correct values to the various model parameters. In solid mechanics for example laboratory testing is performed using fairly large specimens which however, do not immediately provide information about all the parameters in the material model (e.g. elastic modulus, yield stress etc) at the scale of interest, particularly in heterogeneous materials. The present paper is concerned with the problem of identifying parametric variability at various length scales from disparate and noisy measurements and experimental observations using computational models.

In existing methods, deterministic (i.e. optimization-based) or statistical, the representation of the unknown field has coincided with the forward model's. In cases where finite elements are used as forward solvers, the property of interest is assumed constant within each element and therefore the vector of unknowns is of dimension equal to the number of elements. This offers obvious implementation advantages but also poses difficulties as it can lead to over-fitting if the actual scale of variability is larger than the grid.

We propose a nonparametric, Bayesian model which is independent of the grid of the forward solver and is reminiscent of non-parametric kernel regression methods. The unknown parametric field is approximated by a superposition of kernel-type functions centered at various locations. The cardinality of the representation, i.e. the number of such kernels, is treated as an unknown to be inferred in the Bayesian formulation. This gives rise to a very flexible model that is able to adapt to the problem and the data at hand and find succinct representations of the material properties of interest. If prior information on the scale of variability is available, it can be directly introduced in the model.

The determination of the posterior distribution is carried out using Sequential Monte Carlo samplers which represent a set of flexible simulation-based methods for sampling from a sequence of probability distributions; each distribution being known up to a normalizing constant. In this work the sequence of distributions is defined by considering forward models operating at different resolutions. This results in a hierarchical, multi-resolution formulation that reduces the computational cost associated with the identification process and makes near-optimal use of the computational resources available. It also provides confidence metrics that can direct further refinement of the forward model in regions that would be most informative for the identification of the material property of interest. The algorithm is directly parallelizable and capable of identifying multiple modes in the posterior distribution i.e. several significantly different hypothesis that are nevertheless consistent with the data. Several examples on linear and nonlinear problems are presented with emphasis on quantifying predictive uncertainty.

REFERENCES

Del Moral, P., Doucet, A., & Jasra, A. 2006. Sequential monte carlo for bayesian computation (with discussion). *Bayesian Statistics 8*. Oxford University Press.

Holloman, C.H., Lee, H., & Higdon, D. 2006. Multiresolution genetic algorithms and markov chain monte carlo. *Journal of Computational and Graphical Statistics* 861–879.

Safety, Reliability and Risk of Structures, Infrastructures and Engineering Systems – Furuta, Frangopol & Shinozuka (eds)
© 2010 Taylor & Francis Group, London, ISBN 978-0-415-47557-0

Strength variability in brittle materials with randomly occurring flaws

L. Graham-Brady & C. Zingale
Johns Hopkins University, MD, USA

ABSTRACT

Numerical models of failure in brittle materials can exhibit idiosyncratic mesh dependencies that lead to predictions that are more dependent on numerical discretization than on physical reality. A successful technique for reducing this mesh dependency is to introduce random fluctuations in the material properties from element to element in the finite element mesh. While generally successful, this approach requires a physically reasonable method for assigning probability distributions to these random fluctuations. In other words, the random local fluctuations in the material properties must properly reflect fluctuations in the local microstructure.

Paliwal and Ramesh (2009) have developed a computational model for analyzing the strain-rate dependent constitutive relationship in ceramic materials with penny-shaped flaws, given a crack growth law, flaw density, and flaw size distributions. Simultaneously, recent experimental research has addressed estimating flaw size and flaw density distributions in SiC materials, including ultrasound, digital image characterization, and three-dimensional subsample porosity measures. Using the global flaw size distributions and global flaw densities obtained from the experimental observations, local flaw size distributions and flaw densities can be derived, which are a function of the observed bulk flaw size distribution, the flaw density, and the size of the local region. These are the distributions found in local regions of the material, which may vary significantly from the global flaw size distribution.

In the current work, stochastic simulation techniques and analytical approaches are used to establish the variations in the local flaw density. Using these results in conjunction with the analytical models to establish the strain-rate dependent behavior, some measure of variability in strength is established. The output from these analyses gives a basis for assigning random variations of material properties in a larger scale finite element analysis. Results show that as the finite element mesh is refined, the variability of strength from element to element is expected to increase. Furthermore, a higher strain rate tends to lead to a higher variability in the strength calculated for the elements, indicating that this approach is particularly important in the high-strain-rate regime.

REFERENCES

Paliwal, B. and K.T. Ramesh. (2009). "An Interacting Micro-Crack Damage Model for Failure of Brittle Materials under Compression". Journal of the Mechanics and Physics of Solids, In Press

Safety, Reliability and Risk of Structures, Infrastructures and Engineering Systems – Furuta, Frangopol & Shinozuka (eds)
© 2010 Taylor & Francis Group, London, ISBN 978-0-415-47557-0

Reliability analysis for fracture modelled by the partition of unity method

E.C. Schimmel

Faculty of Aerospace Engineering, Delft University of Technology, Delft, The Netherlands

M.A. Gutiérrez

Faculty of Mechanical, Maritime and Materials Engineering, Delft University of Technology, Delft, The Netherlands

J.J.C. Remmers

Department of Mechanical Engineering, Eindhoven University of Technology, Eindhoven, The Netherlands

C.V. Verhoosel

Faculty of Aerospace Engineering, Delft University of Technology, Delft, The Netherlands

ABSTRACT

The geometric reliability method is often used to approximate the probability of failure in solid mechanics. Further to this quantitative result, valuable insight in the failure mechanisms of a body can be obtained. In this contribution geometric reliability is applied to specimens consisting of a quasi-brittle material in which discrete cracks rather than failure process zones are considered. The discrete cracks are modelled by the partition of unity method (PUM).

In PUM the discontinuity introduced by the crack in the displacement field is modelled by an additional displacement field containing a Heaviside step function. Extra degrees of freedom accompany this additional displacement field and are added to only those nodes that are in the support of intersected elements. A constitutive model describes the relation between traction and opening of the crack.

The random fields of fracture strength and fracture toughness are discretised by the Karhunen-Loeve expansion and the local β-points along the limit-state surface are identified. Importance sampling is used to increase the accuracy of the computed probability of failure. The required sensitivity of the solution with respect to the random variables is computed through direct differentiation. Special attention is paid to the fact that the equilibrium solution of a partition of unity model also depends on the course of the crack. A variation of the random variables leads to a change of this course, which needs to be taken into account for the proper computation of the derivatives.

Discrete analysis of failure requires an accurate prediction of the propagation behaviour of the crack. The instant as well as the direction of the propagation are determined on the major principal stress at the tip. Propagation takes place when it surpasses the fracture strength, while the direction is normal to it. Use of the local stress at the tip gives an unsatisfactory prediction of the propagation behaviour. It has been shown by (Jirásek 1998) that use of a non-local stress gives better results for the crack path, however, it delays the propagation due to an underestimation of the tip stress. In this paper, a method is proposed that gives better prediction of the instant as well as the direction of propagation.

The presented algorithm is applied to a three-point bending experiment. Although the dominant mode at the limit-state surface corresponds to a centred crack, variability of the strength along the beam allows for off-centre cracks to appear with significant probability. Therefore, a dedicated method is applied to appropriately capture the shape of the limit-state surface.

REFERENCES

Jirásek, M. (1998). Embedded crack models for concrete fracture. *Euro C 1998 Computational Modelling of Concrete Structures 1*, 291–300.

Organized Session (OS26) Safety Assessment of Offshore and Marine Structures

Safety, Reliability and Risk of Structures, Infrastructures and
Engineering Systems – Furuta, Frangopol & Shinozuka (eds)
© 2010 Taylor & Francis Group, London, ISBN 978-0-415-47557-0

Reliability-based calibration of design code for bulk carriers with combined loading

T. Moan

Centre for Ships and Ocean Structures, Norwegian University of Science and Technology, Trondheim, Norway

H. Amlashi

Department of Marine Technology, Norwegian University of Science and Technology, Trondheim, Norway

ABSTRACT

Rational ship design rules should be based on explicit limit states and direct analysis of the load effects and resistance as well as a recognition of the inherent uncertainties. This calls for a reliability-based code. The recent IACS CSR (JBP, 2006) effort to establish harmonized rules for tankers and bulk carriers is a step in the right direction.

A reliability study of bulk carriers that fulfil semi-probabilistic design criteria for ultimate hull girder strength, based on a limit state function for combined global and local loads that was recently established by Amlashi and Moan (2008) is carried out with reference to a North Atlantic environment. Global and local still-water load models are reviewed. The definition of characteristic still-water and wave-induced loads and their inherent uncertainties are discussed. The effect of the design format, in terms of characteristic values and load factors, on the reliability level is investigated. The base case for load factors for wave- and still-water loads are taken to be 1.2 and 1.0, respectively, while the strength factor is chosen to be 1.1. The effect of possible load control, implying a truncation in still-water load model, has not been investigated in this paper, but it would affect both characteristic values of the design equation as well as the reliability model and results.

A load combination factor for still-water- and wave-induced loads that was recently derived, is considered in the reliability analysis, and not in the design format. The correlations between global and local loads for both still-water and wave loads are accounted for.

Calculation of the hull girder strength by using linear elastic bending stress combined with ultimate strength check of longitudinal elements implies a model bias of about 1.15 for well balanced hulls with longitudinal bulkheads, while a Smith type method would correspond to a bias close to 1.0. For bulk carriers with significant double bottom bending,

"modified" simplified approaches are needed. The limit state formulation for hull girders under combined global and double bottom bending is established based on an interaction formula.

The systematic uncertainties addressed above together with the random uncertainties determine the reliability level for a given design formulation. A bias in strength or load effect has the same effect as a resistance or load factor. The definition of characteristic value is also directly related to the partial safety factor.

It is found that the interaction formula results in a consistent safety level. However, the model uncertainty in the prediction of the ultimate pressure should be more thoroughly assessed.

The safety level should be compared with that of other ship types and a target level established in view of the failure consequences.

In this paper various issues relating to a rational development of semi-probabilistic design codes for bulk carriers have been addressed. The crucial issue is the characterisation of uncertainties in load effects and resistance. Efforts are still needed especially to assess the uncertainty of still-water loads for bulk carriers. It would also be of interest to shed some more light on the effect of load control on the distribution of the still-water loads. Further work is also needed to estimate the effect of heavy weather avoidance on wave loads as well as to assess the ultimate strength of the double bottom structure.

REFERENCES

Amlashi, H.K.K. & Moan, T. 2008. Ultimate Strength Analysis of a Bulk Carrier Hull Girder under Alternate Hold Loading Condition, Part 2: Stress Distribution in the Double Bottom and Simplified Approaches. *Marine Structures*, On-line, 25.02.09.

JBP (2006); "Common Structural Rules for Bulk Carriers", Technical backgrounds, IACS, London

*Safety, Reliability and Risk of Structures, Infrastructures and
Engineering Systems – Furuta, Frangopol & Shinozuka (eds)
© 2010 Taylor & Francis Group, London, ISBN 978-0-415-47557-0*

Probabilistic assessment of collision-induced ship hull failure

H. Jia
CeSOS, NTNU, Trondheim, Norway

T. Moan
CeSOS/Department of Marine Technology, NTNU, Trondheim, Norway

ABSTRACT

Risk may be defined as the probability of failure times
the failure consequences. Estimate of risk may serve
as a basis to establish design and operational criteria
for ships. This paper deals with assessing the probability of collision-induced ship hull failure, which could
be extended to make a part of collision risk analysis.
Jia and Moan (2008) proposed to estimate the failure
probability for the damaged vessel conditioned upon
sea state Hs, loading condition LC, collision location
DL and damage properties DP. Assume that LC, DL
and DP are mutually independent events and Hs is
uncorrected with them. Then the failure probability
after damage is given as follows:

$$P_{f|D} = \sum_i \sum_j \sum_k \sum_l (P_{f|Hs,LC,DP,DL}(Hs_i, LC_j, DP_k, DL_l) \cdot$$
$$P_{Hs}(Hs_i) \cdot P_{LC}(LC_j) \cdot P_{DP}(DP_k) \cdot P_{DL}(DL_l)) \quad (1)$$

Based on the equation above, the analysis of sensitivity of the total failure probability to the sea state and
damage is performed. Damage properties are generally
characterized by damage center, length, penetration
and vertical height etc. In this paper the concerned
parameters for the damage are longitudinal location,
damage length and damage vertical extent and height
and damage penetration. Furthermore, penetration of
the outer skin is considered and the mean vertical
height according to HARDER project (2000~2003)
is used.

In this paper, the conditional failure probability in a
3-hour period is calculated for a double hull oil tanker
at a full load condition. Wave heading is assumed to
be uniformly distributed between 0° and 360°. The
analysis of sensitivity of the total failure probability
to the sea state is based on a certain damage while
using a distribution of the significant wave height from
both HARDER and DNV (2000). The effect of peak
period is taken into account. The analysis of sensitivity to damage is based on one year wave event.
Up to three tanks damage and six vertical damages
are considered. For each damage condition, the conditional failure probability is calculated based on the
most critical transverse section. It is assumed that the
longitudinal properties of the damage are uncorrelated
with the vertical properties. The statistical database for
collision damage in HARDER project is adopted to
assess the probability of damage.

REFERENCES

DNV. 2000. Environmental conditions and environmental
loads, *DNV Classification Notes 30.5.*
Jia, H. & Moan, T. 2008. Comparative reliability analysis of
ship hull girders under vector-load processes", submitted
for publication.
HARDER, 2000~2003. Harmonisation of Rules and Design
Rational, DNV, Oslo, Norway

Safety, Reliability and Risk of Structures, Infrastructures and Engineering Systems – Furuta, Frangopol & Shinozuka (eds)
© 2010 Taylor & Francis Group, London, ISBN 978-0-415-47557-0

Development of quantitative coating condition monitoring procedure

N. Yamamoto & K. Terai
Research Institute of ClassNK, Chiba, Japan

A. Nakayama & K. Amaya
Tokyo Institute of Technology, Tokyo, Japan

ABSTRACT

In order to secure a safe and stable operation of ships throughout the ships' lifetime, it is necessary to maintain the effectiveness of the corrosion protection system. Currently, the confirmation of coating condition is made by visual judgment (IACS 2006). However, the uncertainty might exist in a subjective judgment. Because the judgment of timing and scale of the maintenance work of deteriorated coating is a major issue for a ship like LNG ship which plans a long-term operation while maintaining the high quality, the development of a method of judging the coating condition by the quantitative index is demanded.

The procedure to judge the deterioration of the paint coating of a ballast tank of ship by an objective index was developed (Nakayama et al. 2008). This procedure is the one in which the judgment is done by detecting the potential distribution change in a ballast tank. To obtain the surface resistance which relates the coating deterioration condition, an inverse problem approach was applied to the results of potential difference measurements. Figure 1 shows the measurement model in ballast tank.

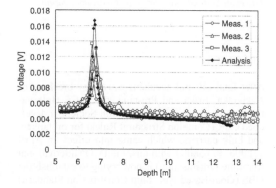

Figure 2. Measured and evaluated voltage distributions in the case that the optional anode is installed at 1.6 m from the seawater surface. (surface resistance of the tank walls is 500 Ωm^2).

In order to verify the feasibility of the developed procedures, potential measurements by the proposed method were made in a ballast tank. Figure 2 shows the measured voltage distributions in the compartment of the tank with the numerically calculated voltage distribution. According to these results, it is confirmed that the proposed method to judge the degree of paint coating deterioration quantitatively in the ship's ballast tank is feasible.

Further, in order to put this procedure into practical application for the coating condition monitoring of actual ship such as LNG ship, the developed coating condition monitoring system was designed and implemented to the LNG ship on trial. The continuous coating condition monitoring will be made for the coming one year.

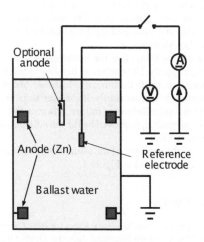

Figure 1. Potential measurement model in ballast tank.

REFERENCES

IACS (2006), Guidelines for coating maintenance & repairs for ballast tanks and combined cargo/ballast tanks on oil tankers, IACS Recommendation 87.

Nakayama, A., Kenji Amaya and Yamamoto N. (2008), Computer Simulation and Potential Measurement of the Cathodic Protection System in Ballast Tank, Proc. of OMAE2008, Estoril, Portugal, OMAE2008-57651.

Safety, Reliability and Risk of Structures, Infrastructures and
Engineering Systems – Furuta, Frangopol & Shinozuka (eds)
© 2010 Taylor & Francis Group, London, ISBN 978-0-415-47557-0

Change in safety level against collapse of ship's hull girder in longitudinal bending according to change in design criteria

Kazuhiro Iijima & Masahiko Fujikubo
Department of Naval Architecture and Ocean Engineering, Osaka University, Suita, Japan

Yasunari Fujii
Ship Design Department, Sanoyas Hishino Meisho Corporation, Kurashiki, Japan

Tetsuya Yao
Tsuneishi Shipbuilding Company, Fukuyama, Japan

ABSTRACT

After fatal oil spill accidents caused by breaking of
Nakhodka, Erika and Prestage, the IACS (International Association of Classification Societies) developed CSR (Common Structural Rules) for oil tankers
and bulk carriers, which came into effect on April
1, 2006. In parallel with this, the IMO (International
Maritime Organisation) is now trying to establish the
GBS (Goal-Based New Ship Construction Standard)
to ensure the safety of ships.

The development of GBS started in a prescriptive
approach limiting the ship types to oil tankers and bulk
carriers. A little later, a safety level approach has also
started. In the safety level approach, a quantitative goal
for safety in terms of risk is first set, and then the design
is verified against the goal. This will allow more freedom for designers to achieve the goal. As an alternative
to set an appropriate safety level as a goal, the safety
level of existing ships that conform with the rules has
to be assessed. In this regard, it is worth while knowing
how the safety level has been affected by design rules
in different times.

In the present paper, four Panamax-size bulk carriers and four Aframax-size oil tankers are chosen for
comparison. Three of the bulk carriers and three of the
tankers were built in accordance with pre-CSR, while
the latest ones were designed in accordance with CSR,
Their safety levels are assessed from the viewpoint of
ultimate hull girder strength in longitudinal bending.
For the reliability analysis, both hull girder capacity
and the extreme load acting on the hull girder have to
be evaluated to get their statistical characteristics or
uncertainty models. A series of progressive collapse
analysis is performed applying the Smith's Method to
evaluate the ultimate hull girder strength and its sensitivities with respect to design parameters. The mean
value and the standard deviation of the ultimate hull
girder strength are evaluated. On the other hand, a time-
dependent nonlinear ship motion analysis is performed
to obtain the time history of the wave-induced bending

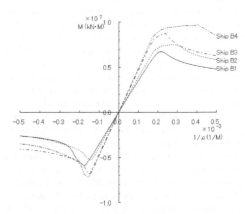

Figure 1. Moment-curvature relationships of cross-sections.

moment applying the nonlinear strip method considering the influence of large motion. Then, the mean
value and the standard deviation of the extreme wave-
induced bending moment are evaluated. Utilising these
results of calculations, reliability indices and failure
probability are calculated applying FORM (First Order
Reliability Method).

It has been found that the newer ships have larger
bending capacity. The maximum bending moment also
becomes larger with the times due to the advancements
in operability. Consequently, the newer ships tend to
have higher probabilities of failure, however, the safety
level is, in general, improved by the introduction of the
new rule, CSR.

One of the example of the ultimate strength assessments are presented in Figure 1.

REFERENCES

IACS. 2006. Common Structural Rules for Bulk Carriers.
IACS. 2006. Common Structural Rules for Double Hull Oil
    Tankers.

Safety, Reliability and Risk of Structures, Infrastructures and
Engineering Systems – Furuta, Frangopol & Shinozuka (eds)
© 2010 Taylor & Francis Group, London, ISBN 978-0-415-47557-0

Load combination factors suitable for probabilistic and semi-probabilistic design of ocean-going ships

W. Huang
Centre for Ships and Ocean Structures, Norwegian University of Science and Technology, Norway
College of Space and Civil Engineering, Harbin Engineering University, P.R. of China

T. Moan
Centre for Ships and Ocean Structures, Department of Marine Technology, Norwegian University of
Science and Technology, Norway

ABSTRACT

In this paper, Poisson square wave processes are employed to model both still water and wave loads, which are have, respectively, a normal and Weibull parent distribution, see Guedes Soares & Moan (1988) and Huang & Moan (2008). Then, based on Poisson load models (Wen 1990), the extreme value distribution of global longitudinal still water and wave bending moments (or loads) in a given design lifetime are determined. Correspondingly, the characteristic extreme value with an exceeding probability α is derived for both loads.

As for combination of two loads, based on Poisson models, the distribution of the extreme value $M_{w,tv}$ of wave loads in a given voyage is obtained firstly. Then, in a given voyage t_v, the maximum $M_{c,tv}$ of the combined bending moment is the sum of $M_{w,tv}$ and corresponding still bending moment M_{sw}, see Figure 1. Therefore, by probabilistic convolution of the distribution of $M_{w,tv}$ and M_{sw}, the distribution of the combined extreme value in one voyage is derived. The combined extreme value in one voyage follow the same time variation as still water loads for one voyage to another, i.e. it also follow a Poisson model. Hence, the distribution of the extreme value of the combined load in a design lifetime can be derived as that of still water loads. By doing so, a consistent solution to the combined extreme value of still water and wave loads is obtained, that is, it will never exceed the sum of extreme values of two individual loads.

Moreover, by approximating a general Weibull distribution with an exponential one and using Turkstra's combination solution (Turkstra 1970), an approximate analytic solution to the mean up-crossing rate of the combined process is derived. Further, a simple analytical formula for a general combined characteristic value with a given exceeding probability is derived. Correspondingly, a simple analytical formula for load combination factors which is based on the companion method is established. Numerical examples show that both numerical and analytical results agree very well. Finally, based on the developed methods, load

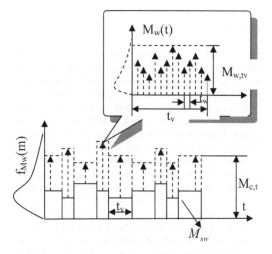

Figure 1. Combination model of still water and wave loads.

combination factors suitable for expressing the combined loads on ships by random variables, suitable for a reliability evaluation as well as semi-probabilistic design, are established.

In addition, the load combination factors can be determined based on an arbitrary given exceeding probability α. Therefore, they would be more useful than those based on the return period, see Huang and Moan (2008).

REFERENCES

Guedes Soares, C. & Moan, T. 1988. Statistical analysis of still water load effects in ship structures. SNAME 96: 129–156.

Huang, W. & Moan, T. 2008. Analytical method of combining global longitudinal loads for ocean-going ships. Probabilistic Engineering Mechanics 23: 64–75.

Turkstra, C.J. 1970. Theory of structural design decisions. Study No.2; Solid Mechanics Division, University of Waterloo, Ontario, Canada.

Wen, Y.K. 1990. Structural load modelling and combination for performance and safety evaluation. Amsterdam: Elsevier.

*Safety, Reliability and Risk of Structures, Infrastructures and
Engineering Systems – Furuta, Frangopol & Shinozuka (eds)
© 2010 Taylor & Francis Group, London, ISBN 978-0-415-47557-0*

Study on prediction model for deterioration of RC structure received chloride induced damage based on reliability theory

M. Matsushima
KAGAWA University, Takamatsu, Japan

H. Nakagawa & M. Yokota
Shikoku Research Institute, Takamatsu, Japan

ABSTRACT

The method to predict the damage level of deteriorated structure received chloride induced damage is proposed in this paper. The deterioration phenomenon of structures in actual environment is varied widely. Therefore, the stochastic method is carried out in order to consider the actual structure with uncertainty in nature. Three parameters, Equivalent diffusion coefficient, Error of cover thickness and Threshold chloride ion density is modelled as stochastic parameters.

A probabilistic model is developed to represent the deterioration of concrete structures using deterioration process model as shown in Figure 1. The parameters involved in the model are expected to vary extensively since the phenomenon takes place in the natural environment. The deterioration process is regarded as consisting of three phases; The first is the period t_s until depassivation of the reinforcement occurs after completion of the structure, and the second is the period t_{cr} commencing from the moment of depassivation and including the development of corrosion at a perceptible rate until the limit state is attained when cracking appears in the surface concrete. The former is called the incubation period and the latter the progressive period. After cracking of surface concrete is called the accelerative period. In this paper, three parameters, Equivalent diffusion, Error of cover thickness and Threshold chloride ion density is modelled as stochastic parameters.

Target structure is the slab of landing pier located at sea side in South Japan. 30 years has passed since the landing pier is constructed.

The results computed using above method show Figure 7 with the results of actual data. The vertical axis indicates the corrosion rate and the lateral axis indicates the elapsed year in figure. The solid line in figure means the mean value and mean value ± standard deviation in computation. Corrosion rate of steel bars indicate approximately zero before 10 years and increases exponentially with elapsed year after 10 years in computation. The variation of corrosion

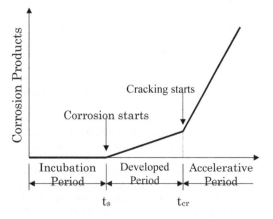

Figure 1. Deterioration process.

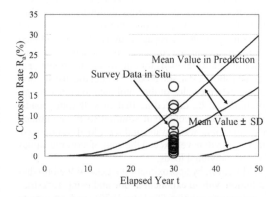

Figure 7. Comparisons of Survey Data and Prediction.

rate also increases with elapsed year in computation. Mark ○ indicates the survey data of corrosion rate of steel bar obtained from target structure. Since the survey data of 85% consists in the mean value ± standard deviation in computation at 30 elapsed year, the accuracy of this simulation model can be judged to be proper.

General Session (Fatigue Reliability)

Safety, Reliability and Risk of Structures, Infrastructures and Engineering Systems – Furuta, Frangopol & Shinozuka (eds)
© 2010 Taylor & Francis Group, London, ISBN 978-0-415-47557-0

Fatigue properties of corroded parallel wire cable

C.M. Lan & H. Li
School of Civil Engineering, Harbin Institute of Technology, Harbin, China

J.P. Ou
School of Civil Engineering, Harbin Institute of Technology, Harbin, China
School of Civil and Hydraulic Engineering, Dalian University of Technology, Dalian, China

ABSTRACT

The corrosion degree and the fatigue life of wires which were obtained from dissection of actual parallel wire cables used on Tianjin Yonghe cable-stayed bridge were presented. The stay cables were used about 18 years. The percent of wire cross-section loss obeyed lognormal distribution with mean 2.66% and standard deviation 3.01%. The discreteness of the corrosions was much larger, local corrosions were serious. It could be seen that corrosions made the surface roughness of wires and stresses concentrated at the deformed location, for this reason the birth of a microcrack generated relative easier, and fatigue life decreased sharply.

The fatigue life of individual wire was described by Weibull distribution considered some useful parameters such as, stress range, mean stress, mean static strength and length effects (Faber, *et al.* 2003). The fatigue life predictions of cable are based on the weakest link model (see Figure 1.) (Stallings and Frank, 1991). Based on the fatigue life of individual wire, the fatigue properties of corroded parallel wire cable were investigated by Monte Carlo simulation.

The calculated results showed that the fatigue life of a cable was controlled by a small fraction of the cable wires with the shortest fatigue lives. The fatigue life

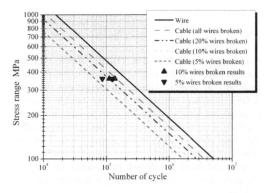

Figure 2. Comparison results between calculation and test.

of a cable was significantly lower than the mean life of wire with equal length. Cable life decreased as the variability of the lives of the cable wires decreased. This result indicates that wires of high quality for used in parallel wire cable should be not only a long mean life, but also low variability of fatigue life. Cable fatigue life decreased with increasing cable length. The variability in cable lives decreased as the number of wires increased. For practical cable sizes, the mean cable life was insensitive to the number of wires in the cable. Consideration to economy and safety of a parallel wire cable, it was reasonable to define the fatigue failure of 10% of the cable wires as the useful life of a cable.

Finally, comparisons between calculated results and test results were shown in Figure 2. The reasons of difference of two results were stated in the paper.

REFERENCES

Faber, M.H. Engelund, S. & Rackwitz, R. 2003. Aspects of parallel wire cable reliability. *Structural Safety* 25:201–225

Stallings, J.M. & Frank, K.H. Stay-cable fatigue behavior. *Journal of Structural Engineering*. ASCE, 117(3): 936–950

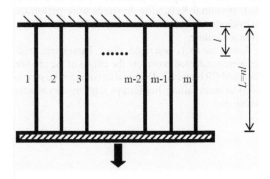

Figure 1. Model of parallel wire cable.

Safety, Reliability and Risk of Structures, Infrastructures and Engineering Systems – Furuta, Frangopol & Shinozuka (eds)
© 2010 Taylor & Francis Group, London, ISBN 978-0-415-47557-0

Performance-based engineering framework for buckling-restrained braced frames

B.M. Andrews
Wiss, Janney, Elstner Associates, Inc., Northbrook, IL, USA
University of Illinois, Urbana, IL, USA

J. Song & L.A. Fahnestock
University of Illinois, Urbana, IL, USA

ABSTRACT

Buckling-restrained braces (BRBs) have recently become popular in the United States for use as primary members of seismic lateral-force-resisting systems. A BRB is a steel brace that does not buckle in compression but instead yields in both tension and compression. Concentrically-braced frames incorporating BRBs are known as buckling-restrained braced frames (BRBFs). Although design guidelines for BRB application have been developed, procedures for assessing performance and quantifying reliability are still needed.

This paper proposes a performance-based engineering framework (PBEF) for a BRBF subjected to seismic loads (the architecture for the proposed framework is shown in Figure 1). The proposed framework quantifies the risk of BRB failure due to low-cycle fatigue fracture of the BRB core.

Using the Input Module, input ground acceleration records were randomly generated from power spectrum models and modulated with envelope functions (to account for non-stationary). The generated time records were used as input excitations to a single-degree-of-freedom lumped-mass BRBF System Model. The BRB hysteretic behavior was modeled using a Bouc-Wen model. Non-linear Dynamic Simulation was performed to obtain BRB core deformation time history records.

This study uses the BRB CPD Capacity Models developed by Andrews et al.

Given the BRB core deformation history as an input, these models predict the remaining CPD capacity of the brace, where values less than zero indicated failure.

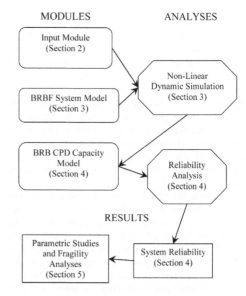

Figure 1. Proposed PBEF Architecture.

Given BRB demand (i.e. core deformation histories generated from the dynamic analyses) and supply (i.e. remaining capacity predicted by the CPD models), structural Reliability Analyses were performed to evaluate the System Reliability and the probability of brace failure.

Using the tools described above, Parametric Studies were conducted to explore the effects of the seismic load and BRB and BRBF characteristics on the probability of brace failure. In addition, fragility curves were developed.

Safety, Reliability and Risk of Structures, Infrastructures and Engineering Systems – Furuta, Frangopol & Shinozuka (eds)
© 2010 Taylor & Francis Group, London, ISBN 978-0-415-47557-0

A simulation on fatigue crack propagation in welded joints

T. Hanji, K. Tateishi & K. Tsuchiya
Nagoya University, Nagoya, Japan

ABSTRACT

A lot of fatigue cracks have been observed in steel bridges throughout the world (Miki, 2000). One of the main reasons to trigger fatigue cracks in the bridges is secondary stress, which is caused by three dimensional deformation of the bridge and not estimated in design phase (Fisher, 1984).

The crack propagation manner caused by the secondary stress is sometimes rather complicated because of intricate stress field around the crack. The cracks often propagate in various manners even though they initiate from the similar locations in the similar joints. This variation can result from the difference of the stress field around the crack. In other words, the crack propagation path may provide a great hint to estimate the stress field around the crack, which can be useful information for determining an appropriate repairing method.

In this study, fatigue crack propagation simulations on T-shape welded joints were carried out by Zencrack (Zentech, 2005), which is a commercial software for three dimensional crack propagation analysis based on fracture mechanics and interfaced to the finite element codes MSC/Marc. The simulation model is shown in Figure 1. Main plate of the joint was supported at ends and tensile and bending loads were applied to an attached rib. An initial crack of 0.5 mm in length was given at weld toe on the main plate. We picked up five kinds of stresses as parameters, which were nominal bending stress, membrane stress and average shear stress in the main plate (σ_b, σ_m and τ), nominal bending stress and membrane stress in the attached rib (σ_{rb} and σ_{rm}). We performed the crack propagation simulation with changing these parameters and investigated the crack propagation path in the thickness direction.

Simulation results demonstrated that the crack direction at the beginning (θ) was influenced by these parameters, particularly the average shear stress in the main plate. This result was supported by the fatigue tests conducted under the same conditions as the simulation. Finally, we presented the correlation between the average shear stress and the crack direction at the beginning as shown in Figure 2.

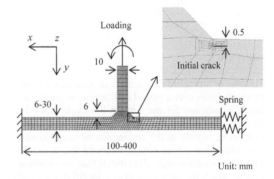

Figure 1. Simulation model.

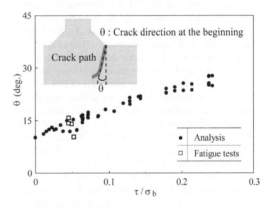

Figure 2. Relationship between θ and τ/σ_b.

REFERENCES

Fisher, J.W. 1984. *Fatigue and Fracture in Steel Bridges: Case Study*. New York: John Wiley & Sons.

Miki, C. 2000. Fatigue and fracture of welds in civil structures. *Science and Technology of Welding and Joining* 5(6): 347–355.

Zentech International Ltd. 2005. *User Manual for ZEN-CRACK 7.4*.

Safety, Reliability and Risk of Structures, Infrastructures and
Engineering Systems – Furuta, Frangopol & Shinozuka (eds)
© 2010 Taylor & Francis Group, London, ISBN 978-0-415-47557-0

A study on the equivalent initial flaw size of Al 2024-T3 riveted specimens under variable load

Jung-Hoon Kim, Goang-Seup Zi & Jung-Sik Kong
Department of Civil, Environmental and Architectural Engineering, Korea University, Seoul, Korea

Min-Seong Kim
Agency for Defense Development, Daejeon, Korea

ABSTRACT

The loss of strength in an aircraft structure as a result of cyclic loads over a period of life time is an important phenomenon for the aircraft life-cycle analysis. Service loads are accentuated at the areas of stress concentration, mainly at the connection of components. The initial flaw influencing the service life was estimated by using the equivalent initial flaw size method (EIFS) which has been proven as a useful design tool for life prediction of aircraft structures. The equivalent initial flaw size (EIFS) concept was developed nearly 30 years ago in attempt to consider the initial quality of structural details. In this study, the fatigue test on small single-rivet specimens made of Al 2024-T3 was conducted. The value of EIFS is calculated by using the back extrapolation technique and the Paris law of fatigue crack growth. The EIFS is used to predict the fa-tigue life of single-rivet specimens subject to nonuniform loads. To predict fatigue life from initial flaw size distribution, the equivalent stress intensity factor (SIF) and a modified Wheeler model by Huang (Huang & Moan 2007) were used. And we analyzed an example problem considering EIFS distribution in multiple site damage specimen. An example problem was calculated by the extended finite element method (XFEM) to analyze multiple site damage (MSD) problem under same EIFS distribution in each cracks.

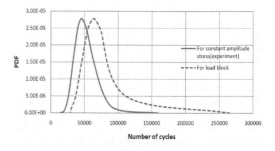

Figure 2. Comparison of PDF at crack length 11mm for constant amplitude stress and load block.

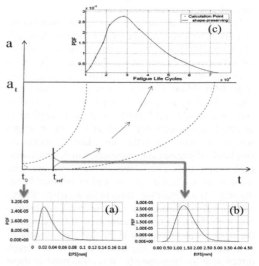

Figure 3. (a) EIFS distribution PDF (b) EIFS distribution PDF in reference time (c) Fatigue life cycles of MSD specimen at faure.

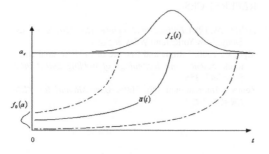

Figure 1. The process showing the compatibility between TTCI and EIFS distribution.

REFERENCE

Fawaz SA. Equivalent initial flaw size testing and analysis; 2000.

Safety, Reliability and Risk of Structures, Infrastructures and
Engineering Systems – Furuta, Frangopol & Shinozuka (eds)
© 2010 Taylor & Francis Group, London, ISBN 978-0-415-47557-0

Fatigue design, evaluation and inspection for steel bridges on the Honshu-Shikoku expressways

I. Yamada, A. Moriyama & K. Fumoto
Honshu-Shikoku Bridge Expressway Company Limited, Japan

1 INTRODUCTION

Recently, fatigue failures on steel bridges have been reported in Japan. Especially, on urban-area expressways and on heavy-traffic highways, the fatigue failures of steel girders or orthotropic steel decks have been increasing for a few decades.

On the other hands, no fatigue failure are presently reported on the three routes of the Honshu-Shikoku Expressways, which were completed by the end of the last century, mainly consisting of long-span bridges between two major islands in Japan. The latest fatigue design was applied to the long-span bridges. However, the structural safety can not be necessarily ensured in the future of the Honshu-Shikoku Expressways.

This paper describes the fatigue design for long-span bridges, including the fatigue tests with full full-scaled specimens. The paper also describes fatigue evaluation and inspection on steel bridges of the Honshu-Shikoku Expressways against the future fatigue damages.

2 FATIGUE DESIGN

In the design and construction stage of the Honshu-Shikoku Expressways, the latest research and development on the fatigue durability were made. For example, the various fatigue tests were carried out, using full-sized specimens for steel truss/box girders in the long-span bridges. Based on the results of fatigue tests, the structural details of steel truss/box girders were improved.

In addition, the advanced specifications for fatigue design were made and applied to the superstructure design and fabrication of steel girders on the long-span bridges, including the suspension bridge and the cable-stayed bridge.

However, general steel bridges, including the plate girder bridge, were designed without any specific consideration to the fatigue design.

3 FATIGUE EVALUATION

In the operation stage of the Honshu-Shikoku Expressways, various technical efforts for the fatigue durability were made. For example, the fatigue assessments were carried out in order to judge the status of the Honshu-Shikoku Expressways, as compared with the other expressways and highways.

The cumulative fatigue damage and the fatigue life of steel bridges were estimated and assessed, carrying out the field measurement on a long-span bridge and the structural analysis based on the Fatigue Design Manual for Steel Highway Bridges.

The fatigue assessments show that the steel bridges on the Honshu-Shikoku Expressways are healthy at present and in the near future.

4 FATIGUE INSPECTION

Since the early detection is required in the inspection for structural safety, the fatigue hazard figure for bridge inspection, which shows important details to inspect against fatigue failures in the future, were developed in order to conduct efficient and effective inspections for steel bridges.

In addition, the application test of the advanced sensing technology, using the fatigue detecting sensor, is under way on a plate girder bridge in order to obtain a more accurate estimation of fatigue damage and to optimize the preventive maintenance.

5 CONCLUSION

The conclusions are summarized as follows,

1) In the design and construction stage, the latest studies on the fatigue durability were made for the Honshu-Shikoku Expressways, especially for long-span bridges.

2) In the operation stage, the fatigue assessments were conducted for steel bridges on the Honshu-Shikoku Expressways, resulting in the sufficient safety in the near future.

3) In order to conduct efficient and effective inspections, the fatigue hazard figures were developed for steel bridges and the application test of fatigue detecting sensor is conducted on a steel bridge on the Honshu-Shikoku Expressways.

The future efforts are required to prevent fatigue failures on steel bridges and to enhance the structural safety and reliability of steel bridges.

Mini-Symposia (MS12) Monte Carlo Simulations Applied to Civil Engineering

Safety, Reliability and Risk of Structures, Infrastructures and Engineering Systems – Furuta, Frangopol & Shinozuka (eds)
© 2010 Taylor & Francis Group, London, ISBN 978-0-415-47557-0

Stochastic modeling and simulation of transient processes

Lijuan Wang
Technip USA, Inc. Houston, TX, USA

Ahsan Kareem
NatHaz Modeling Laboratory, University of Notre Dam, Notre Dame, IN, USA

ABSTRACT

This study focuses on data-driven modeling and simulation of non-stationary transient events such as earthquake ground motions, extreme wind and ocean wave profiles characterized by time-varying amplitude and frequency features. The evolutionary characteristics of transient events like earthquake ground motion, gust front winds, and transient wave fronts are first modeled utilizing a time-frequency domain framework. In this scheme, the stationary wavelet transform is first introduced to decompose a sample of a non-stationary random process into a set of mono-component processes. These components are then transformed to analytic signals using the Hilbert transform which yields their instantaneous amplitude and frequency. The simulation is carried out by introducing the phase angle differences as normally distributed processes. Without the customary assumption of piece-wise stationarity, or an assumed modulation function, this method helps to simulate non-stationary random processes based on a given sample. The uni-variate simulation scheme is extended to multi-variate cases by introducing the proper orthogonal decomposition (POD) of the covariance matrix of the instantaneous frequency. Several quantitative criteria are employed to assess the quality of non-stationary simulations. Also included this study is a framework for the conditional simulation of non-Gaussian, non-stationary, random processes. Focusing on time-frequency description of non-stationary random fields, the evolutionary cross-correlation structure of non-stationary random fields is established analytically in terms of wavelet coefficients. By extending conventional kriging technique to the time-frequency domain, the proposed method facilitates conditional simulation of non-stationary random fields with known time-dependent correlation structure. Numerical examples concerning the conditional simulation of downburst wind velocities are presented to demonstrate the accuracy and efficacy of the proposed scheme. The proposed general scheme has immediate applications to earthquake engineering, wind engineering and ocean engineering in simulating additional time histories based on measurements at limited locations.

Safety, Reliability and Risk of Structures, Infrastructures and
Engineering Systems – Furuta, Frangopol & Shinozuka (eds)
© 2010 Taylor & Francis Group, London, ISBN 978-0-415-47557-0

Stochastic interpolation of the underground contamination by Monte Carlo filter

O. Maruyama
Tokyo City University, Tokyo, Japan

M. Shinozuka & S.R. Chaudhuri
University of California-Irvine, Irvine, CA, USA

ABSTRACT

There are numerous contaminated sites in local and regional scales due to accidental spills, and improper disposal. These contaminants can pose lasting and costly risk to humans, other living receptors, and natural environment to the extent that the human society becomes no longer sustainable.

Over the last decades, concern grew over the impact of environmental contaminations, and several environmental regulations have been enacted worldwide to control the release of contaminants at active sits and remediate the sites that were contaminated in the past. In this context, the source identification and interpolation of the underground contamination has special significance.

This paper focuses on stochastic interpolation of the fate of underground contaminants and quantitative estimation by the Monte Carlo filter technique.

We are concerned with the interpolation of a spatial random field at discrete points. Let $C = [c_1, c_2, \cdots c_i, \cdots, c_n]^T$ be a spatial random field with pdf $p(C)$, where c_i is a random quantity at the ith discrete spatial point, and the posterior field with pdf $p(C|Y)$ will be estimated, after an observation $Y = [y_1, y_2, \cdots, y_\ell]^T$ is carried out on C. The observation is not necessarily made on every corresponding c_i. In addition, the observation Y is not necessarily the same physical quantity as C, and they are generally functionally related, irrespective of linearity or nonlinearity.

Under this condition, we may represent an observation equation by

$$Y = H(C) + w \qquad (2)$$

where $w =$ observation noise vector of $\ell \times 1$ with pdf $r(w)$.

It is noted that $H(C)$ are derived from Eq.(1) and are not generally linear functions. C and w are not necessarily Gaussian vectors.

In order to obtain an optimal estimate on the spatial field C after the observation of $Y = [y_1, y_2, \cdots, y_\ell]^T$, the following conditional probability $p(C|Y)$ for updating or filtering after processing the observation data become a basic formulas. For the updating or filtering, the conditional probability law gives

$$p(C|Y) = \frac{p(Y|C)p(C)}{\int p(C)p(Y|C)dC} \qquad (3)$$

Equation (2) is the basis for fundamental discussions on updating or filtering of stochastic fields. Equation (2) indicates that the conditional pdf $p(C|Y)$ can be obtained if $p(C)$ is known, and if $p(Y|C)$ is evaluated. But it might be difficult to derive $p(Y|C)$ based on Eq.(2), because of non-linearity of $H(C)$ and non-Gaussian property of w.

Under the above mentioned difficulties, as a versatile tool to update such system, Monte Carlo filter is focused that is an algorithm to estimate $p(C|Y)$ experimentally based on generation of a set of m sample realizations of $C, s^{(i)} = [c_1^{(i)}, c_2^{(i)}, \cdots, c_n^{(i)}]$, $i = 1, 2, \cdots, mp(X)$ for a priori field $p(C)$ and a set of m sample realizations of conditional $C, f^{(i)} = [c_{1|Y}^{(i)}, c_{2|Y}^{(i)}, \cdots, c_{n|Y}^{(i)}]$, $i = 1, 2, \cdots, m$ for a posteriori field $p(C|Y)$ or filtering, bypassing the integration of Eq.(3). These are general frameworks of MCF. In other words, the generation of $s^{(i)}$ and $f^{(i)} i = 1, 2, \cdots, m$ is the major key in the algorithm.

Safety, Reliability and Risk of Structures, Infrastructures and Engineering Systems – Furuta, Frangopol & Shinozuka (eds)
© 2010 Taylor & Francis Group, London, ISBN 978-0-415-47557-0

Processing acceleration simulations

A. Zerva

Drexel University, Philadelphia, PA, USA

S. Liao

Rosenwasser/Grossman Consulting Engineers, New York, NY, USA

ABSTRACT

The safety of any structure against earthquake loads cannot be reliably assessed unless the seismic excitation is correctly characterized. Because of the underlying physical causes (instrument noise and/or ground tilt), the need to process recorded seismic data has been recognized essentially since records have become available. Contrary to recorded data, however, the processing of simulated ground motions has not received significant attention. Just as recorded data, simulated accelerations also require processing, but, unlike recorded data, the reasons for processing simulated motions are purely numerical, as, for example, large, numerically generated amplitude variability in the low frequency range. Unprocessed acceleration simulations will lead to unrealistic velocity and displacement waveforms.

This paper presents an integrated approach for the processing of simulated seismic acceleration time series. It adopts the six step approach proposed by Boore et al. (2002), namely:

1. Subtract the mean value from the entire acceleration time series;
2. Apply a short cosine taper function (usually 50–100 steps depending on the specific application) to the acceleration time series to set their initial values to zero;
3. Integrate accelerations to velocities, and apply a quadratic function fitted to the velocity time series in the least-squares minimization sense;
4. Apply a high-pass filter to the acceleration time series;
5. Remove the derivative of the velocity quadratic fitting function from the accelerations;
6. Integrate the accelerations to get velocities and displacements.

Response-spectrum compatible acceleration time series were first generated by means of the spectral representation scheme and then processed. In Step 5, the present approach utilizes a 4-th order, acausal Butterworth filter, and concentrates on the establishment of the high-pass filter's corner frequency. For this purpose, it examines the behavior of error functions

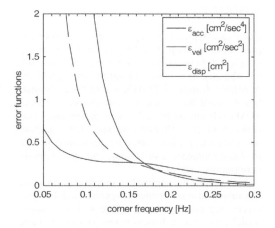

Figure 1. Error functions depicting the squared differences of acceleration, velocity and displacement time series processed with consecutive corner frequencies.

resulting from acceleration, velocity and displacement time series, indicated by ε_{acc}, ε_{vel} and ε_{displ}, respectively, that were processed with consecutive corner frequencies. Figure 1 illustrates the behavior of the error functions as they start "stabilizing", i.e. their slope starts tending to zero. It can be seen from the figure that there exists a narrow corner frequency range (0.16–0.19 Hz) where the three lines intersect. The average value of 0.175 Hz can then be viewed as the lowest corner frequency for which the error functions of all quantities of interest start tending to constant values, and, hence, as the value to be used in the processing of the simulations. It is cautioned, however, that this selection criterion needs to be validated with the influence of the choice of the corner frequency on the structural response.

REFERENCE

Boore, D.M., Stephens, C. & Joyner W. 2002. Comments on baseline correction of digital strong-motion data: examples from the 1999 Hector Mine, California, earthquake. *Bulletin of the Seismological Society of America* 92: 1543–1560.

Safety, Reliability and Risk of Structures, Infrastructures and Engineering Systems – Furuta, Frangopol & Shinozuka (eds)
© 2010 Taylor & Francis Group, London, ISBN 978-0-415-47557-0

Reliability analysis of offshore structures by reduced order models

B.J. Leira

Department of Marine Technology, Norwegian University of Science and Technology, Trondheim, Norway

M.D. Grigoriu

Cornell University, NY, USA

ABSTRACT

ROM's are approximate representations of stochastic processes consisting of a relatively small number m of samples of these processes occurring with probabilities obtained as solutions of optimization problems. A notable feature of ROM's is that their construction does not involve input-output analyses. Once a ROM has been constructed, system performance evaluation involves m deterministic input-output analyses and elementary calculations are performed in oder to obtain moments, distributions, and other response performance metrics.

Such models are applied to a simple offshore structure subjected to an exponentially correlated stochastic process representing the wave forces. The structural model represents a bottom-based gravity structure. The deck displacement and its statistical properties are considered. A refined model of the soil foundation is subsequently introduced in order to represent nonlinear behaviour. It is illustrated how the simulated sample functions of the deck displacement then will exhibit a much larger variation for a given set of load histories. A comparison of the average cumulative sample functions for the deck displacement based on the linear versus the nonlinear soil model is shown in Figure 1 below.

Assessment of the efficiency of the present method is made as compared to pure Monte-Carlo simulation. Application of the same approach to other types of offshore structures is also addressed.

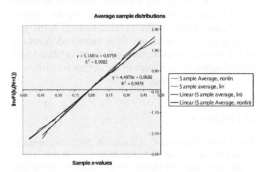

Average sample distributions

$y = 5,1881x + 0,0759$
$R^2 = 0,9982$

$y = 4,4979x + 0,0686$
$R^2 = 0,9978$

InvFi(K/(N+1))

— Sample Average, nonlin
— Sample average, lin
— Linear (Sample average, lin)
— Linear (Sample Average, nonlin)

Sample x-values

Figure 1. Comparison between average sample distribution functions for linear versus nonlinear soil model. 10 samples.

REFERENCES

M. Babuska, R. Temptone, and E. Zouraris (2004). "Galerkin Finite element approximations of stochastic elliptic partial differential equations". *SIAM Journal of Numerical Analysis*, 42(2): 800–825.

R.O. Duda, P. E. Hart, and D. G. Stork (2001) *"Pat tern Classification"*, John Wiley & Sons, Inc., New York, second editionn.

R.G. Ghanem and P. D. Spanos (1991) *"Stochastic Finite Elements: A Spectral Approach"* Springer Verlag, New York.

M. Grigoriu (1995) *"Applied Non-Gaussian Processes: Examples, Theory, Simulation, Linear Random Vibration, and MATLAB Solutions"*, Prentice Hall, Englewoods Cliffs, NJ.

M. Grigoriu (2002) *"Stochastic Calculus. Applications in Science and Engineering"*, Birkhauser, Boston.

M. Grigoriu (2006) "Evaluation of Karhunen-Loeve, spectral, and sampling representations for stochastic processes", *Journal of Engineering Mechanics, ASCE*, 132(2):179-189.

M. Grigoriu (2008) "Reduced order models for random functions. Application to stochastic problems", Applied Mathematical Modelling, in press.

Karunakaran, D.; Leira, B.J.; Spidsøe, N. and Hoen,C. (1994): "Extreme Nonlinear Dynamic Response of a Monotower GBS", Proc. Behaviour of Offshore Structures (BOSS), MIT, Boston.

Leira, B.J.; Karunakaran, D. and Hoen, C.: "'Nonlinear behaviour and extreme dynamic re sponse of the Troll gravity platform", Proc. OMAE 1994, Houston, USA.

MacCamy, R.C. and Fuchs, R.A.(1954): "Wave forces on piles; A dif fraction theory", Beach Board Technical Memo No. 69.

A.C. Miller and T. R. Rice (1983) "Discrete approximations of probability distributions" M*anagement Science*, 29(3):352–362. P.B. Nair and A.J. Keane (2002) "Stochastic reduced basis methods", *AIAA Journal*, 40(8): 1653–1664.

S.I. Resnick (1998) *"A Probability Path*. Birkhauser, Boston.

E. Rosenblueth (1981) Point estimates for probability moments. *Applied Mathematical Model ling*, 5: 329–335.

D. Stoyan, W. S. Kendall, and J. Mecke (1987) *"Stochastic Geometry and Its Applications"* John Wiley & Sons, New York.

J.P. Walser (1999) *"Integer Optimization by Local Search"*, Springer, New York.

Safety, Reliability and Risk of Structures, Infrastructures and
Engineering Systems – Furuta, Frangopol & Shinozuka (eds)
© 2010 Taylor & Francis Group, London, ISBN 978-0-415-47557-0

Stochastic forecast analysis of the service life of reinforced concrete structures subjected to chloride penetration

K.G. Papakonstantinou, S.J. Kwon & M. Shinozuka
Department of Civil and Environmental Engineering, University of California-Irvine, Irvine, CA, USA

ABSTRACT

Durability of reinforced concrete (RC) structures is a critical issue in their management. Durability of structures can be defined as their capacity to withstand the influence of actions in the course of time. Deterioration processes can greatly affect durability and thus the service life of a structure, which can be defined as the period during which the structure is able to satisfy specified requirements without repairs. Deterioration due to chloride penetration, the effects of which are often observed in but not limited to sea-shore structures, is analyzed in this study.

Chloride penetration can lead to steel corrosion of RC structures and capacity reduction. For this reason, prevention of steel corrosion has to be taken into consideration for the whole duration of a structure's service life. Apart from the essential quality controls at the construction site, probabilistic tools aiming at the design of durable structures are also needed. Conventional deterministic approaches for the solution of this problem are inadequate, due to the large amount of uncertainties involved in the problem.

Probabilistic methods including Fick's second law and Monte Carlo simulations are employed in this study to model chloride penetration. The parameters of the probabilistic distributions used were obtained from field investigation and specifications. Time dependence and independence of chloride diffusion coefficient is considered respectively in two different models and the results are compared. Moreover, models of sound and early-aged cracked concrete, as well as models whose variables vary spatially and non-spatially, have been conducted and compared.

Spatial variability of the parameters involved is taken into account and modeled by the generation of non-Gaussian, two-dimensional, uni-variate random fields by the spectral representation method. To the best of the writers' knowledge, this is the first time that this method is used in durability problems dealing with chloride intrusion. Therefore, the simulation procedure presented in this work is in and of itself a valuable tool in predicting the service life of RC structures subjected to chloride penetration.

In addition, evaluation of this study's obtained results can suggest that when accurate information is available about quality of concrete in RC structures, models with apparent diffusion coefficient being time-dependent should be used, for both sound and early-cracked structures. However, when information about concrete quality is insufficient, models with apparent diffusion coefficient being time-independent could preferably be used in terms of safety. Furthermore, it can been shown that the combination of a universal criterion for corrosion, that does not take into account localized effects, together with the spatially uneven corrosion rate, that describes spatially correlated models, results in mean value prediction similarity and mean value variation prediction dissimilarity, between non-spatially and spatially correlated models. Thus, according to the limit state function used and the level of accuracy that is desired, proper selection can be made between a non-spatially correlated model and the somewhat more complicated spatially correlated one.

REFERENCES

Kwon, S.J., Na, U.J. & Song, H.W. 2009. Service life prediction for cracked RC structures using probability approach: chloride attack through field investigations. *ICOSSAR 2009* (submitted).

Shinozuka, M. & Deodatis, G. 1996. Simulation of multi-dimensional Gaussian stochastic fields by spectral representation. *ASME, Applied Mechanics.* 49(1): 29–53.

Tang, L. & Joost, G. 2007. On the mathematics of time-dependent apparent chloride diffusion coefficient in concrete. *Cement and Concrete Research.* 37: 589–95.

Yamazaki, F. & Shinozuka, M. 1988. Digital generation of non-Gaussian stochastic fields. *ASCE, Journal of Engineering Mechanics.* 114(7): 1183–97.

*Safety, Reliability and Risk of Structures, Infrastructures and
Engineering Systems – Furuta, Frangopol & Shinozuka (eds)
© 2010 Taylor & Francis Group, London, ISBN 978-0-415-47557-0*

Statistical, sensitivity and reliability analysis using software FReET

D. Novák, M. Vořechovský & R. Rusina

Brno University of Technology, Faculty of Civil Engineering, Brno, Czech Republic

ABSTRACT

The objective of the paper is to present methods and
software for efficient statistical, sensitivity and reliability assessment implemented in FReET software
(Novák at al. 2008). The attention is given to those
techniques that are developed for analyses of computationally intensive problems like nonlinear FEM.
Sensitivity analysis is based on nonparametric rank-
order correlation. Statistical correlation is imposed by
the the simulated annealing (Vořechovský & Novák,
2003). The most interesting applications of FReET
software are referenced.

State-of-the-art probabilistic algorithms are implemented in FReET to compute the probabilistic
response and reliability. FReET is a modular computer system for performing probabilistic analysis
developed mainly for computationally intensive deterministic modeling such as FEM packages, and any
user-defined subroutines. The main features of the
software are (version 1.5):

Response/Limit state function

- Closed form (direct) using implemented Equation
  Editor (simple problems)
- Numerical (indirect) using user-defined DLL function prepared practically in any programming
  language
- General interface to third-parties software using
  user-defined *.BAT or *.EXE programs based on
  input and output text communication files
- Multiple response functions assessed in same simulation run

Probabilistic techniques

- Crude Monte Carlo simulation
- Latin Hypercube Sampling (3 alternatives)
- First Order Reliability Method (FORM)
- Curve fitting
- Simulated Annealing
- Bayesian updating

Stochastic model (inputs)

- Friendly Graphical User Environment (GUE)
- 30 probability distribution functions (PDF), mostly
  2-parametric, some 3-parametric, two 4-parametric

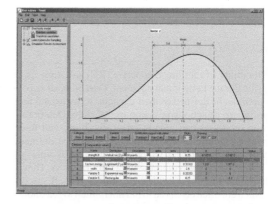

Figure 1.   Window "Random variables".

(Beta PDF and normal PDF with Weibullian left
tail), Figure 1.
- Unified description of random variables option-
  ally by statistical moments or parameters or a
  combination
- PDF calculator
- Statistical correlation (also weighting option)
- Categories and comparative values for PDFs
- Basic random variables visualization, including statistical correlation in both Cartesian and parallel
  coordinates

REFERENCES

Novák, D., Vořechovský, M., and Rusina, M. 2008.
    "FREET v. 1.5–program documentation", User's and
    Theory Guides, Brno/Cervenka Consulting, Prague,
    http://www.freet.cz.
Vořechovský, M. & Novák, D. 2003. Statistical correlation in
    stratified sampling. Proc. of *9th Int. Conf. on Applications
    of Statistics and Probability in Civil Engineering – ICASP
    9*, San Francisco, USA, Rotterdam Millpress: 119–124.

Mini-Symposia (MS13) Smart Structural Health Monitoring
and Safety Assessments
[Session Organized on Behalf of IASSAR Subcommittee SC4]

Mini-Symposia (MS) (3) Smart Structural Health Monitoring and Safety Assessments

[Session Organized on Behalf of IASSAR Subcommittee SC-4]

Safety, Reliability and Risk of Structures, Infrastructures and Engineering Systems – Furuta, Frangopol & Shinozuka (eds)
© 2010 Taylor & Francis Group, London, ISBN 978-0-415-47557-0

Development of an advanced robot system for automated bridge inspection and monitoring

J.S. Lee, I. Hwang, H.S. Lee & S.H. Hong

Department of Civil and Environmental Engineering., Hanyang University, Ansan, Korea

ABSTRACT

Conventional bridge inspection involves the physical positioning of an inspector by the hydraulic telescoping boom of a "snooper truck" thereby providing visual access to bridge components. The process is time consuming, hazardous, and may be affected by various environmental conditions. Therefore, it is of great interest that an automated and/or teleoperated inspection robot be developed to replace or complement the manual inspection practice.

This paper describes an advanced bridge inspection robot system developed at the Bridge Inspection Robot Development Interface (BIRDI) at Hanyang University. The robot is consisted of three major components, namely, the robot platform and motion control system, the machine vision system, and the informatics and bridge management system (BMS), which are described in detail. A field test is performed and the results are shown to demonstrate its applicability. The promise and limitation of the developed system is discussed in conjunction with its field deployment and commercialization.

REFERENCES

Chen, L.C., Shao, Y.C., Jan, H.H., Huang, C.W. & Tien, Y.M. 2006. Measuring system for cracks in concrete using multi-temporal images. *Journal of Surveying Engineering* 132(2): 77–82.

Fujita, Y., Mitani, Y. & Hamamoto, Y. 2006. A method for crack detection on a concrete structure. *ICPR* Hong Kong: 901–904

Huston, D.R., Pelczarski, N., Esser, B., Gaida, G., Arms, S. & Townsend, C. 2001. Wireless Inspection of Structures Aided by Robots. *SPIE Symposium on Smart for Bridges, Structures, and Highways* Newport Beach CA, 4330–09.

Iyer, S. & Sinha, S.K. 2006. Segmentation of pipe images for crack detection in buried sewers. *Computer–Aided Civil and Infrastructure Engineering* 21(6): 395–410.

Lee, J.S., Hwang, I. & Lee, H.S. 2007. Advanced Robot System for Automated Bridge Inspection and Monitoring. *IABSE Symposium Weimar* A-0715.

Lorenc, S.J., Handlon, B.E. & Bernold, L.E. 2000. Development of a Robotic Bridge Maintenance System. *Automation in Construction* 9: 251–258.

Moghadam, M.E. & Jamzad, M. 2007. Linear motion blur parameter estimation in noisy images using fuzzy sets and power spectrum. *EUROSIP Journal on Advances in Signal Processing* 2007: 1–8.

Pack, R.T., Iskarous, M.Z. & Kawamura, K. 1996. Climber Robot. *US Patent* No. 5551525.

Skibnrewski M.J. 1988. *Robotics in Civil Engineering.* Ashurst Lodge: Computational Mechanics Publications.

Yamaguchi T. & Hashimoto S. 2006. Image processing based on percolation model. *IEICE Transactions on Information and Systems* E89-D (7): 2044–2052.

Advanced crane-truck & Robot platform

Machine Vision System

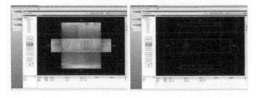

Data Management System

Figure 1. Process of the bridge inspection robot system.

Safety, Reliability and Risk of Structures, Infrastructures and Engineering Systems – Furuta, Frangopol & Shinozuka (eds)
© *2010 Taylor & Francis Group, London, ISBN 978-0-415-47557-0*

Application of robot navigation technology for civil robot development

Hyun Myung, Seungmok Lee & Haemin Jeon
Department of Civil and Environmental Engineering, KAIST, Korea

Bum-Joo Lee
Department of Electrical Engineering, KAIST, Korea

ABSTRACT

In implementing autonomous navigation systems, localization and map building of mobile robots are key technologies. Owing to many efforts in the past decade it has proved that simultaneous localization and map building (SLAM) is possible. Features or landmarks obtained from sensor measurements are registered and associated in SLAM if they are sustainable and correctly matched ones. To have such features it is essential to use a sensor system that provides precise range measurements. Typical sensors for range measurement include the laser range finder (LRF), ultra sonic, or vision sensor. A LRF provide very accurate and long range measurements while measurements from ultrasonic sensors are shorter and coarse due to cone shaped beam patterns. In this paper, a structured light system which combines laser and vision is considered for use with structural displacement measurement. We propose multiple dual laser-vision system to be used as a structural displacement measurement. The proposed dual laser-vision system is quite similar to the structured light system but it solves accurate relative pose instead of range map of the environment. Some preliminary simulation was done to show the effectiveness of the system for the long span structural displacement monitoring. Another good application example such as LBS (Location-Based Service) for U-City (ubiquitous city) will be introduced in the later part. Without initially specifying the locations of ubiquitous sensors, SLAM technique simultaneously estimates the locations of beacons, robot, and even humans.

ACKNOWLEDGMENT

This research was supported by a grant (07High Tech A01) from High tech Urban Development Program funded by Ministry of Land, Transportation and Maritime Affairs of Korean government.

REFERENCES

Dissanayake, G., Newman, P., Clark, S., Durrant-Whyte, H.F., and M. Csorba. 2001. A solution to the simultaneous localisation and map building (SLAM) problem *IEEE Transactions of Robotics and Automation*, 17(3): 229–241.

Montemerlo, M. July 2003. *FastSLAM: A Factored Solution to the Simultaneous Localization and Mapping Problem with Unknown Data Association* Doctoral dissertation, tech. report CMU-RI-TR-03-28, Robotics Institute, Carnegie Mellon University.

Myung, H., Jung, M.-J., Hong, S.-G., Park, D., Lee, H.-K., and Bang, S. 2004. Evolutionary simultaneous localization and map-building (SLAM) in home environments using structured light 2d range finder. In *Proc. of Simulated Evolution And Learning (SEAL)*. Busan, Korea. on CD-ROM.

Smith, R., Self, M., and Cheeseman, P. 1990. Estimating uncertain spatial relationships in robotics. *Autonomous Robot Vehnicles.*

Djugash, J., Singh, S., and P.I. Corke, July, 2005. Further Results with Localization and Mapping using Range from Radio In *Proc. of International Conference on Field and Service Robotics (FSR '05).*

Kurth, D., May 2004. Range-only robot localization and SLAM with radio Master's thesis, Robotics Institute, Carnegie Mellon University, Pittsburgh, PA, tech. report CMU-RI-TR-04-29.

Safety, Reliability and Risk of Structures, Infrastructures and Engineering Systems – Furuta, Frangopol & Shinozuka (eds)
© 2010 Taylor & Francis Group, London, ISBN 978-0-415-47557-0

Structural health monitoring system for reliable seismic performance evaluation of infrastructures

J.-H. Yi
Korea Ocean Research & Development Institute, Korea

D. Kim & S.H. Ko
Kunsan National University, Korea

M.Q. Feng
University of California, Irvine, CA, USA

ABSTRACT

Structural identification techniques can be classified as Level 1 for damage alarming, Level 2 for damage locating, Level 3 for damage quantifying and Level 4 for performance evaluaion. Many studies have been carried out for purpose of Level 1, 2, and 3 identifications, and they were successfully verified through a number of numerical simulations and laboratory-scale experiments, and the structural health monitoring (SHM) systems are popularly instrumented on newly constructed large-scaled civil infrastructures with dense sensor array. However, SHM systems still need to be improved for more practical tool because there are many obstacles (such as measurement noise and modeling errors, environmental effects on structural response, and complexity and redundancy of structures) for successful monitoring. To enlarge the applicability of instrumented SHM system, SHM system is applied for more reliable seismic performance evaluation of civil infrastructures, and a systematic procedure for evaluation of structural performance based on measured vibration data is proposed in this study. The procedure includes (1) identification of modal properties based on ambient vibration test, (2) updating of a preliminary finite element (FE) model using the identified modal properties (Feng *et al.* 2004) and (3) evaluation of the structural performance utilizing the updated FE model (Opensees 2004 and Shinozuka *et al.* 2002) (see Figure 1). A seismic fragility was chosen as an index of structural performance. One of the three instrumented bridges, the Jamboree Road Overcrossing, was used to demonstrate the procedure (see Figure 2). A preliminary FE model was constructed based on the bridge design drawings and then updated based on modal parameters identified from the measured vibration data. Finally the structural seismic fragility curves were developed based on both the preliminary analysis model and the updated model.

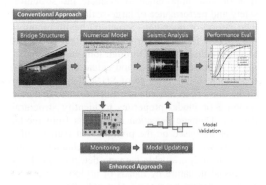

Figure 1. Description of Jamboree Road Overcrossing.

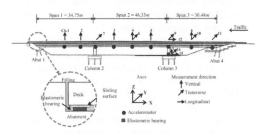

Figure 2. Description of Jamboree Road Overcrossing.

REFERENCES

Feng, M.Q., Kim, D.-K., Yi, J.-H., and Chen, Y. 2004. Baseline models for bridge performance monitoring. Journal of Engineering Mechanics, ASCE. 130(5), 562–569.

OpenSees. 2004. Open system for earthquake engineering simulation, http://opensees.berkeley.edu.

Shinozuka, M., Feng, M.Q., Kim, H.K. and Ueda, T. 2002. Statistical Analysis of Fragility Curves. Technical Report at Multidisciplinary Center for Earthquake Engineering Research, NY, USA.

Safety, Reliability and Risk of Structures, Infrastructures and
Engineering Systems – Furuta, Frangopol & Shinozuka (eds)
© 2010 Taylor & Francis Group, London, ISBN 978-0-415-47557-0

Probability-based structural assessment of Tsing Ma Bridge deck sections using in-service monitoring data

Y.Q. Ni, H.W. Xia & J.M. Ko
Department of Civil and Structural Engineering, The Hong Kong Polytechnic University, Hung Hom,
Kowloon, Hong Kong

K.Y. Wong
Bridge and Structures Division, Highways Department, The Government of The Hong Kong Special Administrative
Region, Hong Kong

ABSTRACT

It is of vital importance to evaluate the structural safety and performance of large-scale bridges during their period of operation. Structural health monitoring (SHM) technique provides an effective means to achieve this objective through long-term monitoring of structural response and actual environmental conditions. In the practice of SHM, the majority of studies have been carried out using measured dynamic response or modal properties to identify structural damage. Damage detection algorithms from modal analysis are based on the premise that damage in a structure will cause changes in modal characteristics. Most currently available damage detection methods are global in nature by means of a global structural analysis of input-output or output-only data. However, experimental verification of damage detection algorithms using vibration data from large-scale structures showed that modal characteristics might be insensitive to localized structural damage.

While global vibration data remains valuable, the strain or stress response is probably the most important data as it directly indicates the safety reserve of structural component. The strain data acquired from a SHM system is a result of multi-path combination of loadings and environmental effects. Uncertainty and variation are intrinsic for strain or stress response of civil infrastructure under operational environment. Probability-based assessment methods incorporating monitoring data from SHM systems enable inclusion of the uncertainty and variation. Also, the usage of

site-specific monitoring data for structural assessment allows the elimination of a substantial portion of modeling inaccuracy in live load characterization and structure itself, and therefore leads to a more accurate evaluation.

A probability-based method for structural assessment of bridge deck making use of in-service strain monitoring data is proposed and applied to the Tsing Ma Bridge (TMB) which has been instrumented with a long-term SHM system. The proposed method consists of the structural assessment at two levels: (i) deck structural components, and (ii) deck cross-sections. The in-service strain monitoring data from bridge deck is mainly due to three effects: traffic (highway and/or railway), wind, and temperature. However, the temperature-caused strain, although considerably large, contributes little to the stress as the majority of temperature-caused strain is released as the movement of the bridge deck is accommodated at the expansion joints and bearings. The trend ingredient due to temperature variation should be eliminated from the total measurement. A wavelet-based filtering method is proposed to eliminate the low-frequency trend ingredient in the monitoring strain data due to temperature variation. The filtered strain data is then used to evaluate the probability density functions (PDFs) of the stress under different loading conditions and to assess the safety indices of the bridge deck components and the deck cross-sections. One-year strain monitoring data from a deck cross-section of the TMB is used to demonstrate the proposed method.

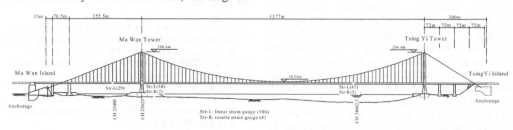

Figure 1.   Layout of Tsing Ma Bridge and sections with strain gauges.

Organized Session (OS25) Monitoring Performance under Uncertainty

Safety, Reliability and Risk of Structures, Infrastructures and Engineering Systems – Furuta, Frangopol & Shinozuka (eds)
© 2010 Taylor & Francis Group, London, ISBN 978-0-415-47557-0

Reliability assessment of reinforced concrete slab bridges strengthening

D. Novák
Faculty of Civil Engineering, Brno University of Technology, Brno, Czech Republic

W. Lawanwisut
IMMS Co., Ltd., Bangkok, Thailand

W. Kijawatworawet
Department of Civil Engineering, Kasetsart University, Nakhon Pathom, Thailand

S. Sirinaranun & S. Issarangkul Na Ayuthaya
College of Engineering, Rangsit University, Pathumthani, Thailand

ABSTRACT

One of various techniques of bridge strengthening is to change the static scheme of bridge from simple span (simply supported beam) to be continuous beam using steel plates fixed by bolts near supports. This topic is the subject of a project being carried out by the Department of Rural Roads (DOR), Ministry of Transportation, Thailand, on methods of bridge strengthening (IMMS 2005).

The aim of the paper is to study and evaluate the existing bridge behavior before and after strengthening of the slab type reinforced concrete bridge with the typical span length of 10 meters. Diagnostic load tests for both static and dynamic are performed to evaluate and assess the bridge capacity. The modification of the bridge from simple span to be continuity has shown the capacity to serve the legal vehicle loads by increasing from 21 tons to be 25 tons with the required reliability.

The bridge was calculated by the nonlinear finite element fracture mechanics to model first experiment and then real structure (Cervenka & Pukl 2005). As many uncertainties are involved in the problem, mainly concerning the knowledge of material parameters (both concrete and reinforcing steel parameters), a statistical approach was used to assess the influence of uncertainties on structural response. It is based on a randomization of nonlinear finite element analysis of concrete structures. The scatter of ultimate load capacity was obtained using Monte Carlo type simulation called Latin hypercube sampling, statistical

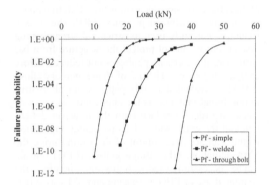

Figure 2. Reliability increase for simply supported beam and two alternatives of strengthening.

correlation was also considered. Reliability measures could be quantified in a parametric way at prescribed loading before and after strengthening. The approach utilizes advanced nonlinear techniques and advanced statistical and reliability techniques implemented in software FReET (Novák et al. 2008).

The computational model bridge (Fig. 1) was randomized, the scatter of limit load – resistance has been obtained. The failure probabilities were estimated as a probability that resistance – maximum jack load will be smaller than a certain level of load, Figure 2. We can see clearly a reliability increase – from simply supported beams to best alternative – steel plate with through bolt.

REFERENCES

Červenka, V. & Pukl, R. 2005. *ATENA Program documentation.* Cervenka Consulting, Czech Republic.
IMMS Co., Ltd. 2005. *Bridge interactive behavior between beam and column,* Final Report, Bureau of Materials, Research and Development, Department of Rural Roads, Ministry of Transport, Thailand.
Novák, D., Vořechovský, M. & Rusina, R. 2008. *FREET – Feasible Reliability Engineering Efficient Tool, User's and Theory guides. Version 1.5,* Brno/Cervenka Consulting, Czech Republic, http://www.freet.cz

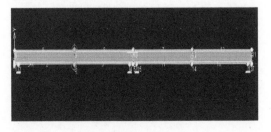

Figure 1. Computational model (ATENA).

*Safety, Reliability and Risk of Structures, Infrastructures and
Engineering Systems – Furuta, Frangopol & Shinozuka (eds)
© 2010 Taylor & Francis Group, London, ISBN 978-0-415-47557-0*

Monitoring with FBG to control cracking of r.c. structures before retrofitting

F. Daghia, A. Giammarruto, R. Carli & G. Pascale
DISTART, University of Bologna, Italy

ABSTRACT

The structural reliability of concrete beams can be strongly reduced by several cracking patterns, because of the rebar corrosion due to moisture. External bonding of fiber reinforced polymer (FRP) sheets is a technique currently used as structural strengthening system, but it can not protect the structure from bar corrosion, if the crack opening is not reduced before applying the sheets. The use of shape memory alloys (SMA) is evaluated in this paper as an alternative to traditional systems, as the application of upwards loads, in order to reduce both the midspan deflection and the crack width. This particular application requires an accurate control of the strain during the SMA actuation. Fiber Bragg grating (FBG) sensors can be successfully used with this aim. Therefore, the combined use of FBG as sensing and SMA as actuation systems is evaluated. Small scale smart concrete beams with FBG sensors and SMA actuators were designed and tested. FBG were used to monitor the deformation in the SMA wires and to determine the beam critical conditions (Fig. 1).

As Fig. 2 shows, once the SMA wires are actuated the midspan displacement gradually starts to decrease, until a minimum is reached once the wires have developed their maximum recovery stress. Upon cooling, part of the recovered displacement is gained again as the SMA wires relaxe.

As for the SMA actuators capability in reducing the cracks opening, the SMA actuation was able to partly close the developed cracks.

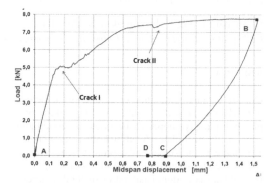

Figure 2. Load/displacement diagram.

The feasibility of monitoring the SMA deformation via FBG sensors has been shown, although further investigation is required to identify the best combination of SMA actuation temperatures and glue for the FBG sensor.

REFERENCES

Daghia F. Active fibre-reinforced composites with embedded shape memory alloys. PhD Thesis 2008, University of Bologna, DISTART department.

Deng Z, Li Q, Sun H. Behavior of concrete beam with embedded shape memory alloy wires. Engineering Structures 2006; 28: 1691–1697.

Arduini M, Bonfiglioli B, Manfroni O, Pascale G. New Applications of Fiber Optic Sensors for Structural Monitoring. In: Proceedings of IABSE Symposium "Structures for the Future- The Search for Quality", August 25–27, 1999, Rio de Janeiro, Brazil, p. 491–498.

Ansari F. Fiber optic sensors for construction materials and bridges. Technomic: Lancaster, USA, 1998.

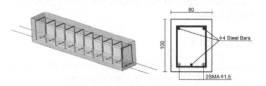

Figure 1. Geometry and reinforcement the specimens.

Safety, Reliability and Risk of Structures, Infrastructures and Engineering Systems – Furuta, Frangopol & Shinozuka (eds)
© 2010 Taylor & Francis Group, London, ISBN 978-0-415-47557-0

Fatigue reliability assessment of steel bridges based on monitoring and FE modeling

K. Kwon & D.M. Frangopol
Lehigh University, Bethlehem, PA, USA

ABSTRACT

This paper focuses on fatigue reliability assessment of steel bridges by using prediction models at several sensor locations based on monitoring data and finite element modeling (FEM). Bridge performance that may have been deteriorated due to fatigue has to be steadily assessed and predicted. Bridge fatigue reliability assessment can be performed more efficiently if long-term monitored data at all critical locations are regularly provided. Continuous monitoring of bridges during their entire service lives, however, may not be possible because of various restrictions, including budgetary constraints. For this reason, it may be necessary to estimate fatigue life based on limited time monitoring under uncertainty and finite element analysis.

The monitored data related to fatigue deterioration process are used to predict the number of cycles and stress ranges under potential displacements. The original measured data are used in a reliability assessment. In this study, stress range histograms are created by random number generation techniques based on information obtained from both stress results computed from FEM and monitored data. The FE models are validated by comparing the analytical results from the FEM with the field measurements. The ranges of applied relative displacements are determined based on actual monitored data.

The basic formula, $N = A/\Delta F^{1/3}$, used to estimate the number of stress cycles, N, is provided in the AASHTO Specifications (2002). The fatigue details coefficient, A, associated with the defined category is herein treated as a random variable, and the stress range, ΔF, is determined by using the ratio of the measured stress from monitoring to the computed stress from FEM at the sensor locations.

Consequently, the fatigue reliability assessment is based on simulation by using the AASHTO Specifications, monitoring, and FEM. The proposed approach is illustrated on an existing steel highway bridge. As a method of the assessment for the retrofit, monitoring programs were performed. The monitored data obtained on this bridge offer the opportunity to perform studies regarding fatigue reliability assessment using prediction monitoring.

REFERENCES

AASHTO Guidelines 2002. AASHTO Standard Specification for Highway Bridges. Washington, D.C.

ABAQUS 2007. ABAQUS Version 6.7-1: Users manual. Hibbitt, Karlsson & Sorensen, Inc., Pawtucket, RI.

Alampalli, S. & Lund, R. 2006. Estimating Fatigue Life of Bridge Components Using Measured Strains. Journal of Bridge Engineering 11(6).

Connor, R.J. & Fisher, J.W. 2002. Report on Field Inspection, Assessment, and Analysis of Floorbeam Connection Cracking on the Birmingham Bridge Pittsburgh PA. Lehigh University's Center for Advanced Technology for Large Structural Systems (ATLSS), Bethlehem, PA.

Connor, R.J., Fisher, J.W., Hodgson, I.C. & Bowman, C.A. 2004. Results of Field Monitoring Prototype Floorbeam Connection Retrofit Details on the Birmingham Bridge. Lehigh University's Center for Advanced Technology for Large Structural Systems (ATLSS), Bethlehem, PA.

Frangopol, D.M. & Estes, A.C. 1997. Bridge Maintenance Strategies Based on System Reliability. Structural Engineering International, Journal of IABSE 7(3): 193–198.

Frangopol, D.M. & Estes, A.C. 1998. RELSYS: A Computer Program for Structural System Reliability Analysis. Structural Engineering and Mechanics, Techno-Press 6(8): 901–919.

Frangopol, D.M. & Liu, M. 2007. Maintenance and Management of Civil Infrastructure Based on Condition, Safety, Optimization and Life-Cycle Cost. Structure and Infrastructure Engineering 3(1): 29–41.

Frangopol, D.M., Strauss, A. & Kim, S. 2008. Bridge Reliability Assessment Based on Monitoring. Journal of Bridge Engineering, ASCE 13(3): 258–270.

Liu, M., Frangopol, D.M. & Kwon, K. 2008. Fatigue Reliability Assessment of Retrofitting Distortion-induced Cracking in Steel Bridges based on Monitored Data. (submitted for publication).

Miner, M.A. 1945. Cumulative Damage in Fatigue. Journal of Applied Mechanics 12(3): 159–164.

Pourzeynali, S. & Datta, T. K. 2005. Reliability Analysis of Suspension Bridges against Fatigue Failure from the Gusting of Wind. Journal of Bridge Engineering 10(3): 262–271.

Safety, Reliability and Risk of Structures, Infrastructures and Engineering Systems – Furuta, Frangopol & Shinozuka (eds)
© 2010 Taylor & Francis Group, London, ISBN 978-0-415-47557-0

Assessment of structural performance considering spatial variability of material properties

R. Pukl
Červenka Consulting, Prague, Czech Republic

M. Vořechovský & D. Novák
Institute of Structural Mechanics, Faculty of Civil Engineering, Brno University of Technology, Czech Republic

ABSTRACT

Reliability assessment based on the non-linear finite element analysis represents an innovative tool for predicting performance and safety of civil engineering structures and for supporting their life-cycle management and maintenance. In order to obtain realistic results of the non-linear analysis (structural response, damage) uncertainty and variability of structural properties should be considered in a realistic way. In particular, the spatial variability and in-homogeneity of material properties, which occur in the real structures, are important phenomena causing random occurrence of damage (e.g. cracks in concrete) and should be accounted for appropriately. Methodology for introducing spatial variability of material properties into the non-linear finite element system is presented together with a brief description of selected applications.

A high level and also very natural (physical) technique of uncertainties modeling is their representation by *random fields*. This paper focuses on representation of material properties (namely mechanical properties of concrete and steel) in advanced numerical models.

The nonlinear finite element analysis employs advanced constitutive models for concrete based on damage mechanics, nonlinear fracture mechanics and plasticity theories with smeared crack approach. It is a proven tool for computer simulation of reinforced concrete structures including failure mechanism and post-peak behavior. It enables to evaluate response of the structure (load-deflection curve, stresses, deflections, crack widths etc.) to external action (forces, settlements of supports, volumetric or environmental effects). It was extended by encapsulating of the existing material models to incorporate the spatial variability of material properties based on the random fields approach.

Appropriate reliability methods are used to calculate stochastic properties of the response variables and structural resistance from random inputs, and consequently to assess structural safety. For the time-intensive calculations like nonlinear fracture

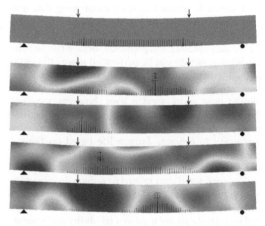

Figure 1. Four-point bending tests – random fields of concrete strength and crack patterns.

mechanics of concrete, the small-sample simulation techniques based on stratified sampling of Monte Carlo type represent a rational compromise between feasibility and accuracy. Therefore, Latin hypercube sampling method (LHS) was selected as a key fundamental technique. Random fields based on this approach describe the random distribution of a structural (material) property over the region representing the structure based on the prescribed correlation length and the autocorrelation function.

The presented technology enables to model uncertainties and inhomogeneities in the nonlinear finite element solution and consequent assessment of structural performance. It allows accounting for the spatial variability of material properties in the non-linear finite element framework. Random occurrence of structural damage (cracks) can be simulated even in a homogeneous stress state, such as the bending span in Fig. 1, where different realizations of local concrete strength field lead to different crack patterns.

Safety, Reliability and Risk of Structures, Infrastructures and
Engineering Systems – Furuta, Frangopol & Shinozuka (eds)
© 2010 Taylor & Francis Group, London, ISBN 978-0-415-47557-0

Structural reliability assessment using sensors and Bayesian updating

D.M. Frangopol
Lehigh University, Bethlehem, PA, USA

A. Strauss
University of Natural Resources and Life Sciences, Vienna, Austria

S. Kim
Lehigh University, Bethlehem, PA, USA

ABSTRACT

Probabilistic design and assessment methods for new and existing structures are now commonly used. These methods require information under uncertainty with respect to various parameters related to structural performance. In general, such information is not fully provided by traditional inspection methods. For this reason, monitoring systems, in combination with traditional inspection methods, take a greater significance. Monitoring systems used in structural engineering can provide essential data for reliability assessment and maintenance planning. However, an extensive storage and evaluation effort is necessary due to the huge amount of data. Therefore, the proper handling of the continuously provided monitoring data is necessary. The design of long-term monitoring systems needs important decisions on the necessary number of sensors and their locations, the data management, and the adjustment of prediction functions, among others. An example of predicting reliability index profile with respect to steel yielding based on daily extreme values indicated by a sensor placed on an existing bridge is provided in Fig. 1.

The purpose of this paper is to review, present and illustrate (a) methods for the effective incorporation of monitored data in the reliability assessment of structures and structural components, (b) acceptance based methods for the determination of the necessary number of sensors, and (c) methods for the adjustment of prediction functions incorporating Bayesian updating.

REFERENCES

Ang, A.H.-S. & Tang, W.H. 2007. *Probability Concepts in Engineering Planning*, 2nd Edition, John Wiley.
Connor, R.J. & McCarthy, J. 2006. *Report on Field Measurements and Uncontrolled Load Testing of the Lehigh River Bridge (SR-33)*. Lehigh University's Center for Advanced Technology for Large Structural Systems (ATLSS), ATLSS Phase II Final Report 06-12.
Connor, R.J. & Santosuosso, B. 2002. *Field Measurements and Controlled Load Testing on the Lehigh River Bridge (SR-33)*. Lehigh University's Center for Advanced Technology for Large Structural Systems (ATLSS), ATLSS Report 02-07.
Frangopol, D.M., Strauss, A. & Kim, S. 2008a. Bridge Reliability Assessment Based on Monitoring. *J. Bridge. Engrg.*, ASCE, 13 (3): 258–270.
Frangopol, D.M., Strauss, A. & Kim, S. 2008b. Use of monitoring extreme data for the performance prediction of structures: General Approach, *Engineering Structures*, Elsevier, 30 (12): 3644–3653.
Mahmoud. H.N, Connor. R.J. & Bowman C.A. 2005. *Results of the Fatigue Evaluation and Field Monitoring of the I-39 Northbound Bridge over the Wisconsin Rever*. Lehigh University's Center for Advanced Technology for Large Structural Systems (ATLSS), ATLSS Report 05-04.
Strauss, A., Frangopol, M. & Kim, S. 2008. Use of monitoring extreme data for the performance prediction of structures: Bayesian updating. *Engineering Structures*, Elsevier, 30(12): 3654–3666.

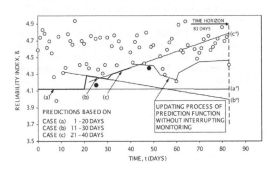

Figure 1. First order polynomial prediction functions obtained by using the monitored extreme data provided by a sensor on an existing bridge.

Safety, Reliability and Risk of Structures, Infrastructures and
Engineering Systems – Furuta, Frangopol & Shinozuka (eds)
© 2010 Taylor & Francis Group, London, ISBN 978-0-415-47557-0

Monitoring of road traffic load as basis for the development of traffic models for weight restricted bridges based on the Eurocode 1

D. Proske

University of Natural Resources and Applied Life Sciences, Vienna, Austria

S. Loos

Ingenieurbüro Lopp, Weimar, Germany

ABSTRACT

Road traffic weight restricted historical bridges are common in Germany. However often it is unclear in which amount road users follow this restriction and how the road weight distribution looks like. Therefore in the German city Dresden a weight my motion for road traffic was installed on a weight restricted bridge. This weighting machine included a software package for identification of single vehicles based on the measured axle loads. This package was running several months/years and data was collected.

Since currently the Eurocode does not include a model for weight restricted bridges, the data collected has been used to extend the current road traffic load concept of the Eurocode and the German DIN-reports to such bridges. This has been done by using the measured data do provide statistical information for a Monte Carlo Simulation. The Monte Carlo Simulation resulted in statistical distributions of the internal forces due to the road loading. The load calibration factor α has then been determined for comply with this simulated data. Furthermore load calibration factors for other weight restrictions have been estimated based on a mixture of measured and estimated load distributions. Finally a comparison with the unrestricted Eurocode road traffic model and the Auxerre load has been carried out to control the applied technique.

Table 1 summarizes the computed load calibration factors. The data in Table 1 depend on the roadway quality. Usually this property is not given in codes, however, here it is assumed that for country roads with weight restrictions lower roadway quality can be found and this has impacts on the chosen α-factor due to the dynamic properties. Here the model from Merzenich & Sedlacek (1995) has been applied.

According to the knowledge of the authors this is the first correct adaptation of the Eurocode road traffic load model to weight restricted bridges in contrast to sometimes found corrections (Vockrodt 2005). However other adaptations are known dealing with country roads (Novák et al. 2007).

Table 1. Factors for recalibration classes.

Bridge class	Roadway quality	Lane 1 α_{Q1}	α_{q1}	Lane 2 α_{Q2}	α_{q2}
3/3*	Average	0.10	0.22		
6/6*	Average	0.20	0.24		
9/9*	Average	0.25	0.26		
12/12	Good	0.30	0.28	0.20	1.00
	Average	0.30	0.30	0.25	1.00
16/16	Good	0.35	0.30	0.35	1.00
	Average	0.35	0.40	0.45	1.00
30/30	Good	0.55	0.70	0.50	1.00
	Average	0.60	0.70	0.80	1.00
Simulation Auxerre	Good	1.0	0.90	1.00	1.00
Load model	1 DIN 101	0.80	1.00	0.80	1.00

* First drafts

REFERENCES

Merzenich G & Sedlacek G 1995 Hintergrundbericht zum Eurocode 1 – Teil 3.2: Verkehrslasten auf Straßenbrücken. Forschung Straßenbau und Straßenverkehrstechnik. Bundesministerium für Verkehr, Heft 711.

Novák B, Brosge B, Barthel K & Pfisterer W 2007 Anpassung des Verkehrslastmodells des DIN FB-101 für kommunale Brücken, Beton- und Stahlbetonbau 102, Heft 5, pp 271–279.

Vockrodt H-J 2005 Instandsetzung historischer Bogenbrücken im Spannungsfeld von Denkmalschutz und modernen historischen Anforderungen. 15. Dresdner Brückenbausymposium, Technische Universität Dresden, pp. 221–241.

Mini-Symposia (MS10) Structural Safety and Reliability by Means of Structural Health Monitoring

Safety, Reliability and Risk of Structures, Infrastructures and Engineering Systems – Furuta, Frangopol & Shinozuka (eds)
© *2010 Taylor & Francis Group, London, ISBN 978-0-415-47557-0*

Quantitative deployment of long-term force monitoring devices in cable-stayed bridges. A Life Cycle Cost (LCC) approach

A. Ladysz
Wroclaw University of Technology, Wroclaw, Poland

J.R. Casas
Technical University of Catalonia, Barcelona, Spain

ABSTRACT

The paper describes two possible device-alternatives of Structural Health Monitoring (SHM) as a solution for the inspection and maintenance of stays of cable-stayed bridges. It presents Monitoring as an important investment that brings comfort of the maintenance during assumed service life. Especially, it covers the problem of quantitative deployment of the stay-force measuring devices and foreseen benefits from their application. Deliberations are conducted on the basis of a newly designed cable-stayed bridge in Wroclaw, Poland (Fig. 1), equipped with a long term, remote SHM (Biliszczuk 2007).

Under consideration are taken two possible monitoring solutions, according to the state-of-the-art in the field of the force measurements: measurement of the forces by means of accelerometers and related vibrating chord theory (Casas 1994) or measurement of the forces by means of elasto-magnetic sensors based on the ferromagnetic properties of the steel (Jarosevic 2008). It presents the selection of the monitoring devices and the optimal number to be used, based on the minimization of the total expected cost of the bridge, including the cost of monitoring. The main objective of the example is to mature these methods and highlight their influence on the total cost of

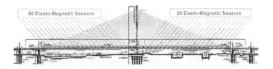

Figure 2. 160 elasto-magnetic sensors (1 per cable).

the bridge, considering a Life-Cycle Cost Analysis (LCCA) approach.

The benefit B of installing a monitoring system can be expressed as:

$$B = C_{et} - C_{et}^m$$

C_{et} – Expected total cost, without monitoring
C_{et}^m – Expected total cost, including monitoring

In the above total cost, the following components are included: expected cost of the design and erection, routine inspection with maintenance, expected cost of a failure, cost of the monitoring application and its maintenance.

SHM application is reasonable only if B is greater than zero. The type of the devices and their quantity is also related to the maximization of the expected benefit B. In the paper, different arrangements of the SHM system are assessed. The optimal solution is presented in figure 2. Prices of the sensors, their accuracy in the measurement of cable forces, labor costs, materials for the reparation and miscellaneous services are the input data considered in the problem.

REFERENCES

Biliszczuk, J. 2007. Podwieszony most przez Wisłe w Płocku", *DWE, Wroclaw*.
Casas, J.R. 1994. A combined method for measuring cable forces: the cable-stayed Alamilo bridge, Spain. *Structural Engineering International* 4 (4): 235–240.
Jarosevic, A. 2008. *Magnetoelastic method of stress measurements in steel* www.dynamag.com, Accessed: 2008.

Figure 1. Visualisation of the newly designed Cable-Stayed bridge in Wroclaw.

Safety, Reliability and Risk of Structures, Infrastructures and
Engineering Systems – Furuta, Frangopol & Shinozuka (eds)
© 2010 Taylor & Francis Group, London, ISBN 978-0-415-47557-0

Service life prediction of masonry under high loading: Modelling and probabilistic evaluation

E. Verstrynge, L. Schueremans & D. Van Gemert
Department of Civil Engineering, KULeuven, Heverlee, Belgium

ABSTRACT

During the past decennia, several collapses of historical masonry structures throughout Europe have drawn
our attention to the vulnerability of masonry constructions under high loading. Monumental masonry
structures, such as tall belfries and medieval towers, are subjected to high dead weight, which causes
time-dependent deformations. The combination of
this time-dependent deformation behaviour with accu-
mulating weathering phenomena has caused several
historical masonry structures to collapse rather unexpected and without an immediate visible cause such as
fire or thunderstorms.

This paper discusses the service life prediction of
masonry under high loading. Therefore, two key elements are addressed: How to model the masonry's
time-dependent deformation behaviour? And how to
simulate this long-term process in laboratory conditions to obtain parameters for the model? An attempt
was made during an extensive test program including
accelerated creep tests and incorporating the uncertainties and scatter on the parameters of the model by
performing a probabilistic analysis. The different tests
involved in the research program are:

   short-term compressive tests;
   three different types of short-term creep tests:
   accelerated creep tests (ACT), cyclic accelerated
   creep tests (CACT) and accelerated creep tests with
   an additional small cyclic loading (ACT + C);
   long-term creep tests.

The long-term deformation behaviour of historical masonry is being modelled using a rheological
model. Damage accumulation is incorporated within
the model by means of damage variables (D) (Papa &
Talierco, 2005). The model parameters and the evolution of the damage variables are obtained from the
experimental tests (Verstrynge, 2008).

A probabilistic analysis is adopted in order to consider the scatter on the parameters of the model. The
sensitivity analysis indicates that the scatter on the
compressive strength, f_c, and the Young's modulus, E^M,
together with the evolution of the damage parameter,

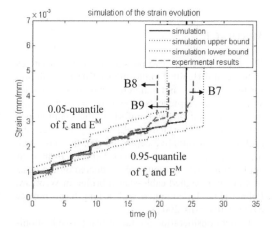

Figure 1. Simulation and experimental results of ACT's.

D^V, have the highest influence on the failure probability. Comparison of the model simulations with the
experimental results from the short-term creep tests
indicates a rather good fit between both (see Fig. 1).

A Monte Carlo (MC) analysis is used to investigate
the failure probability at the end of the short-term creep
tests, which demonstrates a good correlation with the
experimental results. Additionally, MC and MC + VI
sampling are used to obtain a reliability index for the
masonry's structural integrity after a period of 300
years at different stress levels.

The proposed framework provides a powerful tool in
decision making, which will be extended towards the
assessment of real structures and realistic timeframes.

REFERENCES

Papa E. & Talierco A. 2005. A visco-damage model for brittle
materials under monotonic and sustained stresses. *International journal for numerical and analytical methods in
geomechanics*, Vol 29 (3): 287–310.
Verstrynge, E., Ignoul, S., Schueremans, L., Van Gemert, D.
2008. Modelling of damage accumulation in masonry subjected to a long-term compressive load. Proc. of the 6th
Int. Conf. on structural analysis of historical constructions.
Bath 2–4 July 2008, p. 525–532.

Safety, Reliability and Risk of Structures, Infrastructures and
Engineering Systems – Furuta, Frangopol & Shinozuka (eds)
© 2010 Taylor & Francis Group, London, ISBN 978-0-415-47557-0

Updating future reliability of nonlinear systems with low dimensional monitoring data using short-cut simulation

Jianye Ching

Department of Civil Engineering, National Taiwan University, Taipei, Taiwan

Yi-Hung Hsieh

Department of Construction Engineering, National Taiwan University of Science and Technology, Taipei, Taiwan

ABSTRACT

This paper proposes a novel stochastic simulation method of updating future reliability of a nonlinear system with high dimensional uncertainties when the monitoring data is low dimensional. The novelty of the proposed framework is to bypass the most difficult part of the problem: drawing samples of uncertain variables conditioning on the low dimensional monitoring data. This part can be extremely challenging and even prohibitive when the dimension of the uncertain variables is high. This research proposes a short-cut simulation approach: instead of drawing samples of possibly high dimensional uncertain variables conditioning on the monitoring data, it is shown that the problem can be solved by drawing samples of the low dimensional monitoring data conditioning on the future failure event. The latter action turns out to be quite straightforward. Moreover, as long as the probability distribution of the uncertainties and the mathematical model of the target system are given, the entire functional relationship between the updated future failure probability and the monitoring data can be obtained prior to the monitoring process.

The goal of regular reliability analyses is to estimate future failure probability given the probability distribution of the uncertainties in the target system and the mathematical model M of the system. When monitoring data D is available, it is essential to incorporate it to reduce the uncertainties and to update future reliability because D may contain much information about system parameter Θ. Therefore, it is desirable to develop a methodology to update future reliability based on these measurements.

To illustrate the problem of updating future reliability, let us consider the schematics in Figure 1, where $X^{current}$ denotes the uncertain excitation during the (current) monitoring process, while X^{future} denotes the future uncertain excitation. Note that $X^{current}$ and X^{future} can be completely of different type of excitation. It is assumed in this study that the system parameter Θ stays constant from the time instant of monitoring

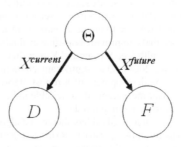

Figure 1. The graphical model for $X^{current}$, X^{future}, Θ, D and F

to future excitation and that $X^{current}$ and X^{future} are independent.

A brief procedure of estimating $P(F)$, $p(d|F)$ and $p(d|F^C)$ is presented in the following.

(1) Estimating $P(F)$
(2) Drawing samples from $p(\theta|F)$ and $p(\theta|F^C)$
(3) Drawing samples from $p(d|F)$ and $p(d|F^C)$
(4) Estimating $p(d|F)$ and $p(d|F^C)$

Given the samples from $p(d|F)$ and $p(d|F^C)$, these two PDFs can be estimated using suitable density estimation methods, e.g.: histograms and kernels. Note that most density estimation approaches are not robust against D dimension, so this step is not robust against D dimension. Together with the estimated $P(F)$, estimation of $P(F|d)$ can be made based on the following equation, which is re-written here for convenience:

$$P(F|d) = [p(d|F) \bullet P(F)]/\{p(d|F) \bullet P(F) + p(d|F^C) \bullet [1 - P(F)]\}$$

Because the entire probability density functions of $p(d|F)$ and $p(d|F^C)$ are obtained, the $P(F|d)$ is estimated as a function of d.

Safety, Reliability and Risk of Structures, Infrastructures and Engineering Systems – Furuta, Frangopol & Shinozuka (eds)
© 2010 Taylor & Francis Group, London, ISBN 978-0-415-47557-0

Damage identification on a shear story structure using multivariate autoregressive time series modeling

A. Cheung & A.S. Kiremidjian
Stanford University, Stanford, CA, USA

Development of a robust structural health monitoring (SHM) system for evaluation of civil structures has the potential to increase life safety and facilitate disaster response. Advances in wireless sensor technology have made it feasible to create an easy-to-install sensor network that can provide real-time evaluation of a structure's health (see Straser and Kiremidjian, 1998, Lynch et al., 2004). One significant challenge that remains is the development of a robust algorithm to connect sensor measurements to reliable indicators of damage location and magnitude.

Recent research in SHM has shown that algorithms based on using autoregressive (AR) modeling of vibration time series as the feature extraction step of a pattern classification framework are capable of detecting structural damage to varying degrees (Sohn et al., 2001, Nair et al., 2006, Noh et al., 2009). However, a comprehensive algorithm capable of detecting and localizing damage robustly has not been fully developed. This paper presents a multivariate AR model for damage identification on a multistory shear structure, and links parameters of the model to the stiffness terms of the structure, in order to justify their selection as a damage sensitive feature (DSF). Next, a new damage identification algorithm using the AR model is presented.

A multivariate AR model for damage identification has been developed by Monroig and Fujino (2006), which is derived from the equation of motion of a shear story structure and finite difference equations. This paper follows a similar approach starting with the equation of motion. However, a new AR model is proposed based on the Tustin transformation, a method of approximating an ordinary differential equation with a linear difference equation, which is more accurate than the finite difference equations used to derive the previous model. It is shown that certain AR coefficients are functions of the stiffness terms of the structure. Consequently, changes in the structural stiffness due to damage should result in corresponding migration of the AR coefficients.

A new damage identification algorithm based on this model is proposed, which uses the AR coefficients as the DSFs. Feature discrimination is done using a simple two sided t-test (Nair et al., 2006), and the *t* statistic itself is used as a damage measure for damage quantification.

The damage identification algorithm using both the previous and the new model is then evaluated using numerically simulated data from the ASCE Phase I benchmark structure for SHM. Both the previous and the new model are shown to be capable of detecting and localizing major damage patterns with approximately equal success. However, for minor damage patterns, only the Tustin transformation model is capable of detecting damage. Furthermore, the Tustin transformation model is able to quantify and locate the approximate region of damage.

Further development of this algorithm for possible use in real structures involves the following steps. First, the algorithm needs to be verified on experimental data, preferably with an extensive reference case collected under varying environmental conditions in order to test algorithm robustness. Second, wireless sensors with the damage identification algorithm embedded on a microprocessor should be developed, in order to determine the feasibility of implementing this algorithm in practice. Third, additional physical models should be developed and fit to AR models in order to expand the scope of this algorithm beyond shear story structures.

Safety, Reliability and Risk of Structures, Infrastructures and
Engineering Systems – Furuta, Frangopol & Shinozuka (eds)
© 2010 Taylor & Francis Group, London, ISBN 978-0-415-47557-0

Substructure identification for shear structures: Power spectral density method

D. Zhang & E.A. Johnson

Sonny Astani Department of Civil and Environmental Engineering, University of Southern California, Los Angeles, CA, USA

ABSTRACT

A shear structure is widely used to model the dynamic behavior of building structures; thus, accurately identifying the parameters of a shear structure plays a vital role in structural health monitoring (SHM) and damage detection for the buildings.

In previous studies by authors (Zhang & Johnson, 2006, 2009a), an innovative substructure identification method, based on a Fourier transform of responses, for shear structures was proposed. By using the dynamic equation of motion of each floor, a series of substructure identification problems can be formulated, from which all structural parameters can be estimated from top to bottom in an inductive manner. In each step of the identification, the Fourier transform of two or three adjacent floors acceleration responses are utilized to formulate a substructure identification problem in which the stiffness and damping coefficient of a certain story are identified. Repeating this procedure, the stiffness and damping coefficients of the whole structure can be identified from top to bottom in an iterative manner. However, due to the noisy nature of the acceleration measurements, this Fourier transform based method can provide accurate results only when the measurement noise is not too large. To improve identification accuracy under large noise, an improved substructure method using transfer functions was subsequently developed by authors (Zhang & Johnson, 2009b). However, the implementation of this new method requires several strict constraints that prevent its wide application.

In this paper, a new substructure identification method, using power spectrum densities, is derived from the differential equations governing the structural random responses. This identification method can overcome the previous constraints required in the transfer function method. A reference response, which is jointly wide sense stationary (WSS) with all structural responses, is introduced and the cross power spectral density between this reference response and structural accelerations, calculated by averaging long wide sense stationary responses in the frequency domain, are used to formulate the new substructure identification, which greatly improves the identification accuracy by reducing the effect of measurement noise through an averaging technique. An identification error analysis is performed to reveal how the uncertainty in the identification process will affect the parameter identification errors. Based on this result, a smart selection mechanism is designed to choose the best reference response candidate which can significantly reduce the effect of measurement noise and, thus, further improve the identification. A 5-story shear building structure is used to demonstrate the effectiveness of the proposed substructure identification method. The simulation results show that this new method can provide very accurate parameter estimates (e.g. the relative root-mean-square-errors of stiffness estimates for most stories are a fraction of a percent) under the disturbance of very large measurement noise (40% measurement noise in root-mean-square sense). Furthermore, compared with randomly selected reference response, the optimal reference response chosen by the proposed selection mechanism does make the identification much more accurate.

REFERENCES

Zhang, D, and Johnson, E.A. 2006. "Substructure Parameter Identification Method for Shear Type Structure," *4th World Conference on Structural Control and Monitoring*, San Diego, July 11–13, 2006, paper 4WCSCM-371.

Zhang, D, and Johnson, E.A. 2009a. "Substructure Identification for Shear Structures I, identification method," *Structural Control and Health Monitoring*, submitted.

Zhang, D, and Johnson, E.A. 2009b. "Substructure Identification for Shear Structures, transfer function method," *Structural Control and Health Monitoring*, submitted.

Safety, Reliability and Risk of Structures, Infrastructures and
Engineering Systems – Furuta, Frangopol & Shinozuka (eds)
© 2010 Taylor & Francis Group, London, ISBN 978-0-415-47557-0

Comparison of reliability of two on pile-supported wharves from structural monitoring data

H. Yáñez-Godoy
OXAND S.A., Avon-Fontainebleau, France

F. Schoefs, A. Nouy & M. Chevreuil
GeM Nantes Atlantic University, Nantes, France

ABSTRACT

During these last decades, developments focus on
the modeling of materials including the updating
from inspections (Rouhan & Schoefs 2003, Faber &
Sorensen 2002). The survey of structures (displacement) gives additional information but the updating
of the modeling is a great challenge when a lot of
influencing factors are involved. Thus, intrusive structural monitoring of complex structures is actually the
only way for reaching as close as possible their real in-
service behavior. This is also of first importance for
their re-analysis.

We present here a reliability analysis of pile-
supported wharves whose containers cranes are submitted to extreme storm loading. These wharves were
built with same materials but different method of
building. The modeling of in-service behavior of these
structures in conjunction with structural monitoring
allows assessing the level of loading and provides more
realistic models.

The behavior of pile-supported wharves is conditioned by several hazards in particular because of the
difficult conditions of building (Yáñez-Godoy et al.
2008) and uncertainties on extreme loadings (storms).
Storm conditions play a dominant role as they act on
cranes; they are of major importance for re-analysis
of old structures which were designed without taking
these situations into account. The updating of wind
speed due to climate changes increases the need of suitable models. Reliability analysis gives efficient tools
to perform such computations as long as boundary
conditions of the structural model are known with a
given uncertainty.

A stochastic modeling takes into account the large
scatter of measured quantities (see Figure 1). It lets
to identify main random fields influencing mechanical behavior. A decomposition on polynomial chaos of
random identified fields and wind loading acting on a
container crane is then selected. A performance criterion, based on wharf displacement and in line with the
so-called Service-Limit-States is suggested for reliability analysis. We make use of non-intrusive methods, particularly projection method and Monte-Carlo
simulations. Projection method has a highly accuracy

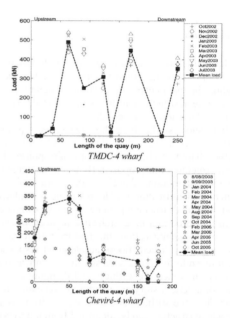

Figure 1. Medium-term loading evolutions in each wharf.

in representing the response distribution tails allowing
to post-evaluate the probability of failure. A comparison with a great number of Monte-Carlo simulations is
then accomplished to verify the accuracy. Finally, sensitivity studies are then performed to rank the set of
basic variables and to analyze the effect of the choice
of the critical value of the performance criterion.

REFERENCES

Rouhan A., Schoefs F., 2003, Probabilistic modeling of
inspection results for offshore structures, *Journal of
Structural Safety, 25: 379–399.*

Faber M.H., Sorensen J.D., 2002, Indicators for inspection
and maintenance planning of concrete structures, *Journal
of Structural Safety, 24.*

Yáñez-Godoy H., Schoefs F., Casari P., 2008, Statistical Anal-
ysis of the Effects of Building Conditions on the Initial
Loadings of On-piles Quays, *Journal Structural Health
Monitoring, 7(3): 245–263.*

Organized Session (OS27) Hoshiya Memorial

Safety, Reliability and Risk of Structures, Infrastructures and Engineering Systems – Furuta, Frangopol & Shinozuka (eds)
© 2010 Taylor & Francis Group, London, ISBN 978-0-415-47557-0

Seismic reinforcing design for an existing bridge pier

T. Koike & Y. Watanabe
Musashi Institute of Technology, Tokyo, Japan

T. Imai
JFE Engineering Corporation, Yokohama, Japan

ABSTRACT

Many existing infrastructures constructed prior to 1980 in Japan were designed for a particular seismic load which is smaller than the Level 2 ground motion caused by the maximum considered earthquake (MCE). Those existing structures must be retrofitted to comply with the seismic requirement for the level 2 ground motion to be newly designated after 1995 Hyogoken-Nanbu earthquake. For such pile-supported structures, a new seismic retrofitting approach is proposed by introducing additional reinforing piles. The effectiveness for this approach will be discussed based on the seismic performance-based design method.

The basic concept of seismic design for such structure is based on two purposes; (1) both the structure and pile-supported foundation are in the elastic state corresponding to a Level 1 ground motion, while (2) the structure moves into the inelastic state for a Level 2 ground motion before a plastic hinge is formed at a critical point in a pile.

If the structure is retrofitted with excessive reinforcement for strong seismic effect like Level 2 ground motion, the piles supporting the structure will create plastic hinge or local buckling. On the other hand adding excessive piles to increase the original strength of the foundation will cause an unexpected failure at the weakest portion of the structure.

In this study, a simplified design method is proposed for seismic design of reinforcing piles supporting an existing structure. Based on the reliability analysis of an existing bridge structure after possible future earthquakes, the present study discusses the optimal combination of the structural strength and the additional pile reinforcement that are necessary to obtain the effective maintenance strategies of deteriorating structures under seismic risks.

Discussions are devoted on (1) the definition of seismic performance level and its probability of damage states, (2) seismic response of existing structure with pile foundation, especially in stressing the effect of structural characteristics coefficient, and (3) numerical studies based on various parameters.

Fig. 1 is a schematic illustration of the structure-pile-foundation system. The group piles are allocated

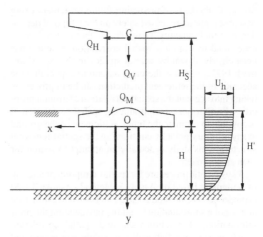

Figure 1. Schematic profile of a bridge pier supported by pile foundation system.

in the 4, × 3 system. The additional reinforcing piles are applied along the outskirt of the original pile allocation. The foundation is also enlarged to make a space for additional pile driving.

When the structural system is reinforced with m additional piles, the major damage mode of the pile foundation and its exceeding probability of major damage mode of the foundation can be defined as

$$D_a^{B*} \equiv \left[\bigcap_{j=1}^{n+m} \left\{ u_2^B(x_j) < u_a^B(x_j) \right\} \right] \bigcap \left[\theta_2^B < \theta_a^B \right]$$

$$\bigcap \left[\bigcap_{j=1}^{n+m} \left\{ S_{V2}^B(x_j) < R_a^B(x_j) \right\} \right], \quad p_{fa}^{B*} = P\left[\overline{D}_a^{B*} \right] \tag{1}$$

Using the damage mode of D_a^{B*}, the exceeding probability of damage mode after the reinforcement can be formulated as follows.

$$p_{fo}^{S*} = P\left[\overline{D}_o^S \middle| D_a^{B*} \right] \cdot P\left[D_a^{B*} \right] = p_{fo}^{S|B*} \cdot \left(1 - p_{fa}^{B*} \right)$$

$$p_{fa}^{S*} = P\left[\overline{D}_a^S \middle| D_a^{B*} \right] \cdot P\left[D_a^{B*} \right] + P\left[\overline{D}_a^{B*} \right] = p_{fa}^{S|B*} \cdot \left(1 - p_{fa}^{B*} \right) + p_{fa}^{B*} \tag{2}$$

Safety, Reliability and Risk of Structures, Infrastructures and
Engineering Systems – Furuta, Frangopol & Shinozuka (eds)
© 2010 Taylor & Francis Group, London, ISBN 978-0-415-47557-0

Settlement prediction by spatial-temporal random process

P. Rungbanaphan & Y. Honjo
Gifu University, Gifu, Japan

I. Yoshida
Musashi Institute of Technology, Tokyo, Japan

ABSTRACT

So far, all methods of predicting future settlement using past observations are based solely on the temporal dependence of their quantity. However, the fact that soil properties tend to exhibit a spatial correlation structure has been clearly shown by several studies in the past (Vanmark 1977, etc.). It is therefore natural to expect that the accuracy of the settlement prediction can be improved by taking into account the spatial correlation of ground properties. Furthermore, by introducing spatial correlation, it is possible to estimate the future settlement of the ground at any arbitrary point by considering the spatial-temporal structure. This study is actually an attempt to search for such an approach.

A systematic procedure for spatial-temporal prediction of settlement based on Asaoka's Method (Asaoka 1978) is proposed. The method is based on Bayesian estimation and Asaoka's formulation by taking into account the prior information, observation data, and spatial correlation structure. Auto-correlation distance of the parameters and the observation-model error are also estimated simultaneously based on Bayesian estimation using the observed data. The Kriging method (Wackernagel 1998, etc.) is considered to be a suitable approach for estimating the predicted settlement at any arbitrary location and time based on the estimated parameters.

Several case studies are carried out using simulated data with the assumed observation layout (for example, Fig. 1). It is concluded that, with relatively strong spatial correlation, the estimation of the model parameters and the final settlement can be significantly improved

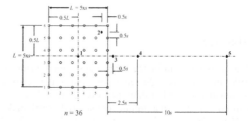

Figure 1. Layout plan of the observation points and the point to be estimated (for $n = 36$).

by taking into account the spatial correlation structure in comparison to the case of ignoring spatial correlation structure (Table 1). The proposed approach also gives the rational prediction of the settlement at any location at any time with quantified uncertainty. It was also found that the accuracy of settlement prediction is relatively insensitive to the changes of auto-correlation distance. Therefore, it can be concluded that the proposed method is practical for the settlement prediction.

REFERENCES

Asaoka, A. 1978. Observational procedure of settlement prediction. *Soil and Foundations* 18(4): 87–101.
Vanmarcke, E. H. 1977. Probabilistic modeling of soil profiles. *Journal of the Geotechnical Engineering Division, ASCE* 103(GT11): 1227–1246.
Wackernagel, H. 1998. *Multivariate Geostatistics: An Introduction with Applications.* 2nd ed. Germany: Springer–Verlag Berlin Heidelberg.

Table 1. Comparison of error of final settlement estimation between considering and ignoring spatial correlation structure for estimation at the 50th time step ($T_v = 0.424$).

| n | s/η | Error of y_f estimation | | | | | |
| | | Ignoring spatial corr. | | considering spatial corr. | | Improvement |
		Mean (%)[1]	Bias (%)	Mean (%)[2]	Bias (%)	(%)*
16	0.5	5.696	−0.344	4.788	−0.570	15.937
	0.25	5.631	−0.398	4.082	−0.580	27.498
	2	5.695	−0.548	5.662	−0.586	0.576
36	1	5.612	−0.596	5.320	−0.620	5.200
	0.5	5.703	−0.517	4.799	−0.621	15.853
	0.25	5.611	−0.314	3.975	−0.584	29.165
64	0.5	5.680	−0.477	4.678	−0.409	17.641
	0.25	5.673	−0.249	3.845	−0.386	32.213

*Improvement (%) = [(1) − (2)] × 100/(1)

Safety, Reliability and Risk of Structures, Infrastructures and
Engineering Systems – Furuta, Frangopol & Shinozuka (eds)
© 2010 Taylor & Francis Group, London, ISBN 978-0-415-47557-0

Frequency and damage analysis of industrial accidents

S. Hanayasu & K. Sekine
Center for Risk Management and Safety Sciences, Yokohama National University, Yokohama, Japan

W.H. Tang
Hong Kong University of Science and Technology, Hong Kong SAR, China

ABSTRACT

This paper deals with the risk analyses of various
industrial accidents in Japan. The major concern on
these accidents involves the frequency of accidents
and the damage consequences associated with them.
The objective of this study is to establish appropriate
probabilistic models for characterizing the industrial
accidents with emphasis on the frequency and damage
relations. Among different kind of damages caused
by industrial accidents, the number of injured workers and fatalities involved in a serious labor accident
were employed as yardsticks of damage magnitude
in this study. An extensive database covering various serious industrial accidents from 1977 to 1990 in
Japan was prepared for the analyses. Conclusions are
summarized as follows:

1. Frequency distribution of accident occurrences of
   serious industrial accidents can be modeled by
   Poisson or Negative Binomial distribution.
2. Frequency distribution of damage consequences
   due to serious labor accidents such as the number of
   injuries or fatalities involved in a serious accident
   can better be modeled by Geometric distribution
   rather than Poisson distribution.
3. Hence, sum of injuries or fatalities due to plural
   accidents can be modeled as the sum of Geometric
   distribution, i.e. Negative Binomial distribution.
4. The probability distribution of the total number of
   injuries or fatalities within a fixed interval of time
   can be obtained by adding up the products of probabilities of frequency of accident occurrences and
   that of conditional damage distribution associated
   with accidents.
5. Generalized Poisson-Poisson and Generalized
   Poisson-Geometric distributions having a certain
   lower damage size limitation have been proposed.
   Generalized Poisson-Geometric distribution had a
   good agreement with the actual total damage distributions. Figure 1 shows the frequency distribution
   of total number of injured workers within one
   month involved in a serious labor accident over 14
   years classified by structure collapse accidents with
   the lower damage limit of $h_c = 3$ in each accident.

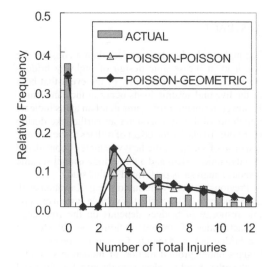

Figure 1. Frequency distribution of total monthly injuries
due to structure collapses ($h_c = 3$, $k = 0$ and $k \geq 3$).

6. The effectiveness of safety counter measures
   against industrial accidents in terms of preventing
   occurrence of accidents is evaluated by change of
   frequency parameter λ, while that of reducing damages is assessed by change of damage parameter τ.
   Risk reduction of industrial accidents is attained by
   balance of frequency and damage measures.

REFERENCES

Hanayasu S. & Tang W.H. 2000. On Generalized Frequency
Distribution of Accidents Considering Their Damage
Magnitude. *Proceedings of the 5th International Confer-
ence on Probabilistic Safety Assessment and Management
(PSAM-5) in Osaka*, 2149–2154, Universal Academy
Press.
Hanayasu S. & Tang W.H. 2001. On Generalized Damage
Distribution of Accidents Considering Their Size Limi-
tations. *Proceedings of the 8th International Conference
on Structural Safety and Reliability (ICOSSAR'01) in
Newport Beach*, A.A.Balkema Publishers.

Safety, Reliability and Risk of Structures, Infrastructures and Engineering Systems – Furuta, Frangopol & Shinozuka (eds)
© 2010 Taylor & Francis Group, London, ISBN 978-0-415-47557-0

Effect of traffic loads to seismic responses of highway viaducts under functional evaluation earthquakes

Mitsuo Kawatani, Chul-Woo Kim & Rie Kitaura
Department of Civil Engineering, Kobe University, Kobe, Japan

ABSTRACT

The highway bridge design codes do not consider the live load in the seismic design of highway bridges because of the low probability of the event that both of the live and seismic loads occur at the same time. However, frequent traffic jams in urban areas indicate a high possibility to encounter an earthquake during rush hour. To clarify the effect of traffics to the seismic response of viaducts, the heavy vehicle is considered as a dynamic system and a three-dimensional seismic response analysis is performed in this study.

Previous studies by the authors (e.g. Kawatani et al. 2008) indicate that the effect of vehicle system on seismic response of bridges depends on the frequency characteristics of ground motions: the vehicles on the bridge act as dampers for seismic response of bridges under ground motion of moderate soil site; on the other hand no clear damper effect is observed under the ground motion of stiff soil site. Usually the ground motion of stiff soil sites has dominant frequency higher than those of moderate soil sites. It indicates that the relation between vehicles effect on seismic responses of bridges and frequency characteristic of ground motion need to investigate. The effect of vehicle's dynamic system to the seismic response of the bridge is examined through the investigation using sinusoidal inputs firstly. A dump truck is assumed to be represented sufficient by a discrete rigid multi-body system with 12 degrees of freedom (DOFs), which considers sway, yawing, bouncing, pitching and rolling motions of the dump truck. Observations obtained from actual ground motion input are explained based on the result obtained from sinusoidal inputs.

The plot of RMS values taken from seismic responses of the bridge under sinusoidal inputs is shown in Figure 1. For scenarios of disregarding vehicles and considering vehicles as additional mass, the input frequency providing the peak RMS value coincides with the first natural frequency of the viaduct: i.e. 1.64 Hz for disregarding vehicles; and 1.59 Hz for considering vehicles as additional mass.

A notable point is that the response of bridge considering vehicles as additional mass is greater than that

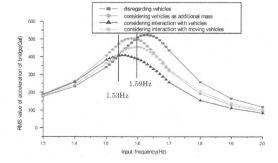

Figure 1. RMS values of seismic response of bridge according to sinusoidal inputs.

of the bridge disregarding vehicles under the condition of sinusoidal input lower than 1.59 Hz. On the other hand, above 1.59 Hz the considering vehicles as additional mass gives smaller response than that of disregarding vehicles. Another interesting point is that considering vehicle as a dynamic system tends to decrease bridge response. One of the reasons for the decrease is that, dynamic system of vehicles on the bridge acts as a damper for seismic response of the bridge. Considering interaction with moving vehicles also results decrease of seismic response of the bridge, even though the amount of decrease is smaller than that of stationary vehicle loadings because of comparatively fewer vehicles on the bridge. Considering vehicles as additional mass gives greater responses than the case considering vehicles as dynamic system.

REFERENCE

Kawatani, M., Kim, C. W., Iwashita, K. & Yasui, K. 2008. Seismic response of highway viaducts incorporating bridge-vehicle interaction, Journal of Structural Mechanics and Earthquake Engineering, JSCE, 64(4): 678–691. (*in Japanese*)

Organized Session (OS05) Vulnerability and Robustness of Structures

Safety, Reliability and Risk of Structures, Infrastructures and
Engineering Systems – Furuta, Frangopol & Shinozuka (eds)
© 2010 Taylor & Francis Group, London, ISBN 978-0-415-47557-0

Estimation of robustness about structures based on contaminated distribution

Xinjian Kou, Chong Zhou & Jimian Song
Shanghai Jiao Ttong University, Shanghai, China

Structual reliability is affected by many of uncertain factors which come from difeerent sources such as structual design, construction, manufacture and so on . In additon to these ramdon deviations, Human mistake is the most important factor influencing strucural safety because its effects ofern exceed the ramdon deviations. Therefore, uncertain factors may be classified into three categories based on their characterristics: systimatic error, random error and gross error.

Human mistake is the major causes of gross errors. Large numbers of facts have shown that structural failures, in normal load conditions, may be caused by the gross errors due to human mistakes. Gross errors may form the defects in a structure. These defects ofen can not be tolerated.

Many experts have proposed that the roubustness or vulnerability of structure should be considered to ensure structural safety. In study of vulnerability, the major purpose is focused upon the structural capability resistenting extreme events such as erthquakes or explodes. However, the largest risk is the defects hidden in structures, so a structure should possess the property with which the structure can be able to stand

gross errors on a specific level. For the large scale structure, gross errors may lead the structure failure entirely. In this case, structure may failure even though no extreme loads act on it.

In analysis of structural reliability, the random variables are assumed that they come from a specific distribution. However, random errors and gross errors may coexsit in the every stages of construction. In order to describe the influence of the two kinds of errors acting on structures, the authors employ the contaminated distribution, which was introduced by Huber (1964), as follows

$$p(x) = (1 - \varepsilon)f_1(x) + \varepsilon f_2(x)$$

In which, $f_1(x)$ is the main distribution that express the statistical properties about the random variables and $f_2(x)$ is the interference distribution that express the statistical properties of the gross errors. In this paper the distribution is applied to estimate the reliabilities of structures. The purpose of the study is to analysis the safety of structure when there are gross errors contained in the structure.

Safety, Reliability and Risk of Structures, Infrastructures and Engineering Systems – Furuta, Frangopol & Shinozuka (eds)
© *2010 Taylor & Francis Group, London, ISBN 978-0-415-47557-0*

Vulnerability assessment of structural systems

J. Agarwal
University of Bristol, UK

G.N. Liu
WSP, London, UK

ABSTRACT

An analysis of demands and capacities has worked well for most engineering systems. However, systems have grown in complexity and new methods have been developed to deal with complexity. For example, reliability theory is used to address the complexity arising as a result of uncertainties in system parameters. Risk calculations are performed using the notional probabilities of failures for the most likely failure scenarios. However, such calculations alone are not sufficient in themselves to produce robust systems. A complex system can fail in many different ways, some of which may lead to disproportionate consequences.

The issue of robustness of structures has been debated more intensely in recent years but there is yet no theory of robustness. However, an insight into the lack of robustness can be gained through an identification of potential vulnerabilities in a structure. These may arise due to an improper form of the structure or could be due to the operating environment. The aim of this paper is to examine the nature of vulnerabilities and their potential measures with a view to developing a framework for the vulnerability assessment of structural systems. A structure that is vulnerable may continue to perform satisfactorily as long as there is no action to exploit it. But it is important to be able to quantify vulnerability of a system to damage so that appropriate remedial measures could be taken. In the literature, different measures of vulnerability have been suggested. Some of these are based on the probability of failure for known demands. However, the damage may come from known or unknown sources; hence the measures based on the form of a system could lead to a better insight into the lack of robustness.

Structural vulnerability analysis is primarily concerned with the form of the system and it leads to the identification of vulnerable failure modes. It is also possible to include the effect of loads and non-linearity in the analysis. The changes in the orthogonal modes for some external actions give a measure of the topological changes in the system. An assessment of vulnerability to natural hazards is usually based on an analysis of damage resulting from a specific hazard of increasing intensity. In the context of social systems, vulnerability indicators are also used. It is proposed that vulnerability assessment should examine both the form of a structural system and the associated processes. The combination of vulnerability and threats can produce high risks. An analysis of vulnerability would help reduce risks which may arise due to unknown events.

Safety, Reliability and Risk of Structures, Infrastructures and Engineering Systems – Furuta, Frangopol & Shinozuka (eds)
© 2010 Taylor & Francis Group, London, ISBN 978-0-415-47557-0

Evaluation of robustness from building failures

S. Thelandersson & E. Frühwald
Lund University, Structural Engineering, Lund, Sweden

ABSTRACT

A database consisting of 127 cases where failure occurred in timber structures is utilized to investigate robustness. The investigated failures are related to events implying potential risks for human lives. The primary causes for failures were determined in the original investigation and classified into five main groups as shown in Table 1. Further details are given in Frühwald et al. (2007) and Thelandersson and Frühwald (2007).

Similar to the results in other failure surveys, human errors dominate among causes behind failure events. More than half of the cases occurred due to errors in design, which may be explained by a generally lower competence among engineers about timber structures.

In this paper, the failure cases in the database are investigated with focus on the relations between initial failure events, secondary failure events and consequences. The degree of progressiveness of the failure events and nature of warning before collapse were also assessed.

62% of the investigated cases were classified as *collapse* and the rest as *no collapse*. As an example, all cases where collapse occurred were investigated with respect to progressiveness, i.e. the extent of secondary damage after primary failure. The result from this exercise is shown in Figure 1. A relatively high proportion (41%) of the cases, where collapse occurred, was classified as large secondary damage.

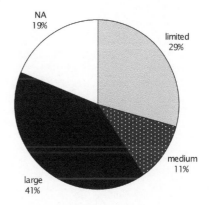

Figure 1. Large, medium or limited extent of secondary effects after primary failure (79 cases).

The failure events were also assessed against the criteria that the consequences should not be disproportionate the cause. It was found that in about one third of all cases the consequences could be classified as "very disproportionate".

A general (subjective) assessment of robustness showed that the systems investigated were to a comparatively large degree classified as having low robustness. Special methods should therefore be developed and implemented to design this class of structures to improve robustness. Although databases of this type never can be seen as representative for structural systems in general, it was concluded that it is valuable to check methods to assess robustness against this type of "empirical" information.

REFERENCES

Frühwald, E., Serrano, E., Toratti, T., Emilsson A., Thelandersson, S. 2007. *Design of safe timber structures –How can we learn from structural failures in concrete, steel and timber.* Report TVBK-3053, Div. of Struct. Eng. Lund University.

Thelandersson S., Frühwald, E. 2007. *Design of safe timber structures – How can we learn from structural failures.* Paper 40-15-7. CIB-W18, Timber Structures, Meeting 40, Bled, Slovenia, Aug. 2007.

Table 1. Distributions of errors causing failures for the cases investigated

Cause	% of 127 cases	Comments
Materials/manufacturing of products	11	
Construction on site	27	
Design	53	
Deficiencies in building code	4	Snow loads exceeded code values
Other/unknown	5	
Total	100	

Mini-Symposia (MS02) NDT Reliability and Its Use for Maintenance Applications

Safety, Reliability and Risk of Structures, Infrastructures and Engineering Systems – Furuta, Frangopol & Shinozuka (eds)
© 2010 Taylor & Francis Group, London, ISBN 978-0-415-47557-0

Development of posterior probability models using the Metropolis-Hastings algorithm and Laplace approximations

P.N. Thodi, F.I. Khan & M.R. Haddara

Faculty of Engineering and Applied Science, Memorial University, St. John's, Canada

ABSTRACT

The Bayesian posteriors find extensive applications in probabilistic risk analysis (PRA). The uncertainty and variability in data is modeled by prior probability distributions. These prior distributions need to be updated to posteriors with likelihood functions, which are the evidence supporting priors, using Bayes theorem. Bayes theorem encapsulate a process of learning, by which best inferences about the posterior parameter can be made based on any prior knowledge and the evidence at hand, reserving the right to revise the present knowledge continuously with new information. Mathematically, Bayes' theorem states how to update the prior probability distribution, $p(\theta)$ with a likelihood function, $l(x/\theta)$ to obtain the posterior distribution, $p(\theta/x)$:

$$p(\theta / x) = \frac{p(\theta)l(x / \theta)}{\int p(\theta)l(x / \theta)d\theta}$$

Since certain non-conjugate priors one comes across in PRA, like the Weibull, Lognormal and Extreme Value distributions, do not belong to the conjugate pair of exponential family, the conventional posterior estimation, in closed form, is not possible.

The performances of two Bayesian posterior development methods are discussed in this paper. One is a rejection sampling based Metropolis-Hastings (M-H) algorithm and second one is the analytical Laplace approximation method. The M-H algorithm is used to generate a sequence of samples from a probability distribution that is difficult to sample directly. This sequence is used in Markov chain Monte Carlo (McMC) simulations to approximate a distribution. In Bayesian applications, the normalization factor is difficult to compute, so the ability to generate the posterior samples without knowing this constant of proportionality is a major virtue of this algorithm. Laplace method is used for approximating the parameters of the posterior densities when direct estimations are difficult. It is a handy tool when a normal approximation

to posterior is reasonable. The basic idea is to carry out a Taylor series expansion around the maximum likelihood estimate value, ignore the negligible terms, and normalize.

In order to judge the applicability of these two computational methods, results of the M-H algorithm and Laplace approximations are compared with that of conjugate pairs. Errors are computed using the known conjugate pair estimates as the true values. The M-H algorithm suggests better results and hence recommended for the posterior model development. Also, it is observed that the change in threshold parameter from prior to posterior is insignificant. Although the Laplace method produces reasonable approximation to posterior mean, it fails to estimate variances accurately. Moreover, it is computationally intensive while working with distributions having more than two parameters.

Furthermore, the developed procedure is applied to a case study involving the stochastic degradation of process assets. The priors and likelihoods were of 3P Weibull, 3P Lognormal and Type1 Extreme value distributions. The estimated posteriors were observed to follow the same form as that of priors and likelihoods. The developed posteriors can be used in the risk assessment of asset degradations.

REFERENCES

Chib, S. & Greenberg, E. 1995. Understanding the Metropolis- Hastings algorithm. *The American Statistician*, 49 (4): 327–335.

Tanner, M.A. 1996. *Tools for Statistical Inferences*. Springer Series in Statistics, 3rd Ed., Springer-Verlag New York Inc., 175 Fifth Avenue, New York -10010, USA.

Thodi, P. N., Khan, F.I. & Haddara, M.M. 2008. The selection of corrosion prior distributions for risk based integrity modeling, *Jr. of Stochastic Environmental Research and Risk Assessment* (in press: DOI 10.1007/s00477-008-0259-x).

Tierney, L. & Kadane, J. B. 1986. Accurate approximations for posterior moments and marginal densities, *Jr. of American Statistical Association*, 81 (393): 82–86.

Safety, Reliability and Risk of Structures, Infrastructures and
Engineering Systems – Furuta, Frangopol & Shinozuka (eds)
© 2010 Taylor & Francis Group, London, ISBN 978-0-415-47557-0

Effect of the shape of ROC on risk based inspection: A parametric study

F. Schoefs
GeM, Nantes Atlantic University, Nantes, France

J. Boéro
OXAND S.A., Avon, France

ABSTRACT

Reassessment of existing structures generates a need
for up-dated materials properties. In a lot of cases,
on-site inspections are needed and in some cases
visual inspection is not sufficient. For example Non
Destructive Testing (NDT) tools are required for the
inspection of coastal and marine structures where
marine growth acts as a mask or immersion area
gives harsh condition of inspection. In these fields,
the cost of inspection can be prohibitive and an accurate description of the on-site performance of NDT
tools must be provided. Inspection of existing structures by a NDT tool is not perfect and it has become
a common practice to model their reliability in terms
of probability of detection (PoD), probability of false
alarms (PFA) and Receiver Operating Characteristic
(ROC) curves. These results quantities are generally
the main inputs needed by owners of structures in view
to achieve Inspection, Maintenance and Repair plans
(IMR) (Sheils et al., 2008). The assessment of PoD and
PFA is even deduced from inter-calibration of NDT
tools or from the modelling of the noise and the signal.

Theoretical aspects coming from detection theory
and probabilistic modelling of inspections results in
view to provide inputs in the computation of mathematical expectation of RBI cost models are described
in Schoefs et al. (2008).

First, let us focus in this paper on the benefit of
the combination of multiple Non-Destructive Testing
(NDT) and the role of expert judgement in this process.
In this case, expert judgement acts at two levels:

– The knowledge of ageing laws to provide the probability of defect existence that is needed when
  computing likelihood of events that govern the cost
  expectation;
– The way to address the decision after obtaining
  results from the two inspections.

The process is illustrated for the RBI of steel
harbour structures.

Then, the effect of the shape of ROC curve on the
decision process is highlighted. To this aim, a parametric study is performed to analyze the influence of
the polar coordinates of the best performance point of

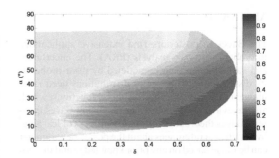

Figure 1. Mapping of cost of no detection $E(C)_{nd}$ in polar
plan for probability of defect existence $\gamma = 0.9$.

the NDT tool (NDT-BPP) on the expectation of the
cost of detection and the cost of no detection (see Figure 1). NDT-BPP is defined as the closest point of the
ROC curve to the best performance point BBP of coordinates (PFA = 0, PoD = 1) and the NDT-BPP polar
coordinates are defined by:

– The radius δ_{NDT} equal to the performance index
  (NDT-PI) (distance between the best performance
  point and the ROC curve) (Schoefs, 2007);
– The angle α_{NDT} between axis (PFA=0) and the line
  (BBP, NDT-BPP).

REFERENCES

Schoefs, F., Clément, A., Boéro, J; Capra, B. 2008. Expert
judgement for combining NDT tools in RBI context:
Application to marine structures, *4th ASRANet Collo-
quium, 25–27 June 2008, Athens, Greece.*
Schoefs, F., Clément, A., Memet, J.B., and Nouy, A. 2007.
Spatial dependence of Receiver Operating Characteristic
curves for Risk Based Inspection of corroded structures:
application to on-pile wharf, *Proc. of 10th International
Conference on Applications of Statistics and Probability in
Civil Engineering, (I.C.A.S.P'07), July31-August 3 2007,
Tokyo, Japan.*
Sheils, E., O'Connor, A., Breysse, D., Schoefs, F., and
Yotte, S. 2008. Development of a two stage inspection
process for the assessment of deteriorating bridge struc-
tures, *Journal of Bridge Engineering (Publication of the
American Society of Civil Engineers)*, to be published.

Safety, Reliability and Risk of Structures, Infrastructures and
Engineering Systems – Furuta, Frangopol & Shinozuka (eds)
© 2010 Taylor & Francis Group, London, ISBN 978-0-415-47557-0

Holistic statistical analysis of structural defects inspection results

S.A. Timashev & A.V. Bushinskaya
Science and Engineering Center "Reliability and Safety of Large Systems and Machines"
Ural Branch Russian Academy of Sciences, Ekaterinburg, Russia

ABSTRACT

The paper discusses current possibilities and drawbacks of in-line inspection (ILI) and direct assessment (DA) in detecting, identifying, locating and sizing of all types of defects in oil & gas pipelines, offshore platform structures and the like. A holistic methodology is presented that extracts maximum value from the ILI and DA measurements of defects and verifying their results.

Currently ILI/ DA are the preferred technologies to evaluate the condition of the pipe wall and assure pipeline integrity. A high-resolution and deformation ILI (geometry, corrosion, and crack) tool can inspect piggable pipeline sections and detect, locate, measure, and display irregularities in the pipe wall. These irregularities may represent geometric deformations (dents, gouges, ovalities, wrinkles, ripples, and buckles), laminations, pittings, local corrosion, cracks and other defects. However, as demonstrated by recent accidents on hazardous liquid and natural gas pipeline systems throughout the world, some pipelines that were inspected by all types of ILI devices continue to fail relatively soon (6 mo – 1 year) after the pipelines were inspected and pipeline cleared for further safe operation.

According to the sample OPS DOT USA data the most common causes of failures of pipelines soon after their inspection are: 1) non-detection (51% of all cases), 2) underreporting of defect depth (33.3%) and 3) misidentification (15.7%). The root causes of such types of failure may be different from case to case. But one thing is obvious: pipeline operators in general do not have a consistent, standardized process for evaluating and assessing data extracted by ILI/DA devices, and for providing guidance to ILI vendors, contract field inspection personnel, and company personnel on how to assess the ILI/DA data. Because of this, there is a general public concern about the performance issues associated with in-line inspection tools and how the data from these devices is being integrated with other information on the pipeline systems.

In this paper a new holistic approach to defect assessment is developed based on unconventional statistical analysis of inspection measurements data. The format of the resulting output is in full compliance with the reliability and safety methods that are currently used to assess structural safety. This allows creating adequate statistical models of defect sizes, and of crack/defect growth in time. The main emphasis is put on building a comprehensive, consistent and precise stochastic model of measurements and methods of their verification. This model allows for constructing consistent and unbiased assessments of the true *immeasurable* sizes of defect parameters and their variances for the case when the needed information about the ILI/DA tool and the verification instrument VI are obtained from the field and lab measurements.

A comprehensive and consistent methodology is presented for assessing the "*in the field*" statistical properties of the measurement errors of ILI/DA and verification tools (for the case "*one measurement by each tool*"). A method for calibrating the inspection tool is presented, which allows assessment of the true values of defect parameters.

Results of analysis of real ILI data (obtained on an oil pipeline with a MFL tool) and simulated data for a gas pipeline using the above methodology provide an illustration of the main algorithms of the holistic approach. A new approach is offered to assess the necessary number of verification measurements. The methodology of holistic statistical analysis of ILI data outlined in the paper permits assessing the components of the total variance of the ILI technology, including attribution of measurement error variance to the ILI tool, verification tool and the diagnostician. The algorithm outlined above is implemented in a user friendly software package PRIMA, which permits remote usage over the internet and provides visualization of all results.

Organized Symposia (OS14) Non-Destructive Testing of Concrete Structures

Safety, Reliability and Risk of Structures, Infrastructures and
Engineering Systems – Furuta, Frangopol & Shinozuka (eds)
© 2010 Taylor & Francis Group, London, ISBN 978-0-415-47557-0

Accurate crack inspection on concrete structure surface images

T. Yamaguchi & S. Hashimoto
Department of Applied Physics, Waseda University, Tokyo, Japan

ABSTRACT

As a diagnostic application of automated image analysis, crack inspection based on image processing and
measurement is proposed. Crack inspection in concrete surfaces during the maintenance and diagnosis
of concrete structures is important to ensure the safety.
The crack width is an important data in order to evaluate the durability and degradation of concrete surfaces.
Although many applications and products have been
developed in this field [Architectural Institute of Japan
2003], it is not easy to measure cracks accurately
with exact scale for practical use. In our previous
study, we performed two approaches individually to
achieve each problem. We proposed a percolation-
based image processing to detect cracks [Yamaguchi
and Hashimoto 2006]. This method was demonstrated
to be more accurate than the conventional methods.
On the other hand, to cope with measurement of

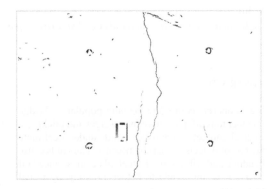

Figure 2.   Result of crack detection.

the exact size of crack width, we attached the crack
scale on the concrete surface in the step of image
acquisition because the crack scale has already been
written the size per scale on the crack scale. Then, we
used the brightness of width per the crack scale and
compared it with one of the detected cracks [Yam-
aguchi and Hashimoto 2007]. This paper presents a
high-accurate crack inspection method by combined
our detection and measurement approaches. Moreover, in order to build automatic inspection system,
we introduce automatic crack scale detection which
is manually detected in previous work as shown in
Figure 1. Through the experiment, we conduct proper
evaluations for the performance of the system in order
to indicate the reliability for practical use as shown in
Figure 2.

REFERENCES

Architectural Institute of Japan. 2003. Shrinkage cracking in
    reinforced concrete structures – Mechanisms and practice
    of crack control, AIJ.
Yamaguchi, T. & Hashimoto, S. 2006. Image processing based
    on percolation model, *IEICE Trans. Info. and Sys.* E89-
    D(7): 2044–2052.
Yamaguchi, T. & Hashimoto, S. 2007. Practical Image Mea-
    surement of Crack Width for Real Concrete Structure,
    IEEJ Trans. On Electronics, Information and Systems
    127(4): 605–614.

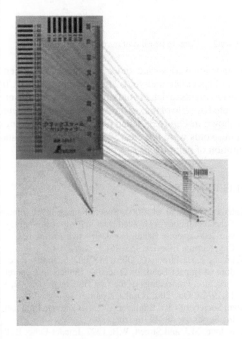

Figure 1.   Automatic crack scale detection.

Safety, Reliability and Risk of Structures, Infrastructures and
Engineering Systems – Furuta, Frangopol & Shinozuka (eds)
© 2010 Taylor & Francis Group, London, ISBN 978-0-415-47557-0

Impact elastic wave method for detecting voids under steel plate in steel concrete composite

T. Watanabe & C. Hashimoto

Department of Civil and Environmental Engineering, The University of Tokushima, Tokushima, Japan

K. Nishioka

Nippon Koei Co., Ltd, Tokyo, Japan

Y. Ito

Ishikawajima construction materials Co., Ltd, Tokyo, Japan

ABSTRACT

Steel concrete composite become popular as bridge member to reinforce and reduce weight and thickness. In steel concrete composites, voids under steel plate are one of defects in cast-ing process. The voids cause bending and vibration of steel plate in service. Of course it is impossible to detect the voids by visual inspection.

In order to detect the voids under steel plate in steel concrete composite, impact elastic wave method are employed. Impact elastic wave method is a non-destructive evaluation for detecting defects in concrete structures analyzing elastic wave due to a mechanical impact. Elastic wave is generated on a surface of concrete structures and propagation and reflection wave are detected by a sensor on surface. It is clarified that the procedure is effective to detect and visualize an ungrouted duct of a pre-stressed concrete beam (Watanabe and Ohtsu, 2000) and a void of a concrete slab (Watanabe, Hashimoto and Ohtsu, 2002).

In order to detect vibration due to flexural mode of steel plate on voids, impact tests were carried out at the surface of steel concrete composite specimens. Steel concrete composite specimens were mode and

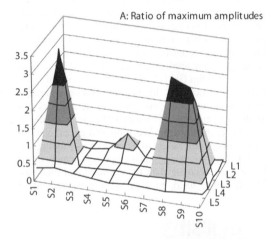

Figure 2. Contour of ratio of maximum amplitude.

some voids were varied under steel plate. The dimension of specimen and voids are shown in Figure 1. In measurements to detect voids in specimens, voids were detected by ratio of maximum amplitudes of parameter as shown in Figure 2. When plate thickness becomes 6 mm, voids were not clearly detected by energy and duration of parameters.

REFERENCES

Watanabe T, Ohtsu M. 2000. Spectral Imaging of Impact Echo Technique for Grouted Duct in Post-tensioning Pre-stressed Concrete Beam: Nondestructive Testing in Civil Engineering, Elsevier, 453–461.

Watanabe T, Hashimoto C, Ohtsu M. 2002. Scanning Procedure of Impact Echo for Detecting Defects in Concrete Structure, Proceedings of the first fib Congress 2002, 8, 15,19–28, Oct. 2002, Osaka, Japan.

Japan Society of Civil Engineers (2004), Concrete Engineering Series 61: 48–49 (in Japanesse)

Sansalone, M.J. and Streett, W.B. 1997. Impact-Echo. Bullbrier Press, Ithaca, N.Y.

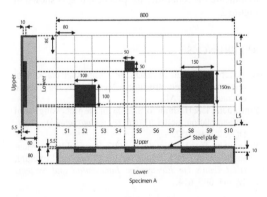

Figure 1. Concrete steel composite specimens.

Safety, Reliability and Risk of Structures, Infrastructures and
Engineering Systems – Furuta, Frangopol & Shinozuka (eds)
© 2010 Taylor & Francis Group, London, ISBN 978-0-415-47557-0

Characterization of damage status for concrete structures by means of stress wave techniques

T. Shiotani & H. Ohtsu
Kyoto University, Kyoto, Japan

D.G. Aggelis
University of Ioannina, Ioannina, Greece

S. Momoki
Tobishima Corporation, Chiba, Japan

ABSTRACT

Ageing infra-structures are increasing worldwide and the preservation of their structural integrity is a crucial issue. The present tax revenue shortage due to economic recession and demographic change resulted in the reduction of construction investment, leading to much attention to the maintenance work prior to the construction of new structures. For repair program, some theoretical models to manage the structures, namely infra-asset management, have been proposed, in which based on the current status the forthcoming deterioration process is estimated, and a proper timing of maintenance is assumed in consideration of life cycle cost (LCC).

Specifically the present damage amount of structures, which would be employed to make priority of structure to repair, would be assumed by the amount of loss and damage indices through the monitoring.

Due to the complexity of deterioration process as well as various kinds of structures, any decisive investigation technique has been established so far. Acoustic emission (AE) and ultrasonic techniques (UT), non-destructive techniques using stress wave, could be applied for the infrastructures diagnosis.

In the paper, a 45 m long PC bridge, showing no deterioration throughout the conventional survey, is monitored nondestructively: sensitive monitoring of AE was conducted in order to investigate the early damage, followed by UT aided by tomography.

Firstly the relatively damaged area for the longitudinal direction was extracted with lineally placed 28-AE sensor during mobile loading over the bridge. Based on the resulted AE activity the part of structure being most likely deteriorated is further investigated with a dense array of AE sensors, allocated in 2D. Using this 2D arrayed AE sensors, a surface ultrasonic examination is also performed to obtain a tomogram of wave velocities. Through these monitoring, following findings were obtained.

A most likely deteriorated area (hereafter 'area of interest') was successfully extracted from AE activity

emerged always in a particular area from linearly placed AE sensors (see Figure 1). As in Figure 2, followed 2D AE monitoring, conducted in the area of interest, showed the remarkable number of AE sources, corresponding well to the area of low velocity. It was confirmed that the area indicated by AE activity was actually deteriorated, suggesting the potential of AE testing both for as global- and local-monitoring.

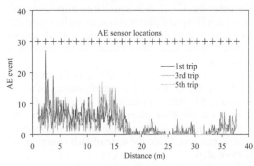

Figure 1. 1D AE sources along with repeated mobile loads.

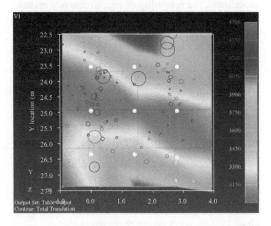

Figure 2. Identified 2D AE sources along with mobile load and velocity tomogram obtained from surface UT measurement.

*Safety, Reliability and Risk of Structures, Infrastructures and
Engineering Systems – Furuta, Frangopol & Shinozuka (eds)*
© *2010 Taylor & Francis Group, London, ISBN 978-0-415-47557-0*

Eigenfrequency estimation of RC slab based on impact sound recorded by moving vehicle

Y. Oshima, K. Sugiura & H. Kawano
Kyoto University, Kyoto, Japan

ABSTRACT

Herein we verify the concept of easy and simple
method to evaluate eigenfrequency obtained by impact
sound which is generated and recorded by moving
vehicle. In order to attain this goal, noise reduction
in recorded sound must be done since the sound is
recoded by moving microphones with wind noise,
and also Doppler Effect should be confirmed. Therefore, eigenfrequency estimation of beam specimens
with noise was experimentally carried out; denoise
algorithm was based on independent component analysis. Then Doppler Effect was confirmed by numerical
model as well as the experiment using small carrier
moving on the slab, and influence of the effect was
evaluated.

The concept of inspecting vehicle is shown in Figure 1. In order to capture the eigenfrequency of a slab
structure, RC slab is impacted by a hummer from a
moving vehicle. Microphones are also installed on the
vehicle to record the impact sound. This inspecting
vehicle can investigate the state of slab continuously
without stop.

First of all, the eigenfrequency of beam was estimated by impact sound with noise. On the basis of this
experiment, the feasibility of eigenfrequency estimation by impact sound and the influence of noise in the
recorded sound were evaluated. Additionally, the type
of hummer head was also evaluated to simulate the
effect of pavement. As a result of this experiment, more
frequencies can be detected by impact sound than by
the installed sensors, because the installed sensors are
located in the limited area and it is difficult to capture
all the modes, but the sound may include all the modes.
As for noise reduction, ICA based on AR model can
reduce the noise with high frequencies but the accuracy depends on the order of AR model. To obtain
more exact solution, appropriate order of AR model
must be determined. Furthermore, the noise in this
study mainly consisted of high frequencies, which can
be separated by a low-pass filter. Thus for noise reduction, both methods should be applied as the situation
demands.

Next, in order to confirm the influence of Doppler
Effect, concrete slab was impacted and the sound
was recorded by moving microphones. In this experiment, the slab of a residential building was used and
microphones were installed on the manual carrier.

As a result, we found that several peaks were found
only when the microphones were moving, but many
other peaks were almost identical even for different
velocity of the microphones. This may be attributed to
the fact that the velocity of carrier in this experiment
was not so large to affect the frequency. Now the vehicle moves at 6 km/h, the difference between the static
and fast case becomes 1.6%, which can be ignored. But
if the inspecting vehicle moves at 60 km/h, the difference is 4.1%. Thus Doppler Effect must be considered
when the velocity is high.

Finally, the inspecting vehicle proposed herein is
feasible on the basis of experimental results in this
study, but the concept must be verified by real bridge
slab and vehicle.

REFERENCE

M. Nitta and K. Sugimoto (2006). ICA Based Blind Iden-
tification fia Exact Parameterization, 17th International
Symposium on Mathematical Theory of Networks and
Systems, 402–407.

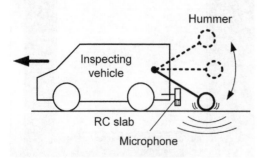

Figure 1. Concept of the inspecting vehicle.

Safety, Reliability and Risk of Structures, Infrastructures and
Engineering Systems – Furuta, Frangopol & Shinozuka (eds)
© 2010 Taylor & Francis Group, London, ISBN 978-0-415-47557-0

Evaluation method of mechanical performance of RC beams with corroded reinforcing bars by FEM utilizing results from nondestructive test

H. Minezawa, T. Kamada & S. Uchida
Osaka University, Osaka, Japan

S. Miyazato
Kanazawa Institute of Technology, Ishikawa, Japan

I. Kuroda
Defense Academy of Japan, Kanagawa, Japan

ABSTRACT

Generally, concrete inspections are applied to determine the current strength and durability of a structure. However, it is difficult to determine the strength and performance of a structure from inspection results, therefore currently indirect estimation from visual observation of cracks, corrosion levels from nondestructive test among others are used. Therefore, developing a method to directly determine the current conditions of a structure in a standardized manner will enhance rationalization of management of structures.

In this study, to develop a method directly determine the current conditions of a structure, various non-destructive tests were performed on RC beams whose rebar has been corroded through electricity. And according to the results of nondestructive tests, deterioration maps were made. Then, the results were input into finite element method (FEM) analysis and the mechanical performance of the specimens investigated. The validity of the analysis results were confirmed by conducting flexural loading tests on the RC beams. The evaluation method used is shown in Figure 1.

In experiment, RC beam specimens were made. And electrolytic corrosion was carried out for 14 and 28 days in order to have specimens with varying levels of corrosion. Before and after electrolytic corrosion, 4 types of NDT were performed to estimate mechanical performance of specimens. Firstly, crack propagation was confirmed through visual investigation. Thereafter, the half cell potential was measured to determine the corrosion of rebar. Moreover, wave velocity was measured to evaluate the condition of the concrete surface layer and rebar. Furthermore, to determine the rate of corrosion, polarization resistance was concurrently measured. Then, damage maps were made which reflect the results of nondestructive tests. Thereafter, flexural loading tests were conducted.

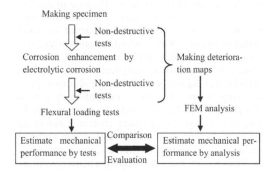

Figure 1. Evaluation method.

While, FEM analysis based on the results of NDT was performed. Results of NDT were reflected to the corrosion of rebar and adhesion between rebar and concrete. The analysis was done in a 2 dimensional, plane stress condition.

From both results, the analysis model accurately predicted adhesive levels between concrete and rebar and rebar mass reduction rate, therefore proving to be an appropriated evaluation method for the mechanical properties of RC beams whose rebar have undergone corrosion.

REFERENCES

ASTM C876. 1999. *Standard Test Method for Half-Cell Potentials of Uncoated Reinforcing Steel in Concrete*: Annual book of ASTM Standards.
Japan Society for Civil Engineers Specification. 2007. *JSCE-E601*: JSCE
Japan Information Processing ServiceCo.,Ltd. 2002. *DIANA7 User's Manual Release7*: JIP

Safety, Reliability and Risk of Structures, Infrastructures and
Engineering Systems – Furuta, Frangopol & Shinozuka (eds)
© 2010 Taylor & Francis Group, London, ISBN 978-0-415-47557-0

Influence of conformity control on the strength distribution of concrete and the safety level of concrete columns

R. Caspeele & L. Taerwe

*Magnel Laboratory for Concrete Research, Department of Structural Engineering,
Ghent University, Ghent, Belgium*

ABSTRACT

Besides quality verification, conformity control of concrete production also has a filtering effect on the offered strength distribution. Because of the fact that conformity control rejects certain concrete lots, the outgoing strength distribution (indexed 'o') has a higher mean and a lower standard deviation in regard to the incoming strength distribution (indexed 'i'). This effect can be quantified analytically using Bayesian updating techniques and leads to a closed-form expression only in case of one type of conformity criteria (Rackwitz 1979, Taerwe 1985). In order to investigate more complex conformity criteria, a numerical algorithm is developed based on Bayesian updating techniques and numerical Monte Carlo simulations. With this model, also autocorrelation between consecutive test results can be taken into account. The filter effect of some different types of conformity criteria are quantified, among which the filter effect of the conformity criteria for continuous production control mentioned in the European standard EN 206-1.

The filtering effect of conformity control also has an influence on the safety level of concrete structures.

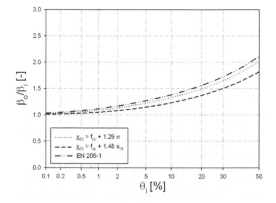

Figure 2. Safety index ratio for concrete structures, designed according to Eurocode 2, taken into account different conformity criteria (based on autocorrelated observations).

This effect is investigated for structures, designed according to Eurocode 2, for which the concrete compressive strength is a sensitive variable in regard to the safety index. A general approach was used, without any strength model or distributional assumptions for loads. It was found that the safety level of these type of concrete structures increases significantly when conformity control is taken into account. Finally, also an example is provided regarding the influence of conformity control on the safety level of a specific design situation.

REFERENCES

Rackwitz, R. 1979. Über die Wirkung von Abnahmekontrollen auf das Verteilungsge-setz von normalen Produktionsprozessen bei bekannter Standardabweichung. In: *Materi-alprüfung* 21(4): 122–124.

Taerwe, L. 1985. *Aspects of the stochastic nature of concrete strength including compliance control (in Dutch).* PhD thesis, Ghent University, Ghent, Belgium.

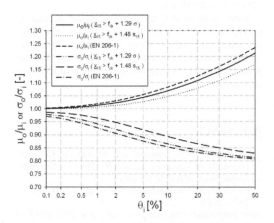

Figure 1. Filter effect corresponding to different conformity criteria (based on autocorrelated observations).

General Session (Geotechnical Engineering)

Safety, Reliability and Risk of Structures, Infrastructures and
Engineering Systems – Furuta, Frangopol & Shinozuka (eds)
© 2010 Taylor & Francis Group, London, ISBN 978-0-415-47557-0

Bayesian statistical framework for the estimation of soil properties and deformations during supported excavations

J.-K. Park, P. Gardoni & G. Biscontin
Zachry Department of Civil Engineering, Texas A&M University, College Station, TX, USA

ABSTRACT

Construction of supported excavation systems inevitably causes horizontal wall deflections and ground movements including surface settlement. The observational method of design in geotechnical engineering has proven to be a valuable tool for addressing soil and structural uncertainty during subsurface construction projects. In the observational method, project design and construction sequences are evaluated and revised as necessary based on the comparison between observed and predicted responses.

This paper presents a Bayesian statistical framework (Box & Tiao 1992) to assess soil properties to better predict excavation-induced horizontal deformations and surface settlement. The soils properties and the models parameters are updated after each of excavation stage. The updated parameters are then used to develop new and more accurate predictions of the deformations in the subsequent stages until the end of excavation project. As an application, the proposed framework is used to assess the moduli of elasticity of multiple soil layers for two examples, using both deformation data at different depth locations and surface settlement data for four incremental excavation stages. The first example illustrates the proposed methodology for a simple idealized case. In the second example, actual excavation data recorded during a supported excavation project in Evanston, Illinois, are used (Finno & Roboski 2005).

Since the model $\mathbf{D}(\boldsymbol{\theta}; \mathbf{z})$ is nonlinear in the unknown parameters $\boldsymbol{\theta}$, a closed-form solution is not possible. In this case, computation of the posterior statistics, as well as the normalizing constant κ, is not a simple matter, as it requires multifold integration over the Bayesian kernel. In this paper, a Markov Chain Monte Carlo (MCMC) algorithm was used for computing the posterior statistics (Robert & Casella, 2005).

In two application examples, the standard deviation of unknown parameters $\boldsymbol{\Theta}$ gradually decreased as excavation steps increase. This indicated that the uncertainty can be reduced by proposed probabilistic approach. The small values of Mean Absolute Percent

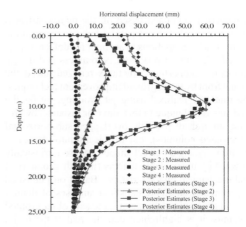

Figure 1. Comparison of measured and predicted horizontal displacement based on posterior estimates for Lurie case.

Error (*MAPE*) indicate that the proposed probabilistic models are accurate. The predicted values capture accurately the horizontal displacement (as shown in Figure 1), the settlement profile and the location of the maximum surface settlement.

The developed approach can be used for the design of optimal revisions for supported excavation systems. By applying a Bayesian approach to reliability-based design of geotechnical engineering projects, engineers can combine the adaptive advantages of the observational method with the advantages of probabilistic design methods.

REFERENCES

Box, G. E. P. and Tiao, G. C. 1992. *Bayesian Inference in Statistical Analysis*. New York: Willy.
Finno, R. J. & Roboski, J. F. 2005. Three-Dimensional Responses of a Tied-Back Excavation through Clay. *Journal of Geotechnical and Geoenvironmental Engineering* 131(3): 273–282.
Robert, C. P. and Casella, G. 2005. *Monte Carlo Statistical Methods*. Springer.

Safety, Reliability and Risk of Structures, Infrastructures and
Engineering Systems – Furuta, Frangopol & Shinozuka (eds)
© 2010 Taylor & Francis Group, London, ISBN 978-0-415-47557-0

Effect of defects on structural safety of jet grouted umbrellas in tunneling

G.P. Lignola & G. Manfredi

University of Naples Federico II – Department of Structural Engineering, Naples, Italy

A. Flora

University of Naples Federico II – Department of Hydraulic, Geotechnical and Environmental Engineering, Naples, Italy

ABSTRACT

A numerical approach based on the matrix stiffness method to predict the structural behavior and on the Monte Carlo simulations to account for defects is presented. The aim of this work is to assess and design temporary supporting structures realized ahead the tunnel front by using ground improvement techniques. are routinely realized using jet grouting, by far the best suited technique to this aim (for instance, by creating an arch of partially overlapped sub-horizontal jet grouted columns), but this leads into structures far from having a perfect shape because intrinsically affected by unavoidable defects (in both geometrical and mechanical characteristics). Based on published experimental data and statistical analyses of diameter and centroid position along the axis of jet grouted columns, it is highlighted that the real shape of jet grouted supporting structures is not that of a regular frustum of cone. Columns overlapping decreases along the span, thus there is a critical length after which structural continuity is difficult to obtain, depending on the statistical characteristics of the jet grouted columns. Defects in axis orientation play the major role on this critical length.

In this paper non-closed tunnel supporting structures are considered. Even though tunnel excavation is a fully 3D problem, it is typical in tunnel lining design to adopt a simpler 2D approach on simplified arch schemes, with reference to the part of the supporting structure in which continuity is guaranteed.

A numerical code was developed to analyze these structures considering the effective geometry of the supporting structure (affected by geometrical defects, namely partial overlapping of sub horizontal jet grouted columns decreasing along the span, defects in diameter and position) and to account for the horizontal active loads that are different from the vertical ones. The developed code is able also to analyze the soil–structure interaction (SSI) in detail.

In the paper a series of Monte Carlo simulations were carried out to find the great influence of possible defects, usually overlooked by practitioners, within the column in the static performance of these supporting structures. The variability and statistical distributions of the geometrical and mechanical parameters were

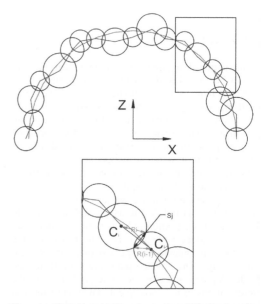

Figure 1. Typical umbrella cross section, differing from the ideal shape because of defects.

considered and they are based on the large experimental evidence collected in field trials by the authors and published in open literature.

It is pointed out that the design of such structures may hide unforeseen risks and it could happen that the structure is unable to carry soil loads with the desired factor of safety. As a consequence, this is the typical case in which sophisticated numerical analyses may just give the illusion of being refined, if possible defects are not correctly taken into account.

Safety checks for soilcrete arches have been established accounting for four different failure modes.

A case study highlights the role of defects and SSI on static performances. For the chosen set of statistical parameters, starting from 3/4 span, the failure probability, P_f, reaches 100%, while in the case of ideal geometry (without defects), no failure is detected along the entire umbrella. If SSI is neglected the P_f would be 100% starting from the very first arch, both for an ideal or real geometry.

Safety, Reliability and Risk of Structures, Infrastructures and
Engineering Systems – Furuta, Frangopol & Shinozuka (eds)
© 2010 Taylor & Francis Group, London, ISBN 978-0-415-47557-0

Evaluation of the seismic reliability for the stability of cantilever retaining wall

F.K. Huang

Department of Water Resources and Environmental Engineering, Tamkang University, Taipei County, Taiwan

G.S. Wang

Department of Construction Engineering, Chaoyang University of Technology, Taichung County, Taiwan

ABSTRACT

Despite advances in geotechnical engineering, it is common to find retaining walls experiencing near or complete failure during strong earthquakes (Kramer 1996). Effects of earthquakes on retaining walls often include large translational and rotational displacements, settlement of backfill soils, and bearing capacity failure under the toe. Traditionally, factors of safety (FS) have been used to evaluate the seismic stability of retaining walls in analysis and design practice. Factors of safety are normally selected empirically, i.e. based on past experience or experience with similar engineering structures, but a formal relationship does not exist between the factor of safety and the probability of failure. The inherent empiricism in the factor-of-safety approach may lead to instability when uncertainties are greater than anticipated (Duncan 2000). Accordingly, the reliability-based design (RBD) or performance-based design (PBD) approach is preferred in the evaluation of seismic stability of retaining walls.

In this study, the seismic reliability for the stability of cantilever retaining wall is evaluated by the Monte Carlo simulations. The effects of variability and uncertainty of evaluation parameters are all included in the analysis model accompanied by the hazard-consistent ground motion of Taiwan area. The reliability for the stability of the wall is expressed by the annual probability of factor of safety that less than a specified code value. The results thus obtained give the risk-based failure potential of wall which is consistent with the seismic hazard of the site, and constitute a common standard for comparing failure modes among different structures. The decision of countermeasures against failure of wall is also easy to make by the methods established here.

From this study, it is shown that when both uncertainties of the wall-soil system and the site effects are included, the factors of safety against sliding (FS_s) and bearing capacity failure (FS_b) are all likely to be governing factors of design, especially for FS_b

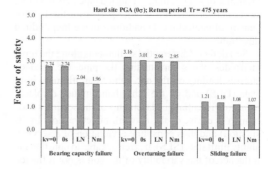

Figure 1. Comparison of factor of safety for 475 return period (hard site condition; PGA: median value).

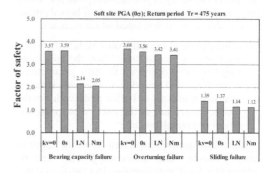

Figure 2. Comparison of factor of safety for 475 return period (soft site condition; PGA: median value).

(as shown in Figs 1–2). Appropriate countermeasures against failure must be taken to ensure the reliability of the wall.

REFERENCES

Kramer, S.L. 1996. *Geotechnical Earthquake Engineering.* New Jersey: Prentice Hall, Inc.

Duncan, J. M. 2000. Factor of safety and reliability in geotechnical engineering. *Journal of Geotechnical and Geoenvironmental Engineering* 126(4):307–316.

Safety, Reliability and Risk of Structures, Infrastructures and Engineering Systems – Furuta, Frangopol & Shinozuka (eds)
© *2010 Taylor & Francis Group, London, ISBN 978-0-415-47557-0*

Modelling the structural reliability of buried networks in a heterogeneous and aleatory environment

S.M. Elachachi & D. Breysse
University of Bordeaux 1, Ghymac, Talence, France

INTRODUCTION

Dysfunctions and failures of buried pipe networks like sewer networks result in large part from the heterogeneity of geotechnical conditions in the longitudinal direction.

Soil defects (differential settlements along the pipe, landslides, voids surrounding the pipe, etc.) induce stresses leading on the one hand to an ultimate limit state (ULS) and on the other hand induce displacements (differential displacements of the section's ends bring a decrease of the network effective flow by its siltation) and thus constitute a violation of a serviceability limit state (SLS). It is remarkable to note that the influence of the variability of the soil is not reflected in current European standards.

That is why it is necessary to focus on the impact of the spatial variability of geotechnical properties on the longitudinal behaviour of a network of buried pipes.

A model has been developed which includes:

(a) a description of the soil spatial variability, within the frame of geostatistics, where the correlation length of soil properties is the main parameter;
(b) a mechanical description of the soil–structure interaction of a set of buried pipes with flexible connections resting on the soil by a two parameter model (Pasternak model).

Reliability analysis is performed on the sewer by using a Monte Carlo simulation (MCS), with the reliability index β being calculated for two limit states:

– Serviceability limit state, corresponding to a two large "counter-slope" in a given pipe (slope = 4% is taken as an average limit value), which can prevent the normal flow of fluids,

– Ultimate limit state, corresponding to a too large bending moment, thus bending stress, which can cause crack in the pipes (2 MPa is taken as an average limit value for concrete pipe for example).

Several conclusions are drawn:

(a) Soil heterogeneity induces effects (differential settlements, bending moments, stresses and potential cracking) that cannot be predicted if homogeneity is assumed.
(b) The magnitude of the induced stresses depends mainly on four factors:

– The magnitude of the soil variability (i.e. its coefficient of variation);
– A soil-structure stiffness ratio,
– A structure-connection stiffness ratio (relative flexibility),
– A soil-structure length ratio, which combines the soil fluctuation scale and a structural characteristic length (buried pipe length).

A worst value, corresponding to the value leading (from a statistical point of view) to the (statistically) largest effects in the structure, can be found. The principal benefit of such an approach is to provide some new approaches for better considering phenomena such as the geometrical irregularities in the longitudinal profile during the control of soil compaction of sewer trench filling. This kind of approach can also give experts new tools for better calibration of safety in soil-structure interaction problems, when the soil variability is an influential parameter.

*Safety, Reliability and Risk of Structures, Infrastructures and
Engineering Systems – Furuta, Frangopol & Shinozuka (eds)
© 2010 Taylor & Francis Group, London, ISBN 978-0-415-47557-0*

Probabilistic optimization methodology for secant pile wall watertightness

T. Micic

City University, London, UK

ABSTRACT

Secant pile walls are, increasingly, a standard piling technique for construction of deep basements and
earth retaining structures, especially in urban areas.
Construction of the secant pile wall includes, drilling
for, usually, un-reinforced primary (female) piles at
a fixed distance that is smaller then pile diameter.
Shortly after casting and while the concrete is still
fresh secondary (male) reinforced piles are drilled
between primary piles. If there is a sufficient overlap
the pile wall will be watertight. However, the secant
pile walls are being used for ever deeper basements
and limitations for this piling technique have started
to emerge.

Secant pile wall watertightness is significantly
influenced by construction aspects such as tolerances,
limited verticality monitoring, influence of adjacent
buildings and services, guide wall construction, concrete mix design and development of early strength,
drilling method, piling platform and orientation of
the piling rig, Suckling et al. Furthermore, contractor expertise and specific construction method, wall
layout (straight or curved), drilling method, orientation
of the rig, ground conditions obstructions and required
depth have significant influence on water-tightness of
the secant pile wall. All these aspects bring uncertainties and implementation of probabilistic methods,
Melchers 1999, to improve the management of risk
is investigated here. Probabilistic modelling is, at
present, restricted to definition of probability density functions for dimensions and spatial positions.
The sufficient overlap area between a pair of primary and secondary piles, as shown in Figure 1, has
been assumed to be 75% of the design overlap, thus
$\lambda = 0.75$. As the insufficient overlap is considered a
failure we focus on finding the likelihood of that event.

The paper proposes that the standard FOSM is
enhanced by imbedding it within an optimization
procedure. Namely, rather then evaluating notional
reliabilities a continuous search for a minimum
notional reliability index for the increasing depth
and variable soil conditions is carried out. It is

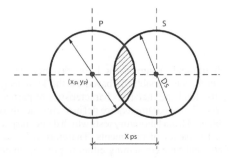

Figure 1. Primary/secondary segment of the straight secant pile wall.

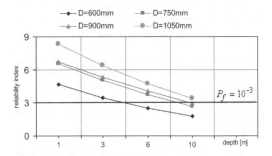

Figure 2. Reliability index variation as a function of depth for different pile diameters.

demonstrated that FOSM methods provide a rational
analytical tool for risk monitoring during construction
of secant pile walls.

REFERENCES

T.P. Suckling, C. J. Wren, V. M. Troughton; Secant Pile
Walls – A Consistent Approach to Risk Management; in
Proc. of the 30th Annual Conference on Deep Founda-
tions, Chicago, Illinois, USA, 2005

Melchers, R.E.; Structural Reliability, Analysis and Prediction, Ellis Horwood Ltd, Chichester & Halsted Press (John
Wiley & Sons), 1999

Safety, Reliability and Risk of Structures, Infrastructures and
Engineering Systems – Furuta, Frangopol & Shinozuka (eds)
© 2010 Taylor & Francis Group, London, ISBN 978-0-415-47557-0

Computer simulation of cyclic-load triaxial testing with respect to uncertainties in input parameters

A. Florian & J. Pencik
Department of Structural Mechanics, Brno University of Technology, Brno, Czech Republic

S. Karaskova
Department of Road Structures, Brno University of Technology, Brno, Czech Republic

The traditional design method for pavement structures in Czech Republic is based on the knowledge of traffic load and material characteristics of particular layers. These parameters are defined empirically or are obtained from laboratory tests which do not respect the real behavior of pavements. Therefore it is very difficult and time consuming either to put new materials into practice or to change traditional thickness of structural layers.

Cyclic-load triaxial test is an innovative laboratory test method that is presently implemented in Czech Republic. By simulating both the vertical loading and the matching horizontal pressure caused by individual crossings of vehicles, the real traffic loading and stress conditions in corresponding layers of the structure can be obtained. The success of the proposed method in a practical use is dependent on the knowledge of loading effects. The horizontal stress can not be obtained experimentally by long-term measurements on real pavements.

To obtain necessary input data the computer simulation of triaxial test taking into account uncertainties in input variables is proposed. The plane (2D) FEM numerical model of a pavement structure is created. The contact between individual layers of a structure is modeled. The model enables to determine stress and deformation conditions in the specimen and/or in the particular material layers.

The reliability analysis takes into account the uncertainties in basic input variables – thickness and material properties of individual layers. The source of uncertainties in the problem analyzed is the fact that input variables are generally random ones, and also the lack of representative and correct data about these variables, our incomplete knowledge about behavior of the structure, and the accuracy of analytical model.

The statistical and sensitivity analysis of horizontal stresses caused by the axle load is performed for three variants of road type 9,5 m with different thickness of bituminous layer. As result of the statistical

analysis, the estimates of mean value, standard deviation, coefficient of variation, skewness, and minimal and maximal values of the horizontal stresses are obtained. In addition, the deterministic analysis with input variables set to their nominal (mean) values is performed.

In the sensitivity analysis, the influence of random variability of input variables on the random variability of stresses is evaluated. To measure the relative influence of each input variable on the output, the sensitivity coefficient based on the Spearman rank correlation coefficient is proposed, Florian and Navratil (1993).

The modern simulation technique Updated Latin Hypercube Sampling, Florian (1992), with 50 simulations is used for reliability analysis. It is an improved variant of Latin Hypercube Sampling, McKay et al. (1979). The method keeps the methodology of Latin Hypercube Sampling, but uses the improved strategy of generating input samples based on specially modified tables of random permutations of rank numbers. The modified tables consist of random permutations that are mutually statistically independent. Using of Updated Latin Hypercube Sampling generally results to the further increase of accuracy, quality and reliability of the results obtained from reliability analysis.

REFERENCES

Florian, A. 1992. An Efficient Sampling Scheme: Updated Latin Hypercube Sampling. *J. Probabilistic Engineering Mechanics* 7(2): 123–130.
Florian, A. & Navrátil, J. 1993. Reliability Analysis of the Cable Stayed Bridge in Construction and Service Stages. *6th International Conference on Structural Safety and Reliability ICOSSAR'93*, Innsbruck, Austria: 869–877.
McKay, M., Beckman, R.J. & Conover, W.J. 1979. A Comparison of Three Methods for Selecting Values of Input Variables in the Analysis of Output from a Computer Code. *Technometrics* 2: 239–245.

General Session (Offshore and Marine Structures)

Safety, Reliability and Risk of Structures, Infrastructures and Engineering Systems – Furuta, Frangopol & Shinozuka (eds)
© *2010 Taylor & Francis Group, London, ISBN 978-0-415-47557-0*

Maintenance planning for the decks of bulk carriers and tankers

Y. Garbatov & C. Guedes Soares

Centre for Marine Technology and Engineering (CENTEC), Instituto Superior Técnico, Technical University of Lisbon, Lisboa, Portugal

ABSTRACT

A probabilistic maintenance and repair analysis of the desk plates of bulk carriers and tankers subjected to general corrosion wastage is presented here. The decisions about when to perform maintenance and repair on structure that may reach a failed state are studied. Different practical scenarios are analyzed and optimum repair intervals are proposed. The optimum repair age and intervals are based on a Weibull analysis. The total cost is calculated in normalized form to avoid the difficulties in costing. A comparative analysis of the maintenance planning for different types and lengths of ship is also provided.

Modern methodologies, such as reliability analyses, make full use of data acquired from non-destructive testing and can gather, filter and process information about the condition of ageing structures providing an estimate of the parameters defining the actual structural strength. Decisions about maintenance problems, such as defining when to perform maintenance on structures that is subjected to deterioration needs information about when the structure is expected to reach a failed state.

The set of corrosion data, deck plates of ballast of tankers provided by American Bureau of Shipping (ABS, 2002) is analyzed here. The set includes 1168 measurements of deck plates from ballast tanks with original nominal thickness varying from 13.5 to 35 mm on ships with lengths between perpendiculars in the range of 163.5 to 401 m. The second set analyzed includes a survey of corrosion data on deck plates of three 38000 DW bulk with a length between perpendiculars $L = 172$ m with data consisting of 1244 measurements of corrosion thickness collected from four inspections of service life of ships. The design thickness of measured deck plates varies between 10 and 30 mm. The bulk carriers have been subjected to similar corrosion environment along the entered ship

life including environment condition related to temperature, humidity, chlorides and sulphur dioxide, oxygen and same cargo. The corrosion data of this set was initially presented for tankers by Garbatov et al, 2007 and for bulk carriers by Garbatov & Guedes Soares, 2008.

The corrosion wastage of deck plates of ballast and cargo tanks was analyzed by Garbatov et al, 2007 is based on the non-linear corrosion model proposed by Guedes Soares & Garbatov, 1999. This model can describe an initial period without corrosion as a result of the presence of a corrosion protection system, a transition period with a nonlinear increase of wastage up to a steady state of long-term corrosion wastage.

An effort has been made to establish realistic decisions about when to perform maintenance on structure that will reach a failed state. Different scenarios are analyzed and optimum interval and age are proposed. The optimum age and intervals are based on Weibull analysis and some assumptions about the inspection and the time required for repair in the case of failure have been considered here.

REFERENCES

ABS, 2002, Database of Corrosion Wastage for Oil Tankers, American Bureau of Shipping RD 2002-07

Garbatov, Y., Guedes Soares, C. and Wang, G., 2007, Non-linear Time Dependent Corrosion Wastage of Deck Plates of Ballast and Cargo Tanks of Tankers, *Journal of Offshore Mechanics and Arctic Engineering*, Vol. 129, No. 1, pp. 48–55.

Garbatov, Y. and Guedes Soares, C., 2008, "Corrosion Wastage Modeling of Deteriorated Ship Structures", *International Shipbuilding Progress*, Vol. 55, pp. 109–125.

Guedes Soares, C. and Garbatov, Y., 1999, Reliability of Maintained, Corrosion Protected Plate Subjected to Non-linear Corrosion and Compressive Loads, *Marine Structures*, Vol. 12, pp. 425–445.

Safety, Reliability and Risk of Structures, Infrastructures and Engineering Systems – Furuta, Frangopol & Shinozuka (eds)
© 2010 Taylor & Francis Group, London, ISBN 978-0-415-47557-0

Uncertainty effects on dynamic response of offshore platform with wind energy production

K. Kawano, Y. Kimura, K. Ito & N. Tanaka
Kagoshima University, Kagoshima, Japan

Keywords: *offshore platform, dynamic forces, uncertainty, reliability*

It is expected that development of the offshore wind energy production would be very effective to carry out the reduction of greenhouse gas emissions. The uncertainty effects on the dynamic response evaluation of the idealized offshore platform model on offshore wind energy production subjected to wave force and seismic force are carried out in the present study. The steady wind force, which can be provided to generate the wind energy production, would be obtained in the coast area with water depth about 50 m. The offshore structure is expected to have important roles on development of wind energy production system. The environmental condition of the offshore structure is more severe than the land structure. If the offshore structure is located in the seismic activity area, it is essential to perform the dynamic response estimation on the wave force and seismic force in order to carry out the reliable design of the structure.

The wave and seismic force are usually estimated by relevant expressions with some parameters of uncertainties such as the significant wave height and the maximum acceleration of seismic motion. Since uncertain parameters on the wave and seismic force have very different characteristics on the dynamic response evaluations, it is suggested that the second moment approach could become an effective evaluation for the maximum response characteristics of the offshore platform with wind energy production. Applying the Monte Carlo Simulation (MCS) method to the dynamic response of the offshore platform, it

is efficiently carried out the dynamic response estimations for these uncertainties. Since the offshore platform of the wind energy production is the structure with the top heavy type, which has the properties to be very susceptive to seismic force, it is very important to clarify the uncertainty effects on the seismic response evaluation. The reliability evaluation of the offshore platform with the MCS simulation can be provided the available evaluations on the offshore platform with wind energy production subjected to dynamic forces with considerably different characteristics. It is demonstrated that contributions of uncertain parameters on the dynamic response evaluations of the offshore platform with wind energy production can be evaluated with the reliability index by applying the simulation results.

It is shown that in order to perform the reliability evaluation of the offshore platform with the wind energy production, it is significant to carry out the available estimation to the wave force, wind force, and seismic force with uncertainty.

REFERENCES

K. Kawano, Y. Kimura, M. Park, T. Iida(2007): *Performance based evaluation of offshore structure to wave and seismic forces with uncertainties*, Proce. of the seventeenth interernational offshore and polar engineering Conference, pp. 3554–3561

Safety, Reliability and Risk of Structures, Infrastructures and Engineering Systems – Furuta, Frangopol & Shinozuka (eds)
© 2010 Taylor & Francis Group, London, ISBN 978-0-415-47557-0

Ultimate strength of ship hull girder under random initial imperfections

S.C. Vhanmane
Research and Rule Development Division, Indian Register of Shipping, Mumbai, India.

B. Bhattacharya
Department of Civil Engineering, Indian Institute of Technology Kharagpur, Kharagpur, India

ABSTRACT

Due to poor workmanship, improper handling of cargo during loading and unloading, improper use of grab, and due to slamming during voyages, etc. the ship hull is subjected to various damages which collectively are termed as "initial imperfections." Available statistics show that these initial imperfections are random in their magnitudes. Consequently, the strength of the ship hull girder is random as well. It is important to Classification Societies as well as ship owners that the effect of such random initial imperfections on the safety of ship structures be assessed accurately.

Here we look at two common forms of initial imperfection: the initial deflection and residual welding stresses in the plating between longitudinals. This paper builds up on our earlier work on determining the ultimate strength of hull girders subjected to initial imperfections (initial deformation of plating between stiffeners and initial welding residual stresses). A simplified analytical incremental-iterative method given in IACS Common Structural Rules (for Tankers and Bulk Carriers) has been used to estimate the hull girder

Table 1. Effect of random initial imperfections alone (100 cases).

Ship Type	Condition	Mean (MN-m)	c.o.v. (%)
Bulk Carrier	Sagging	3575.0	1.63
	Hogging	5341.7	0.84
Tanker	Sagging	20133.0	1.00
	Hogging	25025.0	0.61

Table 2. Effect of combined random yield strength and random initial imperfections (100 cases).

Ship Type	Condition	Mean (MN-m)	c.o.v. (%)
Bulk Carrier	Sagging	3747.9	2.53
	Hogging	5767.5	2.35
Tanker	Sagging	21018.0	1.90
	Hogging	26793.0	2.46

ultimate strength under these initial imperfections; randomness in the magnitude of initial imperfections has been included for the first time. Also, the effect of randomness in the material yield stress, with moderate statistical dependence between those of stiffeners is also studied.

The above analysis can provide the modeling uncertainty in the hull girder strength when fabrication and service related damages are taken into account. This additional uncertainty may be used to modify the design equation for new ships through an "uncertainty factor" in its early design stages; it can also lead to a quick and economical means for early assessment of the effects of initial imperfections, perhaps in an LRFD-type equation format.

This methodology is applied to a bulk carrier and a double hull VLCC tanker, both modelled after two existing ship structures. Three primary cases are considered: effect of random yield strength (alone), effect of random initial imperfections (alone) and effect of combined random yield strength and random initial imperfections.

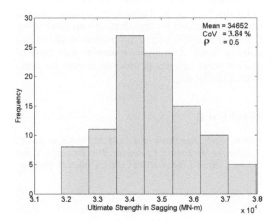

Figure 1. Random yield strength (alone) for Tanker – hogging (100 cases).

Safety, Reliability and Risk of Structures, Infrastructures and Engineering Systems – Furuta, Frangopol & Shinozuka (eds)
© 2010 Taylor & Francis Group, London, ISBN 978-0-415-47557-0

Aspects of reliability assessment of tidal stream turbines

D.V. Val

Heriot-Watt University, Edinburgh, UK

ABSTRACT

Growing demand for energy in the world along with depletion of fossil fuels and global initiative to achieve substantial cuts in greenhouse gas emissions have led to an increasing interest in renewable energy technologies. Many governments set targets to significantly increase the amount of energy from renewable sources in the next few decades, e.g., the UK Government has a target for a 10% contribution from renewable sources to UK electricity supply by 2010 and 20% by 2020. Tidal streams are one of the sources of renewable energy, which according to recent estimates can supply up to 5% of the UK energy demand.

Since extraction of tidal stream energy is a new technology there is significant uncertainty associated with the prediction of the performance of energy converting devices (i.e., tidal stream turbines), which causes stakeholders to be more cautious and less willing to invest into this area. One of the aims of the EPSRC-funded SUPERGEN Marine Energy Consortium is to develop a methodology for reliability assessment of such devices, which will enable to reduce this uncertainty and eventually make investment decisions more attractive. The paper presents preliminary results of this on-going research.

It starts with a brief description of the technology and the main types of existing (or proposed) tidal stream turbines. A generic turbine used in reliability studies is defined. This is a horizontal axis free-stream turbine, which is fixed to or gravitationally mounted on the seabed. It is also assumed that the generic turbine has a single indirect drive-train with a mechanical gearbox. The turbine is without a yawing mechanism and has variable pitch blades. On the whole, the selected generic tidal turbine is very similar to commonly used wind turbines. This provides an opportunity to use relevant data on failures of wind turbines, which have been collected during the last 15 years.

A methodology for the turbine reliability analysis is then described. A tidal stream turbine is a complex system consisting of structural, mechanical and electronic parts; therefore, the methodology should involve methods of both structural and system reliability. A simple model of the generic tidal turbine for system reliability analysis is presented. Since this a new technology the main problem with the application of system reliability methods is a lack of statistical data on failures of subsystems/components of tidal stream turbines. Depending on the subsystem/component the following approaches to estimating the failure rates are proposed:

(i) direct use of data on failure rates of similar components from other industries;
(ii) modification of the base failure rates (i.e, the failure rates derived from data, which have been collected at known operational and environmental conditions) to the condition of a tidal turbine by using influence factors (NSWC 2007); and
(iii) evaluation of the failure rates by direct reliability analysis of the turbine subsystem/component.

Each of the approaches is then described in more detail. In particular, differences in operational and environmental conditions of tidal stream and wind turbines are discussed, including their effect on the applicability of the failure rate data collected for wind turbines to reliability analysis of tidal stream turbines.

Finally, the concept of availability in relation to tidal stream turbines and factors affecting it are considered. It is shown that subsystems/components of tidal stream turbines should have higher reliability than those of onshore wind turbines in order to achieve the same level of availability.

REFERENCE

NSWC. 2007. *Handbook of Reliability Prediction Procedures for Mechanical Equipment*. West Bethesda, Maryland: Naval Surface Warfare Center, Carderock Division.

Safety, Reliability and Risk of Structures, Infrastructures and Engineering Systems – Furuta, Frangopol & Shinozuka (eds)
© *2010 Taylor & Francis Group, London, ISBN 978-0-415-47557-0*

A level-1 reliability-based design method for caisson-type breakwaters in view of sliding displacement

Takeshi Yoshioka
Electric Power Development Co., Ltd., Tokyo, Japan

Yoichi Moriya
Coastal Development Institute of Technology, Tokyo, Japan

Takashi Nagao
National Institute for Land and Infrastructure Management, Yokosuka, Japan

Following the adoption of ISO 2394, research on application of the reliability-based design method to civil engineering structures is progressing in various areas. Japanese technical standards for port and harbor facilities were revised in 2007 based on the recent research and the level-1 reliability-based design method (partial factors method) was introduced as a standard design method from the viewpoint of harmonization with international standards.

Focusing on the external stability of caisson-type breakwaters, the standard reliability-based design method takes into consideration of the equilibrium condition of horizontal forces against sliding. Meanwhile, another method that evaluates the expected cumulative sliding displacement during the design lifetime by Monte Carlo simulation was proposed by Shimosako & Takahashi (1999). The method is more advanced compared with the standard method, however, the adoption of the method to the technical standards was postponed considering the fact that calculations are time-consuming and discussion of a system reliability considering other failure modes is difficult.

The objectives of this research are to construct a practical calculation formula for sliding displacement of caisson-type breakwaters, and to construct a level-1 reliability-based design method by determining partial factors based on a reliability analysis.

Firstly, the practical calculating formula for sliding displacement of caisson-type breakwater caused by the design wave height was constructed. It gives approximately the same sliding displacement as that obtained by numerical integration of equation of motion for caisson assuming a time-series model of wave force proposed by Tanimoto et al. (1996). Using the sliding displacement S given by the formula and an allowable value S_a (30 cm assumed in the study), the

performance function Z is defined by $Z = S - S_a$. Secondly, first-order reliability analysis was conducted and reliability indices and sensitivity factors of random variables were evaluated. As a result, partial factors method was constructed to control the failure probability P_{f1} considering the target failure probability P_{f1T}. We used 33 cases collected from design datum for caisson-type breakwaters throughout Japan in this study.

The method well evaluates the failure probability by the single wave action, however, we also have to establish the method for the evaluation of the failure probability P_{f50} that the cumulative sliding displacement ΣS exceeds S_a during the design working life. We found that the P_{f50} greatly depends on the probability distribution of wave heights. Therefore, we established the method to evaluate the P_{f50} by using the failure probability by the single wave action.

The proposed method is a level-1 reliability-based design method which can control the failure probability against the cumulative sliding displacement of the caisson-type breakwater during the design lifetime with high accuracy.

REFERENCES

Shimosako, K. & Takahashi, H. 1999. Application of deformation-based design for coastal structures, *Proceeding of the international conference on coastal structures '99.*

Tanimoto, K., Furukawa, K. & Nakamura, H. 1996. Fluid resistant forces during sliding of breakwater upright section and estimation model of sliding under wave forces, *Proceedings of coastal engineering*, JSCE, Vol. 43, pp. 846–850, (in Japanese).

Safety, Reliability and Risk of Structures, Infrastructures and Engineering Systems – Furuta, Frangopol & Shinozuka (eds)
© 2010 Taylor & Francis Group, London, ISBN 978-0-415-47557-0

A comparison of safety levels of sheet pile quay wall based on minimization of life cycle cost and expected total cost

T. Nagao
National Institute for Land and Infrastructure Management, Yokosuka, Japan

R. Ozaki
Chuo Fukken Consultants, Co., Ltd., Osaka, Japan

ABSTRACT

The technical standards for port and harbor facilities in Japan were revised in 2007 and the reliability-based design method using partial factors was introduced. Partial factors for the ordinal condition were decided on the basis of minimization of the expected total cost (hereafter, ETC, Ozaki et al. 2005). As the standard design working life for port facilities is fifty years in Japan, the section modulus of the sheet pile is set considering fifty years' corrosion in the previous study. This means that the ETC is of one year and not of the whole design working life. Although the partial factors are rounded off to the conservative side from that of minimum ETC, this might arise the discussion: are the partial factors appropriate from the viewpoint of minimum life cycle cost (hereafter, LCC)?

We conducted reliability analyses changing the section modulus of the sheet pile from initial value to that after fifty years and evaluated the reliability indices according to the decrease of section modulus. It should be noted that the evaluation of LCC is not an easy task such as the sum of multiplication of failure loss cost and failure probability for each section modulus. The difficulty in the evaluation of LCC derives from the condition that the section modulus becomes initial value after the restoration work. It was necessary to consider about 563 trillion failure event cases altogether in order to evaluate LCC.

Figure 1 shows an example of LCC curve about water depth of −11 m with cumulative failure probability $P_{f\Sigma 50}$ during design working life of fifty years. This figure also shows an example of ETC curve under the section modulus considering the fifty years' corrosion. Table 1 summarized the failure probabilities corresponding to the minimum LCC and ETC, which also shows the failure probabilities by the level-1 reliability-based design results using the partial factors in the technical standards.

The failure probabilities for minimum ETC are larger than those for minimum LCC. However, the failure probabilities by the level-1 reliability-based design are smaller than those for minimum LCC. It was shown that the partial factors in technical standards are set to

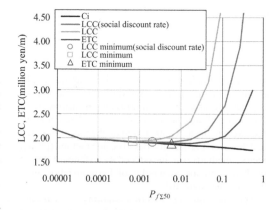

Figure 1. Example of the LCC curve.

Table 1. Comparison of failure probability.

Design	Evaluation of economic loss cost*	Section modulus of sheet pile m³/m	Diameter of tie rods mm	$P_{f\Sigma 50}$
LCC minimum	$\mu + \sigma$	0.00119	38	6.898E-04
	$\mu + 3\sigma$	0.00119	38	6.898E-04
ETC minimum	$\mu + \sigma$	0.00119	38	6.898E-04
	$\mu + 3\sigma$	0.00119	38	6.898E-04
L1 design**	$\beta = 2.7$	0.00119	38	6.898E-04
	$\beta = 3.6$	0.00134	38	1.628E-04

* Two amounts of the economic loss cost were considered similar to the past work (Ozaki et al. 2005).
** In L1 design, $\beta = 2.7$, 3.6 are corresponding to the economic loss cost of $\mu + \sigma$ and $\mu + 3\sigma$, respectively.

the conservative side from the viewpoint of minimum LCC.

REFERENCE

Ozaki R., Nagao T. & Shibasaki R. 2005. Level-1 reliability-based design method for port and harbor facilities under ordinal conditions for minimization of expected total cost, Proceedings of ICOSSAR2005. CD-ROM

General Session (Reliability-Based Design and Regulations & Reliability-Based Optimization and Control & Reliability Theory)

Safety, Reliability and Risk of Structures, Infrastructures and
Engineering Systems – Furuta, Frangopol & Shinozuka (eds)
© 2010 Taylor & Francis Group, London, ISBN 978-0-415-47557-0

Reliability-based code calibration: Case study of Brazilian structural design codes

A.T. Beck & A.C. Souza Junior
Structural Engineering Department, University of São Paulo, São Carlos, SP, Brazil

ABSTRACT

This paper addresses the reliability-based calibration
of partial safety factors for Brazilian design codes
NBR8681:2003 (actions and safety of structures) and
NBR8800:2008 (design of steel structures). The general format of these codes is very similar to the new
EUROCODES:

$$S\left[\gamma_D \cdot D_n + \gamma_i \cdot Q_{ni} + \sum_{j \neq i}^{n} \psi_j \cdot \gamma_j \cdot Q_{nj} \right] \leq R\left[\frac{r_k}{\gamma_R} \right] \quad (1)$$

Brazilian structural design codes have never been
subject to reliability-based calibration of partial safety
and load combination factors. The present calibration
effort is based on actual data for wind loads in Brazil,
but uses mainly international data for other parameters.
So far, the investigation is limited to structural steel
members.

A preliminary calibration run involving only dead
load revealed the optimum partial factors $\gamma_R = 1.1$ and
$\gamma_D = 1.35$, for a target reliability index of $\beta_T = 3.0$.
Fixing these values and running the calibration process again, for the variable load combinations, resulted
in the partial and combination factors presented in
Table 1.

Table 1 shows significant differences in the partial
factors currently in use and the factors found through

Table 1. Brazilian code (NBR) and calibrated partial
factors.

Factor	NBR8800 NBR8681	Calibration for $\beta_T = 3.0$
γ_R	1.10	1.10
γ_D	1.35	1.35
γ_L	1.50	1.65
γ_W	1.40	1.70
ψ_L	0.70	0.30
ψ_W	0.60	0.30
$\gamma_L \cdot \psi_L$	1.05	0.50
$\gamma_W \cdot \psi_W$	0.84	0.51

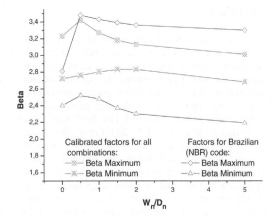

Figure 1. Bounds on reliability indexes for Brazilian codes
and for calibrated partial factors, $\beta_T = 3.0$.

the calibration process. It is observed that the calibrated factors are larger for the principal loads (γ_L
and γ_W) but the combination factors are significantly
smaller, resulting in significantly smaller values of the
accompanying variable actions (products $\gamma \cdot \psi$).

Reliability indexes resulting from use of these partial factors are compared graphically in Figure 1. The
figure shows lower and upper bounds on reliability
indexes obtained for all designs (load ratios) considered, with β given as the minimum between all
limit state (load combination) equations. The bounds
show the uniformness (or lack of) of reliability indexes
obtained using the different sets of partial factors.

It can be clearly seen that the calibrated combination factors lead to larger and more uniform reliability
indexes, when compared to their Brazilian design code
counterparts. This is also consequence of larger partial
factors applied on dead and principal variable loads.

The set of calibrated partial factors is an *optimum*
(or, at least, a better) set, in comparison to what is
currently used in Brazilian design codes. This conclusion is dependent on further confirmation, given the
limitations of the present calibration effort.

Safety, Reliability and Risk of Structures, Infrastructures and
Engineering Systems – Furuta, Frangopol & Shinozuka (eds)
© 2010 Taylor & Francis Group, London, ISBN 978-0-415-47557-0

Reliability-based design specifications for simply supported steel beams exposed to fire

S. Iqbal & R.S. Harichandran

Department of Civil and Environmental Engineering, Michigan State University, East Lansing, MI, USA

ABSTRACT

Until recently, steel beams exposed to fire were designed using prescriptive approaches that are very simplistic and do not account for actual loading conditions and real fire scenarios. In the last decade, performance-based codes which allow more rational engineering approaches for the fire design of steel members have been promoted. The 2005 AISC Steel Construction Manual (AISC 2005) now allows steel beams to be designed against fire using room temperature design specifications and reduced material properties. According to the AISC manual, the same resistance factors used in room temperature design should be used for fire design. However, most other codes suggest that a capacity reduction factor of 1.0 be used (e.g., in the Eurocode 3, the partial safety factor γ_M is 1.0 for fire design). This recommendation is based on arguments that the probability of fire occurrence and the strength falling below the design value simultaneously is very small, and that the fire design is based on the most likely expected strength (Buchanan 2001). However, no substantial work has been done to develop a strength reduction factor based on rigorous reliability analysis.

A methodology is proposed for developing the capacity reduction factor for the flexural strength of simply supported steel beams exposed to fire. Fire load, ventilation factor, ratio of floor area to the total area of fire compartment, thermal inertia of compartment enclosure, thickness, density and thermal conductivity of fire protection material, yield strength of steel, plastic section modulus, dead load, and live load are taken as random variables. The chosen statistics of the live load, fire load, and ratio of floor area to total surface area of the fire compartment are representative of typical office buildings in the U.S. The FERUM (Finite Element Reliability Using Matlab) software is used to account for the true distributions of the stochastic fire design parameters. Instead of using the approximate mean and variance of the steel temperature, the steel temperature is included in the

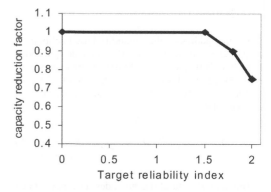

Figure 1. Capacity reduction factor vs. target reliability index.

performance function in terms of the basic random variables. The effect of active fire mitigating systems (e.g., sprinklers, smoke/heat detectors, fire brigade) in reducing the probability of occurrence of a severe fire is included.

The probability of failure, and correspondingly the target reliability index, should vary depending on the presence of active fire mitigating systems because these measures affect the occurrence of a severe fire. Consequently, the capacity reduction factor should vary for each design situation. The plot of the capacity reduction factor versus the target reliability index found from preliminary analyses is shown in Figure 1. The capacity reduction factor for flexural strength is not constant for all design situations as suggested in design specifications, and varies depending on the presence of active fire mitigating systems in a building.

REFERENCES

AISC. 2005. Steel construction manual. 13th edition, American Institute of Steel Construction, Inc.

Buchanan, A.H. 2001. *Structural design for fire safety*. Chichester: John Wiley& Sons Ltd.

Safety, Reliability and Risk of Structures, Infrastructures and
Engineering Systems – Furuta, Frangopol & Shinozuka (eds)
© 2010 Taylor & Francis Group, London, ISBN 978-0-415-47557-0

A simulation-based framework for robust stochastic design of base isolation systems

A.A. Taflanidis
Civil Engineering and Geological Sciences Department, University of Notre Dame, Notre Dame, IN, USA

ABSTRACT

Over the last decades, there has been a growing interest in the application of base isolation techniques to civil structures. Of the many relevant research topics, the efficient design of additional dampers (passive or semi-active), to operate in tandem with the isolation system, has emerged as one of the more important. One of the main challenges of such applications has been the explicit consideration of the nonlinear behavior of the isolators or the dampers in the design process. Another challenge has been the efficient control of the response under near-field ground motions. Such motions frequently include a strong longer-period pulse that has important implications to base-isolated systems.

The current work proposes a framework for design of base isolation systems which address both these challenges. A probability logic approach is adopted for addressing the uncertainties about the structural model as well as the variability of future excitations. In this approach, probability can be interpreted as a means of describing the incomplete, i.e., missing, information about the system under consideration. This is established by characterizing the relative plausibility of different properties of the system and future excitations by probability models. In this stochastic setting, a realistic model (Taflanidis et al., 2007) for the description of near-field ground motions is discussed. This model establishes a direct link, in a probabilistic sense, between our knowledge about the characteristics of the seismic hazard in the structural site and future ground motions. The design objective is then defined as the maximization of structural reliability, quantified as the probability that the structural response will not exceed acceptable performance bounds. A simulation-based approach is implemented for evaluation of the stochastic performance of the base isolated structure. This approach explicitly takes into account all non-linear characteristics of the structural response in the design process and allows for complex characterization for the system and excitation models.

An efficient framework (Taflanidis and Beck, 2008) is discussed for performing the associated challenging design optimization and for selecting values of the controllable system parameters (design variables) that optimize the system reliability. This approach also allows for a sensitivity analysis for the influence of the uncertain model parameters on the stochastic performance (Taflanidis and Beck, 2009).

The methodology is illustrated through application to a base isolated building with lead-rubber bearings and nonlinear viscous dampers. Uncertainty is included in both the structural and excitation models. The properties of the viscous dampers correspond to the design variables in this case. The effect of their maximum forcing capabilities is also investigated. The sensitivity analysis established for the model parameters shows that the stochastic excitation characteristics, in particular, the moment magnitude, epicentral distance and peak ground velocity, have the greater importance for the structural performance. The results also shows that the addition of the optimally designed dampers provides a significant improvement for the seismic performance of the isolated structure and that nonlinearities of the damper behavior are appropriately addressed in the context of the proposed framework.

REFERENCES

Taflanidis, A.A. & Beck, J.L. 2008. An efficient framework for optimal robust stochastic system design using stochastic simulation, *Computer Methods in Applied Mechanics and Engineering*. 198(1): 88–101.

Taflanidis, A.A. & Beck, J.L. 2009. Stochastic Subset Optimization for reliability optimization and sensitivity analysis in system design. *Computers and Structures*. 87: 318–331.

Taflanidis, A.A., Scruggs, J.T. & Beck, J.L. 2007. Smart base isolation design including model uncertainty in ground motion characterization, *4th International Conference on Earthquake Geotechnical Engineering*, Thessaloniki, Greece, June 25–28.

Safety, Reliability and Risk of Structures, Infrastructures and Engineering Systems – Furuta, Frangopol & Shinozuka (eds)
© 2010 Taylor & Francis Group, London, ISBN 978-0-415-47557-0

CAE-based sequential topology optimization and reliability analysis

K.W. Liao

Tamkang University, Tamsui, Taiwan

ABSTRACT

Sizing, shape and topology optimization problems address different aspects of the structural design problem. This study will focuses on the topic of topology optimization only. Many algorithms for Deterministic Topology Optimization (DTO) have been proposed, such as: (1) Genetic Algorithms (GA) which explore the entire design space and are less likely to be trapped in local minima but additional computational expense is expected; (2) gradient search methods which is one of the most popular approaches but may converge to a local optimum and usually intermediate densities are obtained, which provide no physical meaning; (3) Evolutionary Structural Optimization (ESO) that do not have the problem of intermediate density and is based on the concept of gradually removing redundant material to achieve an optimal design.

DTO does not consider the randomness/uncertainty in design variables/parameters while optimizing the structure. Once considered, the optimal topology will be affected by the deviation of the design variables/parameters. Because many engineers use the results of topology optimization as their starting point in the development of product design, the integration of the reliability or safety criteria with DTO by introducing the reliability constraints is necessary. Reliability-Based Topology Optimization (RBTO) incorporates probabilistic analysis into topology optimization process to take into account the randomness of design parameters in physical models. Earlier researches have shown the optimal topologies are different between RBTO and DTO. Researches also have proved RBTO can provide a design having a great chance of staying in the feasible design space.

A typical approach of RBTO is a nested double-loop approach where the outer loop conducts optimization and the inner loop deals with probabilistic analysis. The computational cost for this double-loop procedure is prohibitive for most practical problems. Moreover, topology optimization inherently possesses a large number of design variables making the problem more difficult. To overcome the high computational cost and build a suitable methodology for practical engineers, this study utilizes the concept of single-loop approach to RBTO problem. The single-loop approach is well developed in RBDO, and its main idea is that the random fields at different design points (e.g., different topologies in RBTO) in the single optimization iteration are identical so that one can decouple the nested RBDO into a sequential analysis procedure and has been proved to be effective by many researches.

Application of the single-loop technique on the RBTO has not drawn much attention at this point. Only few researches have investigated this topic and recognized the efficiency of the single-loop approach. For example, Kharmanda et al (2004) have applied the single-loop algorithm into RBTO to consider the randomness in design parameters. The design domain in Kharmanda's study is not prescribed; an alternative single-loop approach, in which the design domain is prearranged, is proposed and its effectiveness is investigated thoroughly in this study.

In practical, engineers are more familiar with the operation of CAE-based software than the theory behind. The single-loop algorithm is readily to take advantage of this trend if a platform incorporating commercial software can be built. That is, engineers can achieve the jobs of RBTO using the same tools already known.

Several numerical examples are provided to show the efficiency and accuracy of the proposed approach.

REFERENCE

Kharmanda G, Olhoff N, Mohamed A and Lemaire M (2004) Reliability-based topology optimization. Struct Multidisc Optim 26:295–307

Safety, Reliability and Risk of Structures, Infrastructures and Engineering Systems – Furuta, Frangopol & Shinozuka (eds)
© 2010 Taylor & Francis Group, London, ISBN 978-0-415-47557-0

The study on reliability with an improved truncated normal distribution

Zhangmiao. Li, Jimian Song & Xinjian. Kou
Shanghai Jiao Tong University, Shanghai, China

Normal distribution has been extensively applied in different fields as well as structural reliability, but there are a mount of documents suggest that normal distribution was not according with the actual situations. Because of the defect of normal distribution, a modified normal distribution has been used to model the stochastic structure of random variables.

The modified normal distribution is derived by intercepting certain breadth of normal distribution at two sides of mean and named truncated normal distribution. The breadth is determined by engineering practice. Truncated normal distribution has catastrophe where it is truncated, the modification is not flawless because catastrophe at two-tail is not matched with actual situations. Therefore there is also a difference between calculate results used truncated normal distribution model and theory results. It is necessary to introduce a new probability distribution that has no trails and no catastrophe.

Three-parameter truncated normal distribution is derived from normal distribution (see figure 1). Let there a normal distribution with mean μ and variance σ^2, firstly, shift ax x of normal distribution upward to new ax x' to make function values at two endpoints of normal distribution probability density function curve are zero. Then magnify ax y in order to make the area under the density curve between $\mu - k\sigma$ and $\mu + k\sigma$ in the new coordinate system $x'oy$ is 1. The 'density curve between $\mu - k\sigma$ and $\mu + k\sigma$ in the new axis is the very new probability density curve. Obviously the new probability density function is more reasonable than normal distribution and truncated normal distribution. The breadth of intercepted interval is different from reality, so introduce new parameter k named breadth intercepted parameter. The new probability distribution function is determined by the mean μ, the standard deviation σ and the breadth intercepted parameter k completely. Then the new probability distribution can be named three-parameter truncated normal distribution. This new probability distribution has no trails and no catastrophe. The problem is that whether difference between this new distribution model and theory is smaller than difference between normal distribution model (or truncated normal distribution) and theory.

Firstly, probability density function of three-parameter truncated normal distribution must be determined by mathematic method. Then the numerical aspects and linear combination of three-parameter truncated normal distribution can be calculated. The point estimation of three parameters of new probability distribution is completed by a new method introduced in paper. Then theory studies of three-parameter truncated normal distribution is finished, the application of three-parameter truncated normal distribution in engineering reliability is simulated by computer and the simulation results can suggest that the three-parameter truncated normal distribution model is more approximate to reality than normal distribution(or truncated normal distribution).

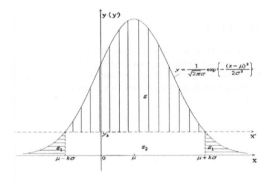

$$y = \frac{1}{\sqrt{2\pi}\sigma} \exp\left\{-\frac{(x-\mu)^2}{2\sigma^2}\right\}$$

Figure 1. Three-parameter truncated normal distribution.

Safety, Reliability and Risk of Structures, Infrastructures and
Engineering Systems – Furuta, Frangopol & Shinozuka (eds)
© 2010 Taylor & Francis Group, London, ISBN 978-0-415-47557-0

A Monte Carlo based method for system reliability

A. Naess

Centre for Ships and Ocean Structures & Department of Mathematical Sciences
Norwegian University of Science and Technology, Trondheim, Norway

B.J. Leira

Department of Marine Technology, Norwegian University of Science and Technology, Trondheim, Norway

O. Batsevych

Centre for Ships and Ocean Structures, Norwegian University of Science and Technology, Trondheim, Norway

ABSTRACT

The focus of this paper is the development of a Monte Carlo based method for estimating the reliability of structural systems. The use of Monte Carlo methods for system reliability analysis has several attractive features, the most important being that the failure criterion is usually relatively easy to check almost irrespective of the complexity of the system. The flip side of such methods is the amount of computational efforts that may be involved. However, by reformulating the reliability problem to depend on a parameter and exploiting the regularity of the failure probability as a function of this parameter, it is shown that a substantial reduction of the computational efforts involved can be obtained.

A prominent problem in structural reliability is the prediction of the failure probability of a system of failure modes, which is expressed either as a series system or a parallel system, or a combination of the two. In general, such problems are often exceedingly hard to calculate directly by the methods of structural reliability. Therefore, several approximation procedures have been proposed, often in the form of upper and lower bounds on the failure probability (Ditlevsen and Bjerager 1986; Melchers 1999).

In principle, the reliability of complicated systems can be accurately predicted by standard Monte Carlo simulation methods, but the computational burden may be prohibitive. The authors have therefore started the development of a new Monte Carlo based method for estimating system reliability that aims at reducing the computational cost. It exploits the regularity of tail probabilities to set up an approximation procedure for the prediction of the far tail failure probabilities based on the estimates of the failure probabilities obtained by Monte Carlo simulation at more moderate levels. In the paper the usefulness and accuracy of the estimation method will be illustrated by application to two specific examples: A ten-bar truss structure and a series system modelling of the reliability of a portal frame structure.

REFERENCES

Ditlevsen, O. & Bjerager, P. 1986. Methods of structural systems reliability. *Structural Safety 3* 195–229.
Melchers, R. E. 1999. *Structural Reliability Analysis and Prediction* (Second ed.). New York: John Wiley & Sons, Inc.

Mini-Symposia (MS06) Uncertainties in Civil Structures & Infrastructure Engineering

Safety, Reliability and Risk of Structures, Infrastructures and
Engineering Systems – Furuta, Frangopol & Shinozuka (eds)
© 2010 Taylor & Francis Group, London, ISBN 978-0-415-47557-0

Role of modeling uncertainties in assessing vulnerability and risk to civil infrastructure from rare events

Bruce R. Ellingwood

School of Civil and Environmental Engineering, Georgia Institute of Technology, Atlanta, GA, USA

ABSTRACT

Civil infrastructure facilities are designed to withstand demands imposed by their service requirements and by natural environmental events. While their performance under such demands generally has been acceptable, events *beyond* the design envelope, including extreme storms or earthquakes, fires, accidents and malevolent attack, may cause severe damage to a facility or precipitate a catastrophic failure. Changes in design and construction practices over the past several decades have made some modern structural systems vulnerable to such events. Moreover, public awareness of infrastructure performance and safety issues has increased markedly during the past thirty years as a result of well-publicized natural and man-made disasters. Improved building practices to lessen the likelihood of unacceptable damage from low-probability, high-consequence threats now are receiving a high level of interest among engineers and other design professionals, especially in light of the worldwide move toward performance-based engineering of civil infrastructure systems.

The complexities of modern engineering analysis and design give rise to numerous uncertainties. Some of these uncertainties stem from factors that are inherently random (or aleatoric) at the customary scales of resolution in engineering or scientific analysis. Others (epistemic) arise from a lack of knowledge, ignorance, or approximations in modeling. Still others are associated with human behavior – the manner in which individuals, groups and political entities perceive, respond to, and make decisions regarding rare or threatening events. The risk that arises as a result of these uncertainties must be managed in the public interest by engineers, code-writers and other regulatory authorities.

Structural reliability and probabilistic risk analysis provide the theoretical framework for modeling uncertainties associated with hazard occurrences, damage states and consequences, and for optimizing investments in risk reduction for civil infrastructure. Communicating the role of uncertainty and its impact on risk to diverse stakeholders – architects, engineers, urban planners, insurance underwriters, and local governmental agencies and regulatory authorities – is one of the major challenges to risk-informed decision-making. This paper reviews recent advances in uncertainty modeling and risk-based decision tools applicable to performance assessment of civil infrastructure systems, discusses their limitations, and identifies research issues that must be addressed to facilitate further advances in practical risk-informed decision-making for facilities exposed to a spectrum of natural and man-made hazards.

REFERENCES

Ellingwood, B. R. and Y. K. Wen (2005), "Risk-benefit based design decisions for low probability/high consequence earthquake events in Mid-America," *Prog. in Struct. Engrg. and Mat.* 7(2):56–70.

Ellingwood, B.R. (2005), "Risk-informed condition assessment of civil infrastructure: state of practice and research issues" J. Struct. & Infrastruct. Engrg. 1(1):7–18.

Ellingwood, B. R. (2007), "Strategies for mitigating risk to buildings from abnormal load events," Int. J. of Risk Assessment and Mgt. 7(6/7):828–845.

Faber, M. and M.G. Stewart (2003), "Risk assessment for civil engineering facilities: a critical appraisal," Reliability Engrg. & System Safety 80: 173–184.

Safety, Reliability and Risk of Structures, Infrastructures and
Engineering Systems – Furuta, Frangopol & Shinozuka (eds)
© 2010 Taylor & Francis Group, London, ISBN 978-0-415-47557-0

Probabilistic fragility assessment method for structural intervention schemes

S.H. Jeong

Department of Architectural Engineering, INHA University, Korea

Amr S. Elnashai

Department of Civil and Environmental Engineering, University of Illinois at Urbana-Champaign, IL, USA

ABSTRACT

Due to the high uncertainties of structural responses under earthquake loadings, it is desirable to use probabilistic performance targets for decision-making related to intervention methods. The latter is based on the probabilistic performance assessment of structures with various retrofit options and requires extensive computer simulations to account for the randomness in both input motion and response characteristics. An approach whereby a set of fragility relationships with known reliability are derived based on the fundamental response quantities of stiffness, strength and ductility is presented (Jeong & Elnashai 2007). An exact solution for a generalized single-degree-of-freedom system is employed to construct a response database of maximum responses. Once the fundamental response quantities of a wide range of structural systems are defined and the Response Database is constructed, the fragility relationships for various limit states can be constructed without recourse to further simulation.

By virtue of its instantaneous nature, the proposed method is especially useful for practical application of the analytical fragility assessment to the planning of seismic rehabilitation, and regional earthquake mitigation where fast estimation of probabilities of reaching damage states for a large number of structural configurations and different mitigation measures is required. For cases of selection between different retrofitting options, the proposed approach gives rapid estimates of probabilities of various damage levels being inflicted onto the structures under consideration, given only the stiffness, strength and ductility for each alternative retrofitting scheme. Additionally, the presented fragility assessment methodology enables the analyst to practically investigate the vulnerability of large numbers of structural configurations instantly without simulation.

For a given earthquake intensity and a limit state, the fragility is a function of response demand which is directly estimated by stiffness and strength in the

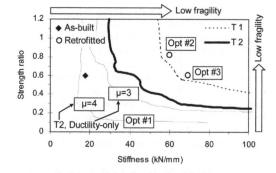

Figure 1. Fragility contour.

proposed method. The latter function can therefore be represented as contours provided that the response estimate is at hand (Fig. 1). The contours of the limit state probability can be considered as probabilistic performance targets of various structural configurations that cover wide range of stiffness and strength.

The fragility contour is an inverse problem of the conventional fragility assessment; finding structural parameters with given probability of exceedance. The inverse function, $(T, SR) = F^{-1}(P_i)$, can be solved only by the Response Database. Since the Response Database is a collection of pre-analyzed responses for a wide range of structures, the response estimation of structures entails only retrieving the relevant value from the latter database. Thus, the contour chart (Fig. 1) which needs estimation of limit state probabilities for all possible response parameters (stiffness and strength) can be obtained instantly.

REFERENCE

Jeong, S.H. & Elnashai, A.S. 2007. Probabilistic Fragility Analysis Parameterized by Fundamental Response Quantities. *Engineering Structures* 29(6): 1238–1251.

Safety, Reliability and Risk of Structures, Infrastructures and
Engineering Systems – Furuta, Frangopol & Shinozuka (eds)
© 2010 Taylor & Francis Group, London, ISBN 978-0-415-47557-0

Story-specific demand models and seismic fragility estimates for low-rise buildings

J. Bai, P. Gardoni & M. Hueste

Zachry Department of Civil Engineering, Texas A&M University, College Station, TX, USA

ABSTRACT

A seismic fragility analysis is conducted to assess the vulnerability of low-rise building structures. Seismic fragility is defined as the conditional probability of attaining or exceeding a certain performance level given a specified earthquake hazard level. The focus of this paper is on the development of seismic demand models for low-rise buildings that reflect the potential for more than one story to be at or near a critical interstory drift limit.

The maximum interstory drift is a convenient parameter to describe overall structural response to lateral loads. Typical demand models used to develop seismic fragility curves relate the maximum interstory drift to a characteristic earthquake intensity measure, such as spectral acceleration, for the corresponding time history of ground motion. For a suite of *n* ground motions, *n* data points are generated using time history response analysis. These data points can be used to develop demand models describing the relationship between these parameters. There are sources of uncertainty associated with demand models including earthquake intensity, validation limits for analytical results, and other assumptions. A probabilistic methodology is used to account for the above uncertainties.

Traditionally, demand models have been developed based on the maximum interstory drift experienced over the entire building height. However, when only the maximum interstory drift of the building is considered, the actual fragility estimates tend to be underestimated, in particular if the interstory drifts for more than one story are close to the maximum value. Therefore, to assess the probability that any interstory drift exceeds a specified limit for a given structural performance level, it is important to evaluate the drift demand for each story within the structure. In this paper, story-specific demand models are developed to consider the maximum interstory drift for each story versus seismic intensity. Correlations among individual story demand models for a given structure are included to properly capture the potential dependence between maximum interstory drifts. Both linear and bilinear models are considered to describe the demand relationships and a Bayesian methodology is used to assess the statistics of the model parameters. In the development of the story-specific demand models, the computed responses that exceed the validation limit of the analytical model are treated as lower bound data. The interstory drift limits that correspond to particular structural performance limits can also differ over the height of a structure. Therefore, capacity limits for each story are determined based on the building response from a story-by-story push-over analysis.

Using the developed story-specific demand models and the corresponding capacity limits, seismic fragility curves are developed using Monte Carlo simulation for a typical low-rise building structure as an application. The building is a two-story reinforced concrete flat slab building typical of construction in the central United States. The developed fragility estimates are compared to those based on demand models describing the overall maximum interstory drift. The proposed methodology provides a refined approach that includes more building response information than typical demand models, allowing for more accurate estimates of seismic fragility for low-rise building structures.

Safety, Reliability and Risk of Structures, Infrastructures and
Engineering Systems – Furuta, Frangopol & Shinozuka (eds)
© 2010 Taylor & Francis Group, London, ISBN 978-0-415-47557-0

Seismic fragility of a bridge on liquefaction susceptible soil

Oh-Sung Kwon
*Department of Civil, Architectural, and Environmental Engineering, Missouri University
of Science and Technology, MO, USA*

Anastasios Sextos
Civil Engineering Department, Aristotle University Thessaloniki, Greece

Amr Elnashai
*Mid-American Earthquake Center, Civil and Environmental Engineering Department,
University of Illinois at Urbana-Champaign, IL, USA*

ABSTRACT

This paper reports an exploratory framework for
including the effect of soil liquefaction in fragility
analysis bridge-foundation-soil system. The framework is demonstrated through an application to the
well-studied and instrumented Melloland Overcrossing Bridge. A powerful computational scheme was
developed for this purpose involving both 3-D and
simplified inelastic finite element models.

Detailed inelastic site response analyses considering liquefaction effects were carried out. An effort is
also made to propose a Global Damage Index accounting for different soil, foundation and superstructure
failure modes. The results indicate that the inelastic
Figure 1. Fragility curves for the four foundation flexibility and excitation cases studied dynamic response of
the investigated bridge-foundation-soil system is significantly affected by the liquefaction of upper sand
layers. In addition to the significance of accounting for
liquefaction, the spatial extent of liquefaction is also
shown to be important. Both liquefaction and its spatial
distribution are influential on determining the characteristics of the input motion and the demand imposed
on foundations and superstructure. The assumptions
made to facilitate the application of the proposed
framework for liquefaction-sensitive fragility analysis
are not integral to the procedure and may be improved
by researchers seeking to quantify the probabilistic
response of soil-foundation-structure systems.

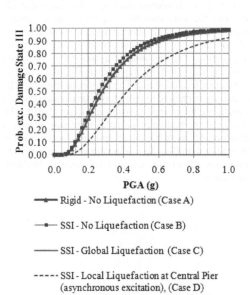

Figure 1. Fragility curves for the four foundation flexibility
and excitation cases studied.

REFERENCES

Bray, J.D. and Stewart, J.P. 2000. Damage patterns and
foundation performance in Adapazari, in Kocaeli, Turkey
Earthquake of August 17, 1999 Reconnaissance Report.
Earthquake Spectra, **16**(Suppl. A), 163–189.
Elgamal, A., Yang, Z. and Parra, E. 2002. Computational
Modeling of Cyclic Mobility and Post-Liquefaction Site
Response, *Soil Dyn. and Earthq. Eng.*, Vol. 22, No. 4,
259–271.
Elnashai, A. S., Papanikolaou, V., and Lee, D. 2002. Zeus
NL – A System for Inelastic Analysis of Structures. *Mid-
America Earthq. Center, University of Illinois at Urbana-
Champaign.*
Kwon, O.-S., Nakata, N., Elnashai, A. and Spencer, B. 2005.
A Framework for Multi-Site Distributed Simulation and
Application to Complex Structural Systems, *J. of Earthq.
Eng.*, 9(5), 741–753.
Kwon, O.-S. Sextos, A and Elnashai, A. 2008. Liquefaction-
dependent fragility relationships of complex bridge-
foundation-soil systems', International Conf. on
Earthquake Engineering and Disaster Mitigation, Jakarta,
Indonesia.
Kwon, O.-S. 2007. Probabilistic Seismic Assessment of
Structure, Foundation and Soil Interacting Systems, Ph.D.
Thesis, University of Illinois at Urbana-Champaign.
McKenna, F., and Fenves, G. L. 2001. The OpenSees com-
mand language manual, version 1.2., *Pacific Earthquake
Engineering Research Center*, Univ. of California at
Berkeley.

Safety, Reliability and Risk of Structures, Infrastructures and Engineering Systems – Furuta, Frangopol & Shinozuka (eds)
© 2010 Taylor & Francis Group, London, ISBN 978-0-415-47557-0

Seismic vulnerability assessment of embankments

Y. Tsompanakis
Department of Applied Sciences, Technical University of Crete, Greece

N.D. Lagaros
School of Civil Engineering, National Technical University of Athens, Greece

Earthquake engineering applications, mainly the geotechnical ones, are characterized by inherent uncertainties related to material properties, geometry, loading conditions, numerical modelling, etc. Nevertheless, deterministic simplifications are employed in geotechnical engineering practice due to the complexity and computational cost required by the probabilistic methods. Since the latter offer a more realistic way to determine the seismic assessment of structures/geostructures they are gaining popularity due to the advances in computational resources and numerical methods. For large-scale geostructures, such as embankments under seismic loading conditions, probabilistic vulnerability analysis can be used to produce fragility curves and assess their safety margins for various demand and capacity levels.

Embankments constitute large-scale geostructures of great importance, the safety of which are directly related to environmental and social-economical issues. The seismic vulnerability of such infrastructures became the subject of systematic research following the Loma Prieta, Northridge and Kobe earthquakes, after which extended investigations took place to examine the failures that occurred in embankments due to seismic actions. In geotechnical earthquake engineering practice, the slope stability of an embankment is most often evaluated utilizing the pseudostatic method, in which the horizontal and vertical pseudostatic inertial forces are included in the safety factor calculations. In contrast, vulnerability analysis of structures is becoming a very useful tool for assessing their performance. The core of contemporary vulnerability analysis is the development of fragility curves. There are two basic approaches to develop a fragility curve: either using the lognormal assumption of the response or via reliability analysis methods. The Monte Carlo Simulation (MCS) method is the most widely applied reliability analysis method. In a recent work by the authors it was found that MCS-based fragility analysis offers a more reliable approach compared to the simplistic lognormal assumption.

Nevertheless, in order to reduce the excessive computational cost of the process, efficient approximation techniques have to be used. New non-conventional data processing and simulation methods such as Artificial Neural Networks (ANN) have been widely applied. In the present work efficient ANN-based metamodels are used for estimating the safety factor required for the slope stability analysis under pseudo-static conditions of a typical embankment, and subsequently, are applied in the framework of MCS-based reliability analysis in order to develop the fragility curves of the geostructure. Randomness related to the soil mechanical properties and the geometry of the embankment as well as the seismic demand is considered.

The employed ANN schemes are trained and tested utilizing available information generated from selected pseudostatic analyses of the geostructure. Subsequently, the trained ANN is used within MCS to predict the safety factor against slope instability of the geostructure, replacing the conventional analysis procedure. For this purpose two ANN-based approaches are implemented. It is demonstrated that the use of ANN metamodels for the numerical generation of MCS fragility curves can be very effective both in terms of time and accuracy. The use of ANN can practically eliminate any limitation on the sample size and lead to extremely accurate estimates of the fragilities without using any simplifying assumptions. Conclusively, it is evident that by incorporating efficient ANN-based metamodels into the time consuming process of fragility curves construction it is possible to increase the MCS sample size and predict more accurately the safety margins of a geostructure within a fraction of time compared to the conventional procedure. Thus, ANN offer a precise and efficient way to assess the seismic vulnerability of a geostructure for multiple hazard levels and vulnerability states, in the viewpoint of the Performance-based Earthquake Engineering (PBEE).

Mini-Symposia (MS13) Smart Structural Health Monitoring and Safety Assessments [Session Organized on Behalf of IASSAR Subcommittee SC4]

Safety, Reliability and Risk of Structures, Infrastructures and Engineering Systems – Furuta, Frangopol & Shinozuka (eds)
© *2010 Taylor & Francis Group, London, ISBN 978-0-415-47557-0*

Evaluation of the applicability of surface wave method on the sloping surface

J.T. Kim, H.J. Park & D.S. Kim
Department of Civil and Environmental Engineering, KAIST, Daejeon, Korea

S.W. Kim
Korea Railroad Corporation, Daejeon, Korea

ABSTRACT

The evaluation of shear modulus profile of a site is very important in the various fields of geotechnical engineering. To obtain shear wave velocity profile, various in-situ seismic methods using surface waves have been developed. During the rainy season, softened slope causes landslide occur, and it is very important to know stiffness variation and bedrock depth of the slope. For seismic design of earth dam, the shear wave velocity profile of each zone needs to be evaluated, but little data have been obtained for the sloping rockfill zones. So, it is required that reliable dynamic properties have to be evaluated in slope by in-situ test.

Spectral Analysis of Surface waves (SASW) method has many advantages for determining Vs profile of the ground. It is non-intrusive, rapid and economical. However SASW method has an assumption that the evaluated system consists of a set of homogeneous layers of constant thickness, but in reality, the ideal conditions cannot be obtained. Several researchers have already studied the effects of inclined layers, but the effect of sloping ground surface on Vs profile has not been studied.

The purpose of this study is an application of SASW method to the sloping ground surface. This paper is composed of numerical simulation, model test and field application. FE method has been used to assess effects of the sloping surface on wave propagation and evaluation of dispersion curve. SASW method has been simulated by numerical program. Numerical simulation and model test are performed varying several parameters such as slope angle, direction of wave propagation, setup of a source, and stiffness contrast. Consequently, effects of the sloping surface on experimental dispersion curves can be compared with theoretical ones. Finally, field tests were performed to estimate the reliability of surface wave test method at sloping surface at highway cut slope and slope of rock-fill dam.

When a source was hit to the slope, wave fronts of surface wave were normal to the surface. Therefore particle movements normal to the slope must be required. When a vertical source was applied, propagation shape was asymmetric. Theoretical and experimental dispersion curves of both directions were different. Because reflected body wave distorted the phase information of normal signals obtained from receivers. In case of a normal source, waves propagated symmetrically and theoretical and experimental dispersion curves of both directions were identical. For applying SASW method to the sloping ground surface, source must be hit normally to the slope and receivers have to obtain normal signals. Consequently a unique Vs profile of the slope could be derived. It is also verified through model test and field application. Through field applications and comparison with other test results, the good accuracy and applicability of surface wave test on the sloping surface were verified.

REFERENCES

ABAQUS/Standard Version 6.5 User Manual, Finite element software package. Hibbit, Karlsson & Sorenson, Inc.
Gabriels, P. & Snider, R. & Nolet, G., 1987. In situ measurements of shear wave velocity in sediments with higher-mode Rayleigh waves. *Geophysical Prospecting.* Vol 35: 187–196.
Gucunski, N. et al., 1996, Effects of Soil Nonhomogeneity on SASW testing, Department of Civil and Environmental Engineering, Rusters University.
Joh, S. H., 1992, *WinSASW User's Guide*, The Univ. of Texas at Austin.
Ludovic B. et al., 2004, Effects of Dipping Layers on Seismic Surface Waves Profiling : A Numerical Study, *Laboratoire Central des Ponts et Chaussees*, Nantes, France
McMechan, A. George. & Yedlin, J. Mathew., 1981. Analysis of dispersive waves by wave field transformation. *Geophysics.* Vol 46 (No. 6). 869–874.
Nazarian, S. & Stokoe, KH, 1984. In situ shear wave velocities from spectral analysis of surface wave. *Proc., 8th Conf On Earthquake Eng*, S.Francisco, 31–38.
Park, C.B.& Miller, R.D. & Xia, J., 1999. Multi-channel analysis of surface waves. *Geophysics*, 64(3), 800–808.
Park, H.C. & Kim, D.S., 2001. Evaluation of the Dispersive Phase and Group Velocities using Harmonic Wavelet Transform. *NDT&E International*, 34(7), 457–467.
Yoon, J. G., 2000, Application of the SASW method in an Inclined Layer – Numerical study, Dissertation, Master, *KAIST*

Safety, Reliability and Risk of Structures, Infrastructures and
Engineering Systems – Furuta, Frangopol & Shinozuka (eds)
© 2010 Taylor & Francis Group, London, ISBN 978-0-415-47557-0

Monitoring techniques for lab-scaled underwater landslides

Q. Hung Truong, Hyung-Koo Yoon & Jong-Sub Lee
Civil, Environmental and Architectural Engineering, Korea University, Korea

ABSTRACT

Underwater landslides include rotational and translational slides, debris flow, mud flow and avalanches.
Most of the underwater landslides are triggered by
earthquake, sea wave loading, tides, sedimentation,
natural gas, glaciations, and man-made civil works.
Therefore, the underwater landslides may pose hazards for costal and offshore living environment as well
as civil facilities (pipeline, cables, etc.). Several of the
under-sea landslides have generated even the tsunami.
In this work, two monitoring techniques, ultrasonic
reflection imaging and pressure-based method, are
used to investigate 1g lab-scaled underwater landslides. The simulation of lab-scaled landslide is carried in a $360 \times 540 \times 314$ mm container by a gate
system. Soil is ideally modeled by glass-bead and silica flour. In the high resolution reflection method,
two 500 kHz focal-type transducers, which work as
source and receiver, are fixed on a designed horizontal translation system. The reflection images are
obtained from the geo-material interfaces due to the
mechanical impedance mismatches. Piezoelectric film
sensors are used in the pressure-based method with a
designed circuit to capture the changes of height of soil
column.

The experimental results show that ultrasonic
reflection techniques not only obtained landslide
images but also assessed the spatial distribution of
the specimens with submillimetric resolution. The
pressure-based method enhanced the detection of
under-water landslide results by simply using the voltage ratios. The pressure-based technique using flexible
force sensors can examine the landslide fast and
continuously. These two methods can provide the complementary information. Although this study focuses
on laboratory scale specimens at 1 g, these two geophysical techniques suggested in this study also can
be effectively used for monitoring in geotechnical
centrifuge models at high g. The ultrasonic reflection imaging, moreover, can be done on the sea water
surface.

(a)

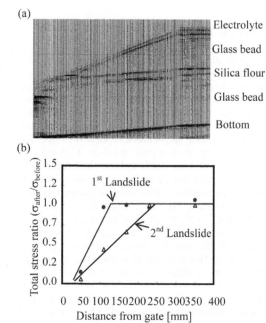

(b)

Figure 1. Underwater landslide monitoring: (a). ultrasonic
scanning images after 2nd landslides; (b). Ratios of voltage
values obtained from piezoelectric film sensors before, and
after 1st and 2nd landslide.

REFERENCES

Lee, J.S. and Santamarina, J.C., 2005. P-wave reflection
imaging. Geotechnical Testing Journal, 28, 197–206.
Locat, J. and Lee, H.J., 2002. Submarine landslides: advances
and challenges. Canadian Geotechnical Journal, 39,
193–212.
Ward, S.N. and Day, S. (2002). Suboceanic Landslide,
2002 Yearbook of Science and Technology, McGraw-Hill,
349–352.
Santamarina, J. C. in collaboration with Klein, K. A., and
Fam, M. A. (2001). Soils and waves – Particulate materials
behavior, characterization and process monitoring, Wiley,
New York.

Safety, Reliability and Risk of Structures, Infrastructures and Engineering Systems – Furuta, Frangopol & Shinozuka (eds)
© *2010 Taylor & Francis Group, London, ISBN 978-0-415-47557-0*

Wireless sensor network based slope monitoring system considering variability of matric suction

Seung Rae Lee, Jung Chan Choi, Young Hun Song, Hyun Myoung & Yun Ki Kim
Department of Civil & Environmental Engineering, KAIST, Daejeon, Republic of Korea

ABSTRACT

Rainfall-induced slope failures are one of the most frequent natural hazards in heavy rainy season in many countries. Generally these kinds of failures are triggered by reduced soil strength which is induced by rainfall infiltration. Thus measuring the hydraulic soil properties in a soil slope and reliable slope stability assessment are important for effective slope maintenance.

A wireless sensor network based slope monitoring system considering variability of matric suction was proposed in this study for reliable slope maintenance. A wireless sensor module was developed for slope monitoring and reliability based slope stability assessment method which is based on AFORM(Advanced First Order Reliability Method) was suggested for real-time slope stability assessment.

The WSN based slope monitoring system is composed of a wireless module, gateway, tensiometer, ECH2O-10 and rain gage as shown in Figure 1. Wireless module transmits data wirelessly, and gateway receives it. Tensiometer measures matric suction, ECH2O-10 measures water content and rain gage measures rainfall events. V-Link (Microtrain Inc, 2007) was adopted as a wireless module.

The reliability index of slope considering the measured matric suction was proposed based on AFORM approach. Cohesion c', friction angle ϕ', and matric suction $(u_a - u_w)$ were considered as random variables and assumed that the values have normal distribution.

Figure 2. Model slope.

The proposed system was applied to a model slope as shown in Figure 2.

The results show that the WSN based slope monitoring system and the reliability index can be used for efficient slope management by quantifying the risk of slope in real time.

REFERENCES

Cho, S. E. and Lee, S. R. 2001. Instability of unsaturated soil slopes due to infiltration. *Computers and Geotechnics* 28(3): 185–208.

Fredlund, D.G., Morgenstern, N.R., and Widger, R.A. 1978. The shear strength of unsaturated soils. *Can. Geotech. J.* 15: 313–321.

Hasofer, A. M., and Lind, N. C. 1974. Exact and Invariant Second-moment Code Format. *J. Eng. Mech. Div. ASCE* 100(1): 111–121.

MicroStrain, Inc. 2007. V-Link 2.4 GHz Wireless Voltage Node Quick Start Guide.

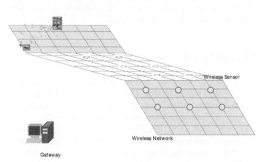

Figure 1. WSN based slope monitoring system.

Safety, Reliability and Risk of Structures, Infrastructures and
Engineering Systems – Furuta, Frangopol & Shinozuka (eds)
© 2010 Taylor & Francis Group, London, ISBN 978-0-415-47557-0

Measurement of load transfer between anchor and grout using optical FBG sensors embedded smart anchor

Y.S. Kim, J.M. Kim, D.N. Suh & H.J. Sung
Department of Civil & Environmental Engineering, Chonnam National University, Yeosu, Republic of Korea

S.R. Lee
Department of Civil and Environmental Engineering, Korea Advanced Institute of Science and Technology, Daejeon, Republic of Korea

ABSTRACT

Despite of increasing popularity of 7-wire strand for a cable-stayed bridge and geo-reinforcement system – e.g., rock or soil anchor, no accurate or simple method is available for directly measuring the stress (force) in those systems. FBG Sensor, which is smaller than conventional strain gauge and has better durability and does not have a noise from electromagnetic waves, was embedded into the center king cable of 7-wire strand as shown in Figure 1.

A smart anchor was developed by using a smart tendon to measure the distribution of strain along the wire strand and the change of anchor force. A series of pullout tests were performed to verify the feasibility of smart anchor and find out the load transfer mechanism around the steel wire fixed to rock with grout. Figure 2 show the pullout test results. Distribution and transfer of shear stresses during loading-unloading phase at steel wire-grout interface is assessed effectively by the optical fiber sensors.

It is found that FBG sensors, which are located at the different depth of center king cable, can measure the strain change of the 7-wire strand at each location effectively during loading–unloading–reloading processes. Secondly strain distribution at the interface of smart tendon and grout was measured, which can be used to develop the equation for estimation of stress distribution during loading. Finally progressive failure due to load transfer phenomenon can be monitored successfully by smart tendon at the interface between wire strand and grout.

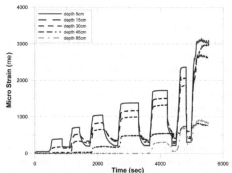

(a) time-strain curves measured at each FBG sensor during pull out test

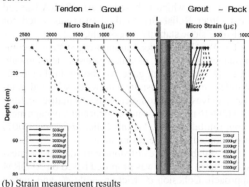

(b) Strain measurement results

Figure 2. Strain measurements of interface between 7-wire strand and grout(left), strain measurements of interface between grout and rock body(right).

REFERENCES

Wang, M.L., Wang, G. & Yim, J.S. 2006. Smart Cables for Cable-Stayed Bridge. *Proceedings of US-Korea Workshop on Smart Structures Technology for Steel Structures*, 16~18 November 2006, Seoul, Korea, pp.25~31.
Kim, Y.S., Suh, D.N., Kim, J.M. & Lee, S.R. 2008. Development and Application of A Smart Anchor with Optical FBG Sensors. *KGS Spring National Conference*, March 28~29, Seoul, Korea. (in Korean)

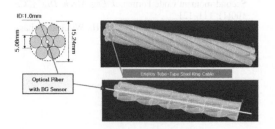

Figure 1. Conceptual figure of smart tendon.

Organized Session (OS25) Monitoring Performance under Uncertainty

Safety, Reliability and Risk of Structures, Infrastructures and Engineering Systems – Furuta, Frangopol & Shinozuka (eds)
© *2010 Taylor & Francis Group, London, ISBN 978-0-415-47557-0*

Assessment methods for highway bridges

D.L. Allaix & G. Mancini
Department of Structural and Geotechnical Engineering, Politecnico di Torino, Torino, Italy

A.C.W.M. Vrouwenvelder
Faculty of Civil Engineering and Geosciences, TU Delft, Delft, The Netherlands

ABSTRACT

The reliability assessment of highway bridges is a relevant topic in the management of the roadway system of the European countries, since a large part of the goods transport travels by road. Several studies showed a noticeable increment of the goods travel in the past few decades and an increasing trend is expected also in the future. At the same time, the existing structures undergo deterioration processes. Consequently, the governments and highway agencies need detailed information about the actual situation and future scenarios in order to select a proper repair strategy. The reliability assessment of such highway structures is therefore a key-point for any further decision (Allaix 2008). The research advances in this field point out significantly the attention on the probabilistic-based methodologies for the reliability assessment ((BRIME 1999), (COST 2004)), suggesting the formulation of suitable probabilistic models for the actions effects and the structural resistances based on actual measurements. While a large number of documents about the modelling of structural properties is available (JCSS 2001), only few guidelines are available for the modelling of the traffic loads, which represent the most important variable action. Therefore, the paper is focused on the analysis of traffic loads acting on highway bridges. A refined traffic load model based on measurements done in The Netherlands is presented. The traffic flow composition is derived from the observed passages of vehicles and representative types of vehicles are modelled subsequently. The static and the dynamic properties of the trucks are described by random variables, whose statistical parameters are estimated from the observed data. Within the static properties, the gross weight requires particular care in the modelling of the upper tail of the distribution, which is dominant for the ultimate limit states. The traffic load model is employed to estimate the distribution of the extreme traffic load effects in selected longitudinal locations of two highway bridges. The coupling of the Monte Carlo method with a structural solver is used to simulate the traffic flow approaching the bridges and to evaluate the traffic load effect by static and dynamic analyses. A comparison with the design load model LM1 of the Eurocode EN1991-2 (CEN 2003) is also presented.

REFERENCES

Allaix, D.L. 2008. Bridge reliability analysis with an up-to-date traffic load model. PhD thesis, Politecnico di Torino.
BRIME 1999. Development of models for material strength and traffic loading. Internet publication, www.trl.co.uk/brime/deliver.htm
CEN 2003. Eurocode 1 – Actions on structures – Part 2: Traffic loads on bridges.
COST 2004. COST 345 – Procedures required for assessing highway structures, work packages 4 and 5 report, Liubjana.
JCSS 2001. *Probabilistic Model Code*. Joint Committee on Structural Safety, Internet publication, www.jcss.ethz.ch

Safety, Reliability and Risk of Structures, Infrastructures and Engineering Systems – Furuta, Frangopol & Shinozuka (eds)
© 2010 Taylor & Francis Group, London, ISBN 978-0-415-47557-0

Measurement and identification of constructional tolerances in connections between steel and concrete

P. Spyridis, K. Bergmeister & A. Strauss
Institute of Structural Engineering, University of Natural Resources and Applied Life Sciences, Vienna, Austria

D.M. Frangopol
Department of Civil and Environmental Engineering, ATLSS Center, Lehigh University, Bethlehem, PA, USA

ABSTRACT

Dimensional deviations of structural components are always present in construction. In order to deal with these deviations in the execution of a structure, constructional tolerances are intended and standardized in a way to facilitate the assembly of components. However, the synthesis of tolerances may affect the performance of the final structure to greater or smaller extend. In the case of fastenings to concrete and particularly those placed close to a free edge and loaded in shear towards the edge, tolerances between the anchors and the steel base plate are critical for the capacity of the system. Due to installation tolerances, anchors are randomly configured within the clearance holes, which results in random distribution of loads to each fastening point and therefore uncertainty on the actual shear resistance of the anchorage.

As a demand for the assembly of steel components in fastening systems, the diameter of the holes on the fixture of the group is typically larger than that of the anchors to be installed. Due to this fact, the distribution of a shear force to each individual anchor in a group is not predictable; each anchor may be randomly positioned in the hole of the base plate, so initially not all the anchors in the group will be in contact with the plate. According to the random configuration of the anchors in the holes, various distances between the plate and the anchors will exist. Therefore the participation of each anchor to the group's performance and the final group's capacity remains unknown. In the case of a fastening close to a free edge subjected in shear towards the edge, each anchor's capacity is dependent on the distance from the edge, which introduces one more factor of complexity in the analysis of the system.

Actual configurations of the anchors in the holes is valuable information. Riha (2006) has held important work on the identification of actual imperfections and randomness in the configurations of fasteners. Based on the experience gained from this study, new concepts of monitoring tools for tolerances are being

developed in order to enrich this information. This procedure could help to update information on the anchor's locations.

Once the information on the anchor's configurations is acquired, the estimation of a probabilistic distribution of the anchors' positions and the group's shear capacity as a random variable can follow. In order to calculate the shear capacity of a fastening to concrete near to an edge with a specific anchor configuration, an algorithm has been developed. Furthermore, this algorithm can serve as a stochastic model for random positions of the anchors in order to estimate the distribution of the anchorage's capacity.

Based on this work, relevant suggestions for a designer of fastenings can be provided.

REFERENCES

Frangopol, D.M., Strauss, A. & Kim, S. 2008a. Bridge reliability assessment based on monitoring. *Journal of Bridge Engineering* 13(3): 258–270

Frangopol, D.M., Strauss, A. & Kim, S. 2008b. Use of monitoring extreme data for the performance prediction of structures: General Approach. *Engineering Structures* 30: 3644–3653

Riha, A. 2006. *Montageungenauigkeiten von Befestigungen in Beton.* (Installation imprecisions of fastenings in concrete) Diploma thesis. University of Natural resources and Applied Life Sciences of Vienna (in German)

Spyridis, P. 2008. Behavior of anchor groups under shear loads. In Eligehausen, R & Gehlen, C. (eds) *7th fib PhD Symposium in Civil Engineering, Part 3: Fastening,* Stuttgart, September 2008, Stuttgart: Ibidem

Spyridis, P., Unterweger, A., Mihala, R. & Bergmeister, K. 2008. Querbeanspruchung von Anker-gruppen in Randnähe – Stochastische Studien (Behavior of Anchor Groups Close to an Edge under Shear Loads – Stochastic Studies) *Beton- und Stahlbetonbau* 103: 617–624 (in German)

Strauss, A., Frangopol, D.M. & Kim, S. 2008. Use of monitoring extreme data for the performance prediction of structures: Bayesian Updating. *Engineering Structures* 30: 3654–3666

Safety, Reliability and Risk of Structures, Infrastructures and
Engineering Systems – Furuta, Frangopol & Shinozuka (eds)
© 2010 Taylor & Francis Group, London, ISBN 978-0-415-47557-0

Automated speed detection of moving vehicles from remote sensing images

Wen Liu & Fumio Yamazaki
Graduate School of Engineering, Chiba University, Chiba, Japan

Tuong Thuy Vu
Royal Institute of Technology, Stockholm, Sweden

ABSTRACT

A new method is developed for both vehicle extraction and speed detection from two consecutive digital aerial images and a pair of QuickBird (QB) panchromatic (PAN) and multi-spectral (MS) images.

From a high resolution aerial image or a QB PAN image, a vehicle has a clear shape and can be extracted by an object-based method. Several global and local parameters of gray values and sizes are examined to classify the objects in the image. The vehicles and their associated shadows can be discriminated by removing the big objects such as the roads. The procedure is tested using digital aerial images. The environment surrounding roads affects the result accuracy, however, 97% of vehicles can be extracted.

The resolution of a QB MS image in about 2.4 m, and it is difficult to extract the accurate edge and position of vehicle. Thus, an area correlation method is proposed to estimate the location of vehicles from a MS image in a sub-pixel level. A template is selected from PAN image using the location information of vehicle extraction. The reference image is selected with the same center as the template but bigger than that. Then the position of the vehicle in the MS image with the highest probability is obtained by calculate the cross-correlation coefficient between the template and the reference image. The template and the reference image are converted to 0.24 m/pixel by cubic convolution to be matched in a sub-pixel level.

Since two consecutive digital aerial images taken by UltramD have 80% overlap, the speed of moving vehicles in the overlap can be detected by the movement between two images and the time lag (about 3 sec). The vehicle database of each image is developed after the object-based vehicle extraction process from two consecutive aerial images. Then the vehicles in the two databases are linked by the order, moving direction, size and distance. Using the positions of the same vehicle in two images, the speed can be computed. 71% of vehicles' speeds were detected from the aerial images

Figure 1. Result of automated speed detection from two consecutive aerial images of Tokyo.

Figure 2. Result of the sub-pixel based automated speed detection from the QB image of Bangkok.

for Tokyo, Japan (Figure 1). The standard deviation for the difference of speed between the automated and visual detections is 0.83 km/h.

QB images also can be used to detect the speed of moving vehicles, by the time lag (about 0.2 sec) between PAN and MS images. The location of vehicles in a QB's PAN image can be extracted by the object-based method. And the most possible location of a vehicle in the MS image can be obtained by the area correlation method. Then speed of vehicles is computed by the location change between the PAN and MS images with the time lag about 0.2 s. The approach is tested on one QB scene covering the central Bangkok, Thailand (Figure 2).

The result of this study may be used in assessing traffic conditions of large urban areas without using ground-based traffic sensors.

Safety, Reliability and Risk of Structures, Infrastructures and Engineering Systems – Furuta, Frangopol & Shinozuka (eds)
© 2010 Taylor & Francis Group, London, ISBN 978-0-415-47557-0

Multifunctional materials: From manufacturing to application, planning for the integration of reliability and risk concepts

T.B. Messervey & D. Zangani
D'Appolonia, S.p.A., Genova, Italy

D.M. Frangopol
Lehigh University, Bethlehem, PA, USA

ABSTRACT

Materials with embedded sensing capabilities have great promise for the construction industry, the management of civil infrastructure, and for the promotion and use of structural health monitoring methods. The data collected from these materials can be utilized to ensure safe construction conditions, to assess in-service performance, to schedule inspection, maintenance, and repair activities, to conduct performance-based design, and eventually to provide quantitative information that enables the improvement of existing codes and guidelines (Frangopol & Messervey, 2008). However, despite the potential of these products and the need for them, there are some very practical issues that must be solved. Across different applications, the type of measurement and how it relates to reliability and risk will be unique. Algorithms, data-processing techniques, methodologies, and tools to communicate useful information to owners and infrastructure managers must be developed. Equally as important, the manufacturing procedures and machinery modifications to produce sensor-embedded materials that are rugged, durable, and affordable must be created. Lastly, guidelines for on-site installation, handling, and maintenance are required.

This paper begins to investigate solutions to some of these challenges by describing work ongoing in the European funded research project "Polyfunctional Technical Textiles against Natural Hazards (POLY-TECT)." This goal of this four year project is to provide reinforcing strength and monitoring capability for geotechnical and masonry applications through the design and industrial production of multifunctional technical textiles. Now entering its third year, project activities are focusing on field test opportunities. Two field tests in particular are discussed, the monitoring of an unstable slope at a coal mine excavation site, and the reinforcement and monitoring of a mid-scale masonry structure subjected to shaking table tests. For each case, how to relate monitoring data to reliability and risk concepts is different. For the coal mine, information from the owner to estimate the impact of shutting down mine operations due to road failure becomes critical as well as the conduct of a slope stability analysis. For the masonry structure, a probabilistic method that can quantify the decrease in risk associated with structural retrofit (increase in strength and ductility) is required. For this application, the PEER (Pacific Earthquake Engineering Research Center) approach is selected (Cornell & Krawinkler, 2000). For both upcoming field applications, this paper describes how reliability and risk concepts are being included in the planning phase to maximize the utility of these tests and to eventually construct simple tools and examples that can assist project industrial partners in building a market for these materials.

REFERENCES

Casciati, S., Faravelli, L., 2008. Vulnerability for Medieval Civic Towers. *Structures and Infrastructure Engineering.* (accepted for publication)

Cornell, A. & Krawinkler, H., 2000. Progress and challenges in seismic performance assessment. *PEER Center News* 3(2): 1–3.

Frangopol, D.M., Messervey, T.B., 2008. Maintenance principles for civil structures. Chapter 89 in *Encyclopedia of Structural Health Monitoring.* Ed. Christian Boller, Fu-Kuo Chang, and Yozo Fujino. John Wiley & Sons, Ltd., UK, pp. 1533–1562.

Glaser, S.D., Li, H., Wang, M.L., Ou, J., & Lynch, J, 2007. Sensor technology innovation for the advancement of structural health monitoring: a strategic program of US-China research for the next decade. *Smart Structures and Systems*, 3(2), pp 221–244

Glisic, B. & Inaudi, D., 2007. Fibre Optic Methods for Structural Health Monitoring. John Wiley & Sons, LTD, England.

Safety, Reliability and Risk of Structures, Infrastructures and
Engineering Systems – Furuta, Frangopol & Shinozuka (eds)
© *2010 Taylor & Francis Group, London, ISBN 978-0-415-47557-0*

Structural vulnerability analysis for capacity assessment and enhancement of truss bridge

G. Yu, Z. Sun & L.M. Sun

State Key Laboratory on Disaster Reduction of Civil Engineering, Tongji University, Shanghai, China

ABSTRACT

For a structural system, its vulnerability can be defined as structural performance susceptibility to local damage. This paper presents a structural vulnerability analysis method based on the theory of plastic limit analysis. By applying the mechanism generation method, the ultimate loading factor and the corresponding failure mode of a structural system can be obtained. The ultimate loading factor is then used to measure the performance of a structural system. The variation of this factor to different damage scenarios due to vehicle collision is studied.

To illustrate the proposed method, a case study on a truss bridge (as shown in Figure 1) is performed. Structural vulnerabilities to all possible damage scenarios caused by the vehicle collision are studied.

Figure 2 shows the susceptibilities of the normalized structural performance measures to local damage severities for these four damage scenarios. For different damage scenarios, two types of drawings can be observed. Figure 2a presents a curve without a plateau. This figure tells that the structure is of no damage tolerance to the considered damage scenarios. The structure is thus highly susceptible to this kind of local damage scenario. Figure 2b presents a curve with a plateau. This figure tells that structural performance will not

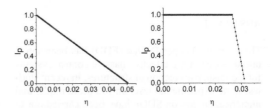

Figure 2. Susceptibility of limit load factor to local damage. (a) The 1st damage scenario. (b) The 2nd damage scenario.

decrease if the corresponding damage scenario is in a certain range. In another word, the structure is of robustness to this kind of local damage scenario. The range is called the damage tolerance of the structural system to this specified damage scenario. Since only the 1st damage scenario is of no damage tolerance, it is regarded to be the most vulnerable damage scenario for this structure to vehicle collision.

Based on the result of structural vulnerability analysis, structural capacity under different damage scenarios can be easily evaluated according to component level damage evaluation result. Moreover, in order to reduce structural vulnerability to vehicle collision damage, the local enhancement strategy can be efficiently made. The enhancement of the components should transfer the weakest part of the structural system to the position where the extreme events are not easy to approach.

The results of case study show that the structure capacity may not be influenced by some damage scenarios; it depends on the damage location and severity of the given damage scenarios. Knowing this quantitatively, structural capacity assessment and enhancement after damage can be conducted more efficiently.

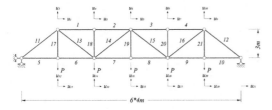

Figure 1. The basic geometry of truss bridge.

Safety, Reliability and Risk of Structures, Infrastructures and Engineering Systems – Furuta, Frangopol & Shinozuka (eds)
© *2010 Taylor & Francis Group, London, ISBN 978-0-415-47557-0*

Seismic performance of new Friction Damper Device under near field earthquake excitation

M. Kohrangi, T. Taghikhany & M. Tehranizadeh
Department of Civil Engineering, Amirkabir University of Technology, Tehran, Iran

Keywords: Friction Damper; Near field Earthquake; Energy Dissipation

ABSTRACT

The Friction Damper Device (FDD) has been introduced as one of the effective passive control systems for the seismic protection of structures. In recent studies, some experimental tests on full scale model and numerical studies on SDOF have been carried out to endorse its efficiency. This paper focuses to assess the degree of improvement in performance achieved through use of FDD device in steel structures within near field earthquakes, and evaluates the suitability of optimum design of this system. Accordingly, the nonlinear dynamic response of five, eight and twelve stories steel moment-resistant frames with and without Friction Damper Device (FDD) under seismic excitation is obtained using nonlinear analysis program, OPENSEES. Optimum design philosophy of frames was based on significantly reducing maximum displacements and base shear demand. Moreover the overall seismic performance and complete response history can be described by energy dissipation of total input energy. Different governing parameters such as stiffness of frames and bracings and amount of prestressing force in bracings were identified and their influences on response of the frames investigated. The numerical studies reported clearly demonstrate that passive response control systems based on the new FDD present a viable alternative to the conventional ductility-based earthquake-resistant design both for new construction and for upgrading existing structures.

Mini-Symposia (MS10) Structural Safety and Reliability by Means of Structural Health Monitoring

Safety, Reliability and Risk of Structures, Infrastructures and
Engineering Systems – Furuta, Frangopol & Shinozuka (eds)
© 2010 Taylor & Francis Group, London, ISBN 978-0-415-47557-0

GPS based structural health monitoring: Uncertainty of the measurements

F. Casciati & C. Fuggini
Department of Structural Mechanics, University of Pavia, Pavia, Italy

ABSTRACT

Sensors placement, structural identification, diagnostic processes are different sources of uncertainty in the acquired measurements. Within any Structural Health Monitoring (SHM) process, in particular sensors placement is the topic of this study, where SHM is based on a Global Navigation Satellite System (GPS, GLONASS and the future GALILEO).

The effectiveness of such a Structural Health Monitoring process, given the uncertainties of the measurements obtained by GPS network, is studied with reference to a standard industrial building (Figure 1). The aim is to asses if, and in which per cent, the uncertainty of GPS data influences the response displacements of the building during similar strong wind events, occurred in the period from November 2007 to October 2008.

Different GPS units locations on the roof of the building, at different hours during the day, are considered. One also investigates the possibility of detecting any loss of signal and/or any dilution of precision (DOP). The displacement time histories, in the two directions North and East, as recorded by the GPS sensors at different sampling rates are processed. The application of a suitable filtering schemes leads one to distinguish the part of the GPS signal, which refers to oscillations induced by the monitored wind from the noise component (i.e., the background component). An estimation on how the accuracy with which the dynamical response of the building is detected by GPS is pursued.

REFERENCES

Casciati, F. Fuggini, C. & Bonanno, C. 2007. Dual frequency GPS receivers: reliability of precision of the measures, *Proceedings of 4th C2I, Nancy, 15–17 October 2007*, 604–612.

Casciati, F. & Fuggini, C., Measuring the displacements of a steel structure by GPS units, *Proceedings of 1st International Symposium on Life-Cycle Engineering, Varenna, 11–14 June 2008*, 501–506

Casciati, F. & Fuggini, C., Monitoring an Industrial Steel Building by GPS Receivers, *Proceedings of the 4th European Workshop on Structural Health Monitoring, Krakow, 2–4 July 2008*, 219–226

Casciati, F. & Fuggini, C., GPS Based Structural Identification and health monitoring, *Proceedings of the Fourth European Conference on Structural Control, St. Petersburg, 8–12 September 2008*, 133–140

Dana, P. H. 1997. Global Positioning System (GPS) Time Dissemination for Real-Time Applications, Real-Time Systems. *The International Journal of Time Critical Computing Systems 12*, 9–40.

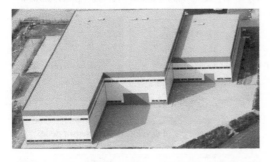

Figure 1. Bird flight view of the industrial building under investigation.

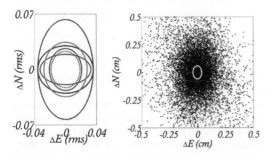

Figure 2. The rms ellipses computed at different hours during the day. The largest is also drawn in the plot on the right side of the figure.

Safety, Reliability and Risk of Structures, Infrastructures and
Engineering Systems – Furuta, Frangopol & Shinozuka (eds)
© 2010 Taylor & Francis Group, London, ISBN 978-0-415-47557-0

Autonomous smart sensor nodes for hybrid health monitoring of steel plate-girder railroad bridges

J.T. Kim, D.D. Ho, D.S. Hong, J.H. Park & S.Y. Lee
Pukyong National University, Busan, Korea

ABSTRACT

SHM systems are widely adopted to monitor the structural responses, to detect damage, and to assess the effect of damage on the structural integrity. Many researchers have developed novel sensing technologies and damage monitoring techniques for the practical SHM applications. The SHM system for long-span bridges mainly include a number of sensors, a huge amount of signal transmitting wires, data acquisition (DAQ) instruments, and one or more centralized data storage servers. The stored data in the centralized servers are handled for off-line signal and information analysis for damage monitoring and safety evaluation.

The high costs associated with wired SHM systems can be greatly reduced through the adoption of wireless sensors. One of great advantages for using wireless sensors is that the autonomous operation for SHM can be implemented by embedding advanced system technologies. The new paradigm by adopting smart sensor nodes offers an automated, cost-efficient, SHM system.

In this study, autonomous smart sensors are developed for hybrid SHM of steel plate-girder railroad bridges as shown in Figure 1. Also, the feasibility of the autonomous hybrid health monitoring using the two smart sensors are evaluated on a laboratory-scaled steel plate-girder railroad bridge model as shown in Figure 2.

The system is consisted of an acceleration-based smart sensor (Acc-SSN), an impedance based smart

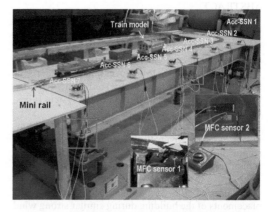

Figure 2. Test setup for model steel plate-girder railroad.

(a) Acc-SSN   (b)Imp-SSN

Figure 3. Prototype of developed smart sensor node.

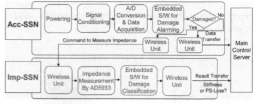

Figure 4. Schematic of SHM system using smart sensors.

sensor (Imp-SSN) and a main control server as shown in Figure 3 and 4. Each smart sensor autonomously performs the SHM such as data acquisition, damage alarming and classification, and wireless transmission/receipt. The main control server receives and stores only damage index values.

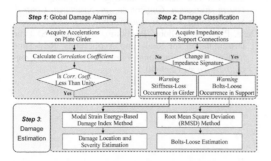

Figure 1. Hybrid health monitoring scheme for steel plate-girder railroad bridges.

Safety, Reliability and Risk of Structures, Infrastructures and Engineering Systems – Furuta, Frangopol & Shinozuka (eds)
© 2010 Taylor & Francis Group, London, ISBN 978-0-415-47557-0

Laboratory studies to explore uncertainty quantification for systems-based structural health monitoring

F. Necati Catbas & H. Burak Gokce
University of Central Florida, Orlando, FL, USA

Dan M. Frangopol
Lehigh University, Bethlehem, PA, USA

ABSTRACT

The use of objective data from a monitoring system is expected to improve the reliability estimation of structural systems. In real structures, it is possible to obtain a variety of different measurements by continuously monitoring the loads and responses to characterize the performance and safety. It is also important to define and quantify uncertainties associated with the loading, structural responses and structural health monitoring devices such as sensor characteristics. While aleatory uncertainty due to random nature of the data cannot be reduced, structural reliability can be better assessed with continuous monitoring data. The epistemic uncertainty can be reduced by better modeling. In this study, the writers are conducting extensive laboratory studies on a steel structure using various response measurements.

The component and system reliability estimates based on different measurements are investigated for the quantification of the uncertainties. These fundamental studies serve as a necessary prelude to real life implementation of structural health monitoring systems for estimating reliability.

Figure 2. Sunrise Boulevard Bridge in Florida.

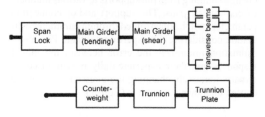

Figure 3. Simplified system model for the Sunrise Bridge.

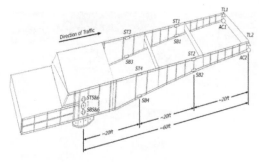

Figure 4. Instrumentation for the structural health monitoring of the Sunrise Bridge.

Figure 1. Laboratory test setup for concept demonstration and calibration studies.

Safety, Reliability and Risk of Structures, Infrastructures and
Engineering Systems – Furuta, Frangopol & Shinozuka (eds)
© 2010 Taylor & Francis Group, London, ISBN 978-0-415-47557-0

Modeling and boundary parameter estimation of a bridge-like structural steel frame

Yunus Dere
Department of Civil Engineering, Selcuk University, Engineering and Architecture Faculty, Konya, Turkey

F. Necati Catbas
Department of Civil and Environmental Engineering, University of Central Florida, Orlando, FL, USA

ABSTRACT

The computer models of real world structures make many assumptions and idealizations when it comes to support boundaries and element connections. During the service life of steel structures, support deteriorations are commonly observed. The behavior of existing structures can be better understood with updated computer models which consider existing condition by means of experimental data. In this work, one ⊢ shaped steel frame is constructed and tested under static and dynamic loads to explore the integration of experimental data, finite element modeling and parameter updating using optimization. (Fig. 1).

The frame is heavily instrumented with strain gages, displacement gages, accelerometers, load cells and tilt meters. During the tests, the supports of the frame are changed ranging from pin supports to various numbers of elastomeric pads. The support and/or connection parameters will be determined through parameter estimation considering the experimental data. Parameter estimation minimizes the error between the model responses and their experimentally measured counterparts iteratively changing the guessed value of the chosen unknown parameters (Eqn. 1).

Figure 2. The FE solid mesh of the frame created in ANSYS®.

where x is the vector containing displacements, strains, etc. of various points at the structure.

In the Figure 2, the finite element model of the structure is shown.

PARIS (PARameter Identification Software) will be employed for parameter estimation analyses. The updated model will be verified against dynamic experimental data by comparing the natural frequencies and mode shapes of the frame structure.

$$\left\| x_{measured} - x_{mod\,el} \right\| \leq Tolerance \qquad (1)$$

REFERENCES

Catbas, F. N. & Brown, D. L. & Aktan, A. E. 2004. Parameter estimation for multiple-input multiple output modal analysis of large structures. *ASCE Journal of Engineering Mechanics* 130(8): 921–930.

Sanayei, M. & Imbaro, G.R. & McClain, J.A.S. & Brown, L.C. 1997. Structural model updating using experimental static measurements. *Journal of Structural Engineering* 123(6): 792–798.

Sanayei, M. 1998. PARIS© PARameter Identification System, Software Package. Department of Civil and Environmental Engineering, Tufts University, Medford, Massachusetts.

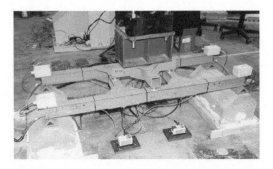

Figure 1. The test setup with all the sensors and the load application.

Safety, Reliability and Risk of Structures, Infrastructures and
Engineering Systems – Furuta, Frangopol & Shinozuka (eds)
© 2010 Taylor & Francis Group, London, ISBN 978-0-415-47557-0

Performance-based design of structural health monitoring systems

Y.Y. Qian & Y.Q. Ni
Department of Civil and Structural Engineering, The Hong Kong Polytechnic University, Hong Kong

K.Y. Wong
*Bridge and Structures Division, Highways Department, The Government of The Hong Kong
Special Administrative Region, Hong Kong*

ABSTRACT

With the development of structural health monitoring (SHM) technology and successful applications to civil engineering structures in the past decade, it can be argued that SHM has matured to the stage where radical principles and detailed guidelines have emerged. SHM system monitors structural loading and response, recognizes structural mechanism, provides standard working model of structures, evaluates the performance/safety of structures and supervises operation management with a deep pool of mixed knowledge: modern sensing technology, network communication, signal analysis and processing, data management, knowledge mining, structural analysis, etc. The function of SHM system is to supervise operation, maintenance and management of the structure, to verify design assumption or coefficients, to alarm abnormal loading or response, and structural damage/deterioration, to provide data support for safety evaluation after disaster, and to evaluate the efficiency of structural overhaul.

The development of structural health monitoring technology for surveillance, evaluation and assessment of existing or newly built bridges has now attained some degree of maturity. On-structure long-term monitoring systems have been implemented on bridges in Europe (Andersen & Pedersen 1994, Casciati 2003), the United States (Pines & Aktan 2002), Canada (Cheung & Naumoski 2002), Japan (Fujino & Abe 2004), Korea (Koh et al. 2003), China (Ko & Ni 2005) and other countries.

In this study, performance-based design of SHM systems is proposed as one future direction in this field, with the target to design and implement SHM systems that are enabled to achieve both structural performance objectives and SHM system performance objectives. Taking the SHM projects in Hong Kong as paradigms, this paper describes the functions and requirements of performance-based SHM design. The SHM system architecture with four integrated systems is suggested and elaborated in order to further advance the concept of performance-based design as a cornerstone in SHM field. The structural performance objectives and the monitoring system performance objectives for performance-based SHM were defined, and limit states and limit events for performance-bases SHM were established. A modular architecture was recommended in the design of a SHM system. The architecture of the structural health monitoring system described consists of six integrated modules: the sensory system, the data acquisition and transmission system, the data processing and control system, the structural health evaluation system, the structural health data management system, and the inspection and maintenance system. The SHM system provides the operation strategy under two processes, the structural health monitoring process and the structural safety evaluation process.

REFERENCES

Andersen, E.Y. & Pedersen, L. 1994. Structural monitoring of the Great Belt East Bridge. In: Krokebogr J, editor. *Strait crossings 94*. Rotterdam: Balkema: 189–195.

Casciati F. 2003. An overview of structural health monitoring expertise within the European Union. In: Wu Z.S. & Abe M., editors. *Structural health monitoring and intelligent infrastructure*. Lisse: Balkema: 31–37.

Fujino Y. & Abe M. 2004. Structural health monitoring—current status and future. In: Boller C, Staszewski WJ, editors. *Proceedings of the 2nd European workshop on structural health monitoring*. Lancaster (PA): DEStech; 3–10.

Ko J.M. & Ni Y.Q. 2005. Technology developments in structural health monitoring of large-scale bridges. *Engineering Structures* 27 (2005):1715–1725.

Koh H.M., Choo J.F., Kim S.K & Kim C.Y. 2003. Recent application and development of structural health monitoring systems and intelligent structures in Korea. In: Wu ZS, Abe M, editors. *Structural health monitoring and intelligent infrastructure*. Lisse: Balkema: 99–111.

Pines D.J. & Aktan A.E. 2002. Status of structural health monitoring of long span bridges in the United States. *Progress in Structural Engineering and Materials* 4(4): 372–380.

Safety, Reliability and Risk of Structures, Infrastructures and
Engineering Systems – Furuta, Frangopol & Shinozuka (eds)
© *2010 Taylor & Francis Group, London, ISBN 978-0-415-47557-0*

Active and semiactive control of a distributed mass damper system that integrates structural and environmental controls in buildings

Tat S. Fu & Erik A. Johnson

Sonny Astani Department of Civil and Environmental Engineering, University of Southern California,
Los Angeles, CA, USA

ABSTRACT

Recent developments of a distributed mass damper (DMD) system integrate structural and environmental control systems for buildings (Fu & Johnson 2008). External shading fins are used as mass dampers such that they can (*i*) control building energy consumption by adjusting the fins and, thus, the amount of sunlight coming into the building and (*ii*) control structural movements by dissipating energy with the dampers during strong motions due to wind or earthquakes (Figure 1). Shading fins are placed along the height of the building, distributing the mass along the building instead of being concentrated in a few locations like traditional tuned mass dampers (TMDs).

This paper focuses on active and semiactive control strategies for these DMD systems.Using a Kanai-Tajimi (Soong & Grigoriu 1993) model of ground motion, this paper presents astudyof control strategies to reduce structural vibration under stochastic earthquake excitations. The active DMD system is analyzed and compared with other active control strategies including an ideal fully active system and an active mass driver (AMD) system. Among all of these control strategies, the AMD and active DMD systems best reduce the structural motions within a reasonable control force range.The active DMD system gives greater benefit (smaller primary system displacements and accelerations) but at larger cost (larger damper displacements and control force) than the AMD system.

Finally, semiactive control strategies are studied to evaluate the effectiveness of the DMD system with semiactive dampers. The actuators required for moving the shading fins may be small and too weak for full active DMD control; semiactive dampers can serve as alternatives/complements. Since only dissipative forces can be applied by semiactive devices, the semiactive DMD system is slightly less effective in reducing responses compared to the active DMD system that can constantly apply control forces on the system. Using gain scheduling, the performance of the semiactive DMD system is successfully improved. The nonlinear semiactive DMD system is first divided into piecewise linear active DMD systems. At each time step, an appropriate linear controller is assigned to apply control force depending on which active DMD system is most suitable fitting the current states of the semiactive DMD system.

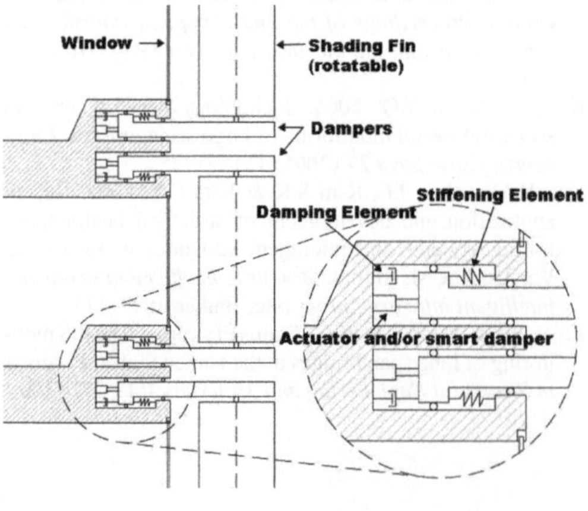

Figure 1. Example shading fin mass damper detail.

REFERENCES

Fu, T.S.& Johnson, E.A. 2008. Distributed Mass Damper System for Integrating Structural and Environmental Controls in Buildings, *Proceedings of the Inaugural International Conference of the Engineering Mechanics Institute*, Minneapolis, MN, May 18–21.

Soong, T. & Grigoriu, M. 1993. *Random Vibration of Mechanical and Structural System*, Prentice Hall, Englewood Cliffs, NJ.

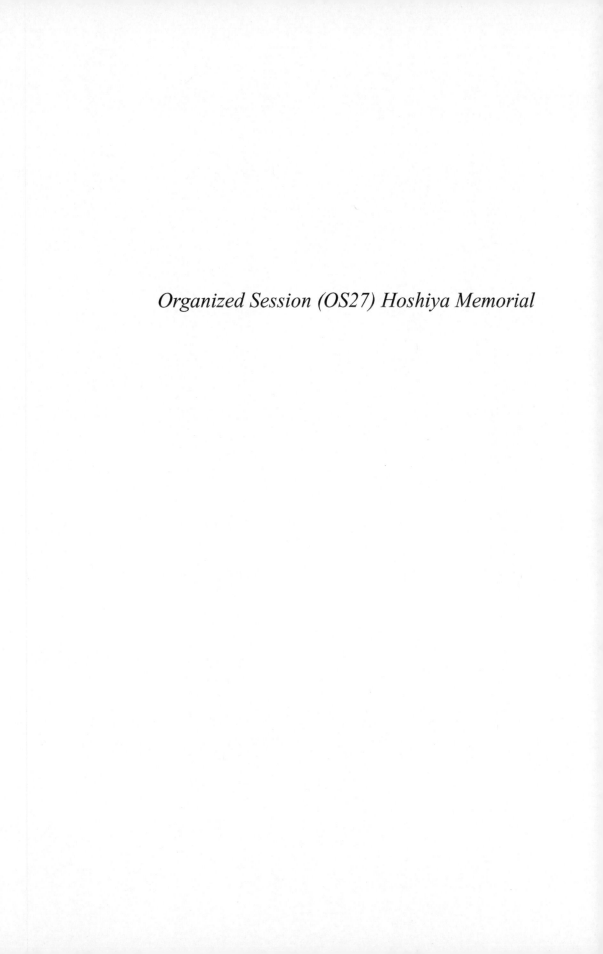

Organized Session (OS27) Hoshiya Memorial

Safety, Reliability and Risk of Structures, Infrastructures and
Engineering Systems – Furuta, Frangopol & Shinozuka (eds)
© 2010 Taylor & Francis Group, London, ISBN 978-0-415-47557-0

Partial factors for the earthquake resistant design on pile-supported wharves against the serviceability limit state

T. Nagao
Ministry of Land, Infrastructure and Transport, Yokosuka, Japan

Y. Sakai, M. Fujita & M. Suzuki
Shimizu Corporation, Tokyo, Japan

T. Sanuki & Y. Yoshinami
Fukken Co. Ltd., Hiroshima, Japan

ABSTRACT

In recent years Japanese structural design codes have been modified so as to comply with the international standard, such as ISO2394 (ISO 1998). One of the fundamental views of the international standard is to carry out the quantitative verification of structural safety with respect to required performances. In line with this trend, the Japanese design code for port and harbor facilities was revised in 2007 (MOLIT 2007) in accordance with the concept of the performance-based design. This revised design code adopts the partial factor format in the safety verification. Authors made an investigation about the reliability level of the wharves designed by the former design code and determined the partial factors for the revised design code (Nagao et al. 2007). This paper reports the content of the investigation for the earthquake resistant design of pile-supported wharves against serviceability limit state and how the partial factors were determined and verified.

The return period of the reference earthquake ground motion is 75 years and the serviceability is required against it. It is known that the steel pile head is the first part damaged by the horizontal seismic motion. So, the limit state is set to be the yield of steel pile head. First, 30 types of wharves with various conditions (ground type, pile length and pile configuration) were designed in accordance with the former design code. Then, the probability distribution functions of the variables (seismic intensity, coefficient of lateral ground reaction and yield strength of steel pipe pile) were determined and the FORM analysis was made to each type of the designed wharf to obtain the reliability index and the sensitivity factor of each variable. Based on the calculated reliability indices, the target reliability index relating to the importance of the wharf was determined by the code calibration method. Using the target reliability index (β_t), the sensitivity factor (α) and the coefficient of variation (COV), the partial factor for each variable was calculated.

Table 1. Partial factors for revised design code.

Grade	Basic variable	γ
B	Yield strength	1.00
	Coef. of lateral subgrade reaction	0.80
	Seismic coefficient	1.23
A	Yield strength	1.00
	Coef. of lateral subgrade reaction	0.72
	Seismic coefficient	1.36
Special	Yield strength	1.00
	Coef. of lateral subgrade reaction	0.66
	Seismic coefficient	1.68

18 types of wharves were designed in accordance with the level 1 reliability method using the partial factor format with the determined partial factors. It was confirmed that the quantities of steel pipe piles designed by the level 1 reliability method was almost the same as those designed by the former design code. It was also confirmed by the FORM analysis that the designed wharves have the reliability indices close to the target reliability. Through these studies, the calculated partial factors were adopted for the revised design code. The partial factors for each wharf grade are shown in Table 1.

REFERENCES

ISO, 1998. General principles on reliability for structures, ISO2394.

MOLIT, 2007. Technical standards and commentaries for port and harbor facilities in Japan (in Japanese), The Japan Port and Harbor Association.

Nagao, T., Kikuchi, Y., Fujita, M., Suzuki, M. and Sanuki, T., 2006. Reliability-based design method for pile-supported wharves against the level-one earthquake ground motion (in Japanese), JSCE, Proc. of Structural Engineering, Vol.52A.

Nagao, T. and Tashiro, S. 2002. Analysical study on earthquake resistant evaluation method for pile-supported wharves (in Japanese). Proc. of JSCE, No. 710/I-60. JSCE

Safety, Reliability and Risk of Structures, Infrastructures and
Engineering Systems – Furuta, Frangopol & Shinozuka (eds)
© 2010 Taylor & Francis Group, London, ISBN 978-0-415-47557-0

Evaluation of outage time for a system consisting of distributed facilities considering seismic damage correlation

T. Shizuma & T. Nakamura
Shinozuka Research Institute, Tokyo, Japan

H. Yoshikawa
Musashi Institute of Technology, Tokyo, Japan

INTRODUCTION

In a seismic risk evaluation for distributed facilities located in wide area, it is important to consider the effect of damage correlation, because a shape of probability mass function of physical loss such as restoration cost of facilities is affected by damage correlation between distributed facilities (Nakamura 2007). This seismic damage correlation relates to correlation of seismic ground motion intensity. Recently, statistical studies of spatial correlation of seismic ground motion intensity have been advanced, and spatial correlation of seismic ground motion intensity depending on relative distance between sites is clarified in detail (Wang & Takada 2007). However, it is applied only to physical loss in the evaluation considering seismic damage correlation.

This paper focuses on an analytical method that is able to consider seismic damage correlation concerning a probability mass function of outage time for a system. This system consists of distributed facilities. The proposed method based on the theorem of total probability and on the principle of simultaneous restoration is practicable from two viewpoints; a lot of computing time is not required, and physical damage correlation between facilities can be considered.

For examination of applicability, Hanshin Expressway series system consisting of RC single piers was adopted in the evaluation of probability mass function of outage time when Uemachi Fault Earthquake occurred. A shape of probability mass function and main descriptors such as mean value and coefficient of variation (COV) were compared in three cases with different levels of damage correlation. In the three cases, damage correlation between RC single piers is classified as "independence", "full correlation" and "correlation depending on relative distance between piers". Main results from Figure 1 are as follows:

(1) A shape of probability mass function of outage time is strongly affected by seismic damage correlation.
(2) If seismic damage correlation between facilities is stronger, mean value of probability mass

Condition of damage correlation	Mean value	COV
Independence ($\rho_G = 0.0$)	182.3	0.91
Correlation depending on relative distance between piers ($\rho_G = 0.66 - 0.8$)	79.3	1.75
Full correlation ($\rho_G = 1.0$)	29.5	3.06

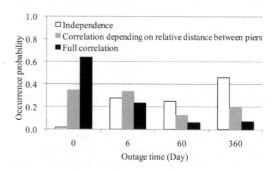

Figure 1. A probability mass function of outage time for Hanshin Expressway system in Uemachi Fault Earthquake occurrence.

function of outage time becomes smaller, while COV becomes larger.

In earthquake prevention measures, decision-making by using only mean value and median, etc. is not preferable because of result(2) above mentioned. Moreover, it suggests the importance of seismic risk evaluation considering the effect of seismic damage correlation between facilities.

REFERENCES

Nakamura, T. 2007. A Portfolio Seismic Loss Estimation Considering Damage Correlation. *Journal of structural and construction engineering Transactions of AIJ*, No.623: 49–56 (in Japanese).

Wang, M. & Takada, T. 2007. Macro-spatial Correlation Model of Seismic Ground Motions. *Proceedings of ICOSSAR'05*, (CD-ROM), Balkema.

Safety, Reliability and Risk of Structures, Infrastructures and Engineering Systems – Furuta, Frangopol & Shinozuka (eds)
© 2010 Taylor & Francis Group, London, ISBN 978-0-415-47557-0

An identification of correlation between demand performances to damage of tunnel lining using AHP

A. Sutoh

Department of Urban & Civil Engineering, Tokyo City University, Tokyo, Japan

T. Sato & H. Nishi

Civil Engineering Research Institute for Cold Region, Public Works Engineering Research Institute, Sapporo, Japan

ABSTRACT

Infrastructure maintenance is becoming increasingly important in Japan. Especially in Hokkaido, about 350 conventional construction method road tunnels have been constructed over the past 40~50 years, of which many tunnels were constructed during the so-called economical developing period. This research project will be developed to the efficient tunnel maintenance system and a quantitative criterion from pictures of tunnel lining concrete using the life cycle cost theory. Firstly, demand performance of tunnel linings in cold region are identified by the Analytic Hierarchy Process (AHP) model (Saaty1980) based on the weights and consistency indices (CI) using the tunnel user, tunnel inspection engineer's and tunnel management engineer's interview.

The AHP is a structured technique for helping people deal with complex decisions. The AHP first decompose their decision problem into a hierarchy of more easily comprehended sub-problems, each of which can be analyzed independently.

Once the hierarchy is built, the decision makers systematically evaluate its various elements, comparing them to one another in pairs. It will be able to make a priority of the optimal combination of the inspection methods to be applied and the demanded performance and condition of tunnel lining automatically on the system (See Figure 1).

Secondly, the outline of conventional construction method road tunnel management system is illustrated and defects and problems to be overcome are clarified. And, many type of cracks and detailed items for evaluation of tunnel linings were set up based on the aforementioned interview using the AHP procedure.

The proposed criterion was employed for some actual conventional construction method read tunnels in order to investigate the applicability. The tunnel lining crack evaluation results by using the proposed criterion were generally correct to results of detailed

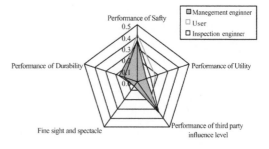

Figure 1. Demand performance of tunnel lining.

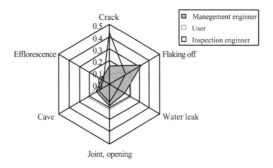

Figure 2. Relationship safe performance to deterioration.

inspection at site. However, the evaluation for exfoliation of concrete was not always corresponding to the detailed (See Figure 2).

Finally, numerical examples are worked out to demonstrate the usefulness of this AHP procedure for non-structural system identification.

REFERENCES

Saaty, T.L, 1980. The Analiytic Hierarchy Process, McGraw-Hill.

General Session (Design Concepts)

Safety, Reliability and Risk of Structures, Infrastructures and Engineering Systems – Furuta, Frangopol & Shinozuka (eds)
© 2010 Taylor & Francis Group, London, ISBN 978-0-415-47557-0

Reliability of water tank reinforced concrete elements under direct tension and flexure

Devdas Menon & D. Navaneeth Kumar
IIT Madras, Chennai, India

ABSTRACT

Reinforced concrete water tank elements, such as walls and slabs, are designed to satisfy strength and serviceability requirements. In general, it is the limit state of cracking that governs the design. Limiting the crack width is necessary to ensure water tightness, for aesthetics and to safeguard the reinforcement against corrosion. The traditional method of design world-wide was to achieve crack control in slabs and walls indirectly by providing relatively thick concrete sections (using the 'uncracked' section concept) and reduced allowable tensile stresses in steel (using the 'cracked' section concept). This is being replaced by more rational limit states design. However, there are significant disparities among international codes relating to design procedures that aim to restrict crack-widths. In this context, the proposed shift in design philosophy in codes such as the proposed Indian Standard, IS 3370, needs to be supported by a reliability study that accounts for the various uncertainties involved.

A detailed study has been done to determine the most appropriate crack-width estimation formula, under direct tension and flexure. In particular, the recommendations given in BS 8007/BS 8110, ACI 224 committee report/ACI 318 and Eurocode 2 have been studied and compared with experimental data reported in the literature. It is inferred that the BS code procedures provide relatively the most accurate predictions that lie in between the Eurocode and ACI predictions. These crack-width formulas, based on BS 8007, applicable for the target crack-width of 0.1 mm and 0.2 mm, under direct tension and flexure, account for the influence of tensile stress in steel, spacing of bars and clear cover to reinforcement.

The reliability analysis is carried out by MonteCarlo simulation and also by first order reliability method. It is seen that the probability of failure (exceeding crack widths of 0.1 mm and 0.2 mm) underlying current codes is fairly high in comparison with the target of 1 in 100 recommended in BS 8007, with reference to the uncertainty data relevant for Indian conditions. Suggestions are made to improve the present design procedures in IS 3370 (under revision) in order to achieve the target reliability.

REFERENCES

Control of cracking in concrete structures, Journal of American Concrete Institute, ACI Committee 224, Vol. 69, No. 12, December 1972: 717–752.

Clark, A.P., Cracking in reinforced concrete flexural members, Journal of the American Concrete Institute, Proceedings, Vol. 52, April 1956: 851–862

BS 8007 (1987) British Standard Code of Practice for structural use of concrete for retaining aqueous liquids, British Standards Institute.

*Safety, Reliability and Risk of Structures, Infrastructures and
Engineering Systems – Furuta, Frangopol & Shinozuka (eds)*
© 2010 Taylor & Francis Group, London, ISBN 978-0-415-47557-0

Analysis of the probability distribution of snow load effects for gable roofs

M. Kasperski

Department of Civil and Environmental Engineering Sciences, Ruhr-University Bochum, Bochum, Germany

ABSTRACT

The specification of the design value for any action requires a thorough analysis of the probability distribution of those action effects which govern the final design. In case of action effects induced by snow on gable roofs, at least the following basic variables are to be considered with their random characteristics:

- the snow load on the ground s_{gr}
- the shape factor η which describes the total amount of snow going to be accumulated on the roof
- the re-distribution factor c_r which takes into account the effects of snow drift.

The required probabilistic models of the shape factor η and the re-distribution factor c_r can be obtained from the recommendations of the Joint Committee on Structural Safety JCCS. The probability distribution of the snow load on the ground can be identified based on meteorological observations. Depending on the actual snow climate, yearly extreme may not be sufficient basis. Then, as further basic random variable the number of independent periods with a permanent snow cover in a twelve month cycle may have to be considered. The paper will show respective examples from the snow climate in Germany.

Since both parameters, the shape factor and the re-distribution factor, depend on the wind conditions during and after the snow fall events, the joint probability distribution of wind and snow has to be analyzed. The paper will present respective results for some typical stations.

Basically, the number of required snow load distributions depends on the number of design-decisive action effects. For each identified design-decisive action effect, the corresponding probability distribution is obtained from a convolution integral over s_{gr} and η and c_r, where the latter two variables will be considered with different weighting coefficients for different action effects.

The analysis of the probability distribution of action effects aims in the specification of the design values for the contributing basic variables. Analogue to the recently adopted strategy in wind engineering, the paper proposes appropriate target values for the exceedance probability of the analyzed action effects with reference to the projected working life. The target values distinguish between four classes of structures from class A – structures with a special post disaster function to class D – structures presenting a low degree of hazard to life and other properties.

REFERENCE

JCCS – Joint Committee on Structural Safety - Probabilistic model code Part II Load Models, Part 2.12 Snow Loads, May 2001, www.jcss.ethz.ch

Safety, Reliability and Risk of Structures, Infrastructures and Engineering Systems – Furuta, Frangopol & Shinozuka (eds)
© 2010 Taylor & Francis Group, London, ISBN 978-0-415-47557-0

Detailed analysis of the safety of design buckling curves for steel members

J. Szalai
KESZ Ltd., Hungary

F. Papp
Budapest University of Technology and Economics, Hungary

ABSTRACT

One of the most important application areas of probabilistic structural analysis is the reliability calibration of standard design methods and moreover the development of the safety concept and background behind the formulas. In the engineering practice the structural design standards provide the most frequently applied tool which guarantees the required safety level for the designer. Accordingly there have been developed many approaches for the calibration of different parts of design standards. These special purpose calibration methods are usually quite varying as they try to adapt to the characteristic signs of the examined task. This paper focuses on the probabilistic calibration procedure of design resistance formulas for steel structures. The unified regulations for the member states of the European Union are collected in the EN 1993. This part of the Structural Eurocodes is a progressively developing code including a great number of new design formulas. At this point some notes should be made in order to understand clearly the current situation around the calibration of these new resistance formulas:

1. In the earlier versions of the Structural Eurocodes the resistance formulas were usually calibrated directly to experimental results, however currently the new formulas developed on the basis of appropriate high level numerical analysis exploiting the power of computer calculations;
2. Only deterministic calibration of the new design mechanical models is executed, a complete probabilistic verification – which shows the appropriate reliability level – is usually missing;
3. There is only one recommended statistical calibration method for the resistance formulas which is rather an evaluation method of experimental test results – but sometimes applied for numerical results as well –, this method obviously does not take advantage of the opportunities implied in numerical calculations.

According to these notes in this paper a simple method is developed for the probabilistic evaluation of standard resistance models based on the results of numerical calculations. The method utilizes the features a design resistance model implies: description of only one failure mode, simple construction of resistance functions, the random variables at the resistance side have usually sufficient measured data for statistical analysis. The method also attempts to unify the accuracy level of the three main components of such calculations: (1) the deterministic model, (2) the probabilistic model, and (3) the set of random variables and their reliable statistics.

As an application example the design buckling curves for column and lateral-torsional buckling are evaluated as the most important basis for the stability design resistance of steel members. These curves are typical examples for the mentioned problems, as they were originally calibrated to tests by the recommended statistical evaluation method, later some modifications were added however without any probabilistic verification. Although, the column buckling is a deeply verified design problem, the widening knowledge of uncertainty data and the progressive methods for analysis of structural behavior allow but also obligate the engineer to reanalyze the standardized methods from time to time by applying the most up-to-date information and techniques. On the other hand it should be emphasized that the lateral-torsional buckling problem – which is far more difficult than the column buckling – has no such strong experimental and theoretical background resulting in a great deviation in the different national code regulations. The example is not sufficiently comprehensive to be eligible for a complete probabilistic verification of the resistance model, but gives a transparent overview about the working of the proposed method and the inconsistencies in the design curves.

Safety, Reliability and Risk of Structures, Infrastructures and Engineering Systems – Furuta, Frangopol & Shinozuka (eds)
© *2010 Taylor & Francis Group, London, ISBN 978-0-415-47557-0*

A comparative method for improving the resistance of designs to failure initiated by flaws

M.T. Todinov

School of Technology, Oxford Brookes University, Oxford, UK

ABSTRACT

Calculating the absolute reliability built in a product is often an extremely difficult task because of the complexity of the physical processes and physical mechanisms underlying the failure modes, the complex influence of the environment and the operational loads, the variability associated with reliability-critical design parameters and the non-robustness of the prediction models. Predicting the probability of failure of loaded components with complex shape for example is associated with uncertainty related to: the type of existing flaws initiating fracture, the size distributions of the flaws, the locations and the orientations of the flaws and the microstructure ant its local properties. Capturing these types of uncertainty, necessary for a correct prediction of the reliability of the component is a formidable task which does not need to be addressed if a comparative reliability method is employed, especially if the focus is on reliability improvement.

The new comparative method for improving the resistance to failure initiated by flaws proposed here is based on an assumed failure criterion, an equation linking the probability that a flaw will be critical with the probability of failure associated with the component and a finite element solution for the distribution of the principal stresses in the loaded component. The algorithm is of linear complexity $O(n)$, where n is the number of finite elements into which the component has been divided. The proposed method is very precise because statistical information is collected from all parts of the stressed component. The method does not rely on a Monte-Carlo simulation and is also very efficient.

Another advantage of the proposed comparative method is that it does not depend on knowledge of the size distribution of the flaws and material properties. This essentially eliminates uncertainty associated with the material properties and the population of flaws.

On the basis of counterexamples we show that even for the simple case of a single group of of identical flaws, the Weibull distribution fails to predict correctly the probability of failure.

We also show that contrary to the common belief, in the case of non-interacting flaws in a stressed volume, the Weibull distribution is not the mathematical formulation of the weakest-link concept. The distribution of the minimum failure stress $\sigma_{min,f}$ for a bar subjected to tension is

$$P(\sigma_{min,f} < \sigma) = 1 - \exp\left(-\bar{\lambda} V F_c(\sigma)\right)$$

where, $\bar{\lambda}$ is the average number density of the flaws, $F_c(\sigma)$ is the probability that the flaw will be critical (will initiate failure if present in the stressed volume) at a loading stress σ and V is the volume of the specimen. It can also be interpreted as a strength distribution of the flaws.

This equation has been verified by Monte Carlo simulations. It is the correct mathematical expression of the weakest-link concept, in the case of failure locally initiated by non-interacting flaws. The equation does not require any assumption concerning the physical nature of the flaws and the physical mechanism of failure and can be applied in any situation of locally initiated failure by non-interacting entities. The flaws present in real materials are rarely simple cracks that satisfy the equations of fracture mechanics. The flaws could be machining flaws, scratches, voids, inclusions, etc. The mechanism of forming unstable cracks from these flaws is very complex and still not very well understood (e.g. it may also involve plasticity effects). The strength distribution $F_c(\sigma)$ however is relevant to each of these types of flaws.

Finally, we propose a method for estimating the stress dependence of the number density of critical flaws $\varphi(\sigma) = \lambda F_c(\sigma)$ from experimental measurements. Knowing the real stress dependence $\varphi(\sigma)$ of the number density of critical flaws, makes unnecessary the assumption of a power law stress dependence. From the real stress dependence of the number density of critical flaws characterizing the material, the reliability of a component with complex shape is obtained by integration throughout the volume of the component.

Safety, Reliability and Risk of Structures, Infrastructures and
Engineering Systems – Furuta, Frangopol & Shinozuka (eds)
© 2010 Taylor & Francis Group, London, ISBN 978-0-415-47557-0

Calculation of load factors for reliability-based bridge evaluation

E.-S. Hwang & H.T. Nguyen

Department of Civil Engineering, Kyung Hee University, Korea

ABSTRACT

The most important factor that affects the design and
evaluation of a bridge is the load effect. Of all loads,
the live load is the dominant factor for most bridges.
Since the strength evaluation of the bridge focuses on
the live load capacity, it is essential to calculate the
live load effect very carefully. The live load effects
used in bridge evaluation are different from those in
bridge design because they must consider evaluation
interval, local effects and so on.

To consider the difference of time effect between
evaluation and design, this study has analyzed truck
weights in the period of 5 years for evaluation and 100
years for design. Figure 1 shows the estimation of 5 and
100 maximum weight based on plot of 20% upper data
on Gumbel probability paper and linear extrapolation.
The results are shown in Table 1. The obtained reduction factor due to the service life is estimated as 0.9.

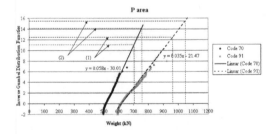

Figure 1. Plots on Gumbel probability paper of 20% truck
weight data in P area.
(1) Estimation of 5-year maximum truck weight.
(2) Estimation of 100-year maximum truck weight.

Table 1. Estimated maximum truck weights (kN) at 6
locations.

	100-year interval		5-year interval	
Truck Weight	Code 70	Code 91	Code 70	Code 91
Average	815	1,182	739	1,049
COV	0.10	0.09	0.08	0.08
Ave. COV	0.10		0.08	

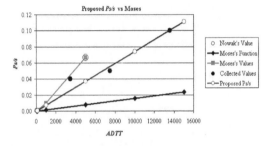

Figure 2. Comparison proposed $P_{s/s}$ with Moses.

Table 2. Estimated maximum truck weights (kN)
within $P_{s/s}$.

	ADTT	100	1,000	5,000	10,000	15,000
Avg.	Code 70	556	614	655	673	683
Weight	Code 91	726	831	903	934	952
Reduction	Code 70	0.85	0.94	1.00	1.03	1.04
Factor	Code 91	0.80	0.92	1.00	1.03	1.05
Avg Factor		0.83	0.93	1.00	1.03	1.05

Table 3. Calculated live load factors for evaluation.

ADTT	100	1,000	5,000	10,000	15,000
γ	1.21	1.36	1.46	1.51	1.54

Figure 2 shows the comparison of probability of
multiple presence depending on ADTT. The linearly
fitted function of ADTT for side-by-side probability
is determined and proposed. Based on the multiple
presence probability, average reduction factor of load
factor are calculated as shown in Table 2.

In the calibration process, the load and resistance
factors are determined from statistical properties of
load and resistance. For the bridge evaluation, the COV
of live load effect is assumed as 15% compared with
20% for design. This difference results the live load
reduction factor of 0.93.

Consequently, the load factors for evaluation of
bridges are proposed as in Table 3 compared with the
design load factor of 1.75.

Organized Session (OS12) Inspection and Analysis of Concrete Bridges in Coastal Environments

Safety, Reliability and Risk of Structures, Infrastructures and Engineering Systems – Furuta, Frangopol & Shinozuka (eds)
© 2010 Taylor & Francis Group, London, ISBN 978-0-415-47557-0

Properties of concrete and reinforcing steel in PC bridge girders severely damaged in coastal environment

I. Iwaki
Nihon University, Koriyama, Japan

ABSTRACT

This paper presents the properties of concrete and reinforcing steel in PC bridge girders severely damaged by chloride induced deterioration in a coastal environment. This bridge was located on the Japan Sea shore in Aomori Prefectural Government (Aomori), the northern end of the main island in Japan. In this study, concrete cores, PC strands, and stirrups were sampled from the girders, and the properties of concrete and reinforcing steels were investigated. Further, due to making comparisons between the results of visual inspection and those of laboratory testing, the usefulness of visual inspection for deteriorated bridges was discussed. Finally, considering mechanical properties of strands, the deterioration of this bridge was compared with that of another bridge, which is known as one of the most severely damaged bridges by chloride induced deterioration.

The main conclusions are as follows;

(1) This bridge was severely damaged by chloride induced deterioration exposed to salty air, sometimes directly to the seawater, in winter time.
(2) The condition state of girders is remarkably different depending on the position of girders. These differences seem to be affected by various conditions, such as climate condition, environmental condition, load condition, construction errors and so on.

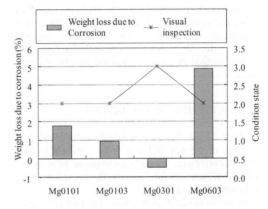

Figure 2. Relation between visual inspection results and testing results.

(3) The visual inspection results with the condition state of girder are comparatively correlated to the weight loss due to corrosion of reinforcing steel, which greatly influences to the load-extension relation of reinforcing steel. This fact seems to indicate that visual inspection can be used as a first and simple method of bridge inspection.
(4) Comparing with the data of another bridge which is known as one of the most severely damaged by chloride induced deterioration in Japan, the deterioration of this bridge is more moderate than that of another one but the process of deterioration is very similar.

REFERENCES

Kawamura, H., Kudo, K., Soma, M., Kawaragi, H. & Kaneuji M. 2008. Effective Bridge Manage-ment using ABMS, Bridge Maintenance, Safety, Management, Health Monitoring and Informat ics, in Seoul, Koh&Frangopol(eds), London: Taylor&Francis Group, 1853–1860.

Matsumura, E., Senoh, Y., Sato, M., Miyahara, Y., Kaneuji, M. & Sakano, M. 2006: Condition eval-uation standards and deterioration prediction for BMS, in: Bridge Maintenance, Safety, Manage-ment, Life-Cycle Performance and Cost, in Cruz, Frangopol & Neves (eds), Proc. of the third IABMAS conference, Porto

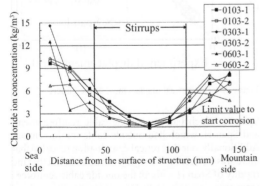

Figure 1. Distribution of chloride ion concentration inside web.

Safety, Reliability and Risk of Structures, Infrastructures and
Engineering Systems – Furuta, Frangopol & Shinozuka (eds)
© 2010 Taylor & Francis Group, London, ISBN 978-0-415-47557-0

Static loading test of a PC bridge damaged in a coastal environment

H. Naito & M. Suzuki

Graduate School of Engineering, Tohoku University, Sendai, Japan

1 INTRODUCTION

The purpose of the present study was to examine the
structural performance of a pre-stressed concrete (PC)
bridge damaged in a coastal environment. Structural
performance includes such factors as rigidity, strength,
ductility, etc. We report here on static loading tests
of the damaged PC bridge using the weight of dump
tracks.

2 DESCRIPTION OF THE BRIDGE

The Shin-Akaishi bridge was a road bridge located on
the coast of the Sea of Japan in the northern part of
Aomori Prefecture. It had been exposed to a severe
coastal environment for 32 years. Bridge information
is shown in Table 1.

We examined 2 different spans, one with a low level
of damage (Span L), and one with a high level of
damage (Span H). Span H was reinforced by outside
cables.

3 EXPERIMENTAL TESTS

We conducted static loading tests of the PC bridge
using dump trucks. As shown in Figure 1, 6 dump
trucks were placed on the road.

The vertical displacement of Spans L and H is
shown in Figure 2. Note that the rigidity of the girders
(vertical displacement) was approximately equal.

4 STRUCTURAL PERFORMANCE

The calculation results shown in Figure 2 indicate that
there was no damage with respect to rigidity in Span
L, because the deformations observed in experimental tests corresponded approximately to the theoretical
values of the undamaged span. However, in Span H, the
rigidity of the girders without outside cable reinforcement was approximately 50% of its rigidity without
damage.

5 CONCLUSION

Based on the results of loading tests of the PC bridge,
there was approximately equal bending rigidity in

Table 1. Information on the Shin-Akaishi bridge

Structure type	PC bridge with T-shaped girders on simple supports
Span	5 @ 31.84 m = 163.0 m
Width of road	11.1 m
Construction date	January, 1976

Figure 1. Loading tests with dump trucks.

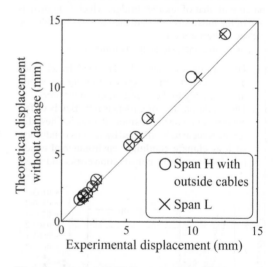

Figure 2. The vertical displacement of the girders.

Span L, which had no outside cable reinforcement,
and Span H, which had outside cable reinforcement.
Additionally, our theoretical examination revealed that
there was no decrease in rigidity in Span L, but that the
rigidity of Span H without the outside cable reinforcement was approximately 50% of its rigidity without
damage.

Safety, Reliability and Risk of Structures, Infrastructures and
Engineering Systems – Furuta, Frangopol & Shinozuka (eds)
© 2010 Taylor & Francis Group, London, ISBN 978-0-415-47557-0

Aerosol chlorides condition and chloride penetration into concrete in seashore area

T. Saeki
Niigata University, Niigata, Japan

ABSTRACT

The purpose of this study is to evaluate the aerosol chloride condition in the coastal area in Japan and the relation between the environmental condition and chloride penetration into concrete.

In this study, the amounts of aerosol chloride were measured at 15 places in the coastal area of Niigata prefecture and the amounts of chloride in mortar and concrete specimens exposed in the same places were measured. From the experimental results, the amount of the aerosol chloride at each investigation point is strongly influenced of individual conditions. And the amount of the chloride penetrated into concrete depended on the amount of the aerosol chloride. Therefore, the chloride penetration into concrete may be predicted by the result of short-term mortar exposure test as shown in Figure 1 and Figure 2.

Based on the above-mentioned results, the 2nd stage research was performed. The amount of chloride ions in the thin disk mortar specimens were measured at 8 places in the coastal area of Niigata, Hokkaido and Okinawa. Four kinds of mix proportions mortar were used.

The amount of chloride ions penetrated into specimen under the same environmental condition can be estimated by the effective diffusion coefficient, which is an index of mass transfer characteristic of mortar (Figure 3). And the relative influence of the effective diffusion coefficient on the amount of chloride penetration is almost same even if exposure places differ. From these results, relation between the amount of aerosol chloride and chloride penetrated into concrete was studied. The prediction method for the amount of chloride penetration under aerosol chloride condition using effective diffusion coefficient and chloride binding capacity of binder was also discussed.

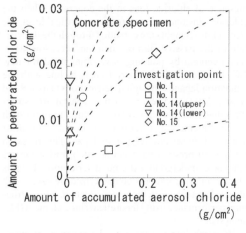

Figure 2. Relation between amount of accumulated aerosol chloride amount of chloride ions penetrated into concrete specimen.

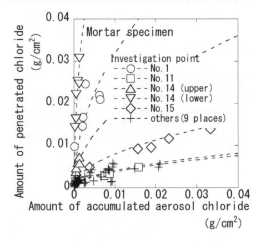

Figure 1. Relation between amount of accumulated aerosol chloride amount of chloride ions penetrated into mortar specimen.

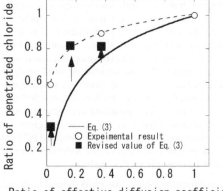

Figure 3. Revised relationship between effective diffusion coefficient and amount of chloride penetration.

Safety, Reliability and Risk of Structures, Infrastructures and Engineering Systems – Furuta, Frangopol & Shinozuka (eds)
© 2010 Taylor & Francis Group, London, ISBN 978-0-415-47557-0

Prediction for structural performance of PC bridge girders damaged by salt attack

A. Kamiharako
Hirosaki University, Hirosaki, Japan

ABSTRACT

Reinforced concrete (RC) structures located in coastal areas undergo deterioration as a result of corrosion to their steel reinforcements. This phenomenon is known as salt attack in RC structures, and is caused by the presence of chloride ions in the concrete. When the chloride ion concentration reaches a critical level, the steel reinforcements are corroded through the destruction of the passive film on their surface. Nowadays, reinforcement by painting is recommended for new RC structures built in coastal areas in line with the regulation Japanese design specifications (Japan Road Assoc. 2002). However, existing coastal RC structures that were built 30 to 50 years ago have no such reinforcement, and suffer deterioration as a result of salt attack in seaside regions of Japan. This research focuses on prestressed concrete (PC) bridge girders that have undergone extreme deterioration by this process in environments characterized by severe salt attack. The author attempts to evaluate residual structural performance using theoretical finite element (FE) analysis.

Research into the theoretical prediction of structural performance as a function of time has already been conducted by many researchers (e.g., Shimomura et. al., 2006). However, in most cases, the prediction of material transport and deterioration are emphasized, and there have been relatively few studies on the prediction of structural performance after material deterioration. Figure 1 shows an outline of this research. The author aimed to develop numerical models of deteriorated PC structures for use in FE analysis, and conducted prediction of structural performance for demolished PC bridge girders deteriorated as a result of salt attack. The input data were determined on the basis of destructive tests consisting of weight loss investigation and tensile testing for corroded PC bars.

Based on above discussion, the following conclusions can be drawn from this study:

(1) The corrosion weight loss ratio showed large values when the rating grade (derived by visual inspection) was less than 2.5. This means that

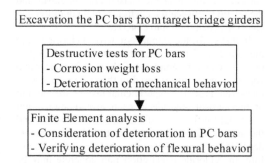

Figure 1. Research outline.

visual inspection is efficient in estimating the corrosion state of steel reinforcements.

(2) The deterioration of mechanical performance in the PC bars, including the maximum load/ elongation and the yielding load, did not depend on the magnitude of the corrosion weight loss ratio because the diameter of the bars was not uniformly distributed due to uneven corrosion.

(3) The deterioration of flexural behavior in the PC girders depended on the loss of prestress in the PC bars.

REFERENCES

Japan Road Association, 2002. Specifications for Highway Bridges, Part III: Concrete Bridges

Shimomura T. et al., 2006. Report of research project on structural performance of deteriorated concrete structures by JSCE-331, Proceedings of the International Workshop on Life Cycle Management of Coastal Concrete Structures: 151–170.

Safety, Reliability and Risk of Structures, Infrastructures and
Engineering Systems – Furuta, Frangopol & Shinozuka (eds)
© 2010 Taylor & Francis Group, London, ISBN 978-0-415-47557-0

Inspection of PC bridge girders severely damaged in coastal environment

H. Minagawa
Tohoku University, Sendai, Japan

Y. Koda
Nihon University, Koriyama, Japan

M. Hisada
Tohoku University, Sendai, Japan

ABSTRACT

In recent period, it becomes important matter that the mechanism of deteriorations is investigated from existing concrete structures, and, it is absolutely necessary to make the best use of the investigation results in the next generation's design of concrete structures. This paper focused on the chloride induced deterioration, which is one of the typical deterioration phenomenon of concrete structures, and reported the investigation results of the existing bridge located along the shoreline. The investigation items and methods in this study are shown in Table 1.

According to the investigation results and JSCE Standard: Standard Specification for concrete

Table 1. Investigation Items and methods.

Items	Methods
Defects of concrete surface	Visual observation Infrared thermography method
Compressive strength	Surface hardness method
Carbonation depth	Phenolphthalein method
Steel rebar corrosion	Half-cell potential method Chipping test

Figure 1. Cracks at the lower-flange of the girders occurred along the bridge axis.

structures-2001, "Maintenance", the grade of appearance and deterioration of structure were assessed as follows.

In this bridge, the steel rebar corrosion can be observed in every girder or slab as shown in Figure 1. After considering the comparison result of the carbonation depth and the cover depth of a girder as well as the environmental condition of this bridge, it is obvious that the reason of the steel rebar corrosion in these girders was chloride induced corrosion.

Many cracks, that direction was along the bridge axis, can be seen at the lower-flange of all main girders, and the steel exposure according to spalling as well as the stain of rust can be observe at the cracks. Moreover, the lack of section of the steel rebar due to corrosion can be observed at the lower-flange of the main girders. From these results, it can be assessed that the steel rebar in the lower-flange of the main girders corresponds to Grade III (deterioration stage).

On the other hand, the existence of some delaminations can be confirmed in a web, the bottom surface of a slab, and a cross girder, and the spalling concrete and the steel exposure were observed partially. From these results and considering the visual observation test and the half cell potential of steel rebar, it can be assessed that the steel rebar in these parts corresponds to Grade II (acceleration stage).

Therefore, it is presumed that the main cause of deterioration of this bridge is the chloride induced deterioration. Moreover, it can be assessed that the grade of appearance and deterioration of this bridge corresponds to Grade II or III (acceleration stage). However, this assessment is based on the limited investigation. Therefore, more elaborate inspections, such as a measurement of a distribution of chloride ion concentration or a loading test, must be needed, in order to conclude to clarify a structural performance of this bridge.

REFERENCES

Japan Society of Civil Engineers, 2001. JSCE Standard: Standard Specification for concrete structures-2001, "Maintenance", JSCE Guidelines for concrete, No. 4, p. 103.

Safety, Reliability and Risk of Structures, Infrastructures and
Engineering Systems – Furuta, Frangopol & Shinozuka (eds)
© 2010 Taylor & Francis Group, London, ISBN 978-0-415-47557-0

Prediction of deterioration in PC bridge girders severely damaged in coastal environment

H. Tsuruta

Kansai University, Osaka, Japan

ABSTRACT

This paper presents the prediction of deterioration in PC bridge girders severely damaged by chloride induced deterioration. This bridge in Route 101 was located on the sea shore of the Sea of Japan in Aomori, northeastern part of Japan. Since the location is subjected to a strong monsoon wind from sea at winter, this bridge was exposed to severe corrosive environment by sea water. Although the bridge had been served for 30 years with repeated repairs, it was dismantled in 2006 due to the severe chloride induced deterioration. The length of bridge was 33m. The bridge was composed of 6 main T-shaped post-tensioned girders.

In this study, the prediction of deterioration was carried out with the probabilistic methodology. And the result was compared with the result by the probabilistic methodology and the actual state of repairing in this PC bridge girder. First of all, in probabilistic methodology, data was collected to make a database from many literatures. The data was the apparent diffusion coefficient of concrete and the chloride ion content on the surface concrete. In the prediction, the chloride ion content on the surface concrete, apparent diffusion coefficient of chloride ions, cover depth of concrete and corrosion rate of steel bar etc were used as parameters and were considered with probability theory. The calculation was carried out with data in database and data from actual bridge. In the calculation, the Monte Carlo method that had 5000 trial frequency was used. The occurring time of the deterioration event was decided by providing the peak of the probability. As a result, three kinds of results of prediction and actual state of repairing were compared each other. The three results were the result by the deterministic methodology, the one by probabilistic methodology using database, and the one by probabilistic methodology using data of actual bridge. The result by the deterministic methodology was calculated by owner of this bridge.

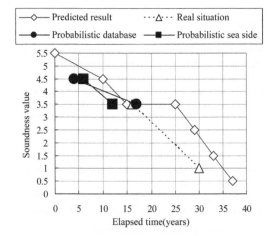

Figure 1. The comparison between results of two prediction methods.

Figure 1 shows about the comparison between results of two prediction methods. The results by the deterministic methodology are corrected using progress state of deterioration in real bridge. Then if the plots of "Predicted result" are right approximately, we can think that probabilistic results are also reasonable and safety results.

The main conclusions are as follows; (1) At the result of predicting deterioration using data from database and data from actual bridge, it is confirmed that prediction by using database enables to predict well approximately. (2) It is confirmed that this probabilistic methodology is good to predict the deterioration in severely damaged coastal environment. (3) The distribution of data is different by the kinds of data, so it is confirmed that the difference of the distribution of data occurs the big difference for result of predicting deterioration.

General Session (Reliability-Based Design and Regulations & Reliability-Based Optimization and Control & Reliability Theory)

Safety, Reliability and Risk of Structures, Infrastructures and
Engineering Systems – Furuta, Frangopol & Shinozuka (eds)
© 2010 Taylor & Francis Group, London, ISBN 978-0-415-47557-0

Safety assessment of composite sinusoidal-web beams

S.M.C. Diniz, R.J. Pimenta, G. Queiroz, R.H. Fakury, A. Galvão & F.C. Rodrigues
Federal University of Minas Gerais, Belo Horizonte, Brazil

ABSTRACT

Steel girders have been successfully used for many
years in the composite construction market. Further
optimization and advances in fabrication technology
have led to a new generation of structural shapes.
One of these developments, – the sinusoidal-web beam
(Fig. 1) –, has been recently introduced in the Brazilian
market. A sinusoidal-web beam is a built-up I-girder
with a thin-walled corrugated web (with a sinusoidal
profile) and flat plate flanges. The sinusoidal corrugation considerably increases the rigidity and the
resistance to shear forces and local effects, reducing,
or even eliminating, the local buckling of flat parts that
exist in trapezoidal corrugation. Therefore, it allows
for the use of thinner web sheets without the need for
transverse stiffeners. It also allows for weight reduction while increasing load capacity, thus leading to
economical benefits.

Due to the high structural efficiency and easy execution, the usage of sinusoidal-web beams has been
increasing significantly in the most diverse segments
of civil engineering construction. At present, composite beams (concrete deck slab over sinusoidal-web
beams, Fig. 2) have not yet been used in floor systems. However, with the increasing interest on these
elements, their potential for economical gains, and the
large volume of research in this area, hopes are that
they eventually will be used in composite construction, especially in large span floor systems. In spite
of the advantages this type of composite construction
may offer, there are no design standards or specifications dealing with the behavior of such beams. As a
result, there is a need to develop design recommendations that properly address the composite flexural
capacity of these elements.

In this paper, design recommendations for flexural capacity of composite sinusoidal-web beams under
positive moments are presented. To this end: (i) finite-
element models for composite connections and composite beams have been developed (ii) finite-element
models were validated by experimental data and other
results reported in the literature. Additionally, the
safety levels implicit in the proposed recommendations are assessed. First, a description of model errors
estimated from both experimental and numerical analyses are presented. Second, statistical descriptions of
the random variables involved in the design process
are investigated. Third, reliability evaluations using
first-order methods and Monte Carlo simulation are
performed. It is shown that the safety levels resulting
from the proposed recommendations are in agreement
with current trends in structural engineering practice.

Figure 1. Sinusoidal-web beams.

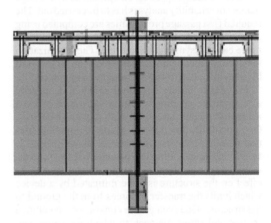

Figure 2. Composite sinusoidal-web beams.

Safety, Reliability and Risk of Structures, Infrastructures and Engineering Systems – Furuta, Frangopol & Shinozuka (eds)
© 2010 Taylor & Francis Group, London, ISBN 978-0-415-47557-0

Optimal probabilistic design of friction-based seismic isolation devices

Christian Bucher
Vienna University of Technology, Austria

ABSTRACT

In earthquake engineering the design of energy dissipating devices play an important role to ensure structural safety and integrity. Such devices one one hand must allow for a sufficiently high level of energy dissipation in order to reduce structural damage, and on the other hand must provide enough stiffness in order to prevent excessive permanent deformations or off-sets. This leads to trade-off considerations which can be dealt with through an optimization process. In the present paper, the earthquake excitation is treated as non-stationary random process. Therefore the optimization has to be based on a probabilistic characterization of the dynamic response, i.e. the first passage-probability of critical response level. In order to cover the design space for an optimal design, a large number of reliability analyses has to be carried out. The required first passage probabilities are computed using a novel efficient Monte Carlo based simulation technique called Asymptotic Sampling. The optimization involves conflicting objectives which can be resolved by applying a Pareto-type optimization approach.

In order to ensure structural safety and integrity in earthquake conditions it may be useful or even necessary to equip structures with protective devices. One possibility for this are energy-dissipating friction devices. The basic scenario is shown in Fig. 1. In this scenario, a structure is subjected to an earthquake represented by the ground acceleration $a(t)$. Its effect on the structure is to be mitigated by a device which limits the transfer of forces from the ground to the structure. One such device consists of a combined friction and spring element, in which the spring can also be replaced by a recentering force due to gravity effects (so-called friction pendulum systems). The device has two characteristic parameters, one is the maximum force transmitted by friction and the other is the re-centering (or restoring) spring stiffness constant. Both parameters should be chosen such as to maximize the protective effect on the structure.

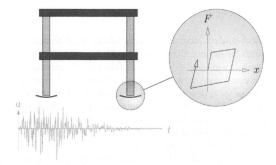

Figure 1.   Structure and friction damping device.

The question of optimal design for stiffness and friction of seismic isolation systems has been addressed previously in the literature. In the present paper, a full nonlinear dynamic analysis will be used as a basis for the computation if the structural response. The ground motion is modeled as a non-stationary random process, hence the response needs to be characterized in probabilistic terms. For the purpose of safety assessment, the appropriate measure is the first passage probability of response quantities of interest such as displacements or internal forces.

The present paper aims at developing a computationally efficient method for probability-based design optimization of the protective device. In this method, the first passage probabilities are computed using a novel efficient Monte Carlo based simulation approach. The resulting probabilities then enter a cost function which can be constructed in such a way as to provide a design compromise between conflicting optimization targets.

Safety, Reliability and Risk of Structures, Infrastructures and
Engineering Systems – Furuta, Frangopol & Shinozuka (eds)
© 2010 Taylor & Francis Group, London, ISBN 978-0-415-47557-0

An efficient algorithm for reliability–based design optimization with the system reliability constraint

H.M. Koh & S.C. Kang
Seoul National University, Korea

J.F. Choo
Konkuk University, Korea

J.O. Kim
Seoul National University, Korea

ABSTRACT

In conventional structural design, deterministic optimization which satisfies codified constraints is performed to ensure safety and maximize economical efficiency. However, uncertainties are inevitable due to the stochastic nature of structural materials and applied loadings. Thus, deterministic optimization without considering these uncertainties could lead to unreliable design. Recently, there has been much research in reliability-based design optimization (RBDO) taking into consideration both reliability and optimization. RBDO involves the evaluation of probabilistic constraints that can be estimated using two different approaches: the reliability index approach (RIA) and the performance measure approach (PMA)(Youn et al, 2004). RIA defines a probabilistic constraint as reliability while PMA defines a probabilistic constraint as performance measure. It is generally known that PMA is more stable and efficient than RIA. These approaches focus on the component limit states only. However, the structural failure does not depend only on the component limit states but also on the system limit state itself. Therefore, RBDO with the system reliability constraint is necessary to represent the structural failure mechanism but presents the disadvantage to be very time consuming for large scale applications.

In this study, we developed an efficient algorithm for RBDO with the system reliability constraint. Previous studies already reported that PMA is difficult to apply to conventional RBDO with the system reliability constraint because PMA only calculates the probabilistic performance measure for a limit state equation and does not evaluate the reliability index or the failure probability (Yi et al, 2008). In order to overcome these difficulties, PMA is introduced in the decoupled algorithm proposed by Aoues et al. (Aoues et al. 2008). In this method, RBDO is sequentially performed with updated target component reliability indices until the calculated system reliability index

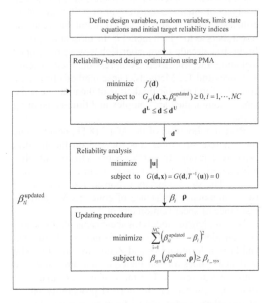

Figure 1. Detailed procedure of the proposed method.

approaches the target system reliability index. The detailed procedure is illustrated in Figure 1. The effectiveness of the proposed method is demonstrated through a numerical example.

REFERENCES

Youn, B.D. and Choi, K.K., 2004. A new response surface methodology for reliability-based design optimization, Computers and Structures, 82: 241–56.

Frangopol, D.M. and Maute, K. 2003. Life-cycle reliability-based optimization of civil and aerospace structures, Computers and Structures, 81: 397–410.

Aoues, Y., Chateauneuf, A. 2008. Reliability-based optimization of structural systems by adaptive target safety – Application to RC frames. Structural Safety, 30: 144–161.

Safety, Reliability and Risk of Structures, Infrastructures and
Engineering Systems – Furuta, Frangopol & Shinozuka (eds)
© 2010 Taylor & Francis Group, London, ISBN 978-0-415-47557-0

Reliability model and calibration of current provisions for shear in R/C beams

P. Paczkowski & A.S. Nowak

University of Nebraska – Lincoln, NE, United States

ABSTRACT

Shear resistance of reinforced concrete beams has been a subject of intensive research for over 50 years. Nevertheless, there is still a disagreement between the researchers as to which model is the most appropriate. In the last decade many approaches were developed including the Modified Compression Field Theory and Strut and Tie Methods. These methods provide a consistent and reliable prediction of the ultimate resistance, though; they are complex and time-consuming in use.

Recent calibration of the ACI 318 [1] design code for concrete structures was focused on the flexural capacity. The objective of this paper is to present reliability based calibration for shear in reinforced concrete beams. The shear capacity is considered as a sum of two components: resistance of concrete, V_c, and the resistance of shear reinforcement, V_s.

Four analytical models for shear capacity for beams are considered: two ACI Code formulas [2], Tureyen and Frosch formula [3], and Zsutty formula [4]. The analytical values for considered models are compared with the test results. The ratio of test vs. analytical values are plotted on the normal probability paper. Statistical parameters of resistance are estimated separately for each formula. ACI 318 formula 11–3 results show the highest degree of variation, particularly it is insensitive to longitudinal reinforcement ratio. Zsutty formula and ACI formula 11–5 show similar variability. Tureyen and Frosch formula has similar variability but it is shifted conservatively so that lower tale of the CDF approaches asymptotically to 1.

The optimum reliability depends on the consequences of failure and incremental cost of safety [5]. Based on [1] target reliability index for the cast in place beams in shear is $\beta = 3.5$.

To assure proper level of reliability without unnecessary conservativeness it is clear that for each formula designer should use different resistance factor. A selection criterion for the resistance factor is closeness to the target reliability index.

For the recommended resistance factors, reliability indices are calculated as a function of dead load to live load ratio and ratio of the shear capacity of transverse reinforcement to overall shear resistance.

The following resistance factors are recommended:

For ACI 318 Eq. 11–3, use $\phi = 0.75$
For ACI 318 Eq. 11–5, use $\phi = 0.8$
For Frosch formula, use $\phi = 1$
For Zsutty formula, use $\phi = 0.85$

REFERENCES

[1] Szerszen, M., Nowak, A.S., 2003, „Calibration of the Design Code for Buildings (ACI 318): Part 2 – Reliability Analysis and Resistance Factors.", *ACI Structural Journal*, May, pp. 383–391

[2] ACI Committee 318, 2008, Building Code Requirements for Reinforced Concrete (ACI 318–08) and Commentary-ACI318RM-08, American Concrete Institute, Detroit,.

[3] Tureyen A. K., and Frosh R.J., 2003,"Concrete Shear Strength: Another Perspective", ACI Structural Journal, Vol. 100, Nr. 5, September–October, pp. 609–615.

[4] Zsutty T.C., 1968, Beam Shear Strength Prediction by analysis of Existing Data. ACI Struct. J., v.65, Nov., p. 943–951

[5] Nowak A.S., Collins K. R., 2000, "Reliability of Structures." McGraw-Hill International Editions, Civil Engineering Series, Singapore, 338 pages.

Safety, Reliability and Risk of Structures, Infrastructures and
Engineering Systems – Furuta, Frangopol & Shinozuka (eds)
© 2010 Taylor & Francis Group, London, ISBN 978-0-415-47557-0

Reliability-based analysis and criteria for aerodynamic instability of cable-supported bridges with the SOSM method

Y.J. Ge
State Key Lab for Disaster Reduction in Civil Engineering, Tongji University, Shanghai, China

Z. Zhou
Chongqing Architectural Design Institute, Chongqing, China

ABSTRACT

The objective of this paper is to make the reliability-based analysis for aerodynamic instability of fourteen cable-supported bridges in China with the second-order second-moment (SOSM) method and to propose probabilistic criteria for the assessment of aerodynamic instability. The probabilistic analysis for flutter instability is carried out with a proposed reliability-based model:

$$g(X) = U_{cr} - U_e = C_w U_f - G_s U_b = 0 \qquad (1)$$

where C_w is the wind conversion factor following a normal distribution, U_f is the basic flutter speed following a log-normal distribution, G_s is the gust speed factor following a normal distribution and U_b is the maximum wind speed following a Gumbel distribution. The stochastic limit state equation, Eq. 1, is solved by the second-order second-moment (SOSM) method after the non-normally distributed variables have to be transformed to the equivalent normal variables at the design point.

Probability-based analysis with the SOSM method and the first-order and second-moment (FOSM) has been applied in 14 cable-supported bridges in China,
including 8 cable-stayed bridges and 6 suspension bridges. With the first and second sets of parameters of basic random variables, the numerical results of reliability index β and failure probability P_f as well as relative errors E are listed and compared in Table 1 due to the SOSM and the FOSM methods.

Having made comparison and contrast among the results in Table 1, as a more precise method with conservative result the SOSM method is recommended in probability-based analysis for aerodynamic instability of cable-supported bridges, and the reasonable values of standard deviations can be assumed in the magnitude of the first set of parameters. Probability-based criteria for aerodynamic instability of cable-supported bridges can be classified into three grades in accordance with failure probabilities or reliability indexes. The top grade has $P_f < 10^{-6}$ or $\beta > 4.7$, which has been met with six bridges among fourteen invested ones, and the middle and the bottom grades have $10^{-6} < P_f < 10^{-5}$ or $4.2 < \beta < 4.7$ and $10^{-5} < P_f < 10^{-4}$ or $3.7 < \beta < 4.2$, met with three bridges, among 14 cable-supported bridges.

The work described in this paper is partially supported by the NSFC 90715039, the MOST 2006AA11Z108 and the MOC 2006-318-494-26.

Table 1. Reliability index and failure probability of investigated bridges due to aerodynamic flutter.

Bridge Type	No.	Bridge Name	The first set of parameters				The second set of parameters			
			β	P_f (SOSM)	P_f (FOSM)	E (%)	β	P_f (SOSM)	P_f (FOSM)	E (%)
Cable	1	Shanghai Kezhushan	4.342	7.08E-06	6.65E-06	6.0	3.311	4.65E-04	4.21E-04	9.4
stayed	2	Hainan Shiji	5.875	2.11E-09	1.94E-09	8.1	4.313	8.03E-06	7.06E-06	12.2
bridges	3	Shanghai Nanpu	4.040	2.68E-05	2.52E-05	5.6	3.116	9.18E-04	8.36E-04	8.9
	4	Hubei Jingsha	6.327	1.25E-10	1.14E-10	8.7	4.608	2.03E-06	1.77E-06	12.9
	5	Shanghai Yangpu	4.989	3.04E-07	2.83E-07	6.9	3.733	9.45E-05	8.45E-05	10.6
	6	Fujian Qingzhou	3.403	3.33E-04	3.17E-04	4.8	2.709	3.37E-03	3.11E-03	7.8
	7	The 2nd Nanjing	7.516	2.83E-14	2.53E-14	10.5	5.376	3.81E-08	3.25E-08	14.8
	8	Jiangsu Sutong	4.792	8.25E-07	7.71E-07	6.6	3.605	1.56E-04	1.40E-04	10.3
Suspension	1	Guangxi Hongguang	4.541	2.80E-06	2.62E-06	6.3	3.441	2.90E-04	2.61E-04	9.8
bridges	2	Guangdong Humen	4.037	2.71E-05	2.55E-05	5.6	3.114	9.23E-04	8.41E-04	8.9
	3	Hubei Yichang	5.421	2.97E-08	2.75E-08	7.5	4.016	2.96E-05	2.62E-05	11.4
	4	Jiangsu Jiangyin	4.569	2.45E-06	2.29E-06	6.3	3.459	2.71E-04	2.44E-04	9.9
	5	Jiangsu Runyang	3.444	2.87E-04	2.73E-04	4.9	2.735	3.12E-03	2.88E-03	7.8
	6	Zhejiang Xihoumen	3.960	3.74E-05	3.53E-05	5.5	3.065	1.09E-03	9.95E-04	8.8

Mini-Symposia (MS06) Uncertainties in Civil Structures & Infrastructure Engineering

Safety, Reliability and Risk of Structures, Infrastructures and
Engineering Systems – Furuta, Frangopol & Shinozuka (eds)
© 2010 Taylor & Francis Group, London, ISBN 978-0-415-47557-0

Optimal restoration scheduling considering multiple group work under uncertain environments

Koichiro Nakatsu & Hitoshi Furuta
Department of Informatics, Kansai University, Japan

Yasutoshi Nomura
Organization of Advanced Science and Technology, Kobe University, Japan

Ken Ishibashi & Hiroshi Hattori
Graduate School of Informatics, Kansai University, Japan

ABSTRACT

In resent years, serious earthquakes have frequently occurred in Japan. In the near future, there is the fear that large earthquakes occur. Therefore, it is necessary to develop a synthetic disaster prevention program based on the recognition that lifeline systems may unavoidably suffer from damages when big earthquakes occur. Several researches are studied in order to obtain the optimal restoration schedule. However, road networks after earthquake disasters have an uncertain environment, that is, the restoring works are not progressing on schedule. Therefore, it is necessary to obtain the restoration schedule which has the robustness. In addition, in order to make the restoration more effective, the restoration schedule is planned with considering the cooperating work by multiple groups. Furthermore, the cooperating works are expected to make the restoration schedule more robust. In order to obtain applicative solutions, an attempt is made to develop a decision support system of the optimal restoration scheduling by using the improved Genetic Algorithm (GA) to consider uncertainty.

Table 1 presents the result of solution obtained by Simple GA by 1,000 times simulations, which can check the robustness of restoration schedule. The simulation has two kinds of uncertainties; increase of damage and delay of restoring days. In Table 1, "Impossible number" means the number of occurrence of impossible restoration. From Table 1, when there is unexpected increase of damage, "Impossibility number" is shown as about 30 percents.

Because information of devastated areas is ever-changing, the damage quantity of the devastated area might change significantly. Furthermore, if the schedule with considering uncertainty is searched by Simple

GA, a lot of calculation time is necessary to obtain the schedule.

GA Considering Uncertainty (GACU) is an improved GA to obtain the robust solution against uncertainties in the problem. In this study, Improved GACU (IGACU) is applied in order to obtain a robust restoration schedule. IGACU can search for global solutions and local solutions properly with the density of individuals. The proposed method can maintain the diversity of individuals which have equal evaluations due to selecting the individual which is in thin density area.

In Table 1, the evaluation of IGACU's solution is worse than Simple GA's. The solutions of Simple GA have better evaluation because Simple GA aims at the early restoration only and does not consider the uncertainties. On the other hand, IGACU aims at the early and robust restoration. So, the proposed method can obtain the solutions which have the robustness and the practicality. The impossible restoration does not occur on the schedule of the proposed method. This reason is that the restoring group which has no restoring equipment and facility necessary for the large damage work is assigned the works which have the possibilities to change the large damage state as the cooperating work. If a low ability group can not restore a work due to the uncertainties, the impossible restoration does not occur since another group which is cooperating with that group restores that work.

In this study, an attempt was made to develop a new method to obtain a robust restoration schedule considering the multiple group work under uncertain environments after earthquake. The proposed method was able to obtain the robust and practical solutions under uncertain environment. Furthermore, the features of robust restoration schedule were clarified.

Table 1. Effects of uncertainties by 1000 simulations.

Considered Uncertainty Method	Only Increase of Damage		Only Delay of Restoring Day		Both Uncertainties	
	Simple GA	IGACU	Simple	IGACU	Simple GA	IGACU
The Worst Evaluation	∞	9.87	10.05	11.26	∞	12.57
The Best Evaluation	6.76	8.27	6.45	8.1	7.11	8.72
The Longest Delay of Restoring Day	∞	6 days	13 days	12 days	∞	13 days
The Shortest Delay of Restoring Day	1 day	1 day	1 day	no delay	3 days	2 days
Average Delay of Restoring Day	3.50 days	2.99 days	4.51 days	4.02 days	6.72 days	5.98 days
Impossibility Number	264/1000	0/1000	0/1000	0/1000	273/1000	0/1000

Safety, Reliability and Risk of Structures, Infrastructures and
Engineering Systems – Furuta, Frangopol & Shinozuka (eds)
© 2010 Taylor & Francis Group, London, ISBN 978-0-415-47557-0

Nonlinear global seismic reliability analysis of buildings

P.Y. Song, D.G. Lu, X.H. Yu & G.Y. Wang
School of Civil Engineering, Harbin Institute of Technology, Harbin, China

ABSTRACT

To overcome the shortcomings of the conventional
failure mode approach (FMA) in structural system reliability theory, a new trend in which structural systems
reliability is approximately calculated by using global
limit states based on nonlinear structural analysis techniques recently has been increasingly of interest in
many different communities (Zhao & Ono, 1998;
Onoufriou & Forbes, 2001; Li et al., 2002, 2004, 2006;
among others).

In this paper, a global performance function for
nonlinear seismic reliability of structural systems is
firstly built up, and then, a random pushover analysis approach based on an improved point estimation
method (PEM) is developed to efficiently estimate the
first fourth statistical moments of the global performance function. The global seismic reliability index
and failure probabilities of the buildings are computed
by second-moment, third-moment and fourth-moment
methods proposed by Zhao & Ono (2001), respectively. The developed methodology is applied to a
two-dimension RC frame building considering nonlinear material and geometric effects. The semi-analytical
approach combining FORM and random pushover
analysis, and Monte Carlo simulation, are also used
to verify the accuracy and efficiency of the proposed
methodology. Through the comprehensive study in this
paper, it is found that the method of system reliability analysis based on global load-carrying capacity is
simple, practical and efficient. On the one hand, this
method can overcome many difficulties of conventional system reliability theory; on the other hand, it
can be linked with the current design codes so that
the static reliability method can solve the difficult
dynamic seismic reliability problems.

REFERENCES

Li, G.Q. & Li, J.J. 2002. A semi-analytical simulation method
for reliability assessments of structural systems. *Reliabil-
ity Engineering and System Safety* 78(3): 275–281.
Li, G.Q., Liu, Y.S. and Zhao, X. 2006. Advanced Analysis
of Steel Frames and System Reliability Based Design.
Beijing: China Architecture & Building Press.
Li, J.J. & Li, G.Q. 2004. Reliability-based integrated design
of steel portal frames with tapered members. *Structural
Safety* 26(2): 221–239.
Liu, P.L. & Der Kiureghian, A. 1986. Multivariate distribution models with prescribed marginals and covariances.
Probabilistic Engineering Mechanics 1(2): 105–112.
National Standard of China P.R. 2001. Seismic Design Code
of Buildings (GB50011-2001). Beijing: China Architecture & Building Press.
Onoufriou, T. & Forbes, V.J. 2001. Developments in structural
system reliability assessments of fixed steel offshore platforms. *Reliability Engineering and System Safety* 71(2):
189–199.
Ou, J.P., Duan, Y.B. and Liu, H.Y. 1994. Random seismic action and its statistics. *Journal of Harbin Building
Engineering and Architecture* 27(5): 1–10.
Ou, J.P. & Duan, Y.B. 1995. Seismic reliability analysis and
optimum design of high-rise building structures. *Earthquake Engineering and Engineering Vibration* 15(1):
1–13.
Rosenblueth, E. 1975. Point estimates for probability
moments. *Proceedings of the National Academy of Science* 72(10): 3812–3814.
Zhao, Y.G. & Ono, T. 1998. System reliability evaluation of ductile frame structures. *Journal of Structural
Engineering* 124(6): 678–685.
Zhao, Y.G. & Ono, T. 2000. New point estimates for probability moments. *Journal of Engineering Mechanics* 126(4):
433–436.
Zhao, Y.G. & Ono, T. 2001. Moment methods for structural
reliability. *Structural Safety* 23(1): 47–75.

Safety, Reliability and Risk of Structures, Infrastructures and Engineering Systems – Furuta, Frangopol & Shinozuka (eds)
© 2010 Taylor & Francis Group, London, ISBN 978-0-415-47557-0

Probabilistic seismic assessment of concrete dams

A. Lupoi
Department of Structural and Geotechnical Engineering, Sapienza University of Rome, Italy

C. Callari
Facoltà di Ingegneria, University of Molise, Italy

ABSTRACT

Dam engineering has a long tradition in Europe where hundreds of dams have been built in the first half of the last century. In Italy, for example, the average age of concrete dams is over 50 years. Many of these dams were designed neglecting or underestimating the seismic action and employing out-of-date analysis methods.

In view of the relevant economic losses due to eventual limitations of reservoir operation or dam dismission, the seismic assessment of existing dams, complementary to the ordinary maintenance and preservation of such old structures, is a topic of great interest and urgency.

Motivated by the above considerations, a methodology for the seismic assessment of existing dams employing state-of-the-art methods is investigated in the present work. Two main issues need to be tackled: the physical complexity of the system and the large uncertainties affecting the problem.

In fact, a dam system consists of at least three components: dam body, foundation rock mass and reservoir water, all contributing to the dynamic response of the system. The interaction between dam, foundation and water, whose influence on the dynamic behavior of the dam structure is not negligible, can be properly taken into account employing finite element discretizations of all the three aforementioned components (Callari and Abati, 2008).

For what concerns the second issue, the primary source of uncertainties in any seismic assessment is the variability of the ground motion. In the case of dams, the height of the water reservoir is a further source of uncertainty which stems from the action side. Dams are often characterized also by significant uncertainties of structural nature, typically due to lack of data regarding for example dam geometry and site geology as well as to inaccuracies in models describing capacities.

This work presents the main results of the application in the seismic assessment of dams of a probabilistic method based on the approach developed for building and bridge structures by Lupoi et al. (2006). The seismic response of the structure is estimated from a reduced number of dynamic time-history analyses; the system fragility curves are then obtained from a standard Monte Carlo simulation procedure. The procedure is able to account for uncertainties both in external (seismic) action and in system properties. To the author's best knowledge, few other examples of probabilistic seismic analysis of dams are available in literature (Ellingwood and Tekie 2003).

The methodology has been applied to the case of Kasho Dam, a 46.4 m high concrete gravity dam which experienced the Western Tottori earthquake without serious damage. The assessment has been carried out with respect to an operational limit state; for which the "critical" failure mechanisms have been identified and numerically evaluated, both in terms of demand definition and capacity evaluation.

Available monitoring data have been used to validate numerical analyses and the definition of the critical mechanisms (cracking/sliding at dam foundation, cracking at dam upstream face, excessive drift deformation, etc.).

Results of the seismic assessment are finally presented in terms of fragility curves for the whole system and for each failure mechanism.

REFERENCES

Callari C., Abati A. (2008), Finite element methods for unsaturated porous solids and their application to dam engineering problems, *Computers and Structures*, in press.

Lupoi G., Franchin P., Lupoi A. and Pinto P.E. 2006. Seismic Fragility Analysis of Structural Systems, *Journal of Engineering Mechanics ASCE*, 132(4): 385–395.

Ellingwood B. and Tekie P.B. 2003. Seismic fragility Assessment of concrete gravity dams, *Earthquake Engineering and Structural Dynamics*, 32: 2221–2240.

Safety, Reliability and Risk of Structures, Infrastructures and
Engineering Systems – Furuta, Frangopol & Shinozuka (eds)
© 2010 Taylor & Francis Group, London, ISBN 978-0-415-47557-0

Structural reliability analysis using subset simulation and neural networks

D.G. Giovanis, V. Papadopoulos, N. Lagaros & M. Papadrakakis
Institute of Structural Analysis & Seismic Research, National Technical University of Athens, Athens, Greece

ABSTRACT

This paper examines a methodology for computing the probability of structural failure by combining neural networks (NN) and Subset Simulation (SS). A methodology is proposed which exploits the capability of a NN to approximate the structural performance with acceptable precision, allowing the computation of response parameters at a very low computational cost. A better estimation of the failure region is been succeeded, with no additional computational cost, which leads to a more precise estimation of the failure probability.

1 INTRODUCTION

In the present study a methodology for computing the probability of structural failure by combining neural networks (NN) and Subset Simulation (SS) is being proposed. The goal is to determine the failure region more precisely with the aid of neural networks. The proposed procedure where NN are incorporated into the framework of the SS method enhances the computational performance of the method. The basic concept is to train the neural networks according to the samples generated within the Markov Chains. Once properly trained, a NN allows the determination of the structural performances with a very small number of operations and at a fraction of the cost of the corresponding structural analysis. The proposed methodology takes advantage of the special characteristics of the SS methodology in order to train effectively the NN network, allowing this way for very accurate predictions of the failure region and consequently of the failure probability.

The NN-based SS method is demonstrated by means of a numerical example, where the efficiency of the NN implementation is being showed. The performance function $f(x)$ is a mathematical one. A failure event occurs if the performance function exceeds a certain value. The partial failure probabilities are predefined to specific values for each conditional SS level. The proposed methodology takes advantage of the special characteristics of the SS methodology in order to train effectively the NN, allowing this way for very accurate predictions of the failure region and consequently of the failure probability. It is shown that with the proposed approach, the accuracy of a MCS estimate of the probability of failure is reached with orders of magnitude less required samples.

REFERENCES

Schuëller G.I and Pradlwarter H.J., 2007, Benchmark study on reliability estimation in higher dimensions of structural systems – An overview, *Struct Saf* 29 , pp. 167–182.
Au SK, Beck JL, 1999, A new adaptive importance sampling scheme, *Struct Safety* 21: 135–58, (1999)
Au SK, Beck JL, 2001, Estimation of small failure probabilities in high dimensions by subset simulation. *Probabilistic Engineering Mechanics*, 16(4), 263–277.
Papadrakakis M., Papadopoulos V. and Lagaros N.D., 1996, Structural reliability analysis of elastic-plastic structures using neural networks and Monte Carlo simulation, *Comp. Meth. Appl. Mech. Eng.* 136, 145–163.

Safety, Reliability and Risk of Structures, Infrastructures and
Engineering Systems – Furuta, Frangopol & Shinozuka (eds)
© 2010 Taylor & Francis Group, London, ISBN 978-0-415-47557-0

Reliability analysis of idealized dam and power intake structure

M.C. Westberg

Department of Structural Engineering, Lund University, Lund, Sweden

ABSTRACT

International Commission on Large Dams (ICOLD) defines a safe dam as "a dam with appropriate reserves, taking into account all reasonably imaginable scenarios of normal utilization and exceptional hazard which it may have to withstand during its life".

Concrete dams are, in Sweden as well as world wide, designed based on safety factors (Level 0 according to JCSS). A research project aiming at introducing structural reliability analysis for concrete dams is in progress. This paper describes the failure modes, the random variables defining them and gives two examples of application.

The limit states functions analyzed are sliding and overturning. Since the headwater level is usually kept constant close to retention water level (rwl), and only in exceptional situations exceed this, each failure mode has to be analyzed for two cases; at rwl and above rwl separately. To get the reliability of the structure the resulting four failure modes are combined as a series system.

The target reliability used is 4.8/year and is taken from the Swedish design guideline for structures, BKR.

The random variables are defined as in a licentiate thesis (Westberg), and a short summary is given in the paper.

– Large uncertainties are associated with the shear strength (described by cohesion and friction angle). A sensitivity analysis is performed for friction angle and cohesion.
– Concrete density is defined according to JCSS.
– When the headwater level exceeds rwl it is assumed possible to describe by an exponential distribution and the parameters are found by information of water levels and return periods. This information is given by design flood calculation.
– The uplift pressure is found by using a "geostatistical approach" where the hydraulic conductivity beneath the dam is assumed to have certain mean value, variance and spatial range. Statistical distributions of uplift force and moment is derived from approximately 1000 realizations of the random field with these properties, using a FE analysis. This method has proved itself useful.
– Ice load, concrete volume, base area, density of soil and earth pressure are also random variables.

Two examples will be given of reliability analysis of concrete dams. Both examples are analyzed using computer software COMREL.

The first example is an idealized dam, where dimensions are chosen so that the structure fulfils the Swedish dam safety requirements (RIDAS). This case is analyzed for different assumptions of drain effectiveness to investigate the safety index of a "safe" dam in different conditions. The results show that when designed without drains, sliding is the most likely failure mode and zero cohesion (assumed in the Swedish dam safety guideline) gives very low safety index. For a dam designed with drains, where the drains are malfunctioning, the most likely failure mode is overturning if the cohesion is medium-high. Cohesion and friction angle dominates the uncertainties, and research should be focused on describing these parameters as well as the drain effectiveness that is also of importance.

The second example is an intake structure in a Swedish river where rock anchors had to be installed to fulfil the stability requirements. The structural reliability analysis was mainly performed to demonstrate the methodology. The analysis shows that the stability is sufficient if the cohesion is high. The uplift pressure is of importance, but the largest sensitivity values are related to cohesion and friction angle. The tendons (installed to increase safety and decrease non-compressive stresses) gives only a small contribution to the safety, but significantly reduce the base area in "non-compression". Failure will, most likely, occur due to sliding at normal water levels. The safety of this intake structure is probably sufficient, but some more information of shear strength is necessary.

The conclusions are that there are problems that need to be solved to use structural reliability analysis for concrete dams, but the examples shows that the analysis is useful and gives important information.

The use of risk analysis and assessment for dams is increasing and structural reliability analysis may give input to quantitative assessment procedure.

Mini-Symposia (MS07) Vulnerability Assessment and Risk-Based Life-Cycle Analysis and Design

Safety, Reliability and Risk of Structures, Infrastructures and
Engineering Systems – Furuta, Frangopol & Shinozuka (eds)
© 2010 Taylor & Francis Group, London, ISBN 978-0-415-47557-0

Vulnerability curves to compare retrofit solutions of a medieval civic tower

S. Casciati
ASTRA Department, University of Catania, Siracusa, Italy

L. Faravelli
Department of Structural Mechanics, University of Pavia, Pavia, Italy

ABSTRACT

The results reported in (Faravelli and Casciati, 2008)
are based on the implementation of a numerical model
for the medieval civic tower shown in Figure 1, which
was also the object of an experimental campaign. The
next step in the study of the tower structural behaviour
consists of using the experimental data for damage
identification and localization, which also relies on
visual inspections.

Several design situations, including different
retrofitting policies and/or devices, are then considered. By repeating the analyses on the modified

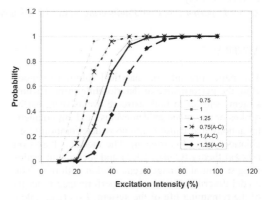

Figure 2. Vulnerability curves obtained with the excitation
in one of the two orthogonal directions, either B-D or A-C
with the von Mises stress criterion as limit state: damaged
model.

numerical models including the retrofitting effects,
vulnerability curves are built for all the considered
design situations. The amount of translation to the left
(or, alternatively, the amount of clockwise rotation)
with respect to the original curves represents an index
of the retrofit effectiveness and provides the basis for
a cost-benefit optimization.

The ultimate goal of comparing vulnerability estimates (see Figure 2) justifies the adoption of a linear
model which is not representative of the damage phenomenon, but it is sufficient to classify the damaged
state of the structure as either severe, moderate, or
light. On the other hand, the uncertainties related
to the constitutive law of ancient structures prevent
one from elaborating a sophisticated nonlinear model,
which is instead replaced by an equivalent homogenized material. This choice, together with providing
a simple to handle numerical tool, provides a preliminary understanding of the structural behaviour that
can be afterwards confirmed by repeating the analyses on a model that includes the crack discontinuities
identified from the visual inspections.

REFERENCES

Faravelli, L and Casciati, S. 2008, Vulnerability assessment
for medieval civic towers, accepted for publication in
Structure and Infrastructure Engineering.

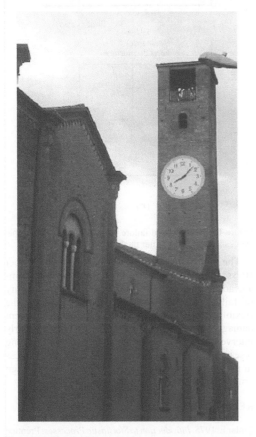

Figure 1. The Soncino civic tower

Safety, Reliability and Risk of Structures, Infrastructures and
Engineering Systems – Furuta, Frangopol & Shinozuka (eds)
© 2010 Taylor & Francis Group, London, ISBN 978-0-415-47557-0

Probabilistic modeling of deteriorating systems subject to extreme events

Department of Civil and Environmental Engineering, Universidad de Los Andes, Bogotá, Colombia

G.-A. Klutke

Department of Industrial and Systems Engineering, Texas A&M University, College Station, TX, USA

D.V. Rosowsky

Zachry Department of Civil Engineering, Texas A&M University, College Station, TX, USA

ABSTRACT

This paper presents an analytical solution to modeling the performance of deteriorating infrastructure systems subject to both extreme events (e.g., earthquakes or any other sudden events) and environmental actions that cause continuous deterioration (e.g., chloride-induced corrosion). The model is based on renewal theory (Çinlar, 1975) and focuses on evaluating structural damage accumulation with time. The model describes the structural performance in terms of the remaining life of the system, $V(t)$ (e.g., inter-story drift). The probability of exceeding a predefined limit performance value, for one life-cycle, can be written as:

$$P(V(t+dt) \leq s^* \,|\, V(t) > s^*) =$$

$$\sum_{n=1}^{\infty} \left(\int_{V_1(t,s^*,n)}^{\infty} \lambda(t)dG(y) \right) P(N=n,t) \tag{1}$$

where $\lambda(t)$ is the shock process occurrence rate, N is the number of shocks; s^* is the threshold value (in remaining life units) that defines the need for an intervention; and $dG(y)$ describes the probability of having a shock size between y and $y+dy$. The lower integral limit in equation 1 is defined by:

$$V_1(t,s^*,n) =$$

$$\int_0^{u_0-s^*-A_p(t)} (u_0 - s^* - A_p(t) - y)dG^{(n)}(y) \tag{2}$$

where u_0 is the remaining life at the beginning of the cycle; $A_p(t)$ is a progressive deterioration function; and $dG^{(n)}(y)$ is the n-th convolution of $G(y)$ with itself. For structures that have been in operation for a long time, an asymptotic solution can be found averaging equation 1 with respect to the length of a cycle.

The paper presents the derivation of these equations and compares the solution for the cases of failure after a single or multiple shocks; and with and without progressive deterioration. The results are summarized in Figure 1 where the importance of including deterioration caused by both extreme events and environmental factors is shown.

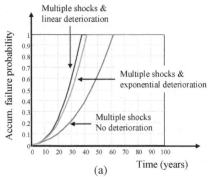

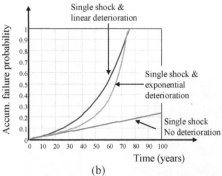

Figure 1. Accumulated failure probability for (a) Multiple shocks and (b) failure after the first shock.

The illustrative example discussed in the paper stresses the importance of taking into account damage accumulation in the estimation of the probability of failure. Furthermore, the paper emphasizes the flexibility of the model to accommodate different damage limit states and to define policies for early intervention (e.g., preventive maintenance). Ongoing research is currently directed to include randomness in progressive deterioration, defining optimal design criteria and intervention policies.

REFERENCES

Çinlar E. 1975. *Introduction to Stochastic Processes*. Prentice Hall. New Jersey.

Safety, Reliability and Risk of Structures, Infrastructures and Engineering Systems – Furuta, Frangopol & Shinozuka (eds)
© 2010 Taylor & Francis Group, London, ISBN 978-0-415-47557-0

Life-cycle cost optimal design of passive dissipative devices for seismic risk mitigation

A.A. Taflanidis
Civil Engineering and Geological Sciences Department, University of Notre Dame, IN, USA

J.L. Beck
California Institute of Technology, Division of Engineering and Applied Sciences, CA, USA

ABSTRACT

The cost effective performance of structures has long been recognized to be an important topic in the design of civil engineering systems. This design approach requires proper integration of (i) methodologies for treating the uncertainties related to natural hazards and to the structural behavior over the entire life-cycle of the building, (ii) tools for evaluating the performance using socioeconomic criteria, as well as (iii) algorithms appropriate for stochastic analysis and optimization.

A complete probabilistic framework is presented in this paper for detailed *estimation* and *optimization* of the life-cycle cost of earthquake engineering systems. The focus is placed on the design of passive dissipative devices. The framework is based on a knowledge-based interpretation of probability (Jaynes, 2003), which leads to a realistic framework for formulating the design problem, and on an efficient novel approach to stochastic optimization problems (Taflanidis and Beck, 2008). The latter facilitates an efficient solution of this design problem and thus allows for consideration of complex models for describing structural performance.

A comprehensive methodology is initially discussed for earthquake loss estimation; this methodology uses the nonlinear time-history response of the structure under a given excitation to estimate the damages in a detailed, component level. A realistic probabilistic model is then presented for describing the ground motion time history for future earthquake excitations. This model establishes a direct link between the probabilistic seismic hazard description of the structural site and the acceleration time history of future ground motions. In this setting, the life-cycle cost is given by an expected value over the space of the uncertain parameters for the structural system, performance evaluation and excitation models. Because of the complexity of these models, calculation of this expected value by means of stochastic simulation techniques is adopted. This approach, though, involves an unavoidable estimation error and significant computational cost, features which make the associated optimization challenging. An efficient framework, consisting of two stages, is presented for the optimization in such stochastic design problems. The first stage implements a novel approach, called Stochastic Subset Optimization (SSO), for efficiently exploring the sensitivity of the objective function to both the design variables as well as the model parameters. Using a small number of stochastic analyses SSO iteratively identifies a subset of the original design space that has high plausibility of containing the optimal design variables and additionally consists of near-optimal solutions. The second stage, if needed, adopts some other stochastic optimization algorithm to pinpoint the optimal design variables within that subset. All information available from the first stage is exploited in order to improve the efficiency of the second optimization stage.

An example is presented that considers the retrofitting of a four-story reinforced concrete office building with viscous dampers. Complex system, excitation and performance evaluation models are considered, that incorporate all important characteristics of the true system and its environment into the design process. The results illustrate the capabilities of the proposed framework for improving the structural behavior in a manner that is meaningful to its stakeholders (socio-economic criteria), as well as its capabilities for computational efficiency and the treatment of complex analysis models.

REFERENCES

Jaynes, E.T. 2003. *Probability Theory: The logic of science.* Cambridge University Press, Cambridge, UK.
Taflanidis, A.A. & Beck, J.L. 2008. An efficient framework for optimal robust stochastic system design using stochastic simulation, *Computer Methods in Applied Mechanics and Engineering*, 198(1): 88–101.

Safety, Reliability and Risk of Structures, Infrastructures and Engineering Systems – Furuta, Frangopol & Shinozuka (eds)
© 2010 Taylor & Francis Group, London, ISBN 978-0-415-47557-0

Vulnerability assessment for cable-stayed bridges under blast loads

D. Yan & C.C. Chang

Department of Civil and Environmental Engineering, Hong Kong University of Science and Technology, Hong Kong

ABSTRACT

Vulnerability of a structure under terrorist attack can be regarded as the study of its strength and robustness against blast-induced loads. A structure is vulnerable if a small damage can trigger disproportionately large consequence and lead to a cascade of failure events or even collapse. The performance of structural vulnerability depends upon factors such as external loading condition and structural properties. As many of these factors are random in nature, it is necessary to develop a vulnerability assessment technique in the probabilistic domain. In this study, one such assessment framework is proposed for cable-stayed bridges. The framework consists of two stages of analysis: determining the probability of direct damage and assessing the subsequent probability of collapse. In the first stage assessment, damage of bridge component is defined as the exceedence of a predefined limit state such as displacement or yielding. The damage probability is obtained through a stochastic finite element analysis and the first-order second-moment reliability method. The second stage assessment further calculates the probability of collapse due to direct damage of some

component via an event tree approach. The proposed assessment methods are illustrated on a hypothetical single-tower cable-stayed bridge. It is seen that the proposed methods provide a quantitative tool for analyzing the vulnerability performance of cable-stayed bridges under terrorist attack.

REFERENCES

Agarwal, J., Blockley, D., and Woodman, N. 2003. Vulnerability of structural systems. *Structural Safety*, 25(3): 263–286.

Elnashai, A. S., Borzi, B., and Vlachos, S. 2004. Deformation-based vulnerability functions for RC bridges. *Structural Engineering and Mechanics*, 17(2): 215–244.

Florek, J. R., and Benaroya, H. 2005. Pulse-pressure loading effects on aviation and general engineering structures — review. *Journal of Sound and Vibration*, 284(1-2): 421–453.

Pandey, A. K., Kumar, R., Paul, D. K., and Trikha, D. N. 2006. Non-linear response of reinforced concrete containment structure under blast loading. *Nuclear Engineering and Design*, 236(9): 993–1002.

Safety, Reliability and Risk of Structures, Infrastructures and
Engineering Systems – Furuta, Frangopol & Shinozuka (eds)
© 2010 Taylor & Francis Group, London, ISBN 978-0-415-47557-0

Simplified fragility analysis methods for minimum life-cycle cost seismic design of generic structures

D.G. Lu, X.H. Yu, M.M. Jia & G.Y. Wang
School of Civil Engineering, Harbin Institute of Technology, Harbin, China

ABSTRACT

Recently, life-cycle cost (LCC) analysis of civil engineering structures and infrastructure systems, and the design optimization based on LCC, have become the key ingredients of life-cycle civil engineering (LCCE). Meanwhile, minimization of the expected life-cycle cost under low-probability and high-consequence seismic hazards based on benefit-cost criteria is forming into the ultimate goal of performance-based seismic design.

LCC of civil engineering structures under earthquakes includes initial cost and the total loss expectation under multi-level damage states. It is needed to estimate the failure probabilities of multi-level seismic damage states in estimation of the loss expectation, thus how to effectively calculate these probabilities is one of the key issues in successfully applying life-cycle cost design theory to real-world civil engineering practice.

In this paper, the theoretical foundations of seismic fragility theory are built for evaluation of seismic damage state probabilities of structures, in which the concepts and definitions of limit state fragility and damage state fragility are presented. Two practical analysis approaches for evaluation of seismic damage state probabilities of structure are analyzed from the viewpoint of seismic fragility theory. It is found that the first approach is a semi-theoretical and semi-empirical method, which considers the uncertainty in the macro-scope seismic capacity of structures; and that the second approach is a simplified seismic fragility analysis method, which doesn't consider the uncertainty in the macro-scope seismic capacity of structures. The two approaches are compared based on the computed results of seismic damage state probabilities of structures, and the suggestions of selection of two simplified seismic fragility methodologies for the optimal fortification level decision-making and minimum life-cycle cost design of structures are given.

REFERENCES

Lu, D.G., et al. 2006. Finite element reliability method in structural seismic fragility analysis. *Jounal of Application Base and Engineering Science* 15(suppl): 264–272.

Lu, D.G., et al. 2008. Minimum life-cycle cost design and optimal earthquake intensity decision-making of structures based on finite element reliability analysis. In Biondini F & Frangopol DM (Eds.), Life-Cycle Civil Engineering: *Proceedings of the International Symposium on Life-Cycle Civil Engineering* (IALCCE'08), Varenna, Lake Como, Italy, June 10–14, CRC Press – Taylor & Francis Group, 361–366.

Lu, D.G., Li G., and Wang, G.Y. 2007. Optimal fortification load decision-making and life-cycle cost design for structures according to Chinese codes. In Cho, H.-N., et al. (Eds.), Life-Cycle Cost and Performance of Civil Infrastructure Systems: *Proceedings of the 5th International Workshop on Life-Cycle Cost Analysis and Design of Civil Infrastructure Systems* (LCC5), Seoul, Korea, Oct. 16-18, 2006. London: Talyor & Francis: 239–248.

Lu, D.G. & Wang, G.Y. 2001. Decision-making method of optimal fortification level for structures based on damage performance. *China Civil Engineering Journal* 30(6): 12–19.

Lu, D.G., Zhang, P., and Wang, G.Y. 2002. Minimum total life cycle cost design of structures: Principle and method. *The Second China-Japan-Korea Joint Symposium on Optimization of Structural and Mechanical Systems* (CJK-OSM2), Busan, Korea, TS3-2: 129–134.

Wang, G.Y. 1997. Optimal fortification load and reliability for disaster resistant engineering. *China Civil Engineering Journal* 34(1): 44–49.

Wang, G.Y., et al. 1999. *The Applicable Methods of Engineering System Optimal Design.* Beijing: Chinese Architecture & Building Press.

Wang, G.Y., Ji, T.J., and Zhang, P. 2003. An optimal design for total lifetime cost of structures. *China Civil Engineering Journal* 36(6): 1–6.

Wang G.Y., & Lu, D.G. 2001. Optimal fortification load and reliability of structures. In Spencer and Hu, (Eds.), *Earthquake Engineering Frontiers in the New Millennium,* Swets and Zeitlinger, Lisse, The Netherlands: 371–376.

Wang, G.Y, Lu, D.G., and Gu, P. 1999. *Optimal Fortification Load Decision-Making for Disaster Resistant Engineering.* Beijing: Science Press.

Safety, Reliability and Risk of Structures, Infrastructures and
Engineering Systems – Furuta, Frangopol & Shinozuka (eds)
© 2010 Taylor & Francis Group, London, ISBN 978-0-415-47557-0

A procedure for performance-based wind engineering

M. Ciampoli, F. Petrini & G. Augusti
Sapienza – Università di Roma, Italy

ABSTRACT

A modern approach to wind engineering must consider performances as key objectives of structural design. Hence optimal design procedures should be developed in the framework of Performance-Based Design or, better, Performance-Based Wind Engineering (PBWE): the relevant structural performance requirements should be satisfied with a sufficiently high probability throughout the whole life-cycle.

Many studies have already been carried out on this topic. In Italy, Augusti and Ciampoli (2006) suggested the extension of the PEER approach (Porter 2003) into PBWE; Petrini *et al.* (2008 a, b) applied the risk assessment procedure to the assessment of (a) the fatigue damage of the hangers and (b) the serviceability and ultimate performances of a proposed long span suspension bridge.

The goal of this paper is to give some further contribution towards the development of a probabilistic procedures for the application of Performance-Based Design concepts to wind engineering.

Namely, a PBWE procedure is proposed that can be summarized in intermediate steps aimed at:

- defining the Aeolian hazard at the site, in terms of wind intensity and/or parameters of the wind velocity field;
- analyzing the structural response;
- defining and evaluating indicators of the structural damage (identified with unacceptable performances);
- defining the decisional variables that are appropriate to quantify the performances required of the structure, in terms of damages;
- evaluating the structural risk by the probabilistic characterization of the decision variables;
- optimizing design, i.e. minimizing risk, by appropriate techniques of decision analysis.

The decision variables that quantify the performances are distinguished between those corresponding to possible consequences on structural and personal safety (low performances), and to effects on serviceability and comfort (high performances).

As an example, the proposed procedure is applied to the assessment of the high- and low-performance risks of the same long-span suspension bridge examined in Petrini *et al.* (2008a, b).

For high performances three response parameters have been considered, namely vertical acceleration, rotational velocity and longitudinal acceleration of a typical cross section. "Avoiding the flutter instability" has been the low-performance criterion, and this has been recognized through the damping of the vertical displacement of the mid-span section. The probabilistic characterizations of all the relevant parameters have been obtained by Monte Carlo simulations.

The performed analytical and numerical analyses prove that PBWE is feasible, but to make it more reliable it is essential to improve the probabilistic description of the parameters of the wind field at the site and the phenomena that represent the interaction between the wind actions and the structure. This will require much further research work.

REFERENCES

Augusti, G. & Ciampoli, M. 2006. First steps towards Performance-based Wind Engineering. In: *Performance of Wind Exposed Structures: Results of the PERBACCO project* (G. Bartoli, F. Ricciardelli, A. Saetta, V. Sepe eds.), Firenze University Press; 13–20.

Petrini, F., Bontempi, F. & Ciampoli, M. 2008a. Performance-based wind engineering as a tool for the design of the hangers in a suspension bridge. *Proc. Fourth International ASRANet colloquium*, Athens, Greece, Abstract Volume p. 25, paper in CD-ROM.

Petrini, F., Ciampoli, M. & Augusti, G. 2008b. Performance-based Wind Engineering: risk assessment of a long span suspension bridge. *Proc. IFIP WG7.5 Conference*, Toluca, Mexico, (in press).

Porter, K.A. 2003. An Overview of PEER's Performance-Based Engineering Methodology. *Proc of the Ninth International Conference on Applications of Statistics and Probability in Civil Engineering* ICASP9. San Francisco, CA, USA; Millpress Rotterdam.

Mini-Symposia (MS15) System Identification and Structural Health Monitoring
[Session Organized on Behalf of IASSAR Subcommittee SC5]

Safety, Reliability and Risk of Structures, Infrastructures and
Engineering Systems – Furuta, Frangopol & Shinozuka (eds)
© 2010 Taylor & Francis Group, London, ISBN 978-0-415-47557-0

Probability logic, model uncertainty and robust predictive system analysis

James L. Beck & Sai Hung Cheung
California Institute of Technology, Pasadena, CA, USA

ABSTRACT

Probability can be viewed as a multi-valued logic
that extends binary Boolean propositional logic to the
case of incomplete information. The key idea is that
the probability P[b/c] of a proposition (statement) b,
given the information in proposition c, is a measure
of how plausible b is based on c. Boolean logic deals
with the special case of complete information where
the truth or falsity of b is known from c. R.T. Cox
was the first to derive the probability logic axioms
from those of Boolean logic but this interpretation of
probability has a long history and has received increasing interest recently, especially due to E.T Jaynes
who emphasized its connection with the information-
theoretic ideas stemming from C.E. Shannon's work..
It is consistent with the Bayesian point of view that
probability represents a degree of belief in a proposition. A.N. Kolmogorov's statement of the probability
axioms, which are neutral with respect to the interpretation of probability, can be viewed as a special case
where the statements refer to uncertain membership of
an object in a set. Probability logic provides a rigorous
unifying framework for treating modeling uncertainty,
along with excitation uncertainty, when using models
to predict the response of a real system.

An overview is given of the foundations of probability logic and its application to quantifying uncertainty
for robust predictive analysis of systems. A key concept is a *stochastic system model class*, which consists
of a set of probabilistic input-output predictive models for a system together with a chosen probability
distribution, the *prior*, over this set that quantifies the
initial relative plausibility of each predictive model in

the set. These probability models are viewed as representing a state of knowledge about the real system,
conditional on the available information, and not as
inherent properties of the system. Any set of dynamic
models for a system can be used to construct a model
class through *stochastic embedding* in which Jaynes'
Principle of Maximum Information Entropy plays an
important role in establishing the fundamental probability models. A model class can be used to create
both *prior* (initial) and *posterior* (updated using system data) *robust predictive models* based purely on the
probability axioms. Since there is always uncertainty
in choosing a model class to represent a system, one
can choose a set of candidate model classes. The probability axioms then lead naturally to *prior* and *posterior*
hyper-robust predictions that combine the predictions
of all model classes in the set. The posterior probabilities of the candidate model classes also give a measure
of their relative plausibility based on the data and so
can be used for model class selection.

In applications, integrals over high-dimensional
parameter spaces are usually involved that cannot
be evaluated in a straight-forward way. Useful computational tools are Laplace's method of asymptotic approximation and Markov Chain Monte Carlo
(MCMC) methods such as the Metropolis-Hastings,
Gibbs sampler and Hybrid Monte Carlo algorithms.
Parameter estimation, which selects just one predictive model in the model class, is justified via Laplace's
method if the class is globally identifiable based on
the data; if not, a posterior robust predictive analysis
should be performed using an MCMC method that
combines the probabilistic system predictions from
each model in the model class.

Safety, Reliability and Risk of Structures, Infrastructures and Engineering Systems – Furuta, Frangopol & Shinozuka (eds)
© 2010 Taylor & Francis Group, London, ISBN 978-0-415-47557-0

Comparison of different model classes for Bayesian updating and robust predictions using stochastic state-space system models

Sai Hung Cheung

Institute of Computational Engineering and Sciences (ICES), The Center for Predictive Engineering and Computational Sciences (PECOS), University of Texas at Austin, Austin, TX, USA

James L. Beck

Division of Engineering and Applied Science, California Institute of Technology, Pasadena, CA, USA

ABSTRACT

A stochastic system-based framework for Bayesian model updating of dynamic systems was presented in Beck and Katafygiotis (1998). One key concept in this framework is a *stochastic system model class* which consists of probabilistic predictive input-output models for a system together with a prior probability distribution over this set that quantifies the initial relative plausibility of each predictive model. Past applications of this framework focus on model classes which consider an uncertain prediction error as the difference between the real system output and the model output and model it probabilistically using Jaynes' Principle of Maximum Information Entropy.

In this paper, in addition to these model classes, we also consider an extension of such model classes to allow more flexibility in treating modeling uncertainties when updating state space models and making robust predictions; this is done by introducing prediction errors in the state vector equation in addition to those in the system output vector equation. The extended model classes allow for interactions between the model parameters and the prediction errors in both the state vector equation and the system output equation to give more robust predictions at unobserved DOFs. Bayesian model class selection is used to evaluate the posterior probability of model classes for the comparison of the extended model classes and the original one. To make predictions robust to model uncertainties, Bayesian model averaging is used to combine the predictions of these model classes. State-of-the-art algorithms (Cheung & Beck 2007, 2008; Ching & Chen 2007) are used to solve the computational problems involved. The importance and effectiveness of the proposed method is illustrated with examples for robust reliability updating of structural systems.

REFERENCES

Beck, J.L. & Katafygiotis, L.S. 1998. Updating models and their uncertainties. I: Bayesian statistical framework. *Journal of Engineering Mechanics* 124(4): 455–461.

Beck, J.L. & Yuen, K.V. 2004. Model selection using response measurements: A Bayesian probabilistic approach. *Journal of Engineering Mechanics* 130(2): 192–203.

Cheung, S.H. & Beck, J.L. 2007. Algorithms for Bayesian model class selection of higher-dimensional dynamic systems. *Proc. ASME 2007 Intl Design Engineering & Computers and Information in Engineering Conferences (DETC/CIE 2007)*, Las Vegas, Nevada, USA, September 4–7, 2007.

Cheung, S.H. & Beck, J.L. 2009. Bayesian model updating using Hybrid Monte Carlo simulation with application to structural dynamic models with many uncertain parameters. *Journal of Engineering Mechanics* 135(4): 243–255.

Ching, J. & Chen, Y.J. 2007. Transitional Markov Chain Monte Carlo method for Bayesian model updating, model class selection and model averaging. *Journal of Engineering Mechanics* 133(7): 816–832.

Safety, Reliability and Risk of Structures, Infrastructures and
Engineering Systems – Furuta, Frangopol & Shinozuka (eds)
© 2010 Taylor & Francis Group, London, ISBN 978-0-415-47557-0

Robust ground vibration predictions based on SASW tests

Mattias Schevenels, Sayed Ali Badsar, Geert Lombaert & Geert Degrande
Department of Civil Engineering, K.U. Leuven, Belgium

ABSTRACT

Ground vibrations in the built environment are an important issue, as they cause nuisance to people, disturbance of sensitive equipment, and damage to buildings. They are caused by a variety of sources, such as earthquakes, road and rail traffic, construction activities, and industrial machinery. Several numerical methods have been developed for the prediction of ground vibrations [1]. These methods can be used to assess the efficiency of vibration reduction measures in new or existing situations. The soil is typically modelled as a layered elastic halfspace. The dynamic properties of the soil layers have to be determined by means of in situ tests or laboratory tests.

The Spectral Analysis of Surface Waves (SASW) test is a commonly used in situ test to determine the dynamic shear modulus and the material damping ratio of shallow soil layers [2]. It is based on an in situ experiment where surface waves are generated by means of a falling weight or an impact hammer. The resulting wave field is recorded by a number of sensors at the soil's surface and used to determine the dispersion and attenuation curves of the soil. An inverse problem is finally solved to identify the corresponding soil profile. However, the dispersion and attenuation curves are insensitive to variations of the soil properties on a small spatial scale or at a large depth. The information on the soil properties provided by these curves is therefore limited. As a result, the solution of the inverse problem is non-unique: the soil profile obtained from the inversion procedure is only one of the profiles that fit the experimental data [3].

The present paper focuses on the impact of this non-uniqueness on the prediction of ground vibrations. As an example, the prediction of the vibrations due to a hammer impact on a small surface foundation is considered. First, a Bayesian approach is followed to determine an ensemble of soil profiles that fit the experimental data obtained from an SASW test. These profiles are subsequently used in a Monte Carlo simulation for the prediction of ground vibrations. Based on the variability of the results, the frequency range where the SASW test allows for robust vibration predictions is determined. The methodology developed in this paper allows for an optimization of the SASW test: it can be used to determine the sensor configuration that maximizes the robustness of vibration predictions in the frequency range of interest.

REFERENCES

[1] G. Lombaert, G. Degrande, J. Kogut, and S. François. The experimental validation of a numerical model for the prediction of railway induced vibrations. Journal of Sound and Vibration, 297(3–5): 512–535, 2006.

[2] S. Nazarian and M.R. Desai. Automated surface wave method: field testing. Journal of Geotechnical Engineering, Proceedings of the ASCE, 119(7): 1094–1111, 1993.

[3] M. Sambridge and K. Mosegaard. Monte Carlo methods in geophysical inverse problems. Reviews of Geophysics, 40(3): 1–29, 2002.

Safety, Reliability and Risk of Structures, Infrastructures and
Engineering Systems – Furuta, Frangopol & Shinozuka (eds)
© 2010 Taylor & Francis Group, London, ISBN 978-0-415-47557-0

Stochastic model of tunnel concrete degrading process

O. Maruyama & A. Sutoh
Department of Urban & Civil Engineering, Tokyo City University, Japan

T. Sato & H. Nishi
*Cold-Region Construction Engineering Research Group, Civil Engineering Research Institute for Cold Region,
Public Works Engineering Research Institute, Japan*

ABSTRACT

A resurgence of attention to and interest in system identification techniques has recently been observed among engineers in the field of structural engineering in conjunction with the rehabilitation of existing structures possibly damaged by past earthquakes and other loads. The load resisting capacity of these structures may also degraded due to aging. As a structure deteriorates or approaches its design life, the existing condition may be quite different from that of the original system. In this regard, the field of system identification has special significance in the connection with the asset management of the existing structure.

Especially, tunnel concrete degradation has become serious social problem since tips of concretes fell off in concrete structures. So far, the monitoring and maintenance of concrete structure has been done by visual inspections. Then, system identification techniques must exhibit analytical stability and numerical efficiency in identifying significant parameters indicative of deteriorating process of tunnel concrete.

The present problem is the identification of the degrading process of the tunnel concrete represented by the Ito stochastic differential equation as follows.

$$dX(t) = \beta X(t)dt + \sigma X(t)dW_1(t) \qquad (1)$$

where, β: the constant drift parameter, σ: the constant volatility parameter, $W_1(t)$: the wiener process.

The maximum likelihood estimation is employed to identify the parameters of Ito stochastic diffrential equation based on the monitoring data from existing tunnels. Then parameters in eq. (1) are obtained to maximize the following criteria.

$$\log p(X_{t0}, X_{t1}, \cdots, X_{tN}) = \log p(y_{t0}) + \log \prod_{n=0}^{N-1} p(y_{tn+1} \mid y_{tn})$$

$$(2)$$

in which, $p(X_{t0}, X_{t1}, \cdots, X_{tN})$: probability density function of inspected degrading ratio and $y_t = \ln(X_t)$.

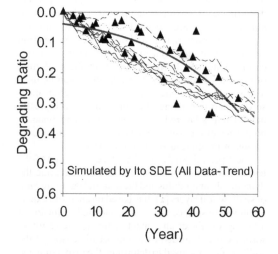

Figure 1. Simulated degrading process.

Inspected data from existing tunnels in Hokkaido Island used for the identification of Ito stochastic differential equation. Inspection has been carried out last several decades, such as crack width, crack length and crack expanse of forty-five tunnels. The results summarized as degrading ratio from 0.0(no damage) to 1.0(critical damage).

For example, figure 1 shows simulated degrading process in which inspected data obtained from existing aged tunnels indicated by symbol: Δ, also indicated reproduced degrading process by Ito stochastic differential equation. The result shows the efficiency and potential of this method.

*Safety, Reliability and Risk of Structures, Infrastructures and
Engineering Systems – Furuta, Frangopol & Shinozuka (eds)*
© 2010 Taylor & Francis Group, London, ISBN 978-0-415-47557-0

Stochastic analysis of vehicle trajectories in bend: Toward a risk indicator construction

A. Koita & D. Daucher

Université Paris-Est, LEPSIS, INRETS/LCPC (Laboratoire Central des Ponts et Chaussées), Paris, France

ABSTRACT

Most of the road safety studies and research works are based on statistics of accidents, and mainly of fatalities. However, this approach has many limitations: (i) accident on a given road section or for a given vehicle is a rare event with a very low probability, (ii) accidents mostly result of a series of causes including human errors, (iii) therefore the statistics of accident are rather poor to explain the real causes of accident related to the infrastructures and to prevent them. Therefore, it was decided to investigate more in details the concept of vehicle trajectories.

Trajectories of vehicles are defined not only as the path followed on the surface of the road, but as realizations of stochastic process of the following type:

$$\widetilde{U} : T \times \Omega \rightarrow \mathrm{IR}^6$$

$$(t, \omega) \mapsto \widetilde{U}(t, \omega) = (x, y, \dot{x}, \dot{y}, \ddot{x}, \ddot{y}, \ddot{z})(t, \omega)$$

These trajectories constitute a good indicator of the effective behaviours of the vehicles and drivers in their context and in interaction with the infrastructure. They can also reveal the use of the infrastructures by the vehicles.

In the literature, the trajectories modeling works are on the one hand in the geometrical model based on the interpolation formalism, on the other hand, in the dynamic models of vehicles which are based on the representation in system of state of the vehicle equations motion. Many simplifying assumptions are carried out on these models, because the mechanical system is complex, much of difficulties of interactions modeling between the vehicle and the road.

This paper presents a stochastic analysis of experimental trajectories in bend within a suitable mathematical framework.

After a first part on proposed analysis of the problem, we define vehicle's trajectory as a vectorial stochastic process. Then, we used classification (or clustering) methods on the experimental observed trajectories in order to identify several families (class) of trajectories. Afterwards, a functional filter (or distance) judiciously chosen are used to project on IR the

Figure 1. Sample of a figure caption.

set of the observed trajectories in order to get a stationary scalar process. For that, Mahalanobis distance and Euclidean distance are used. Lastly, we characterize a process governing each class (the process of the deviation to the average of each class).

This work is a precondition to the construction of an indicator of risk. This work is supported by the French Ministry of Transport.

REFERENCES

Brossard J. P. & al. 2006. Vehicle dynamic and complex system modelling. INSA Lyon. France

Daucher, D. & Fogli, M. and Clair, D. 2006. Modelling of complex dynamical behaviours using a state representation technique based on a vector ARMA approach. Probabilistic Engineering Mechanics, Volume 21: pages 73–80

Kree, P. and Soize, C. 1983. Mathematics of random phenomena. Dordrecht

Lemaire, M. and al. 2005. Fiabilité des structures, Couplage mécano-fiabiliste statistique.

Nakache J. P. and Confais J. 2005. Approche pragmatique de la classification, Technip

Mini-Symposia (MS16) Risk Assessment of Lifeline Networks and Decision Support

Safety, Reliability and Risk of Structures, Infrastructures and
Engineering Systems – Furuta, Frangopol & Shinozuka (eds)
© 2010 Taylor & Francis Group, London, ISBN 978-0-415-47557-0

Seismic topology optimization of lifeline systems

Jie Li & Wei Liu
Department of Building Engineering, Tongji University, Shanghai, China
State Key Laboratory of Disaster Reduction in Civil Engineering, Tongji University, Shanghai, China

ABSTRACT

With the development of modern cities, lifeline systems, including water distribution, gas supply and power networks, play more and more important roles to urbanites' everyday life. The investigations of many previous earthquakes indicate that the seismic performances of lifeline systems will directly determine not only the property losses and casualties during the disasters, but also the recovery speed afterwards. However, nearly all of the lifeline systems suffered serious damages during previous strong earthquakes. For example, in 1995 Kobe earthquake, the main gas supply network suffered extensive damages. The number of leaks or breaks was as high as 5190. As a result, about 857,000 customers encountered gas supply stoppage and the secondary disaster caused by fire led to even higher losses. It took about three months for the gas supply network to be fully recovered.

Consequently, many researchers focus on the seismic reliability analysis of lifeline networks, including elements seismic analysis and networks seismic analysis. However, it should be noted that the seismic analysis is not the ultimate goal of lifeline engineering, whereas the analysis-based design and update is. As lifeline systems are usually distributed as large networks, the network topology optimization provides a useful tool for this purpose.

In this paper, a procedure to obtain the least-cost topology of lifeline system is presented. Taking system cost as the optimization objection and system seismic reliability as the optimization constraint, a topology optimization model is established where the optimization parameter is network topology structure. In order to speed up the optimization process, element probabilistic and investment importance factor is introduced based on recursive decomposition algorithm. As solving this model is a typical combinatorial optimization problem, three approaches, genetic algorithm (GA), simulated annealing algorithm (SAA) and simulated annealing genetic algorithm (SAGA), are used to solve this optimization model. When GA is used for this problem, a generation including many genes is initially created with each gene representing a network. Then by using selection, crossover and mutation operators, a new generation of networks is evolved. The fitness of each gene determines whether it will survive or not. After a number of iterations or when some criteria are met, a near-global optimal solution could usually be found. SAA takes a network topology as its current solution and produce a new solution by perturbing. If the perturbation result is an improved solution, it is accepted and the current solution is updated accordingly. Otherwise, it can also be accepted at a probability that is determined by the corresponding parameters. The perturbations and updates repeat until some criteria are met. Replacing the mutation operator in GA with perturbations and updates in SAA, SAGA is established to solve this topology optimization model. In order to compare the efficiency of these algorithms, two fictitious and a real network are evaluated.

REFERENCES

Li Jie. 2005. Lifeline earthquake Engineering-Basic Method and Application (in Chinese). *Beijing: Science Press.*

Li Jie, He Jun, 2002. A recursive decomposition algorithm for network seismic reliability evaluation, *Earthquake Engineering and Structural Dynamics*; 31: 1525–1539

Li J, Liu W Bao Y.F. 2008. Genetic algorithm for seismic topology optimization of lifeline network systems, *Earthquake Engineering and Structural Dynamics.*

Holland JH, 1975. Adaptation in neural and artificial systems, *Ann Arbor: University of Michigan Press*

Kirkpatrick S, Gelatt Jr C D, Vecchi M.P. 1983. Optimization by simulated annealing. *Science*, 220: 671 ∼ 680.

Safety, Reliability and Risk of Structures, Infrastructures and Engineering Systems – Furuta, Frangopol & Shinozuka (eds)
© 2010 Taylor & Francis Group, London, ISBN 978-0-415-47557-0

Methodology for identifying near-optimal interdiction strategies for a power transmission system

V.M. Bier, N. Haphuriwat & W. Magua
Department of Industrial and Systems Engineering, University of Wisconsin-Madison, Madison, WI, USA

E.R. Gratz
General Electric – Technology Infrastructure, Wauwatosa, WI, USA

K.R. Wierzbicki
Department of Electrical and Computer Engineering, University of Wisconsin-Madison, Madison, WI, USA

ABSTRACT

Previous methods for assessing the vulnerability of complex systems to intentional attacks or interdiction have either not been adequate to deal with systems such as electricity transmission, in which flow readjusts in response to the interdiction, or have been complex and computationally difficult. We propose a relatively simple, inexpensive, and practical method ("Max Line") for identifying promising interdiction strategies in such systems. The method is based on a greedy algorithm in which, at each iteration, the transmission line with the highest load is interdicted. We apply this method to sample electrical transmission systems from the Reliability Test System developed by the Institute of Electrical and Electronics Engineers, and compare our method and results with those of other proposed approaches for vulnerability assessment. We also study the effectiveness of protecting those transmission lines identified as promising candidates for interdiction. These comparisons shed light on the relative merits of the various vulnerability assessment methods, as well as providing insights that can help to guide the allocation of scarce resources for defensive investment.

*Safety, Reliability and Risk of Structures, Infrastructures and
Engineering Systems – Furuta, Frangopol & Shinozuka (eds)
© 2010 Taylor & Francis Group, London, ISBN 978-0-415-47557-0*

Efficient sampling techniques for seismic risk assessment of lifelines

N. Jayaram & J. W. Baker

Department of Civil and Environmental Engineering, Stanford University, CA, USA

ABSTRACT

Seismic risk assessment of lifelines is considerably
more complicated than that of a single structure on
account of the geographical spread of lifelines. Lifeline risk assessment requires knowledge about ground-
motion intensities at multiple sites. Further, the link
between the ground-motion intensities and lifeline performance is typically not available in closed form.
These complications render analytical risk assessment
tools insufficient for a probabilistic study of lifeline
performance. The current study proposes an efficient
simulation-based lifeline risk assessment framework
based on importance sampling (IS). In this framework,
'important' ground-motion fields are preferentially
sampled, and their impacts on lifeline performance
are studied. Important ground-motion fields are generated by preferentially sampling large magnitude
earthquakes and above-average ground motions corresponding to the sampled earthquakes. The study
proposes IS density functions that can be used for such
preferential sampling. The study also suggests techniques that can be used to estimate the parameters of
these sampling densities.

The proposed IS framework is used to evaluate the
seismic risk of an aggregated form of the San Francisco bay area transportation network. The accuracy
of the proposed method is demonstrated by showing
that the risk estimates obtained using the IS framework
match those obtained using random Monte Carlo simulation (MCS). It is also shown that the IS approach
can produce risk estimates that are comparable in
accuracy to results from the MCS approach while
using roughly one-hundredth the number of realizations . Finally, the study shows that the uncertainties
in the ground-motion intensity and the spatial correlations between ground-motion intensities at various
sites must be modeled in order to obtain unbiased
estimates of lifeline risk.

Safety, Reliability and Risk of Structures, Infrastructures and Engineering Systems – Furuta, Frangopol & Shinozuka (eds)
© *2010 Taylor & Francis Group, London, ISBN 978-0-415-47557-0*

Empirical evidence from Hurricane Katrina on the failure potential of bridges under storm surge loading

J.E. Padgett & A. Spiller
Rice University, TX, USA

ABSTRACT

Susceptibility of coastal bridges to damage during hurricane induced storm surge has been evidenced along the Gulf Coast of the United States in 2005 and throughout historical record. This poses a significant threat to the safety of nation-wide transportation systems, effectiveness of post-event emergency response and recovery activities, and socio-economic stability further afforded by functioning transportation infrastructure. Moreover, effects of climate change coupled with the locale of population growth and urbanization will likely enhance future threats to coastal communities. This punctuates the need to quantify and understand risks to bridge infrastructure from storm surge events to inform smart development and management of these systems. While recent research targets associated challenges of predicting the hazard itself or deterministically estimating bridge deck loading for capacity-demand checks, probabilistic bridge vulnerability assessment has not been addressed to date. Nationwide risk and loss assessment packages lack any reliable input models of bridge fragility to assess the risk to the transportation infrastructure posed by hurricane induced storm surge. However, these tools are essential for reliability comparisons with different design details, consequence risk assessment, and loss estimation.

As a first step toward the development and validation of quantitative models of bridge reliability under storm surge loading, evidence from the 2005 Hurricane Katrina is used in this paper to analyzed trends in bridge damage and estimate failure probability for coastal bridges. Reconnaissance following Katrina revealed that 44 bridges were damaged along the Gulf Coast in Louisiana, Mississippi, and Alabama, where considerable storm surge elevations were suffered (TCLEE, 2006; Padgett et al, 2008). Several of the failed bridges are major water crossings having tens to hundreds of simply supported spans, such as the I-10 Twin spans over Lake Pontchartrain suffering 54 unseated spans and 473 shifted spans (Figure 1). Analysis of empirical data collected following Hurricane Katrina will provide the unique opportunity to

Figure 1. Bridge damage from Hurricane Katrina: (top) US-90 Bay St. Louis Bridge in Mississippi, (bottom) I-10 Twin Spans in New Orleans, Louisiana.

develop several levels of observed statistics which can be compared to predicted statistics in future models and used for refinement or validation purposes.

REFERENCES

Padgett, J., et al., (2008) *Bridge damage and repair costs from Hurricane Katrina.* Journal of Bridge Engineering. **13**(1): p. 6–14.

TCLEE, (2006) *Hurricane Katrina: Performance of Transportation Systems,* in *ASCE Technical Council on Lifeline Earthquake Engineering Monographs,* R. DesRoches, Editor. 2006.

Safety, Reliability and Risk of Structures, Infrastructures and Engineering Systems – Furuta, Frangopol & Shinozuka (eds)
© 2010 Taylor & Francis Group, London, ISBN 978-0-415-47557-0

Systems modeling of large transportation networks exposed to multiple hazards

M. Sánchez-Silva
Department of Civil and Environmental Engineering, Universidad de Los Andes, Bogota, Colombia

R. Castro & D.V. Rosowsky
Zachry Department of Civil Engineering, Texas A&M University, College Station, TX, USA

ABSTRACT

The impact and the extent of damage to infrastructure networks exposed to a hazard usually go beyond the physical bounds of the event. Within this context, this paper presents a model that combines a systems approach with strategies for detecting "community" structures in networks (Newman, 2004) in order to make better estimates of the hazard impact. The proposed model will assess the form (topology) of the network in order to develop a hierarchical description of the system. The hierarchical structure is obtained by unraveling the system progressively following state of the art network clustering algorithms (Van Dongen, 2000). In the proposed approach, the network performance is not modeled as a result of a collection of separate elements but rather as a dynamic hierarchically structured functional unit. The focus is not only on the components but also on their interaction and dependencies, which define the system's emergent properties at every level in the hierarchy. The model allows using information more efficiently for decision making.

The output of the model is an impact index that contributes to understand better the relationship between local and global effects. For instance, if the assessment is taken to the upper levels, the interconnection of components is more redundant and the emergent properties are more complex. Then, the impact of the event on the system is defused leading to smaller index values and adding more elements to the set of affected nodes. The impact index-based maps do not indicate the level of expected damage nor the size of the threat. They indicate the degree to which every part of the system if affected as a result of the network interactions.

The model was illustrated using data obtained from the impact of hurricane Ike in Texas in September 2008. Figure 1 presents the impact index maps developed for flooding and high wind speeds at level 2 (in the hierarchical representation of the system). Estimates of population affected and loss of productivity are presented in the paper along with the importance of considering network connectivity. One of the main advantages of the model is its potential to estimate indirect losses which are strongly related with the

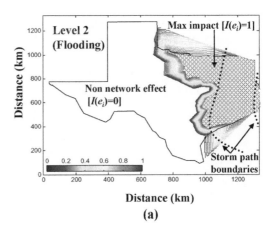

(a)

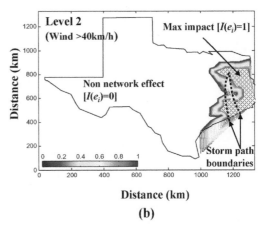

(b)

Figure 1. Network impact index spatial distribution for storm IKE considering only flooding.

emergent properties of the network (connectivity, flow, etc.) and not to the direct physical response to the event.

REFERENCES

Newman M.E.J. 2004. Detecting community structure in networks, *Eur. Phys. J. B*, 38, 321–330.

Van Dongen S. 2000, *Graph Clustering by Flow Simulation*. PhD thesis, University of Utrecht, May.

Mini Symposia (MS17) Earthquake Engineering and Engineering Seismology

Safety, Reliability and Risk of Structures, Infrastructures and Engineering Systems – Furuta, Frangopol & Shinozuka (eds)
© 2010 Taylor & Francis Group, London, ISBN 978-0-415-47557-0

Study on dynamic response of slightly compressible fluid-solid coupling system by ALE finite element method

Jingkui Zhang, Liaojun Zhang & Shuhe Wei

College of Water Conservancy and Hydropower Engineering, Hohai University, Nanjing, Jiangsu, China

ABSTRACT

At present, Arbitrary Lagrangian Eulerian (ALE) finiteelement method has been widely used to solve fluid-solid coupling problems (Nakayama & Washizuk 1987, Ramasuwamy & Kawahara 1987, Texieira & Awruch, 2005). The ALE finite element method with alternative algorithm format was studied in literature (Wang, 1998) and used to simulate fluid-solid coupled problem in fast reactor. The dynamic problems of flexible storage tank induced by fluid inertia and liquid sloshing with free surface during earthquake were studied based on the ALE finite element method (Liu, 1981). The ALE fractional steps finite element method was used to analyze fluid-rigid container interaction in Tuned Liquid Damper (Yue, 2000). These studies have obtained remarkable achievement.

Liquid referred in fluid-solid coupling system was usually treated as incompressible fluid in the past. However, compressibility in so-called "incompressible" fluid plays key role in some situations. For example, when a fluid is subject to a sudden load of disturbance and huge dynamic pressure, or the fluid domain is enclosed by deformable boundaries,or the propagation of a pressure wave is to be calculated, the compressibility should be taken into consideration, thus, the results could be exacter. Furthermore,it is difficult to calculate low Mach number flows, such as water and oil. So the slightly compressible fluid model by Professor Charles C. S.Song is adopted to consider fluid compressibility in seismic response calculation of liquid-filled container.

Based on the continuity equation and Navier-Stokes equation in Arbitrary Lagrangian Eulerian system, the discretized finite element equations are obtained by scheme of fractional steps in time domain and the Galerkin weighted residual method in space domain. Taking fluid-solid coupling condition into account, this paper establishes coupled finite element equations of coupling system solved by iteration method, thus, new ALE fractional steps finite element method which could solve nonlinear slightly compressible fluid-solid coupling system is presented in this paper.

REFERENCES

Nakayama, K. & Washizu, K. 1987. The boundary element method applied to the analysis of two-dimensional nonlinear sloshing problems. International Journal for Numerical Methods in Engineering 17: 1631–1646.

Ramasuwamy, B. & Kawahara, M. 1987. Arbitrary Lagrangian-Eulerian finite element method for unsteady convective incompressible viscous free surface fluid flow. Int. J.Numer. Methods fluids 7: 1053–1075.

Texieira, P.R.F. & Awruch, A.M. 2005. Numerical simulation of fluid-structure interaction using the finite element method. Computer & Fluids 34: 249–273.

Wang, Jian-jun. 1998. Strong Fluid-structure Analysis of main vessel of fast Breeder Reactor. In Tsinghua University(ed.).

Liu, W.K. 1981. Finite element procedure for fluid-structure interactions and application to liquid storage tanks. Nuclear Engineering and Design 65:221–238.

Yue, Bao-zeng. 2000. Numerical study of the threedimensional free surface in cylindrical tank. Chinese Journal of Computational Mechianics 19(4): 1–4.

Safety, Reliability and Risk of Structures, Infrastructures and Engineering Systems – Furuta, Frangopol & Shinozuka (eds)
© *2010 Taylor & Francis Group, London, ISBN 978-0-415-47557-0*

Sensitivity study on the seismic performance of container cranes

B.D. Kosbab, L. Schleiffarth, R. DesRoches & R.T. Leon
Georgia Institute of Technology, Atlanta, GA, USA

ABSTRACT

Past earthquakes have shown that container cranes are vulnerable to damage due to seismic loading (Kanayama et al. 1998). This study aims to identify some of the key parameters which negatively affect the seismic response of jumbo container cranes. For a representative jumbo crane, as shown in Figure 1, a deterministic sensitivity study is used in which the probability distributions of uncertain input parameters are assumed. This study employs a tornado diagram approach (Porter et. al 2002), in which the "swing," or relative importance of each input parameter, is ranked and graphically depicted. The input parameters chosen for this study are the ground motion profile, earthquake intensity in terms of spectral acceleration at the fundamental mode, mass, structural strength and stiffness, damping, and boundary condition friction coefficient.

The importance of each parameter is quantified based on the difference between evaluations of the engineering demand parameter at the lower and upper bounds of each parameter. A ductility capacity-based damage index is chosen as the engineering demand parameter, which is calculated from dynamic nonlinear time history analyses of a detailed finite element model. Results are reported graphically in Figure 2.

It is found that ground motion uncertainties are significantly more influential than the structural properties considered, and that an idealistic boundary model is sufficient. The results suggest the need for improved understanding of the ground motion characteristics, and for the development of performance-based design

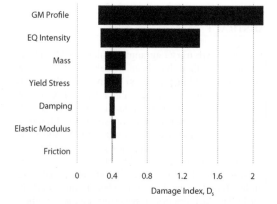

Figure 2. Results of the sensitivity study. The figure, called a tornado diagram, shows how the damage index is affected by variations in the input parameters. The swing of the damage index for a given parameter is represented by the damage index calculated at its lower (10th percentile) and upper (90th percentile) bounds. Parameters are shown in decreasing importance. The vertical line at Ds = 0.40 gives the damage index when all parameters are taken at their 50th percentile values.

criteria that take advantage of an ultimate ductility evaluation procedure.

REFERENCES

Kanayama, T., Kashiwazaki, A., Shimizu, N., Nakamura, I., & Kobayashi, N. 1998. Large Shaking Table Test of a Container Crane by Strong Ground Excitation. *Proc. Of Pressure Vessels and Piping Conference*, ASME PVP 364: 243–248.
Porter, K.A., Beck, J. L., & Shaikhutdinov, R.V. 2002. Sensitivity of building loss estimates to major uncertain variables. *Earthquake Spectra*, 18(4): 719–43.

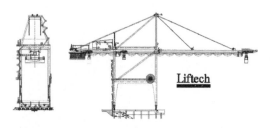

Figure 1. Geometry of a representative jumbo container crane, courtesy of Liftech, Inc.

Safety, Reliability and Risk of Structures, Infrastructures and
Engineering Systems – Furuta, Frangopol & Shinozuka (eds)
© 2010 Taylor & Francis Group, London, ISBN 978-0-415-47557-0

Nonlinear analysis of reinforced concrete frames, utilizing new joint element

S.SH. Hashemi, A.A. Tasnimi & M. Soltani
Tarbiat Modares University, Tehran, Iran

ABSTRACT

In this paper a new method for nonlinear analysis of
two dimensional reinforced concrete frames is proposed. Each frame is divided into two types of joint and
beam-column elements. The effect of bond-slip is considered in the formulation of beam-column element.
The formulation is based on displacement method
and derives from the principle of stationary potential
energy (Fig. 1). Joint elements are formulated upon
major behaviors including Pull-out of embedded longitudinal bars, shear and flexural deformation of joint
panel and shear-slip in interface section between joint
and neighboring element. Four types of joint elements
have been generated depend on their position in the
frame as an exterior, corner, interior or footing connection and are illustrated in Figure 2. Each element
type has been modeled based on major behaviors of
that through the combination of one or more defined
mechanisms and sub-elements. The sub-elements are:
a concrete and a reinforced concrete thick beam. In
which the effects of shear and flexural deformations
has been considered based on Timoshenko beam theory. The two considered mechanisms are: Pull-out of
beam or column longitudinal bars embedded in the
joint, and shear-transfer at the side of joint. The number of degrees of freedom in each side of joint element
is compatible with the degrees of freedom in the ends
of beam-column elements that are in the neighboring of the joint element. Finally, the reliability of the
proposed numerical method is assessed through the
comparison of experimental and analytical results.

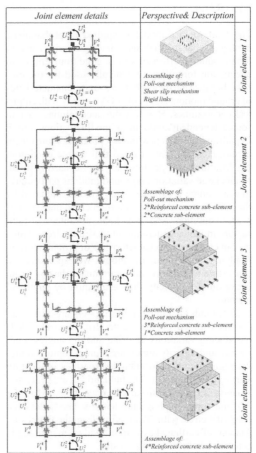

Figure 2. Joint element types.

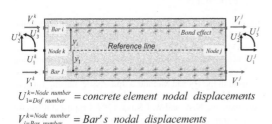

$U_{1=Dof\ number}^{k=Node\ number} = concrete\ element\ nodal\ displacements$

$V_{i=Bar\ number}^{k=Node\ number} = Bar's\ nodal\ displacements$

Figure 1. Reinforced concrete beam-column element.

*Safety, Reliability and Risk of Structures, Infrastructures and
Engineering Systems – Furuta, Frangopol & Shinozuka (eds)
© 2010 Taylor & Francis Group, London, ISBN 978-0-415-47557-0*

Study on probabilistic earthquake hazard analysis applying the fault rupture model

T. Kagawa

Tottori University Graduate School of Engineering, Tottori, Japan

ABSTRACT

The fault rupture model is applied to probabilistic earthquake hazard analysis of back ground seismicity for introducing effect of source condition. Empirical attenuation formulae are generally used for the purpose, because the methods give average characteristics of ground motion; e.g. PGA, PGV quickly with empirical uncertainties. However the uncertainties contain source dependent randomness and fluctuations caused by path and site effects without separation. From recent earthquakes, we have learned that source condition; e.g. locations of asperities and rupture starting point, results large variation on ground motions particularly in near source region. Ground motions estimated by the fault rupture model naturally include the effect of source condition. And remarkable advantage of fault rupture model is generation of waveform from which we can estimate any characteristics of ground motion. Applying the fault rupture model, we expect that we can estimate more natural earthquake hazard from background seismicity than estimation by empirical method.

Thousands of fault rupture models are randomly generated based on the recipe (Irikura et al. 2004) introducing variations of strike and dip angles, fault rupture and asperity areas, stress drop, rupture velocity, rupture duration, locations of asperities and rupture starting point and so on. The fault models are generated around a target site with variable magnitudes (Mw 5.5–7.0) and corresponding fault lengths. The faults exist inside the circle with 50km radius from the target site. Amount of earthquakes with individual magnitude is controlled by introducing b-value.

Strong ground motions from the models are calculated by using stochastic Green's function method (Kagawa 2004) with considering frequency dependent radiation pattern and response of layered structure. Minimum shear wave velocity of supposed layered structure is 350 m/s. Peak ground accelerations of the resulting ground motions are distributed between 20 to 940 cm/s². Response spectra of calculated ground motions are successfully compared with those by an empirical attenuation technique (Abrahamson & Silva 1997) especially in short period range. However, in

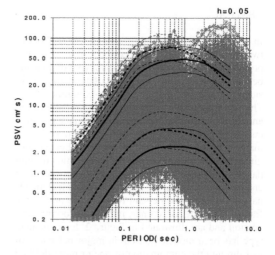

Figure 1. Comparison of response spectra by the fault rupture model (back gray lines) and those by empirical method with uncertainty (dense lines and dashes).

period range longer than one second, ground motions by the proposed method have larger variation than empirical method because of rupture directivity effect (see Figure 1).

The advantage and disadvantage for applying fault rupture model to probabilistic earthquake hazard analysis are discussed using the data base of calculated ground motions.

REFERENCES

Irikura, K et al., 2004. Recipe for predicting strong ground motion from future large earthquake, Proc. 13WCEE: 1371.

Kagawa, T., 2004. Developing a Stochastic Green's Function Method having more accuracy in long period range to be used in the Hybrid Method, J. Jpn. Assoc. Earthq. Eng., 4: 21–32.

Abrahamson, N. A. and W. J. Silva, 1997. Empirical response spectral attenuation relations for shallow crustal earthquake, Seism. Res. Lett., 68: 94–127.

Safety, Reliability and Risk of Structures, Infrastructures and
Engineering Systems – Furuta, Frangopol & Shinozuka (eds)
© 2010 Taylor & Francis Group, London, ISBN 978-0-415-47557-0

Shallow S-wave seismic reflection survey around the K-NET Anamizu site

Nozomi Kobayashi
Engineering Department, Kyoto University, Japan

Hiroyuki Goto & Sumio Sawada
Disaster Prevention Research Institution, Kyoto University, Japan

Koji Yamada
Hanshin consultants co., LTD, Japan

ABSTRACT

Around the K-NET Anamizu site, some residences are damaged during the 2007 Noto Hanto Earthquake, Japan. Hayashi *et al.* [2007] reported relations between the residences damages and the subsurface structure estimated from surface-wave survey and so on. The subsoil boundary of estimated velocity structure is expected to become clear by applying shallow S-wave seismic reflection survey. We develop the seismic reflection survey system to reduce the survey costs by devising equipment and materials.

Shallow S-wave seismic reflection survey is performed in Omachi West Children's Park where K-NET Anamizu site is located. Total profile length is 40 m from south to north and 10 m from west to east. The number of participants is two. S-wave source is excited by a wooden hammer stroke on one side of a wooden placed perpendicular to the seismic line. The signal is enhanced through 10 times stacks of the hammer stroke on both sides. 24 geophones of natural frequency of 28 Hz receive the excited horizontal ground motions. Each receiver observes one component. Data logger (ES-3000) in this survey supports only 8 channel data sets. To use 24 receivers, three sets of observations with 8 receivers at same source point are conducted. To select 8 receivers as one set easily, we made a switch box (see Figure 1). Receiver interval is set to be 1.0 m.

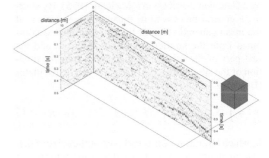

Figure 2.   3D time section.

After data processing, we estimated the 3D time section profile. The shallow S-wave seismic reflection survey gives detailed structures of the surface layers shallower than 20 m (see Figure 2).

In addition, this survey also observed waveforms in Line L among 0 m–24 m. To calculate the theoretical travel time of refraction waves, we conducted a modeling of simple two-layer structure and comparing to the theoretical refracted travel time.

REFERENCES

Hayashi, K. & Tamura, M. & Hirade, T. & Yu, S. & Muraoka, M. & Kikuti, Y. 2007. Site Investigation by Geophysical Exploration and Sounding Around the K-NET Anamizu Observation Station. *PROGRAMME AND ABSTRACTS THE SEISMOLOGICAL SOCIETY OF JAPAN 2007, FALL MEETING* A12-07: 11.

Vidale, J. 1988. Finite-Difference calculation of travel times, *Bull. Seism. Soc. Am.* 78: 2062–2076

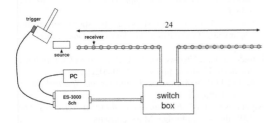

Figure 1.   Survey system.

Safety, Reliability and Risk of Structures, Infrastructures and Engineering Systems – Furuta, Frangopol & Shinozuka (eds)
© 2010 Taylor & Francis Group, London, ISBN 978-0-415-47557-0

Stochastic sensitivity of ground motion parameters to structural response

A. Yazdani

Department of Engineering, University of Kurdistan, Sanandaj, Iran

T. Takada

Graduate School of Engineering, University of Tokyo, Tokyo, Japan

A theoretical relation is presented between the seismological Fourier amplitude spectrum and the mean squared value of the elastic response which is defined by a Gaussian distribution (Newland, 1975). By shifting a general process to its mean value, spectrum of the mean squared value of the displacement is computed from the Fourier amplitude spectrum and the real part of the relative displacement transfer function of the single-degree-of-freedom elastic oscillator (Brune 1970, Boore 2003):

$$\sigma_Y^2 = \int_{-\infty}^{\infty} S_{YY}(\omega)d\omega = \int_{-\infty}^{\infty} |H_D(\omega;\omega_n,\xi)|^2 S_{AA}(\omega)d\omega \quad (1)$$

In which $|H_D(\omega;\omega_n,\zeta)|$ is real part of the relative displacement transform, $S_{YY}(\omega)$ and $S_{AA}(\omega)$ are the power spectral density functions of the output process and input ground motion respectively. The power spectral density of input ground motion is defined as:

$$S_{AA}(\omega) = |F(\omega)|^2 / T \quad (2)$$

Where $F(\omega)$ is Fourier spectrum of a ground motion acceleration and T is earthquake ground motion duration. In fact, the amplitude Fourier spectrum has been, so far, the most widely used from of specifying ground-motion characteristics in engineering seismology. Brune (1970) assumes that the far-field accelerations on an elastic half space are band-limited, finite-duration, white Gaussian noise, and that the sources spectra are described by single corner-frequency model which corner frequency depend on earthquake size. The Fourier amplitude spectrum, $F(\omega)$, used in a seismological model (Nigam, 1983) can be broken into contributions from earthquake source model, and typical geometric, anelastic whole path and upper crust attenuation, and site functions, so that:

$$|F(\omega)| = C.E(\omega).G.An(\omega).P(\omega).A(\omega) \quad (3)$$

Where C is a scaling factor, $E(\omega)$ is source spectrum, G is the geometric attenuation factor. The $An(\omega)$ factor, by definition, includes all the losses which have

not been accounted for by the geometrical attenuation factor. The attenuation, or diminution, operator $P(\omega)$ in Equation (3) accounts for the path independent loss of high-frequency in the ground motions. $A(\omega)$ is the upper crust amplification factor and it is a function of shear-wave velocity vs. depth.

Variations in moment magnitude, focal distance, geometric and material attenuation relations, and site condition affect the uncertainty in structural response. In order to study the stochastic response, these variations were modeled as random variables. The Monte Carlo method, which it uses randomly generated samples of the input variables for estimating response, is used to evaluate the mean and variance of displacement.

The relative contributions of variance in random variables to the variance in the maximum structural response are demonstrated. The results reveal that magnitude, focal distance and site amplification are the main sources of uncertainty affecting the probabilistic the mean squared value of the displacement of the structures. The contributions of focal distance in near distances are smaller than that for far distances. The amplification for very hard rock is substantially below those for generic rock site and the site effect amplification in generic rock is more important than very hard rock.

REFERENCES

Boore, D.M. 2003. Prediction of ground motion using the stochastic method. *Pure Appl. Geophys.*, Vol. 160, pp: 635–676.

Brune, J.N. 1970. Tectonic stress and the spectra of seismic shear waves from earthquake. *Journal Geophys. Res.*, Vol. 75, pp: 4997–5009.

Karadeniz, H. & Vrouwenvelder T. 2003. *Overview reliability methods*. TU-Delft press, the Netherlands.

Newland, D.E.1975. *Random Vibrations and Spectral Analysis*, Longman, London,.

Nigam, N.C. 1983. *Introduction to Random Vibrations*, The MIT press, Cambridge, Massachusetts.

Mini-Symposia (MS08) Time-Dependent Reliability Methods and Their Applications

Safety, Reliability and Risk of Structures, Infrastructures and Engineering Systems – Furuta, Frangopol & Shinozuka (eds)
© 2010 Taylor & Francis Group, London, ISBN 978-0-415-47557-0

Seismic reliability analysis of corrosive pipelines

Wei Liu & Jie Li
Department of Building Engineering, Tongji University, Shanghai, China
State Key Laboratory of Disaster Reduction in Civil Engineering, Tongji University, Shanghai, China

ABSTRACT

Buried pipelines are commonly used in many lifeline engineering systems, such as water distribution, gas and oil supply systems. Due to the possible serious consequence of lifeline system failure, the pipeline performance under earthquake has been intensively studied. During recent decades, many approaches have been presented to predict the pipeline seismic response, such as elastic foundation beam approach, shell model, finite element method, and so on. Although so many methods have been proposed, none of above studies has ever considered the effect of corrosion on pipeline. However, the pipeline with serious corrosion will generally suffer more significant damages under earthquake than those with no corrosion or slight corrosion. Investigations show that almost half of buried pipelines are made of metals, which might be corroded gradually by the soil around them or the materials conveyed inside them during working time. As the result, the seismic reliability of pipeline decreases gradually as the working time increases.

Therefore, in order to accurately evaluate the pipeline performance under earthquake, corrosion could not be ignored. However, because of the uncertainty of the soil around pipeline and the materials conveyed inside pipeline, the pipeline corrosion should been stochastically described. In this paper, homogeneous Markov process is used to simulate the corrosion occurrence of pipeline. Combining with linear corrosion development model, i.e. the corrosion width and depth are the linear functions of time interval, the probability density evolution of pipeline area corrosion percentage as in-service time increases is derived. Also, elastic foundation beam model is employed to express the pipeline seismic displacement and stress as a function of pipeline area which is viewed as basic random variable. Combining with the free boundary and the continuity of displacement and axial force at the interface between adjacent pipe segments, a equation set is established to evaluate the uncertain constants in seismic response expression. As random variables exist in the equation set, random perturbation approach is employed to solve this equation set and the mean and variance of the pipeline seismic response are given. Based on Chinese code for gas pipeline, a limit state equation about the corrosive pipeline under earthquake is established. First order method is employed to calculate the seismic reliability of corrosive pipeline. In order to validate the proposed approach, a 200-meter long pipeline is investigated in detail. The results indicate that the proposed corrosion model can efficiently simulate the area evolution of buried pipeline with the increase of the working time. Also, the seismic reliability of an actual network located in China is given.

REFERENCES

Li Jie. 2005. Lifeline Earthquake Engineering-Basic Method and Application (in Chinese). *Beijing, Science Press*

M. Ahammed. 1998. Probabilistic estimation of remaining life of a pipeline in the presence of active corrosion defects. *International Journal of Pressure Vessels and Piping*; 75:321~329

Lu Dajin. 1986. Stochastic Processes and its Applications (in Chinese). *Beijing: Tsinghua University Press.*

Liu Limin, Yu Jianxin et,al. 2002. Evaluation of Erosion Remaining Life on Submarine Pipelines (in Chinese). *China Offshore Oil and Gas (Engineering)*; 14(3):42–44

Qu Tiejun, Wang Qianxin. 1993. Series Solutions for the Axial Vibration of Buried Pipeline under Earthquake Excitation (in Chinese). *Earthquake Engineering and Engineering Vibration*; 13(4):39–45

Li Jie. 1996. Stochastic Structural System-Analysis and Model (in Chinese). *Beijing, Science Press.*

Safety, Reliability and Risk of Structures, Infrastructures and Engineering Systems – Furuta, Frangopol & Shinozuka (eds)
© 2010 Taylor & Francis Group, London, ISBN 978-0-415-47557-0

Decision of the optimal interval check time of deterioration structure system

Chao Jia, Xingfang Liu & Yafei Li
School of Civil Engineering, Shandong University, Jinan, Shandong ,China

ABSTRACT

Functions of a constructed project will deteriorate as time goes by and its failure probability increases. During the operation, normal operation state goes into abnormal state and then malfunction. If maintenance is not performed, normal function would be affected and finally used performance would even be lost totally till some point. To avoid that, repair and maintenance are necessary in order to improve the structure's performance and prolong its used life. Therefore, timely examination is needed for constructed and operating projects to find out potential problems as early as possible and avoid possible hazards. However, if the interval check time is too long, the structure would stay in malfunction for a long time which would definitely affect its function and cause unnecessary loss. If the interval check time is too short, the repair time and cost are increased while the structure's inherent reliability is affected. Thus, to find out the optimal interval check time becomes necessary.

Traditional method of determining the maintenance time of a structure is to consider it as a non-deterioration system. Actually, the whole life-cycle of structures fits the characteristics of deterioration system, so establishing the maintenance strategy of deterioration system has a significant engineering meaning.

Based on previous researches at home and abroad, this paper takes project structures as deterioration system. By combing the reliability theory with structure usability, the maintenance strategy which takes minimum cost in life circle as the objected function is established to determine structures' optimal interval check time.

Take the Heihushan Reservoir Dam in Qingzhou, Shandong Province for example. Based on the reliability theory, the reservoir dam's stability against sliding, overtopping functions and the objected function are established.

The objective function takes check availability p and interval check time τ as independent variables. Figure 1 is the relationship curve of them.

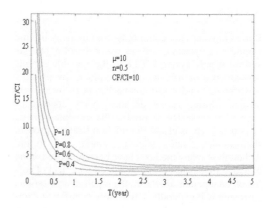

Figure 1. Relation of cost and check availability, interval check time.

As the Fig. 1 that when check availability is certain, the shorter the interval check time is, the more cost it takes, that is, check frequency increases check cost. For the same interval check time, more detailed check costs more. Based on the least cost, there is one optimal interval check time τ_{opt} corresponding to different check availability.

To conclude, the method in this paper can be applied to other fields, like the maintenance time decision of mechanical facilities,military facilities and other projects.

REFERENCES

Ang, A.H-S & W.H. Tang. 1990. Probability Concepts in Engineering Planning and Design, Volume II - Decision, Risk, and Reliability. New York: John Wiley & Sons.

JIA Chao, LIU Ning, CHEN Jin. Studying of engineering structure maintenance time decision. Engineering Journal of Wuhan University, 2003, 36(6): 1–4.

Safety, Reliability and Risk of Structures, Infrastructures and Engineering Systems – Furuta, Frangopol & Shinozuka (eds)
© 2010 Taylor & Francis Group, London, ISBN 978-0-415-47557-0

Time-dependent reliability analysis using finite difference method in the framework of directional simulation

M.R. Moarefzadeh

Civil Engineering Department, Imam Hossein University, Tehran, Iran

ABSTRACT

Time dependent reliability analysis normally requires the initial probability of failure (the probability of structural failure at $t = 0$) and the out-crossing rate both to be known. While the first is usually estimated by the use of well-established techniques (i.e. First/Second Order Reliability Methods, FORM/SORM, or simulation), the latter needs more sophisticated approaches in which time plays a significant role. Different precise solutions have been given in the literature to treat varieties of problems the structural reliability analysis is being involved (i.e. for example see Melchers, 1999 and Ditlevsen & Madsen, 2005). They, however, restrain to many constraints which impose limitations for their implementation. For example, stationarity and Gaussianity of the involving stochastic processes and linearity of the existing limit state functions are of usual provisions which are required for the close-form solutions to be directly used. Even-though a large amount of efforts have been made in recent years to relax the limitations and to give more general solutions, still better and less sophisticated techniques are desired. One of the efforts carried out recently to propose a more convenient treatment of time-dependent reliability problem is to use a parallel system reliability formulation for computing the out-crossing rate by the use of basic time-invariant reliability means such as FORM/SORM methods. This idea was first presented by Hegan and Vdeit (1991) and then more developed by many others. One of the efforts made in this area is the method so-called PHI2 (due to Andrieu-Renaud et.al; 2004). In this method, the time-parameter is only considered in the computations indirectly and also, by dividing the structural life-time into some time-increments (within which the involving processes may be seen stationary), non-stationary problems could also be taken into account.

Despite its considerable abilities, the method is mostly dependent to the time-increment (i.e. Δt) taken in the analysis. This is the major drawback from the original work noted above. Further, since the method uses the conventional FORM method, it involves all undesired limitations originated in this method (e.g. need to work in the standard Gaussian space rather than original space, need to have almost linear limit state functions and also need to use techniques such as bounds to evaluate structural system reliability).

In this paper attempt is made to use the main idea given above, but, in the space of $X(t) - \dot{X}(t)$ (i.e. instead of the space $X(t) - X(t + \Delta t)$). This makes the analysis to be almost independent from the value of Δt, taken to carry out the analysis. The idea is applied first to the PHI2 method and then is used in the framework of directional simulation. The latter application enables the analysis to achieve more precise results when the limit state(s) are highly nonlinear and also relaxes the requirement to transform the original existing space to equivalent standard normal space.

REFERENCES

Andrieu-Renaud C.B., Sudret B. and Lemaire M. 2004. The PHI2 Method: a Way to Compute Time-variant Reliability, Reliability Engineering & System Safety, 84, 75–86.

Ditlevsen O. and Madsen H.O. 2005. Structural Reliability Methods, Chichester, J. Wiley and Sons,

Hagen, O. and Tvedt, L. 1991, Vector Process Out-crossing as Parallel System Sensivity Measure, Journal of Engineering Mechanics, 117(10), 2201–2220,

Melchers, R.E. 1999. Structural Reliability, Analysis and Prediction. Chichester, J. Wiley and Sons,

Sudret B. 2005. Analytical Derivation of the Out-Crossing Rate in Time-Variant Reliability Problems. Proceedings of the 12th IFIP WG7.5 Working Conference, (Aalborg, May 2005) (Ed. J.D. Sorensen and D.M. Frangopol), 125–133.

Safety, Reliability and Risk of Structures, Infrastructures and Engineering Systems – Furuta, Frangopol & Shinozuka (eds)
© 2010 Taylor & Francis Group, London, ISBN 978-0-415-47557-0

Time variant analysis of structural reliability – numerical comparison of various approaches

M. Sykora

Klokner Institute, Czech Technical University in Prague, Prague, Czech Republic

ABSTRACT

Civil engineering structures are often exposed to combinations of time-variant loads. In probabilistic reliability analysis, random load fluctuations within a working life may be described by rectangular wave renewal processes with intermittencies. It was indicated by Rackwitz (1998), Melchers (1995) and Sykora (2005) that the renewal processes provide a simple model with a relatively small number of input parameters.

Since closed-form solutions of the failure probability related to the working life are available for trivial cases only and simulation-based computations may be cumbersome, bounds on the failure probability are commonly used to estimate a reliability level. In particular an upper bound on the failure probability is of a great importance in practical applications.

The present paper is aimed at comparison of various techniques proposed by Rackwitz (1998), Melchers (1995) and Sykora (2005). In addition estimates based on the rule by Turkstra (1970) are also provided. Differences among the approaches are briefly discussed and reliability of simple structural members and structures is then analyzed.

It follows that the upper bound by Rackwitz (1998) and its modification by Sykora (2005) yield nearly the same results for always present processes. Considerably lower estimates of the failure probability are

Table 1. Reliability index β.

Technique	Study A	Study B
Melchers (1995)	2.36	1.42
Sykora (2005)	2.28	1.33
Turkstra's rule	2.58	1.63
"Exact" solution	2.45	1.50

obtained for spike-like processes using the bound by Rackwitz (1998) as shown in Figure 1.

Other numerical studies (see Table 1) indicate that results based on the technique by Melchers (1995) are in a good agreement with those obtained using the bound by Sykora (2005). Turkstra's rule (1970) leads to similar results as the techniques by Melchers (1995) and Sykora (2005) and seems to provide sufficiently accurate estimates of the failure probability for civil engineering applications.

It is, however, emphasized that these conclusions are indicative only and more numerical studies are needed to generalize these findings. In particular future research should be focused on further comparisons of the techniques by Melchers (1995) and Sykora (2005) with Turkstra's rule that is of a great practical importance due to its simplicity.

Reliability index

Probability of "on"-state of the leading action

Figure 1. Variation of the reliability index with probability of "on" – state of the leading action.

REFERENCES

Melchers, R.E. 1995. Load Space Reliability Formulation for Poisson Pulse Processes. *Journal of Engineering Mechanics* 121(7): 779–784.

Rackwitz, R. 1998. Computational techniques in stationary and non-stationary load combination - A re-view and some extensions. *Journal of Structural Engineering* 25(1): 1–20.

Sykora, M. 2005. Load combination model based on intermittent rectangular wave renewal processes. In *Proc. 9th Int. Conf. on Structural Safety and Reliability – ICOSSAR 2005, Rome, 19–23 June 2005*. Rotterdam: Millpress.

Turkstra, C.J. 1970. *Theory of Structural Design Decisions*, SM Studies Series No. 2. Ontario: University of Waterloo.

Safety, Reliability and Risk of Structures, Infrastructures and
Engineering Systems – Furuta, Frangopol & Shinozuka (eds)
© 2010 Taylor & Francis Group, London, ISBN 978-0-415-47557-0

Stochastic fatigue lifetime computation: Evaluation by polynomial chaos and resampling

A. Notin
CETIM, SENLIS, France

N. Gayton & M. Lemaire
*LaMI EA 3867 – FR CNRS 2856, Université Blaise Pascal et Institut
Francais de Mécanique Avancée, Aubière, France*

J.L. Dulong
Laboratoire Roberval, Université de Technologie Compiègne, CNRS FRE No. 2833, Compiegne, France

ABSTRACT

The estimation of fatigue lifetime is critical in the
industry in order to assess and verify safety margins.
In reliability analysis, the failure criterion is defined as
a limit state function which depends on random inputs
and response quantities. Using stochastic finite elements, the response quantity can be written as series
expansion which allows an approximation of the limit
state function. Resampling techniques are a tool to
explore database and to compute confidence intervals
which are applied to reliability analysis in the context
of fatigue lifetime computation. Fatigue is a random
process in essence. The randomness comes from the
loading process and the fatigue resistance of material. Monte Carlo simulation methods are commonly
used to solve this kind of problems. Even if these
methods have strong advantages, they usually become
computer time-consuming as the complexity and the
size of the embedded deterministic models increase.
In order to find an alternative to Monte Carlo simulations, the so called Stochastic Finite Element Method

Table 1. Reliability index and its confidence interval for a fatigue limit at 10^6 cycles and 500 experiments.

	Chaos order 3		
β	Lo. bound 1,0408	Estimation 1,0841	Up. bound 1,1274
	4		
β	Lo. bound 1,0420	Estimation 1,0655	Up. bound 1,0890

has been developed. The response is expanded onto
a particular basis of the probability space called the
polynomial chaos. The polynomial chaos can be seen
as a response surface. On that account, it is build from
a design of experiments containing more or less data
(depends on the case study). But, in the industry, new
data can be very expensive to obtain. The use of resampling techniques permits to explore and evaluate the
variability of results without new computations (using
bootstrap methods): taking the initial design of experiments into account, the validity of the results can be
checked. Confidence intervals are applied to reliability
analysis in the context of fatigue lifetime computation
which leads to a resampling polynomial chaos method
(RPCM). The method is illustrated on the calculation
of the fatigue life of a bogie support fixing.

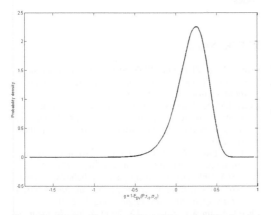

Figure 1. Bogie – Representation of the probability density function of g.

REFERENCES

Dang Van, K. & Griveau, B. & Message, O. 1989. On a
new multiaxial fatigue criterion: theory and application.
M.W. Brown and K.J. Miller (ed.). *Biaxial and Multiaxial
Fatigue*. London: Mechanical Engineering Publications.
DiCiccio, T.J. & Efron, B. 1996. Bootstrap confidence
intervals, *Statistical Science* 11(3): 189–228.
Ghanem, R.G. & Spanos, P.D. 2003. *Stochastic finite ele-
ments - A spectral approach* (Revised Edition). Dover.

Safety, Reliability and Risk of Structures, Infrastructures and Engineering Systems – Furuta, Frangopol & Shinozuka (eds)
© 2010 Taylor & Francis Group, London, ISBN 978-0-415-47557-0

Reliability analysis of structures considering resistance deterioration over time

Wei-Qiu Zhong, Zhi-Wei Zhang & Li-Yuan Xie

School of Civil and Hydraulic Engineering, Dalian University of Technology, Dalian, China

ABSTRACT

Structural resistance changes greatly over time in adverse environment, such as chemical environment and ocean environment. Reliability analysis considering structural resistance change over time is very complex. In the current Chinese building codes, structural resistance change over time is not considered (Zhao, Jin & Gong 2000). In reference (Gong & Zhao 1998), a reliability analysis method considering resistance change over time is given. In the context, it is called the equal resistance method. In the present paper, the minimum resistance method is presented and compared with the other methods. The minimum resistance method is simple in computation and intuitive in concept.

Based on the method of current Chinese building codes, it is known that

$$Z = g(R, G, Q_T) = R - G - Q_T \qquad (1)$$

Z is the performance function and can be solved by first order second moment method.

In reference (Gong & Zhao 1998), the current reliability calculation method was improved and resistance change over time was considered. By discrete resistance $R(t)$, the equal structural performance function considering resistance change over time is achieved.

$$Z = g(R_{eq}, G, Q_T) = R_{eq} - G - Q_T \qquad (2)$$

where R_{eq} is the structural equal resistance and can be expressed by

$$R_{eq} = -\frac{1}{\alpha_T} \ln\left\{ \frac{1}{n} \sum_{i=1}^{n} \exp[-\alpha_T R(t_i)] \right\} \qquad (3)$$

In the minimum resistance method, resistance change over time is considered either. The minimum resistance value in design working life is adopted to calculate the reliability index. In many cases, this point is also the ending point in design working life.

$$Z = g(R_{\min}, G, Q_T) = R_{\min} - G - Q_T \qquad (4)$$

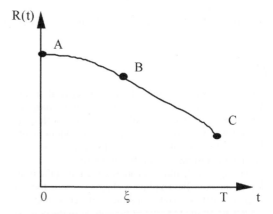

Figure 1. Different resistance at different time adopted by the three methods respectively.

In the minimum resistance method, the minimum value of resistance fading function $R(t)$ in $[0,T]$ which is the terminal point value of resistance in design working life in many cases is considered only. Subsection summation or integration is not needed and the method is conservative, simple in calculation, intuitive in concept and compatible with the current building codes.

REFERENCES

Gong J.X. & Zhao G.F. 1998. Structural reliability analysis considering resistance change over time, *Journal of Building Structures* 19(5): 43–51.

Gong J.X. & Zhao G.F. 2000. Influence of corrosion to reliability of R.C. structures in atmospheric environment, *Journal of Dalian University of Technology* 40(2): 210–213.

Li T. & Liu X.L. 1994. Durability design of concrete structures, *Journal of Civil Engineering* 27(2): 47–55.

Wang G.Y. 1990. Discussion on dynamic reliability and repair theory of structures in service period, *Journal of Harbin Institute of Architectural Engineering* 23(2): 1–9.

Zhao G.F., Jin W.L. & Gong J.X. 2000. *Reliability theory of structures*. Beijing: China Building Industry Press.

Organized Session (OS07) Performance-Based Design for Steel Structures

Safety, Reliability and Risk of Structures, Infrastructures and
Engineering Systems – Furuta, Frangopol & Shinozuka (eds)
© 2010 Taylor & Francis Group, London, ISBN 978-0-415-47557-0

An investigation on dynamics of large scale induced draft fan installation system using finite element method

Kishore K Rathore, Sanjeev Srivastava & R.L. Deshmukh
Aditya Birla Group, Grasim Industries Ltd (Cement Business), Technology & Research Centre, Mumbai, India

ABSTRACT

Over the past decade, the capacities of large process machinery have increased substantially. Subsequently, the concrete structures and pedestals sup-porting these units are becoming taller and larger - resisting higher static and dynamic loads. Moreover, in industrial plants such as, cement, refineries, chemical plants, gas plants and power plants, it is normal practice to support multiple machines on a common platform due to space limitations. For this case, the effect of interaction between units on dynamic responses can be significant and should be considered in the design. Dynamic response has been recognized as one of the significant factors affecting the service life & safety of structure in any process industry. The unacceptable dynamic response in a form of large displacement results in fatigue deterioration and can override any cost efficiency of these systems because of the increased maintenance cost & rehabilitation. Ever since centrifugal fans have been manufactured they have been subject to vibration related problems. Of these, problem related to resonance predominant. Resonance problems are often two fold on large fan assemblies. The first component that has to be addressed is critical speeds. Most fans are designed to operate below first critical speed. If a fan operates above first critical speed then careful attention has to be paid to vibration levels as the fan accelerates up to operating speed and, more importantly, coasts down to a stop from operating speed. Excessive levels of vibration while passing through a critical speed can lead to severe damage.

The second factor, structural resonance, can be much more challenging to predict. If a fan system operates at a structural resonance point that is not corrected it can lead to component and/or system failures. Proper machinery installation is a fundamental element of high equipment reliability and low life cycle cost. Conversely, improper installation ensures machinery will become a chronic source of downtime, poor product, quality, reduced capacity and high cost. Thus, accurate predictions of critical speeds whirl responses and modal characteristics both for the rotor and foundation system are necessary for safe & efficient operations.

In cement plant huge chunks of limestone are crushed & milled to a fine powder blended with other raw materials & eventually moved into a huge rotary kiln. The draft fan draws out hot gas off the rotary kiln & preheat cyclone tower before sending the dust laden air into the bag house for removal of particulates. The operator reported based on the monitored vibration levels at the bearings that the system approaches the maximum vibration trip level set for this installation when operated at or above 850rpm. An initial perception was that the excessive vibration was caused by a rotor imbalance perhaps from uneven buildup of debris on the impeller blades. However, after a routine maintenance and cleaning of the fan the problem persisted even in the absence of significant build up of coating on the impeller blades.

Currently, the traditional approach of cut-and-try in the design phase and physical model testing is being replaced by the finite element analysis. However, literature relating the physical observations from the test rig through condition monitoring to approximations from virtual testing with Finite Element Analyses techniques and on planning the strategies for simulating the physical testing with perfection is rare.

This work is dedicated to development of a convenient & reliable analysis methodology using a finite element technique that can accurately predict the static & dynamic response of the system. The paper demonstrates the used methodology on a large scale induced draft fan installation system of a cement process industry. A substructure procedure, wherein the overall system of a fan is divided into subsystems, namely, rotor, the foundation (base frame, pedestal and concrete foundation) and the dampers are addressed using software ANSYS. The work scope includes determination of the vibration characteristics of a fan system under different operating and design conditions. Dynamic analysis of rotor and its supporting structure are discussed. The effects of mass load on the base frame for rotating machinery in avoiding resonance have been emphasized through several numerical design iterations. Simulation results are validated with the field test data and are observed in good agreement.

Safety, Reliability and Risk of Structures, Infrastructures and Engineering Systems – Furuta, Frangopol & Shinozuka (eds)
© *2010 Taylor & Francis Group, London, ISBN 978-0-415-47557-0*

Characteristics of performance-based design specifications of JSCE for steel and composite structures

T. Yoda

Waseda University, Tokyo, Japan

ABSTRACT

A new performance-based design specification for steel and composite structures has been developed by the Japan Society of Civil Engineers. The new design specification is based mainly on the recent researches and developments in the field of steel and composite structures. The performance-based design specifies the performances required for structures and verification of whether or not the required performances are satisfied, by using the appropriate verification indices.

Important thing is what required performances are to be set for the structure, which leads to the characteristics of the design specifications. The safety and serviceability are performances common to almost all the design specifications while others are roughly classified into the durability, restorability, environmental compatibility, constructability and maintainability, and economy, although such classifications are not universally employed in design specifications.

Selection of design values in a limit state design and assurance of appropriate levels of structural reliability are important tasks made by code writers. The statistical basis of the present design format is the use of the low values of resistance population and the high values of action as design values with the help of partial factors, in which a large volume of good quality data play a key role in the determination of design values related either to resistance or to action.

The purpose of the present paper is to demonstrate the ongoing activities concerning a performance-based design specification for steel structures in Japan from the view point of probability–based limit state

design methodology. Emphasis is given to the method for determining design values of resistance and appropriate levels of structural reliability, by the reason that available statistical information on resistance is limited at present and that the failure probability is sensitive to the tail of distributions. Happily, however, the present design format is similar to most of conventional design rules, and existing statistical information on resistance can be easily incorporated in the selection of design values to attain consistent failure probability. With this format, gradual transition will be an easy task to arrive at an improved design rule in the sense of consistent reliability.

The main verification check is based on the linear analysis in the member level as shown in Table 1. The new design specifications opens up the opportunity for new, more accurate structural analyses. The more powerful and more accurate structural analyses will become readily available in the form of reliable commercial software packages, resulting in the possibility of introducing new materials and new structural forms.

REFERENCES

ISO2394, 1998. General Principles on Reliability for Structures.
Japan Society of Civil Engineers, 2007. Standard Specifications for Steel and Composite Structures-2007.
Japanese Society of Steel Construction, 2001. Guidelines for Performance-Based Design of Civil Engineering Steel Structures, Technical report No.49.
Ministry of Land, Infrastructure and Transport (2002) Basis of Structural Design for Buildings and Public Works.

Table 1. Level of structural analyses.

Level	Verification Check	Structural Analysis
Action Level	$\gamma_d P_d \leq P_u$	Nonlinear Frame Analysis
(Structure Level)	(γ_d: System Reduction Factor)	Nonlinear FEA
Stress Resultant Level	$\gamma_i \frac{S_d}{R_d} \leq 1$	Linear Analysis
(Member Level)	(γ_i: Capacity Reduction Factor)	
Stress Level	$\sigma_{max} \leq \sigma_u / \gamma_s$	Linear FEA
(Structural Detail Level)	(γ_s: Strength Reduction Factor)	

Safety, Reliability and Risk of Structures, Infrastructures and Engineering Systems – Furuta, Frangopol & Shinozuka (eds)
© 2010 Taylor & Francis Group, London, ISBN 978-0-415-47557-0

Assessment of vibration serviceability for bridge using artificial wheel load

Sang-Hyo Kim, Kwang-Il Cho, Moon-Seock Choi & Ji-Young Lim
Yonsei University, Seoul, Korea

ABSTRACT

A recent trend of designing the infrastructures in advanced countries including the United States, Japan and Europe is the expansion of considerations from involving only the life safety and the economic feasibility of construction to involving in addition performance levels for various interests. Among the vast range of lists on the performance needed for a structure, the vibration serviceability is one of the principle elements in that it affects the users in direct ways.

For the assessment of vibration serviceability, it is desirable to analyze with simple load models for the convenience of engineers. In the previous study, a generation method of artificial wheel load has been developed based on a PSD function from stationary stochastic process. The artificial wheel load enables a simpler procedure for engineers to perform a dynamic analysis of a bridge under running vehicles. Specific procedure to apply artificial wheel load using PSD function can be shown in Kim et al., (2009)..

The vibration serviceability of a bridge is assessed by applying certain evaluation criteria. Among existing serviceability criteria, Reiher-Meister curve (Reiher and Meister, 1931) and ISO 2631 (ISO 1997), BS 6841 (BSI 1987) are applied in this study for the assessment of vibration serviceability of a bridge. A five-span continuous steel box girder bridge considered for the evaluation of validity through dynamic analyses using the moving vehicle model (real wheel load) and the artificial wheel load. For all serviceability criteria, the serviceability assessment via artificial wheel load showed more conservative result than one from moving vehicle model. Figure 1 shows the result of each criterion. Because of limited space, serviceability result from Reiher-Meister curve is showed only for road roughness A. For figure 1-(b), WRMS stands for weighted root mean square and WRMQ stands for weighted root mean quad.

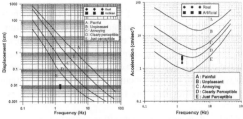

(a) Reiher-Meister curve (Displacement & Acceleration)

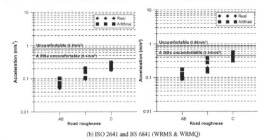

(b) ISO 2641 and BS 6841 (WRMS & WRMQ)

Figure 1. Serviceability assessment result using various evaluation criteria

REFERENCES

BSI. 1987. *BS 6841 British standard guide to measurement and evaluation of human expose to whole-body mechanical vibration and repeated shock,* British Standard Institution.

ISO. 1997. *ISO 2631 Mechanical vibration and shock-evaluation of human exposure to whole-body vibration,* International Organization for Standardization, Switzerland.

Kim. S.-H & Cho. K.-Y. & Choi. M.-S. & Lim. J.-Y. 2009. Development of a generation method of artificial vehicle wheel load to analyze dynamic behavior of bridges. *Advances in Structural Engineering (accepted)*

Reiher. H & Meister. F. J. 1931. Sensitivity of human to vibration. *Forschung auf Gebeitedes Ingenieurwesens,* Vol. 2 (11): 381

Safety, Reliability and Risk of Structures, Infrastructures and
Engineering Systems – Furuta, Frangopol & Shinozuka (eds)
© 2010 Taylor & Francis Group, London, ISBN 978-0-415-47557-0

Estimation of load combination factors based on Monte Carlo simulations

T. Sugiyama

University of Yamanashi, Kofu, Japan

ABSTRACT

Quantitative estimation of the load combination factors for highway steel bridges under the combination with environmental or accidental loads based on Monte Carlo simulation method has been made. Respective load effects are expressed by the rectangular wave model whose occurrence on time-axis, intensity and duration time are taken as random variables.

Obtained load factors for several load combination cases have been tabulated as shown in Table 1 so as to be put into practical use. According to the results obtained here, the values of increment factor for allowable stress under the combination with environmental loads adopted in the current design code for steel highway bridges in Japan are not necessarily inappropriate from the probabilistic viewpoint except for earthquake.

Table 1. Proposed load combination factor values

Load combination	γ_D	γ_L	V_T	γ_T	V_W	γ_W	V_{EQ}	γ_{EQ}	remark*
D+L	0.90	0.90							
D+L+T	0.90	0.60	0.05	0.90					
			0.10	0.85					
			0.15	0.80					
D+L+W	0.90	0.65				0.85			②,③
					0.05	0.80			
					0.10	0.75			①
					0.15	0.70			
D+L+T+W	0.90	0.55	0.10	0.65	0.05	0.80			
					0.10	0.70			④
					0.15	0.65			
			0.05	0.80					
			0.10	0.70	0.10	0.65			②
			0.15	0.65					
D+EQ	0.85						0.10	0.85	
	0.80						0.50	0.70	
	0.75						0.90	0.65	
D+T+EQ	0.85		0.10	0.85			0.10	0.85	⑤
				0.80					⑥
				0.75					⑦
	0.80		0.10	0.80			0.50	0.70	⑤
				0.70					⑥
				0.65					⑦
	0.80		0.10	0.75			0.90	0.65	⑤
				0.65					⑥
				0.60					⑦

* ① D:L:W = 2:2:1 ② D:L:W = 1:1:1 ③ D:L:W = 2:2:3 ④ D:L:T = 1:1:1
⑤ D:T:EQ = 2:2:1 ⑥ D:T:EQ = 1:1:1 ⑦ D:T:EQ = 2:2:1

Safety, Reliability and Risk of Structures, Infrastructures and
Engineering Systems – Furuta, Frangopol & Shinozuka (eds)
© 2010 Taylor & Francis Group, London, ISBN 978-0-415-47557-0

Reduction of design live load for bridges by controlling the heavy traffics

S.-H. Kim, K.-I. Cho & M.-S. Choi
Yonsei University, Seoul, Korea

W.-S. Hwang
Inha University, Incheon, Korea

I.-T. Kim
Busan University, Busan, Korea

ABSTRACT

A recent trend of designing the infrastructures in advanced
countries is the expansion of considerations from involving only the structural safety and the economic feasibility
of construction to involving in addition performance levels for various interests. For this reason, the performance-
based design of structure is the design concept that does
not always follows the prescribed design criterion suggested by design specifications but uses suitable values
which may rational to the specified structure.

In the case of bridge structures, it is irrational to use
the same live load model without considering the practical
traffic condition. In fact, some bridges are suffering safety
problems because of frequent passage of heavy traffics.

In this study, a traffic control system using weigh-
in-motion(WIM) sensors is introduced to control heavy
vehicles passing through a bridge. This system is simply
filters over-weighted vehicles which are restricted to enter
bridge and controls the heavy vehicle to enter bridge more
rationally. To simulate this system, vehicle data including

their running patterns and load characteristics is determined based on their probabilistic model, based on Kim
et al., (1991)'s research.

To simulate the proposed traffic control system, following procedure is proposed.

1) Randomly produce an array of vehicles.
2) Perceive heavy traffic on ahead.
3) Let the following heavy traffic to enter suitable position.
4) After step 3) is finished, the simulator will perceive
   next heavy traffic.

To verify the effectiveness of proposed traffic control
system, maximum load effect analysis for daily traffic volume and annual traffic volume is performed using simply
supported girder bridge. Four bridges are prepared for
this analysis with different span length. Type-I Gumbel
distribution is used for analysis.

The analysis results are shown in Table 1 and 2. The
unit of load effects in table 1 and 2 is standardized unit
where specified live load for DB-24 or DL-24(Korean
ministry of land, transport and maritime affairs, 2005) is
assumed as 1. As shown in the tables, the maximum load
effect is decreased when proposed traffic control system
is activated.

REFERENCES

Kim. S.-H., Lee, J.-H., Park, H.-S. & Yang, Y.-S. (1991) Prob-
    abilistic analysis of bridge design loads, *Korea institute of
    construction technology*, Final report, No. 91-SE-113-2
    (in Korean)
Korean ministry of land, transport and maritime affairs (2005)
    Korean bridge design specification for highway bridges,
    Korean ministry of land, transport and maritime affairs

Table 1. Results of maximum load effect analysis without
any traffic control (95% maximum).

Span Length (m)	Heavy traffic composition rate: 25%		Heavy traffic composition rate: 40%	
	Daily Maximum	Annual Maximum	Daily Maximum	Annual Maximum
30	0.7228	1.0811	0.8302	1.1108
50	0.8641	1.1334	0.9925	1.1646
100	0.5370	0.7340	0.6168	0.7541
150	0.4408	0.5926	0.5063	0.6089

Table 2. Result of maximum load effect analysis with proposed traffic control system (95% maximum).

Span Length (m)	Heavy traffic composition rate: 25%		Heavy traffic composition rate: 40%	
	Daily Maximum (Decrement %)	Annual Maximum (Decrement %)	Daily Maximum (Decrement %)	Annual Maximum (Decrement %)
30	0.4896	0.8664	0.5013	0.8785
50	0.4749	0.8544	0.4862	0.8663
100	0.3687	0.5618	0.3776	0.5696
150	0.2856	0.4809	0.2925	0.4876

General Session (Random Vibration)

Safety, Reliability and Risk of Structures, Infrastructures and Engineering Systems – Furuta, Frangopol & Shinozuka (eds)
© 2010 Taylor & Francis Group, London, ISBN 978-0-415-47557-0

Approximate calculation method for integral of mean square value of nonstationary random response

S. Aoki & A. Fukano

Tokyo Metropolitan College of Industrial Technology, Tokyo, Japan

The response of the structure subjected to nonstationary random process such as earthquake excitation is nonstationary random process. Mean square value of the response is one of the representative values which describe the statistical characteristics of the response. Mean square value of the response is related to energy. Energy of the response is used to evaluate absorbed energy and cumulative damage of the structure. Calculating method for mean square value of the nonstationary random response is complicated. Then, some approximate methods are presented.

In this paper, an approximate method to calculate integral of mean square value of the nonstationary random response using mean square value of stationary random response is proposed. Nonstationary random excitation is assumed to be product of stationary random process and envelop function. The envelop function represents nonstationary amplitude characteristic of excitation. Mean square value of nonstationary random response is approximated as product of mean square value of stationary random response and square of envelop function. Integral of mean square values with respect to time form 0 to infinity are obtained. As an analytical model of the structure, a single-degree-of-freedom system is used.

First, the nonstationary random excitation is given by multiplying stationary white noise by envelop function in order to examine the fundamental effectiveness of the proposed method. In this case, the dynamic characteristics of the ground are not considered. Exact value of integral of mean square value of the response is obtained from autocorrelation function. Approximate values of integral of displacement response and velocity response are obtained for some values of the damping ratio ζ and the natural period T_n of the structure and compared with exact values. It is shown that the proposed method gives exact values of integral of mean square value of the response. Table 1 shows results for some values of the damping ratio.

Second, the nonstationary random excitation is given by multiplying stationary filtered white noise by envelop function. In this case, the dynamic characteristics of the ground are considered. As ground model, Tajimi model is used. Exact value of integral

Table 1. Integral of mean square value of response ($T_n = 0.5$ s).

ζ	Displacement (m²·s)		Velocity (m²/s²·s)	
	Exact	Approximate	Exact	Approximate
0.01	8.44×10^{-1}	8.44×10^{-1}	1.33×10^2	1.33×10^2
0.02	4.22×10^{-1}	4.22×10^{-1}	6.65×10	6.67×10
0.05	1.69×10^{-1}	1.69×10^{-1}	2.65×10	2.67×10
0.10	8.44×10^{-2}	8.44×10^{-2}	1.31×10	1.33×10

of mean square value of the response is obtained from the moment equations. Approximate values of integral are obtained for some values of the damping ratio and the natural period of the structure and the dynamic characteristics of the ground and compared with exact values. It is shown that the proposed method gives exact values of integral of mean square value of the response.

An approximate method to calculate integral of mean square value of the nonstationary random response using mean square value of stationary random response is proposed. Mean square values of stationary random process for abovementioned systems are given as formula. The proposed method is a practical method to obtain integral of mean square value of nonstationary random response.

Safety, Reliability and Risk of Structures, Infrastructures and
Engineering Systems – Furuta, Frangopol & Shinozuka (eds)
© 2010 Taylor & Francis Group, London, ISBN 978-0-415-47557-0

A spectral approach for the analysis of pedestrian-induced vibrations of footbridges

F. Tubino & G. Piccardo
University of Genoa, Genoa, Italy

ABSTRACT

Modern footbridges are very slender structures, often characterized by natural frequencies in the vertical and lateral directions within the range of the dominant walking harmonics; thus, they can be very sensitive to walking-induced vibrations. Their serviceability assessment is based on a comparison between the pedestrian-induced acceleration and a suitably defined limit value (FIB 2005).

Until the nineteen nineties, standard codes (e.g. BS5400 1978, 2001) required the analysis of vertical pedestrian-induced vibrations of footbridges considering a loading model constituted by a single pedestrian, modeled as a deterministic, resonant, moving harmonic load. This loading model is conventional and it is not representative of real pedestrian traffic conditions. From a general point of view, real pedestrian traffic conditions should be probabilistically modeled (Zivanovic et al 2007, Sahnaci & Kasperski 2005), considering several sources of randomness among which pedestrian arrivals, walking frequencies and velocities, force amplitudes and pedestrian weights.

The authors of the present paper have statistically analyzed a realistic scenario modeling pedestrians arrivals as a Poisson random variable and schematizing pedestrian pacing frequencies, walking velocities, force amplitudes and weight as normally distributed (Tubino & Piccardo 2008, Piccardo & Tubino 2009). Furthermore, an equivalent synchronization factor has been introduced as a function of suitable non-dimensional parameters, in order to allow an easy estimate of the maximum dynamic response.

In the present paper, the dynamic response to stationary streams of pedestrians is analyzed through a spectral approach, introducing models for the power spectral density function and for the coherence function of pedestrian-induced forces, considered as a stationary random process. The problem is tackled with the classical methods of linear random dynamics and the maximum dynamic response is determined in closed-form using an approach similar to that used to study dynamic effects of turbulent winds (Brownjohn et al. 2004). Hypotheses validating this kind of approach are briefly discussed. A preliminary comparison with techniques based on equivalent distributions of uniform resonant loads (e.g., Piccardo & Tubino 2009) points out some differences in the evaluation of the structural peak acceleration. A discussion about the choice of scenarios to be studied for footbridge serviceability analyses appears fundamental in order to validate the possibility of using a spectral approach like the one here proposed. In particular, the definition of a conventional pedestrian flow duration (e.g., in wind engineering the gust-excited response of structures is estimated considering time intervals of 10 minutes) is an important, open point to be still discussed.

REFERENCES

Brownjohn, J.M.W., Pavic, A., Omentzetter, P. 2004. A Spectral Density Approach for Modelling Continuous Vertical Forces on Pedestrian Structures Due to Walking. *Canadian Journal of Civil Engineering* 31(1): 65–77.

BS5400. 1978, 2001 *Steel, Concrete and Composite Bridges*. Specification for loads, Part 2 (BD37/01, Appendix B). London: British Standard Institution.

FIB. 2005. Guidelines for the design of footbridges. Bulletin 32. Lausanne: FIB.

Piccardo, G. & Tubino, F. 2009. Simplified procedures for the vibration serviceability analysis of footbridges subjected to realistic walking loads, *Computers and Structures*, submitted for publication.

Sahnaci, C. & Kasperski, M. 2005. Random loads induced by walking. In: Soize C, Schuëller GI, editors. *Proceedings of Structural Dynamics, EURODYN 2005*, Rotterdam: Millpress. 441–446.

Tubino, F. & Piccardo, G. 2008. Deterministic and stochastic approaches in the vibration serviceability assessment of pedestrian bridges, *Proc. VII European Conference on Structural Dynamics, EURODYN 2008*.

Zivanovic, S., Pavic, A., Reynolds, P. 2007. Probability-based prediction of multi-mode vibration response to walking excitation. *Engineering Structures* 29: 942–954.

Safety, Reliability and Risk of Structures, Infrastructures and
Engineering Systems – Furuta, Frangopol & Shinozuka (eds)
© 2010 Taylor & Francis Group, London, ISBN 978-0-415-47557-0

PDF solution to hardening hysteretic systems excited by Gaussian white noise

G.K. Er, H.T. Zhu, V.P. Iu & K.P. Kou
Department of Civil and Environmental Engineering, University of Macau, Macao, P.R. China

ABSTRACT

The Bouc-Wen hysteretic model is popularly adopted
to simulate the dynamic behavior of hysteretic restoring force. It was proposed by Bouc (1967) and further
developed by Wen (1976) and others. It is a smooth
and versatile model described with a differential equation which represents a wide range of hysteretic shapes
corresponding to different parameter values.

Initially Galerkin method (Wen 1976) was applied
to analyze the second moments of the hysteretic system responses. Thereafter, equivalent linearization
(EQL) method was extensively employed to analyze
the hysteretic systems (Hurtado and Barbat 2000). The
associated assumption in the standard EQL method is
made that all the system responses are jointly Gaussian. This assumption can be violated in view that the
hysteretic restoring force is a highly nonlinear function
of displacement and velocity. Cumulant-neglect closure technique was utilized to evaluate the moments
of system responses. This technique is suitable for the
system which responses are nearly Gaussian because
it is based on the assumption that the higher order
cumulants are neglected, which is valid only if the
system responses are nearly Gaussian. In order to estimate the non-Gaussian behavior, some non-Gaussian
PDFs, such as generalized exponential-type distribution, multi-dimensional Edgeworth expansion, truncated bivariate Gram-Charlier series expansion and a
linear combination of Gaussian and Dirac densities,
were introduced to improve the accuracy of the EQL
procedure and the cumulant-neglect technique, respectively. With the non-Gaussian PDFs introduced to the
EQL procedure and the cumulant-neglect closure technique, the accuracy of the obtained second moments
of system responses can be improved. Stochastic averaging method was also investigated in analyzing the
PDF solution of the Bouc-Wen hysteretic systems. It
is known that the tail behavior of the PDF of system
responses has a significant influence on the reliability
analysis of a system and it was less addressed.

In this paper, the exponential-polynomial closure
(EPC) method (Er 1998) is adopted to analyze the PDF
solutions of the stochastic hysteretic systems incorporated with the Bouc-Wen model. The joint PDF of
the system responses and hysteretic restoring force
is assumed to be an exponential function of polynomial in state variables. Special measure is taken such
that the corresponding FPK equation is satisfied in the
weak sense of integration with the assumed PDF. The
problem of determining the unknown parameters in the
approximate PDF finally results in solving simultaneous nonlinear algebraic equations. In order to assess
the effectiveness of the EPC method in this case,
a nonlinear system incorporated with the Bouc-Wen
hysteretic model is studied numerically. The hardening behavior of the hysteretic model is considered in
the cases of either weak or strong excitations. Numerical results show that the PDFs obtained with the EPC
method are in good agreement with those from Monte
Carlo simulation, especially in the tail regions of the
PDFs. The analysis also reveals that the tails of the
PDFs are far from being Gaussian even if the excitation
level is low.

REFERENCES

Bouc, R. 1967. Forced vibration of mechanical systems with
hysteresis. Abstract, Proc. Fourth Conf. on Nonlinear
Oscillation, Prague, Czechoslovakia.
Wen, Y.K. 1976. Method for random vibration of hysteretic
systems. J. Eng. Mech., 102(2): 249–263.
Hurtado, J.E. and Barbat, A.H. 2000. Equivalent linearization
of the Bouc-Wen hysteretic model. Eng. Struct., 22(9):
1121–1132.
Er, G.K. 1998. An improved closure method for analysis
of nonlinear stochastic systems. Nonlinear Dyn., 17(3):
285–297.

Safety, Reliability and Risk of Structures, Infrastructures and Engineering Systems – Furuta, Frangopol & Shinozuka (eds)
© 2010 Taylor & Francis Group, London, ISBN 978-0-415-47557-0

Evaluation of the dynamic characteristics of the coupled system structure and occupants by simulation

E. Agu & M. Kasperski

Department of Civil and Environmental Engineering Sciences, Research Team EKIB, Ruhr-University Bochum, Bochum, Germany

ABSTRACT

Structures like stands in stadia and floors in assembly halls can be excited by dynamic crowd loads to considerable vibration amplitudes which may affect the structural safety as well as the comfort of the audience. The modelling of the crowd has to distinguish between active and passive persons. Active persons can be modelled as external loads. Since the human body forms a complex dynamic system with more than one natural frequency and considerable damping capacities, the coupled system of a structure and the occupants can hardly be described with its basic dynamic characteristics considering only the mass of the occupants. A simple description of the dynamic characteristics of the human body has been obtained by Griffin [L. Wei & M.J. Griffin (1998), Y. Matsumoto & M.J. Griffin (2003)], modelling the human body as a two degree of freedom system. For sitting and standing persons, two natural frequencies and the corresponding masses and damping capacities have been identified. Since the basic dynamic interaction between the structure and the occupants is non-linear, strictly speaking it is not appropriate to model the occupants with their averaged dynamic characteristics.

In a first step, based on the experimental work by Griffin, a separate probabilistic model for the random dynamic characteristics of the human body in sitting and standing posture is developed. The individual parameters are analysed by order statistics. From biomechanical reasons it can be concluded that the probability distributions of the different dynamic characteristics have to be limited to both end, i.e. there should be an upper value which can not be exceeded and there should be no negative values, i.e. the smallest possible value should be 0. The Beta-Distribution offers these basic features and therefore is used to describe the probability distributions of the two natural frequencies f_1 and f_2, the corresponding damping capacities D_1 and D_2 and the corresponding mass ratios m_1/m_2 and m_3/m_{tot} (with m_{tot} being the total body mass). As additional information, the probability distribution of the body weight for male and female persons, as it has been observed in Germany [Robert Koch Institute (1995)] is used.

The paper will present results from simulations for some example structures with different natural frequencies and different participation rates of the audience in the range from 0 (all active) to 1 (all passive). Furthermore, different ratios in regard to standing and sitting posture of the occupants are analysed. The occupants are modelled to have random individual dynamic characteristics. From each simulation run, the dynamic amplification function is obtained. Based on a sufficient large number of independent runs, the 5% and 95% fractile values of the amplification are identified.

REFERENCES

L. Wei & M.J. Griffin Mathematical model for the apparent mass of the seated human body exposed to vertical vibration, Journal of Sound and Vibration (1998), 212 (5), pp 855–874

Y. Matsumoto & M.J. Griffin Mathematical model for the apparent masses of standing subjects exposed to vertical whole body vibration, Journal of Sound and Vibration (2003), 260, pp 431–451

Robert Koch Institut, Berlin, 1995, Datenbank OWDB

Safety, Reliability and Risk of Structures, Infrastructures and
Engineering Systems – Furuta, Frangopol & Shinozuka (eds)
© 2010 Taylor & Francis Group, London, ISBN 978-0-415-47557-0

Application of the tail-equivalent linearization method for stochastic dynamic analysis with asymmetric hysteresis

Salvatore Sessa
University Federico II, Napoli, Italy

Armen Der Kiureghian
University of California, Berkeley, CA, USA

ABSTRACT

The tail-equivalent linearization method (TELM) is used to investigate the stationary response of a system having a highly asymmetric hysteretic behavior and subjected to a discretized white-noise excitation.

TELM, is a non-parametric method for approximate solution of nonlinear stochastic dynamic problems. In this method, the tail-equivalent linear system is defined by equating its tail probability for a specified threshold with the first-order approximation of the tail probability of the nonlinear response for the same threshold. This equality is imposed by matching the so-called design points of the linear and nonlinear responses in the space of a vector of standard normal random variables, which are obtained by discretizing the Gaussian input excitation. Here, the "design point" refers to the point of linearization in the first-order reliability method (FORM) approximation of the tail probability problem. Fujimura and Der Kiureghian (2007) have shown that knowledge of the design point provides complete information about the tail-equivalent linear system in terms of its unit impulse response function. This information is obtained in a numerical form and does not require definition of a parameterized linear model. In this sense, TELM is a non-parametric equivalent linearization method. An important advantage of TELM is that it can accurately capture the non-Gaussian distribution of the nonlinear response in first-order approximation, particularly in the tail (small probability) regions. In particular, if the hysteretic law of the system is asymmetric, the predicted response distribution is not only non-Gaussian, but is also asymmetric. Thus, TELM offers significant advantages over the conventional ELM in nonlinear stochastic dynamic analysis, particularly when the hysteresis law of the structure is asymmetric.

Previous studies have investigated TELM for stationary and non-stationary excitation processes, but always in presence of symmetric hysteresis loops that lead to a zero-mean and symmetric response distribution. In this work, the performance of TELM for response analysis of stochastic dynamical systems with asymmetric hysteretic behavior is investigated. For this purpose, a smoothed version of the generalized Bouc-Wen model (Song and Der Kiureghian 2006) is developed, so that the response is continuously differentiable. It is found that TELM is able to capture the non-Gaussian and asymmetric distribution of both the point-in-time response, as well as the maximum response over a time interval (the first-passage probability distribution) for large thresholds (small exceedance probabilities). The distributions for the positive and negative thresholds are distinctly different, reflecting the difference in the behavior of the system for the two regimes. It is found, however, that TELM is unable to provide good accuracy for thresholds near zero. In particular, the cumulative probability approximation by TELM for a zero threshold is always 0.5, regardless of the shape of the hysteresis loop. This is a characteristic of FORM, which can only be overcome through a higher-order approximation method, such as SORM. Although probability values for small thresholds are not of interest in reliability analysis, work is currently in progress to construct a better approximation for small thresholds by use of SORM.

REFERENCES

Fujimura, K. and Der Kiureghian, A. 2007. Tail-equivalent linearization method for nonlinear random vibration. *Prob. Engrg. Mech.* 22: 63–76.

Song, J. and Der Kiureghian, A. 2006. Generalized Bouc-Wen model for highly asymmetric hysteresis. *J. Engrg. Mech., ASCE* 132: 610–618.

Safety, Reliability and Risk of Structures, Infrastructures and Engineering Systems – Furuta, Frangopol & Shinozuka (eds)
© *2010 Taylor & Francis Group, London, ISBN 978-0-415-47557-0*

Modeling nonlinear systems by Volterra series

L. Carassale
Department of Civil, Environmental and Architectural Engineering, University of Genova, Italy

A. Kareem
NatHaz Modelling Laboratory, University of Notre Dame, IN, USA

INTRODUCTION

The Volterra series is a mathematical tool widely employed for the representation of the input-output relationship of nonlinear dynamical systems. It is based on the expansion of the nonlinear operator representing the system into a series of homogeneous operators. Such operators are completely defined given the Volterra Volterra frequency-response functions (VFRF, Schetzen 1980), which can be reviewed as a generalization of the usual frequency-response function.

Applications of the Volterra series are present in several fields of engineering and physics and can be roughly classified into two distinct classes. In the first case, the Volterra series is used to build a model of an observed dynamical phenomenon and the VFRFs are estimated from experimental or numerically-generated data. This approach is applied with the purpose of realizing mathematical models able to simulate some observed physical behavior (e.g. Koukoulas et al. 1995) or to construct reduced models to reproduce selected features of a complex numerical model (e.g. Lucia et al. 2004). Other applications regard the analysis of dynamical systems that are already represented by an analytical model, for example a differential equation. In this case, the potentiality of the Volterra series representation has been used to investigate the behavior harmonically-excited nonlinear systems (e.g. Worden & Manson 2005) or to calculate the probabilistic response of randomly-excited systems (e.g. Kareem et al. 1995).

When an analytical model of the dynamical system is available, the VFRFs are usually calculated by means of the harmonic probing method, consisting in evaluating analytically the response of the system excited by products of harmonic functions with different frequencies (Bedrosian & Rice 1971). As discussed by Peyton Jones (2007), this operation is straightforward for simple systems, but may reach a prohibitive level of computational complexity when dealing with high-order nonlinear systems or for the calculation of a high-order VFRF.

A new technique concerning an alternative to the harmonic probing approach is presented here with the goal of simplifying the evaluation of the VFRFs of dynamical systems represented by analytical models and to enable the analysis of systems realized by combining analytical models and Volterra systems obtained by a numerical or experimental derivation. This method involves the representation of a complex dynamical system by an assemblage of simple operators for which VFRFs are readily available. According to this concept, the paper provides the VFRFs of some simple dynamical systems, and presents a set of rules for the evaluation of the VFRFs of composite systems. This approach makes the computation of the VFTFs very simple, at least from a formal point of view, enabling the use in symbolic manipulation software.

Two simple examples of wave and wind-excited single-DOF systems are considered for demonstration. Comparisons are made in terms of statistical moments of the structural response.

REFERENCES

Bedrosian, E. & Rice, S.O. 1971. The output of Volterra systems (nonlinear systems with memory) driven by harmonic and Gaussian inputs. *Proc IEEE* 59(12): 1688-1707.

Kareem, A., Zhao, J. & Tognarelli, M.A. 1995. Surge Response Statistics of Tension Leg Platforms Under Wind & Wave Loads: A Statistical Quadratization Approach. *Prob Engrg Mech* 10(4): 225–240.

Koukoulas P, Kalouptsidis N. 1995. Nonlinear system identification using Gaussian inputs. *IEEE Trans. Sign. Proc.* 43: 1831–41.

Lucia, D.J., Beranb, P.S. & Silva, W.A. 2004. Reduced-order modeling: new approaches for computational physics. *Progress in Aerospace Sciences*, 40, 51–117

Peyton Jones, J.C. 2007. Simplified computation of the Volterra frequency response functions of non-linear systems. *Mechanical Systems and Signal Processing*, 23: 1452–1468.

Schetzen, M. 1980. *The Volterra and Weiner theories of nonlinear systems*. New York: John Wiley & Sons, Inc.

Worden, K. & Manson, G. 2005. A Volterra series approximation to the coherence of the Duffing oscillator. *Journal of Sound and Vibration* 286: 529–547.

Organized Session (OS16) Safety Prediction and Evaluation of Wind-Induced Phenomena

Safety, Reliability and Risk of Structures, Infrastructures and Engineering Systems – Furuta, Frangopol & Shinozuka (eds)
© 2010 Taylor & Francis Group, London, ISBN 978-0-415-47557-0

Typhoon hazard analysis in southeast China coastal region with the CE wind field model

Y.F. Xiao & Z.D. Duan
School of Civil Engineering, Harbin Institute of Technology, Harbin, Heilongjiang, China

Y.Q. Xiao
Harbin Institute of Technology, Shenzhen, Graduate School, Shenzhen, Guangdong, China

ABSTRACT

Two aspects of improvements are introduced in doing the typhoon hazard analysis in southeast China coastal region. One is the U.S. Army Corps of Engineers (CE) wind model, which is the full nonlinear solution to the equations of motion for a translating typhoon and well demonstrated through comparison with observed typhoon wind speed records. Additionally, there will be some difficulties in obtaining the data of two important input parameters to the wind field model namely the Holland pressure profile parameter (B) and radius

to maximum winds (R_{max}) with the absence of information about those data of historical records used by previous studies in the China Meteorological Administration (CMA), so the formula presented by Jakobsen et al. (2004) and that by Li et al. (1995) are used to estimate the value of the B and R_{max}, respectively.

Combined with the results described above, modeling the filling model using the exponential function and other parameters using the previous method based on the data of typhoon entering into the site subregion with a diameter of 500 km, the typhoon hazard analysis in southeast China coastal region is done with the assumption that the track of typhoon is a line path. 1000 years of storms are simulated for each city to obtain the annual extreme series, based on which the goodness-of-fits of the cumulative density function (CDF) of both Weibull with tail length parameter $\gamma > 4$ and Gumbel distribution are tested reasonably well for the seven cities.

The results are shown in Table 1 and there is reasonable agreement between the design wind speeds with various return periods in this paper and those in the Load Code for the Design of Building Structures of China (GB 50009-2001) for most of the cities, indicating that the simulation method proposed in this paper is feasible to estimate the design wind speeds for critical structures in the major cities of China.

Table 1. 10-minute mean wind speeds (m/s) at 10m-level in B terrain as a function of return period for the seven cities.

City	Method	Distribution	Return period(years)		
			50	100	200
Hong Kong	*Code**	Gumbel	38.05	39.09	40.01
	This paper	Weibull	33.54	36.79	39.89
		Gumbel	34.22	38.00	41.77
Fu zhou	*Code**	Gumbel	33.61	37.04	39.88
	This paper	Weibull	34.36	37.59	40.66
		Gumbel	34.78	38.64	42.49
Shen zhen	*Code**	Gumbel	34.68	37.99	40.74
	This paper	Weibull	31.60	34.67	37.60
		Gumbel	32.57	36.24	39.89
Guang zhou	*Code**	Gumbel	28.30	31.00	33.25
	This paper	Weibull	27.18	29.66	32.03
		Gumbel	27.60	30.57	33.53
Hai kou	*Code**	Gumbel	34.67	37.98	40.74
	This paper	Weibull	33.21	36.13	38.88
		Gumbel	33.92	37.64	41.34
Ning bo	*Code**	Gumbel	28.30	31.01	33.25
	This paper	Weibull	33.27	36.80	40.20
		Gumbel	33.29	37.24	41.17
Zhan jiang	*Code**	Gumbel	35.83	39.04	41.73
	This paper	Weibull	30.48	32.89	35.14
		Gumbel	31.60	34.95	38.29

* Load Code for the Design of Building Structures of China (GB 50009-2001).

REFERENCES

Jakobsen, F. & Madsen, H. 2004. Comparison and further development of parametric tropical cyclone models for storm surge modeling. *Journal of Wind Engineering and Industrial Aerodynamics* 92(5):375–391.

Li, X.L., Pan, Z.D., and Yu, J. 1995. An adjustment method for typhoon parameters. *Journal of Oceanography of Huanghai & Bohai Seas* 13(2):11–15 (in Chinese).

Ministry of Construction. 2002. The Load code for the design of building structures (GB 50009-2001), *Construction Industry Press of China*, Beijing, China

Thompson, E.F. & Cardone, V.J. 1996. Practical modeling of hurricane surface wind fileds. *Journal of Waterway, Port, Coastal and Ocean Engineering*, ASCE 122(4):195–205.

Safety, Reliability and Risk of Structures, Infrastructures and Engineering Systems – Furuta, Frangopol & Shinozuka (eds)
© 2010 Taylor & Francis Group, London, ISBN 978-0-415-47557-0

Wind-induced lateral-torsional buckling analysis of long-span suspension bridge

H. Yamada, H. Katsuchi & E. Sasaki
Yokohama National University, Yokohama, Japan

D. Ishihara
Pacific Consultant Co. Ltd., Nagoya, Japan

ABSTRACT

Wind effect is one of the most important design concerns for a long-span bridge. Not only wind dynamic effect such as flutter but also wind-induced static instability such as divergence and lateral-torsional buckling should be carefully checked for a very long-span bridge. The static instability is usually believed to be stable more than dynamic instability such as flutter. However, past study showed that much longer span would exhibit the static instability at lower wind speed. The Wind-resistant Design Code for Honshu-Shikoku Bridges provides a calculation formula of the onset wind speed of wind-induced lateral-torsional buckling. However, there are not so many studies on the static instability of a long-span bridge compared with flutter instability.

This study focuses on the static instability of a long-span suspension bridge and investigates the possibility of numerical simulation for the onset of lateral-torsional buckling. A precise FEM bridge model of the Akashi Kaikyo Bridge where all truss members were modeled was developed as shown in Figure 1. Then three-component wind forces were applied to the model, and wind-induced response such as deformation (Figure 2) and possible lateral-torsoinal buck-

ling phenomenon was analyzed with consideration of aerodynamic and structural nonlinearity.

The paper presents main results of the analysis together with analytical procedures. Comparing the result with the past study, the possibility of FEM analysis for the onset of wind-induced lateral-torsional buckling and further improvement is discussed.

Figure 2. Wind-induced deformation at wind speed of 150 m/s.

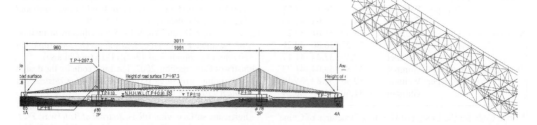

Figure 1. General plan and FEM model of Akashi Kaikyo Bridge.

Safety, Reliability and Risk of Structures, Infrastructures and
Engineering Systems – Furuta, Frangopol & Shinozuka (eds)
© 2010 Taylor & Francis Group, London, ISBN 978-0-415-47557-0

The problem of uncertainty in the measurement of aerodynamic derivatives

C. Mannini & G. Bartoli

CRIACIV-Department of Civil and Environmental Engineering, University of Florence, Florence, Italy

ABSTRACT

Modern long-span bridges are more and more sensitive to wind loads and aeroelastic phenomena, due to designs characterized by light structures and low frequencies of vibration. In particular flutter can lead to structural collapse, therefore an adequate and reliable safety margin with respect to the onset critical wind speed has to be guaranteed. Consequently, the limit of flutter stability is a severe and important constraint in the conception of long-span bridges. At the present state of the art, flutter assessment is based on wind-tunnel tests, either by the direct measure of the instability wind speed on a model that satisfies a certain number of similitude parameters or by the measure of a series of unsteady aerodynamic coefficients, the so-called aerodynamic derivatives (Scanlan & Tomko 1971). These coefficients are also important to predict the response of the bridge to turbulent wind.

Aerodynamic derivatives are usually treated as deterministic functions but it is clearly shown in Figure 1 and in previous publications (Bartoli & Mannini 2005, Mannini 2006) that wind-tunnel measures are affected by significant dispersion. In case the measurements are performed with the common free-vibration technique, the scatter tends to increase with the wind speed and becomes particularly large near the flutter onset. In this paper the probabilistic description of aerodynamic derivatives is outlined on the basis of a wind-tunnel test campaign on a common bridge deck geometry, expressly conceived for this purpose (Mannini 2006). It is found that at a given reduced wind speed the random variables corresponding to the aerodynamic derivatives can be considered as normally distributed. In addition, the correlation between these random variables is investigated. It turns out that at high wind speed and above all near the flutter onset aerodynamic derivatives are correlated but less than expected. As a matter of fact, strong correlation is found between the coefficients called H_1^* and H_2^* or A_1^* and A_2^*. In contrast, no significant correlation is observed, for instance, between lift and moment aerodynamic derivatives.

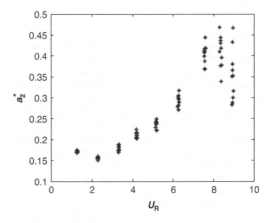

Figure 1. Dispersion of the wind-tunnel measures of the aerodynamic derivative $a_2^* = -KA_2^*$. U_R is the reduced wind speed.

A model t o account for the effects on the flutter boundaries of the uncertainty affecting the aerodynamic derivatives is proposed. The probability distribution of the flutter critical wind speed is obtained via Monte-Carlo simulation and can be used in the framework of the Performance-Based Design approach. At present the improvement of the probabilistic model is underway.

REFERENCES

Bartoli, G. & Mannini, C. 2005. Reliability of bridge deck flutter derivative measurements in wind tunnel tests. In G. Augusti, G.I. Schuëller & M. Ciampoli (eds.), *Proc. IX International Conference on Structural Safety and Reliability, Rome, Italy, 19–23 June 2005.* Rotterdam: Millpress: 1193–1200.

Mannini, C. 2006. *Flutter Vulnerability Assessment of Flexible Bridges.* Ph.D. thesis, University of Florence, Italy – TU Braunschweig, Germany. Saarbrücken: Verlag Dr. Müller (2008).

Scanlan, R.H. & Tomko, J.J. 1971. Airfoil and bridge deck flutter derivatives. *J. Eng. Mech. Div.* 97(6): 1717–1737.

Safety, Reliability and Risk of Structures, Infrastructures and
Engineering Systems – Furuta, Frangopol & Shinozuka (eds)
© 2010 Taylor & Francis Group, London, ISBN 978-0-415-47557-0

Simulation of wind speeds using a hidden Markov chain model

Rwey-Hua Cherng & Shih-Che Kao
Department of Construction Engineering, National Taiwan University of Science and Technology, Taipei, Taiwan

Jian-Ye Ching
Department of Civil Engineering, National Taiwan University, Taipei, Taiwan

ABSTRACT

Determination of design wind speeds and the associated uncertainty is the most essential step in wind-resistant structural designs and in quantatative wind risk assessments. Design wind speeds are usually obtained by analyzing the recorded field wind speeds using various statistical methods. However, recorded field wind data are often quite limited and are not sufficient for reliable statistical analyses, especially when one is predicting long-return-period design wind speeds or directional design wind speeds. Accordingly, the generation of artificial wind speeds whose statistical properties are compatible to those from observed wind speeds is a critical research topic.

This paper proposes a non-stationary auto-regressive model with a hidden Markov chain to analyze field wind speeds and then to simulate synthetic wind speed time series. Compared to the models adopted by previous researches, the chosen model is different in the way that it is a non-stationary model. Moreover, new algorithms are developed to generate synthetic wind speeds based on the model and field wind speeds.

The daily wind speeds at a weather station are collected to demonstrate the implementation of the proposed approach. The long-term trend is first modeled and filtered out of the collected data. The empirical cumulative distribution function derived from the filtered data is modified by a generalized Pareto distribution; the resulting compound distribution function is used for transforming the filtered data into 'standard' data with a Gaussian distribution.

Next, a non-stationary auto-regresive model with random coefficients modeled by a hidden Markov chain is developed for analyzing the 'standard' data; the variation of standard deviations in a year is shown in Figure 1. The maximum-likelihood estimates for the variances of daily winds are computed by an Expectation-Maximization algorithm; the random coefficients are sampled based on a Kalman filter algorithm. It is shown in the current study that the statistical properties (e.g., means, standard deviations, auto-correlation functions as well as exceedance probabilities (Figure 2)) of the

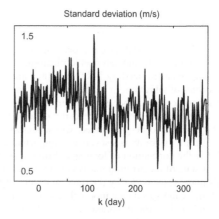

Figure 1. Plot of variation of standard deviations in a year.

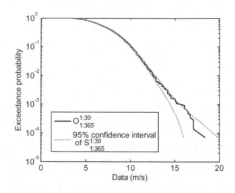

Figure 2. Comparison of exceedance probability curves of $O_{1:365}^{1:39}$ and $S_{1:365}^{1:39}$.

simulated wind speeds are quite close to those of the field wind speeds. In addition, the simulated wind speeds exhibit the similar non-stationarity inherent in the field wind speeds; it reveals that the proposed model is capable of simulating non-stationary wind speeds.

The proposed model can be extended to deal with wind speeds and associated wind directions simultaneously; the results will appear in another paper.

Safety, Reliability and Risk of Structures, Infrastructures and
Engineering Systems – Furuta, Frangopol & Shinozuka (eds)
© 2010 Taylor & Francis Group, London, ISBN 978-0-415-47557-0

Specification of the design wind load considering long term trends in the wind climate

M. Kasperski & N. Aylanc

Department of Civil and Environmental Engineering Sciences, Ruhr-University Bochum, Bochum, Germany

ABSTRACT

Any extreme value analysis of the wind climate suffers from uncertainties due to the confined observation period. Even for a stationary wind climate 55 years are not enough to identify the describing parameters. Then, the evaluation of the statistical significance of trends becomes impossible.

For a consistent theoretical model, at least the intensity of storms and the number of storms per year are required [1]. In the actual study, it is assumed that the intensity of storms follows a Generalized Pareto Distribution [2] with the two describing parameters scale s and shape k. The threshold value v_s is treated as deterministic. Furthermore, it is assumed that the number of storms per year follows a Poisson Distribution. Then, as third parameter the average number of storms per year λ is obtained. Target values for the non-exceedance probability of the design wind speed are specified for four structural classes with reference to the design working life.

In a first step, the new approach estimates the probability that for any triple (s, k, λ) in J observation years the observed triple (s_{obs}, k_{obs}, λ_{obs}) is obtained (figure 1).

Next, for each triple (s, k, λ) a design wind speed is calculated. The obtained design wind speeds are sorted in ascending order together with their corresponding probabilities. The plot of the cumulative probabilities versus the sorted design wind speeds gives the range of possible design wind speeds for the observed triple (s_{obs}, k_{obs}, λ_{obs}). Choosing an appropriate confidence interval than allows the final specification of the design wind speed in terms of a best estimate. Analogue to the recommendations in the Eurocode [3] for 'design by testing' the one-sided 75% confidence is used.

The new method is applied to observation data of the meteorological station at Düsseldorf airport. The straight-forward estimation of the design wind speed based on the observed pair (s_{obs}, k_{obs}) leads to $v_{des} = 23.34$ m/s. The best estimate of v_{des} is about 10% larger.

Small trends are observed for both, the intensity and the number of storms. Each trend has a statistical significance of about 80%, which means that the occurrence of both trends is about 96%.

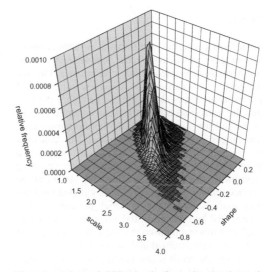

Figure 1. Joint-probability density for getting the observed parameters s_{obs} and k_{obs} from an ensemble with 92 events

Considering the observed trends in the estimation of the design wind speed requires to introduce the start of exposure as further influencing parameter. Basically, increasing trends will lead to an increase of the design wind speed. The effect becomes larger if the starting of exposure is far from the centre of the trends. For the example, the increase of the design wind speed is about 14% for a class B structure with a design working life of 50 years compared to the assumption of a stationary wind climate.

REFERENCES

M. Kasperski 2002 A new wind zone map of Germany, Journal of Wind Engineering and Industrial Aerodynamics (2002),90, pp 1271–1287.

J.D. Holmes & W.W. Moriarty 1999 Application of the generalised Pareto distribution to wind engineering, Journal of Wind Engineering and Industrial Aerodynamics (1999),83, pp 1–10.

Eurocode – Basis of structural design, German version EN 1990:2002

Safety, Reliability and Risk of Structures, Infrastructures and
Engineering Systems – Furuta, Frangopol & Shinozuka (eds)
© 2010 Taylor & Francis Group, London, ISBN 978-0-415-47557-0

Improved spectral analysis of turbulent wind response

Steen Krenk

Department of Mechanical Engineering, Technical University of Denmark, Kgs. Lyngby, Denmark

Abstract

Some basic properties of the spatial covariance structure of gusty wind loads have been extracted from the theory of convected isotropic turbulence. The two principal results concern the transverse coherence of the turbulent wind velocity field. The theory predicts that the transverse coherence must integrate to zero over a full transverse plane, and therefore must contain ranges with negative coherence. This confirms the intuitive observation that a local observation of a positive velocity component does not imply a positive probability of a positive net flow over an infinitely extended plane. This basic result is violated by common analysis procedures, in which the coherence is represented by an exponential function. As a consequence of this result the coherence appears more local that commonly assumed in design calculations.

The second result is concerned with the coherence of the wind field in the low-frequency limit. It is commonly assumed that the coherence can be described in terms of the non-dimensional distance $\omega r/U$, where r is the separation, ω the angular frequency, and U is the mean wind speed. Such a relation would imply that the zero frequency limit would have full coherence over the entire transverse plane. This seeming paradox is explained by the fact that the theoretical coherence for a wind spectrum with a length-scale λ is a function of the combined variable $[\lambda^{-2} + (\omega/U)^2]^{1/2}r$, with the well defined finite lower limit r/λ. This leads to a finite coherence less than unity at any frequency for all points with finite separation. In a design context it is of interest that this leads to a reduction of the low-frequency part of the turbulent wind load on large structures.

Turbulent wind load generally lead to a combination of resonant response at the natural frequency, and a non-resonant response at the frequencies below the natural frequency, generated by the non-white character of the wind spectrum. It is demonstrated via a suitable rational approximation for the wind load spectrum, that the total response variance σ_x^2 can be expressed in the form

$$k^2 \sigma_x^2 \simeq \frac{\pi \omega_0}{2\zeta} S_f(\omega_0) + \sigma_f^2 \left[\frac{(\omega_0 \lambda/U)^2}{1 + (\omega_0 \lambda/U)^2} \right]^2$$

where k is the structural stiffness, $S_f(\omega)$ is the spectral density of the wind load f, and σ_f^2 is its variance. The first term is the classic resonant white noise approximation inversely proportional to the damping ratio ζ. The second term is an approximate form of the response to 'background' turbulence. The magnitude of this term depends on the magnitude of the structure resonance frequency ω_0 to a reference frequency U/λ formed by the ratio of the mean wind speed U to the length scale λ of the turbulence.

The wind field and response results can be combined into an improved explicit design procedure, in which the spatial characteristics of the wind field are represented by a consistent transverse coherence function, and the variance of the response is obtained by a consistent combination of resonant and non-resonant parts.

Organized Session (OS08) Machine Learning in Structural Reliability and Probabilistic Mechanics

Safety, Reliability and Risk of Structures, Infrastructures and Engineering Systems – Furuta, Frangopol & Shinozuka (eds)
© 2010 Taylor & Francis Group, London, ISBN 978-0-415-47557-0

Life cycle monitoring for urban public infrastructure based on classification techniques

G.M. Atanasiu
Department of Civil Engineering, Technical University of Iaşi, Romania

M.H. Zaharia & F. Leon
Department of Computer Science and Engineering, Technical University of Iaşi, Romania

ABSTRACT

This paper presents a methodology for monitoring seismic structural vulnerability, while taking into account the damages revealed during seismic hazard scenarios, aiming at the identification of the seismic serviceability and operability of constructions located in seismic zones.

By using GIS-based visualization of geospatial data, one can develop a useful scenario in order to improve the knowledge on structural seismic vulnerability for city built infrastructure and, at the same time, to provide for stakeholders the needed support for their decisions for the management of possible emerging crises and/or for planning the rehabilitation programs at the local or governmental level.

Modeling, simulation, and analysis scenario based on linear or non-linear dynamic procedures are described, with applications to some different classes of models of buildings from the existing urban infrastructure of Iasi municipality, the second largest city of Romania, located in the North-Eastern region of the country.

An improved methodology is presented for monitoring the urban risk management using the instruments of the information society is presented. The approach is obtained by mixing the intelligent agent based cluster computing technologies, distributed GIS systems with decision support add-ons and the EDRI model.

The main idea was to offer enough computer power, needed for seismic risk analysis, at reasonable prices, and also to create a distributed knowledge database that will automatically make continuous improvements into the local models of the urban structures by direct simulations and also by exchanging information with other related urban computing centers. The idea is feasible for medium developed economies that need urban risk management but cannot afford the creation of dedicated supercomputing centers.

The present urban profile of Iasi municipality permits the identification of different classes of public buildings some of them residential, typical of the urban development of this category of cities in Romania.

A set of classification techniques such as decision trees and elliptical basis axis parallel neural networks are used as tools for the awareness and monitoring of seismic effects taking into account possible future disasters in the urban areas.

REFERENCES

Atanasiu G. and Gâlea D. (eds.), 2005. GIS monitoring of urban seismic risk, Politehnium, Iasi, Romania.

Davidson, R. A., 1997. An urban Earthquake Disaster Risk Index, BLUME-121, Report No. 121, The John A. Blume Earthquake Engineering Center Stanford, California, http://blume.stanford.edu/pdffiles/Tech%20Reports/TR12_Davidson.pdf.

FEMA, 1999. HAZUS, Earthquake Loss Estimation Methodology, Technical Manual, National Institute of Building Sciences for the Federal Emergency Management Agency.

Leon, F. and Atanasiu G. M., 2006. Data Mining Methods for GIS Analysis of Seismic Vulnerability, Proceedings of the First International Conference on Software and Data Technologies, ICSOFT 2006, INSTICC Press, Portugal, vol. 2, pp. 153–156,

Leon, F., Zaharia, M. H. and Atanasiu, G. M., 2008. Adaptive DOTNET Multiagent System Based on JADE and CLIPS for Creating Seismic Risk Scenarios, Management & Marketing.

Sprague, R. H. and Carlson, E. D., 1982. Building effective decision support systems. Englewood Cliffs, N.J., Prentice-Hall.

Safety, Reliability and Risk of Structures, Infrastructures and
Engineering Systems – Furuta, Frangopol & Shinozuka (eds)
© 2010 Taylor & Francis Group, London, ISBN 978-0-415-47557-0

Response surface reliability analysis of steel plates with random fields of corrosion

A.P. Teixeira & C. Guedes Soares

*Centre for Marine Technology and Engineering (CENTEC), Instituto Superior Técnico, Technical University of
Lisbon, Lisboa, Portugal*

ABSTRACT

Reliability methods are widely used in structural
design and, particularly, to assess the implicit levels of safety of structural systems and components.
In many cases of practical importance, particularly
for real structures, the evaluation of each realization of the limit state function often requires the
use of advanced non-linear finite element analysis.
First and Second Order Reliability Methods (FORM/
SORM), developed mainly to deal with explicit limit
state functions, and Monte Carlo simulation methods, are well established techniques that have been
successfully used for reliability assessment of structures, as reviewed by ()Rackwitz, (2001). In these
cases, direct or even advanced Monte Carlo simulation may become not feasible if the deterministic
structural analysis is time-consuming. First and Second Order Reliability Methods (FORM/SORM) can
also be adapted to handle with implicit limit state functions via direct coupling of the finite element code and
the FORM/SORM code. However, the solution of reliability problems in connection with non-linear finite
elements calculations may be not practical, particularly, when all sources of uncertainty are taken into
account including the spatial variability of structural
parameters represented by random fields discretized
into a high-dimensional random vector.

Among the techniques available to cope with such
problems, the Response Surface Method (RSM) has
proved to be an efficient and widely applicable method
in structural reliability analysis. The basic idea in utilizing the Response Surface Method is to replace the
true limit state function by an approximation, the so-
called response surface. The functions are typically
chosen to be first- or second-order polynomials that
substitute the real limit state function only at the
neighborhood of the design points, where their contribution to the total failure probability is more important
((()Bucher and Bourgund, (1990); ()Rajashekhar and
Ellingwood, (1993); ()Kim and Na, (1997); ()Das and
Zheng, (2000)).

Although the response surface method has been traditionally restricted to problems of small dimension of

the uncertainty space it can be extended to larger problems involving random fields as the derivatives of the
real implicit limit state function are avoid.

In this paper, the efficiency of the application of
Response Surface Methods in structural reliability
of steel plates with random fields of corrosion is
assessed. The limit state function is defined in terms of
the ultimate strength of the corroded plate calculated
by nonlinear finite element analysis, as proposed by
()Teixeira and Guedes Soares, (2008). The approach
is first illustrated with an example of the ultimate
strength of plates under in-plane compression with
random imperfections and material properties and the
technique is then applied to plates with random fields
of corrosion discretized using the Expansion Optimal
Linear Estimation method proposed by ()Li and Der
Kiureghian, (1993).

REFERENCES

Bucher, C. G. and Bourgund, U., (1990), A fast and effe-
cient response surface approach for structural reliability
problems, *Structural Safety*, Vol. 7, pp. 57-66.
Das, P. K. and Zheng, Y., (2000), Cumulative formation
of response surface and its use in reliability analysis, *Probabilistic Engeneering Mechanics*, vol. 15, pp.
309-15.
Kim, S.-H. and Na, S.-W., (1997), Response surface method
using vector projected sampling points, *Structural Safety*,
(1), 19, pp. 3-19.
Li, C.-C. and Der Kiureghian, A., (1993), Optimal discretization of random fields, *Journal of Eng. Mechanics*, N. 6,
Vol.119, pp. 1136-1154.
Rackwitz, R., (2001), Reliability analysis - a review and some
perspectives, *Structural Safety*, Elsevier, vol. 23, pp. 365-
395.
Rajashekhar, M. R. and Ellingwood, B. R., (1993), A new look
at the response surface approach for reliability analysis,
Structural Safety, Vol. 12, pp. 205-220.
Teixeira, A. P. and Guedes Soares, C., (2008), Ultimate
strength of plates with random fields of corrosion, *Structure and Infrastructure Engineering*, No. 5, Vol. 4, pp.
363-370.

Safety, Reliability and Risk of Structures, Infrastructures and Engineering Systems – Furuta, Frangopol & Shinozuka (eds)
© 2010 Taylor & Francis Group, London, ISBN 978-0-415-47557-0

Patchwork approximation scheme for reliability assessment and optimization of structures

S. Pannier, J.-U. Sickert & W. Graf
Institute for Structural Analysis, TU Dresden, Dresden, Germany

ABSTRACT

The design of engineering structures is mainly focused on an ideal performance under compliance of high reliability levels for the intended lifetime. The realization of such challenging tasks requires ambitious simulation techniques for optimization and reliability assessment. Naturally, this is always linked to a repetition of the structural analysis. In view of industry-sized problems, such a structural analysis is mostly based on Finite Element analyses of large nonlinear systems. High requirements at the precision of determined results force the computation time up to several hours. This impedes an application of the addressed simulation techniques in its crude form.

Increasing the numerical efficiency is a central demanding issue. In accordance to Simpson et al. 2001 appropriate methods replace the response surface, constituted by results of the structural analysis, by approximations schemes, so called metamodels. Those response surface approximation (RSA) functions reproduce the functional relationship between input and result quantities. Typically, they are defined global, setting the respective design of experiments without any inside into the specific problem. In consequence, the approximation quality may not be sufficient in all parts of interest. This fact contradicts the desire in reliability assessments to determine even small failure probabilities to a high level of accuracy. Likewise, in optimization it is dispensable to approximate the functional relationship in a global manner.

The idea of the proposed approach is to apply the RSA functions in sections to improve the approximation quality of local function features. Hence, a patchwork-like result is obtained.

The design of a patch is not bounded to any requirements but rather user-defined. Thus, it may be best constituted in dependence of the respective problem and the available set of input-output pairs. Varying the patch size patchwork approximations features, as a respective special case, both a pure deterministic approach and a global approximation scheme.

The design of experiments (Sacks et al. 1989) for the patchwork approximation scheme, opposite to global approximation schemes, is not performed in advance but rather adapted to the regions of interest. This sidesteps a generation of information in insignificant regions, e.g., in reliability assessments and optimization of structures. In consequence, under a constant approximation quality the numerical efficiency can be improved.

The approximation of the functional relationship within a patchwork approach may be performed with arbitrary approximation schemes. Generally, the behavior of local parts of the response surface is less complex than the function features of the global response surfaces. Thus the requirements on an approximation scheme of local function features are less rigorous. To preserve a high degree of generality and flexibility an application of sophisticated approximation schemes is reasonable.

The proposed approach is demonstrated by means of examples. Thereby, the respective approximation schemes apply neural networks (Haykin 1999) as surrogate model. The advantages of the patchwork approach in comparison to global approximation schemes are exemplified with benchmarks. Finally, the applicability is demonstrated for an industry-sized problem of a sheet metal forming process.

REFERENCES

Haykin, S. 1999. *Neural Networks: A Comprehensive Foundation.* New York: Prentice Hall.
Simpson, T., Poplinski, J., Koch, P.N. & Allen, J. 2001. Metamodels for computer-based engineering design: Survey and recommendations. *Engineering with computers* 17(2): 129–159.
Sacks, J., Welch, W.J., Mitchell, T.J. & Wynn, H.P. 1989. Design and analysis of computer experiments. *Statistical Science* 4(4) 409–435.

Safety, Reliability and Risk of Structures, Infrastructures and
Engineering Systems – Furuta, Frangopol & Shinozuka (eds)
© 2010 Taylor & Francis Group, London, ISBN 978-0-415-47557-0

Sampling-based Fast Probability Analyzer (FPA) for reliability-based maintenance optimization

Y.-T. Wu
Applied Research Associates, Inc., Raleigh, NC, USA

M. Shiao
FAA William J. Hughes Technical Center, Atlantic City International Airport, NJ, USA

ABSTRACT

Mechanical and structural systems often use scheduled inspections and maintenance to sustain structural integrity throughout the service life. Developing maintenance plans under various uncertainties will benefit from probabilistic analyses of damage accumulations, detections, and mitigation. Given the wide spectrum of maintenance options and the complexities in risk modeling, random sampling-based methods are preferred because of their computational robustness and simulation flexibility. This paper describes an efficient random simulation-based computational approach featuring a two-stage sampling framework and a new Adaptive Stratified Importance Sampling (ASIS) method

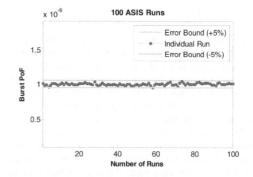

Figure 3. Error-Controlled ASIS Results.

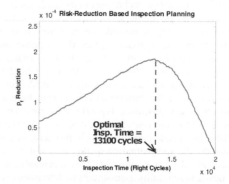

Figure 4. Risk Reduction Plot for Inspection Optimization.

for computing probability of failure with error control and for generating failure samples for maintenance simulations. The new approach has been implemented in a Fast Probability Analyzer (FPA) software. Selected examples are presented to demonstrate the efficiency and accuracy of FPA for Reliability-Based Maintenance Optimization (RBMO) applications.

Using demonstration examples, Fig. 1 compares failure samples generated by MCMC and ASIS. Fig. 2 shows a probability-of-failure convergence history using ASIS.

Fig. 3 shows that the 100 ASIS runs produced results within the specified 5% error. Using failure samples, an optimal inspection time was obtained as shown in Fig. 4.

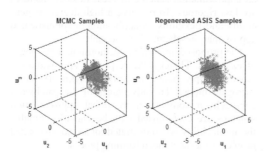

Figure 1. Failure Samples Generated by MCMC and ASIS.

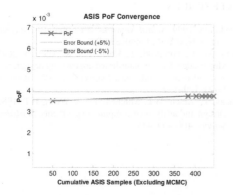

Figure 2. ASIS Probability Convergence History.

Safety, Reliability and Risk of Structures, Infrastructures and
Engineering Systems – Furuta, Frangopol & Shinozuka (eds)
© 2010 Taylor & Francis Group, London, ISBN 978-0-415-47557-0

Safety evaluation of thermal radiation from tank fire in petrochemical plants based on Monte Carlo method

Y. Yamanoi, T. Kurashiki, H. Ueda & M. Zako

Osaka University, Osaka, Japan

ABSTRACT

In a petrochemical plant, there are a lot of combustible
materials in tanks. If fire occurred in the plant, people and structures are exposed. It is important for the
safety and reliability of chemical plants to estimate
the spread of damages which will be caused by these
materials. However, it is very difficult to estimate the
dangerous area caused by the disaster due to complex phenomena, such as ignition, heat radiation from
fire, explosion pressure and so on. If the affected area
caused by disaster can be estimated with numerical
simulation, it will become a very useful tool for safety
and reliability of chemical plants.

In the previous study, we had developed the computer simulation program in order to estimate the
spread of fire, and proposed a probabilistic estimation
approach of the risk based on the developed simulation
(Kurashiki et al. 2005). However, the effect of distribution of wind velocity in the height of tank on the
dangerous area had not been considered in the previous study. Heat radiation is greatly affected by wind
and shielded by structures or slope.

Therefore, a flame model considering wind effects
is important in order to evaluate the heat radiation.
The commonly used models of a flame considering
wind effects have the shape of inclination of cylinder
as shown in Fig. 1(a).

However, in the conventional method for heat radiation from tank fire, the distribution of wind velocity

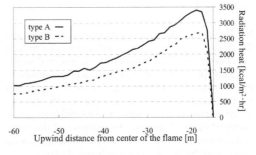

Figure 2. The relation between heat radiation and upwind
distance from tank fire.

in vertical direction and upward plume had not considered completely. Generally, wind velocity increase
toward to upper direction. In the actual tank fire, the
heat radiation is dependent on the wind conditions and
upward plume as shown in Fig. 1(b).

To estimate dangerous distance, the relation
between heat radiation and upwind distance from tank
fire is shown in Fig. 2.
where type A is the numerical result of heat radiation
based on proposed method, type B is the numerical
result of heat radiation based on conventional model.

In case of heat radiation considering upward
plume, the dangerous distance for human body
(2000 kcal/m²hr) is about 35 [m]. On the other hand,
in the conventional method, the dangerous distance
is about 27 [m]. It is clarified that the difference
of heat radiation with/without upward plume, and
distribution of wind velocity is important for the extin-
guishments and prevention of fire that is required
safety estimation.

REFERENCES

Kurashiki, T., Zako, M., and Fumita, M., 2005, A practical
estimation method of safety and reliability for chemical
plants, Proc. of International Conference on Structural
Safety and Reliability.

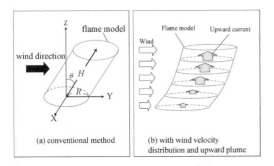

(a) conventional method   (b) with wind velocity
                              distribution and upward plume

Figure 1. Flame models with considering wind effect.

Safety, Reliability and Risk of Structures, Infrastructures and
Engineering Systems – Furuta, Frangopol & Shinozuka (eds)
© 2010 Taylor & Francis Group, London, ISBN 978-0-415-47557-0

Development of simulation system for hazardous area caused by disaster in petrochemical plants

H. Ueda, T. Kurashiki & M. Zako
Osaka University, Osaka, Japan

ABSTRACT

If a big earthquake attacks petrochemical plants and the tanks are broken, disaster will extend due to heat radiation from fire, diffusion of toxic gas, and so on. Recently, phenomena of tank fires and gas diffusion often occurred in petrochemical plants in Japan. So, the estimation of the hazardous area caused by disaster is very important for the safety and reliability of petrochemical plants. Therefore, the purpose of the study is development of a simulation system for hazardous area caused by disaster in petrochemical plants.

In Japan, the High Pressure Gas Safety Institution of Japan had established the estimation procedure of the dangerous zone cased by disaster in a plant based on the guideline (KHK, 1974). However, there are some problems in the guideline, ie., the heat radiation considering the inclination of flame due to the wind had not been estimated because of the difficulty for solving the integral equation of heat radiation. In order to solve the problem, an actual simulation system for hazardous area caused by heat radiation and gas diffusion has been developed.

As the estimation of geometrical view factor based on Monte Carlo method has much CPU time for the calculation, it is not suitable to obtain the calculation results immediately. Therefore, in order to get the situation of hazardous area immediately, the area had been calculated with Monte Carlo method under several conditions of geometries of tanks and wind conditions. And, the results of geometric view factors are stored database. The simulation for heat radiation is connected to the database with GUI. Figure 1 shows the developed simulation system for heat radiation from tank fire. The numerical results are compared with the Yellow Book method (TNO, 2005). There are good agreements with both results. We can get the hazardous area caused by tank fire with considering various condition of wind.

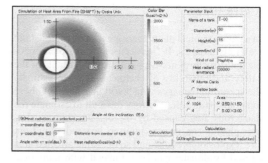

Figure 1. Simulation system for heat radiation from fire.

Furthermore, the simulation system for estimation of gas diffusion considering the combination effect of leakage, evaporation, and diffusion has also been developed. The hazardous area of gas can be estimated conveniently with the developed database and GUI. We believed that the proposed system contributes to the safety evaluation for petrochemical plants.

REFERENCES

KHK(The High Pressure Gas Safety Institution of Japan), 1974, Safety and loss prevention for large scale petrochemical and related industries, KHK E 007-1974

TNO Committee for the Prevention of Disasters, 2005, Methods for the calculation of physical effects resulting from releases of hazardous materials, CPR-14E.

Mini-Symposia (MS06) Uncertainties in Civil Structures & Infrastructure Engineering

Safety, Reliability and Risk of Structures, Infrastructures and Engineering Systems – Furuta, Frangopol & Shinozuka (eds)
© *2010 Taylor & Francis Group, London, ISBN 978-0-415-47557-0*

Gaussian process emulators for dynamical systems with random parameters

F.A. DiazDelaO & S. Adhikari
Swansea University, Swansea, UK

ABSTRACT

Complex engineering systems are studied using computer codes (also known as *simulators*) which might be very expensive to run. An increasingly common solution is to employ a less expensive surrogate model to investigate these systems. One possibility is to construct a statistical approximation to the simulator, known as *Gaussian process emulator*. Such technology is based on the analysis and design of computer experiments (Sacks et al., 1989; Satner et al., 2003), and on concepts of Bayesian statistics. Using this approach, it is possible to ef?ciently make inference about unknown output values by evaluating a fairly limited number of carefully selected points in the in-put domain of the simulator.

Gaussian process emulators work in the follow-ing way: A small set of code runs is treated as training data used to update the prior beliefs about the simulator. Such prior beliefs take the form of a Gaussian stochastic process distribution (Haylock and O'Hagan, 1996; O'Hagan, 2006). After conditioning on the training runs and updating this prior distribu-tion, the mean of the resulting posterior distribution approximates the output of the simulator at any un-tried input, whereas it reproduces the known output of the simulator at each initial input.

This paper compares two strategies for selecting points to run a simulator of the mean frequency re-sponse of a dynamical system with a random param-eter, taking Monte Carlo simulation as a benchmark. The input domain of the simulator is divided in two subdomains: the one corresponding to the frequency domain and the one associated with the random nature of the response. The results obtained con?rm that em-ulation is less computationally expensive than Monte Carlo simulation. More importantly, the comparison of selection strategies helps elucidate whether the ac-curacy and computation time of Gaussian process em-ulators can be improved.

REFERENCES

Haylock, R. and O'Hagan, A. (1996). *Bayesian Statistics 5*, chapter On inference for outputs of computationally expensive algorithms with uncer-tainty on the inputs. Oxford University Press, Ox-ford, UK.

O'Hagan, A. (2006). Bayesian analysis of computer code out-puts: A tutorial. *Reliability Engineering & System Safety*, 91(10-11):1290–1300.

Sacks, J., Welch, W., Mitchell, T., and Wynn, H. (1989). Design and analysis of computer experi-ments. *Statistical Science*, 4:409–435.

Satner, T., Williams, B., and Notz, W. (2003). *The Design and Analysis of Computer Experiments*. Springer Series in Statistics, London, UK.

Safety, Reliability and Risk of Structures, Infrastructures and Engineering Systems – Furuta, Frangopol & Shinozuka (eds)
© 2010 Taylor & Francis Group, London, ISBN 978-0-415-47557-0

Stage critical strength branch and bound algorithm based bridge failure mode identification strategy

Xin Gao

School of Civil Engineering, Harbin Institute of Technology, Harbin, China

Jinping Ou

School of Civil and Hydraulic Engineering, Dalian University of Technology, Dalian, China

ABSTRACT

The assessment of system reliability of real structures by the failure mode approach (FMA) leads to numerous failure modes and requires complex modeling. In most cases, however, only a small fraction of the modes contributes significantly to the overall failure probability of the system. Hence, an important part of system reliability estimation is to identify the subset of significant failure modes. Several different approaches have been developed, and some of them present elegant approaches, but they are seldom applied to the bridges.

The objective of this article is to present a strategy on how to search bridge significant failure modes under traffic load. A stage critical strength branch and bound algorithm is introduced.(Dong & Yang, 1993) and modified to suit for performing the system reliability assessment of bridges.

The main steps of stage critical strength branch and bound algorithm are shown in Figure 1. The strategy includes six concerned problems: a) Definition of bridge system failure; b) Structural modeling and analysis; c) The failure mode analysis method; d) Failure tree search strategy; e) Shortest failure path set search strategy; f) False failure mode updating method. In the following part of this paper, the concerned problems are discussed in details.

The proposed method is implemented in combination with the ANSYS finite element software and MATLAB procedure, and is illustrated with the help of a truss bridge example (as shown in Figure 2 and Figure 3). The proposed algorithm is found efficient and reasonably accurate. The method in general, is applicable to both ductile and brittle behaviors, and overcomes the limitations of the analytical techniques. It also accounts for the contribution of various significant failure modes towards the overall failure probability of the structural system. At the same time, computational effort is not wasted in enumerating a large number of failure modes, most of which may not contribute to the failure probability of the system. This algorithm can be applied to any kind of bridge systems without doing much additional programming for

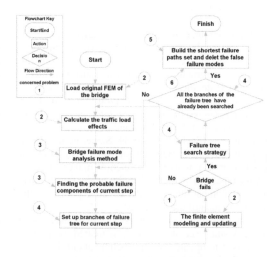

Figure 1. Flow chart of bridge failure mode analysis under traffic load.

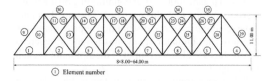

Figure 2. Steel truss bridge FEM.

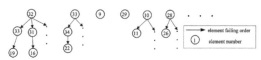

Figure 3. Important failure modes.

which the component failure modes can be defined through limit state equations.

REFERENCES

Dong, C & Yang QX. 1993. Theory and Algorithm of Structural System Reliability under Static Loading. *Structure & Environment Engineering* 02:1-8.

Safety, Reliability and Risk of Structures, Infrastructures and Engineering Systems – Furuta, Frangopol & Shinozuka (eds)
© 2010 Taylor & Francis Group, London, ISBN 978-0-415-47557-0

Depth to bedrock randomness effect on the design spectra

M. Badaoui & A. Mébarkis
Université Paris-Est, Laboratoire Modélisation et Simulation Multi Echelle, MSME FRE3160 CNRS, Marne-la-Vallée, France

M.K. Berrah
Ecole Nationale Polytechnique, El-Harrach, Algiers, Algeria

ABSTRACT

To show the effect of the soil profile height randomness on the dynamic structural response, a seismic analysis of typical reinforced concrete buildings by spectral analysis is carried out in the present study.

Several reinforced concrete buildings (6 and 12 story) are arbitrarily selected and submitted to the response spectra deduced at the top of the multilayered random soil.

The buildings have three different floor plans that are symmetric, monosymmetric, and unsymmetric. Columns and beams are modeled as frame elements, structural walls as shell elements and slabs as rigid diaphragms in each story level. In order to evaluate the seismic response of the buildings, an elastic analysis is carried out by the response spectrum method (modal spectral approach).

This analysis concerns the responses deduced on the surface of a multilayered soil composed of four layers, where the total height and the shear wave velocity are considered as random variables with a lognormal distribution. Monte Carlo simulations are combined with the stiffness matrix method, used herein as a deterministic method.

Figure 1 gives the response spectra for different types of soils and soil height coefficient of variation. For varying Cv_H, the response spectra decrease for fundamental periods and increase elsewhere. The frequency content increases with the soil height variability. The building acts as both a filter on the frequency content and a magnifier on the spectral amplitudes

This variability induces an increase of the base shears for six story buildings, an increase in the

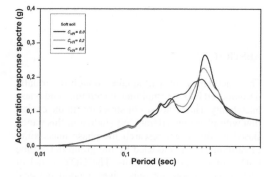

Figure 1. Stochastic design spectra for soft soil.

interstory drifts and story displacements for rock soft soils.

REFERENCES

Badaoui, M., Berrah, M.K. & Mebarki, A. 2006. Soil-structure interaction with random soil height. *Fifth International Conference on Engineering Computational Technology*, Proc. intern. Conf., Las Palmas de Gran Canaria. 12–15 September 2006. Stirlingshire: Civil-Comp Press.

Badaoui, M. 2008. Influence de l'hétérogénéité géologique et mécanique sur la réponse des sols multicouches, thèse de Doctorat, Université Paris-Est, France.

Doğangün, A. & Livaoğlu, R. 2006. A comparative study of the design spectra defined by Eurocode 8, UBC, IBC and Turkish Earthquake Code on R/C sample buildings. *Journal of seismology* 10(3): 335–351.

Manolis, G.D. 2002. Stochastic soil dynamics. *Soil Dynamics and Earthquake Engineering* 22(1): 3–15.

Safety, Reliability and Risk of Structures, Infrastructures and Engineering Systems – Furuta, Frangopol & Shinozuka (eds)
© 2010 Taylor & Francis Group, London, ISBN 978-0-415-47557-0

Analytical solution of structural response under stochastic excitation with phase uncertainty

Lingyi Y. Lu
Southeast university, Nanjing, China

Tadanobu Sato
Kobegakuin University, Kobe Japan

Yoshitaka Murono
Railway Research Institute, Kokubunnji, Japan

ABSTRACT

The uncertainty of earthquake motion is usually decomposed into the amplitude uncertainty and phase uncertainty. The responses of structures to the excitation with phase uncertainty are studied in this paper. Based on the recent research the phase uncertainty of earthquake was modeled by the group delay time (GDT) of the ground motions. The GDT model of earthquake motion not only gives a good physical interpretation for the earthquake phase uncertainty, but also leads to a simple path to the solution. In this paper, an analytical procedure is proposed to calculate stochastic response of structures to the excitation with phase uncertainty. In addition, an explicit analytical solution for the analysis of stochastic response of a SDOF system is presented. The theoretical solutions are compared with numerical simulations. The results show that the approaches proposed in this paper is reliable and efficient.

REFERENCES

Ghanem, R. G. and Spanos, P. D., 1991. Stochastic finite elements, A specific approach, Sringer-Verlag.
Izumi, M and Katukura, H., 1980. Representation of nonstationary nature of seismic waves based on phase (PratI). Proceedings of the International Research conference on Earthquake Engineering, 199–212.
Kameda, H., 1977. On a method to computing evolutionary power spectra of strong motion seismograms. Journal of Structural mechanics and Earthquake Engineering, JSCE, 235, 55–62 (in Japanese).
Kiureghian, A. D., 1981. A response spectrum method for random vibration analysis of MDF systems. Earthquake Engineering and Structural Dynamics, 9, 419–435.
Lin, Y. K., 1967. Probabilistic theory of structural dynamics, McGraw-Hill Inc.
Osaki, Y., kanda, J., Iwasaki, R., Masano, T., kitasa, T., and Sakata, K., 1984. Improved method for generation of simulated earthquake ground motions. Proceedings of the 8th World Conference on Earthquake Engineering, 573–580.
Sato, T., Murono, Y. and Nishimura A., 1999. Modeling of phase spectrum to simulate design earthquake motion. Proceedings of the Optimizing Post-Earthquake Lifeline System Reliability, ASCE, Seattle 1999, 804–813.
Sato, T. and Murono, Y., 2002. Simulation of earthquake motions based on phase information. Proceedings of the 12th European Conference on Earthquake Engineering CD-ROM487.
Sato, T., Murono, Y. and Murakami Y., 2003. Modeling of Phase Spectra For Simulation of Near-Fault Design Earthquake. Proceedings of the advancing mitigation technologies and disaster response for lifeline systems. Technical Council on Lifeline Earthquake Engineering Monograph, 25, 769–778.
Yang, J. N., 1972. Simulation of random envelope process. Journal of Sound and Vibration, 21, No.73.

*Mini-Symposia (MS07) Vulnerability Assessment and
Risk-Based Life-Cycle Analysis and Design*

Safety, Reliability and Risk of Structures, Infrastructures and
Engineering Systems – Furuta, Frangopol & Shinozuka (eds)
© 2010 Taylor & Francis Group, London, ISBN 978-0-415-47557-0

A conceptual model for time varying risk assessment for extreme events

A.S. Kiremidjian
Stanford University, Stanford, CA, USA

ABSTRACT

Risk assessment models and performance based engineering for extreme events such as earthquakes and
hurricanes are now widely used for variety of purposes
including building specific evaluation for retrofit
strategies, portfolio analysis for insurance pricing and
regional disaster planning and mitigation. When these
models are used for disaster planning and mitigation, it
is critical that a reliable estimate of the losses be made
at the time of the event, which may occur at some
future time. Similarly, when large portfolio owners or
insurance companies are interested in their exposure
risk for periods longer than a few years, the changes to
the structures or content of properties in their portfolio
needs to be considered in the risk forecast. Currently,
regional risk assessment studies and portfolio analyses
are performed on annual basis and future risk forecasts
are obtained under the assumption that the annual risk
rate is independent of time thus enabling modelers to
invoke the simple Poisson model for their analyses.

Over the past two decades significant advances have
been made in the development of time dependent models for seismic hazard analysis starting with the early
work of Kiremidjian and Anagnos (1985), and Lutz
and Kiremidjian (1993). Considerable research has
been done since these early developments and many
of them are now implemented in conventional hazard mapping. Time- variant systems have been treated
mostly at the conceptual lever in system reliability
analysis with early work focusing on load combination
over time. Consideration of time variance of structural resistance is discussed in Casiati and Faravelli
(1991). However, little has been done to carry the
time dependence to the risk level. Rosenblueth (1976)
and Esteva (1978) were some of the first researchers
to recognize the problem of time dependence in risk
assessment in relationship to optimal earthquake resistant design. Recently these early developments were

cast within the current performance based earthquake
engineering design framework by Esteva (2008). More
specifically, in that study Esteva presents a model for
optimal seismic design that includes a structure's life
cycle costs.

Expanding on the developments by Esteva, we
present a model for regional or portfolio risk assessment over time that includes the changes in the building or infrastructure attributes over the same period of
time. For specific structures, the changes to the capacity of the structure are considered in the evaluation of
future damage and losses. When a portfolio of structures or infrastructure components that are part of a
system are considered, then the growth of the portfolio
or infrastructure components is also included. The time
dependent risk assessment model focuses on capturing
the temporal variations of the attributes of a building
or other infrastructure components. The changes in the
structure are represented through a generic deterioration model and the inventory changes are represented
through a time series model. Single structure risk
assessment formulation is presented while a simulation procedure is introduced to treat a portfolio with
changing components.

REFERENCES

Casiati, F. & Faravelli, L. 1991. Fragility Analysis of Complex
    Structural Systems, John Wiley and Sons, Inc., NY
Esteva, L. 2008. Life Cycle Optimization in Earthquake Engi-
    neering, in Life-Cycle Civil Engineering, Biondini, F. &
    Frangopol, D. (eds.), Taylor and Francis Group, London,
    pp. 35–46.
Kiremidjian, A. S. & Anagnos, T. 1984, A Stochastic Slip-
    Predictable Model for Earthquake Occurrences, Bull.
    Seism. Soc. Am., Vol. 74, No. 2, 739–755.
Rosenblueth, E. 1976. Optimum Design for Infrequent Dis-
    turbances. ASCE Journal of the Structural Division, Vol.
    102 (ST9). 1779–1796.

Safety, Reliability and Risk of Structures, Infrastructures and
Engineering Systems – Furuta, Frangopol & Shinozuka (eds)
© 2010 Taylor & Francis Group, London, ISBN 978-0-415-47557-0

Risk, vulnerability and cost functions for electrical substation and power transmission towers under strong winds in Mexico

D. De León
Engineering School, UAEM, Mexico

A. López
Institute of Electrical Research, Mexico

ABSTRACT

Electrical substations and transmission towers located on areas exposed to significant wind hazards are important infrastructure facilities where the potential failure involves exceptionally high consequences given the relevance of its contribution to the GNP. Also, the service interruption represents a strong negative impact to extensively populated areas.

Recent damages on towers and substations for electrical power transmission in Mexico derived on the request, by electrical industry managers and operators from CFE, Federal Electrical Commission, to take a closer look and make practical recommendations to improve the current design guidelines of these structures.

As a first step towards an optimal design of these structures, vulnerability, cost and risk curves have been developed by considering typical failure modes, costs and design and construction practices. Also, occurrence probabilities of wind velocities are considered for several sites on both coasts of Mexico as the description of the wind hazard to which the structures must respond.

The acceptable failure probability for each type of structure is calculated, as a function of the cost of consequences, from the minimum expected lifecycle cost.

Conditional annual failure probability curves for typical substations and towers are calculated by examining their responses under assumed mean values of the maximum wind velocities and the unconditional annual failure probability results from the convolution of these conditionals with the occurrence probability of the assumed maximum wind velocities. However,

given that the estimation of the mean maximum wind velocity is subject to error due to scarce data and the imperfection of probabilistic models, epistemic uncertainty is implicit into the failure probability assessment and it requires a specific modeling.

Initial and expected cost functions are obtained for given wind velocities and ranges of maximum reliability/cost ratios are identified for those wind velocities.

Given the scarcity of data and the imperfect prediction models for wind velocity, the epistemic uncertainty is treated as separated from the aleatory uncertainty. The epistemic uncertainty on the mean maximum wind velocity is included by considering the mean as a random variable and, as a consequence, the unconditional failure probability becomes also a random variable. Unconditional relative frequency distributions of the annual failure probability are obtained for several mean values of maximum wind velocity.

Beta distribution functions are fitted to the unconditional annual failure probabilities and estimations of the annual failure probability are made to illustrate the procedure for specific sites and for several given confidence levels. These estimations are valuable for risk averse managers who prefer the use of conservative confidence levels as opposite to the traditional use of mean annual failure probabilities.

The above allows for conservative decisions and constitutes a basis to improve the current design specifications and to justify the investment on wind velocity measurement and recording equipment to protect critical infrastructure. The formulation may also be adapted to protect other infrastructure facilities under wind hazard and under other hazards.

Safety, Reliability and Risk of Structures, Infrastructures and
Engineering Systems – Furuta, Frangopol & Shinozuka (eds)
© 2010 Taylor & Francis Group, London, ISBN 978-0-415-47557-0

System effects on the fatigue reliability of deteriorating riveted railway bridges

B.M. Imam & M.K. Chryssanthopoulos
University of Surrey, Guildford, UK

D.M. Frangopol
Lehigh University, Bethlehem, PA, USA

ABSTRACT

A large proportion of the existing railway bridge infrastructure in Europe and North America is of riveted construction and is already exceeding 100 years of age. Considering their large number, replacement of these structures will be extremely difficult from an economic point of view and is likely to create severe network problems. In terms of maintaining these old bridges and planning any future repair actions, the assessment of their remaining fatigue life is a vital requirement. A key aspect in determining the remaining fatigue life is the estimation of the fraction of the fatigue life that has already been expended, followed by a prediction of the remaining fatigue life.

Building on previous work by the authors, this paper presents a system-based model for the fatigue assessment of old, deteriorating riveted railway bridge connections, which are built up of a number of basic components. A finite element model of a typical, short-span railway bridge is used to convert train loading into probabilistic fatigue load spectra via Monte Carlo train simulations. The spectra are developed for critical hot-spots on different components of the connection, such as holes, rivets, angle fillets. Uncertainties in terms of loading, resistance and modelling are taken into account. These spectra show that the majority of the stress cycles experienced by the hot-spots are below the fatigue limit of the material.

The riveted connection is then treated through generic sub-systems capturing potential damage in identifiable hot-spots, such as rivets, holes and angle fillets. These hot-spots are treated as the elements of a structural system susceptible to fatigue failure. Fatigue damage calculations are based on the Theory of Critical Distances (TCD), which is a recently developed theory that considers the entire stress distribution ahead of any given stress concentration. Reliability profiles and probabilistic remaining fatigue life estimates for the connection are obtained, based on different failures modes of the connection and different system assumptions. The results are shown in Figure 2.

The reliability profiles show that the remaining life of the connection is sensitive to the assumptions made

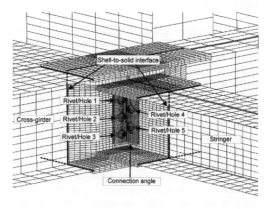

Figure 1. Finite element model of riveted bridge connection.

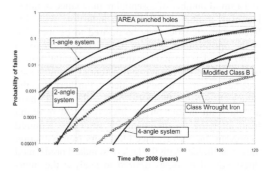

Figure 2. System probability of failure versus time obtained for the connection.

regarding the form of the system considered for the analysis. Low remaining lives were obtained for a number of the connection's hot-spots indicating that fatigue cracking may be imminent or may have already initiated in a number of similar existing bridge connections. Reliability profiles obtained from the TCD method and the system representation has shown a more rapid increase in failure probability of the connection compared to similar results obtained using the traditional nominal S-N approach.

Safety, Reliability and Risk of Structures, Infrastructures and Engineering Systems – Furuta, Frangopol & Shinozuka (eds)
© *2010 Taylor & Francis Group, London, ISBN 978-0-415-47557-0*

Lifetime functions for multi-criteria optimization of life-cycle maintenance programs considering availability, redundancy and cost

N.M. Okasha & D.M. Frangopol
Lehigh University, Bethlehem, PA, USA

ABSTRACT

Lifetime functions are an extremely helpful tool in characterizing the lifetime performance of existing structures and predicting their future reliability. Most importantly, they offer a probabilistic basis to quantify the lifetime survivability of structural components and systems. Lifetime redundancy is another crucial performance that quantifies the availability of warning in the structure if collapse is to occur (Okasha and Frangopol 2010). Redundancy can also be estimated based on lifetime functions (Okasha and Frangopol 2008). Lifetime functions have successfully been used to find optimum maintenance strategies for bridge structures by enumeratively comparing all alternatives (Yang et. al. 2006). This was made possible by the use of only essential maintenance, one performance indicator (survivability) with a predefined threshold, and a small number of essential maintenance options. However, multiple performance indicators with different characteristics may be desired to be included in the optimization process (Okasha and Frangopol 2008). Besides, the maintenance options available may be numerous. Preventive and essential maintenance together may need to be considered. Furthermore, the values chosen for the performance thresholds have a significant impact on the life-cycle cost resulting from applying the maintenance strategies required to prevent the performance indicators from violating these thresholds. With the lack of standard predefined thresholds, it becomes desired to analyze the interaction between the performance thresholds and the life-cycle cost in order to ensure an economical and effective choice of threshold values. Finding trade-off solutions by solving a multi-objective optimization problem that includes the performance thresholds and life-cycle cost as criteria is the best way to achieve this goal. These issues render the non-automatic solution of the problem impossible. Luckily, the advances achieved in the genetic algorithms optimization technology (Deb et al. 2002) provide a robust automatic capability to solve (Neves et al. multi-objective optimization problems 2006). What is left is a proper formulation of the problem that takes the best advantage of this technology and provides the optimum solutions to the problem.

The objective of this study is to present an automated approach that finds trade-offs among optimum maintenance strategies considering their life-cycle cost, lifetime availability and lifetime redundancy through multi-objective optimization and using genetic algorithms. This approach finds the optimum maintenance strategies by automatically selecting the appropriate essential and preventive maintenance actions and determining the optimum times of their application to satisfy a given set of performance thresholds. The approach is able to provide optimum maintenance strategies considering a combination of multiple essential maintenance options and multiple preventive maintenance options applied at regular or irregular time-intervals. The lifetime availability and the lifetime redundancy are calculated using lifetime distribution functions. The proposed approach is illustrated on a highway bridge example.

REFERENCES

Deb, K., Pratap, A., Agrawal, S. & Meyarivan, T. 2002. A fast and elitist multiobjective genetic algorithm: NSGA-II. *Transaction on Evolutionary Computation* 6(2): 182–97.

Okasha, N.M. & Frangopol, D.M. 2008. Redundancy of Structural Systems with and without Maintenance: An Approach Based on Lifetime Functions. (submitted).

Okasha, N.M. & Frangopol, D.M. 2010. Time-variant redundancy of structural systems. *Structure and Infrastructure Engineering* 6(1-3), (in press)

Neves, L., Frangopol, D.M. & Petcherdchoo, A. 2006. Probabilistic Lifetime-Oreinted Multiobjective Optimization of Bridge Maintenance: Combination of Maintenance Types. *Journal of Structural Engineering* 132(11): 1821–1834.

Yang, S-I, Frangopol, D.M. & Neves, L.C. 2006. Optimum maintenance strategy for deteriorating structures based on lifetime functions. *Engineering Structures* 28(2): 196–206.

Mini-Symposia (MS15) System Identification and Structural Health Monitoring
[Session Organized on Behalf of IASSAR Subcommittee SC5]

*Safety, Reliability and Risk of Structures, Infrastructures and
Engineering Systems – Furuta, Frangopol & Shinozuka (eds)
© 2010 Taylor & Francis Group, London, ISBN 978-0-415-47557-0*

Structural model updating using measured dynamic responses with the consideration of optimal sensor placement

H.M. Chow, S.K. Au, H.F. Lam & T. Yin
Department of Building and Construction, City University of Hong Kong, HKSAR, China

ABSTRACT

This paper addresses the problem of optimally placing sensors on structures for the purpose of model
updating. A methodology is presented to identify an
effective way to install a given number of sensors on
the structures to extract as much information as possible for structural model updating. The information
entropy is employed as a measure to quantify the uncertainties of the set of identified model parameters. The
problem of optimal sensor placement is then formulated as a discrete optimization problem, in which the
information entropy measure is minimized, with the
sensor configurations as the minimization variables.
The performance of the optimal sensor placement technique is verified by the results of model updating using
measured acceleration responses of a 4-storey shear
building model under laboratory conditions.

A series of comprehensive numerical case studies
were carried out, for shear buildings with different
number of stories, to illustrate the proposed methodology and verify the robustness of using information
entropy for optimal sensor placement problem. The
results of the case studies conclude that the lower part
of the structure is more suitable for placing sensors
to extract more information from the measurement.
In addition, an experiment on a 4-storey shear building model was carried out to show the importance
of sensor locations in the results of structural model
updating. In order to improve the reliability of the analytical model after model updating, the initial modal
conditions obtained analytically from the identified
modal parameters were employed in the analytical
model. The results show that the set of model parameters updated from different sensor configurations can
be very different. For instance, the updated FE model
using the data from the optimal sensor location is much
more reliable than that using the data from the worst
sensor location. As you can see in Figure 1 which
shows the results of model updating using the data
from the optimal sensor location (the 2nd floor). The
responses of most storeys of the updated FE model are
very close to the measured acceleration, except that of
the first storey. Most probably, the poor matching at
the first storey is due to the small signal-to-noise ratio.

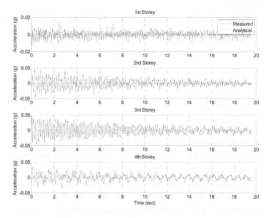

Figure 1. Responses of the updated FE model using data
from the optimal sensor location.

ACKNOWLEDGEMENTS

The work described in this paper was fully supported
by a grant from the Research Grants Council of the
Hong Kong Special Administrative Region, China
(CityU 114706).

REFERENCES

Au, S.K., Ng, C.T., Sien, H.W. & Chua, H.Y. 2005. Modal
   identification of a suspension footbridge using free vibra-
   tion signatures. International Journal of Applied Mathe-
   matics and Mechanics 1(4): 55–73.
Beck, J.L. & Katafygiotis, L.S. 1998. Updating models
   and their uncertainties – Bayesian statistical framework.
   Journal of Engineering Mechanics 124 (4): 455–461.
Jaynes, E.T. 1978. *Where do we stand on maximum entropy?*
   The Maximum Entropy Formalism, Levine, R.D., and
   Tribus, M. (eds), Cambridge: MIT Press.
Papadimitriou, C., Beck, J.L. & Au, S.K. 2000. Entropy-based
   optimal sensor location for structural model updating.
   Journal of Vibration and Control 6 (5): 781–800.

Safety, Reliability and Risk of Structures, Infrastructures and
Engineering Systems – Furuta, Frangopol & Shinozuka (eds)
© 2010 Taylor & Francis Group, London, ISBN 978-0-415-47557-0

Damage classification by probabilistic neural networks based on latent components for a time-varying system

Jian Yuan

College of Aerospace Engineering, Nanjing University of Aeronautics and Astronautics, Nanjing, China

Yan Zhou

Department of Mechatronics, Guangxi Institute of Communication, Nanjing, China

Xin Lu

College of Aerospace Engineering, Nanjing University of Aeronautics and Astronautics, Nanjing, China

ABSTRACT

In this paper, a novel method of damage classification for health monitoring of a time-varying system is presented. The Functional-Series Time-dependant Auto Regressive Moving Average (FS-TARMA) time series model is applied to vibration signal observed in time-varying system for estimating the TAR/TMA parameters and innovation variance, which are the function of time and can be represented by group of projection coefficients on certain functional subspace with specific basis functions. The estimated TAR/TMA parameters and innovation variance are further used to calculate the Latent Components (LCs) as the more informative data for health monitoring evaluation, based on an eigenvalue decomposition technique. The Latent Components (LCs) are then combined and reduced to numerical values as feature sets, which are input to a Probabilistic Neural Networks (PNN) for damage classification. For evaluation of the proposed method, numerical simulations of damage classification for a time-varying system were employed, in which different classes of damage were modeled by the mass or stiffness reductions. It is demonstrated that the method is able to discriminate between the time varying nature of the system parameters and the damages that may happen in the course of operation and cause the change of parameters. As the results will show, by using the proposed method, the success rate of classification considerably enhances in comparison with the other works using non-reduced and ordinary feature extraction methods.

REFERENCES

Serhat S & Emine A. 2003, Feature extraction related to bearing damage in electric motors by wavelet analysis[J]. Journal of the Franklin Institute, 340:125–134.

Huang NE. 1998. The empirical mode decomposition and the Hilbert spectrum for nonlinear and non stationary time series analysis[A]. Proceedings of the Royal Society of London Series[C]. London, A 454, 903–995.

Ettefagh, M.M. & Sadeghi M.H. 2007. Health monitoring of time-varying stochastic structures by latent components and fuzzy expert system[J], Earthquake engineering and engineering vibration, Vol. 7 No. 1 91–106

Poulimenos AG & Fassois SD. 2006. Parametric time-domain methods for non-stationary random vibration modelling and analysis — A critical survey and comparison. Mechanical Systems and Signal Processing, 20:763–816.

Sakellariou JS & Fassois SD. 2006. Stochastic output error vibration-based damage detection and assessment in structures under earthquake excitation. Journal of Sound and Vibration, 297:1048–1067.

Qian S & Chen D. 1996. The Joint time-frequency analysis—methods and applications. Prentice-Hall, Englewood Cliffs, NJ.

Poulimenos AG & Fassois SD. 2003. Estimation and identification of non-stationary signals using functional series TARMA models[A]. Proceedings of the 13th IFAC Symposium on System Identification[C], Rotterdam. The Netherlands, 162–167.

Poulimenos AG & Fassois SD. 2003. Estimation and identification of non-stationary signals using functional series TARMA models. In: Proceedings of the 13th IFAC Symposium on System Identification, Rotterdam. The Netherlands, 162–167.

Poulimenos AG & Fassois SD. 2005. On the estimation of non-stationary functional series TARMA models. In: Proceedings of the 13th European Signal Processing Conference. Antalya, Turkey.

Prado R. 1998. Latent structure in non-stationary time series. PhD thesis, Institute of Statistics and Decision Sciences, Duke University.

West M & Harrison PJ. 1997. Bayesian forecasting and dynamic models (2nd ed.). New York, Springer-Verlag.

West M, Prado R & Krystal AD. 1999. Evaluation and comparison of EEG traces: Latent structure in nonstationary time series. Journal of the American Statistical Association (Applications and Case Studies), 94(446):375–387.

Safety, Reliability and Risk of Structures, Infrastructures and Engineering Systems – Furuta, Frangopol & Shinozuka (eds)
© 2010 Taylor & Francis Group, London, ISBN 978-0-415-47557-0

A novel structural health assessment technique using noise-contaminated limited response information

H. Katkhuda
The Hashemite University, Zarqa, Jordan

R. Martinez-Flores
Talavera Design and Construction, Tucson, AZ, USA

A. Haldar
University of Arizona, Tucson, AZ, USA

ABSTRACT

A time domain finite element-based system identification procedure capable of detecting defects at the local element level using only limited noise-contaminated response measurements and without using any information on the exciting dynamic forces is presented. The procedure tracks the stiffness properties of structural elements and studies the amount or rate of degradation of the stiffness with respect to the "as built" or expected properties, or deviation from the previous values if periodic inspections were conducted. After a repair, it also indicates whether or not all the defects were repaired. Limited experimental verification of the procedure is also presented in the paper.

The method can be used as an inspection-based nondestructive defect evaluation procedure. The structural health can be assessed in normal operating condition or just after natural or man-made events. Defects at the local element level that alter the dynamic response behavior of structures can be detected with the method.

A finite element-based system identification (SI) approach will be ideal to meet the objective since it will locate the defect spot(s). To increase its implementation potential, the procedure should be able to locate defective element(s) using minimum information. To address minimum information required to identify a system, its three components, i.e., (i) input excitation information, (ii) output response information, and (iii) the system to be identified, represented by in an algorithmic form such as finite elements, need further study. Outside the highly controlled laboratory environment, the measurement of excitation information is always error-prone. Furthermore, just after a natural and made-made incidence, the information on the input excitation is expected to be unavailable. Thus, it would be very desirable if a system can be identified without using any information on the excitation.

The required dynamic degrees of freedom (DDOFs) are expected to be very large for real structures. The simultaneous collection of response information at a large numbers of DDOFs may not be practical. Also, measured response information is expected to be noise-contaminated. Thus, the application potential of the nondestructive defect evaluation (NDE) procedure will be significantly improved if the structure can be identified with only limited noise-contaminated response information and without excitation information.

The research team at the University of Arizona is in the process of developing such a method. They proposed a two-stage SI-based procedure for this purpose. The technique is a combination of two methods: (1) the modified iterative least-squares technique with unknown input excitation (MILS-UI) and (2) the extended Kalman filter with a weighted global iteration (EKF-WGI). The authors denote the new method as GILS-EKF-UI.

The procedure was extensively verified using computer generated noise-free and noise-contaminated response information. Then, it was verified using measured response information in the laboratory. A scaled two-dimensional one-bay three-story steel frame was built in the laboratory. The defect-free frame was first successfully identified using the laboratory measured response information. Then, several defects were introduced in the frame. Some of the defects considered are removal of a beam assuming it is completely broken, reduction of cross sectional area of a beam over a finite length, presence of multiple cracks in a beam, and presence of single crack in a beam. In all cases the GILS-EKF-UI method successfully identified the defects and their locations. This will be discussed in more detail during the presentation.

Safety, Reliability and Risk of Structures, Infrastructures and
Engineering Systems – Furuta, Frangopol & Shinozuka (eds)
© 2010 Taylor & Francis Group, London, ISBN 978-0-415-47557-0

Probabilistic damage assessment based on measured vibration data from a progressively damaged beam

G. Lombaert, E. Reynders, M. Schevenels & G. De Roeck
Department of Civil Engineering, K.U. Leuven, Leuven, Belgium

B. Moaveni
Department of Civil & Environmental Engineering, Tufts University, Medford, MA, USA

J.P. Conte
Department of Structural Engineering, University of California, San Diego, La Jolla, CA, USA

ABSTRACT

Vibration-based finite element (FE) model updating has proven to be a powerful methodology to identify (i.e., detect, localize and quantify) structural damage. One of the main challenges in model updating is that the inverse problem is often ill-posed. Therefore, the existence, uniqueness, and stability of the solution with respect to measurement errors are not guaranteed. This issue is dealt with by means of regularization techniques or an appropriate parametrization (Friswell, Mottershead, and Ahmadian 2001; Teughels, Maeck, and De Roeck 2002). Alternatively, a Bayesian inference scheme (Beck and Katafygiotis 1998; Papadimitriou 2004) can be applied to quantify uncertainties in the identification.

Within the frame of the present paper, Bayesian FE model updating is applied for damage identification of a full-scale subcomponent composite beam that has been progressively damaged in several stages through quasi-static loading. At each stage, the modal parameters are derived by means of low-amplitude dynamic excitation. (Moaveni, He, Conte, and de Callafon 2008). A Bayesian FE model updating procedure is used to quantify the uncertainties of the identified bending stiffness distribution, taking into account the uncertain prior information and the uncertain experimental data. A probabilistic model has been used to quantify the uncertainty in the observed prediction error that includes both the measurement and modeling uncertainty. The parameters of this probabilistic model are considered as random variables and are determined from the experimental data as well.

The results show that the informativeness of the modal parameters is largely dependent on the identified probabilistic model for the observed prediction uncertainty. When large values of the observed prediction error are probable, the experimental data are less informative, and the prior uncertainty on the stiffness distribution is only slightly reduced. This is the case when a large measurement uncertainty is present, or when the model does not adequately represent the actual behavior of the system. The estimation of the combined measurement and modeling uncertainty, and the quantification of its effect on the identified parameters is the main benefit of the present approach compared to deterministic procedures. The results of the identification correspond relatively well with the actual damage observed after the test.

REFERENCES

Beck, J. and L. Katafygiotis (1998). Updating models and their uncertainties. I: Bayesian statistical framework. *ASCE Journal of Engineering Mechanics 124*(4), 455–461.

Friswell, M., J. Mottershead, and H. Ahmadian (2001). Finite-element model updating using experimental test data: parametrization and regularization. *Philosophical Transactions of the Royal Society 359*, 169–186.

Moaveni, B., X. He, J. Conte, and R. de Callafon (2008). Damage identification of a composite beam using finite element model updating. *Computer-Aided Civil and Infrastructure Engineering 23*(5), 339–359.

Papadimitriou, C. (2004, July). Bayesian inference applied to structural model updating and damage detection. In *9th ASCE Specialty Conference on Probabilistic Mechanics and Structural Reliability*, Albuquerque, USA.

Teughels, A., J. Maeck, and G. De Roeck (2002). Damage assessment by FE model updating using damage functions. *Computers and Structures 80*(25), 1869–1879.

Safety, Reliability and Risk of Structures, Infrastructures and Engineering Systems – Furuta, Frangopol & Shinozuka (eds)
© 2010 Taylor & Francis Group, London, ISBN 978-0-415-47557-0

Identification of progressive damage in structures using wavelet multi-resolution approximation

Y.F. Shi & C.C. Chang
The Hong Kong University of Science and Technology, Hong Kong, China

ABSTRACT

For the reliability and safety of structures, it is necessary to have the most comprehensive knowledge of the circumstances associated with a possible damage/failure (Farrar & Worden 2007). Most damage identification techniques are often performed separately for different stages (intact and possibly damaged) in which the structural system is assumed to be time-invariant. However, damage usually develops gradually during the monitoring period and is better modeled with time-varying characteristics (Udwadia and Jerath 1980). Hence, it is needed to develop some techniques that can track these time-varying characteristics in a damaged structure.

With its capability in spatial-temporal localization and unconditional function approximation, wavelets have been shown their potential in estimating time-varying systems (Tsatsanis & Giannakis 1993, Wei & Billings 2002, Chang & Shi 2008). In this paper, a time-varying tracking technique using wavelets is investigated to identify progressive damage in the structure. The time-varying parameters which are used to model progressive damage are represented by wavelet multi-resolution approximation. The original time-varying parametric problem is transformed into a nonparametric regression one. One important issue for this technique is to minimize the number of basis functions in the estimation by proper selection of resolution levels for each parameter and the significant basis functions. In this study, the resolution levels are determined using the BIC (Schwarz 1978) model selection criterion. The significant basis functions from the selected resolution levels are determined by the significant term search scheme OFR (Chen at al. 1989) with the AIC (Akaike 1974) as the stopping criterion. Examples using a single-degree-of-freedom structure with Bouc-Wen hysteresis subjected to earthquake excitation are given to illustrate the performance of the proposed technique to track both smooth- and sudden-varying progressive damage.

REFERENCES

Akaike, H. 1974. A new look at the statistical model identification. *IEEE Transactions on Automatic Control* 19(6): 716–723.
Chang, C.C. & Shi, Y.F. 2008. Tracking time-varying properties of hysteretic structure by wavelet multi-resolution analysis. *Proceedings of SPIE* 6932: 693229.
Chen, S., Billings, S.A. & Luo, W. 1989. Orthogonal least-squares methods and their applications to nonlinear system identification. *International Journal of Control* 50(5): 1873–1896.
Daubechies, I. 1992. *Ten lectures on wavelets*. Society for Industrial and Applied Mathematics, Philadephia.
Farrar, C.R. & Worden, K. 2007. An introduction to structural health monitoring. *Philosophical Transactions of the Royal Society A* 365: 303–315.
Schwarz, G. 1978. Estimating the dimension of a model. *Annals of Statistics* 6(2): 461–464.
Tsatsanis, M.K. & Giannakis, G.B. 1993. Time-varying system identification and model validation using wavelets. *IEEE Transactions on Signal Processing* 41(12): 3512–3523.
Udwadia, F.E. & Jerath, N. 1980. Time variations of structural properties during strong ground shaking. *Journal of the Engineering Mechanics Division*, ASCE 106(1):111–121.
Wei, H.-L. & Billings, S.A. 2002. Identification of time-varying systems using multiresolution wavelet models. *International Journal of System Science* 33(15): 1217–1228.

Safety, Reliability and Risk of Structures, Infrastructures and Engineering Systems – Furuta, Frangopol & Shinozuka (eds)
© 2010 Taylor & Francis Group, London, ISBN 978-0-415-47557-0

Coupling of differential evolution algorithm and quadratic approximation for dynamic identification

L. Vincenzi & M. Savoia

DISTART – Structural Engineering, University of Bologna, Bologna, Italy

ABSTRACT

In a FE model updating procedure, the uncertain model properties are adjusted such that the numerical predictions correspond as closely as possible to the measured data [Savoia and Vincenzi, 2008]. To obtain unknown parameters, an optimization problem is solved. Success of the application of the updating method depends on the definition of the optimization problem and the mathematical capabilities of the optimization algorithm.

Global optimization techniques, as Genetic algorithms and Evolution approaches are considered very efficient numerical methods. Unfortunately, they have the disadvantage to require a large number of cost function evaluations since they are based on probabilistic searching without any gradient information. Moreover, before detect global minimum, several number of evaluation must be performed in order to obtain prescribed precision. In FE modal updating, objective function can be often characterized by only one (global) minimum. In this case, sensitivity-based methods are preferred.

Instead of applying the optimization algorithm directly to the objective function, the *Response Surface Methodology* (*RSM*) applies it to an approximate surface of the real objective function. In classical *RSM*, the response surface is obtained by combining first or second-order polynomials fitting the "real" objective function in a set of sampling points. The theory of *Design of Experiments* provides statistical tools to sample the search domain efficiently [Khuri and Cornell, 1996]. Computational effort is very cheap and *RSM* is a very powerful methodology when cost function present only one minimum. The main disadvantage is the fact that due to the use of second-order polynomials, a local minimum is usually reached when the objective function presents several local minima. In Structural identification problems, rarely the cost function shape is known because of it is not explicitly defined.

To improve speed rate of global optimization techniques, the response surface methodology is introduced in Differential Evolution Algorithm to perform dynamic structural identification. Differential evolution (*DE*) algorithms are parallel direct search methods where N different vectors collecting the unknown parameters of the system are used in the minimization process [Storn and Price, 1997]. The vector population is chosen randomly or by adding weighted differences between vectors obtained from the old population.

In the modified algorithm, a new parameter vector can be also defined as the minimum of a second-order polynomial surface which approximate the real cost function. Performance in term of speed rate is improved and higher precision of results is obtained by introducing the second-order approximation in *DE*. Moreover, when objective function presents only one (global) minimum, second-order approximation provide to find the solution with the lowest number of iteration. On the other hand, the global minimum is reach since multiple search points are used simultaneously. *DE* algorithm shows its efficiency when several design parameters must be researched and close to the solution, quadratic approximation gives higher accuracy. *DE* is compare with the modified algorithm to perform a FE modal updating of a steel-concrete bridge. The effectiveness of the algorithm in finding the set of optimization parameters has been verified also using pseudo-experimental input data, obtained by adding a statistic scattering to exact modal data.

REFERENCES

Savoia, M. and Vincenzi, L. 2008. Differential evolution algorithm for dynamic structural identification. *Journal of Earthquake Engineering* 12(5): 800–821.

Khuri, A. and Cornell, J.A. 1996. *Response Surfaces. Designs and Analyses*. Marcel Dekker Inc., New York.

Storn, R. and Price, K. 1997. Differential Evolution – a simple and efficient heuristic for global optimization over continuous spaces. *Journal of Global Optimization* 11:341–359.

*Mini-Symposia (MS16) Risk Assessment of Lifeline
Networks and Decision Support*

Safety, Reliability and Risk of Structures, Infrastructures and Engineering Systems – Furuta, Frangopol & Shinozuka (eds)
© 2010 Taylor & Francis Group, London, ISBN 978-0-415-47557-0

Development of evaluation method of systems consequences due to functional impairment of critical infrastructure in views of seismic disaster risks

G. Shoji
University of Tsukuba, Tsukuba, Japan

N. Kurozumi
InterRisk Research Institute & Consulting, Inc., Tokyo, Japan

ABSTRACT

This study focused on functional impairment of critical infrastructures such as energy supply systems, water treatments, communication networks, and transportation networks, and evaluation method of their consequences on other systems involved in societal and economical activity during a seismic disaster is developed. Among various societal and economical activity, financial business dealing with mutual fund and selling and buying of securities, is selected in developing the method. Based on the proposed method, case studies associated with economical activity by a private company engaged in finance business in a typical office building located at the Tokyo metropolitan area, affected by functional impairment of critical infrastructures, are done anticipating the Tokyo Metropolitan Earthquake. The following results were obtained:

First of all, the framework of evaluation method of subject systems consequences, which are quantitatively measured by cumulative probability distribution function associated with total damage costs in a year, was shown. The cumulative probability distribution function could be computed by adopting the compound Poisson distribution obtained by compounding probability of n_i times occurrences in a unit period corresponding to a consequence $P_i(n_i)$ and probability of an occurrence $g_i(s)$ associated with induced damage costs s.

Second, what is important to prevent subject financial business discontinuity and to reduce catastrophe damage for subject sector, is to take countermeasures associated with the prevention of functional impairment of involved critical infrastructures as well as to reduce physical damage such as 'Causalities of workers', 'Structural damage of building', 'Fire' and 'Overturning of furniture and fixtures' (Figure 1).

Third, it is quite likely that the functional impairment associated with power supply, telecommunication and maintenance of servers becomes crucial to systems consequences, focusing on the functional

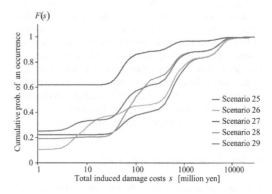

Figure 1. Evaluation of effect of a consequence associated with functional impairment of critical infrastructures on systems consequences.

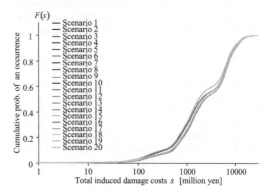

Figure 2. Evaluation of effect of a consequence on systems consequences considering combination of all consequences.

impairment of critical infrastructures for subject financial activity during a seismic disaster (Figure 2). This is a reasonable result which involved stakeholders in subject issue are recognizing qualitatively when they make a plan for the seismic disaster prevention.

Safety, Reliability and Risk of Structures, Infrastructures and
Engineering Systems – Furuta, Frangopol & Shinozuka (eds)
© 2010 Taylor & Francis Group, London, ISBN 978-0-415-47557-0

Risk assessment of shore-based power supply at marine container terminals

D. Pachakis & A. Jacob
Moffatt & Nichol, Long Beach, CA, USA

1 INRODUCTION

In this paper, quantitative risk assessment methodologies are applied on a newly developed ship emissions reduction system known as "Alternative Maritime Power" supply or "Cold Ironing". The main premise of this system is that during the ship's stay at a port, power is provided from the shore so that the ship can turn off its diesel engines and thus reduce air emissions in the atmosphere. Such a system has been currently installed successfully at the OOCL terminal at the Port of Los Angeles and is underway in two more terminals at the ports of Los Angeles and Long Beach.

2 BACKGROUND AND SCOPE

What triggered this study is the request by the electricity provider, which was a public utility company, to be indemnified for any damage that might occur to their power grid due to the operations of this system. The main concern was that, under certain circumstances, ship-based electricity from its generators could flow into the utility power grid and cause damage to other customers.

Moffatt & Nichol (M&N) was assigned the task to analyze the whole shore-power providing system from the terminal to the ship and a) determine the circumstances under which such a power flow reversal could happen b) examine the safety systems responsible for deterring such an event and assess their reliability and c) propose changes in the equipment or procedures that would improve the reliability of the safety systems and reduce the probability of the power reversal to happen.

3 DESCRIPTION OF THE ANALYZED SYSTEM

The shore-based power supply system is comprised of three main sub systems:

1. the land-based infrastructure that get electricity from the utility all the way to the wharf

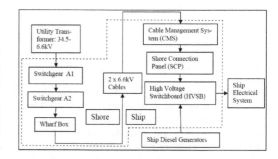

Figure 1. Schematic of the shore-based power supply system. Dashed line shows the boundaries of the analyzed system.

2. the cables that transfers electricity from the wharf to the ship
3. the ship-based infrastructure that synchronizes the shore power with the ship power and performs the switch in the source to the ship's electrical system.

Each of the above sub-systems had been designed by different engineers in the US, Europe and Asia and had not yet been tested in operation. Figure 1 gives a schematic diagram with the electricity flow under normal conditions.

4 SUMMARY AND CONCLUSIONS

Moffatt & Nichol engineers made operational recommendations to the terminal operator on how to increase significantly the reliability of the safety mechanisms with very low cost. Furthermore, they gave policy recommendations regarding the negotiations with the utility company and the extent of risk it would be prudent to them to assume.

Safety, Reliability and Risk of Structures, Infrastructures and
Engineering Systems – Furuta, Frangopol & Shinozuka (eds)
© 2010 Taylor & Francis Group, London, ISBN 978-0-415-47557-0

Seismic risk management for an existing lifeline system

T. Imai
JFE Engineering Corp., Yokohama, Japan

T. Koike
Musashi Institute of Technology, Tokyo, Japan

ABSTRACT

This study proposes a certain measure for decision-making associated with the seismic investment strategy preventing a large-scale malfunction for a existing lifeline system under seismic risks. The most appropriate decision on the seismic investment must be acceptable to all the parties involved whether in the financial or structural side.

When a new construction or an existing retrofitting project are planned, the target performance should be set up which can be accepted by all the sectors including stakeholders, project operation firms, design and construction contractors. In this case, there are two target performances that can be identified. The first target performance (A) would come from stakeholders and the project operating firms and will naturally be in monetary terms. The other target performance (B) is discussed among the engineers of the project operating firms and the contractors in terms of probability of failure or its equivalent measures of safety index, partial safety factor or factor of safety. It is pointed out that the target performance (A) and (B) are not related, unless the target performance should be defined and functionally formulated to include both monetary investment and safety measure variables.

When estimating the feasibility of the project, a final balance of costs and income over the life cycle periods which we call a value index, V_0, should be a key factor for a decision-making.

$$V_0 = B + I - E(C_D + C_M) \tag{1}$$

in which B, I, E, C_D and C_M are the accumulated total amounts of social benefit, income, operational expense, depreciation cost and maintenance cost, respectively, during its life cycle period.

When seismic investment C_S is taken into consideration and serviceability loss due to earthquakes, the value index, V_2, is given by

$$V_2 = B + I - E(C_D + C_M) - (C_R + \Delta B + \Delta I) \cdot 1_{EQ}(t_{EQ}) \tag{2}$$

where,

$$1_{EQ}(t_{EQ}) = \begin{cases} 1: \text{an earthquake occurs at } t_{EQ} \\ 0: \text{an earthquake does not occur at } t_{EQ} \end{cases}$$

in which ΔB and ΔI are losses of social benefit and income, respectively.

In order to decide the seismic investment for a seismic disaster prevention project, several sets of solutions including (1) the content of retrofitting design and restoration plan, (2) the financial requirement for the project and (3) the decision-making information such as the probability of system performance failure p_f^{Target} and the probability of value loss p_{u_0} should be prepared.

$$p_{v_0}(T_p) = P[V_0(T_p) < C_S] = F_{V_0}(C_S) \tag{3}$$

$$p_V^{Target}(T_p, t_{EQ}) = p_{V|EQ_1}^{Target}(T_p, t_{EQ}) \cdot P[EQ_1] + p_{V|EQ_2}^{Target}(T_p, t_{EQ}) \cdot P[EQ_2] + p_{v_0}(T_p)(1 - P[EQ_1]) \cdot (1 - P[EQ_2]) \tag{4}$$

in which T_p, $F_{V_0}(v)$, EQ_1 and EQ_2 are the present point, the probability function, the Level 1 and Level 2 earthquake ground motion, respectively.

Using this stochastic value index, seismic disaster prevention investment for the existing deteriorated lifeline system is discussed.

Several important results can be summarized:

(1) The probability of system performance failure can be related with the probability of value loss through the seismic investment.

(2) Target value on the probability of failure can be determined by the decision maker who can select the target value based on the probability of value loss.

(3) The performance-based design method for seismic retrofit strategies of existing lifelines can be carried out through the target probabilities of value loss and structural failure.

*Safety, Reliability and Risk of Structures, Infrastructures and
Engineering Systems – Furuta, Frangopol & Shinozuka (eds)
© 2010 Taylor & Francis Group, London, ISBN 978-0-415-47557-0*

Modeling infrastructure system performance using BN

M.T. Bensi
University of California, Berkeley, CA, USA

D. Straub
Technical University of Munich, Germany

P. Friis-Hansen
Det Norske Veritas, Norway

A. Der Kiureghian
University of California, Berkeley, CA, USA

ABSTRACT

We are currently working towards the development
of a probabilistic decision-support system (DSS) for
near-real time emergency response and recovery following a seismic event, utilizing a Bayesian Network
(BN) methodology. To date, our efforts toward the
development of the probabilistic DSS have focused
on modeling the earthquake demand as a spatially
distributed random field (Straub et al. 2008) and on
investigating methods for modeling the infrastructure system as a BN. In this paper we provide a
brief description of the seismic demand model and
thereafter focus on modeling system connectivity.

Friis-Hansen (2004) has proposed a BN formulation, in which system connectivity is modeled by
exploiting causal relationships between the system
components necessary for its survival. We have found
that this approach can be effective for certain types of
systems, specifically those for which it is easy to identify all minimum link sets (MLSs). We expand upon
the Friis-Hansen (2004) approach by considering five
formulations for modeling binary systems using BN.
These are described in this paper in conjunction with
a simple example system.

The five approaches to modeling a spatially distributed network by a BN are described and compared.
First, we present a formulation we have designated the
naïve BN formulation; one in which the system connectivity is modeled as a direct function of its components. In this formulation all components are parents of
one system node. Second, we define the *minimum link
set BN formulation* as a BN where the system connectivity is expressed directly as a function of the MLSs.
This is done by representing the system as a node
whose parents are the MLSs. In turn, each MLS node
has its constituent components as parents. Third, we
present a formalization of the Friis-Hansen approach,
in which system connectivity is modeled by exploiting
causal relationships between the system components
necessary for its survival. Rather than modeling the

system as a child of its MLS nodes, this formulation
expresses system connectivity using a causal interpretation of the connectivity paths. We denote this as the
explicit connectivity (EC) BN formulation. Fourth, we
define the dual of the MLS formulation as the *minimum cut set (MCS) BN formulation,* in which the
system node is a child of parents representing MCS
nodes, and each MCS node itself is represented as
a child of nodes representing its constituent components. The system node is a series system of all the
MCS nodes, whereas each MCS is a parallel system
of its parent nodes. Finally, we define the dual of the
EC formulation as the *explicit disconnectivity (EDC)
BN formulation.* Rather than tracing paths that ensure
survival of the system, one pursues causal event paths
that ensure failure of the system. This is a less intuitive
approach than the EC formulation.

In addition to presenting the formulations we also
offer some guidance on the system topologies for
which the differing formulations are applicable. However, the development of explicit rules for when to
apply each formulation remains an area for future
research. A case study of a hypothetical transportation
system is presented to (1) demonstrate the modeling of
earthquake loading as a spatially distributed random
field, (2) present a specific application of the formulations presented in this paper, and (3) to illustrate the
value BNs for information updating.

REFERENCES

Friis-Hansen, P. (2004). Structuring of complex systems
using Bayesian Networks (with additional material from
the unpublished extension to the paper, 2008). *Proceedings, Two Part Workshop at DTU.* Technical University of
Denmark.

Straub, D., M. Bensi, & A. Der Kiureghian (2008). Spatial
modeling of earthquake hazard and infrastructure performance through Bayesian Network. *Proc. Inaugural
International Conference of the Engineering Mechanics
Institute, ASCE,* Minneapolis, MN, May 19–21, 2008.

Safety, Reliability and Risk of Structures, Infrastructures and
Engineering Systems – Furuta, Frangopol & Shinozuka (eds)
© 2010 Taylor & Francis Group, London, ISBN 978-0-415-47557-0

Seismic risk management of water supply system with PML index

K. Yamamoto
Engineering & Risk Services Corporation, Tokyo, Japan

O. Maruyama & M. Hoshiya
Musashi Institute of Technology, Tokyo, Japan

H. Take
City of Yokohama, Public Works office, Yokohama, Japan

ABSTRACT

The numerous major earthquakes such as the Great
Hanshin-Awaji earthquake, Nigata-Chuetsu earthquake caused substantial casualties and property
damages. Private companies/public utility enterprises
should provide disaster reduction investment as part
of risk management and prepare a strategic Business
Continuity Plan. The significance of the Business Continuity Plan has been increasingly important in the connection of corporate social responsibility (Hoshiya &
Yamamoto 2005, 2002).

This study focuses on the seismic risk analysis of
the existing water supply system. Water supply system
may be modeled as a network that consists of supply
nodes (purification plants), demand nodes (pump-
ing stations and service reservoirs) and links (buried
pipes). It may be postulated that water can be supplied
to a demand node as long as this node is connected at
least to either one of supply nodes, and each demand
node is connected to and is in charge of the corresponding supply area of a lower network. It is also
postulated that links are subject to seismic risk and
they have probability of two events, failure or success,
whereas nodes are modeled as a system of components such as motors, water pipes, electric systems and
so forth. Each component has probability of failure
or success, and because of the combination of plural
component probabilities, failure of a node system has
different failure modes, and their probabilities must be
evaluated, for example, by the event tree analysis.

In this research, probable maximum loss (PML)
index is employed to evaluate the seismic risk of a
lower network. An alternative method is proposed to
avoid the complicated event tree analysis and evaluate PML index of the high order water supply network
system. Then a numerical analysis is carried out to
demonstrate the efficiency of the proposed method.

REFERENCES

Hoshiya, M. and Yamamoto, K. 2005 Efficient Investment
in Anti-seismic Disaster Measures, ICOSSAR 2005, 248,
Rome, Italy.
Hoshiya, M. and Yamamoto, K. 2002 Redundancy Index of
Lifeline Systems, Jour. of EM Div., Vol. 128, No.9, ASCE,
2002, 961–968.

Safety, Reliability and Risk of Structures, Infrastructures and Engineering Systems – Furuta, Frangopol & Shinozuka (eds)
© 2010 Taylor & Francis Group, London, ISBN 978-0-415-47557-0

Probability models for the reliability metrics of practical power distribution networks

L. Dueñas-Osorio & J. Rojo

Rice University, Houston, TX, USA

ABSTRACT

This study introduces a method to analytically obtain the probability distribution of power distribution system reliability metrics. The proposed method goes beyond traditional analytical tools that only provide moments of the probability distributions. The new methodology uses the conditional probability distribution of failure of each network element, and the probability distribution of external hazard occurrence. The system-level reliability assessment is derived from a combinatorial procedure that relies on the unconditional probabilities of failure of individual elements to produce the exact probability distribution of standard reliability indices. The system average interruption frequency index (SAIFI) is used in this study to illustrate this exact Poisson-Binomial combinatorial approach. Results for a simple radial feeder and a portion of the Roy Billinton test system (RBTS) suggest that the probability distribution of SAIFI is robust to variations in the probability laws of network element fragility and hazard occurrence.

Figure 1 displays the topology of the power distribution systems of this study, and Figure 2 highlights the exact Poisson-Binomial PMF of SAIFI as well as the binomial and Poisson approximations for the linear and simplified bus 6 circuits. The probability distribution of SAIFI and its approximations are insensitive

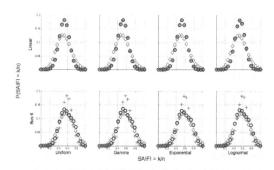

Figure 2. Approximate and exact PMF of SAIFI. Crosses represent the binomial fit to SAIFI data, diamonds the Poisson fit, and circles the exact combinatorial SAIFI distribution.

to the form of the fragility and hazard pairs. In addition, the binomial model fits the linear feeder SAIFI data better than the Poisson probability law, while a contrasting trend is observed in bus 6 where Poisson models fit better the SAIFI data. These closed form approximations to the Poisson-Binomial model enable detailed system sensitivity analyses, including response to extreme hazards, and estimation of the probabilities of attaining regulatory standards.

REFERENCES

Balijepalli, N., S.S. Venkata, and R.D. Christie, 2004a. "Predicting distribution system performance against regulatory reliability standards," *IEEE Transactions on Power Delivery*, vol. 19, no. 1, pp. 350–356.

Billinton, R., and W. Wangdee, 2006. "Delivery point reliability indices of a bulk electric system using sequential Monte Carlo simulation," *IEEE Transactions on Power Delivery*, vol. 21, no. 1, pp. 345–352.

Brown, R.E., 2002. *Electric power distribution reliability*. New York: Marcel Dekker, Inc., pp. 75–114.

Warren, C. A., 2005. "IEEE Reliability indices standards," *IEEE Industry Applications Magazine*, pp. 16–22, Jan./Feb.

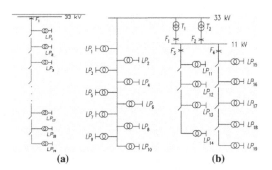

Figure 1. Single line diagram with 19 load points of (a) ideal linear feeder, and (b) simplified RBTS bus 6.

Mini-Symposia (MS17) Earthquake Engineering and Engineering Seismology

Safety, Reliability and Risk of Structures, Infrastructures and Engineering Systems – Furuta, Frangopol & Shinozuka (eds)
© 2010 Taylor & Francis Group, London, ISBN 978-0-415-47557-0

A method for accurate calculation of seismic structural failure probability using multi point estimate

M. Nakajima
Central Research Institute of Electric Power Industry, Abiko-sh, Japan

H. Morikawa
Tokyo Institute of Technology, Yokohama, Japan

ABSTRACT

In performing seismic response analysis of non-linear strcture systems using a group of earthquake ground motions X which are generated to follow a probability distribution, the structural response value Y can be generally represented as a non-linear function $Y = g(X)$. Hence, it can be assumed that the Y is stochastic variable in a case where non-linear structure models are considered.

Although the probability density function $f_Y(y)$ is required in computing seismic structural failure probabilities using structural realibility theory, it is usually difficult to specify the shape of the $f_Y(y)$.

To calculate a failure probaility of structures against the considered limit state, it is necessary to estimate $f_Y(y)$ efficiently and precisely. Therefore, we improve multi point estimate method in order to calculate seismic structural failure probability more efficiently and more precisely. Then, we discuss about the precision of the improved method.

In the current paper, we adopt spring-mass model whose reponse is non-linear and sliding-block model so that we may conduct numerical simulation for the demonstration. In our simulation, the PDF $f_Y(y)$ is modeled in two ways as follows:

(model 1) Y follows normal distribution which is determined using the first and the second moment of g(X), that corresponds to the conventional point estimate method (Rosenblueth[1]),

(model 2) Y follows a probability distribution whose density function which is approximated by Gram-Charlier expansion that is one of the orthogonal polynomials espansion methods. In this simulation, moments of Y are calculated to the fourth order although the higher order moments can be considered.

Conducting the seimic response analysis, the distribution of Y = g(X) is obtained for the model 1 and model 2, respectively. The PDF $f_Y(y)_1$ is statistically estimated from the histogram, which represent the relationship between the response value and its frequency, for the model 1, and the PDF $f_Y(y)_2$ is from the histogram for the model 2.

We compare the PDF $f_Y(y)$ which is obtained from Monte-Carlo Simulation (model 3) with the PDF $f_Y(y)_1$ and $f_Y(y)_2$, first. Furthremore, the failure probaities for the considered limit state are compared among the model 1, model 2 and model 3.

On the basis of the numerical simulation results, we discuss about an appropriate method to calculate failure probability of non-linear structure models using multi point estimate from the viewpoint of numerical precision.

REFERENCES

Rosenblueth, E. : Point estimates for probability moments, Proc.Nat.Acad.Sce, USA, Vol.72, No.10, pp.3812–3814, 1975.

Safety, Reliability and Risk of Structures, Infrastructures and Engineering Systems – Furuta, Frangopol & Shinozuka (eds)
© 2010 Taylor & Francis Group, London, ISBN 978-0-415-47557-0

Seismic safety assessment of concrete gravity dams

J. Huang & A. Zerva
Drexel University, Philadelphia, PA, USA

1 ABSTRACT

The safety of concrete gravity dams in seismic environments has been a subject of significant concern. The evaluation of their response to earthquakes requires particular attention due to the complexities of the dam-reservoir-foundation system. These include: nonlinear behavior of the concrete and foundation rock, interaction between the dam, the reservoir and the foundation, unbounded size of the reservoir and foundation domain, and spatial variation of seismic ground motions over distances comparable to the base dimensions of the dam. This paper addresses some of the aforementioned issues by developing a computational procedure for the rational seismic safety evaluation of concrete gravity dams using the finite element method and conducting sensitivity analyses that investigate the modeling assumptions. Features of the numerical model include: (i) the concrete and rock masses are assumed to be impervious material represented, respectively, by the concrete damaged plasticity model (Lee & Fenves, 1998; Lubliner et al., 1989) and the Mohr-Coulomb model. The models are permitted to respond either in the linear or the nonlinear range. (ii) The unbounded foundation rock is divided into two sub-domains, i.e. the near-field region of interest and the far-field unbounded medium represented, respectively, by finite elements and infinite elements. (iii) The dam-reservoir dynamic interaction is emulated with two approaches: either via Westergaard's added mass technique (Westergarrd, 1933) or Darbre's two-parameter model (Darbre, 1998). (iv) The dam-foundation interaction is modeled using contact surfaces with a "hard" contact and a Coulomb friction model for the interaction normal and tangential to the surface, respectively (Abaqus, 2007). For comparison purposes, the model is also considered to be fixed at its base. (v) The input excitation is considered to be either uniform or propagating from the upstream to the downstream face of the dam, and is applied at the foundation surface of the numerical models.

A preliminary study is conducted to investigate the effects of dam-reservoir hydrodynamic interaction and spatial variation of earthquake ground motions on the response of the Koyna Dam in India. The findings of this research work indicate that: (i) The different models for representing the hydrodynamic pressure effects produce different response with Darbre's (1998) model yielding a lower response than Westergaard's (1933) approach, due to the energy absorption by the dampers of the former model. (ii) The consideration of the infinite foundation domain and the contact interaction at the dam-foundation interface can have a considerable impact on the dam response, as the peak stress distribution and crack formation in the body of the dam differs significantly from the corresponding response values obtained when the dam is considered fixed at its base. (iii) Cracks at the neck of the dam and contact opening at the heel of the dam-foundation interface form under both uniform and nonuniform seismic excitations. For both excitation cases, slipping occurs along the entire dam base with a slightly more pronounced response at the heel of the dam. However, the spatially variable seismic excitation tends to magnify the dam response and may pose a threat to the dam safety.

REFERENCES

Abaqus. 2007. Analysis user's and theory manual, Version 6.7, Simulia.

Darbre, G.R. 1998. Phenomenological two-parameter model for dynamic dam-reservoir interaction. *Journal of Earthquake Engineering* 2(4): 167–176.

Westergaard, H.M. 1933. Water pressures on dams during earthquakes. *Transactions of the American Society of Civil Engineers* 98: 418–433.

Lee, J. & Fenves, G.L. 1998. Plastic-damage model for cyclic loading of concrete structures. *Journal of Engineering Mechanics* 124(8): 892–900.

Lubliner, J., Oliver, J. , Oller, S. and Oñate, E. 1989. A plastic-damage model for concrete. *International Journal of Solids and Structures* 25(3): 229–326.

Safety, Reliability and Risk of Structures, Infrastructures and Engineering Systems – Furuta, Frangopol & Shinozuka (eds)
© 2010 Taylor & Francis Group, London, ISBN 978-0-415-47557-0

Stochastic analysis of floor response spectra in consideration of structural uncertain properties

T. Mochio
Kinki University, Wakayama, Japan

ABSTRACT

In this paper, instead of the Monte Carlo simulation, an attempt was made to evaluate the stochastic seismic response of the structure having uncertain properties by a technique capable of analytical evaluation employing the random vibration theory and statistics technique.

As a practical item of analysis, this paper deals with the floor response spectrum, which is one of the most important quantities in earthquake-resistant design. First of all, an earthquake wave is modeled as product of a non-stationary envelope function and a stationary Gaussian random process of which power spectral density is equal to the Kanai-Tajimi spectrum. Next, applying the transition matrix method with state space to the secondary system mounted on main structure, the non-stationary standard deviation of absolute acceleration response relating to the secondary system is obtained under the assumption of structural properties being certain values. In addition, by introducing the peak factor based on the extreme value theory, the expected maximum value of system concerned can be given with no fluctuation of properties. Finally, application of FOSM(First Order Second Moment) method to the total representation theorem produces statistical values regarding the maximum acceleration response of the secondary system in consideration of randomness for input wave and uncertainty for structural properties. Thus, the floor response spectrum can be stochastically obtained.

The validity of the analytical method is estimated by the Monte Carlo simulation.

REFERENCES

Benjamin, J.R. & Cornell, C.A. 1970. Probability, Statistics, and Decision for Civil Engineers. McGraw-Hill.
Ditlevsen, O. 1981. Uncertainty Modeling. McGraw-Hill.
Tajimi, H. 1960. A Standard Method of Determining the Maximum Response of a Building Structure During an Earthquake. *Proc. Second World Conf. Earthquake Eng., Tokyo.*

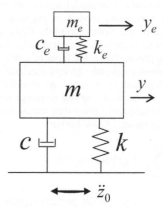

Figure 1. Mathematical model for generating response spectrum.

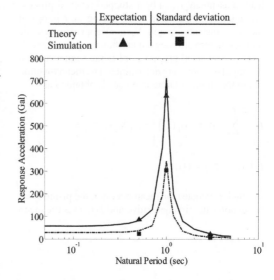

Figure 2. Comparison of analytical results with simulations.

Safety, Reliability and Risk of Structures, Infrastructures and
Engineering Systems – Furuta, Frangopol & Shinozuka (eds)
© *2010 Taylor & Francis Group, London, ISBN 978-0-415-47557-0*

Stochastic seismic linear response via the random train of impulses

A.I. Daniliu
Romanian Association for Earthquake Engineering, Botosani, Romania

ABSTRACT

The random pulse-train model was one of the earliest stochastic earthquake models in engineering literature. Based-on the application of the first mean value theorem, in some author's own papers it has been mathematically proved that at the input of a linear system the continuous earthquake ground acceleration $\ddot{U}(t)$ may be replaced by a series of modulated Dirac impulses. For a given impulse response function of the linear system and a given time instant t we have in the sense of theory of distribution:

$$\ddot{U}(\tau) = \sum_{i=1}^{N_\alpha(t,\tau)} \bar{U}(\tau)\delta(\tau-\alpha_i(t))$$

where $N_\alpha(t, \tau)$ is a point process representing the number of Dirac impulses at time τ for a fixed value of parameter t, the random time instants $\alpha_i(t)$ are not unique determined between two successive zero-crossing times of $\ddot{U}(t)$ and $\bar{U}(\tau)$ is the overall impulse magnitude function. Physical considerations lead to assume $N_\alpha(t, \tau)$ be a stopped renewal process dependent on parameter t and having unit jumps at the time instants α_i and $N_\alpha(t, \tau)$ approximates the number of zero crossings of earthquake ground acceleration. Statistical properties of Dirac sequences and the impulsive noise are determined. The mean function of random train of Dirac impulses is obtained as

$$E\left[\sum_{i=1}^{N_\alpha(t,\tau)} \delta(\tau-\alpha_i(t))\right] = E[N'_\alpha(t,\tau)] =$$

$$E'[N_\alpha(t,\tau)] = \lambda(t,\tau)$$

in which E denotes the mean operator, the prime symbol denotes the first derivative and $\lambda(t, \tau)$ is the mean

impulse arrival rate. The mean and autocorrelation functions of the impulsive noise are

$$E\left[\sum_{i=1}^{N_\alpha(t,\tau)} \bar{U}(\tau)\delta(\tau-\alpha_i)\right] = E[\bar{U}(\tau)]\lambda(t,\tau)$$

$$R_2(\tau_1,\tau_2;t) = E[\bar{U}^2(\tau_1)]\lambda(t,\tau_1)\delta(\tau_2-\tau_1) +$$

$$E[\bar{U}(\tau_1)\bar{U}(\tau_2)]\lambda(t,\tau_1)\lambda(t,\tau_2)$$

in which $\lambda(t, \tau_1)$ approximates the instantaneous rate of zero-crossings per unit time of earthquake ground acceleration. For a linear system having the impulse response function $h(t - \tau)$ the mean and variance functions of the response $X(t)$ are

$$E[X(t)] = \int_0^t E[\bar{U}(\tau)]\lambda(\tau;t)h(t-\tau)d\tau$$

$$E[X^2(t)] = \int_0^t E[\bar{U}^2(\tau_1)]\lambda(\tau_1,t)h^2(t-\tau_1)d\tau_1 +$$

$$+\int_0^t\int_0^t E[\bar{U}(\tau_1)\bar{U}(\tau_2)]\lambda(\tau_1,t)\lambda(\tau_2,t)\cdot$$

$$h(t-\tau_1)h(t-\tau_2)\,d\tau_1\,d\tau_2.$$

where $\bar{U}(\tau_1) \neq \bar{U}(\tau_2)$. The impulse magnitude function and sign oscillation function are introduced to describe the amplitude variations and the frequency content of earthquake ground motions.

Safety, Reliability and Risk of Structures, Infrastructures and
Engineering Systems – Furuta, Frangopol & Shinozuka (eds)
© 2010 Taylor & Francis Group, London, ISBN 978-0-415-47557-0

Database-assisted seismic analysis of tall buildings subjected to Vrancea earthquakes

M. Iancovici
Technical University of Civil Engineering of Bucharest, Romania

C. Gavrilescu
Kohn, Pedersen, Fox Architects (KPF), London, UK

ABSTRACT

The seismic hazard generated from the Vrancea source
is the major concern for practitioners, especially due to
medium-soft soil conditions in the Bucharest-city. Tall
buildings therefore might have the highest exposure to
significant damage as well as economical and human
losses

Large databases of (i) earthquake ground motion
records and (ii) structural properties are now available and can be used thanks to large computational
capabilities.

The *database-assisted seismic analysis* is an integrated structural analysis format that allows (1)
performing the seismic response analysis, energy
balance-based analysis, fragility and seismic risk analyses, by using the time series of real/scaled/simulated
multi-directional earthquake ground motions and
the induced effects (stresses, strains, displacements,
velocities, accelerations, efforts, forces, energies),
(2) the design of the structural elements and connections in a straightforward and transparent manner
and (3) getting higher performance, safer and cost-
effective structures. This format becomes much more
important in the case of irregular structures, having
plan and elevation complex shape, in which the main
directions of motion are not obvious.

The objective of the paper is to investigate the
applicability of *database-assisted seismic analysis*
approach and to emphasize the main features of this
format in the practical seismic design of large scale
structures.

Parametric analyses of a 60 story model subjected to long predominant period ground motions are
conducted and the main results are highlighted.

The *database-assisted structural analysis and
design concept* was initially developed for low-rise
buildings subjected to wind (Whalen *et al.*, 2000).
Later, this approach was extended to tall buildings
(Simiu *et al.*, 2003; Iancovici *et al.*, 2003) and the
concept can be successfully adopted for the seismic
design also.

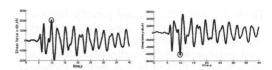

Figure 1. Time-histories of shears in the $x-$ and $y-$
directions in a typical column.

For Vrancea 4 March 1977 record (Bucharest
INCERC station), considering the induced effects, say
in terms of shear forces, the time-histories in the $x-$
and $y-$ directions, corresponding 30° directivity angle
of the orthogonal input ground motions, in a typical
column at the 10th floor, are given in the figure 1.

In a similar way, all the time-histories of the induced
effects are obtained. Demand peak shear forces are
obtained by common maxima superposition rules (e.g.
30% rule, SRSS rule etc.). These are compared with
the peaks obtained by the "exact approach", obtained
directly from the time-series. The time-histories of the
induced efforts (axial forces and bending moments)
are directly used then in design equations in the form
of interaction formulae, provided by various standards
(AISC-LRFD, EC 3 etc.).

Detailed studies are necessary, based on a large
amount of input motions and structural configurations
for setting-up transparent criteria's for design.

REFERENCES

Iancovici, M., Gavrilescu, C., 2008. Seismic energy distribution in high-rise buildings subjected to Vrancea earthquakes. 14 World Conference on Earthquake Engineering,
October 12–17, Beijing, China
Iancovici, M., Riley, M.A., Sadek, F., Simiu, E., 2003. Wind
effects on high-rise buildings: database-assisted design
versus the high-frequency force-balance technique. In:
Proc. 11th Int. Conf. on Wind Engineering, Lubbock, TX,
2003.
Eurocode 3-2005. Design of Steel Structures 1993-1-1, 2005
Manual of steel construction, 2001: load and resistance
factor design. 3rd Edition, American Institute of Steel
Construction (AISC), Chicago, IL, 2001.

Safety, Reliability and Risk of Structures, Infrastructures and Engineering Systems – Furuta, Frangopol & Shinozuka (eds)
© 2010 Taylor & Francis Group, London, ISBN 978-0-415-47557-0

Analytical study on seismic verification method of steel columns subjected to bi-directional horizontal loadings

N.G. Kulkarni & A. Kasai

Department of Civil Engineering, Nagoya University, Nagoya, Japan

ABSTRACT

The hollow steel columns and partially filled with concrete steel columns are popularly used in urban areas of Japan because of their rapid construction and good sustainability during earthquake. After observing disastrous effect of high magnitude earthquake on the bridges, the seismic design codes were revised regularly. However, Kobe earthquake in 1995 again pointed out drawbacks of former specification hence in 1996 revision of code included intensive earthquake motion observed at Kobe (Kawashima 2004) and 2002 version of specification accepted performance-based design concept for the purpose to respond the international harmonization of design codes and flexible application to new structures and new construction methods (Japan Road Association 2002). However, the current specification suggests performing separate seismic analysis in longitudinal and transverse directions of the structures. Whereas, in actual conditions earthquake generates 3D vibrations and the response of the structure to this type of vibration is quite different than uni-directional earthquake motions (Okazaki et al. 2003), hence it is needed to consider bidirectional loading effect on the bridge columns.

Concerned with thin walled steel columns subjected to bi-directional loading, research has been carried out (Watanabe et al. 2000; Goto et al. 2006) to find strength and ductility. However, application of results obtained from the bi-directional loading analysis in the seismic verification process was not proposed until Tsuboi et al. (2007) developed displacement based verification method. After conducting dynamic analysis Tsuboi et al. found that displacement based verification gives slightly over safe seismic design when compared with ultimate strain value. Hence, the requirement of developing strain based verification method for bi-directional earthquake is the aim of present study.

It has been proved that bi-directional circular loading is severe for circular steel columns, hence loading patterns considered here are gradually changes from uni-directional to elliptical shape and then fully circular to observe effect on the column. The thin walled circular steel column is analyzed by FEM shell element model and simplified beam element model. The local buckling effect can be seen in shell element model whereas in beam element model it is ignored. In case of shell element models, a simple method to find out equivalent strain based on local buckling is defined whereas in beam element model average compressive strain is calculated. As a result, after comparing both types of strains, empirical formulas are predicted to propose a strain based verification method for bi-directional earthquake motion.

REFERENCES

Japan Road Association. 2002. Design Specification of Highway bridges, part V Seismic Design.

Kawashima, K. & Unjoh, S. 2004. Seismic Design of Highway Bridges. Journal of Japan Association for Engineering 4(3): 174–186.

Goto, Y., Jiang, K.. & Obata, M. 2006. Stability and Ductility of Thin-walled Circular Steel Columns under Cyclic Bidirectional Loading. Journal of Structural Engineering 132(10): 1621-1631.

Okazaki, S., Usami, T. & Kasai, A. 2003. Elasto-plastic dynamic analysis of steel bridge piers subjected to bidirectional horizontal earthquakes. Journal of Structural and Earthquake Engineering 27: (CD-ROM). (In Japanese)

Tsuboi, H, Torii, J., Kasai, A. & Usami, T. 2007. The seismic method for pipe-section steel bridge piers subjected to bidirectional earthquake motions. Journal of Structural and Earthquake Engineering, JSCE 29: 529–538. (In Japanese)

Watanabe, E., Sugiura, K. & Oyawa, W. 2000. Effect of multidirectional displacement paths on the cyclic behavior of rectangular hollow steel columns. Journal of Structural and Earthquake Engineering, JSCE 17(1): 69–85.

Mini-Symposia (MS08) Time-Dependent Reliability Methods and Their Applications

Safety, Reliability and Risk of Structures, Infrastructures and
Engineering Systems – Furuta, Frangopol & Shinozuka (eds)
© 2010 Taylor & Francis Group, London, ISBN 978-0-415-47557-0

Flexural reliability of corroding segmental post-tensioned bridges

R.G. Pillai, P. Gardoni, M.D. Hueste, K. Reinschmidt & D. Trejo
Zachry Department of Civil Engineering, Texas A&M University, College Station, TX, USA

ABSTRACT

The presence of air-voids, moisture, and chlorides inside the tendons was cited as a reason for the early age (8, 13, and 16 years, respectively, after construction) strand corrosion and failure in the Mid-bay, Sunshine Skyway, and Niles Channel post-tensioned (PT) bridges in Florida, USA. These incidents call for frequent inspection and structural reliability assessment of PT bridges. Hence, bridge management authorities worldwide need time-variant structural reliability models to assess long-term performance to optimize inspection, repair, and maintenance programs of PT bridges. This paper presents modeling and assessment of time-variant structural reliability of segmental posttensioned (PT) bridges and their application to a typical PT bridge exposed to various exposure conditions.

The reliability is estimated as the probability that the moment demand attains or exceeds the corresponding moment capacity. The moment demand is modeled using a probabilistic formulation based on HL93 loading and field data (Nowak and Collins 2000). A probabilistic model for the moment capacity is formulated based on probabilistic models for the tension capacity of strands developed by Trejo et al. (2009), the nonlinear stressstrain formulation for concrete developed by Todeschini (1964), the stress-strain formulations for strands provided in AASHTO LRFD (2007) specifications, and principles of engineering mechanics. The proposed reliability model is applicable to different environmental conditions and accounts for the uncertainties in the damage of and voids in the PT systems.

As an application of the developed reliability model, Monte Carlo simulation is used to determine the reliability of a typical PT bridge (Figure 1) at 0, 25, 50, 75, and 100 years of service. The selected PT bridge

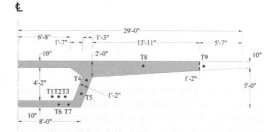

Figure 1. Semi-cross-section at midspan of a typical segmental, PT box girder.

girder has a simply supported span of approximately 30 meters (100 feet) and 18 tendons. The effects of three cases of exposure (i.e., a wetdry exposure condition in 0, 2, and 4 tendons, respectively) are assessed. At 100 years of exposure, the reliability indices for these three cases are determined to be 3.2, 3.1, and 0.1, respectively.

REFERENCES

AASHTO LRFD. 2007. *LRFD Bridge Design Specifications.* American Association of State Highway and Transportation Officials, Washington D.C., USA.

Nowak, A.S. & Collins, K.R. 2000. *Reliability of structures.* McGraw Hill Higher Education, USA.

Todeschini, C.E. Bianchini, A.C. & Kesler, C.E. 1964. Behaviour of concrete columns reinforced with high strength steels. *ACI Structural Journal,* 61(6):701–716.

Trejo, D. Hueste, M.D. Gardoni, P. Pillai, R.G. Reinschmidt, K., Kataria, S. Im, S, Gamble, M. & Hurlebaus, S. 2009. *Effects of voids in grouted, post-tensioned, concrete bridge construction.* Report No. 04588. Texas Transportation Institute, Texas Department of Transportation. Austin, Texas, USA.

Safety, Reliability and Risk of Structures, Infrastructures and Engineering Systems – Furuta, Frangopol & Shinozuka (eds)
© 2010 Taylor & Francis Group, London, ISBN 978-0-415-47557-0

Sum-of-disjoint-products technique for seismic reliability assessment of bridge networks

Z.H. Lu & Y.G. Zhao
Kanagawa University, Yokohama, Japan

X.G. Zhang
Shenzhen University, Shenzhen, China

ABSTRACT

Highway transportation networks, which mainly consists of individual bridges and roads that interconnect these bridges to form the network, are expected to continue to function during and after the occurrence of natural disaster such as an earthquake, so that lifelines can continue to provide emergency services and minimize loss of life and economic distress. Unfortunately, as evidenced in past earthquakes, highway transportation is often disrupted as a direct result of damage and collapse of highway system components. Of all the components of highway systems, bridges are usually considered to be the most vulnerable to earthquake induced damage and collapse. A typical case is the 1995 Hyogo-ken Nanbu (Kobe) earthquake, in which the destructive damage and collapse of bridges resulted in paralyzing of the traffic for months (Ministry of Construction 1995). For this reason, it can be assumed that highway bridges are only vulnerable elements of a highway network, i.e., roads that link any two bridges never fail and therefore, the highway networks is called as a bridge network. Due to the uncertainties in the description of loads especially the earthquake load and the resistances of the bridges, there is a significant need to evaluate the connectivity reliability of bridge networks, i.e., the probability of maintaining the connection between the origin and destination locations, under seismic excitations.

The seismic reliability assessment of a bridge network encompasses two basic steps, i.e., first to compute the seismic reliability of individual bridges and then to conduct the connectivity reliability analysis of the bridge network. The seismic reliability of individual bridges has been the focus of research in the area of dynamic reliability theory for bridge engineering in the past decades and therefore it will not be discussed in detail herein. On the contrary, there are few studies on the connectivity reliability of a bridge network (e.g., Augusti et al. 1994; Liu & Frangopol 2005).

The currently used method for reliability assessment of bridge networks such as the series-parallel path models and the event (fault) tree analysis may fail as the number of individual bridges increases since this computation is a typical NP-hard problem, i.e., the number of paths in the series-parallel path models or the number of branches on the event tree will correspondingly increase in a non-polynomial way. It is necessary to introduce more effective methods for evaluating the connectivity reliability of bridge networks.

The paper attempts to introduce the sum-of-disjoint-products technique (SDPT) (e.g. Fratta & Montanari 1973; Aggarwal et al. 1975) for seismic reliability analysis of bridge networks. It is expected that accurate prediction of connectivity reliability of bridge networks in future earthquakes based on the presented methodology can assist the related administrative departments in achieving a cost-effective strategy in the management of bridge networks.

REFERENCES

Aggarwal, K. K., Misra, K. B., & Gupta J. S. 1975. A fast algorithm for reliability evaluation. *IEEE Transaction on Reliability* 24 (1): 83–85.

Augusti, G., Borri, A., & Ciampoli, M. 1994. Optimal allocation of resources in reduction of the seismic risk of highway networks. *Engineering Structures* 16(7): 485–497.

Fratta, L., & Montanari, U. G. 1973. A Boolean algebra method for computing the terminal reliability in a communication network. *IEEE Transaction on Circuit Theory* 20 (3): 203–211.

Liu, M., & Frangopol, D. M. 2005. Time-dependent bridge network reliability: a novel approach. *Journal of Structural Engineering* 131(2): 329–337.

Ministry of Construction. 1995. Report on the damage of highway bridges by the Hyogo-ken Nanbu earthquake. *Committee for Investigation on the Damage of Highway Bridges Caused by the Hyogo-ken Nanbu Earthquake.*

Safety, Reliability and Risk of Structures, Infrastructures and Engineering Systems – Furuta, Frangopol & Shinozuka (eds)
© 2010 Taylor & Francis Group, London, ISBN 978-0-415-47557-0

Probabilistic approach of concrete durability design in marine environment

F. Deby, M. Carcassès & A. Sellier
LMDC – Laboratoire Matériaux et Durabilité des Constructions – Université de Toulouse – UPS, INSA, – Toulouse, France

ABSTRACT

We often observe that there is generally a random variation in the physical properties of concrete due to variability of fabrication on building sites. The actual design of concrete cover is very empirical and does not take this variability into account explicitly. The aim of this work is to propose an approach that integrates the composition of concrete and the variability of the physical properties associated with chloride ingress, in order to make an objective and probabilistic prediction of its durability. To this end, we need to combine a diffusion model for chloride and a probabilistic method.

First, the diffusion model in a saturated medium is explained. As the study is limited to concrete immersed in sea water, we use a non-linear model of chloride diffusion in a saturated medium where the chloride penetration can be modeled by Fick's law, which is a good com-promise between empirical models and multi-species models.

In this methodology, and for simplicity of use, only the composition of the concrete, the chemical composition of the cement and the measurement of water porosity p are necessary. Two elementary models are proposed to estimate the effective coefficient D_e and the chloride binding isotherm $C_b(c)$ for OPC concrete and OPC concrete with silica fume. The associated errors, considered as random variables, are exposed and their distribution laws assessed from experiments on OPC and OPC with silica fume (see Figure 1).

Finally, four random variables remain: the porosity p, the errors on the effective diffusion coefficient Err_D or Err_{Dsf}, the errors on the bound chlorides Err_C or Err_{Csf}, and the concrete cover thickness e. To update

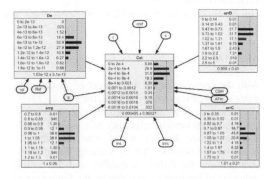

Figure 2. Bayesian network for chloride diffusion.

these distribution laws with new experimental results, a bayesian network has been built. The update is done for the elementary models errors and new findings can be associated with the effective diffusion coefficient or a chloride concentration. The network is presented on Figure 2.

Then the probabilistic method is proposed to evaluate the probability of reaching the critical chloride concentration for a given period of exposure. This is the probability (P_f) that the quantity of chlorides on the first concrete reinforcement bars reaches a critical chloride concentration value. It's estimated through the Hasofer-Lind reliability index β, using the Rackwitz-Fiessler algorithm.

A practical application to a concrete submerged in sea water is then proposed. For instance, the variation of the reliability index β with exposure time is presented for an OPC concrete and the same OPC+SF concrete. The threshold value $\beta_{SLS} = 1.5$ ($P_f \approx 7\%$) recommended by the Eurocodes for Serviceability Limit States at 50 years is compared.

To finish, the probabilistic method is necessary to take into account the distribution laws of the input parameters of the chloride ingress model in order to assess the probability of corrosion initiation for a given period of exposure. We would highlight the general character of this probabilistic methodology, which could offer a new tool to the most chloride diffusion modeling for designing structures under durability constraints.

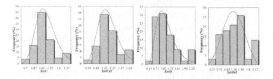

Figure 1. Histograms of elementary models errors.

Safety, Reliability and Risk of Structures, Infrastructures and
Engineering Systems – Furuta, Frangopol & Shinozuka (eds)
© 2010 Taylor & Francis Group, London, ISBN 978-0-415-47557-0

Seismic reliability assessment of thin-wall concrete aqueduct with method of moments

J.G. Xu
Department of Environment and Hydraulic Engineering, Zhengzhou University, Zhengzhou, China

B. Wang & H. Chen
Department of Civil Engineering, Zhengzhou University, Zhengzhou, China

Z.H. Lu & Y.G. Zhao
Department of Architecture and Building Engineering, Kanagawa University, Yokohama, Japan

ABSTRACT

With the construction of the project of transporting water from south to north in China, the aqueduct seismic security arose more and more attention. At present, the seismic dynamical reliability of aqueduct is seldom researched. The method of moments, which is based on the premise that by finding the relationship between the failure probability and the central moments of the limit state function, the failure probability can be assessed, is proposed to the load and resistance faction. Compared with the method of common, the load and resistance factors can be determined even when the probability distributions of random variables are unknown, and the method of moment needs neither iterative calculation of derivatives nor any design points. Based on the thin-wall dynamic analytical model of aqueduct with fluid-structure coupling, the seismic reliability of aqueduct is studied by used of the method of moment. In this model the influence of the bend-torsion coupling, water-structure coupling and nonlinear rubber bearing isolation are all taken into consideration. Based on the type of site soil, the artificial seismic waves are created. In accordance with the method of moment and the measured data of structural internal force under earthquake, the reliability distribution function and the aqueduct reliability are all calculated. Based on Housner theory, the fluid and aqueduct body's dynamical interaction include fluctuation pressure and convection pressure under the transverse earthquake loading. The creations of $[M]$, $[K]$ and $[C]$ could consult document.

The acceleration time-history $a(t)$ is calculated utilizing Fast Fourier's Transform after the amplitude spectrum calculation.

$$a(t) = \sum_{k=1}^{N} A_k \cos(\omega_k t + \varphi_k) = \sum_{k=N_1}^{N_2} A_k e^{i(\omega_k t + \varphi_k)}$$

The 3M reliability index can be written as follow

$$\beta_{3M} = (3 - \sqrt{9 + \alpha_{3G}^2 - 6\alpha_{3G}\beta_{2M}})/\alpha_{3G}$$

The 4M reliability index can be written as follow

$$\beta_{4M} = \frac{p}{D} - D + \frac{l_1}{3k_2}$$

The PDF of z_s can be expressed as follow

$$f(z_s) = K(z_s - r_2)^{\frac{-ar_2 - b}{\sqrt{\Delta}}} (r_1 - z_s)^{\frac{ar_1 + b}{\sqrt{\Delta}}}, \Delta > 0, d < 0$$

$$f(z_s) = K(c + bz_s + ez_s^2)^{-\frac{a}{2e}} \exp[\frac{ab - 2be}{e\sqrt{-\Delta}} \tan^{-1}(\frac{b + 2ez_s}{\sqrt{-\Delta}})], \Delta < 0$$

Utilizing 500 artificial seismic waves, the aqueduct structural response under two cases are calculated. From the analysis of structural internal force under strong earthquake, the bracket base section always entrance the plastic failure period firstly. So the section of the bracket base is the failure control section. The calculation results show that the failure probability of empty aqueduct (case1) is 0.0129, 0.02018, 0.02043 with the $\beta_{2M}, \beta_{3M}, \beta_{4M}$ respectively. The failure probability of design water level is 0.03253, 0.04138, 0.04095 with the $\beta_{2M}, \beta_{3M}, \beta_{4M}$ respectively.

The PDF equation is expressed as

$$f(z_s) = 6.77462e15(14.3697 - 2.10422z_s + 0.408434z_s^2)^{-17.0468}$$

$$\exp[-16.9114 \tan^{-1}(-0.509648 + 0.197847z_s)]$$

$$f(z_s) = 8.82930e - 7(4.806877 + z_s)^{6.43441}(2.58706 - z_s)^{3.00121}$$

REFERENCES

Zhao Guofan. 1996. *Reliability Theory and Its Applications for Engineering Structures.* Dalian: Dalin University of Technology Press. (in Chinese)

Zhao Yangang & Lu Zhaohui, 2006. A simple third-moment for structural reliability. *Journal of Asian Architecture and Building Engineering* 5(1): 129–136.

Zhao Yangang & Lu Zhaohui, 2007. Fourth-moment standardization for structural reliability assessment. *Journal of Structural Engineering* 133(7): 916–924.

Safety, Reliability and Risk of Structures, Infrastructures and Engineering Systems – Furuta, Frangopol & Shinozuka (eds)
© 2010 Taylor & Francis Group, London, ISBN 978-0-415-47557-0

Time-dependent reliability of R.C. structures considering uncertain parameters with unknown CDFs

Y.G. Zhao & Z.H. Lu
Kanagawa University, Yokohama, Japan

W.Q. Zhong
Dalian University of Technology, Dalian, China

1 INTRODUCTION

Some R.C. structures maybe situated in corrosive environment, such as severe ocean environment, chemical industry environment. And they may be attacked by several corrosive factors, such as carbon dioxide, chloride ions, sulfate or sulfuric acid. How to evaluate the safety of such R.C. structures is a necessary and important problem. Time-dependent reliability for such R.C. structures is generally evaluated using system reliability theory in which the correlation among failure mode have to be computed and the random variables are generally assumed with known distribution. In reality however, the distribution of many parameters are unknown since the limitation of the number of data. In the present paper, Dimension reduction integration is developed to evaluate the first few moments of the performance function of a structure with inclusion of the uncertain parameters with known distributions.

2 COMMON CORROSION FAILURE MODES OF R.C. STRUCTURES

The most common corrosion failure modes of R.C. structures are rebar corrosion caused by concrete carbonization, partial corrosion of rebar caused by invasion of chloride ions and concrete damage caused by sulfate or sulfuric acid. In this section, the limit states for failure caused by carbonization, failure caused by invasion of chloride ions and bearing capacity failure are established.

3 DIMENSION REDUCTION INTEGRATION FOR SYSTEM RELIABILITY

Since the system performance function $G(\mathbf{X})$ will not be smooth although the performance function of a component is smooth, it is difficult to obtain the sensitivity of the performance function even for a series system, derivative- based FORM would not be applicable. In the present section, Dimension Reduction Integration (DRI) for system reliability evaluation is investigated. The first few moments of the system performance function are obtained by DRI, from which the moment-based reliability index based on the fourth moment standardization function and failure probability can be evaluated without Monte Carlo simulations. The procedure does not require the computation of derivatives, nor the determination of the design point and computation of the mutual correlations among the failure modes; thus, it should be computationally effective for time-dependent reliability under several corrosive attacks. A numerical example is given to illustrate the application of the method. A specific R.C. structure that was drilled and sampled every 20 years over its service period of 60 years in corrosive environment is investigated.

4 PRINCIPAL CONCLUSIONS

Dimension Reduction Integration (DRI) for time-dependent reliability of reinforced concrete structures under chlorination and carbonation is suggested. A computationally effective method for probability density approach of performance function is proposed and examined; for both series and non-series systems. The method directly calculates the reliability indices (and associated failure probabilities) based on the first few moments of the system performance function of a structure.

Since the system performance function are not smooth, the histograms have good behaviors and the PDF of the system performance function should be smooth, and the histograms can be approached by the PDF defined by the first four moments.

The accuracy of results, including the first four moments of the performance function, and the probability of failure, obtained with the proposed method has been thoroughly examined by comparisons with large sample Monte Carlo simulations (MCS).

*Safety, Reliability and Risk of Structures, Infrastructures and
Engineering Systems – Furuta, Frangopol & Shinozuka (eds)
© 2010 Taylor & Francis Group, London, ISBN 978-0-415-47557-0*

Column over-design factor requirements for ensuring beam failure mechanism with specific reliability levels

W.C. Pu

Tokyo Institute of Technology, Yokohama, Japan

Y.G. Zhao

ABSTRACT

In seismic design it is desirable to have plastic hinges form in beams rather than in columns during an earthquake event. For ensuring a beam-hinging failure pattern, the sum of the columns at a joint is required to possess greater moment capacity than that of the adjoining beam. The "column over-design factor", which is defined as the ratio of column strength to beam strength at beam-column node, is generally used to fulfill such strong-column-weak-beam design.

Most of the previous studies, including experiment examination and numerical analysis, were performed using one or small quantity of specific earthquake ground motions as input. However, the ground motions are of great randomness, and the properties of earthquakes still cannot be predicted accurately. The obtained results by this way are consequently hard to represent the potential response to the next unknown ground motion. Zhao et al. proposed a COF evaluation method with the uncertainties of member strength and earthquake load in consideration, but the dynamic properties of structures are not considered. For a better understanding of the seismic behavior of structures, the dynamic evaluations based on statistical properties are more desired.

Basically, two approaches are available for computing the statistical properties of structural response. The one is to conduct time history analysis using a large quantity of ground motion records, well known as the Monte Carlo simulation. The alternative way is to conduct stochastic analysis based on the mean earthquake response spectrum to obtain the statistical indices directly. Comparably, the stochastic analysis is more efficient. Usually, some linearization techniques are involved in stochastic analysis, which makes it just an approximation method. Despite of this, the stochastic analysis has been well developed and popularly used because it provides helpful information about structural seismic behaviors. Jiang & Lu[11] developed the stationary stochastic analysis of linear system to apply for non-stationary random vibration, and proposed a method to compute the statistical properties of responses of multi degree-of-freedom system. It was demonstrated that the proposed method can be used to simulate the structural seismic behavior with good accuracy.

This paper conducts the stochastic analysis using the aforementioned method to evaluate the effect of COF on the statistical properties of maximum structural responses. The fishbone model is incorporated to account for the interaction between beams and columns. The distribution of the mean and standard deviation of the maximum story drift, beam rotation angle, and ductility ratio are investigated, and the relationship between the COF and those responses are analyzed.

REFERENCES

Jiang J.R. & Lu Q.N. 1984. Stochastic seismic response analysis of hysteretic MDF structures using mean response spectra. Earthquake Engineering and Engineering Vibration. 4(4), 1–13.

Zhao Y.G., Ono T. & Idota H. 1999. Response uncertainty and time-variant reliability analysis for hysteretic MDF structures. Earthquake Engineering and Structural Dynamics. 28: 1187–1213

Organized Session (OS15) Challenging Technology for Condition Screening of Bridge Structures

Safety, Reliability and Risk of Structures, Infrastructures and
Engineering Systems – Furuta, Frangopol & Shinozuka (eds)
© 2010 Taylor & Francis Group, London, ISBN 978-0-415-47557-0

Application of laser peening on steels for structures

Y. Sakino & Y.-C. Kim
Joining and Welding Research Institute, Osaka University, Osaka, Japan

Y. Sano
Power and Industrial Systems Research and Development Center, Toshiba Corporation, Yokohama, Japan

ABSTRACT

Laser peening is an innovative surface enhancement technology to introduce a compressive residual stress in metallic materials. Fundamental process of laser peening is summarized as follows. When an intense laser pulse is focused on the material, the surface absorbs the laser energy and a submicron layer of the surface evaporates instantaneously. Water confines the evaporating material and the vapor is immediately ionized to form plasma by inverse bremsstrahlung. The plasma absorbs subsequent laser energy and generates a heat-sustained shock wave, which impinges on the material with an intensity of several gigapascals, far exceeding the yield strength of most metals. The shock wave loses energy as it propagates to create a permanent strain. After the shock wave propagation, the surface is elastically constrained to form a compressive residual stress on the surface. X-ray diffraction study showed that the compressive residual stress, nearly equal to the yield strength, was imparted to the surface of the material. Laser peening was effective to prevent the initiation and propagation of stress corrosion cracking (SCC). Taking advantage of the inertia-less process of laser peening over mechanical treatment, a remote-controlled process system has been developed and applied to nuclear power reactors as a preventive maintenance measure against SCC.

Laser peening changes tensile residual stress to compressive. So it seems that laser peening will be very effective in enhancing the fatigue strength, because tensile residual stress is one of the most important factors to reduce fatigue strength.

In this study, laser peening conditions for steels for structure were examined by residual stress measurement and Vickers hardness measurement. Four grades of steels, Low-yield steel, $400\,N/mm^2$ grade steel, $490\,N/mm^2$ grade steel and $780\,N/mm^2$ grade

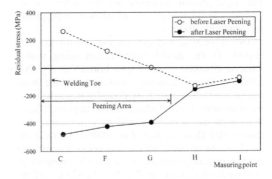

Figure 1. Residual stress distribution of welding toe (σ_y).

steel, were used as specimen. Moreover residual stress and Vickers hardness of the fillet weld zone after laser peening was investigated by comparing it with that before laser peening. Mild steel ($490\,N/mm^2$ grade steel) was used for the plates and fillet weld was used to join the ribs to the plates. X-ray diffraction was used to measure the residual stress.

Main results are summarized as follows. 1) Laser peening affects steel hardness up to the depth of $0.6\,mm$ in case of the over $400\,N/mm^2$ grade steels. 2) In the case of over $400\,N/mm^2$ grade steels, material strength of the steels for structure did not affect the compressive residual stress imparted by laser peening. 3) Laser peening can change tensile residual stress to large compressive residual stress in the welding zone. The nearer to the welding toe, the larger this effect by laser peening became. (Figure 1) 4) The hardness values near the welding toe increase about 100Hv compared to the unpeened area. But all laser-peened area was not hardened with uniformity. The nearest point to the welding toe was more hardened than the other laser peened points.

Safety, Reliability and Risk of Structures, Infrastructures and Engineering Systems – Furuta, Frangopol & Shinozuka (eds)
© 2010 Taylor & Francis Group, London, ISBN 978-0-415-47557-0

Challenge for a drive-by bridge inspection

C.-W. Kim & M. Kawatani

Department of Civil Engineering, Kobe University, Kobe, Japan

ABSTRACT

This paper presents an on-going research for a drive-by bridge health monitoring, which is even adaptable for short span bridges.

Two theoretical methods for the scheme are discussed: 1) the modal identification using traffic induced vibration of short span bridges as a global condition screening method; and 2) the bridge damage identification based on drive-by measurements using an inspection car which has two important functions such as an actuator and data acquisition (or processing) (Kim & Kawatani 2008). A prototype wireless sensor node, which is developed aiming to meet specific requirements of the drive-by monitoring, is also presented (Kim et al. 2008). Feasibility of the scheme is investigated by moving vehicle laboratory experiments.

Observations demonstrate that comparing the identification results between intact and damage girders such as pattern change of identified parameters encourages the use of the first-stage drive-by inspection method for long term health monitoring even for short span bridges as shown in Figure 1. The experimental study also demonstrates that the damage location and severity are well detectable using the proposed second-stage drive-by inspection method (see Figure 2).

The feasibility investigation of the wireless sensor board for the drive-by monitoring clarifies that many

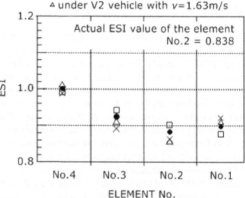

Figure 2. Identified element stiffness index (ESI).

technical tasks are remained unsolved: enhancing wireless communication performance; development of a middleware for on-board computing of bridge's current health condition; and energy harvesting for the system.

The most important and fundamental result obtained through this study is the feasibility of the method for addressing real-world problems. Feasibility investigations for actual bridges will be a strong motivation for the next challenge.

REFERENCES

Kim, C.W. & Kawatani, M. 2008. Pseudo-static approach for damage identification of bridges based on coupling vibration with a moving vehicle, *Structure and Infrastructure Engineering* 4(5): 371–379.

Kim, C.W., Kawatani, M., Tsukamoto, M. & Fujita, N. 2008. Wireless sensor node development for bridge condition assessment, *Advances in Science and Technology*, Trans Tech Publications 56:573–578.

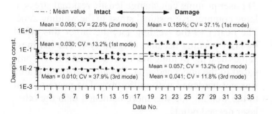

Figure 1. Identified modal damping constants using the first-stage drive-by inspection.

Safety, Reliability and Risk of Structures, Infrastructures and
Engineering Systems – Furuta, Frangopol & Shinozuka (eds)
© 2010 Taylor & Francis Group, London, ISBN 978-0-415-47557-0

On-line computation of dynamic displacements from measured strains through mode superposition

S. Shin & B. Yoon
Inha University, Incheon, Korea

ABSTRACT

A simple algorithm is proposed for computing dynamic displacements from measured strain data. The algorithm can be applied to any beam-type structure including a plate-girder bridge and a long span suspension or cable-stayed bridge. Modal superposition using generalized coordinates are applied to relate the curvature and displacement based on the beam theory.

Simulation studies, laboratory test, and field measurements were carried to examine the proposed algorithm. Figure 1 shows the view of laboratory tests with moving vehicles over a 6 m long simple span model bridge. Model vehicles moved over the bridge in a constant speed between 1 m/s and 6 m/s along the center of the section. FBG sensors were placed at the bottom of side flanges in an equal distance. A LVDT was placed at the bottom of the deck in the middle of the span. Various load cases were applied with one or two model vehicles in varying speeds.

A field test data obtained from a bridge for maglev trains algorithm as shown in Figure 2 was also used to examine the algorithm. FBG sensors were placed at the bottom of the girder bridge. A LVDT was set at the bottom of the middle span. The train ran from 10 *km/hr* to 50 *km/hr* over the girder bridge.

Vibrational displacement was computed and compared with that measured from LVDT in the speed of 50*km/hr* in Figure 3. Except some minor discrepancy around the center when the vehicle passed over it, the estimated displacements are in a quite good agreement with the measured data. The errors in all the train

Figure 2. Field test on a bridge for maglev trains.

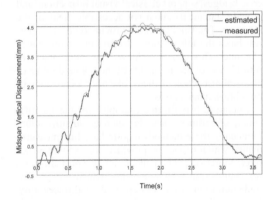

Figure 3. Comparison of estimated and measured displacements at the midspan with the train speed of 50 km/hr.

speeds are small enough to allow engineers to use the proposed algorithm reliably.

REFERENCES

Cho, J.Y., Choi, K.K., and Shin, S. (2005), Development of Algorithm for Estimating Moving Loads Using Measured Acceleration Time History, *Proceeding of the Korea Institute for Structural Maintenance Inspection Symposium*, 469–474, May 11. (in Korean)

Shin, S., Han, M.J., Jo, J.Y., Lee, H.J., and Jung, B.S. (2007), Identification of moving forces by measuring bridge dynamic responses, *The 3rd Int. Conf. on SHM of Intelligent Infrastructure*, paper 165, 1–6, November 13–16, Vancouver, Canada.

| | |
| (a) model bridge | (b) model vehicle |

Figure 1. Laboratory test with model vehicles.

Safety, Reliability and Risk of Structures, Infrastructures and
Engineering Systems – Furuta, Frangopol & Shinozuka (eds)
© 2010 Taylor & Francis Group, London, ISBN 978-0-415-47557-0

Techniques for bending stiffness system identification for RC-structures

S. Hoffmann
DYWIDAG Systems International, Munich, Germany

R. Wendner, A. Strauss & K. Bergmeister
University of Natural Resources and Applied Live Sciences, Vienna, Austria

ABSTRACT

The Project AIFIT – user oriented identification for engineering structures – claims to develop and supply specifically a support for the user in a sense of the expert in charge for the maintenance of a bridge structure. Various methods described in detail in earlier publications (Hoffmann 2008, Wendner 2009, Strauss et al. 2007, Ralbovský 2007, Lehký et al. 2007) shall allow an "inside view" into reinforced concrete bridges, support classical visual inspections and provide statements about not accessible areas of the structure. The final aim is to provide on an economical basis the information allowing the maintenance planning. This can be achieved by more objective information or the advantage of more aimed usage of classical inspection methods, like prior localization of possibly damaged areas.

Literature lists diverse examples for the identification of stiffness used for damage identification. Frequently these methods require base line measurements, which cannot be provided for existing structures. Moreover such methods can give only the basis for engineering type statements, which are in contradiction to the claim of objectivity and traceability. Based on theoretical considerations, numerical studies and especially expert interviews the efforts of the project were drawn to the identification of stiffness distribution of reinforced concrete bridges. These provide independency as far as possible at relatively high significance at the same time, especially in the case that damages embodies by bending cracks. The analyzed methods cover new approaches, all in the context of a practical application of such system identification methods.

Besides the practical application of the analyzed methods, including all necessary measurements at the structure, a comparison requires the assessment of their capabilities for the aimed information. In order to provide this comparison extensive laboratory tests have been conducted, which are described in detail in Hoffmann 2008 and Wendner 2009.

Experiences gained in these laboratory tests have been realized in the field while considering an application of system identification as practical as possible. For this a representative structure was analyzed and its dynamic and quasi static behavior recorded. In this way obtained information about the stiffness distribution of the structure shall provide finally a support for the inspection and maintenance planning of the bridge structure.

The implementation of the experiences in field and the analysis of the real capabilities on the basis of a field test are described in this contribution.

REFERENCES

Hoffmann, S. 2008. System identification by directly measured influence lines. Ph.D. thesis at the University für Bodenkultur. Vienna.

Hoffmann, S.; Wendner, R.; Strauss, A. 2007. Comparison of Stiffness Identification Methods for Reinforced Concrete Structures, In: *6th International Workshop on Structural Health Monitoring*, p. 354, Stanford.

Lehký, D.; Novák, D.; Frantic, P. et al 2007. Dynamic damage identification based on artificial neural networks, SARA – Part IV. In: *Third conference on structural health monitoring of intelligent infrastructure*, p. 183. Vancouver.

Ralbovský, M. 2007. Damage detection in concrete structures using modal force residual method, Ph.D. thesis at the Techniká Univerzita v Bratislave.

Strauss, A.; Bergmeister, K.; Frangopol, D. 2007. Inverse statistical FEM analysis for the assessment of existing structures. In: Jun Kanda, Tsuyoshi Takada, Hitoshi Furuta (Eds.), *Applications of Statistics and Probability in Civil Engineering*, Taylor & Francis, 0th International Conference on Applications of Statistics and Probability in Civil Engineering, ICASP10, 31 July – 3 August 2007, Tokyo, 331–332; ISBN: 978-0-415-45211-3.

Steinhauser, P., Steinhauser, W. 2007. Neue Möglichkeiten zur Erschütterungsuntersuchung durch das VibroScan advanced technology Verfahren, 10. Bauwerksdynamik Symposium, Zürich, 10pp. (in German)

Wendner, R. 2009: Modale Steifigkeitsidentifikation zur Zustandserfassung von Stahlbetonkonstruktionen. Ph.D. thesis at the University für Bodenkultur. Vienna (in German).

Safety, Reliability and Risk of Structures, Infrastructures and Engineering Systems – Furuta, Frangopol & Shinozuka (eds)
© *2010 Taylor & Francis Group, London, ISBN 978-0-415-47557-0*

Estimation of bridge eigenfrequencies based on vehicle responses using ICA

Y. Oshima, K. Yamamoto & K. Sugiura
Kyoto University, Kyoto, Japan

T. Yamaguchi
Osaka City University, Osaka, Japan

ABSTRACT

Herein in order to estimate the vibration component of a bridge from the vehicle response, independent component analysis (ICA) was applied to bridge-vehicle coupled vibration. Generally, ICA can be applied to static problem where signals are mixed simultaneously. But the bridge-vehicle vibration is one of the convolution phenomena. Thus in this study, system model and AR model with state equation were used for ICA to transform the convolution problem into the static problem. To verify this concept, numerical simulation was conducted to produce input signals for estimation analysis.

In this study, a bridge and passing vehicle are assumed as an input-output system. First of all, the vehicle moving on the subgrade is assumed to be a subsystem subject to road roughness profile. In this model, independent component is the road profile, but the other signals are dependent outputs including the sprung and unsprung mass. Then the total system was also assumed including vehicle and bridge interaction as the second step. Figure 1 shows the assumed model for subsystem and total system.

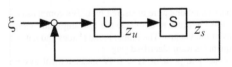

(a) Subsystem model for subgrade

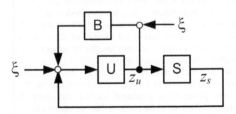

(b) Total system model for bridge and vehicle

Figure 1. System models, where S and U denote sprung and unsprung mass, respectively, and B denotes bridge.

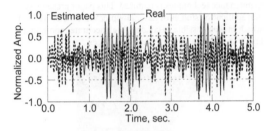

Figure 2. The estimated bridge response for 20 km/h.

The following equation was assumed on the basis of this system.

$$z_u(k) = \sum_{i=1}^{p+t} G_i \xi(k-i) + \sum_{i=1}^{q} C_i z_s(k-i) + \sum_{i=2}^{p+s} F_i z_u(k-i) + M_2 \xi(k) \tag{1}$$

where z_s and z_u are the acceleration of sprung and unsprung mass, ξ is the acceleration of road roughness profile, and G_i, C_i and F_i are the AR coefficients. M_2 is the restoring element for ICA. Then the bridge response can be expressed by

$$y(k) = \sum_{i=1}^{s} A_i z_u(k-i) + \sum_{i=1}^{t} E_i \xi(k-i) \tag{2}$$

where $y(k)$ is the response of the bridge along with the vehicle position, A_i and E_i are also the AR coefficients, which can be given by G and F. Figure 2 shows the estimated signals for the bridge response. Unfortunately the estimated signals do not agree well with the real signals which are directly obtained by simulation. This disagreement may occur due to the inappropriate orders of AR model to represent the bridge response, thus several models should be evaluated in the future.

REFERENCES

Y.B. Yang, C.W. Lin, J.D. Yau: Extracting bridge frequency from the dynamic response of a passing vehicle, *Journal of Sound and Vibration*, Vol. 272, 471–493, 2004.

Safety, Reliability and Risk of Structures, Infrastructures and Engineering Systems – Furuta, Frangopol & Shinozuka (eds)
© 2010 Taylor & Francis Group, London, ISBN 978-0-415-47557-0

Application of laser doppler vibrometers to bridges

T. Miyashita
Nagaoka University of Technology, Nagaoka, Japan

ABSTRACT

A visual inspection of bridges mainly has been conducted as conventional maintenance technique. The results may depend on subjectivity. Therefore, the development of quantitative maintenance techniques is required. This paper presents several application studies of Laser Doppler Vibrometers (LDV), which makes possible to conduct high accurate, non-contact and long distance measurement, to bridges.

Figure 1. Application to a Shinkansen steel box girder bridge.

1) At a Shinkansen (Bullet Train) railway steel box girder bridge, fatigue cracks were observed at the bottom end of vertical stiffeners. The objective of this study is to clarify the dynamic behavior of the bridge by vibration measurement using convectional sensors and LDVs (Fig. 1).
2) A new measurement system using LDVs for monitoring Shinkansen RC viaducts will be presented. Ambient and

Figure 2. Application to Shinkansen concrete viaducts.

Figure 3. Application to tensile force measurement of cables.

Figure 4. Remote non-contact measurement system.

impact vibrations were measured by LDVs compensated for noise in order to eliminate the low amplitude vibrations of the viaduct. The global mode shapes and natural frequencies were identified (Fig. 2).
3) Compact and portable type LDV measurement system was developed and applied to tensile force measurement of cables in the Tatara Bridge. The LDV used in this study integrates sensor head and controller into a body and drives by a battery. Measurement system using the LDV was developed in order to be possible to handle by a laptop PC. As the result, the number of working persons for measurement becomes few due to very excellent movability of the system (Fig. 3).
4) looseness-1When a LDV is far away from measured points, it is especially difficult to confirm the location of laser points on a structural surface. Adjusting the position of the laser to the measured points becomes time-consuming task during field measurements. Therefore, a non-contact measurement system combining LDV with Total Station for long distance measurements, which has the ability of high accurate positioning and automated measurement, was developed (Fig. 4).

Organized Session (OS06) Nonlinear Stochastic Dynamics of Structures with Uncertain Mechanical and Geometric Properties
[Session Organized on Behalf of IASSAR Subcommittee SC1]

Safety, Reliability and Risk of Structures, Infrastructures and Engineering Systems – Furuta, Frangopol & Shinozuka (eds)
© *2010 Taylor & Francis Group, London, ISBN 978-0-415-47557-0*

Critical excitation method for moment-resisting frames subjected to two-directional horizontal ground inputs

Kohei Fujita & Izuru Takewaki

Department of Urban & Environmental Engineering, Graduate School of Engineering, Kyoto University, Kyoto, Japan

A new model is proposed of bi-directional horizontal ground motions (2DGM) (see Fig. 1). It is shown that, in comparison with the Penzien-Watabe model (1975) (see Fig. 2), the cross power spectral density (PSD) function between 2DGM along the building structural axes (BSA) can be treated in a more relaxed manner by using an extended Penzien-Watabe model proposed in this paper. The auto PSD functions of 2DGM along BSA are assumed to be given and the cross PSD function between these 2DGM is treated as a complex variable. A critical excitation problem is then considered for a moment resisting three-dimensional frame (see Fig. 3). The objective function is the corner-fiber stress at the column-end and the worst cross PSD function of the 2DGM is searched for the maximum corner-fiber stress at the column-end. It is shown that the real part (co-spectrum) and the imaginary part (quad-spectrum) of the worst cross PSD function can

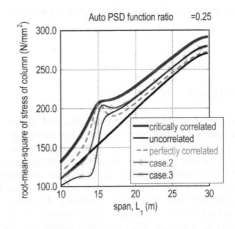

Figure 4. Response to the critically correlated 2DGM and responses to other inputs.

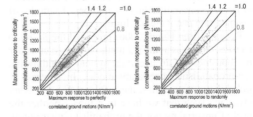

Figure 5. Responses to critically correlated 2DGM, perfectly correlated ones and randomly correlated ones (1000 samples).

be obtained by an algorithm interchanging the double maximization procedure in the time and cross PSD domains. It is concluded that the critically correlated 2DGM are inevitable for reliable design of important structures (Fig. 4, Fig. 5).

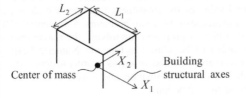

Figure 1. Moment resisting three dimensional frame and the definition of building structural axes.

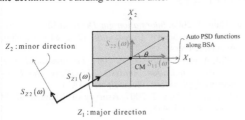

Figure 2. Penzien-Watabe model (1975).

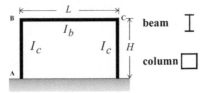

Figure 3. One-story one-span plane frame.

REFERENCES

Fujita, K., Yoshitomi, S., Tsuji, M. and Takewaki, I. (2008). Critical cross-correlation function of horizontal and vertical ground motions for uplift of rigid block, *Engrg. Struct.*, 30(5), 1199–1213.

Penzien, J., and Watabe, M. (1975) Characteristics of 3-dimensional earthquake ground motion. *Earthq. Engrg. Struct. Dyn.*, 3, 365–374.

Safety, Reliability and Risk of Structures, Infrastructures and
Engineering Systems – Furuta, Frangopol & Shinozuka (eds)
© 2010 Taylor & Francis Group, London, ISBN 978-0-415-47557-0

Semidefinite programming for robustness analysis of structures under large uncertainties

Yoshihiro Kanno
Department of Mathematical Informatics, University of Tokyo, Tokyo, Japan

Izuru Takewaki
Department of Urban and Environmental Engineering, Kyoto University, Kyoto, Japan

ABSTRACT

The robust structural design has received increasing attention, which may decrease the sensitivities of mechanical performances with respect to various uncertain parameters. Roughly speaking, there exist two frameworks for considering the uncertain property of a structural system, i.e. probabilistic and non-probabilistic uncertainty modeling. Probabilistic robust design approaches require information on stochastic variation of the uncertain parameters, e.g. parameters of the probability density function. However, it is often difficult to estimate those parameters accurately, especially when the number of samples of the uncertain parameters is limited.

Recently, Ben-Haim (2006) proposed the *info-gap decision theory* as a unified methodology to consider the robustness of an uncertain system including the non-probabilistic uncertainty model. In contrast to probabilistic approaches, the info-gap theory has an advantage such that we have to estimate neither the level of uncertainty nor the probabilistic distribution of uncertain parameters.

In the info-gap theory, the *robustness function* plays a key role which represents the greatest level of uncertainty at which any failure cannot occur. This implies that the robustness function can be regarded as a quantitative measure of robustness, i.e. the constraint conditions can be violated only at great level of uncertainty in a structure with a large robustness function, while they can be violated at small level of uncertainty in a structure with a small robustness function.

Unfortunately, in general, it is difficult to compute the robustness function exactly. In this paper, we first propose a tractable reformulation of the robustness function of a structure under non-probabilistic uncertainties. In particular, based on the S-lemma (Pólik & Terlaky, 2007), we propose an efficient numerical algorithm for computing a lower bound of the robustness function. At each iteration of the algorithm, we solve a *semidefinite programming* (SDP) (Ben-Tal & Nemirovski, 2001) problem.

We next consider two structural optimization problems dealing with uncertainties: (i) a minimization problem of the structural volume over robust constraints, and (ii) a structural optimization problem which attempts to find a structure with the maximal value of the robustness function. We show the equivalence of those two problems. A sequential SDP algorithm is proposed for solving the maximization problem of the robustness function. Numerical experiments are presented for a truss in order to demonstrate the validity of evaluation and maximization of the robustness function.

REFERENCES

Ben-Haim, Y. 2006. *Information-gap Decision Theory: Decisions under Severe Uncertainty*, (2nd ed.). London: Academic Press.
Ben-Tal, A. and Nemirovski, A. 2001. *Lectures on Modern Convex Optimization: Analysis, Algorithms, and Engineering Applications*. Philadelphia: SIAM.
Pólik, I. and Terlaky, T. 2007. A survey of S-lemma. *SIAM Review* 49: 371–418.

GDEE-based reliability evaluation of nonlinear stochastic dynamical systems

J. Li & J.B. Chen

State Key Laboratory of Disaster Reduction in Civil Engineering, School of Civil Engineering,
Tongji University, Shanghai, P.R. China

ABSTRACT

The approaches for dynamic reliability and system reliability evaluation of nonlinear stochastic dynamical systems based on the generalized density evolution equation (GDEE) are outlined.

If $Z(t)$ is a physical quantity of the system, the first-passage reliability is usually defined as

$$R(T) = \Pr\{Z(t) \in \Omega_s, \ t \in [0,T]\} \qquad (1)$$

An absorbing boundary condition can be imposed on the generalized density evolution equation (GDEE) (Chen & Li, 2005). An alternative approach is through the integral of extreme value distribution. For instance, if the safety domain is $\Omega_s = [-b, a]$, $a > 0, b > 0$, then we can define the extreme value variable (Li et al, 2007; Li & Chen, 2009) as

$$Z_{ext}(T) = \max_{0 \le t \le T} \left| Z(t) - \frac{a-b}{2} \right| \qquad (2)$$

Thus the reliability is given by

$$R(T) = \int_0^{\frac{a+b}{2}} p_{Z_{ext}}(z,T) dz \qquad (3)$$

Our task is now to obtain the extreme value distribution $p_{Z_{ext}}(z,T)$. To this end, construct a virtual stochastic process $Y(\tau) = \phi(Z_{ext}, \tau)$ satisfying $Y(0) = 0$, $Y(\tau_c) = Z_{ext}$. A generalized density evolution equation (GDEE) governing the evolution of the joint PDF of (Y, Θ), $p_{Y\Theta}(y, \theta, \tau)$, can be obtained (Li & Chen, 2008; Chen & Li, 2009)

$$\frac{\partial p_{Y\Theta}(y,\theta,\tau)}{\partial \tau} + \dot{Y}(\theta,\tau)\frac{\partial p_{Y\Theta}(y,\theta,\tau)}{\partial y} = 0 \qquad (4)$$

Solving this equation will yield $p_{Y\Theta}(y, \theta, \tau)$ and then give $p_Y(y, \tau)$. Thus there is $p_{Z_{ext}}(z,T) = p_Y(y = z, \tau_c)$. Hence, the dynamic reliability evaluation is transformed to the integral in Equation 3.

The ideas can also be extended to system reliability evaluation. In this case, an equivalent extreme value can be constructed. For instance, consider

$$R(T) = \Pr\left\{ \bigcap_l^m \left(|Z_l(t)| \le b_l\right), \ t \in [0,T] \right\} \qquad (5)$$

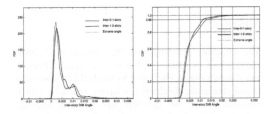

Figure 1.   PDFs and CDF of the maximum inter-story drift angles.

Likewise, when an equivalent extreme value is defined as $Z_{ext}(T) = \max_{0 \le t \le T} \left(\max_{1 \le l \le m} |Z_l(t)/b_l| \right)$ (Li et al, 2007), then the ideas elaborated above can be applied here. The PDF of $Z_{ext}(T)$ could be obtained and the system reliability can be finally evaluated through a one-dimensional integral

$$R(T) = \int_0^1 p_{Z_{ext}}(z,T) dz \qquad (6)$$

In Figure 1 shown are the PDFs and CDFs of the maximum inter-story drifts and the reliability (coordinate of the CDF), demonstrating the effectiveness of the proposed method.

REFERENCES

Chen, J.B. & Li, J. 2005. Dynamic response and reliability analysis of nonlinear stochastic structures. Probabilistic Engineering Mechanics, 20(1): 33–44.

Chen, J.B. & Li, J. 2009. A note on the principle of preservation of probability and probability density evolution equation. Probabilistic Engineering Mechanics, 24(1): 51–59.

Li, J., Chen, J.B. & Fan, W.L. 2007. The equivalent extreme-value event and evaluation of the structural system reliability. Structural Safety, 29: 112–131.

Li, J. & Chen, J.B. 2008. The principle of preservation of probability and the generalized density evolution equation. Structural Safety, 30: 65–77.

Li, J. & Chen, J.B. 2009. Stochastic Dynamics of Structures, John Wiley & Sons.

Safety, Reliability and Risk of Structures, Infrastructures and Engineering Systems – Furuta, Frangopol & Shinozuka (eds)
© 2010 Taylor & Francis Group, London, ISBN 978-0-415-47557-0

Stochastic nonlinear dynamic response of frames with spatially varying non-Gaussian material properties

G. Stefanou & M. Fragiadakis

Institute of Structural Analysis & Seismic Research, National Technical University of Athens, Greece

ABSTRACT

The efficient prediction of the nonlinear dynamic response of structures with uncertain system properties poses a major challenge in the field of computational stochastic mechanics. Current research efforts concentrate on the development of new and the improvement of existing methods, however they are usually limited to linear elastic analysis considering only monotonic loading. In order to investigate realistic problems of structures subjected to transient seismic actions, a novel approach is introduced in this paper. This approach is used to assess the stochastic response of a 3-storey steel moment-resisting frame in the framework of Monte Carlo simulation (MCS), using nonlinear response history analysis. The structure is modeled with a mixed fiber-based, beam-column element, whose kinematics is based on the natural mode method (e.g. Argyris et al. 1988). The adopted formulation leads to the reduction of the computational cost required for the calculation of the element stiffness matrix, while increased accuracy compared to traditional displacement-based elements is achieved (Papaioannou et al. 2005). Thus a cost-effective stochastic formulation of a distributed plasticity element is presented. The uncertain parameters considered are the Young modulus and the yield stress, both described by homogeneous non-Gaussian translation stochastic fields that vary along the element. A non-Gaussian translation field is defined as a nonlinear memory-less transformation of some underlying Gaussian field with known second order statistics (Grigoriu 1998). The frame is subjected to natural seismic records that correspond to three levels of increasing seismic hazard as well as to spectrum-compatible artificial accelerograms.

Under the assumption of a pre-specified power spectral density function of the stochastic fields that describe the two uncertain parameters, the response variability (mean, coefficient of variation (COV) and skewness) is computed using MCS. Probabilities of the demand, expressed in terms of the maximum inter-storey drift θ_{max} exceeding given threshold values that correspond to preset limit-states level, are calculated. Moreover, a parametric investigation is carried out providing useful conclusions regarding the influence of the non-Gaussian distribution as well as of the correlation length of the stochastic fields on the response variability. In contrast to the static case where the response variability shows always the same trend, starting from small values for small correlation lengths corresponding to wide-banded stochastic fields up to large values for large correlation lengths (Stefanou & Papadrakakis 2004), the COV of θ_{max} is found to vary significantly not only with the correlation length but also in many different ways among the records of the same intensity level. Finally, a large magnification of uncertainty is observed in some cases, leading to response COV values that are 1.4-2.3 times greater than the corresponding input COV.

REFERENCES

Argyris, J., Tenek, L. & Mattsson, A. 1988. BEC: A 2-node fast converging shear-deformable isotropic and composite beam element based on 6 rigid-body and 6 straining modes. *Computer Methods in Applied Mechanics and Engineering* 152: 281–336.

Papaioannou, I., Fragiadakis, M. & Papadrakakis, M. 2005. Inelastic analysis of framed structures using the fiber approach. Proceedings of the 5th International Congress on Computational Mechanics (GRACM 05), Limassol, Cyprus, 29 June-1 July 2005, Vol. I, pp. 231–238.

Grigoriu, M. 1998. Simulation of stationary non-Gaussian translation processes. *Journal of Engineering Mechanics (ASCE)* 124: 121–126.

Stefanou, G. & Papadrakakis, M. 2004. Stochastic finite element analysis of shells with combined random material and geometric properties. *Computer Methods in Applied Mechanics and Engineering* 193: 139–160.

Safety, Reliability and Risk of Structures, Infrastructures and
Engineering Systems – Furuta, Frangopol & Shinozuka (eds)
© 2010 Taylor & Francis Group, London, ISBN 978-0-415-47557-0

Transient stability analysis of an aeroelastic problem with random fluid and structure properties

C.V. Verhoosel, T.P. Scholcz & S.J. Hulshoff
Faculty of Aerospace Engineering, Delft University of Technology, Delft, The Netherlands

M.A. Gutiérrez
*Faculty of Mechanical, Maritime and Materials Engineering, Delft University of Technology,
Delft, The Netherlands*

ABSTRACT

Fluid-structure interactions play an important role in many fields. Since unstable interactions can cause structural failure, prediction of their occurrence is of primary importance in the design of aircraft, bridges and many other structures. Fluid-structure interactions have been studied extensively, which has resulted in several analytical and numerical methods for their prediction (Bisplinghoff 1955; Dowell 2004). Generally, these methods presume that the properties of the considered structure and loading conditions are exactly known. In practice, these uncertainties are taken into account indirectly by using conservative safety factors. Direct incorporation of uncertainties in the analysis of fluid-structure interactions should therefore result in more efficient designs.

The direct incorporation of uncertainties in both the structure and fluid can be considered using stochastic finite element methods (SFEM). This approach has recently been used to study the asymptotic stability of a prototypical fluid-structure interaction problem (Verhoosel et al. 2009). Although asymptotic stability is a necessary condition for structural safety, it is not sufficient as certain loading conditions can still lead to untolerable transient responses in asymptotically stable configurations. This occurs in configurations with non-normal dynamical operators, which are typical in fluid-structure interactions. In this contribution, methods to determine the probability of structural failure due to transient responses are developed and assesed.

The problem considered is a panel interacting with a supersonic exterior flow. The panel is modelled structurally as a beam and is discretised using finite elements. The modulus of elasticity of the panel is described by a stationary lognormal random field. The fluid flowing over the panel is described using piston theory, which in combination with transpiration boundary conditions yields an analytical expression for the pressure variations over the plate. To simulate the effect of a turbulent flow field, a stationary decaying and convecting random field of fluctuations with an experimentally measured autocorrelation kernel is superimposed onto the variations. Karhunen-Loeve expansions are used for the discretisations of both random fields.

The probability that a failure measure (e.g. maximum stress) exceeds a specified threshold during a given time interval is characterised by the first-passage probability. A simplified approach to compute this probability is to assume that the stochastic excitations are the only source of randomness. It is, however, well known that structural uncertainties have a significant effect and can therefore not be neglected (Lutes and Sarkani 2004). Here, the effect of these structural uncertainties are instead taken into account using the perturbation method.

Results from numerical experiments are compared with Monte-Carlo simulations. The influence of both structural uncertainties and stochastic excitations are studied.

REFERENCES

Bisplinghoff, R. L. (1955). *Aeroelasticity.* Cambridge, Mass.: Addison-Wesley Pub. Co.
Dowell, E. H. (2004). *A modern course in aeroelasticity.* Dordrecht: Kluwer Academic Publishers.
Lutes, L. D. and S. Sarkani (2004). *Random vibrations.* Elsevier Butterworth-Heinemann.
Verhoosel, C., T. Scholcz, S. Hulshoff, and M. Gutiérrez (2009). Uncertainty and reliability analysis of fluid-structure stability boundaries. *AIAA journal 47*(1), 91–104.

Safety, Reliability and Risk of Structures, Infrastructures and Engineering Systems – Furuta, Frangopol & Shinozuka (eds)
© 2010 Taylor & Francis Group, London, ISBN 978-0-415-47557-0

Statistical properties of third-order nonlinear random waves and wave-induced offshore structural responses

X.Y. Zheng & T. Moan

Center for Ships and Ocean Structures, Department of Marine Technology, Norwegian University of Science & Technology, Trondheim, Norway

ABSTRACT

The linear random wave (LRW) theory has long been adopted in the design of offshore structures. To more accurately calculate wave loadings with strong wave interactions taken into account, nonlinear random wave (NRW) theories need to be employed. Some works have been carried out to study the effects of 2nd-order NRW on the non-Gaussian response of slender-member platforms. The present paper aims at looking into the difference by including 3rd-order NRW.

Based on the 3rd-order NRW model (Ohyama et al. 1995), mean and variance of wave kinematics are derived. Zero mean is found. Valuation of variance relies on the cubic transfer matrices (CTM) of wave interactions.

Simulation of 3rd NRW is a practical concern since 3-fold summations of CTM about frequencies demand considerable computational efforts. Based on earlier proposed 2D-FFT procedures for 2nd NRW, a 3D-FFT scheme is proposed. This scheme is realized by operating FFTs or IFFTs along the three dimensions of CTM respectively.

Time-domain Monte-Carlo simulations are conducted to obtain the total wave force on an idealized jack-up platform. A narrow JONSWAP wave spectrum is considered for simulating the short-term statistics of platform deck sway response. The SDOF vibration system of deck sway has two typical frequencies, $f_n = 2f_p$ and $3f_p$. Spectral analysis shows that NRW promotes the response at the resonant frequency f_n such that the response variance is increased to some extent. The gained amplification at f_n due to 3rd-order NRW is comparable to that due to 2nd NRW. Also, in response spectrum, the peak amplitude at f_n is comparable to that at the dominant wave frequency f_p.

In addition, NRW increases the higher-order statistics of wave force and structural response as well. The skewness of force and response is controlled by 2nd NRW while the increase in kurtosis involves the contribution from both 2nd and 3rd NRW. As a result, the nonlinear behavior of wave force become stronger and

Table 1. Statistics of deck sway.

		mean $(10^{-4}\,\text{m})$	variance $(10^{-3}\,\text{m}^2)$	skewness	kurtosis excess	3 h extreme (m)
$f_n = 3f_p$	LRW	0.03	0.728	−0.007	4.205	0.210
	2nd NRW	−4.42	0.794	−0.221	5.674	0.242
	3rd NRW	−4.49	0.835	−0.226	6.012	0.269
$f_n = 2f_p$	LRW	0.17	5.292	0.009	2.513	0.477
	2nd NRW	−9.64	5.663	0.020	3.315	0.535
	3rd NRW	−9.81	6.163	0.016	3.490	0.587

the averaged 3h maximum deck sway is larger. Compared with LRW case, the maximum deck sway under NRW earns some 20% increase. A half of this increase is sourced from 3rd NRW.

Compared with the system of $f_n = 3f_p$, the system of $f_n = 2f_p$ exhibits more significant dynamic resonances and the deck sway response is more Gaussianized. Its kurtosis excess is much smaller, but variance is larger. Therefore the 3h extreme response is more than doubled.

REFERENCES

Ohyama, T., Jeng, D.S., Hsu, J.R.C. 1995. Fourth-order theory for multiple-wave interaction. *Coastal Eng.*, Vol. 25: 43–63.

Stansberg, C.T. 1998. Non-Gaussian extremes in numerically generated second-order random waves on deep water. *Proc. 8th ISOPE Conf.*, Montreal Canada, (III): 103–110.

Tayfun, M.A. 1986. On narrow-band representation of ocean waves. *J. Geophysical Res.* Vol. 91: 7743–7752.

Winterstein S.R., Torhaug R. 1996. Extreme jack-up response: simulation and nonlinear analysis methods. *ASME J. of offshore mech. and arctic Eng.* Vol. 118 (No. 2): 103–108.

Zhang, J., Chen, L. 1999. General third-order solutions for irregular waves in deep water. *ASCE J. of Eng. Mech.* Vol. 125 (No. 7): 768–779.

General Session (Stochastic Computational Mechanics &
Stochastic Finite Elements)

Safety, Reliability and Risk of Structures, Infrastructures and
Engineering Systems – Furuta, Frangopol & Shinozuka (eds)
© 2010 Taylor & Francis Group, London, ISBN 978-0-415-47557-0

Stochastic seismic response analysis of nonlinear structures

J.B. Chen
*State Key Laboratory of Disaster Reduction in Civil Engineering, School of Civil Engineering, Tongji University,
Shanghai, P.R. China*

Z.J. Liu
College of Civil & Hydroelectric Engineering, China Three Gorges University, Yichang, P.R. China

J. Li
*State Key Laboratory of Disaster Reduction in Civil Engineering, School of Civil Engineering, Tongji University,
Shanghai, P.R. China*

ABSTRACT

The coupling of randomness and nonlinearity should be
rationally taken into account in nonlinear stochastic systems. To this end, an orthogonal expansion of stochastic
ground motion incorporated into the probability density
evolution method is outlined.

An orthogonal expansion for stochastic ground
motions can be adopted (Liu & Li, 2008). Let $\varphi_k(t)$'s be
standard orthogonal basis functions, then the stochastic
process $X(t)$ can be expanded as

$$X(\xi,t) = \sum_{k=1}^{\infty} \xi_k \varphi_k(t), \quad \xi = \sum_{j=1}^{N} \zeta_j \sqrt{\lambda_j} \psi_j \qquad (1a,b)$$

where ξ_k's are random variables and $\zeta = (\zeta_1, \zeta_2, \ldots,
\zeta_N)^{\mathrm{T}}$ are the uncorrelated basic random variables trans-
formed from ξ through the decomposition of the covariance matrix. Substituting Equations 1b in 1a will yield

$$\hat{X}(\zeta,t) = \sum_{k=1}^{N}\sum_{j=1}^{N} \zeta_j \sqrt{\lambda_j} \phi_{jk} \varphi_k(t) = \sum_{j=1}^{N} \zeta_j \sqrt{\lambda_j} f_j(t)$$

where ϕ_{jk} is the k-th component of the eigenvector ψ_j.
Further, for stochastic ground motions, the acceleration
time history can be expressed as

$$\ddot{X}_{\mathrm{g}}(t) = \sqrt{2S_0} \sum_{j=1}^{10} \zeta_j \sqrt{\lambda_j} F_j(t) \qquad (2)$$

where $F_j(t) = -\sum_{k=1}^{300} (2k\pi/T_{\mathrm{s}})^2 \eta_{k+1} \varphi_{j,k+1} \phi_k(t)$. The
Hartley functions $\mathrm{cas}(t) = \cos(t) + \sin(t)$ are
employed, i.e. $\phi_k(t) = 1/\sqrt{T}\mathrm{cas}(2\pi kt/T), k = 0, 1,
2, \ldots$.

Generally, the number s of the basic random variables
is nearly 10. Figure 1 is the comparison between the PSDs
and the shape and the response spectrum of a sample.

Consider an n-DOF nonlinear structure, of which the
equation of motion is

$$\mathbf{M}\ddot{\mathbf{X}} + \mathbf{C}\dot{\mathbf{X}} + \mathbf{f}(\mathbf{X}) = -\mathbf{MI}\ddot{X}_{\mathrm{g}}(t) \qquad (3)$$

where $\ddot{X}_{\mathrm{g}}(t)$ is the stochastic ground motion acceleration
given by Equation 2.

Denote the physical quantities of interest by $\mathbf{Z}(t) =
(Z_1(t), Z_2(t), \ldots, Z_m(t))^{\mathrm{T}}$. According to the principle of
preservation probability (Li & Chen, 2008, 2009), we can

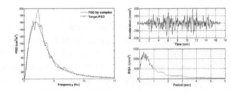

Figure 1. Comparison between PSD and the typical generated time history and the response spectrum.

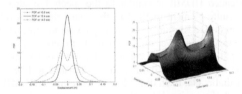

Figure 2. PDF and CDF of the top displacement at typical
time instants.

finally obtain the generalized density evolution equation
(Chen & Li, 2009)

$$\frac{\partial p_{z\Theta}(\mathbf{z},\boldsymbol{\theta},t)}{\partial t} + \sum_{j=1}^{m} \dot{Z}_j(\boldsymbol{\theta},t)\frac{\partial p_{z\Theta}(\mathbf{z},\boldsymbol{\theta},t)}{\partial z_j} = 0 \qquad (4)$$

Figure 2 shows the PDF and CDF of the top displacement
of a nonlinear structure subjected to stochastic ground
motion. The investigations show that the proposed method
is feasible for MDOF nonlinear stochastic systems.

REFERENCES

Chen, J.B. and Li, J. 2009. A note on the principle of
preservation of probability and probability density evolution equation. Probabilistic Engineering Mechanics 24(1):
51–59.

Li, J. and Chen, J.B. 2008. The principle of preservation of probability and the generalized density evolution
equation. Structural Safety 30: 65–77.

Li, J. and Chen, J.B. 2009. Stochastic Dynamics of Structures.
John Wiley & Sons.

Liu, Z.J. and Li, J. 2008. Orthogonal expansion of stochastic
processes for earthquake ground motion. Journal of Tongji
University 36(9): 1153–1159. (in Chinese)

Safety, Reliability and Risk of Structures, Infrastructures and Engineering Systems – Furuta, Frangopol & Shinozuka (eds)
© 2010 Taylor & Francis Group, London, ISBN 978-0-415-47557-0

HDMR based stochastic finite element analysis for random field

Rajib Chowdhury
School of Engineering, Swansea University, Swansea, UK

B.N. Rao & A. Meher Prasad
Indian Institute of Technology Madras, Chennai, India

ABSTRACT

A generic high dimensional model representation (HDMR) method for approximating the limit state/performance function in terms of functions of lower dimensions which has been previously proposed for problems dealing only with random variables is extended in this paper for problems in which physical properties exhibit spatial random variation and may be modeled as random fields. The formulation of the extended HDMR is similar to the spectral stochastic finite element method in the sense that both of them utilize Karhunen–Loeve expansion to represent the input, and lower order expansion to represent the output. However, it is recognized that lower order function expansion, in HDMR, do not translate to linear, quadratic, etc. Each of the low order terms in HDMR is sub-dimensional, but they are not necessarily low degree polynomials. This study helps us to decouple the finite element and stochastic computations, and the finite element code can be treated as a black box, as in the case of a commercial code.

The method appears to be efficient, requiring only conditional responses at selected sample points to accurately compute solution statistics. In comparison, full scale MCS may require thousands of realizations or more for converged statistics. Thus the proposed method substantially reduces the computational effort while maintaining the desired accuracy. However, K–L expansion requires for discretizing the input random field with many terms, which may in turn increase the dimensionality of the problem and thus the computational effort of the SFEM. It can be noticed that, the curse of dimensionality does not arise in the proposed approach, as it discretizes the whole problem into a number of sub-problems; it is a problem in other currently available methods as well.

Also, the proposed method is independent of the structural analysis and treats the finite element model as a black-box. If higher loads introduce nonlinearity in the model response, then the appropriate nonlinear analysis should be used, and the underlying response surface would automatically be different. Thus the proposed method is applicable to the analysis of linear as well as nonlinear structures, provided the right 'black-box' is used. Good agreements are observed comparing the numerical results between extended HDMR and full scale MCS. However, following observations can be made easily from this article:

(1) *Computational effort*: HDMR based approach requires conditional responses at selected sample points and the sample points are chosen along each of the variable axis. It is found in authors' previous work that $n = 5$ or 7 works well for most the problem. Due to this fact, results for $n = 5$ or 7 are used in this paper to construct HDMR based metamodels.

(2) *Handling spatial variability*: HDMR can accurately handle random fields with moderate to large coefficients of variation unlike other methods such as perturbation and Neumann expansion methods.

REFERENCES

Chowdhury, R. and Rao, B. N. 2009. Assessment of high dimensional model representation techniques for reliability analysis. *Probabilistic Engineering Mechanics*, 24(1): 100–115.

Ghanem, R., and Spanos, P.D. 2002. Stochastic Finite Elements: A Spectral Approach. Dover Publications Inc., New York.

Rabitz, H., and Alis, O.F. 1999. General foundations of high dimensional model representations. *Journal of Mathematical Chemistry*, 25(2–3): 197–233.

Chowdhury, R. and Adhikari, S. 2009. High Dimensional Model Representation for Stochastic Finite Element Analysis. *Applied Mathematical Modelling* (In review).

Sudret, B. and Der Kiureghian, A. 2000. Stochastic Finite Element Methods and Reliability: A State-of-the-Art Report. Tech. Rep. *UCB/SEMM-2000/08*, University of California, Berkley, USA.

Safety, Reliability and Risk of Structures, Infrastructures and
Engineering Systems – Furuta, Frangopol & Shinozuka (eds)
© 2010 Taylor & Francis Group, London, ISBN 978-0-415-47557-0

Recent advances in the modelling of space-variant structural properties

D.L. Allaix, V.I. Carbone & G. Mancini
Department of Structural and Geotechnical Engineering, Politecnico di Torino, Torino, Italy

ABSTRACT

The agreement of the research community about the modelling of the random spatial variability of the structural properties by means of the random field theory is well established nowadays. The contributions of the research in the civil engineering field allow to formulate probabilistic models able to take into account in the probabilistic investigations of the structural performance the space-variant characterization of material properties, geometrical dimensions and actions (JCSS 2001). The computational applications regarding the reliability assessment of new ((Araujo 2001), (Vasconcellos Real et al. 2003), (Lee et al. 2004)) and existing structures ((Defaux et al. 2006), (Stewart et al. 2007)) point out the problem of achieving an accurate discretization of the involved random fields. The random fields are represented in this paper by the Karhunen-Loeve series expansion and the discretization is obtained by truncation of the series expansion. As result, the quality of the approximation of the random field depends on the truncation order of the series expansion. It is clear that the accuracy of the discretization is a key point for any further probabilistic investigation of the structural behaviour. Therefore, it is meaningful to quantify the discretization accuracy by means of one or more analytical discretization error estimators related to the relevant properties of the random field being discretized. Moreover, it is reasonable to require that the truncation order of the series expansion has to be chosen in such way the discretization error is less than a prescribed target accuracy. In the applications involving space-variant properties of the concrete structures, the terms of the Karhunen-Loeve series expansion have to be evaluated by numerical treatment of the a Fredholm integral equation. A Galerkin procedure, involving the finite element discretization of the structural domain, is applied for the purpose. As consequence, the discretization accuracy can be linked to the properties of the finite element mesh and the random field approximation can be improved iteratively by a refinement of the finite element mesh. Therefore, starting form a coarse mesh, an accurate random field discretization can be achieved with a limited computational effort on the basis of the spatial distribution of the discretization error. The effectiveness of the procedure is remarkable in the case of two-dimensional random fields defined over regular and irregular domains. The applications example concern the modelling of the random spatial variability of the concrete properties in two-dimensional structures described by homogeneous random fields with reference to the most advances models available in literature.

REFERENCES

Araujo, M.J. 2001. Probabilistic analysis of reinforced concrete columns. Adv. Eng. Softw., 32: 871–879.

Defaux, G., Pendola, M. & Sudret B. 2006. Using spatial reliability in the probabilistic study of concrete structures: the example of a RC beam subject to carbonation inducing corrosion. Journal de Physique, 136: 243–253.

Ghanem, R.G. & Spanos, P.D. 1991. Stochastic finite elements: a spectral approach. New York: Springer-Verlag.

JCSS 2001. *Probabilistic Model Code*. Joint Committee on Structural Safety, Internet publication, www.jcss.ethz.ch

Lee, T.H. & Mosalam K.M. 2004. Probabilistic fiber element modeling of reinforced concrete structures. Comput. Struct., 82: 2285–2299.

Stewart, M.G. & Mullard J.A. 2007. Spatial time-dependent reliability analysis of corrosion damage and timing of first repair for RC structures. Eng. Struct., 29: 1457–1464.

Vasconcellos Real, M., Campos Filho, A. & Maestrini S.R. 2003. Response variability in reinforced concrete structures with uncertain geometrical and material properties. Nucl. Eng. Des., 226: 205–220.

Safety, Reliability and Risk of Structures, Infrastructures and Engineering Systems – Furuta, Frangopol & Shinozuka (eds)
© 2010 Taylor & Francis Group, London, ISBN 978-0-415-47557-0

Sensitivity analysis of nested multiphysics models using polynomial chaos expansions

B. Sudret

Phimeca Engineering, Centre d'Affaires du Zénith, Cournon, France
IFMA & Université Blaise Pascal, Laboratoire de Mécanique et Ingénieries, France

T. Yalamas, E. Noret & P. Willaume

Phimeca Engineering, Centre d'Affaires du Zénith, Cournon, France

ABSTRACT

Uncertainty quantification in computational mechanics has received much attention in the past few years. In this context, global sensitivity analysis (GSA) aims at quantifying which random input parameters (or combinations of parameters) are the most important in order to explain the variability of the model response. Various measures of importance have been proposed including partial correlation coefficients (PCC), standard regression coefficients (SRC) and Sobol' indices [4]. The latter, which are usually computed using Monte Carlo simulation, allow the analyst to rank efficiently the input random variables whatever the complexity of the model [5].

Recently, it has been shown that polynomial chaos expansions (PCE) [3] may be used in order to efficiently evaluate the Sobol' indices *analytically* [6]. The computational cost reduces to that related to computing the PCE coefficients, for which efficient *non intrusive* methods have been proposed [2].

In this paper we are interested in multiscale modeling in which several hierarchical models are nested, *i.e.* the response of one model corresponds to the input of another model and so on. This type of nested models may be represented by a network, in which each node is a model (having its own input and output vector) and some connections with the other nodes. They appear for instance when micromechanical models of materials at various scales are considered (*i.e.* a homogenized material property obtained at one scale is used as the input parameter of a homogenization procedure at the next scale). The concept of nested models appears more generally in all multiphysics models.

When considering nested models from the point of view of sensitivity analysis, it is desirable to establish importance measures (such as Sobol' indices) at various levels, *i.e.* the sensitivity of any response quantity at any level with respect to each variable of lower level (which may be itself the response of a lower level model).

In this paper, the technique developed in [6] is extended to multiscale nested models. It is illustrated on a numerical model of concrete elastic behavior [1].

Cement-based materials microstructure may be represented by a four level nested model. At the lowest level we find the C-S-H matrix that forms at early ages. At the second level the cement paste is composed from C-S-H, C-H, aluminates, cement clinker inclusions and water. The third level refers to mortar, which is made of sand particles embedded in a homogeneous cement paste matrix. Finally level IV corresponds to concrete, *i.e.* aggregates with an interfacial transition zone embedded in an homogeneous mortar matrix. In terms of homogenization schemes, levels I, II and IV are based on the Mori-Tanaka scheme and Level III is based on the self-consistent scheme.

Uncertainties in the material properties, hydration degree, volume fraction, etc. are introduced at each level and the sensitivity of the resulting elastic properties of concrete are finally computed by the proposed method.

REFERENCES

[1] O. Bernard, F.-J. Ulm, and E. Lemarchand. *A multi-scale micromechanics-hydration model for the early-age elastic properties of cement-based materials.* Cement Conc. Res., 33(9), 2003.

[2] M. Berveiller, B. Sudret, and M. Lemaire. *Stochastic finite elements: a non intrusive approach by regression.* Eur. J. Comput. Mech., 15(1-3), pp. 81–92, 2006.

[3] R.G Ghanem and P.D Spanos. *Stochastic finite elements – A spectral approach.* Springer Verlag, 1991. (Reedited by Dover Publications, 2003).

[4] A. Saltelli, K. Chan, and E.M. Scott, editors. *Sensitivity analysis.* J. Wiley & Sons, 2000.

[5] I.M. Sobol'. *Sensitivity estimates for nonlinear mathematical models.* Math. Modeling & Comp. Exp., 1, pp. 407–414, 1993.

[6] B. Sudret. *Global sensitivity analysis using polynomial chaos expansions.* Reliab. Eng. Sys. Safety, 93, pp. 964–979, 2008.

Safety, Reliability and Risk of Structures, Infrastructures and
Engineering Systems – Furuta, Frangopol & Shinozuka (eds)
© 2010 Taylor & Francis Group, London, ISBN 978-0-415-47557-0

Implementation of a polynomial chaos toolbox in OpenTURNS with test-case applications[1]

I. Dutka-Malen
Sinetics department, EDF-R&D, Clamart, France

R. Lebrun
EADS Innovation Works, Sys. Engineering – Applied Mathematics, Suresnes, France

B. Saassouh
LaMSID, EDF-R&D CNRS, Clamart, France

B. Sudret
Phimeca Engineering S.A., Paris, France

ABSTRACT

OpenTURNS is an Open source software to Treat
Uncertainties, Risks'N Statistics in structured industrial
approach (www.openturns.org) jointly developed by EDF,
EADS and Phimeca Engineering. This paper presents the
implementation steps of the so-called *polynomial chaos
expansion* (PCE) toolbox in the software. A comparative
study is presented to show the performance of the implemented method and the way it can be used for distribution,
moment and reliability analyses. The propagation of the
uncertainties in the input parameters gathered in a random
vector $\underline{X}$ (of size M) through the model function $\mathcal{M}$ leads
to a random response of the system, say $Y = \mathcal{M}(\underline{X})$, which
is supposed to be scalar here for the sake of simplicity.
Provided Y has a finite variance, it can be expressed in an
orthonormal polynomial basis as follows (Ghanem(1991)
& Sudret(2007)):

$$M(X) = \sum_{\alpha \in IN^M} a_\alpha \psi_\alpha(X)$$

Where $\Psi_\alpha(X) =$ the orthonormal polynomial basis
functions & $a_\alpha =$ real coefficients (the *coordinates* of Y
in the PC basis) to be determined by suitable algorithms.

Considering the recent progress in computer science
and software engineering, an object-oriented language
was used to ensure the robustness of the code. Therefore, the UML modeling approach was applied at the
conception stage. At this stage, and due to expert judgment, some modeling choices were done to ensure a
quick convergence of the calculation. Usually the input
random variables of the system are transformed into
independent standard normal variables by using an isoprobabilistic transform, leading to a Hermite polynomial
chaos representation. However theoretical studies show
that a generalized chaos basis provides a better convergence rate. In fact, some orthonormal families fit better
with some specific random variable for the same order of
truncation. For example, if X is Gaussian than the best

corresponding family is the Hermite polynomial family.
Accordingly, the solution retained in OT is the use of
some *isoprobabilistic transformations*, such as the *Nataf*
or *Rosenblatt* transform. In this context, the original input
random vector X is recast as a random vector whose margins are independent and has specified PDFs with their
associated orthonormal families presented below:

– Hermite family for Gaussian variables $N(0, 1)$
– Legendre family for Uniform variables $U(-1, 1)$
– Laguerre family for Gamma variables $G(k, 1, 0)$
– Jacobi family for Beta variables $B(r, t, -1, 1)$.

The present implementation in OT deals with three
different cases for the input random vector:

– an input random vector X with independent margins. Such case is treated by using M univariate
  isoprobabilistic transformation.
– an input random vector with Gaussian copula. The
  most efficient transformation for this case is the Nataf
  transformation (accurate with cheap numerical cost).
– for all remaining cases (without special special depen-
  dancy) the Rosenblatt transformation is used.

The last section of the article presents various applications examples where the polynomial chaos approach
is benchmarked with respect to classical methods of
computational stochastic mechanics.

REFERENCES

Ghanem R. & Spanos P. 1991. Stochastic finite elements –
A spectral approach, Springer Verlag. (Reedited by Dover
Publications 2003).
Sudret, B. 2007. Uncertainty propagation and sensitivity
analysis in mechanical models – Contribution to struc-
tural reliability and stochastic spectral methods, Habili-
tation à diriger des recherches, Université Blaise Pascal,
Clermont-Ferrand, France.

---

[1] Authors are cited by an alphabetical order

Safety, Reliability and Risk of Structures, Infrastructures and
Engineering Systems – Furuta, Frangopol & Shinozuka (eds)
© 2010 Taylor & Francis Group, London, ISBN 978-0-415-47557-0

SFE analysis of railway tracks

N. Rhayma, Ph. Bressolette, P. Breul & M. Fogli
LaMI, Blaise Pascal University, Aubière, France

G. Saussine
I&R SNCF, Paris, France

ABSTRACT

The behaviour of track structures (Fig. 1) is strongly
dependent on numerous uncertain parameters related
to operating (drainage, settlements...) and maintenance conditions of the tracks. A realistic description
of track behaviour requires to take into account variability of the parameters using a model based on a
probabilistic approach.

This work propose a method based on the use of
Stochastic Finite Element Methods (SFEM) (Baroth &
al 2007), aiming to evaluate the effect of uncertainties
of geometrical and mechanical random parameters on
the model response.

To reach this goal, the Stochastic Collocation
method using Lagrange polynomials interpolation is
used to evaluate statistical moments of the output
model parameters. The uncertain parameters taken into
account are modelized by a vector of independent
Lognormal random variables (r.v.).

The first step of this study was to develop a
2D multi-layer FE model of a railway track; then
the model was validated comparing to experimental
and numerical results. Geometrical (thickness) and
mechanical (Young modulus) uncertain parameters are
then selected. Their statistical representations has been
deduced from in-situ measurement realized on various
sections of railways tracks (Sol Solution, 2008).

The control parameters selected are indicators of the
track behaviour: sleeper deflection and acceleration,
rail deflection and a heterogeneity indicator called *NL*.

Then a preliminary convergence study of the proposed SFEM have shown that 4 collocation points

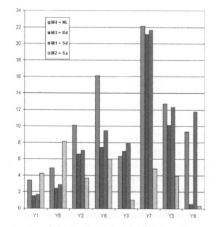

Figure 2. ΔCV_{M_j} for each random parameter Y_i.

is a good compromise between precision and CPU
time, for the mean (m_{Y_i}) and the standard deviation
(σ_{Y_i}), with geometrical and mechanical random input
parameters.

At the end, a uncertainty propagation study have
been conducted with one uncertain input parameter at
a time. The influence of the selected parameters has
been evaluated. In order to illustrate the influence of
random parameters, we computed the increase of the
coefficient of variation for each control parameters
$\Delta CV_{M_i} = \sigma_{M_i}/m_{M_i}$.

Obtained results are summarized in figure 2. They
highlight the influence of the mechanical parameters
(Young modulus) of subgrade layers (form-layer and
sub-ballast layer) on most of the response parameters. The characteristics of the ballast layer show an
important influence on sleepers accelerations.

REFERENCES

Baroth J., Bressolette Ph., Chauvière C., Fogli M. (2007).
An efficient SFE method using Lagrange polynomials:
application to non-linear mechanical problems with uncertain parameters. *Comp. Meth. In Appl. Mech. Engrg*, 196,
45–48: 4419–4429..
Sol Solution (2008). Railway investigation Innotrack project:
Classical line Chambery/Saint Pierre d'Albigny PK
157.000 to PK 151.100 Lane 1, lane 2 and track side lane
2. Geotechnical project.

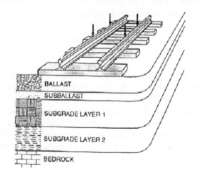

Figure 1. Railway track Structure.

General Session (Simulation Methods)

Safety, Reliability and Risk of Structures, Infrastructures and Engineering Systems – Furuta, Frangopol & Shinozuka (eds)
© *2010 Taylor & Francis Group, London, ISBN 978-0-415-47557-0*

Adaptive directional stratification – An adaptive directional sampling method on a stratified space

M. Munoz Zuniga & J. Garnier
Laboratoire de Probabilités et Modélisation Aléatoire, Université Paris VII, Paris, France

E. Remy & E. de Rocquigny
EDF R&D, Chatou, France

ABSTRACT

In an industrial context, there are different ways to assess the reliability of a structure from physical considerations. One of them is to use a probabilistic approach, that includes the whole probability distribution of the random input variables of the deterministic model representing the physical state of the structure considered.

Many stochastic tools already exist in the literature to estimate a failure probability, but we will essentially focus on the probabilistic simulation methods. The aim of the paper is to present a new accelerated Monte-Carlo simulation method, we named ADS – Adaptive Directional Stratification – and we designed to overcome the main real industrial constraints: robustness of the estimation of the (potentially very low) structural failure probability, limited computational resources and complex (and potentially monotonous) physical models.

The problem of estimating a failure probability from a physical model leads to evaluate a multivariate integral. A standard solution to evaluate this integral is Monte-Carlo simulation method. It is surely consistent and controllable thanks to classical probabilistic theorems which allow to give a precise statistical measure of the estimation error. Nevertheless, the required number of simulations with this method is at least 100 times P_f^{-1} in order to get a relative error of the order of 10%, which is computationally expensive in the case of a rare failure.

So we turn our attention to accelerated Monte-Carlo simulation methods, more particularly to the stratified sampling and the directional simulation. On the one hand, the stratified sampling offers us the opportunity to carry out an adaptive strategy and the possibility to take advantage of an eventually monotonous hypothesis of the physical model. On the other hand, the use of the directional sampling method enables us to keep a good "precision/calculation time" ratio. That's why we combined these two methods and their advantages to develop our ADS technique.

Hence, we first consider our ADS method with two steps in the adaptation phase (2-ADS). We theoretically study the asymptotic properties of the failure probability estimator and get the asymptotic estimator variance expression. Then, we study the dichotomic iterative algorithm required in the directional sampling method to take into account the numerical error propagation to our failure probability estimator: thus, we imagine a stop criterion taking heed of this error. Finally, we generalize our method with L steps in the adaptation phase (L-ADS). We perform the same robust theoretical study to control at the best the results.

To conclude our article, we test our 2-ADS and L-ADS methods on academic examples: we compare their results with the ones obtained by the standard directional simulation method. Lastly, we apply them to an EDF nuclear structure to concretely demonstrate their interest on a practical case.

REFERENCES

Madsen, H.O. and Ditlevsen, O. 1996. Structural reliability methods. Wiley.

Melchers, R.E. 1999. Structural reliability analysis and prediction. Wiley.

Rubinstein, R.Y. and Kroese, D.P. 2007. Simulation and the Monte Carlo Method (second edition). Wiley.

Safety, Reliability and Risk of Structures, Infrastructures and
Engineering Systems – Furuta, Frangopol & Shinozuka (eds)
© 2010 Taylor & Francis Group, London, ISBN 978-0-415-47557-0

Reliability analysis for crack growth of turbine disk by advanced subset simulation

S.F. Song, Z.Z. Lu & X.K. Yuan
School of Aeronautics, Northwestern Polytechnical University, Xi'an, China

ABSTRACT

The estimation of failure probability has attracted a lot
of attention over the years. Typically for the structure
with implicit limit state, the evaluation of structural
reliability remains a challenging problem. As small
failure probability is associated, the accurate reliability analysis becomes more difficult due to a heavier
computational burden. For this kind of difficulty, a
novel technique called Subset simulation (Subsim)
was proposed. Subsim is an efficient simulation to
perform the reliability analysis in a progressive manner. Introducing a set of intermediate failure events,
Subsim separates the original probability space into
a sequence of subsets, and then the small failure
probability can be expressed as a product of larger
conditional failure probabilities. Markov Chain Monte
Carlo (MCMC) simulation is employed for generating conditional samples to estimate these conditional
failure probabilities. However, these conditional samples generated by MCMC are dependent in general.
These samples are used for statistical averaging as if
they are independent and identically distributed (i.i.d.)
with some reduction in efficiency. And the variance
of the failure probability estimator can not be evaluated by an approximate value but only by an upper
limit. Truncated Importance Sampling (TIS) procedure is employed for efficiently generating conditional
samples that correspond to specified levels of failure
probabilities by constructing the TIS density function
gradually. In this paper, TIS in conjunction with the
Subsim, denoted as Subsim/TIS, is developed. The
concept and the implementation of the presented Subsim based TIS are explained in detail. The Subsim/TIS
is employed to estimate the reliability of the crack
growth of powder metallurgy (PM) turbine disk, which
is characterized by finite element model and with very
small failure probability.

The low cycle fatigue (LCF) life of PM alloys is
commonly reduced by non-metallic inclusions which
are introduced during the manufacturing process.
These small particles are well known as initial small
cracks and responsible for fatigue damage of the turbo-
engine disks. An alternative way to evaluate the effect
of inclusions on LCF life seems to be a probabilistic approach. The physical data about inclusions in
PM superalloys, such as type, size, location, etc., have
probabilistic features. Since the shape of the initial
crack was found to have a small effect on LCF life, the
failure probability of the crack growth can be assumed
as a function of the location and the size of these inclusions. The failure probability of crack growth is too
small for Monte-Carlo simulation to calculate, thus
Subsim/TIS is used to analyze the reliability of LCF
life of PM alloys effectively. Based on the residual
strength interference model, the failure probabilities
due to surface, sub-surface and internal inclusion are
calculated, and they indicate that the sub-surface inclusion has the largest effect on the failure probability of
the crack growth among the surface, sub-surface and
internal inclusions. And the risk of failure increases
with the increase of the number of cycle load.

REFERENCES

Au, S.K. and Beck, J.L., 2001. Estimation of small failure
probabilities in high dimensions by subset simulation.
Probab Eng Mech, 16(4): 263–277.
Yuan, X.K. Lu, Z.Z. and Song, S.F., 2007. Improving further
importance sampling method for failure probability estimation. Journal of Northwestern Polytechnical University,
25(5): 752–756.
Lu, Z.Z., Liu, C.L. and Yue, Z.F., 2005. Probabilistic safe
analysis of the working life of a powder-metallurgy turbine
disk. Material Science and Engineering, 395: 153–159.
Grison, J. and Remyl, L., 1997. Fatigue failure probability
in a powder metallurgy Ni-base superalloy. Engineering
Fracture Mechanics, 57(1): 41–55.
Yan, X.J., Li, H.Y. and Nie, J.X., 2004. A probabilistic
model for prediction of LCF life of PM alloys. Aircraft
Engineering and Aerospace Technology, 76(3): 286–292.

Safety, Reliability and Risk of Structures, Infrastructures and
Engineering Systems – Furuta, Frangopol & Shinozuka (eds)
© 2010 Taylor & Francis Group, London, ISBN 978-0-415-47557-0

A quasi ideal importance sampling simulation for structural reliability estimation

M. Yonezawa
Department of Mechanical Engineering, Kinki University, Osaka, Japan

S. Okuda
Kinki University Technical College, Kumano, Japan

H. Kobayashi
Interdisciplinary Graduate School of Science and Engineering, Kinki University, Osaka, Japan

ABSTRACT

This paper describes a quasi ideal importance sampling simulation combined with the conditional expectation method in the simulation-based structural reliability estimation.

It has been a very important issue in the simulation-based reliability estimation to improve the simulation efficiency by adopting various variance reduction techniques (VRT) such as the importance sampling and the conditional expectation. The primary matter of concern in the importance sampling procedure is how to construct the importance sampling probability density function ($p.d.f.$).

In this study, a quasi ideal importance sampling joint $p.d.f.$ is defined on the basis of the ideal importance sampling concept and the respective marginal $p.d.f.$s of the quasi ideal importance sampling joint $p.d.f.$ are so formulated as to be proportional to

the expectation of the conditional failure probability multiplied by the $p.d.f.$ of the respective sampling variable. The marginal $p.d.f.$s are constructed numerically by the simulations based on the conditional expectation and partly by the piecewise integrations.

The importance sampling simulations combined in the conditional expectation are executed to estimate the failure probabilities of structures with multiple failure surfaces. The samples of the basic random variables are generated by applying the inverse transformation to the cumulative distribution functions corresponding to the respective quasi ideal importance sampling marginal $p.d.f.$s constructed in the proposed method.

Numerical examples to estimate the failure probabilities of structures with multiple failure surfaces are presented to illustrate that the proposed method gives accurate estimations with smaller sample size and shorter processing time.

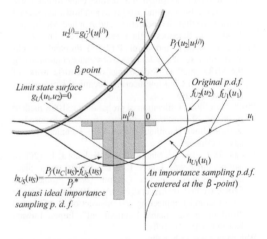

Figure 1. An image of a quasi ideal importance sampling $p.d.f.$ in a two dimensional reliability problem.

REFERENCES

Ayyub, B. M. and Haldar, A. (1984), "Practical structural reliability techniques", *Journal of Structural Engineering*, ASCE, 110, 8, 1707–1724.

Bucher, C. G. (1988). "Adaptive sampling – An iterative fast Monte Carlo procedure", *Structural Safety*, 5, 119–126.

Karamchandani, A. and Cornell, C. A. (1991), "Adaptive hybrid conditional expectation approaches for reliability estimation," *Structural Safety*, 11, 59–74.

Rubinstein, R. Y. (1981), *"Simulation and The Monte Carlo Method"*, John Wiley & Sons, 114–157.

Schuëller, G. I. and Stix, R. (1987), "A critical appraisal of methods to determine failure probabilities," *Structural Safety*, 4, 293–309.

Yonezawa, M., Okuda, S. and Kobayashi, K. (2007), "Structural Failure Probability Estimation Based on Quasi Ideal Importance Sampling Simulation", *Journal of School of Science and Engineering*, Kinki University, 43, 9 (in Japanese).

Safety, Reliability and Risk of Structures, Infrastructures and
Engineering Systems – Furuta, Frangopol & Shinozuka (eds)
© 2010 Taylor & Francis Group, London, ISBN 978-0-415-47557-0

Probabilistic reliability analysis of a continuous reinforced concrete frame using the SBRA method

D. Pustka & R. Čajka
Faculty of Civil Engineering, Technical University of Ostrava, Ostrava, Czech Republic

P. Marek
Institute of Theoretical and Applied Mechanics, Czech Academy of Sciences, Prague, Czech Republic

ABSTRACT

Reinforced concrete structural systems, composed of
eccentrically loaded slender members (e.g. columns,
piers, frame legs, etc.), are used in modern buildings
and bridges very often. Reliability assessment of these
systems should be based on assumption of geometrical
and physical nonlinearity. Significant part of this calculation is determination of critical load of considered
system. This load represents on deformed structure
unstable equilibrium of internal and external forces.
In the construction of bridges there are frequently
used slender columns, which are fixed in the foot and
mutually connected by pin-jointed horizontal beam.
Continuous frames with slender legs are also often
used. For buildings there are used multi-storey frames
with slender columns. In all these structural systems,
the vertical members are slender columns or piers.

Rapid development of computer technology in the
last decades enabled development of fully probabilistic methods of reliability assessment of structures.
Today's codes, based mostly on Partial Factor Design
method, have many limitations, where among the most
important belongs: (a) impossibility of direct quantification of structure reliability, (b) problems with determination of critical combination of multi-component
load effects, and (c) difficulties with consideration of
various time-dependent load effects influencing durability of structure. By these and others shortages of
Partial Factor Design method do not suffer fully probabilistic methods. One of these methods is SBRA
(Simulation-Based Reliability Assessment) – see e.g.
Marek et al. (1995), (1998), (2003), based on Monte
Carlo simulation technique.

Direct quantification of reliability of evaluated
structure is possible, if individual random input quantities are expressed by corresponding histograms. The
SBRA method uses preferably bounded histograms,
which usually better approximate real random quantities, than parametric ones. Potential of this method in
reliability assessment of reinforced concrete structural
system, consisted of slender members, is shortly
indicated in the text of the full paper.

REFERENCES

CEB (1977). Design Manual on Buckling, Bulletin
d'Information No 123, Decembre, Paris.
CEB-FIP (1978). Model Code for Concrete Structures.
Bulletin d'Information No 125-E, Paris.
Gustar, M. (2002). Modern methods of simulation of sys-
tems – SBRA method. Dissertation work. VSB – Technical
University of Ostrava.
Janda, L., Kristek, V., Kvasnicka, M. and Prochazka, J. (1983).
Slender reinforced concrete compressed columns. Prague:
SNTL.
Marek, P., Gustar, M. and Anagnos, T. (1995). Simulation-
Based Reliability Assessment for Structural Engineers,
CRC Press, Inc., Boca Raton, Florida.
Marek, P., Gustar, M. and Bathon, L. (1998). Tragwerks-
bemessung. Von deterministischen zu probabilistischen
Verfahren. Academia, Prague.
Marek, P., Brozzetti, J., Gustar, M. and Tikalsky, J. P. (2003).
Probabilistic Assessment of Structures using Monte Carlo
Simulation. Background, Exercises and Software. Second
Edition. Institut of Theoretical and Applied Mechanics,
Academy of Sciences of the Czech Republic, Prague.
Pustka, D., Cajka, R. and Marek, P. (2008). Probabilistic reli-
ability analysis of a system made of slender columns using
the SBRA method. Proceedings of scientific works of
the Faculty of Civil Engineering, Technical University of
Ostrava, Ostrava.
Pustka, D. (2008). Utilization of probabilistic SBRA method
in design of reinforced concrete structures exposed to
static loadings. Habilitation work. VSB-Technical Univer-
sity of Ostrava.
Pustka, D. Cajka, R., Marek, P. and Kalocova, L. (2008).
Multi-Component Loads Effect Analysis on a Slender Reinforced Concrete Column Using Probabilistic
SBRA Method. In: Eleventh East Asia-Pacific Conference
on Structural Engineering & Construction (EASEC-11)
"Building a Sustainable Environment". Taipei, Taiwan,
November 19–21, 2008.
Web page www.anthill-sbra.com

Safety, Reliability and Risk of Structures, Infrastructures and
Engineering Systems – Furuta, Frangopol & Shinozuka (eds)
© 2010 Taylor & Francis Group, London, ISBN 978-0-415-47557-0

Performance of correlation control by combinatorial optimization for Latin Hypercube Sampling (LHS)

M. Vořechovský

Brno University of Technology, Brno, Czech Republic

ABSTRACT

The objective of this paper is a study of performance of correlation control of recently proposed procedure for sampling from a multivariate population within the framework of Monte Carlo simulations [10–12], especially Latin Hypercube Sampling. In particular, we study the ability of the method to fulfill the prescribed marginals and correlation structure of a random vector for various sample sizes. Two norms of correlation error are defined, one very conservative and related to extreme errors, other related to averages of correlation errors. We study behavior of Pearson correlation coefficient for Gaussian vectors and Spearman rank order coefficient (as a distribution-free correlation measure).

The paper starts with theoretical results on performance bounds for both correlation types in cases of desired uncorrelatedness. It is shown that, under some circumstances, a very high rate of convergence can theoretically be achieved. These rates are compared to performance of other previously developed techniques for correlation control, namely the Cholesky orthogonalization as applied by Iman and Conover [3,4]; and Owen's method [8] using Gram-Schmidt orthogonalization. We show that *the proposed technique* based on combinatorial optimization [10–12] *yields much better results* than the other known techniques.

When correlated vectors are to be simulated, the proposed technique exhibits nearly the same excellent performance as in the uncorrelated case provided the desired vector exists. It is shown that the technique provide much wider range of acceptable correlations than the wide-spread Nataf [7] model [5] (known also as the Li-Hammond model [6] or the NORTA model [1]) and that it is also much more flexible than the Rosenblatt model [2,9].

REFERENCES

[1] Ghosh, S., Henderson, S. G. (2003) Behavior of the NORTA method for correlated random vector generation as the dimension increases, *ACM Transactions on Modeling and Computer Simulation* 13 (3), 276–294.

[2] Hohenbichler, M. Rackwitz, R. (1981) Non-normal dependent vectors in structural safety, *Journal of Engineering Mechanics*, ASCE 107 (6), 1227–1238.

[3] Iman, R. C., Conover, W. J. (1980) Small sample sensitivity analysis techniques for computer models with an application to risk assessment, *Communications in Statistics: Theory and Methods* A9 (17), 1749–1842.

[4] Iman, R. C., Conover, W. J. (1982) A distribution free approach to inducing rank correlation among input variables, *Communications in Statistics B11* (1982) 311–334.

[5] Liu, P., Der Kiureghian, A. (1986) Multivariate distribution models with prescribed marginals and covariances, *Probabilistic Engineering Mechanics* 1 (2) (1986) 105–111.

[6] Li, S.T. and Hammond, J.L. (1975) Generation of pseudo-random numbers with specified univariate distributions and correlation coefficients, *IEEE Transactions on Systems*, Man, Cybernetics 5, pp. 557–560.

[7] Nataf, A. (1962) Détermination des distributions de probabilités dont les marges sont donnés, *Comptes Rendus de L'Académie des Sciences* 225, 42–43.

[8] Owen, A. B. (1994) Controlling Correlations in Latin Hypercube Samples, *Journal of the American Statistical Association* (Theory and methods) 89 (428) (1994) 1517–1522.

[9] Rosenblatt, M. (1952) Remarks on multivariate analysis, *Annals of Statistics* 23, 470–472.

[10] Vořechovský, M. (2007) *Stochastic computational mechanics of quasibrittle structures*. Habilitation thesis presented at Brno University of Technology, Brno, Czech Republic.

[11] Vořechovský, M., Novák, D. (2002). *Correlated random variables in probabilistic simulation*. In: Schießl, P. et al. (Eds.), 4th International Ph.D. Symposium in Civil Engineering. Vol. 2. Millpress, Rotterdam. Munich, Germany, pp. 410–417. Awarded paper. ISBN 3-935065-09-4.

[12] Vořechovský, M., Novák, D. (2009) Correlation control in small sample Monte Carlo type simulations I: A simulated annealing approach, *Probabilistic Engineering Mechanics*, in press, doi: 10.1016/j.probengmech.2009.01.004.

Wednesday Evening (WEE) Sessions

Mini-Symposia (MS15) System Identification and Structural Health Monitoring
[Session Organized on Behalf of IASSAR Subcommittee SC5]

Detection of an obstructed crack on a plate utilizing the two-dimensional spatial wavelet transformation

H.F. Lam, T. Yin & H.M. Chow
Department of Building and Construction, City University of Hong Kong, HKSAR, China

ABSTRACT

The main purpose of this paper is to extend the works of Lam at al. (2005) and Loutridis at al. (2005) to form a practical crack detection method being suitable for plate-type structures when cracks are in an obstructed area (see Fig. 1). The value of the proposed method lies in its ability to detect obstructed crack when measurement at or close to the cracked region is not possible.

When part of the plate is obstructed, the complete two-dimensional spatial wavelet transform of the deflection surface cannot be obtained from laser or optical measurements. In such a situation, a crack detection method would be much more valuable if it can detect cracks even when it's not possible to take measurement in the neighborhood of the cracked region. It is the first time for this paper to reveal that it is possible to use one of the spatial wavelet coefficients in identifying the crack characteristics even when the crack is obstructed. (see Fig. 2).

Furthermore, in order to explicitly handle the uncertainties associated with the crack detection results due to the incomplete measurement and measurement noise, the proposed methodology follows the Bayesian statistical identification framework (Beck & Katafygiotis 1998, Katafygiotis & Beck 1998). Rather than pinpointing the crack location and extent, the proposed methodology aims at calculating the posterior (updated) PDF of the uncertain crack parameters (e.g. crack location, length, and depth) and some system parameters (such as the damping ratio) for a given

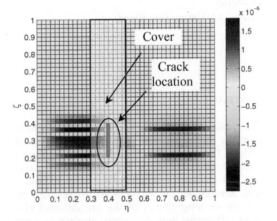

Figure 2. Detail coefficients in vertical direction of the 2D spatial wavelet transform on the deflection surface of the cracked plate at time $t = 0.05$s with cover range being $[0.3, 0.5]$.

set of measured vibration data. The numerical case studies indicate that the proposed method can successfully identify the crack on an obstructed area of a plate-structure utilizing the noised measurement data.

ACKNOWLEDGEMENT

The work described in this paper was fully supported by a grant from CityU (7002240). This generous support is gratefully acknowledged.

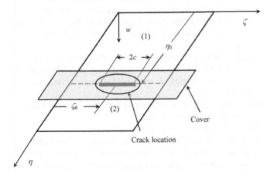

Figure 1. Rectangular plate with a part-through finite-length crack obstructed by a cover being parallel to the crack.

REFERENCES

Beck, J.L. & Katafygiotis, L.S. 1998. Updating models and their uncertainties. I: Bayesian statistical framework. *Journal of Engineering Mechanics*, ASCE 124(4): 455–461.

Katafygiotis, L.S. & Beck, J.L. 1998. Updating models and their uncertainties. II: Model identifiability. *Journal of Engineering Mechanics*, ASCE 124(4): 463–467.

Lam, H.F., Lee, Y.Y., Sun, H.Y., Cheng, G.F., & Guo, X. 2005. Application of the spatial wavelet transform and Bayesian approach to the crack detection of a partially obstructed beam. *Thin-Walled Structures* 43(1): 1–12.

Loutridis, S., Douka, E., Hadjileontiadis, L.J., & Trochidis, A. 2005. A two-dimensional wavelet transform for detection of cracks in plates. *Engineering Structures* 27(9): 1327–1338.

Safety, Reliability and Risk of Structures, Infrastructures and
Engineering Systems – Furuta, Frangopol & Shinozuka (eds)
© 2010 Taylor & Francis Group, London, ISBN 978-0-415-47557-0

Application of Bayesian logic to the condition assessment of a smart precast element

D. Zonta, M. Pozzi & H.Y. Wu
DIMS, University of Trento, Italy

D. Inaudi
Smartec SA, Manno, Switzerland

ABSTRACT

This paper introduces a concept of smart structural
elements for the real-time condition monitoring of
bridges. These elements are prefabricated reinforced
concrete beams embedding a permanent sensing system and capable of self-diagnosis when in operation. Sensors are not just applied to the member,
but are an integral part of the prefabricated element, influencing its design criteria, performance and
detailing. The technology particularly targets new
bridges and includes strain and environmental sensors.
The real-time assessment is automatically controlled
by a theoretical response model and by a numerical algorithm based on Bayesian logic (Bayes 1763,
Papadimitriou 1997, Beck 1998, Gregory 2005, Sivia
2006). The idea is to divide the domain of the possible structural response into a mutually exclusive
and exhaustive set of scenarios, each defining the
structural behaviour in a specific condition (e.g. no
damage, concrete cracking, reinforcement corrosion,
strand failure and so on). The structural response
in the i-th scenario is controlled by a certain number of parameters (e.g. cracking location, cracking
extent). Based on the prior distribution of these parameters and on the real-time recording, the method
assigns posterior probability to each scenario as well
as updated probability distributions for each damage
parameter. To verify the effectiveness of the technology, a reduced-scale prototype Prestressed Reinforced
Concrete beam (size $3.8 \times 0.3 \times 0.5$ m) was produced
with a Dywidag bar, allowing control of the preload
level. The sensor technology selected for this application is a multiplexed version of the standard SOFO
(Surveillance d'Ouvrages par Fibres Optiques) interferometric sensor (www.smartec.ch), where in-line
multiplexing is obtained by separating each measurement field through broadband Fiber Bragg Gratings
(FBGs) (Pozzi 2008). The embeddable part of the
sensing system is prepared in the form of a linear
subassembly, referred to as a smart bar. In this smart
bar, all optical sensors and wires are mounted on a
Fiber Reinforced Polymer (FRP) support. In order to
verify the effectiveness of the damage assessment technology, a 3-field smart bar was produced and built into
the pre-stressed RC element prototype. The sensor system also includes a number of traditional metal-foil
strain gauges. The scope of the experiment was to correlate the response of the embedded sensor to different
damage scenarios, which are artificially induced in the
beam. Different levels of cracking were produced by
vertical loads applied by a hydraulic actuator: the load
protocol included a sequence of load-unload cycles of
increasing amplitude, repeated for different values of
pre-stressing, up to yield of the mild reinforcement.
The method allowed clear recognition of increasing
damage states, simulated on the beam by gradually
reducing the prestressing level.

REFERENCES

Bayes, T. 1763. *An Essay toward solving a Problem in the
doctrine of Chances.* Philosophical Transactions of the
Royal Society of London, 53: 370–418.
Beck, J.L. & Katafygiotis, L.S. 1998. Updating Models And
Their Uncertainties. I: Bayesian Statistical Framework.
Journal of Engineering Mechanics 124(2), 455–461.
Gregory, P. 2005. *Bayesian Logical Data Analysis for the
Physical Sciences.* Cambridge: Cambridge University
Press.
Papadimitriou, C. Beck, J.L. & Katafyogiotis, L.S. 1997.
Asymptotic Expansion for Reliability and Moments of
Uncertain Systems. *Journal of Engineering Mechanics.*
123(12): 380–391.
Sivia, D.S. 1996. *Data Analysis: a Bayesian Tutorial.* Oxford
: Oxford University Press.
Smartec SA. http://www.smartec.ch/.
Pozzi, M. Zonta, D. Wu, H.Y. & Inaudi, D. 2008. Devel-
opment and laboratory validation of in-line multiplexed
low-coherence interferometric sensors. *Optical Fiber
Technology* 14: 281–293.

*Safety, Reliability and Risk of Structures, Infrastructures and
Engineering Systems – Furuta, Frangopol & Shinozuka (eds)
© 2010 Taylor & Francis Group, London, ISBN 978-0-415-47557-0*

Structural damage detection of transmission tower based on dynamic model reduction using ambient vibration measurements

T. Yin, H.F. Lam & H.M. Chow

Department of Building and Construction, City University of Hong Kong, HKSAR, China

ABSTRACT

The collapse of transmission towers due to continuously accumulated damage is not uncommon, especially after typhoons and earthquakes. This has led to
a growing interest in the structural safety and reliability of transmission towers (Yasui et al. 1999). Many
studies have been focused on the vibration behavior of
transmission towers under wind action, but the assessment of the health status of this type of structure is
seldom emphasized in the literature.

The buckling of secondary members (bracings) is
the most common type of damage to transmission
and communication towers, especially after typhoons
and earthquakes. The main objective of this paper
is to develop a practical structural damage detection method that is tailor-made for the detection of
damaged secondary members of large-scale three-
dimensional tower structures.

The proposed methodology contains two phases.
The first phase identifies the modal parameters from
the ambient vibration response of the target transmission tower following the NExT-ERA technique. In
the second phase, the "equivalent" damage extent for
each sub-structure is calculated based on the dynamic
reduction technique (Kidder 1973) using the results
from the first phase. The proposed methodology converts the damage-detection problem into a set of
implicit nonlinear equations. A simple but computationally efficient iteration algorithm is proposed to
solve this set of nonlinear equations.

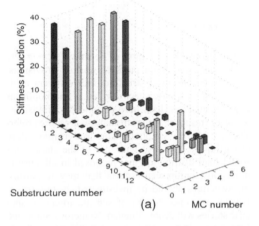

Figure 2. Identified damages with modeling error using
information of different mode combinations (MCs) utilizing
identified modes.

A typical three-dimensional transmission tower
model (with 324 DOFs, see Fig. 1) is employed to
illustrate and verify the proposed methodology under
both single and multiple damage cases in the presence of both measurement noise and modeling error.
The obtained results (see Fig. 2) are very encouraging,
showing that the proposed methodology can identify
the damaged sub-structure by estimating the 'equivalent' stiffness reduction even in the presence of both
measurement noise and modeling error.

ACKNOWLEDGEMENT

The work described in this paper was fully supported
by a grant from CityU (7002240). This generous
support is gratefully acknowledged.

REFERENCES

Kidder, R.L. 1973. Reduction of structural frequency equa-
tions. *AIAA Journal* 11(6): 892.
Yasui, H., Marukawa, H., Momomura, Y., & Ohkuma, T. 1999.
Analytical study on wind-induced vibration of power
transmission towers. *Journal of Wind Engineering and
Industrial Aerodynamics* 83(1–3): 431–441.

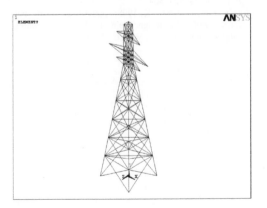

Figure 1. Three-dimensional FE model of the transmission
tower in ANSYS.

Safety, Reliability and Risk of Structures, Infrastructures and
Engineering Systems – Furuta, Frangopol & Shinozuka (eds)
© 2010 Taylor & Francis Group, London, ISBN 978-0-415-47557-0

Location of first plastic hinges in the moment-resisting frames

Ali Reza Khaloo, Mostafa Tazarv & Yousef Javid
Department of Civil Engineering, Sharif University of Technology, Tehran, Iran

ABSTRACT

The horizontal distribution of seismic force is one
of the most important parts of seismic provisions.
For seismic design of structure, four different analyses, static, response spectrum, linear and nonlinear
dynamic analyses, have been provided by provisions.
Static and response spectrum analyses are more preferable due to their simplicity. Due to dynamic nature
of earthquake and also wide use of the static horizontal load distribution that proposed in all seismic
provisions, unforeseen aspects that may jeopardize
the structures should be investigated. In this paper,
location of concentration of seismic forces in specific stories will be investigated. Numerous Static and
Linear time history analyses were carried out for 5,
10 and 15 story moment-resisting frames. Pushover
analysis also carried out to investigate the nonlinear
concentrations of plastic hinges in specific stories. A
new property of moment-resisting frame will be proposed that is the concentration of seismic demands
in the stories between $0.15\,h$ to $0.3\,h$ from the base.
In linear analysis, static lateral load distribution has a
good compatibility with the linear dynamic analyses to
predict force concentrations in these stories. Nonlinear analyses also indicate a deformation concentration
in these stories. Linear dynamic analyses as well as
pushover analysis proves that the first plastic hinge
will occur in these stories. For non-designed structures, static lateral load pattern predict greater forces
for beams in stories between $0.15\,h$ and $0.3\,h$. Therefore, by designing the structures, greater sections and
more rigidity are desirable for these elements. Nevertheless, these stories experience nonlinearity in beams
sooner than the other stories in pushover analysis. It
is worth mentioning that this property of moment-
resisting structures was implicitly observed by other
researches, but they overlooked this property. Foutch &
Yun (2002) noticed a concentration of plastic deformation in 3rd story of 20-story steel moment-resisting
structure. They mentioned that any nonlinear static
procedure that relies on global roof drift for a static
pushover analysis is highly questionable. While in this
paper it will be proved that it is property of moment
structures and has nothing to do with nonlinear static
procedure. Munshi & Ghosh (1998) shown that 12-
story moment structure has a critical and biased drift
concentration. A 20-story moment frame was analyzed
by Filippou & Fenves (Bozorgnia & Bertero 2004) that
its maximum plastic hinge rotation taken place in 6th
story. Plastic hinge rotation is also considerable is 3rd
to 5th story.

REFERENCES

Bozorgnia, Y., Bertero, V. V. 2004, *Earthquake Engineer-*
ing: From Engineering Seismology to Performance-Base
Design, CRC Press LLC, Chapter 6–60.
Foutch, D., Yun, S. 2002, Modeling of steel moment
frames for seismic loads, *Journal of Constructional Steel*
Research, 58, 529–564.
Munshi, J. A., Ghosh, S. K. 1998, Analyses of seismic per-
formance of a code designed reinforced concrete building,
Engineering Structures, Vol. 20, 608–616.

Safety, Reliability and Risk of Structures, Infrastructures and Engineering Systems – Furuta, Frangopol & Shinozuka (eds)
© 2010 Taylor & Francis Group, London, ISBN 978-0-415-47557-0

A consideration on the deterioration of tunnel lining based on actual inspection data

A. Sutoh & O. Maruyama
Department of Urban & Civil Engineering, Tokyo City University, Tokyo, Japan

T. Sato & H. Nishi
Civil Engineering Research Institute for Cold Region, Public Works Engineering Research Institute, Sapporo, Japan

ABSTRACT

This paper proposes deterioration methods based upon actual inspection data in order to carry out strategic maintenance and to rationalize life cycle cost analysis for tunnel structures (Hass and Hudson 1988). The resistance of deteriorating tunnel structures is non-stationary stochastic processes, and reliability problems of such structures are essentially different of time-independent reliability problems.

While the forecasting of deterioration is one of the central takes in infrastructural asset management, it is often the cases where are few data stocks available for estimating the deterioration forecasting model.

Firstly, using deterioration rates, the methodology of predicting of deterioration is discussed to model the deterioration of tunnel lining concrete.

The present problem is the estimation and/or identification of the degrading process of the tunnel lining concrete represented by the Ito stochastic differential equation as follows (Baxter and Rennie 1996).

$$dX(t) = \beta X(t)dt + \sigma X(t)dW_1(t) \qquad (1)$$

where, β: the constant drift parameter, σ: the constant volatility parameter, $W_1(t)$: the wiener process.

And, assume that the resistance recovers immediately after each repair process, so that $Z_i(t)$ becomes discontinuous and returns to vertically at a repair time (Madanat, S. 1997).

$$dZ(t) = \beta Z(t)dt + \sigma Z(t)dW_1(t) + \sum_{i>1} \{Z_1^* - Z_2^*\} l(t - t_1^*) \qquad (2)$$

where, l: the Dirac measure.

In the second place, for the prediction of the individual structure of tunnel lining concrete, a probabilistic approach using the distribution of deterioration rates and its own historical inspection data is proposed.

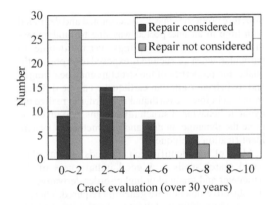

Figure 1. Index of lining Damage (Repair considered).

The validity of these methods is verified through the actual visual inspection data of tunnel lining concrete. And, the applicability of the methodology presented in this paper is examined against the real data concerning the deterioration on the road tunnel lining concrete in Hokkaido (See Figure1).

In addition, the average deterioration curves, variance and distribution density of time history, are obtained using the visual inspection data of tunnel lining concrete, which was considered the repaired process of the each tunnel lining.

REFERENCES

Hass, R. and Hudson, W. R. 1988: Pavement Management Systems.
Baxter, M. and Rennie, A. 1996: Financial calculus: An Introduction to Derivative Pricing, Cambridge University Press.
Madanat, S. 1997. Predicting Pavement Deterioration, ITS Review, Vol.20, Institute of transportation Studies, University of California.

Safety, Reliability and Risk of Structures, Infrastructures and Engineering Systems – Furuta, Frangopol & Shinozuka (eds)
© 2010 Taylor & Francis Group, London, ISBN 978-0-415-47557-0

Hybrid genetic algorithm to identification of real buildings

Grace S. Wang

Department of Construction Engineering, Chaoyang University of Technology, Taichung County, Taiwan

F.K. Huang

Department of Water resources and Evironmental Engineering, Tamkang University, Taipei County, Taiwan

ABSTRACT

Structure properties may be deteriorated and degraded with time in an unexpected way due to randomness in the environment and loadings over its lifetime. In particular, when a structure is exposed to strong earthquake, the properties of the structure may be changed and its behavior after an earthquake can be different from that before the earthquake. In order to realize the dynamic behavior of structural systems, we can determine the dynamic models and parameters by system identification techniques.

Most of the identification methods are calculus-based search method. A good initial guess of the parameter and gradient or higher-order derivatives of the objective function are generally required. There is always a possibility to fall into a local minimum. On the other hand, genetic algorithms (GAs) are optimization procedures inspired by natural evolution. They model natural processes, such as selection, recombination, and mutation, and work on populations of individuals instead of a single solution. In this regard, the algorithms are parallel and global search techniques that search multiple points, and they are more likely to obtain a global solution. Many GA applications have been performed on a variety of optimization problems in engineering area. However, relatively few applications have been on structural identification. Koh et al. (2003) proposed a hybrid strategy of exploiting the merits of GA and local search operator. Two local search methods were studied: an existing SW method and a proposed method called the MV method. The numerical study showed that the hybrid strategy performs better than the GA alone. The author (Wang & Lin 2005) applied the real-coded GA to structural identification problems. The validity and the efficiency of the proposed GA strategy were explored for the cases of systems with simulated input/output measurements. Moreover, the strategy was also applied to the real structure. Genetic algorithms (GAs) are global search techniques for optimization. However, GAs are inherently slow, and are not good at hill-climbing. In order to accelerate the convergence to the optimal solutions, a hybrid GA identification strategy that employs Gauss-Newton method as the local search technique is proposed and verified in this paper.

In this study, the Hybrid Genetic algorithms are used to identify the modal parameters in the time domain. There are two kinds of systems to be identified. The first one identified is 3-story steel frame built in the National Center for Research on Earthquake Engineering. The modal characteristics of this system are identified using the accelerograms collected from the shaking table test when subjected to El Centro earthquake. The second one identified here is the Civil-Environment Building of Chung-Shin University with torsionally-coupled effect. The modal characteristics of this building are identified with one of the recorded accelerograms obtained from the aftershocks of Chi-Chi earthquake. Finally, the estimated parameters enable us to predict the structural response under seismic loading. The comparison is made between the predicted acceleration and the measured one for each case.

REFERENCES

Koh, C.G., Chen, Y.F.& Liaw, C.Y. 2003. A hybrid computational strategy for identification of structural parameters. *Computers and structures* 81: 107–117.

Wang, G.S. & Lin, H.H. 2005. Application of genetic algorithm to structural dynamic parameter identification. *Journal of the Chinese Institute of Civil Engineering and Hydraulic Engineering* 17(2): 281–291.

Mini-Symposia (MS16) Risk Assessment of Lifeline Networks and Decision Support

Safety, Reliability and Risk of Structures, Infrastructures and
Engineering Systems – Furuta, Frangopol & Shinozuka (eds)
© *2010 Taylor & Francis Group, London, ISBN 978-0-415-47557-0*

Time sequential evolution of interdependent lifeline systems

I. Hernández-Fajardo & L. Dueñas-Osorio

Civil and Environmental Engineering Department, Rice University, Houston, TX, USA

ABSTRACT

Realistic models of critical service infrastructure systems must consider interactions in order to evaluate the influence of multi-system emergent response in the performance of individual systems. This paper presents a new methodology for the assessment of the performance of coupled service networks modeling their interdependence as a discrete, unidirectional relationship between individual components in separated networks. The proposed interdependence model is characterized by a set of rectangular matrices, and a standalone variable measuring the strength of the interdependence bond. These two parameters provide a clear identification of the location and strength of the interdependence between two interconnected systems; furthermore, the strength parameter is modeled as a random variable; a feature that introduces a factor on uncertainty hardly discussed in previous research efforts.

In order to model the evolution of the response of coupled networks, the proposed interdependence model is integrated into a novel simulation scheme for the propagation of damage among networks. The algorithm for this procedure separates the external perturbation's triggering role from the continued disruption induced by interdependence links. This characterization portrays interdependence as an inherent source of potential damage created by networks' dependence relationships.

The simulation of fragility evolution attempts to further explore the characteristics of networks' response to perturbation. The evolution algorithm tracks the response of the networks along progressing levels of readjustment from perturbation leading to a final stable network behavior. This final stabilization stage can be achieved through total collapse of the infrastructure of systems, or localized damage absorption by their components.

The simulation of evolution and the inclusion of interdependence obey the objective of obtaining probabilistic measures for the fragility of coupled networks. For this purpose, numerical simulation is needed. Enough simulations of triggering events are executed and once they are concluded, the damage records are analyzed to generate complementary statistical insights on the networks' response in terms of fragility evolution plots and systemic interdependent fragility curves. On one hand, fragility evolution plots describe how the fragility of networks propagates until stabilization for different levels of initial excitation and interdependence. On the other hand, interdependent fragility curves display interdependence influence on the fragility of networks for different performance levels as a function of increasing external excitation magnitudes.

The procedure used for the analysis of the fragility of interdependent networks was tested with a power system and a water network from a real location in the United States. The networks were subjected to the action of seismic hazard characterized by peak ground acceleration values. The Connectivity Loss metric was used for measuring network response performance. The routine was repeated for 5000 simulations, and statistical analysis of the damage records revealed increases of final stable fragilities up to 94% of the original non-evolved fragility estimation when higher interdependence strength values were considered. Also noticeably, the degree and characteristics of interdependence's influence were limited by the localization of interdependence links encapsulated in the interdependence matrices.

The results from this research support the inclusion of detailed interdependence models in complex networks' studies, as well as the tracking of interaction evolution influence on final fragility estimations, a novel feature introduced in this paper. Future work in this area includes the study of spatial variability in coupling strength, evaluation of the mechanisms of damage propagation among systems, use of functional descriptive metrics for the networks performance, and the study of the role of interdependence study in the analysis and prevention of damage propagation among networks.

Safety, Reliability and Risk of Structures, Infrastructures and Engineering Systems – Furuta, Frangopol & Shinozuka (eds)
© 2010 Taylor & Francis Group, London, ISBN 978-0-415-47557-0

Modeling infrastructure interdependencies: Theory and practice

S. Peeta & P. Zhang

School of Civil Engineering, Purdue University, West Lafayette, IN, USA

ABSTRACT

Recent extreme events have suggested that critical interdependencies exist among the infrastructure systems such as transportation, telecommunication, energy, electric power and water. The interdependencies have significant security, engineering and economic implications, which can be revealed in different scenarios: (i) with the current trend of urbanization, new infrastructure systems are being planned, expanded and built; (ii) existing aging infrastructure systems are constantly being maintained and upgraded; (iii) large-scale natural or manmade disturbance may amplify the interdependencies, which may cause common failure and/or cascading catastrophe under extreme situations.

Under these scenarios, the consideration of interdependencies could have great implications on the robustness, effectiveness and efficiency of the infrastructure systems. The key aspects need to be addressed include: (i) identify the sources and characteristics of various interdependencies; (ii) capture the phenomena and consequences that are impossible or difficult to model when the systems are treated in isolation; (iii) evaluate the implications, significance and benefits of a holistic view for relevant practical problems.

In order to address these needs, new generation of modeling frameworks that are capable of incorporating multiple systems seamlessly on the same platform and analyzing different types of infrastructure dependencies simultaneously are needed. Based on a previous work by the author (Zhang, 2008), this paper aims to introduce such a generalized modeling framework using multilayer infrastructure networks (MIN) concept and spatial computable general equilibrium (SCGE) approach. The key advantages of such framework include: (i) The use of economics-based methodology avoids the complexities caused by the disparate physical, institutional, operational,

and network characteristics of the various infrastructure systems. (ii) The use of market-based approach, more specifically, the nested elasticity of substitution functions and network-based approach allows the capture of different types of interdependencies, including the functional, market/economic, physical and budgetary interdependencies, under the same framework. (iii) Both equilibrium and disruption analyses can be conducted using the framework. (iv) Both static and dynamic aspects of the problem can be addressed. (v) Real-world data for calibrating the model parameters is available from sources of different levels. (iv) The framework allows what-if analysis for decision making, policy evaluation and other application problems.

This paper primarily focuses on the implementation and computation aspects of the proposed modeling framework. The basic concept and structure of static equilibrium MINSCGE model are introduced first, followed by the discussion on the various implementation issues such as the choice of function forms, data requirements, calibration issues, and computational tools or approaches. Through a series of illustrative numerical experiments, the paper also demonstrates that the proposed modeling frame-work is able to satisfy the key requirements in the infrastructure interdependency analysis. The systematically designed examples will illustrate the capability of the proposed models, capture of cascading effect, illustrate the importance of holistic decision making over all systems, and the role of infrastructure interdependency in real-world problems.

REFERENCES

Zhang, P. 2008. *A Generalized Modeling Framework to Analyze Interdependencies among Infrastructure Systems.* Purdue University, West Lafayette, IN, USA.

Safety, Reliability and Risk of Structures, Infrastructures and
Engineering Systems – Furuta, Frangopol & Shinozuka (eds)
© 2010 Taylor & Francis Group, London, ISBN 978-0-415-47557-0

Seismic risk and hierarchy importance of interdependent lifeline systems using fuzzy reasoning

Maria N. Alexoudi
University of Macedonia, Thessaloniki, Greece

Kalliopi G. Kakderi & Kyriazis D. Pitilakis
Civil Engineering Department, Aristotle University of Thessaloniki, Greece

ABSTRACT

Strong dependence on lifeline systems is one of the distinctive characteristics of modern urban regions. Unfortunately, lifelines are subjected to several hazards including earthquakes that are closely related to physical damage of the sub-components of the same system (intra-dependency) and of other related systems (inter-dependency). The inherited complexity between the urban environment and lifeline components makes the assessment of inter-dependent lifeline systems' performance a difficult task. A major concern for the quantification of lifeline elements' interactions is the description of the typology and the functioning of systems involved, the nature of the reciprocal influence, the time when the specific dependence is evolved (normal, crisis, recovery period) and the importance of the link (slight/strong) between components and systems.

The methodology presented herein is a combination of decision making and fuzzy linguistic preference relations in order to estimate adequate interdependency indices between different lifeline systems. The main target is to quantify the influence of each lifeline or infrastructure in the performance of complex systems for the normal, crisis and recovery period proposing a hierarchy factor describing the importance of inter-dependences.

The proposed approach involves: (1) the construction of a structural model of interrelationship of all systems under consideration, (2) the quantification of expert opinions using group decision making with uncertain additive linguistic preference relations to illustrate the importance of each lifeline system or element compared to another and (3) the conversion of the opinions into the form of a cross impact matrix composed of elements representing degree of corresponding impact.

The applicability of the proposed methodology is illustrated using the complex system of port facilities. The multiple interactions that exist between the different lifelines systems within port facilities are described for the three periods. The lifelines and infrastructures

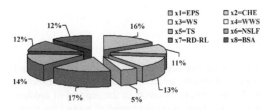

Figure 1. Operability importance of lifelines & infrastructures in port facilities (crisis period).

that were examined are: Electric Power System, Cargo Handling Equipment, Water System, Waste Water System, Telecommunication System, Natural Gas & Liquid Fuels, Roadway/Railway, Buildings, Ships/Wharves & Auxiliary Services.

For example, in crisis period, following the occurrence of an earthquake event, the most important system for the operability of the port is "Telecommunication System" and the least one is "Waste-Water System".

Based on the produced results and for the example of port facilities, pre-earthquake mitigation actions could be undertaken in order to enhance the robustness of the systems that have primary importance (highest level of dependency) in the three periods (electric power and telecommunication systems). A more detailed analysis of their seismic vulnerability taking into account interdependencies could also be performed. Finally, post-earthquake strategies should primarily account for the above systems in order to reduce the restoration time when the human resources are limited.

REFERENCES

Xu, Z.S. 2004. A Method Based on Linguistic Aggregation Operators for Group Decision Making with Linguistic Preference Relations. *Information Sciences* 166: 19–30.

Xu, Z. 2006. A Direct Approach to Group Decision Making with Uncertain Additive Linguistic Preference Relations. *Fuzzy Optimization and Decision Making* 5: 21–32.

*Safety, Reliability and Risk of Structures, Infrastructures and
Engineering Systems – Furuta, Frangopol & Shinozuka (eds)
© 2010 Taylor & Francis Group, London, ISBN 978-0-415-47557-0*

Seismic risk assessment of complex interacting infrastructure using matrix-based system reliability method

Y. Kim
EQECAT, Inc., Oakland, CA, USA

J. Song, B.F. Spencer & A.S. Elnashai
University of Illinois, Urbana, IL, USA

ABSTRACT

Urban infrastructures such as electric power, gas and
water distribution, telecommunication and transportation networks are essential in the support of modern
societies. These systems consist of numerous structural components that are spatially distributed over a
large area, and they are often highly interdependent in
that damage of a particular system are propagated to
other systems and result in additional system impairment due to cascading failures. Therefore, considering
the interdependencies of the complex urban infrastructures is essential to better estimate their reliability
under natural and man-made hazards and to establish
effective response and recovery plans. This paper proposes a new reliability methodology to assess system
reliability of the complex interdependent infrastructures under seismic hazards. First, a probabilistic
model to characterize system interdependencies is
introduced. Second, the matrix-based system reliability method, a recently developed analytical system
reliability framework, is employed to assess the probability of system failures (events). Third, a case study
is presented to demonstrate the proposed method. System reliability of interacting systems with and without
consideration of interdependencies is computed under
various levels of seismic demands. The conditional
failure probability of each system given the failure of
a component in the other system is also presented to
quantify the relative importance of the components.

SELECTED REFERENCES

Amin, M. 2001. Towards self-healing energy infrastructure
systems. *IEEE Computer Applications in Power* 14(1):
20–28.
Chang, L. & Song, J. 2007. Matrix-Based System Reliability Analysis of Urban Infrastructure Networks: A Case
Study of MLGW Natural Gas Network. *The fifth China-
Japan-US Trilateral Symposium on Lifeline Earthquake
Engineering. Haikou.*
Dueñas-Osorio, L., Craig, J.I., & Goodno, B.J. 2007. *Toler-
ance of Interdependent Infrastructures to Natural Haz-
ards and Intentional Disruptions*. Urbana: Mid-America
Earthquake Center
Kang, W.H., Song, J., & Gardoni, P. 2008. Matrix-based
system reliability method and applications to bridge
networks. *Reliability Engineering & System Safety* 93:
1584–1593.
Kim, Y., Spencer, B.F., Elnashai, A.S., Ukkusuri, S. & Waller,
S.T. 2006. Seismic performance assessment of highway
networks. *The 8th National Conference for Earthquake
Engineering. San Francisco, U.S.A.*
Kim, Y, Spencer, B.F., Song, J., Elnashai, A.S. & Stokes,
T. 2007. *Seismic Performance Assessment of Interdepen-
dent Lifeline Systems*. Urbana: Mid-America Earthquake
Center
Li, J. & He, J. 2002. A recursive decomposition algorithm
for network seismic reliability evaluation. *Earthquake
Engineering and Structural Dynamics* 31(8): 1525–1539.
Peerenboom, J., Fisher, R. & Whitfield, R. 2001. Recovering
from Disruptions of Interdependent Critical Infrastruc-
tures. *Prepared for CRIS/DRM/IIIT/NSF Workshop on
"Mitigating the Vulnerability of Critical Infrastructures
to Catastrophic Failures" Lyceum, Alexandria, Virginia,
September 10–11*
Rinaldi, S.M., Peerenboom, J.P., & Kelly, T.K. 2001. Critical
infrastructure interdependencies. *IEEE Control Systems*
21(6):11–25.
Song, J. & Der Kiureghian, A. 2003. Bounds on system reliability by linear programming. *J. of Eng. Mech., ASCE*
129(6): 627–636.
Song, J., and Kang, W.-H. 2009. System reliability and sensitivity under statistical dependence by matrix-based system
reliability method. *Structural Safety* 31: 148–156.
Zimmerman, R. 2001. Social Implications of Infrastructure
Network Interactions. *Journal of Urban Technology* 8(3):
97–119

Safety, Reliability and Risk of Structures, Infrastructures and Engineering Systems – Furuta, Frangopol & Shinozuka (eds)
© 2010 Taylor & Francis Group, London, ISBN 978-0-415-47557-0

Post-hazard flow capacity of bridge transportation network considering structural deterioration

Y.-J. Lee & J. Song
University of Illinois at Urbana-Champaign, Urbana, IL, USA

P. Gardoni
Texas A&M University, College Station, TX, USA

ABSTRACT

The flow capacity of a transportation network can be reduced significantly by structural damage of its constituent bridges, which are often considered as "weakest links" due to their vulnerability to natural or man-made hazards despite of their importance in transportation networks. It is therefore essential to predict the post-hazard flow capacity of the network accurately for risk-informed decision making on hazard mitigation and response. However, it is challenging to estimate the post-hazard flow capacity of a network due to the uncertainty in hazards and structural damages, and the complex nature of the network flow analysis. Moreover, the bridge structures deteriorate over time, which requires time-dependent network reliability analysis. For this post-hazard flow capacity analysis, many researchers used sampling-based approaches that may prevent rapid risk assessment and make it difficult to obtain the parameter sensitivity or component importance measures that are useful for risk-informed decision making.

This paper proposes a new non-sampling-based approach to estimate the time-varying post-hazard flow capacity of a bridge transportation network efficiently and accurately. The proposed approach evaluates the probabilities of structural damage scenarios using the matrix-based system reliability (MSR) method (Kang et al. 2008; Song & Kang 2008) and computes the corresponding flow capacities using a maximum flow capacity analysis algorithm. The matrix-based framework integrates these results to obtain the probabilistic distributions and statistical moments of the network flow capacity (see Figure 1). It also enables computing the conditional mean and standard deviation of flow capacity given structural damages and component importance measures conveniently that facilitate risk-informed decision making. In the proposed approach, probability calculation and network flow analysis are separately performed. This

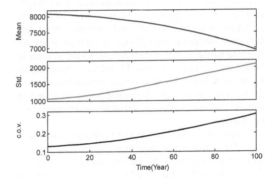

Figure 1. Mean, standard deviation, and coefficient of variation of network flow capacity with deterioration.

"separation of two tasks" in the matrix-based framework requires performing only one set of deterministic network flow analysis to render the time-dependent post-hazard flow analysis efficient. The proposed approach is successfully demonstrated through application to a numerical example based on Sioux Fall network (Ukkusuri 2005; Kim 2007).

REFERENCES

Kang, W.-H., Song, J. & Gardoni, P. 2008. Matrix-based System Reliability Method and Applications to Bridge Networks. *Reliability Engineering & System Safety*. 93: 1584–1593.

Kim, Y.S. 2007, *Seismic Loss Assessment and Mitigation of Critical Urban Infrastructure Systems*, Ph.D. Thesis, University of Illinois at Urbana-Champaign, Urbana, IL.

Song, J. & Kang, W.-H. 2009. System Reliability and Sensitivity under Statistical Dependence by Matrix-based System Reliability Method. *Structural Safety*, 31(2), 148–156.

Ukkusuri, S. 2005. Accounting for Uncertainty, Robustness and Online Information in Transportation Networks. Ph.D. Thesis, University of Texas at Austin.

Safety, Reliability and Risk of Structures, Infrastructures and
Engineering Systems – Furuta, Frangopol & Shinozuka (eds)
© 2010 Taylor & Francis Group, London, ISBN 978-0-415-47557-0

Sewer assets maintenance management: Comparison by two approaches, Markovian process and probabilistic neural networks

S.M. Elachachi & D. Breysse
University of Bordeaux 1, Ghymac, Talence, France

1 INTRODUCTION

Decisions on sewer assets management have large, long-lasting consequences and the decisions have to be taken under uncertainty. Information about the hydraulic, structural and tightness conditions is uncertain due to the fact that this infrastructure system is buried and of uneasy access. Therefore, estimating of the sewer network's performances is problematic and the investments can involve considerable "risk" when one designs with uncertain variables under imperfect knowledge.

In order to be able to sort out prioritary sewer sections for maintenance operation accounting for the specific requirements of the stakeholder, regarding the system itself and/or its urban or socio-economical environment, two approaches were considered: one based on Markovian Processes (MP) and the other one based on Probabilistic Neural Networks (PNN).

The condition state of each section of the sewer asset can not be checked each year (budget limits). This leads to estimate its dysfunction functions from available information (older inspections) or with the support of the elaborated evolution models (when they exist).

The performance indicators relate to three levels of a causal chain, linking defects to their impacts:

- Defects (e.g. cracks) characterize the actual physical state of the facilities and consist of the deviations, present with respect to the current state of the art,
- Dysfunctions (e.g. blockage) represent the consequences of defects on facility operations. Some dysfunctions may be observed using CCTV inspection or alternative types of investigation procedures, while others can only be estimated on the basis of defects.
- Impacts (e.g. pollution of surface water) reflect the degree to which dysfunctions induce noxious effects, depending on the context (vulnerability factors).

The Markov Process approach (MP) is based on Markov chain model. Markovian transition probabilities are estimated in order to characterize the deterioration progress between consecutive states. The followed methodology, allows relating the relative transition frequency of the condition states to section's factors (urban parameters and environmental conditions). The deterioration process is formulated using an exponential hazard model. This approach has the ability to model a multi state deterioration even if it considers that the transition probabilities are independent of the age of the section.

The Probabilistic Neural approach (PNN) becomes commonly used when the relationship between the inputs (here the factors and detected defects of each section) and the outputs (here the section's performance) is complex and weakly known. A (PNN) is an implementation of a statistical algorithm (kernel discriminant analysis) in which the operations are organized into a multilayered feedforward network with four layers. It applies a comprehensive mapping strategy derived from a Bayesian decision rule and from non-parametric estimators of probability density functions. The main goal of the Bayesian decision rule is to assign a section 's' to a class 'c' to which this section is most likely to belong.

A comparison between the two approaches was led, following a parametric study where were compared the ranking of the prioritary sewer sections for each of the approaches and by some aspects, the computational costs (which that conditions the attractiveness of the approach). It was shown that sorting out prioritary sections is sensitive to the method used even without taking account uncertainties in external inputs. For developing accurate models, it is necessary to have panel data that spans over multiple time periods.

Mini-Symposia (MS17) Earthquake Engineering and Engineering Seismology

Safety, Reliability and Risk of Structures, Infrastructures and Engineering Systems – Furuta, Frangopol & Shinozuka (eds)
© 2010 Taylor & Francis Group, London, ISBN 978-0-415-47557-0

A case study of seismic risk management of school buildings based on life cycle cost and safety of students

Y. Murachi & H. Namita
KOZO KEIKAKU Engineering Inc., Tokyo, Japan

ABSTRACT

After the Hanshin-Awaji Great Earthquake, on January 17, 1995, Seismic retrofitting of old school buildings and gymnasiums are facilitated in Japan. However, there are still a lot of reports of damage of those buildings by earthquakes, such as the Geiyo Earthquake (on March 24th, 2001), the Mid Niigata Earthquake (on October 23rd, 2004), and the Off Mid-Niigata Earthquake (on July 16th, 2007). For example, vertical braces of gymnasiums were buckled, or the ceiling panels were falling down.

In Japan, school buildings and gymnasiums are provided as the evacuation center for the people whose houses are collapsed or lifeline utilities are stopped. Even if school buildings and gymnasiums are suffered minor or moderate damage, those buildings cannot be used as the evacuation center. Because there are traumatized evacuees and it is possible to be suffered major damage by aftershocks.

In order to facilitate seismic retrofitting of old school buildings and gymnasiums, local governments set priorities by evaluating those seismic capacities. However, there are two problems to fix an order of priorities. The first problem is how to take account the difference of structural type. Evaluating method of seismic capacities is different between reinforced concrete structure and steel structure. Then seismic performances are different between reinforced concrete structure and steel structure, even if the results of seismic capacities are same value. Consequently, we cannot set priorities by easy comparison of the seismic capacity values. The second problem is how to take account the usability of old buildings. Targets of seismic retrofitting are more than 27-years old buildings. It is necessary to consider the benefit-cost relationship between rebuilding and retrofitting.

For multiple industrial buildings, one of the authors, Murachi et al. (2006), proposed a case study in seismic risk management, and demonstrated that the expected life-cycle cost is useful to explain a reduction/increment of physical and economical damage.

In this paper, the methodology is extended to the seismic risk management of rebuilding and

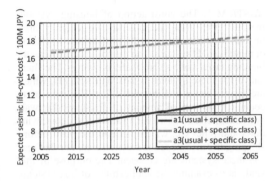

Figure 1. Expected seismic life-cycle cost for each alternative.

retrofitting. As a case study, we deal with two school buildings owned and operated by a local government. Both of them are built by steel structure, and one is for the usual class, the other is for the specific class. Dealing with a school building that accommodates many students, it is also significant to consider their safety. Thus this paper focuses on the computation of the probabilities that the building incurs severe damage states, as the primary indexes of students' safety. In this case study, three alternatives are evaluated for each building: a_1 is the existing building before the seismic upgraded, a_2 is the up-graded one and a_3 is rebuilt one. The estimated results, e.g. damage probabilities of the buildings, restoration time and life-cycle cost, will assist the manager of the local government to decide on the selection of alternatives upon its limited budget.

REFERENCES

Y. Murachi, M. Tsubota, Y. Takahashi: "A case study of seismic risk management of industrial buildings based on safety of employees", The 5th Computational Stochastic Mechanics Conference, June 18–21, 2006, Rhodes, Greece

Safety, Reliability and Risk of Structures, Infrastructures and
Engineering Systems – Furuta, Frangopol & Shinozuka (eds)
© 2010 Taylor & Francis Group, London, ISBN 978-0-415-47557-0

The reasons for choosing box housing systems despite of traditional systems after the earthquake areas in Turkey

Ö Eren
Mimar Sinan Fine Arts University, Turkey

ABSTRACT: The construction of housing systems that built after earthquake in Turkey are generally three or four storey height concrete prefabricared systems or tunnel molds systems. The purpose is complete the building faster and also built safer buildings. The alternative for these systems is box unit systems, which are faster, better quality and safety. The box units contain a substantial amount of finish works, wall and finish, electrical wiring and fixtures, doors, windows and pipes etc. which made in the plant before shipping the module to the erection area. For that reason these systems are choosen. The purpose of this study is to examine the applicability of box systems housing in earthquake region.Modular buildings are pre-fabricated portable structures. The modular buildings are typically transported in large pre-assembled sections and then connected at the building site. In these system the standardization reduces fabrication costs and the factory storage space required to keep a large number of prefabricated components, and also allows for a wide variety of end use customization through userselected components (Eşiz,2001;Orton 1996).

Housing built after the earthquake must meet the following requirements:

- Different options concerning lifestyle and intensity of use should be accommodated by a flexible, multipurpose design that makes maximum use of minimum space.
- The design of the individual components should provide for savings in time, space, manpower, materials and energy consumption, and facilitate scheduling of the construction phases and operations, as well as the assembly and disassembly process
- Once the designated use of the temporary site to be erected has expired, it should provide the options of re-use, reproduction or recycling, and the permanent and temporary structures, consisting of a fixed substructure and variable superstructure components, should be executed accordingly
- The individual units should be adaptable to variable terrain conditions, and the amount of debris left behind on the site should be reduced by way of recycling.

The main advantage of this type of building is speed of construction, as for example the groundwork's can be proceeding at the same time as the building takes shape in the factory. Site assembly quickly produces a weatherproof shell, so progress is rarely delayed by bad weather. Speed of construction also leads to considerable cost savings when compared to conventional building methods. Today in a market potentially worth in excess of a billion pounds, modular and portable buildings continue to fill the gap and the industry is no stranger to off site construction – it is what we do and will continue to develop new ideas and technology.

REFERENCES

Eşiz,Ö. 2001. *Integration of High Tech Building Subsystems*, Doctoral Thesis, Mimar Sinan University.
Orton, A. 1996. *The Way We Build Now*, E&FN Spon, London, p.515

Safety, Reliability and Risk of Structures, Infrastructures and Engineering Systems – Furuta, Frangopol & Shinozuka (eds)
© 2010 Taylor & Francis Group, London, ISBN 978-0-415-47557-0

Optimal criterion for seismic design in Mexico

L.E. Pérez Rocha, A. López López & U. Mena Hernández
Instituto de Investigaciones Eléctricas, Cuernavaca, Morelos, México

D. de León Escobedo
Facultad de Ingeniería Civil, Universidad Autónoma del Estado de México, Toluca, México

ABSTRACT

Recent developments and results used to make the seismic regionalization of Mexico are presented in this work. The formulation to quantify the seismic hazard and their main results, corresponding to excedence rates of peak ground acceleration and the ones of several structural periods, are shown. As well, a formulation based on the total cost of construction is applied to determine the optimal design coefficient for the collapse limit state in high seismicity zones. This cost is the sum of the initial cost and the losses occurred during the life of a structure. However, for low seismicity zones, design accelerations obtained by using the optimal criterion are significantly greater than the ones produced by great earthquakes hard to image in these areas. This is why a concept of maximum or extreme earthquakes is introduced to specify a realistic highest level of intensity in these zones. This hybrid approach yields a continuous descriptor of design seismic intensities along the Mexican territory. For several reference sites located in high, medium and low seismicity zones, this descriptor matches very well with the previous one, consisting in four seismic zones. Finally, a set of expected damage maps are illustrated. These maps show an index of losses in a typical structure constructed with different design criteria, but

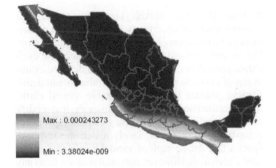

Max : 0.000243273

Min : 3.38024e-009

Figure 2. Annual probability of failure in substations designed for 200 Km/h, under seismic loading.

subjected to the same seismic hazard. This is a way to show how improved is the current criterion based on this hybrid approach. Figure 1 shows a map of optimal Peak Ground Accelerations for design. Figure 2 shows the annual probability of failure in a kind of electrical substations designed with the CFE (2002). The first map will be a part of the Design Handbook for civil structures in Mexico (MDOC-DS, 1993).

REFERENCES

CFE (2002), Especificaciones para diseño de subestaciones JA100-57, Torres autosoportadas J1000-50, and Postes metálicos J6100-54.
MDOC-DS (1993). Capítulo de Diseño por Sismo del Manual de Obras Civiles, Comisión Federal de Electricidad.

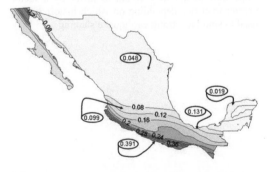

Figure 1. PGA associated to the optimal return periods (as fig. 1) using truncated lognormal distribution ($\varepsilon = 1$).

Safety, Reliability and Risk of Structures, Infrastructures and
Engineering Systems – Furuta, Frangopol & Shinozuka (eds)
© *2010 Taylor & Francis Group, London, ISBN 978-0-415-47557-0*

Quantifying the uncertainty of seismic input motion for dam safety evaluations

G. Bureau

GEI Consultants, Inc., Oakland, CA, USA

ABSTRACT

Professor Bray (U.C. Berkeley) recently stated that the primary source of uncertainty in assessing the performance of dams during earthquakes is the input motion. His comment related to simplified displacement evaluation procedures, but should apply to detailed evaluations of concrete and, especially, embankment dams. Current practice is to select suitable natural earthquake acceleration records (seed histories) and modify these in amplitude and frequency content (spectrum-compatible histories) to closely match the response spectra that define each of various postulated earthquake scenarios. However, potential shaking from any scenario is generally represented by a single set of horizontal and vertical acceleration time histories. Furthermore, embankment dam evaluations often ignore the vertical component of motion because early linear-elastic response studies concluded that it had little significance to such dams.

True nonlinear dynamic response analysis of embankment or concrete dams can simulate the inelastic behavior of soil materials, any buildup of excess pore pressures, and the potential for cracking, base uplift and joint opening. However, they require consideration of both horizontal and vertical motions. While a single set of input acceleration histories seemed acceptable to represent specified seismic criteria for linear-elastic or equivalent-linear dam analyses, this is no longer the case for nonlinear analyses. To quantify the uncertainty in the computed response, the author evaluated an embankment dam and a concrete dam

through nonlinear parametric analyses, using four sets of horizontal and vertical acceleration histories that closely matched the specified spectra, peak ground acceleration and duration of shaking. Three different ground motion levels were specified. To further reduce the uncertainty associated with such shaking, each input history and its "opposite" (all values multiplied by -1) were successively considered. Therefore, the simulation of each of the three earthquake scenarios included sixteen separate analyses with input histories matching the specified horizontal and vertical response spectra.

For the concrete dam, simultaneous occurrence of positive or negative, horizontal or vertical acceleration pulses influence calculated peak stresses and may trigger base sliding or joint opening. Linear-elastic and nonlinear (base uplift and sliding) analyses were successively performed. Calculated stresses and the extent of base uplift of the example gravity dam were influenced by each set of input histories. Peak tensile stresses could differ by a factor greater than two, depending on which set of input histories (representing the same earthquake scenario) had been used.

Downward inertia forces increase transient confining stresses and pore pressures within saturated earthfill or foundation materials. Upward inertia forces reduce the forces resisting failure, hence increase potential slope deformations. For the study embankment dam, computed maximum crest settlements and slope movements were found to differ by a factor greater than two depending on which histories were used to simulate strong earthquake shaking.

Safety, Reliability and Risk of Structures, Infrastructures and
Engineering Systems – Furuta, Frangopol & Shinozuka (eds)
© 2010 Taylor & Francis Group, London, ISBN 978-0-415-47557-0

Analytical fragility models for California box girder bridges

C.S. Yang & R. DesRoches
Georgia Institute of Technology, Atlanta, GA, USA

J.E. Padgett
Rice University, Houston, TX, USA

ABSTRACT

Fragility curves are increasingly being used in probabilistic seismic risk assessment of highway bridges.
Fragility curves, which are conditional probability
statements of a bridge's vulnerability as a function of
ground motion intensity, have been developed using
expert opinion, empirical data from past earthquakes,
and analytical methods (Jernigan & Hwang 2002,
Mander & Basoz 1999). The current fragility curves
used for seismic risk assessment in California are
based on simplified analyses or empirical data from
recent earthquakes. The goal of this paper is to use
detailed nonlinear analytical models of a typical California box girder bridge to develop fragility curves
which can be used in seismic risk assessment of
transportation networks in the region [Hazards US –
Multi-Hazard (HAZUS-MH)]. Because both expert-
based and empirically based fragility curves have
some inherent limitations, analytical methods have
been extensively studied, including spectral analysis,
nonlinear static analysis, and nonlinear time history
analysis.

In this study, analytical fragility curves for California box girder bridges are developed. First, a detailed
review of the California Department of Transportation
Bridge Inventory is conducted to identify the basic
characteristics of typical box girder bridges, including
geometric and material properties. Using this information, a detailed three-dimensional nonlinear analytical
model, which accounts for the nonlinear behavior of
the column, box girder, and abutments, is developed in
the OpenSEES platform. The analytical bridge model
is further calibrated using responses recorded during
the 1979 Imperial Valley earthquake (Fig. 1). Subsequently, the bridge model is used in conjunction with a
suite of eighty recorded ground motions representative
of the seismic hazard in southern California. Using
a set of appropriate limit states, the bridge fragility
curves are developed. The fragility curves for the reference bridge indicate that the bridge is robust compared
to many other bridge types found in other parts of the
United States (Fig. 2).

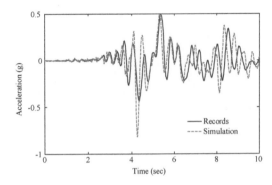

Figure 1. Time history of transverse acceleration at midspan
of the box girder.

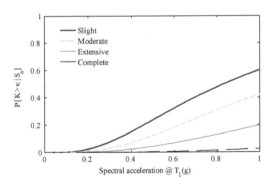

Figure 2. Fragility of the bridge with intensity of spectral
acceleration at the fundamental period.

REFERENCES

Jernigan, J.B. & Hwang, H. 2002. Development of Bridge
Fragility Curves. *7th US National Conference on Earthquake Engineering, Boston, Mass.* EERI.

Mander, J.B. & Basoz, N. 1999. Seismic Fragility Curve Theory for Highway Bridges. *5th US Conference on Lifeline
Earthquake Engineering, Seattle, WA, USA.* ASCE.

Mini-Symposia (MS08) Time-Dependent Reliability Methods and Their Applications

Safety, Reliability and Risk of Structures, Infrastructures and Engineering Systems – Furuta, Frangopol & Shinozuka (eds)
© *2010 Taylor & Francis Group, London, ISBN 978-0-415-47557-0*

Stochastic modelling for gradual deterioration, maintenance and monitoring

A. Barros, M. Fouladirad & Antoine Grall
Institut Charles Delaunay, Université de Technologie de Troyes, CNRS FRE 2848, Troyes, France

ABSTRACT

The aim of this paper is to propose an adequate modelling method and an appropriate maintenance policy for a deteriorating system. The considered system is functioning without any sign of deterioration. In an unknown time the system begins to deteriorate and continues to deteriorate with the same speed. The system suddenly undergoes a change of mode of deterioration at an unknown time. First the aim is to develop an appropriate model that takes into account all the properties of the system and which is easy to manipulate. Afterward, based on the given model a suitable maintenance/detection method that takes into account the presence of unknown change times should be proposed.

In the framework of systems studied in this paper, three typical phases of evolution for the deterioration can be enumerated. In the first phase the system is functioning an no deterioration is visible. At an unknown time the system begins to deteriorate slowly, the system is considered to be in the second phase. In the last phase the deterioration speed grows and the system deteriorates faster and this situation leads to the failure of the system. As a typical example of deteriorating systems studied in this paper, stress corrosion cracking (SCC) can be considered (Blain, Barros, Grall, and Lefebvre 2007). SCC that appears on different elements of nuclear power plants is a very important subject for safety and preventive maintenance planning. SCC is characterised by three main steps. The first step consists in incubation during which no crack appears but the material is in a mechanical and chemical favourable environment for SCC. The second step is the step of initiation during which some cracks appear and are slowly growing.

The last step is the propagation phase during which the largest crack reaches a faster propagation kinetics (main crack) and leads to the rupture of the component.

In the first phase, the random variable of interest is the length of time that the system is functioning without any sign of deterioration. The deterioration process has not started yet. After the first phase, one consider a stochastically deteriorating process described by a scalar ageing variable that summarises the condition of the system (Gamma process). The ageing variable increases with the system deterioration and the failure occurs as soon as the system state crosses a known fixed threshold L called failure threshold. Estimation methods are proposed to treat with unknown model parameters on the basis of operating feedback data.

The system is periodically inspected. The parameters of the deterioration process can change (according to the operating environment for example) at an unknown time T_0. A detection procedure based on the on-line change detection algorithm estimates the instant of change of mode. A maintenance versus detection policy based on the control limit decision rule can be proposed. To avoid a period of unavailability preventive actions can be made if the system state exceeds an alarm threshold A, lower than failure threshold L. The cost of a preventive maintenance action is lower than a corrective maintenance action. The aim is to minimise the global maintenance cost.

REFERENCES

Blain, C., A. Barros, A. Grall, and Y. Lefebvre (2007, Jun.). Modelling of stress corrosion cracking with stochastic processes – application to steam generators. In ESREL 2007, Stavanger, Norway, 25–27 June, Stavanger, Norway.

Safety, Reliability and Risk of Structures, Infrastructures and Engineering Systems – Furuta, Frangopol & Shinozuka (eds)
© *2010 Taylor & Francis Group, London, ISBN 978-0-415-47557-0*

An investigation of COF with variation of load and strength characteristics of material

M. Sharfuddin
Nagoya Institute of Technology, Nagoya, Japan

Y.G. Zhao
Kanagawa University, Yokohama, Japan

ABSTRACT

In the seismic design concept, the frame structures are usually designed with a column over design factor (COF) to ensure that the sum of the plastic moment capacity of column should be greater than that of girder at the joint by some margin. During earthquake this COF plays an important role in determining the building response. Adequate COF value can probabilistically avoid the undesirable story mechanism and can ensure preferable entire beam hinging failure mode. But the nature of the change of the failure modes of the structures as well as COF has not been investigated adequately with the change of the coefficient of variation of material property and earthquake load till now. In the present paper initially the failure modes of structure are investigated with the variation of load and strength characteristics of materials. After comprehensive investigation this paper then investigates the change of the target value of COF with the variation of load and strength characteristics of materials for which the occurrence probability of undesirable story mechanism is limited within a given tolerance. It is found that the coefficient of variation of the member strength merely affects the failure probability of the structure and the target COF while that of earthquake has dominant effect.

The reliability level $\beta_T = 2$ and $\beta_T = 3$ have been taken into consideration. The investigation has been conducted for three storied to six storied two bay frames. The story failure modes are defined and classified before the probabilistic evaluation so that the investigation can be carried out in each type step by step. In this study, the story failure modes are classified into three patterns: upper story failure pattern, middle story failure pattern and lower story failure pattern, which depends on the location of the failure stories. It has been found that failure probability of each collapse type is more sensitive to the change of the co efficient of variation (COV) of earthquake load than that of material strength. For the determination of column overdesign factor (COF) only the most likely failure modes have been taken into consideration. It is to be mentioned here that all the lower story collapse

Table 1. Target COF with COV of earthquake load ($\beta_T = 2$).

COV	6 story	5 story	4 story	3 story
0.6	2.0	1.72	1.49	1.35
0.7	1.9	1.64	1.42	1.31
0.8	1.8	1.57	1.37	1.30
0.9	1.7	1.50	1.33	1.27

modes and the upper story collapse modes with highest failure stories are the most likely failure modes.

After failure mode analysis, the change of target value of COF is investigated. Firstly, COV of earthquake load is kept constant and COV of material strength is varied and secondly COV of material strength is kept constant and COV of earthquake load is varied. It is found that the target column overdesign factor (COF) is insensitive to the co efficient of variation (COV) of the material strength but it is sensitive to that of earthquake load.

Table 1 shows the target COF for three to six storied frame for COV of earthquake load ranging from 0.6 to 0.9 considering reliability level $\beta_T = 2$.

REFERENCES

Kawano, A. Matsui, C. and Shimizu, R. 1998. Basic properties of column overdesign factors for steel reinforced concrete frames. *Journal of Structural and Construction Engineering* 505: 153–159.

Melchers, R.E. 1999. *Structural Reliability Analysis and Prediction.* New York: Wiley & Sons.

Ono, T. Zhao, Y.G. and Ito, T. 2000. Probabilistic evaluation of column overdesign factor for frames. *Journal of Structural Engineering, ASCE* 126(5): 605–611.

Zhao, Y.G. Ono, T. Ishi, K. and Yoshihara, K. 2002. An investigation on the column overdesign factors for steel framed structure. *Journal of Structural and Construction Engineering* 558: 61–67.

Zhao, Y.G. Pu, W.C. Li, H. and Ono, T. 2007. Basic and optimum column overdesign factor avoiding story mechanism for frame structure. *Journal of Key Engineering Materials* 340-341: 1405–1410

Safety, Reliability and Risk of Structures, Infrastructures and
Engineering Systems – Furuta, Frangopol & Shinozuka (eds)
© 2010 Taylor & Francis Group, London, ISBN 978-0-415-47557-0

The time-dependent safety and durability of deteriorating structures

A. Kudzys

Palffy & Associates Inc. Architects & Planners, Tokyo, Japan

O. Lukoševičiené

KTU Department of Construction and Architecture, Kaunas, Lithuania

ABSTRACT

The probabilistic analysis and prediction of the time-
dependent safety and durability of deteriorating structures subjected to recurrent extreme service and climate actions is discussed. The strategy of this prediction is based on the concept that not only a performance
but also a safety margin of deteriorating members of
load-carrying structures are time-dependent random
variables. Thus, the safety margin of design particular
members of deteriorating structural members (beams,
columns, walls) may be defined as the function:

$$Z(t) = g[X(t), \theta] = \theta_R R(t) - \theta_g S_g - \theta_q S_1(t) - \theta_q S_2(t) \quad (1)$$

where $X(t)$ and θ are the vectors of basic and additional
variables, representing respectively random components (resistances and action effects) and their model
uncertainties.

The safety margin process (1) may be treated as
finite decreasing random sequence written as:

$$Z_k = R_{ck} - S_k, \ k=1, 2, \ldots, \ n-1, n, \quad (2)$$

where $R_{ck} = \theta_R R_k - \theta_g S_g$ is the conventional resistance
of members; $S_k = \theta_q S_{1k} + \theta_q S_{2k}$ is the action effect
caused by two independent annual extreme loads
(Kudzys 2007).

The instantaneous and long-term survival probabilities of particular members may be assessed by
the simplified but fairly exact method of transformed
conditional probabilities as:

$$P_k = P\{R_{ck} > S_k \exists k \in [1, n]\} = \int_0^\infty f_{R_{ck}}(x) F_{S_k}(x) dx \quad (3)$$

$$P_t = P_i\{T \geq t_n\} = P\left\{\bigcap_{k=1}^{n}(Z_k > 0)\right\} =$$

$$= \prod_{k=1}^{n} P_k \left[1 + \rho_{n,n-1\ldots1}^{x_n}\left(\frac{1}{P_{n-1}} - 1\right)\right] \times \ldots$$

$$\times \left[1 + \rho_{k,k-1\ldots1}^{x_k}\left(\frac{1}{P_{k-1}} - 1\right)\right] \times \ldots \times \left[1 + \rho_{21}^{x_2}\left(\frac{1}{P_1} - 1\right)\right] \quad (4)$$

where $f_{R_{ck}}(x)$ is the density function of member
resistance R_{ck} and

$$F_{S_k}(x) = \exp\left[-\exp\left(\frac{S_{km} - x}{0.7794 \times \sigma S_k} - 0.5772\right)\right] \quad (5)$$

is the cumulative distribution function of the action
effect S_k; P_k is the instantaneous survival probability
by (2) at the cut k of this sequence.

The correlation factor of deteriorating series system
elements or decreasing sequence cuts:

$$\rho_{k,k-1\ldots1} = \left(\rho_{k,k-1} + \ldots + \rho_{k1}\right)/(k-1) \quad (6)$$

Its indexed value is $\rho_{k,k-1\ldots1}^{x_k}$, where the bounded index
"x_k" may be expressed as:

$$x_k \approx [(4.5 + 4\rho_k)/(1 - 0.98\rho_k)]^{1/2} \approx [8.5/(1 - 0.98\rho_k)]^{1/2} \quad (7)$$

The effect of two coincident recurrent extreme actions
on the survival probability and durability of structures
is analyzed. It is assessed that a failure of members
may occur not only in the case of their coincidence
but also when the value of one out of two effects is
extreme.

The technical service life t_t, as a quantative durability parameter of deteriorating members is related
with the target value of generalized reliability index
$\beta_T = \Phi^{-1}(P_s)$, where $\Phi(\bullet)$ is the inverse of the
standard normal distribution variable (EN 1990). This
value depends on the reliability and functional working
classes of structures.

REFERENCES

EN 1990. 2002. Eurocode-Basic of structural design. CEN,
 Brussels.
Kudzys, A. 2007. Transformed conditional probabilities in
 analysis of stochastic sequences. Summer Safety and Reliability Seminars, SSARS 2007, Gdansk-Sopot, Poland:
 243–250.

Safety, Reliability and Risk of Structures, Infrastructures and Engineering Systems – Furuta, Frangopol & Shinozuka (eds)
© 2010 Taylor & Francis Group, London, ISBN 978-0-415-47557-0

Treatment of uncertainties in life cycle assessment

Jack W. Baker & Michael D. Lepech
Stanford University, Stanford, CA, USA

ABSTRACT

Spurred by European Union directives such as WEEE (Waste Electrical and Electronic Equipment), ROHS (Reduction of Hazardous Substances in Electrical and Electronic Equipment), and ELV (End of Life Vehicles), along with increased interest among US initiatives such as LEED (Leadership in Energy and Environmental Design), life cycle assessment (LCA) is becoming a common tool used in the measurement and evaluation of environmental performance and overall sustainability. Life cycle assessment is an analytical technique for assessing potential environmental, social, and economic burdens and impacts, encompassing all stages of life cycle, from raw material production through end-of-life management. LCA provides metrics that can be used to measure progress toward sustainability.

As outlined by ISO 14040 series standards, any life cycle assessment requires a number of phases beginning with goal and scope definition, inventory analysis, impact assessment, and interpretation. Each of these phases, along with their associated databases and models, has significant uncertainties associated with them. Decisions made regarding design development and improvement, strategic planning, public policy or making, product marketing without recognizing this uncertainty may potentially be flawed.

A general motivation for quantifying uncertainties is to increase the transparency of LCA data and results. Uncertainty is undeniably present in many aspects of analysis, and treating it explicitly will aid in several ways. This paper provides an overview of sources of uncertainty, methods for quantifying uncertainty, and methods for propagating input and model uncertainties in order to determine their effect on uncertainties in the final estimated environmental impacts. Examples are described to demonstrate how uncertainties in impacts can be used to support improved decision-making by users of LCA tools. Impacts include adding the ability to identify alternate systems whose environmental impacts appear to differ at first glance, but for which the impacts are actually statistically insignificant due to uncertainties in the inputs. A second impact is the ability to identify important uncertainties and so focus uncertainty-reduction efforts in the most critical areas.

After describing the importance of this topic and describing tools for performing relevant analysis, prospects for future progress in this field are considered. There are a variety of practical challenges to consider, but also a variety of areas in which further achievements can be made. By looking at examples where uncertainty has been treated explicitly in LCA assessments, one can see where opportunities exist to further apply the tools of probabilistic modeling to this important and rapidly growing field.

Safety, Reliability and Risk of Structures, Infrastructures and Engineering Systems – Furuta, Frangopol & Shinozuka (eds)
© *2010 Taylor & Francis Group, London, ISBN 978-0-415-47557-0*

Development of life cycle condition profile models of some typical types of bridges

Heungmin Park & Kwangkyun Lee
IAM Corporation, Seoul, Korea

Kwangju Lee & Jungsik Kong
Korea University, Seoul, Korea

ABSTRACT

To derive the optimum maintenance scenario, it is needed to analyze the variation in condition of bridge caused by deterioration. However, standardized condition variation deterioration profile models were not available in general. So under these circumstances, the standardization of condition variation and models are immediately necessary to develop.

In this study, the present status of bridges in Korea has been investigated through arrangement of variation in condition with/without maintenance activity. Also we have developed practicable LCP (Life-Cycle condition Profile : LCP) model classified by some typical type of bridges and members to derive optimum maintenance scenario considering deterioration.

The model has been developed based on statistical regression methods and the database of HBMS(Highway Bridge Management system: HBMS).

We used the process in this study as below.

The determined function of regression analysis is determined by coefficient of determination in general, however, characteristics of LCP curve was took

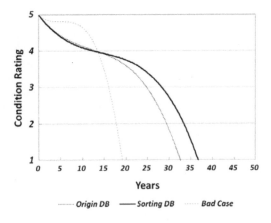

Figure 2. Single Variable Regression Analysis Result (PSCI).

into consideration in determination of final regression analysis model. The feature of LCP curve is as below.

Single variable regression analysis of PSCI girder is as shown in Figure 2.

This study suggested a process to establish LCP model for each bridge and its member using HBMS DB and LCP model by members of bridge structure through single or multi variable regression analysis.

We can see that developed LCP model in this study could be used more efficiently for bridge assessment

REFERENCES

An, Yeong-Gi (2003), Research of Deterioration Prediction Model for Estimation of Life Cycle Cost

Frangopol et al.(2001), Reliability-Based Life-Cycle Management of Highway Bridges.

Furuta, H et al. (2002), Bridge Maintenance System of Load Network Using Life-Cycle Cost and Benefit.

Ito, H et al. (2002), An Optimal Maintenance Planning for Many Concrete Bridges Based on Life-Cycle Cost.

Miyamoto et al.(2000), Bridge Management System and Maintenance Optimization for Existing Bridges.

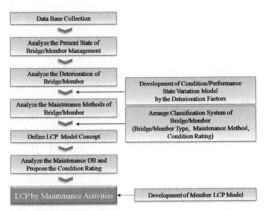

Figure 1. process of this study.

Organized Session (OS15) Challenging Technology for Condition Screening of Bridge Structures

*Safety, Reliability and Risk of Structures, Infrastructures and
Engineering Systems – Furuta, Frangopol & Shinozuka (eds)
© 2010 Taylor & Francis Group, London, ISBN 978-0-415-47557-0*

Prediction of displacement response from the measured dynamic strain signals using mode decomposition technique

Sung-Jin Chang & Nam-Sik Kim
School of Civil and Environmental Engineering, Pusan National University, Busan, Korea

Ho-Kyung Kim
Department of Civil and Environmental Engineering, Seoul National University, Seoul, Korea

ABSTRACT

In this study, a method predicting the displacement response of structures from the measured dynamic strain signal is proposed by using a mode decomposition technique. Evaluation of bridge stability is normally focused on the existing bridges. However, dynamic loadings including wind and seismic loadings could be exerted to the bridges under construction. In order to examine the bridge stability against these dynamic loadings, the prediction of displacement response is very important to evaluate bridge stability. Because it may be not easy for the displacement response to be acquired directly on site, an indirect method to predict the displacement response is needed. Thus, as an alternative for predicting the displacement response indirectly, the conversion of the measured strain signal into the displacement response is suggested, while the measured strain signal can be obtained using fiber optic Bragg-grating (FBG) sensors. As previous studies on the prediction of displacement response by using the FBG sensors, the static displacement has been mainly predicted. For predicting the dynamic displacement, it has been known that the measured strain signal includes higher modes and thus the predicted dynamic displacement can be inherently contaminated by broad-band noise. To overcome such a problem, a mode decomposition technique was used in this study. The measured strain signal is decomposed into each modal component by using the empirical mode decomposition (EMD) as one of mode decomposition techniques. Then, the decomposed strain signals on each modal component are transformed into the modal displacement components. And the corresponding mode shapes can be also estimated by using the proper orthogonal decomposition (POD) from the measured strain signal. Thus, total displacement response could be predicted from combining the modal displacement components. To verity the mode decomposition technique suggested in this study, model experiments were conducted.

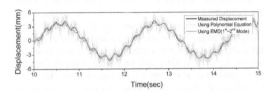

Figure 1. Comparison of displacement responses according to techniques of displacement estimation.

As a result of comparing displacement responses from the polynomial equation with those from the mode decomposition technique, it can be confirmed from Fig. 1 that more noises are removed from the use of mode decomposition technique than that of polynomial equation. The mode decomposition technique can remove noises because it estimated displacement responses using only the desired modes, but the existing method has difficulties in removing noises because it directly uses the measured signals. Therefore, the use of mode decomposition could reduce the noise effect more effectively. Additionally, by applying the FBG sensors with multi-point measurements not influenced by electric noise, it can be expected that the technique would be applicable to infrastructures.

REFERENCES

Nam-Sik Kim, and Nam-So Cho. 2004. "Estimating Deflection of a Simple Beam Model Using Fiber Optic Bragg-grating Sensors", *Experimental Mechanics*. 44(4): 433–439.

N.E. Huang, Z. Shen, S.R. Long, M.C. Wu, H.H. Shih, Q. Zheng, N.-C. Yen, C.C. Tung, and H.H. Liu. 1998. "The empirical mode decomposition and the Hilbert spectrum for nonlinear and non-stationary time series analysis", *Proceedings of the Royal Society of London A*. 903–995.

G. Berkooz, P. Holmes, and J.L. Lumley. 1993. "The Proper Orthogonal Decomposition in the Analysis of Turbulent Flows", *Annual Review of Fluid Mechanics*. 25: 539–575.

Safety, Reliability and Risk of Structures, Infrastructures and
Engineering Systems – Furuta, Frangopol & Shinozuka (eds)
© *2010 Taylor & Francis Group, London, ISBN 978-0-415-47557-0*

Damage assessment of beam by a quasi-static moving vehicular load

C.Y. Wang, C.K. Huang, Y.T. Zeng, C.S. Chen & M.H. Chen
Department of Civil Engineering, National Central University, Chungli, Taiwan, ROC

ABSTRACT

On the development of fast damage assessment techniques for bridge, the idea of using passing vehicle response to assess bridge condition has attracted a great academic interest among researchers. The concept is to use a vehicle moving over a bridge as a message carrier of the properties of the bridge.

In this paper, studies on beams of various damage states are conducted to identify the correlations among damages and the measured responses of the structure. The study uses a quasi-static moving vehicular load to reduce the complicating features arising out of dynamic interactions between vehicle and the bridge. Only deflection at the mid-span of structure is measured while the vehicle crosses the beam. After some investigations on these measured data, it is interested to find that the curvature influence line response due to a quasi-static moving vehicular load can be effectively used to identify the location of damage in the beam. Theoretical, numerical and experimental analyses have been conducted to verify this simple but effective damage assessment technique.

This relatively fast and simple damage assessment technique can be used for both the simply supported and the continuous beams. No previous condition state is required as a reference configuration for the damage assessment. The deflection curvature variations along the span are reliable indicators for damage detection. Different from vibration type damage assessment techniques, the quasi-static type moving load testing method can provide more clean data for the analysis. It also has the potential to reduce the unknown errors due to the incomplete modeling on structure-vehicle interaction effect and environmental factors into damage detection process.

Figure 3. Sensors deployed on the middle span for vehicle moving with constant speed.

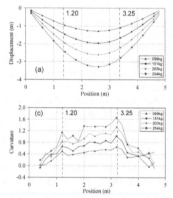

Figure 4. Experimentally measured influence lines and their second derivatives at the middle span of a damaged beam with double cracks of 7 cm depth and 2 mm width located at x = 1.20 m and 3.25 m from the left of the beam sustaining various two-axel wheel loads.

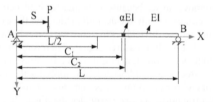

Figure 1. Simply supported beam with damage.

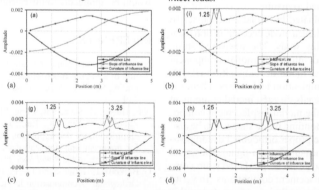

Figure 2. Influence line and its first, second derivatives at the middle span of a beam with cracks located at x = 1.25 m and 3.25 m from the left end of the beam sustaining a two-wheel loading by the numerical analysis. (a) Undamaged beam, (b) Single crack of 8 cm depth, (c) Double cracks of 8 cm and 7 cm depth, (d) Double cracks of 8 cm and 8 cm depth.

Organized Session (OS11) Health Monitoring, Structural Monitoring & Control

Safety, Reliability and Risk of Structures, Infrastructures and
Engineering Systems – Furuta, Frangopol & Shinozuka (eds)
© 2010 Taylor & Francis Group, London, ISBN 978-0-415-47557-0

Fundamental study of damage identification of bridge structures by using AE and sound monitoring data

T. Obata
Department of Environmental Systems Engineering, Osaka Prefectural College of Technology, Osaka, Japan

Y. Sato
Division of Built Environment, Graduate School of Hokkaido University, Sapporo, Japan

ABSTRACT

In recent years, the maintenance and rehabilitation of infrastructures have become a very important problem in all over the world. Also in Japan, the too many infrastructures were constructed for preparation and spread the transportation network from 50's to 70's. The most of infrastructures in Japan are middle-aged structures, also the bridges that more than 50 years active serviced are increase drastically in 2004 and these kind bridges will be over 50,000 in 2021. The developmental spending or investment for preparation of infrastructures such as construction of new transportation network is very difficult in Japan from the problem of economic condition and environmental impact. Therefore it has become a very important subject to aim at the life extension improvements of infrastructures.

The purpose of this study is to perform fundamental investigations to find the damage level and positions by using AE and sound monitoring data of bridges. In this study, steel bridge model and 32 concrete test pieces are prepared and the damage identification experiments are performed for these experimental model. The sound monitoring system in this study is composed of dynamic microphone sensor, digital audio interface and personal computer. This system is able to measure the eight channel's sound of each sensor and to record direct in PC. The sampling frequency is 44.1 kHz and 16 bit digital sound format is used in this study. The analyses for sound data are sound spectrogram and fractal dimension. As the parameters of damage identification, the fractal dimension is calculated from the figure of sound spectrogram. On the other hand, the AE monitoring system is made by tow AE sensors, DAQ card and mobile PC. In this system, the sampling frequency is 200 kHz and the number of sampling data is 32768. The analytical methods of AE data are Fourier spectrum and coherence. The AE sensors are set to near the damaged point on steel bridge model, coherence is calculated between non-damaged data and each of different damaged condition data. For concrete test piece, tow sensors are attached to the opposite side of excitation point symmetrically and coherence is calculated by using these tow data. The results of sound spectrogram experiments of the

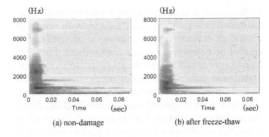

Figure 1. Sound spectrogram (concrete).

(a) non-damage   (b) after freeze-thaw

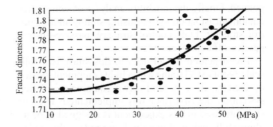

Figure 2. Fractal dimension – Concrete strength.

concrete test pieces are shown in Figure 1. And the relation of compressive strength of concrete test pieces and fractal dimensions are shown in Figure 2. The possibility of application for identifying steel and concrete damages is discussed by these results.

The major conclusions obtained in this study are summarized as follows: From the result of sound spectrogram, it is considered that the change of the frequency response and damping characteristic of sound data can be judged visually and simply. As for the fractal dimension, the significant change has appeared in comparison with non-damaged and damaged condition. The fractal dimension values are decreased, in the case of damages are increasing. The Fourier spectrum results from AE monitoring data is change by link of each damaged level. From the results of coherence, the decrease of the values of coherence in resonant frequency can be confirmed with the increase of damage. Therefore, it is considered that these monitoring methods in this study have possibility of damage identification for infrastructures.

Safety, Reliability and Risk of Structures, Infrastructures and
Engineering Systems – Furuta, Frangopol & Shinozuka (eds)
© 2010 Taylor & Francis Group, London, ISBN 978-0-415-47557-0

Performance and reliability analyses of semi-active equipment isolation system

Chin-Hsiung Loh & Yu-Chen Fan
Department of Civil Engineering, National Taiwan University, Taipei, Taiwan

In this study a combination of a conventional spectral acceleration seismic hazard curve with results of a suite of seismic response analysis of semi-active control of equipment system is examined. First, the hazard curves of floor acceleration spectrum are generated, and then an equipment acceleration response hazard curve is created which accounts for the dispersion on the input ground spectral acceleration. Since the equipment is expected to have control system when subjected to earthquake excitation, performance analysis of equipment isolation system is conducted in which different control strategy of the system is incorporated. Finally, the reliability analyses of semi-active equipment isolation system using sliding mode control strategy, LQR control strategy, passive-on and passive-off control are examined.

The structural model used in this study is a full-scale 3-story steel frame with a light equipment located on the first floor. A semi-active equipment isolation system on the first floor was implemented. The test structure used in this experiment is designed to be an almost full-scale prototype building and is subjected to a one-dimensional ground motion. In this experiment, a single magnetorheological (MR) damper and a friction pendulum-type isolator are installed between the equipment and the first floor. The friction coefficient μ is 0.0001. The natural period of the equipment is 2.77 sec.

Probability assessment of the equipment-isolation system was conducted in conjunction with different control methods. The relationship between the equipment peak response and floor response spectrum is developed through a number of dynamic analysis. Four different control methods are applied to the equipment: passive-off (0 Volt), passive-on (0.2 Volt and 1.2 Volt), and sliding mode control. Finally, the seismic hazard curves of equipment peak response and floor response spectrum were generated with respect to different control method.

REFERENCES

Cornell, C. A. et al. (2002), "Probabilistic basis for 2000 SAC Federal Emergency Management Agency steel moment frame guideline," ASCE, J. of Structural Engineering, 2002, 128(4), 526–533.

Fan, Y.C., Loh, C.H., Yang, J.N. and Lin, P.Y., "Experimental performance evaluation of an equipment isolation using MR-damper," Accept for publication in Earthquake Engineering and Structural Dynamics, July, 2008'

Mysilimaj, B., S. Gamble, D. Chin-Quee, A. Davies, "Base isolation technologies for seismic protection of museum artifacts," The 2003 IMAFA Annual Conference in San Francisco, CA., September 21–24, 2003.

Lin, P.Y., P. N. Roschke and C. H. Loh, "Hybrid base-isolation with magnetorheological dampers and fussy control," Structural Control and Health Monitoring 2007; 3: 384–405.

Loh, C. H., Wu, L.Y., Lin, P.Y., "Displacement control of isolated structures with semi-active control devices," Journal of Structural Control, 2003; 10: 77–100.

Luco, Nicolas and Cornell, C. A., "Effects of random connection fractures on the demands and reliability for a 3-story pre-Northridge SMRF structure," *Proceedings of 6-th US National Conference on Earthquake Engineering*, Seattle, May, 1998.

Ramallo, JC, Johnson EA, Spencer BF, "Smart base isolation systems," Journal of Engineering Mechanics, ASCE, 2002; 128: 1088–1099.

Singh, M. P., "Generation of seismic floor spectra", *Journal of the engineering mechanics division*, ASCE, Vol 106, 1975, 593–607

Yang, J. N., Wu, J. C., Kawashima, K. And Unjoh, S. "Hybrid control of seismic-excited bridge structures," J. of Earthq. Eng. and Structural Dynamics, Vol.24, 1995, 1437–1995.

Yang, J.N., Wu, J.C., Reinhorn, A. M. and Riley, M., "Control of sliding isolated buildings using sliding mode control," J. of Structural Eng., ASCE, Vol. 122, No. 2, 1996, 83–91.

Safety, Reliability and Risk of Structures, Infrastructures and Engineering Systems – Furuta, Frangopol & Shinozuka (eds)
© 2010 Taylor & Francis Group, London, ISBN 978-0-415-47557-0

System identification of non-linear hysteretic rubber-bearings based on experimental tests

Qiang Yin, X.M. Wang & Li Zhou
College of Aerospace Engineering, Nanjing University of Aeronautics & Astronautics, Nanjing, China

Jann N. Yang
Department of Civil & Environmental Engineering, University of California, Irvine, CA, USA

ABSTRACT

Rubber-bearing isolation systems have been used in buildings and bridges. These base isolation systems will become more popular in the future due to their ability to reduce significantly the structural responses subject to earthquakes and other dynamic loads. To ensure the integrity and safety of these base isolation systems, a structural health monitoring system is needed. One important problem in the structural health monitoring is the identification of the system and the detection of damages. This problem is more challenging for the rubber-bearing isolation systems because of their non-linear behavior. In the literature, several hysteretic models for describing the dynamic behavior of rubber-bearings have been proposed, including piecewise-linear hysteretic models, polynomial hysteretic models, curvilinear hysteretic models, etc. Likewise, various system identification techniques in the time domain have been developed for nonlinear structural systems, such as the least-square estimation (LSE), the extended Kalman filter (EKF), the sequential non-linear least-square estimation (SNLSE), etc.

The purpose of this paper is to present the experimental study for the system identification of non-linear hysteretic rubber-bearing isolators. In this study, the Bouc-Wen model is selected to describe the non-linear behavior of rubber-bearings, which has the advantages of being smooth-varying and physically motivated. Further, the extended Kalman filter (EKF) approach (Yang et al, 2006a, b; Zhou et al. 2008) is used to identify the unknown non-linear parameters of the Bouc-Wen model for the rubber-bearing isolators. Experimental tests of a base-isolated structural model on the shake table have been conducted to demonstrate the validity of the proposed approach.

In the experimental tests, rubber-bearings GZN110 were used as the base isolators and the test model consisted of a mass supported by four rubber bearings. Two earthquake records have been used for the excitations, including the El Centro and Kobe earthquakes. During the tests, one acceleration sensor (PCB3701G3FA3G) and one displacement sensor (ASM WS10-250-10V-L10) were installed to measure the responses of the mass. The sampling frequency of all measurements is 200 Hz. The measured acceleration response data and the extended Kalman filter (EKF) approach are used to estimate the unknown parameters of the Bouc-Wen model for the rubber-bearing isolators. Based on the estimated parameters of the Bouc-Wen model, the displacement response of the mass can be estimated. The estimated displacement response is then compared with the measured displacement response data to show the validity and accuracy of the proposed approach. Experimental results obtained in this study demonstrate that: (i) the Bouc-Wen model is quite suitable for describing the non-linear behavior of rubber-bearing isolators, and (ii) the extended Kalman filter (EKF) approach is effective in identifying the non-linear hysteretic parameters.

REFERENCES

Yang, J.N., Lin, S.L., Huang, H.W. & Zhou, L. 2006a. An Adaptive Extended Kalman Filter for Structural Damage Identification. *Journal of Structural Control and Health Monitoring*, 13: 849–867.

Yang, J.N., Pan, S.W., & Huang, H.W. 2006b. An Adaptive Extended Kalman Filter for Structural Damage Identification II: Unknown Inputs. *Journal of Structural Control and Health Monitoring*, 14(3), 497–521.

Zhou L., Wu S.Y., & Yang J.N. 2008. Experimental study of an adaptive extended kalman filter for structural damage identification, Journal of Infrastructure Systems, ASCE, 14(1): 42–51.

Safety, Reliability and Risk of Structures, Infrastructures and
Engineering Systems – Furuta, Frangopol & Shinozuka (eds)
© 2010 Taylor & Francis Group, London, ISBN 978-0-415-47557-0

Live load model for urban highway bridges from BWIM data

M. Kawatani, Y. Nomura & C.-W. Kim
Department of Civil Engineering, Kobe University, Kobe, Japan

M. Nakata
Kobe City, Kobe, Japan

ABSTRACT

In order to establish reliability-based design methods covering the limit state design and the load-resistance factor design (LRFD), a clear grasp of probability characteristics of each load is important. This paper especially focuses on a live load that is a principal load for bridges. The effects of traffic congestion to live load model of bridges are studied. Usually, the characteristics of traffic depend on industrial structure of areas. Moreover the characteristics are different between a toll road such as urban highway bridge and national roadway. This study investigates how

Table 1. Mean and standard deviation of reaction force.

	Reaction force			
	Ordinary congestion		Accidental congestion	
	Mean (tf)	Standard deviation (f)	Mean (tf)	Standard deviation (tf)
Hanshin. Ex	209.6	12.2		
Sakai.Br	260.8	8.8	292.4	12.3
Koushi.Br	344.3	14.4	382.5	17.2

the change of legal live loads affects the extreme value distribution of reaction force. The live loads are constructed based on the available statistical data on truck loads as well as from truck surveys and weight-in-motion measurements of Sakai Bridge in Nagano prefecture and Koushi Bridge in Chiba prefecture (Tamakoshi, et al. 2006). Random traffic data for one month are obtained by means of Monte Carlo simulation using the statistical data of traffic load, and are used for simulating the reaction force at the intermediate support of a three-span continuous steel box girder bridge with the length of 80 m. Probability density function of reaction forces under the ordinary congestion and the accidental congestion is shown in Figure 1. Table 1 depicts the mean and standard deviation of reaction forces under each of traffic congestions. It is observed that the mean value of the reaction force under the accidental congestion is 1.12 times higher than that obtained under the ordinary congestions. And also, reaction force of 356.24tf was obtained in case of using B-live load of Japan Road Association (JRA) code under all the same experimental conditions. It is observed from these results that the reaction force under the accidental congestion obtained from Koushi Bridge is larger than that obtained by using B-live load of JRA code. Observations through the study based on the BWIM data demonstrate that the live load effect varies according to the traffic condition and the area in which the bridge is located.

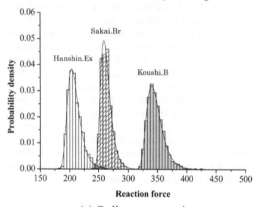

(a) Ordinary congestion

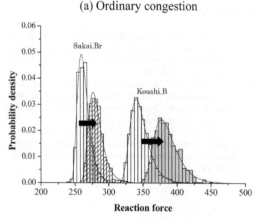

(b) Accidental congestion

Figure 1. Probability density function of reaction force.

REFERENCES

Tamakoshi, T, Nakasu, K and Ishio, M. 2006, Actual data of live loads on highway bridges, Technical note of National Institute for Land and Infrastructure Management No. 295. (in Japanese).

Safety, Reliability and Risk of Structures, Infrastructures and
Engineering Systems – Furuta, Frangopol & Shinozuka (eds)
© 2010 Taylor & Francis Group, London, ISBN 978-0-415-47557-0

Effects of uncertainties on seismic response parameters of reinforced concrete frame

M. Dolsek
University of Ljubljana, Ljubljana, Slovenia

ABSTRACT

Incremental Dynamic Analysis (IDA) has been recently extended in order to quantify also the influence of epistemic uncertainty on the seismic response parameters and not only the record-to-record variability, as applied in the IDA. In addition to the set of ground motion records, which is used in the IDA,

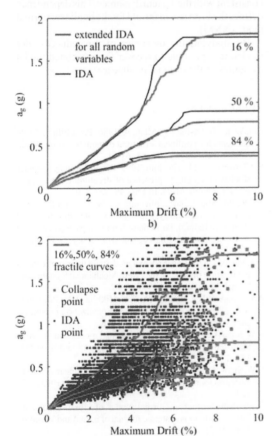

also the set of set of structural models is used in the extended IDA. It is determined by utilizing the Latin Hypercube Sampling technique. In the paper the extended IDA is summarized and demonstrated through an example of a four-storey plane RC frame for which the pseudo-dynamic tests were performed in full scale at ELSA Laboratory in Ispra. Different sources of uncertainty, such as, storey masses, strength of the concrete and that of reinforcing steel, effective slab width, damping and the model for determination of the initial stiffness and ultimate rotation in plastic hinges of beams and columns, are considered in the example.

The results of the extended IDA are the IDA curves, which are presented in Figure 1a together with 16%, 50% and 84% fractile curves, which reflect aleatory and epistemic uncertainty. For comparison, the fractile curves of the extended IDA are compared with fractile curves of the IDA for deterministic model (Fig. 1b). Based on these results, it can be observed that the fractile IDA curves of the extended IDA practically do not deviate from the fractile IDA curves, which are determined by employing the deterministic model. This observation is valid mostly for the 84% fractile curves and for the other curves within a limited range of peak ground acceleration. This is an interesting result, which leads to the conclusion, at least for the presented example, that the epistemic uncertainties do not significantly influence the summarized seismic response parameters, at least in the range near collapse. However, the median collapse capacity is reduced in the case of the extended IDA analysis.

REFERENCES

Vamvatsikos, D., Cornell, C.A., 2002. Incremental Dynamic Analysis. *Earth. Engng. and Struc. Dyn.*, 31: 491–514.
Vorechovsky, M., Novak, D., 2003. Statistical correlation in stratified sampling. In: Der Kiureghian A, Madant S, Pestana JM (Eds.), ICAPS 9, *Proc. Of International conference on Applications of Statistics and Probability in Civil Engineering*, Millpress, Rotterdam, San Francisco, pp. 119–124.
Dolsek M. 2008. Incremental Dynamic Analysis with consideration of modeling uncertainties. *Earth. Engng. and Struc. Dyn.*, doi: 10.1002/EQE.869.

Figure 1. a) the fractile IDA curves, IDA points and collapse points of the extended IDA analysis and b) comparison between the fractile IDA curves and the fractile IDA curves of the extended IDA analysis.

Safety, Reliability and Risk of Structures, Infrastructures and Engineering Systems – Furuta, Frangopol & Shinozuka (eds)
© 2010 Taylor & Francis Group, London, ISBN 978-0-415-47557-0

Dynamics of nonlinear disordered structures by means of nonlinear normal modes

L. Facchini & M. Rizzo
University of Florence, Italy

ABSTRACT

Randomness can affect structural systems by several points of view; in particular, randomness can affect the forcing process that a system undergoes, or might as well influence the structural parameters.

The study of disordered systems, as such structures are usually referred to, is a point of concern in specialized literature: a fair amount of methods to investigate the dynamics of disordered structures can in fact be found.

One of the most common methods is taken into consideration in the present work: the first application is described in Liu et al., 1986, and subsequently enhanced in Chiostrini and Facchini, 1999. It can be classified as a perturbation method and makes use of sensitivity vectors to evaluate the first two moments of the response.

Unfortunately, a severe drawback of the method is that the number of degrees of freedom of the examined structure grows rapidly for increasing number of random parameters, thus leading to the solution of very large (non) linear systems.

The idea that is introduced in the present work is to investigate the possibility to compute and apply the concept of nonlinear modal shapes in order to reduce the number of degrees of freedom. It has recently been developed by Rizzo (2007) in his Ph.D. dissertation.

It is possible to define the normal non-linear (NNMs) modes only as particular synchronous periodic solutions of the non-linear motion equations without looking for any link of such motions to the (linear) principle of superposition.

For free vibration problems one uses system modes to construct reduced order models, and these techniques have been well developed for both linear and nonlinear systems by Vakakis (1997) and by Vakakis et al. (1996).

One such technique, introduced by Shaw and Pierre (1991, 1993, 1993), defines the normal mode of a nonlinear oscillatory system in terms of invariant manifolds in the phase space that are tangent to the linear eigen-modes at the equilibrium point. In such a formulation, a master mode is selected, and the normal mode is constructed by a formulation in which the remaining linear modes of the system, i.e., the slave modes, depend on the master mode in a manner consistent with the system dynamics. This dependence defines the invariant manifold for the nonlinear normal mode (NNM).

By studying the dynamics of the reduced-order model, it is possible to recover the associated modal dynamics of the original nonlinear system.

REFERENCES

Liu, W. K., Belytshko, T., Mani, A. 1986. Probabilistic finite elements for nonlinear structural dynamics. Comp. Meth. Appl. Mech. Engng., 56, pp. 61-81

S. Chiostrini and L. Facchini, 1999. Response analysis under stochastic loading in presence of structural uncertainties. Int. J. for Num. Meth. In Eng. 46, 853-870.

M. Rizzo, Study and application of dynamic reduction of nonlinear structural systems by means of nonlinear modal shapes (in Italian), Ph.D. dissertation, Department of Civil & Environmental Engineering, Univarsity of Florence, 2007.

A.F. Vakakis, Non-linear normal modes and their applications in vibration theory: an overview, Mechanical Systems and Signal Processing 11 (1997) 3–22.

A.F. Vakakis, L.I. Manevitch, Y.V. Mikhlin, V.N. Pilipchuk, A.A. , Zevin, Normal Modes and Localization in Nonlinear Systems, Wiley, New York, 1996.

S.W. Shaw, C. Pierre, Non-linear normal modes and invariant manifolds, Journal of Sound and Vibration 150 (1991) 170–173.

S.W. Shaw, C. Pierre, Normal modes for non-linear vibratory systems, Journal of Sound and Vibration 164 (1993), 85–124.

S.W. Shaw, C. Pierre, Normal modes of vibration for nonlinear continuous systems, Journal of Sound and Vibration 169 (1993) 319–347.

Safety, Reliability and Risk of Structures, Infrastructures and
Engineering Systems – Furuta, Frangopol & Shinozuka (eds)
© 2010 Taylor & Francis Group, London, ISBN 978-0-415-47557-0

Finite element response sensitivity and reliability analysis of structures and/or geotechnical systems

Q. Gu
Department of Structural Engineering, University of California at San Diego, La Jolla, CA, USA

M. Barbato
Department of Civil & Environmental Engineering, Louisiana State University at Baton Rouge, Baton Rouge, LA, USA

J.P. Conte
Department of Structural Engineering, University of California at San Diego, La Jolla, CA, USA

ABSTRACT

In the last decade, Finite Element (FE) reliability methods have emerged as powerful tools to perform probabilistic performance assessment of structures and/or geotechnical systems. In the context of Performance-Based Earthquake Engineering (PBEE), these methods combine FE modeling and seismic response analysis of Soil-Foundation-Structure Interaction (SFSI) systems with state-of-the-art methods in sensitivity and reliability analysis.

FE response sensitivity analysis is a crucial component of FE reliability analysis and is concerned with the relationship between modeling parameters (e.g., geometric, material and discrete loading parameters), and system behavior. FE response simulations of realistic large-scale models of an engineering system must be complemented by extensive studies on response sensitivities to system parameters in order to achieve a broader and deeper under-standing of the system behavior. Accurate and efficient FE response sensitivity computation is required in any application involving gradient-based optimization algorithms. In this paper, Direct Differentiation Method (DDM) for FE response sensitivity computation is adopted and extended to SFSI systems [1, 2]. The DDM consists of differentiating exactly the FE response numerical algorithm with respect to all sensitivity parameters. The DDM-based response sensitivity computation algorithm involves the various hierarchical layers of FE response analysis, namely: (1) structure level, (2) element level, (3) integration point level (or section level), and (4) material level.

In PBEE, structural reliability analysis methods have been employed to estimate the probability of occurrence of various failure events in both structural and SFSI systems (i.e., the failure probability). During an earthquake, the structural/SFSI system usually enters its nonlinear behavior before exceeding a limit-state condition defining failure. Typically, the nonlinear dynamic response of a structural system

cannot be computed in closed-form, particularly when the complex interaction with the foundation and the soil needs to be taken into account. In this case, FE analysis is used to predict the SFSI system nonlinear dynamic response. Modern FE reliability methods are obtained by merging advanced reliability methods and state-of-the-art FE modeling. In this paper, several reliability analysis methods based on the Design Point (DP) search are considered to solve both time-invariant and time-variant reliability problems for SFSI systems, including First- and Second-Order Reliability Method (FORM and SORM), Importance Sampling (IS) method, Orthogonal Plane Sampling (OPS) method, and Mean Out-Crossing Rate (MOCR) analysis based on FORM approximation and on OPS approximation.

This paper considers as benchmark examples both two- and three-dimensional models of an SFSI system consisting of a reinforced concrete frame on layered clay soil. Realistic nonlinear constitutive models are employed for concrete, reinforcement steel and soil materials. FE response sensitivity results are presented and used to evaluate the relative importance of the various material modeling parameters in affecting the system response. Time-invariant and time-variant reliability analyses are performed for the benchmark structure and results obtained from different FE reliability methods are compared in terms of accuracy and computational cost. The use of novel hybrid reliability method, the DP-RS-Sim method, is also illustrated for the time-variant reliability analysis relative to an idealized nonlinear structural model.

REFERENCES

Q. Gu, "Finite element response sensitivity and reliability analysis of soil foundation structure interaction systems", *Ph.D. Dissertation*, UCSD, La Jolla, CA, (2008).

M. Barbato, Q. Gu and J.P. Conte, "Framework for finite element response sensitivity and reliability analyses of structural and geotechnical systems", *Proceedings*, First European Conference on Earthquake Engineering and Seismology, Geneva, Switzerland, (2006).

Safety, Reliability and Risk of Structures, Infrastructures and Engineering Systems – Furuta, Frangopol & Shinozuka (eds)
© 2010 Taylor & Francis Group, London, ISBN 978-0-415-47557-0

Stochastic bridge-vehicle interaction problem with Gaussian uncertainties

S.Q. Wu & S.S. Law

Civil and Structural Engineering Department, Hong Kong Polytechnic University, Hong Kong, People's Republic of China

ABSTRACT

Bridge condition assessment is very popular amongst researchers in recent years, and assessment under operation load is the centre of research in practice. Vehicular axle load is one of the important factors influencing the durability and safety of the bridge structure. When a vehicle passes over a bridge deck, an impact or dynamic amplification factor will be induced which needs to be considered in the response analysis for the assessment.

The dynamic response of a bridge structure subject to moving vehicular loads has been studied for decades. Various bridge-vehicle interaction models are proposed by researchers. The deterministic approach for the analysis can be approximately classified into two kinds: modal superposition technique (Zhu & Law 2003) and finite element method (Henchi *et al.* 1998), with the latter capable of handling a more complex structure. Although most of the methods consider the road roughness as an uncertain factor in the bridge-vehicle interaction problem, yet it is conventionally treated in a deterministic manner, i.e. the road roughness is considered as deterministic samples of irregular profile according to its power spectral density defined in the ISO standard. The bridge-vehicle system often exhibits an inherent randomness. The conventional deterministic analysis generally represents only an "approximation" of the actual reality due to unavoidable uncertainties in the structural properties as well as in the loading processes. Stochastic analysis should be performed instead for the bridge-vehicle interaction problem. Though the topic of dynamic responses of bridge structure under random moving forces has been researched by many engineers (Lin 2006, Seetapan & Chucheepsakul 2006), there is seldom research work (Fryba *et al.* 2003) done on the dynamic analysis of bridge-vehicle interaction problem with random excitation forces applied on the bridge structure which also contains random system parameters. This study is more realistic than the deterministic approach and it itself very important for the subsequent reliability analysis in the structural safety assessment.

A new method for the dynamic analysis of the bridge-vehicle interaction problem considering uncertainties is developed in this paper. The bridge is modeled as a simply supported Bernoulli-Euler beam with Gaussian random elastic modulus and mass density of material with moving forces on top. These forces have time varying mean values and a coefficient of variation at each time instance and are considered as Gaussian random. The mathematic model of the bridge-vehicle system is established using the spectral stochastic finite element method (Ghanem & Spanos 1991). The random Karhunen-Loéve components of the responses can be obtained directly from the Karhunen-Loéve components of the random forces using the Newmark-β method. The response statistics are subsequently derived. Numerical simulation is given to verify the proposed stochastic approach using Monte Carlo simulation with good agreements.

REFERENCES

Fryba, L., Nakagiri, S. & Yoshikawa, N. 2003. Stochastic finite elements for beam on a random foundation with uncertain damping under a moving force. *Journal of Sound and Vibration* 163(1): 31–45.

Ghanem, R. & Spanos, P.D. 1991. *Stochastic Finite Elements: A Spectral Approach.* Springer-Verlag New York Inc.

Henchi, K., Fafard, M., Talbot, M. & Dhatt, G. 1998. An efficient algorithm for dynamic analysis of bridges under moving vehicles using a coupled modal and physical components approach. *Journal of Sound and Vibration* 212(4): 663–683.

Lin, J.H. 2006. Response of a bridge to moving vehicle load. *Canadian Journal of Civil Engineering* 33(1): 49–57.

Seetapan, P. & Chucheepsakul, S. 2006. Dynamic responses of a two-span beam subjected to high speed 2DOF sprung vehicles. *International Journal of Structural Stability and Dynamics* 6(3): 413–430.

Zhu, X.Q. & Law, S.S. 2003. Dynamic behavior of orthotropic rectangular plates under moving loads. *Journal of Engineering Mechanics* 129(1): 79–87.

Safety, Reliability and Risk of Structures, Infrastructures and
Engineering Systems – Furuta, Frangopol & Shinozuka (eds)
© 2010 Taylor & Francis Group, London, ISBN 978-0-415-47557-0

Estimation of uncertain parameters using static pushover methods

M. Fragiadakis

Institute of Structural Analysis & Seismic Research, National Technical University of Athens, Athens, Greece

D. Vamvatsikos

Metal Structures Laboratory, National Technical University of Athens, Athens, Greece

Following recent guidelines (e.g. FEMA-350) seismic performance uncertainty is an essential ingredient for Performance-based earthquake engineering (PBEE). Uncertainty refers to both aleatory uncertainty, associated with the random record-to-record variability, and also to epistemic uncertainty primarily introduced by modeling assumptions or errors. A methodology for the performance-based estimation of the dispersion introduced by parameter uncertainties is developed. The methodology proposed provides an inexpensive alternative to the use of tabulated values, or to performing a series of time-consuming nonlinear response history analyses to obtain parameter uncertainty. As a testbed, the well-known 9-story LA9 2D steel frame is employed using beam-hinges with uncertain backbone properties. The properties of the backbone are fully described by six parameters, which are considered as random variables with given mean and standard deviation values. Using point-estimate methods, first-order-second-moment techniques and latin hypercube sampling with Monte Carlo simulation, the pushover curve is shown to be a powerful tool that can help accurately estimate the uncertainty in the seismic performance. Coupled with the SPO2IDA tool, such estimates can be applied at the level of the results of nonlinear dynamic analysis, allowing the evaluation of seismic capacity uncertainty even close to global dynamic instability. In summary, the method presented can inexpensively supply the uncertainty in the seismic performance of first-mode dominated buildings, offering an estimator of the accuracy of typical performance calculations.

Safety, Reliability and Risk of Structures, Infrastructures and
Engineering Systems – Furuta, Frangopol & Shinozuka (eds)
© 2010 Taylor & Francis Group, London, ISBN 978-0-415-47557-0

Estimation of extreme deflection for a beam with uncertain parameters

A. Naess

Centre for Ships and Ocean Structures & Department of Mathematical Sciences,
Norwegian University of Science and Technology, Trondheim, Norway

Nilanjan Saha

Centre for Ships and Ocean Structures, Norwegian University of Science and Technology, Trondheim, Norway

ABSTRACT

Structures, (e.g., windmills, lighthouse and towers) are
usually subjected to harsh environmental conditions
such as high winds, severe sea states, etc. The response
of such structures under extreme loading conditions
is of prime importance in order to assess the performance of structures on/after such extreme weather
condition. Such structures are highly non-linear and
made up of many structural components. The individual structural components are modelled as jointed
beam or plate elements. Now in traditional analysis
of structural systems, it is usually assumed that the
material and geometrical properties of the system in
question are constant. Although such an assumption
have been a basis for many practical problems, it is
evident that there always exists some amount of randomness in systems properties (e.g. mass density, cross
sectional properties, moduli of elasticity).

In this paper, we calculate the extreme statistical
response of a simply supported beam (modelled as
Euler-Bernoulli one) under stochastic loads and spatially varying random material properties. The loading
on the beam is a concentrated force which follows a
trajectory as the response of a linear oscillator under
white noise. The Young's modulus and mass density of
the beam is assumed to follow a log-normal probability distribution. The beam response (under initial and
boundary conditions) is obtained using central difference method for an ensemble of samples. An extreme
value method is used in obtaining statistical extremes
using time-series obtained through Monte Carlo simulations. The key quantity in such statistical extremes
is the mean upcrossing rates which can be numerically
obtained by counting the number of upcrossings. The
present approach is restricted to cases where the Gumbel distribution is the appropriate asymptotic extreme
value distribution. We then use optimized linear fit and
extrapolation on a double logarithmic scale for prediction of the mean upcrossing rates and thus extreme
response statistics. Numerical analysis shows good
accuracy for the method when applied to structural
components.

General Session (Stochastic Computational Mechanics & Stochastic Finite Elements & Stochastic Processes and Fields)

Safety, Reliability and Risk of Structures, Infrastructures and
Engineering Systems – Furuta, Frangopol & Shinozuka (eds)
© 2010 Taylor & Francis Group, London, ISBN 978-0-415-47557-0

Sensitivity analysis of the tightening of the body-bonnet flange joint in a globe valve by using a non intrusive method

M. Berveiller
EDF R&D Division, Site des Renardières, Moret sur Loing, France

F. Termens
IFMA, Aubière, France

J.P. Mathieu
EDF R&D Division, Site des Renardières, Moret sur Loing, France

A Globe valve is a type of valve used for isolating a piping part inside a circuitry. It generally consists of a movable element and a stationary seat which are relatively moved in order to create flow shutoff. The bonnet is connected to the body and provides the containment of the fluid that is being controlled. The aim of this study is to perform a sensitivity analysis of the effect of tensioning heterogeneities between studs of the body-bonnet flanged joint on the local pressure-tightness repartition between the sealing surfaces.

Sensitivity of a model aims at quantifying the relative importance of each input parameter [1]. Different methods of analysis are available such as the perturbation method, simulation methods and stochastic response surfaces (also called non intrusive finite element methods [2]). The latter is chosen hereand applied together with the Sobol' indices. Stochastic response surfaces allow to expand the response of a system, whose input parameters are random variables, onto a specific polynomial basis, which is called the *polynomial chaos*. Using the non intrusive approach, the expansion coefficients are obtained by evaluating the response of the system for selected values of the input parameters, then applying an analytical post-processing of the corresponding output. In this study a regression method is chosen to evaluate the expansion coefficients of the response: the error between the exact response (obtained with the evaluation of the model) and the approached response (evaluated with the polynomial approximation) is minimized. Using this expansion of the response onto the polynomial chaos, a sensitivity analysis is easily accessible with analytical computations from the expansion coefficients.

In this study, the tightness of the valve is modeled with the finite element code Code_Aster. Different type of tightening of the stud are compared.

REFERENCES

B. Sudret. Global sensitivity analysis using polynomial chaos expansion (2006). Structural Safety, submitted for publication.

M. Berveiller. *Eléments finis stochastiques : approches intrusive et non intrusive pour des analyses de fiabilité*. PhD thesis, Université Blaise Pascal – Clermont Ferrand, 2005.

Safety, Reliability and Risk of Structures, Infrastructures and Engineering Systems – Furuta, Frangopol & Shinozuka (eds)
© 2010 Taylor & Francis Group, London, ISBN 978-0-415-47557-0

A novel hybrid method for the simulation of non-homogeneous non-Gaussian stochastic processes and fields

G. Stefanou, V. Papadopoulos, N.D. Lagaros & M. Papadrakakis
Institute of Structural Analysis & Seismic Research, National Technical University of Athens, Greece

ABSTRACT

The numerical simulation of non-Gaussian stochastic processes and fields has recently received considerable attention in the field of computational stochastic mechanics. However, most of the existing work on non-Gaussian processes and fields has been devoted to the stationary/homogeneous case. Limited work is available for samples with temporally or spatially varying marginal probability density function (PDF) and/or spectral density function (SDF) (e.g. Ferrante et al. 2005). This kind of stochastic processes and fields is very useful e.g. for the simulation of the material properties and microstructure of functionally graded composites or the accurate representation of initial geometric imperfections in shell buckling studies (Ferrante et al. 2005, Papadopoulos & Papadrakakis 2005). The standard translation process concept (Grigoriu 1998) has been extended to the non-stationary case by Ferrante et al. (2005) where it is shown that the upper bound of the target covariance function can be less than 1 and thus the issue of possible incompatibility between the marginal PDF and the correlation structure of a translation process becomes more important in this case.

Recently, an enhanced hybrid method has been proposed for the simulation of homogeneous non-Gaussian stochastic fields with prescribed target marginal distribution and SDF (Lagaros et al. 2005). This approach is based on the translation field concept, but uses an extended empirical non-Gaussian to non-Gaussian mapping for the generation of a non-Gaussian field having the prescribed characteristics. In this way, the possible incompatibility between the marginal distribution and the correlation structure of a translation field is surpassed and an algorithm covering a wider range of non-Gaussian fields is produced. The function fitting ability of Neural Networks (NN) is employed to approximate the power spectrum of the underlying Gaussian field and the target non-Gaussian distribution and SDF are matched with remarkable accuracy even in the case of narrow-banded fields with very large skewness.

In this work, the enhanced hybrid method is extended to the non-homogeneous case i.e. the simulation of non-homogeneous non-Gaussian stochastic processes and fields with prescribed marginal distribution and temporally or spatially varying SDF. The updated algorithm makes use of the spectral representation method coupled with evolutionary spectra theory in order to generate sample functions of the underlying Gaussian stochastic field. The computational efficiency and various features of the algorithm are demonstrated with the simulation of a stochastic field following a highly skewed lognormal distribution.

REFERENCES

Ferrante, F.J., Arwade S.R. & Graham-Brady, L.L. 2005. A translation model for non-stationary, non-Gaussian random processes. *Probabilistic Engineering Mechanics* 20: 215–228.

Grigoriu, M. 1998. Simulation of stationary non-Gaussian translation processes. *J. of Engineering Mechanics (ASCE)* 124: 121–126.

Lagaros, N.D., Stefanou, G. & Papadrakakis, M. 2005. An enhanced hybrid method for the simulation of highly skewed non-Gaussian stochastic fields. *Computer Methods in Applied Mechanics and Engineering* 194: 4824–4844.

Papadopoulos, V. & Papadrakakis, M. 2005. The effect of material and thickness variability on the buckling load of shells with random initial imperfections. *Computer Methods in Applied Mechanics and Engineering* 194: 1405–1426.

Safety, Reliability and Risk of Structures, Infrastructures and Engineering Systems – Furuta, Frangopol & Shinozuka (eds)
© 2010 Taylor & Francis Group, London, ISBN 978-0-415-47557-0

A spectral stochastic finite element method for modal analysis of structures with uncertain material properties

J. Ahmad, Ph. Bressolette & A. Chateauneuf
LaMI, UBP&IFMA, Campus de Clermont-Ferrand, Les Cézeaux, France

ABSTRACT

The problem of handling model uncertainty in the framework of dynamic structural analysis techniques has been a major concern of the literature in the past 15 years (Schueller (2001)). At the origin of this interest lies the unavoidable uncertainty with which the designer has to cope at the model definition stage. In this context, it appears that the handling of the uncertainty in structural models is a natural and necessary extension of present structural analysis techniques.

This paper presents a new technique for the modal analysis of structures with random material parameters. This technique is applied to the calculation of statistics of the dynamic characteristics (e.g. eigenvalues) of mechanical structures. The proposed technique is an extension of the Spectral Stochastic Finite Element Method (SSFEM) (Ghanem & Spanos (1991)). The spectral SFEM is based on a Karhunen-Loève expansion of the random fields involved in the stochastic description of the uncertain model. The truncation of the expansion generates a finite set of orthonormalised random variables the size of which corresponds to the dimension of the stochastic part of the analysis. The required stochastic response is projected on the polynomial chaos obtained with the random variables of the Karhunen-Loève basis. An explicit expression of the random response, analogous to a response surface, is finally obtained and enables the estimation of the statistical moments of the response indicator.

The proposed method is applied to a simple mechanical system composed of rods with random material parameters and the results compared to Monte Carlo simulations. The obtained results reported in Table 1 show good convergence of this approach when increasing the number of terms of the KL decomposition and the polynomial order.

As the frequency of the random fluctuations of the material properties increases, it is clear that additional terms are required in the Karhunen-Loève representation, and the size of the associated chaos expansion grows significantly.

In conclusion, it seems that, for systems whose parameters exhibit high variability, the SSFEM approach provide a suitable representation of the

Table 1. Influence of M and p on the standard deviation of the first eigenvalue (b = 50 m).

M	P	P	σ(SSFEM) $\times 10^3$	Relative error
1	5	6	197.1	3.30 %
2	5	21	203.17	0.32 %
3	3	20	199.31	2.22 %
4	3	70	201.55	1.12 %

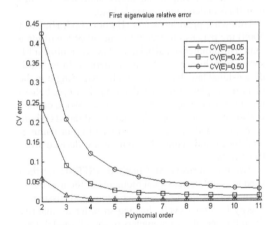

Figure 1. Relative error of CV of the first eigenvalue versus polynomial order for differents values of $CV(E)$.

variability of the system eigenvalues as illustrated in figure 1. Clearly, the higher the standard deviation of the stochastic processes and the shorter the correlation length, the more terms in the polynomial chaos are needed to compute the statistics of the response with accuracy. The problem of the influence of the random variables on the probability distribution function (PDF), required in simulation techniques is also addressed.

REFERENCES

Ghanem, R.G. & Spanos, P.D. 1991. Stochastic finite elements: a spectral approach, Springer, Berlin.
Schueller, G.I. 2001. Computational stochastic mechanics: recent advances, Computers & Structures 79: 2225–2234.

Safety, Reliability and Risk of Structures, Infrastructures and
Engineering Systems – Furuta, Frangopol & Shinozuka (eds)
© 2010 Taylor & Francis Group, London, ISBN 978-0-415-47557-0

SFE analysis of a timber frame undergoing a seismic load sensitivity analysis of a nail plate joint

J. Humbert, J. Baroth & L. Daudeville
Joseph Fourier University, Grenoble, France

C. Faye
FCBA Bordeaux, Bordeaux, France

M. Yasumura
Shizuoka University, Shizuoka, Japan

L. Davenne
ENS Cachan, Cachan, France

ABSTRACT

This paper presents a Stochastic Finite Element (SFE) analysis of a nail plate joint used in prefabricated timber roof-trusses. This work is part of a larger study aiming at achieving a sensitivity analysis on a full roof-truss undergoing a seismic load.

Following previous works (Baroth 2007, Humbert 2008) the paper introduce a FE model of a nail plate joint. The bidimensional behavior law is a nonlinear FE spring with hysteresis (Figure 1) derived from a unidimensional law (Richard 2003) by extending it in the plane of the plates to deal with the non isotropic behavior observed during experimental tests. Wood contact is also taken into account. The law is validated on both monotonic and cyclic test results and input parameters' mean and standard deviation are derived from those tests.

Two non-intrusive SFE methods (SFEM) are then used to compute statistical moments of the maximum strength F of the joint and its cumulated dissipated energy E_d under cyclic loading in order to quantify the influence of the parameters of the FE model. One is based on Lagrange polynomials (Baroth 2007) using 4 collocation points while the second uses the Stroud-3 formulas (Bressolette 2007). Results from the later are not shown here, being rather irrelevant.

Table 1. Influence of uncertainties on the maximal force F of the nail plate joint, using the Lagrange method.

X	σ (X)	μ (X)	C_v (X) %	σ (F) kN	C_v(F) %
K_0 (MN/mm)	109.1	30.3	27.8	0.04	0.14
P_0 (kN)	25.5	3.4	13.4	3.4	12.6
K_1 (kN/mm)	809	207	25.5	0.37	1.39
D_y (mm)	0.4	0.2	50.0	0.04	0.14
P_c (kN)	2.10	0.47	22.4	0.04	0.14

Table 1 shows the influence on the coefficient of variation of the maximal strength F considering uncertainties on chosen parameters of the law modeled with lognormal random variables (r.v.): initial stiffness K_0, tangent stiffness K_1 at peak and associated force P_0 at null displacement, yield displacement D_y, and force at null displacement for cyclic loops P_c. The mean of F remains constant, equal to $\mu(F) = 27.02$ kN. Relevant parameters for F are thus the P_0 force and in a minor extent the slope of the tangent K_1 at peak. Other parameters do not influence the strength in a significant way. Similar results during each traction cycle are then presented for the dissipated energy E_d, which is influenced significantly by P_0, K_0 for small displacements, and P_c for displacements over D_y when damage occurs.

REFERENCES

Baroth, J. et al. 2007. An efficient SFE method using Lagrange Polynomials: application to nonlinear mechanical problems with uncertain parameters. *Comp. Meth. Appl. Mech. Engrg. 196: 4419-4429.*

Bressolette, P. et al. 2007. High-dimensional integration formulas for efficient SFE methods. *The 10th International Conference on Applications of Statistics and Probability (ICASP10), Tokyo, 3 August 2007.*

Richard, N. et al. 2003. Prediction of seismic behaviour of wood-framed shear walls with openings by pseudodynamic test and FE model. *J. Wood Sci. 49: 145-151.*

Humbert, J. et al. 2008. SFE analysis of a timber frame undergoing a seismic load – Sensitivity analysis of a bolted timber joint. *The 10th World Conference on Timber Engineering (WCTE), Miyazaki, Japan.*

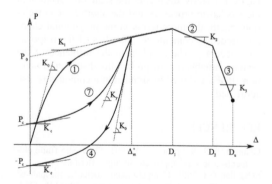

Figure 1. Proposed hysteresis law for nail plate joints.

Safety, Reliability and Risk of Structures, Infrastructures and
Engineering Systems – Furuta, Frangopol & Shinozuka (eds)
© *2010 Taylor & Francis Group, London, ISBN 978-0-415-47557-0*

CDF of the extreme value of nonstationary processes

A.I. Daniliu
Romanian Association for Earthquake Engineering, Botosani, Romania

ABSTRACT

The prediction of the extreme value of stochastic processes is a central topic in the designing in some engineering fields such as earthquake engineering, ocean engineering, wind engineering. An exact expression for the probability law of the extreme value has not been derived. The main objective of the present paper is to derive an approximate closed form for the cumulative probability function of the extreme value of stochastic nonstationary processes. The generalized extreme value distribution method, the exceedances over threshold method and Bayesian approach from extreme value theory are tried in order to obtain mathematical formulas. Let $X(t)$ denote a nonstationary stochastic process with real values and continuous paths and let $X_m = \max{(X(t), 0 \le t \le T)}$ be the largest value of $X(t)$ during an interval of length T. Divide the whole interval $[0, T]$ in an arbitrary way into n small subintervals $[\tau_{i-1}, \tau_i)$ by the points of division and choose the intermediate points $\tau_{i-1} \le t_i \le \tau_i$. The analysis of a nonstationary process with time-varying parameters would be approximately developed by using sequences of dependent random variables $\{X(t_i)\}$. The extreme value distribution is obtained as

$$P\{X_m \le x\} = \lim_{n \to \infty} \int_{-\infty}^{x} .... \int_{-\infty}^{x} f(x_1, ..., x_n) \, dx_1 ... dx_n$$

where $f(x_1, \ldots, x_n)$ is the multivariate joint probability distribution function of the random variables $\{X(t_i)\}$. In the exceedances over threshold method and Bayesian techniques the concept of random time T_m of extreme value is introduced to take into account
the time trend of the nonstationary process. The single random event $X_m \le x$ is assimilated to the union of the mutually exclusive, collectively exhaustive joint events $\bigcup_{i=1}^{n} (X(T_m) \le x, T_m \in [\tau_{i-1}, \tau_i))$. Each sample point of the global maximum value is a local maximum X_p of a particular time history. Firstly the distribution of local maxima $f(x; t)$ occurring during the infinitesimal time interval $(t, t + dt)$ and within $(x, x + dx)$ is determined. Every joint event $(X(T_m) \le x, T_m \in (t, t + dt))$ is equivalent with the joint event $(X(t) \le x, T_p \in (t, t + dt), X(t) \ge u)$ where u is a fixed threshold. The exceedances over threshold method is based on utilizing all peak events of a given time series exceeding a specified threshold. The *CDF* of the global maximum is derived as

$$F_{X_m}(x, T) = P\{X_m \le x\} = \int_0^T \int_u^x f(x; t) \, dx \, dt .$$

Bayesian solution is based on the following estimate

$$F_{X_m}(x, T) = \int_0^T F_{X_m}(x | T_m \in (t, t + dt)) f_{T_m}(t | X) \, dt$$

$F_{X_m}(x | T_m \in (t, t + dt))$ being the conditional probability that $X(T_m) \le x$ given $T_m \in (t, t + dt)$ and $f_{T_m}(t | X)$ denote the posterior density of the parameter T_m given past data X. The results are not entirely equivalent due to different initial hypotheses.

CDF of the extreme value of nonstationary processes

A.J. Torii

ABSTRACT

General Session (Insurance and Management of Risk & Economic Analysis & Probabilistic Risk Analysis & Simulation Methods)

Safety, Reliability and Risk of Structures, Infrastructures and
Engineering Systems – Furuta, Frangopol & Shinozuka (eds)
© 2010 Taylor & Francis Group, London, ISBN 978-0-415-47557-0

Insurer's solvency under catastrophic seismic risk

K. Goda
Department of Earth Sciences, University of Western Ontario, London, Ontario, Canada

H.P. Hong
Department of Civil & Environmental Engineering, University of Western Ontario, London, Ontario, Canada

ABSTRACT

Catastrophic earthquake risk imposes tremendous
financial stress on insurers who underwrite earthquake
insurance policies in a seismic region. To investigate
insurer's solvency under catastrophic seismic risk, a
stochastic model of an insurer's net worth under both
non-catastrophic and catastrophic risks is developed.
The non-catastrophic risks are represented by a diffusion process with an upward drift (Powers & Ren,
2003), whereas the catastrophic risks are modeled
as a jump process by using the seismic risk model
for spatially distributed structures (Goda & Hong,
2009), which takes spatially correlated seismic excitations into account. The developed stochastic model
is applied to an actual building inventory consisting
of 1574 wood-frame buildings located in Richmond,
British Columbia. The relationships between ruin
probability and insurer's asset characteristics, such as
the initial asset, size of non-catastrophic businesses,
and safety loading factor for earthquake insurance cov-
erage, are investigated by considering three correlation
cases: no correlation, partial correlation, and full correlation, where the partial correlation case represents
a realistic situation.

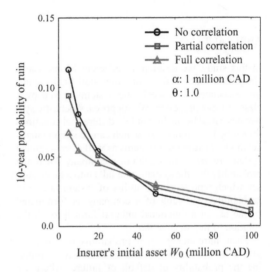

Figure 2. 10-year probability of ruin as a function of the
insurer's initial asset W_0 for the asset growth rate α equal to
1 million CAD and the safety loading factor θ equal to 1.0
by considering three correlation cases.

Analysis results indicate that the spatial correlation of seismic demand affects insurer's earthquake
risk exposure significantly in the upper tail region,
as shown in Figure 1, and that ruin probability for
given insurance policy arrangements can be lowered
by increasing the insurer's initial asset, as shown in
Figure 2. However, the results also suggest that manipulation of the initial asset and the safety loading factor
for earthquake insurance policies alone may not be
sufficient to achieve a diversified insurance portfolio,
depending on the required stability constraint. In such
cases, the use of risk transfer instruments such as rein-
surance is desirable; an extended study on this aspect
will be carried out in the future.

REFERENCES

Goda, K. & Hong, H.P. 2009. Deaggregation of seismic loss
of spatially distributed buildings. *Bull. Earthquake Eng.*
7: 255–272.
Powers, M.R. & Ren, J. 2003. Catastrophe risk and insurer
solvency: A diffusion-jump analysis. *Insurance Risk Management* 71: 239–264.

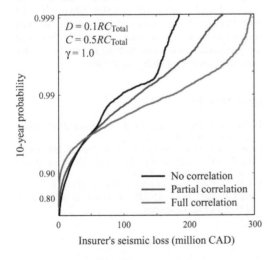

Figure 1. Probability distribution of the insurer's seismic
loss for three correlation cases.

Safety, Reliability and Risk of Structures, Infrastructures and Engineering Systems – Furuta, Frangopol & Shinozuka (eds)
© 2010 Taylor & Francis Group, London, ISBN 978-0-415-47557-0

Efficient monte carlo approach for Lévy processes and its application to risk analyses

Hiroaki Tanaka

Department of Applied Analysis and Complex Dynamical Systems, Kyoto University, Kyoto, Japan

ABSTRACT

In this study, we develop an efficient Monte Carlo simulation scheme for estimating statistical quantities associated with stochastic systems driven by a class of Lévy processes. We suppose a stochastic system described by the Itô random differential equations driven by Lévy processes, which can be a mathematical model of structural systems as well as economical systems representing wealth of a company. Then, the probability that the system state falls into a given subset, which represents probability of system failure or probability of default of a company, is formulated by the use of a functional integral form through the solution of the Itô equation.

Then, introducing a new probability measure, we give a Monte Carlo simulation scheme for estimating the probability of default or failure, where we newly propose a probability measure transformation technique available for Lévy processes. To this end, we propose a systematic method based upon the well-known Lévy decomposition (Lévy 1937) to approximate a class of Lévy processes, where a Lévy process is generally decomposed into an independent sum of a Wiener process and a compound Poisson process. The class of Lévy processes discussed in this study includes gamma processes and variance gamma (VG) processes, which have been frequently used in many fields such as mathematical finance.

Next, using the probability measure transformation method for Wiener processes as well as compound Poisson Processes developed so far (Tanaka 1998, 2002), we give an analytical expression for the Radon-Nikodym derivative, where the Girsanov transformation (Girsanov 1960) and Delbaen-Haezendonck's transformation (Delbaen & Haezendonck 1989) are applied. Further, we newly propose a method to obtain an optimal probability measure realizing effective variance reduction, in which an upper bound of the variance in the Monte Carlo procedure is minimized.

Finally, we give a numerical example in risk analyses for examining our proposed simulation scheme. Here we suppose a well-used wealth model driven

estimated probability of default

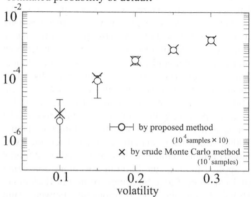

Figure 1. Estimated probability of default comparing with a large scale crude Monte Carlo simulation.

by a VG process. Figure 1 shows estimated probability of default of the wealth for several values of a volatility, which expresses a magnitude of fluctuations of the wealth. Solid circles are results by our proposed method with 10^4 samples, whereas results by the crude Monte Carlo method with 10^7 are also shown to check the accuracy. The results show that the proposed scheme can give good estimations for extremely small probability of default.

REFERENCES

Delbaen, F. & Haezendonck, J. 1989. *Insurance: Mathematics and Economics*, **8**, 269–277.
Girsanov, I. V. 1960. *Theory Probab. Its Appl.* **5**: 285–301.
Lévy, P. 1937. "Theorie de l'Addition des Variables Aleatories", Gauthier-Villars, Paris.
Tanaka, H. 1998. *Proc. of ICOSSAR '97* **1**: 411–418.
Tanaka, H. 2002. *Proc. of ICOSSAR '01*, CD-ROM Paper No. **131**, Balkema, Rotterdam.

*Safety, Reliability and Risk of Structures, Infrastructures and
Engineering Systems – Furuta, Frangopol & Shinozuka (eds)
© 2010 Taylor & Francis Group, London, ISBN 978-0-415-47557-0*

How many containers to inspect to deter terrorist attacks

N. Haphuriwat & V.M. Bier
Department of Industrial and Systems Engineering, University of Wisconsin-Madison, Madison, WI, USA

ABSTRACT

In this paper, we investigate how many containers would need to be screened in order to deter attackers from attempting to smuggle weapons into a defending country in container freight. We hypothesize that with a sufficiently high probability of being detected, attackers might be deterred from smuggling attempts. Thus, our goal is to identify the optimal proportion of containers to inspect in order to minimize the defender's expected loss, using game theory to reflect the fact that attackers are simultaneously trying to maximize their expected rewards. Moreover, our model recognizes that the container-screening policy must simultaneously protect against different types of threats (such as nuclear bombs, dirty bombs, and assault rifles).

Safety, Reliability and Risk of Structures, Infrastructures and Engineering Systems – Furuta, Frangopol & Shinozuka (eds)
© *2010 Taylor & Francis Group, London, ISBN 978-0-415-47557-0*

Seismic fragility estimates for reinforced concrete bridges in M-R-α space

J. Zhong, P. Gardoni & D.V. Rosowsky
Texas A&M University, College Station, TX, USA

ABSTRACT

Typically, the seismic fragility of bridges is estimated conditioning on intensity measures, such as the spectral acceleration, S_a, peak ground displacement, PGD, peak ground velocity, PGV, and peak ground acceleration, PGA. Although these intensity measures have been used broadly in fragility estimation and seismic design, they have two limitations. First, their probability distributions, which are needed to convolve the fragility to estimate the probability of failure, are not easily available. Second, these engineering-oriented intensity measures are not easily understandable by the owners and the public. This can become a hurdle in communicating the outcomes of seismic risk analysis to the public, the policy- and decision-makers. Thus, more intuitive and easily quantifiable intensity measures are needed for the vulnerability assessment of bridges in seismic regions.

In response to these needs, two alternative seismic measures have been proposed to characterize the seismic demand on a structure: the moment magnitude at the source, M, and the source-to-site distance, R (e.g., McGuire & Shedlock 1981, U.S. DOE 1996; U.S. NRC 1997; Bazzurro & Cornell 1999). Most of these studies focus on the probabilistic seismic hazard analysis to compute the hazard level as a function of M and R. Little effort has been made to use M and R in developing fragility estimates. Among these studies, Kafari & Grigoriu (2007) suggested using M and R to develop fragility estimates. However, the fragility estimates are for a simple single-degree-of-freedom (SDF) system and do not account the for the azimuth angle, α, which is found in this paper to be an important variable.

This paper proposes a new methodology to estimate the seismic fragility of reinforced concrete (RC) bridges. The proposed fragility estimates are developed conditioning on more intuitive and easily quantifiable intensity measures: the moment magnitude, M, the epicentral distance, R, and the azimuth angle, α, i.e., in the M-R-α space. In the proposed approach, M captures the amount of slip and energy radiated by the seismic waves of an earthquake, R quantifies the distance that seismic waves travel before hitting a structure, and α introduces the effects of rupture directivity on the seismic demands on a structure.

As an illustrative example, the seismic fragility estimates for an example RC bridge with 11 two-column bents are assessed both at the column and bent levels. It is found that the seismic fragility estimates increase with M and decreased with R and α.

REFERENCES

Kafali, C., & Grigoriu, M. 2007. Seismic fragility analysis: Application to simple linear and nonlinear systems. *Earthq. Engng. Struct. Dyn.*, 36: 1885–1900.

McGuire, R.K., & Shedlock, K.M. 1981. Statistical uncertainties in seismic hazard evaluations in the United States. *Bull. Seism. Soc. Am.*, 71(4): 1287–1308.

U.S. Nuclear Regulatory Commission (U.S. NRC) 1997. Identification and characterization of seismic sources and determination of safe shutdown earthquake ground motion. *Regulations 10 CFR part 100, Regulatory Guide 1.165, Appendix C*, U.S. Nuclear Regulatory Commission, Washington, D.C., 1.165-17/1.165-23

U.S. Department of Energy (U.S. DOE) 1996. Natural phenomena hazards assessment criteria. *DOE-STD-1023-96*, U.S. Department of Energy, Washington, D.C.

Bazzurro, P., & Cornell, C.A. 1999. Disaggregation of seismic hazard. *Bull. Seism. Soc. Am.*, 89(2): 501–520.

Safety, Reliability and Risk of Structures, Infrastructures and
Engineering Systems – Furuta, Frangopol & Shinozuka (eds)
© 2010 Taylor & Francis Group, London, ISBN 978-0-415-47557-0

Structural reliability theory in the wider safety context

R.E. Melchers

Centre for Infrastructure Performance and Reliability, The University of Newcastle, Australia

ABSTRACT

This paper considers the role of structural reliability theory in the wider context of (i) achieving structural safety in practice and (ii) as part of safety assessment for engineering and other projects more generally, including its relationship to achievement of adequate safety in the community generally. It notes that the rules for the design of structures have advanced from 'Factor of Safety' to more rational procedures and these now include the widespread use of structural reliability theory for calibrating design codes. While this is an admirable achievement from a structural engineering viewpoint, and can be considered as a 'stand-alone' matter, increasingly there is an expectation that safety matters in structural engineering must be seen also in a wider perspective, that is in the context of societal expectations for structural engineering. As a result there remain a number of issues for clarification, including how the structural probability measures relate to observations about failures of actual structures. Some of these concerns can be addressed through the adoption of a more clearly defined decision-theoretic framework adapted to structural engineering safety issues. The components for such a framework have recently been outlined and are reviewed herein. It is argued also that structural engineering safety and reliability measures must be compatible with procedures adopted by other potentially hazardous industries.

Safety, Reliability and Risk of Structures, Infrastructures and
Engineering Systems – Furuta, Frangopol & Shinozuka (eds)
© 2010 Taylor & Francis Group, London, ISBN 978-0-415-47557-0

Efficient hit-or-miss Monte Carlo method for structural reliability evaluation

S.A. Matsuho

Anan National College of Technology, Anan, Tokushima, Japan

ABSTRACT

A hit-or-miss Monte Carlo (MC) method is inefficient, but has advantages of easiness of application,
wide applicability, analytical rationality and so on.
Then, this paper considers two methods to improve
the computational efficiency of the hit-or-miss MC
method for the structural reliability evaluation. One
method of them intensively samples limit state events
with low frequency on computer, and is called an
Uncommon-events-sampling (UES) method. Another
method assigns the generated pseudorandom numbers
to basic variables according to their relationship with
which the failure state easily. This method is called an
Assigning-pseudorandom-numbers (APN) method.

In a numerical example, structural reliability of a
steel truss bridge shown in Figure 1 designed based on
Japanese current code (Japan Road Association 1996)
is analyzed by using the proposed MC methods. In a
case of considering strength R and load S only as random variables, it is hard to evaluate reliability of the
truss structure by using the crude MC method, because
not only its R and S but also dimensions of members
are necessary for expression of the limit state function of its member's buckling. In computation, traffic
load model based on statistical observed data on traffic stream of Hanshin (Osaka-Kobe area) Expressway
in Japan is used.

This abstract shows results of the conventional hit-
or-miss MC method and the proposed APN method.
Tables 1 and 2 show estimated results of the probability
P_f by using the conventional hit-or-miss MC method
and the proposed APN method, respectively. The tables
contain results that are computed in cases of deterministic strength and uncertain strength. In these tables, $\hat{P}_f$
is an estimated value, and 'CV' denotes a coefficient

Table 1. System $\hat{P}_f$ estimated by conventional hit-or-miss MC.

Case	Number of tries	$\hat{P}_f$	CV of $\hat{P}_f$
(1) Uncertain strength	1000	3.00e-03	5.77e-01
	10000	2.30e-03	2.08e-01
	100000	2.69e-03	6.09e-02
	1000000	2.54e-03	1.98e-02
	10000000	2.54e-03	6.26e-03
(2) Deterministic strength	1000	1.00e-03	1.00e+00
	10000	1.90e-03	2.29e-01
	100000	2.14e-03	6.83e-02
	1000000	2.01e-03	2.23e-02
	10000000	2.04e-03	7.00e-03

Table 2. System $\hat{P}_f$ estimated by the proposed APN method.

Case	Number of tries	$\hat{P}_f$	CV of $\hat{P}_f$
(1) Uncertain strength	100	1.04e-02	3.09e-01
	1000	2.95e-03	1.84e-01
	10000	2.33e-03	6.55e-02
	100000	2.81e-03	1.88e-02
	1000000	2.56e-03	6.24e-03
(2) Deterministic strength	100	1.04e-02	3.09e-01
	1000	1.47e-03	2.60e-01
	10000	1.78e-03	7.48e-02
	100000	2.31e-03	2.08e-02
	1000000	2.06e-03	6.97e-03

of variation of its estimation. From Tables 1 and 2, we
can see that the coefficients of variation of estimation
in Table 2 are smaller than ones in Table 1. Moreover,
in this example, it is shown that efficiency of the proposed APN method to the conventional hit-or-miss MC
method is about 10.

REFERENCES

Japan Road Association 1996. *Specifications for Highway Bridges*, Part I and II.

Tachibana, Y. 2000. *Bridge Engineering*, 5th Edition, Tokyo: Kyoritsu Publishing Co., Ltd. (in Japanese)

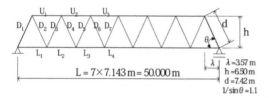

Figure 1.   Main truss (Tachibana 2000).

Thursday Morning (THM) Sessions

Organized Session (OS11) Health Monitoring, Structural Monitoring & Control

Safety, Reliability and Risk of Structures, Infrastructures and Engineering Systems – Furuta, Frangopol & Shinozuka (eds)
© 2010 Taylor & Francis Group, London, ISBN 978-0-415-47557-0

Neural network based technique for on-line bridge diagnostics

S. Beskhyroun, S. Mikami, T. Oshima, Y. Miyamori & T. Yamazaki
Kitami Institute of Technology, Kitami, Japan

ABSTRACT

All load-carrying structures continuously accumulate damage which is usually caused by degradation and deterioration of structural components and connections during their service life. Hence, the need for rapid assessment of the state of critical and conventional civil structures such as bridges, control centers, airports, hospitals among many is essential because undetected damage may lead to structural failure and the loss of human lives. Due to these needs, structural health monitoring (SHM) technology for damage assessment in civil structures has been increasingly receiving great attention in recent years (Ewins 1985).

In this paper, a new damage identification algorithm based on artificial neural network (ANN) is presented. Several research papers (Efstathiadesa et al. 2007) have utilized ANN for damage detection and most of these papers used finite element models for training the network. However, it is extremely difficult to create accurate FEM for complex structures and an inaccurate FEM can degrade or even lead to incorrect result in damage detection. The proposed algorithm is used to detect damage and locate its position using only the structure response data without the need for any modal identification or numerical models. The method is applied to the experimental data extracted from an out of service railway steel bridge after inducing some defects to its members. The damage was introduced to the bridge through the release of some bolts from two stiffeners located on the web of the main girder of the bridge. The obtained results indicate that the current approach can identify the location of small damage effectively. The proposed approach has also shown very good capability of identifying the changes in structure's response produced by damage from those resulted from noise or measurement errors.

One important issue in SHM is obtaining appropriate structural response data. In most cases, one would typically excite the structure with a hammer or shaker, and measure the responses. However, for many structures, using these tools to induce response for SHM purposes may be infeasible or forbidden. For example, the traffic over the bridge must be interrupted in case of using dynamic shakers. Hammers cannot be used for inaccessible members and are time consuming. Moreover, these tools are not convenient for continuous on-line SHM. Many research papers have presented ambient excitation (e.g., microtremor, wind, traffic, etc.) as an inexpensive excitation force that can be used for continuous health monitoring. However, ambient vibrations are very random in nature and it is extremely difficult to obtain two equal excitation forces as many damage identification algorithms require. Additional challenge arise from the fact that damage is typically a local phenomenon and may not significantly influence the lower-frequency global response of structures that is normally measured during vibration tests. In order to overcome these problems, the use of multi-layer piezoelectric actuators (Oshima et al. 2002) as a local excitation force for continuous health monitoring of steel bridges is investigated in this paper. Finally, the paper explains the applicability of the proposed method, combined with the use of piezoelectric actuators, for an online monitoring system of steel bridges.

REFERENCES

Efstathiadesa Ch., Baniotopoulos C.C., Nazarko P., Ziemianski L., Stavroulakis G.E. 2007. Application of neural networks for the structural health monitoring in curtain-wall systems. *Journal of Engineering Structures*. Elsevier 29: 3475–3484.

Ewins, D. J., 1985. *Modal Testing: Theory and Practice*, New York: John Wiley.

Oshima T, Yamazaki T, Onishi K, Mikami S. 2002. Study on damage evaluation of joint in steel member by using local vibration excitation, (In Japanese). *Journal of Applied Mechanics, Japan Society of Civil Engineers (JSCE)* Vol 5: 837–846.

Safety, Reliability and Risk of Structures, Infrastructures and
Engineering Systems – Furuta, Frangopol & Shinozuka (eds)
© 2010 Taylor & Francis Group, London, ISBN 978-0-415-47557-0

Monitoring of prestressed concrete bridge resonated with truck vibration arisen from the road roughness

S. Fukada & Y. Kajikawa
Kanazawa University, Kanazawa, Japan

T. Muroi & Y. Momiyama
West Nippon Expressway Engineering Kansai Co., Ltd., Osaka, Japan

ABSTRACT

Object prestressed concrete bridge with 37.5 m span had vibrated greatly, only when the test truck had ran at the 2nd slow lane. The authors clarified that the suspension spring vibration of the test truck had been arisen from the road roughness with long spatial period nearby the expansion joint of the 2nd slow lane (Fukada et al. 2008). Then monitoring was carried out for a year in order to investigate the dynamic stress of the bridge and the total weight of the running ordinary trucks on the bridge.

This study proposed the new method to evaluate the total weight of running trucks on the bridge. This method measured the vertical displacement of the rotation angles at the shoes. The diurnal maximum total weight of the trucks for a year is shown in Figure 1. 31st August is recorded maximum weight of 1940 kN, and also 1st January is recorded minimum weight of 415 kN. The average of the diurnal maximum is 996 kN.

Figure 2 shows diurnal maximum dynamic stress at each lane for a year. The diurnal maximum dynamic stress in each lane and that of average are 3.0 N/mm² and 2.0 N/mm² respectively.

From the points of view in the design, the tensile stress of the girder by the design live-load using TT-43 load is 6.6 N/mm². The diurnal maximum dynamic stress (3.0 N/mm²) is the half of the tensile stress by the design live-load.

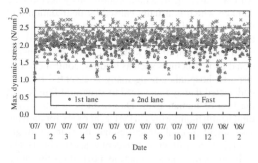

Figure 2. Diurnal maximum dynamic stress.

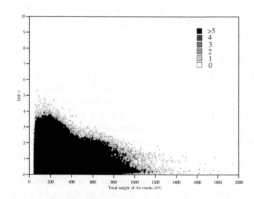

Figure 3. Relation of total weight of trucks and DIF.

Figure 3 shows the relation of the total weight of the trucks and DIF-1.0 at 2nd slow lane by the long term monitoring. It is confirmed that the DIF-1.0 is decreasing as the total weight of the trucks increase.

REFERENCES

S. Fukada, T. Muroi and K. Usui. 2008. Vibration Characteristics of the PC Bridge Resonated to the Vehicles Generated by the Road Roughness nearby the Expansion Joint, *Proceeding of Earth & Space conference 2008*: (on CD-ROM).

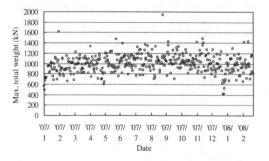

Figure 1. Diurnal maximum total weight of the trucks.

Safety, Reliability and Risk of Structures, Infrastructures and Engineering Systems – Furuta, Frangopol & Shinozuka (eds)
© 2010 Taylor & Francis Group, London, ISBN 978-0-415-47557-0

Stochastic optimal control of a SDOF system subjected to a narrow band excitation

D.V. Iourtchenko
Department of Mathematical Sciences, SPBSPU, S.-Petersburg, Russia

J.-L. Menaldi
Department of Mathematics, Wayne State University, MI, USA

ABSTRACT

The paper considers a problem of stochastic optimal control of a single-degree-of-freedom (SODF) dynamic system, subjected to narrow band random excitation. The problem is handled by the dynamic programming approach. The dynamic programming approach [Bellman, 1957] is one of the possible methods, which serves for solving problems of stochastic optimal control [Fleming & Rishel, 1975]. The method reduces a stochastic optimal control problem to a problem of finding a solution to the deterministic partial differential equation – Hamilton-Jacobi-Bellman (HJB) equation. Thus, the challenge is to solve the Cauchy problem for nonlinear, degenerate equation of parabolic type within the whole state-space. This problem turns out to be far from simple, and a very limited number of solutions available today prove this.

It should be noted that problems of stochastic optimal control of dynamic systems stays apart from other optimal control problems. The reason for that is obvious – the governing equation of motion of any dynamic system, written in a state-space form, does not contain noise in all of the equations. This features leads to the fact that the corresponding HJB equation becomes degenerate. The theorem of existence and uniqueness of a solution in the classical sense for a degenerate parabolic equation, which is yet to be proven. However, certain promises has already been brought by [Lio and Ley, 2006], who has recently proven existence and uniqueness of a solution to such an equation in the viscosity sense.

The only problem of stochastic optimal control that has an exact solution is a Linear Quadratic (LQ) problem. Obviously the assumption of an unbounded in magnitude control force, made in LQ problem, seems unreasonable and unfeasible in many practical applications. Thus, consideration of a bounded in magnitude control force appears to be well justified. In this paper the author considers bounded in magnitude control law, which make the considered problem closer to real practical applications.

The paper considers a narrow band excitation, which is modeled as a filtered Gaussian white noise. Thus, the resulting system of governing equation of motion consists of four stochastic differential equations. The primary goal of the optimal control is to minimize mean system response energy. The corresponding HJB equation is four dimensional with respect to space. The paper uses the newly proposed hybrid solution method for finding a solution to the corresponding HJB equation. The method consists of two steps: first, an exact analytical solution to the corresponding HJB equation found within a certain outer domain. It has been proved that this solution provides an asymptotic behavior of the Bellman function [Bratus et. al., 2006]. Secondly, the obtained analytical solution is used as a boundary condition to solve the corresponding HJB equation within the inner domain, thereby constructing the solution within the entire state-space domain. It is one of the unique cases when a solution to the 4D HJB equation was possible to obtain.

REFERENCES

Bellman R., (1957), Dynamic Programming}, Princenton University Press.
Fleming W. and Rishel R., (1975), Deterministic and Stochastic Optimal Control, Springer-Verlag.
Lio F.D., Ley O. (2006), Uniqueness results for second order Bellman-Isaacs equaton under quadratic growth assumptions and applications. SIAM J. Contr. Optim. 45, N 1, pp.74–106.
Bratus A.S, Iourtchenko D.V., Menaldi J.-L. (2006). Local solutions to the Hamilton – Jacobi – Bellman equation in stochastic problems of optimal control. Doklady Mathematics.74, N 1, pp. 610–613.

Safety, Reliability and Risk of Structures, Infrastructures and Engineering Systems – Furuta, Frangopol & Shinozuka (eds)
© 2010 Taylor & Francis Group, London, ISBN 978-0-415-47557-0

Study on remote vibration monitoring and control of infrastructures by using digital video processing and information technology

T. Obata
Osaka Prefectural College of Technology, Neyagawa, Japan

Y. Miyamori
Kitami Institute of Technology, Kitami, Japan

INTRODUCTION

Vibration measurement of structures is a promising method for structural identification and safety monitoring in infrastructure's management. Many novel methods have been proposed for vibration monitoring based on innovative information and communication technologies for these years. Cost effectiveness and adequate performance are required to such monitoring system. Recent inexpensive semiconductor sensor and computer network may meet such requirements.

In this study, low-cost non-contact vibration monitoring system is developed by using household digital video camera as shown in Photo 1. A low-cost household CCD camera is used for sensor device. The CCD camera connected to a computer takes movement of target mark and digital image processing program calculates displacement of target in real-time (Plat 1991). A telescope can be attached to camera lens for improving resolution of measurement if the target object is remote. The monitoring system saves calculated displacement in the computer and sends to another computer via TCP/IP connection (Miyamori 2003).

The vibration-measuring experiment to pedestrian bridge is performed to evaluate the performance of the system. Conventional monitoring system with strain gage-type accelerometers is also used to evaluate the performance of proposed monitoring system. From experimental results, time history displacement wave and predominant vibration frequencies of lower vibration mode of the bridge are equivalent to measured data of conventional vibration monitoring system.

The displacement monitoring system is then applied to On/Off semi-active structural control. Control system is designed to restrain vibration response of tower structure. A 2.1 m height laboratory model is used for experiment. This experimental model has a control device composed of stiffener, oil damper and electromagnetic connection. The control device is connected to the structure body as amplitude of lower mode vibration increases and structural stiffness increases, and vice versa. In this control system, 2 computers

Photo 1. Displacement measuring system.

are employed; one is a computer of the displacement monitoring system, the other works as a semi-active controller. The semi-active controller switches the state of electromagnet based on the displacement at the top of tower model sent from the monitoring system.

A semi-active structural control test is also performed using this structural control method with this non-contact vibration monitoring system. Vibration response of structure is re-strained by semi-active control. Time delay due to image processing and data communication does not reduce control performance.

The proposed system therefore has applicability for structural monitoring system. And the system has also possibility for control system of smart structures.

REFERENCES

Miyamori Y, Adachi K & Obata T, A consideration of remote vibration control by using digital image processing on steel towers, JSSC Journal of Constructional Steel, Vol. 11, pp.83–90, 2003. (in Japanese)

Pratt, William K, *Digital Image Processing*, J. Wiley, New York, 1991.

Safety, Reliability and Risk of Structures, Infrastructures and Engineering Systems – Furuta, Frangopol & Shinozuka (eds)
© 2010 Taylor & Francis Group, London, ISBN 978-0-415-47557-0

Structural health monitoring accommodating varying environmental and operational conditions

N.H.M. Kamrujjaman Serker, Z.S. Wu & A.P. Adewuyi
Department of Urban & Civil Engineering, Ibaraki University, Hitachi, Japan

ABSTRACT

In last few years, a large number of methods and/or techniques have been proposed in the field of aeronautics, civil and mechanical engineering for the purpose of damage identification. However, many of them are not suitable for the practical application of real life structures due to many reasons [1]. Varying environmental and operational conditions have been considered as the important barriers to develop a reliable and continuous structural health monitoring (SHM) system. Researchers have paid their attention to overcome these barriers and several damage detection methods have been proposed in this research arena. However, a general and widely accepted methodology or strategy is yet to develop especially in the field of civil SHM. In this paper, a new SHM strategy based on the data sensed by long-gage fiber Bragg grating (FBG) sensors is presented. Since confirmation of the presence of any damage is the primary goal of any SHM project, this study is limited on the identification of the structural damage. The damage detection technique is presented graphically in Figure 1. In a recent work, the applicability of the proposed method has been verified using the laboratory experimental test data [2]. In this approach modal macro strain (MMS) [3] is extracted from the measurement over a period of time. Next, the extracted features of a target location are plotted against that of a reference location and to find the regression line. The slope of the regression line will change and shift to a new position as soon as the target location receives damage. The presence of the damage can be identified from the changes of the slope i.e. shifting of the best fit line to a new position as shown in figure 1.

The interesting features of the proposed strategy is that the varying operational and environmental conditions are automatically included in the damage identification process as well as the large volume of monitored data to be gathered can easily be handled. This study discusses the results obtained by the application of real time monitored data from a high-way

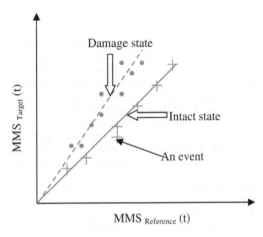

Figure 1. Graphical representation of the proposed damage identification technique.

bridge in Japan which has been instrumented with long-gage FBG sensors. From the visual inspection it was found that the monitored bridge is in the healthy condition and no artificial damage has been introduced. This paper focuses on the novelty and robustness of the proposed SHM strategy to deal with the varying operational and environmental conditions as well as the large volume of monitored data.

REFERENCES

H. Sohn and C.R. Farrar, 2001. Damage diagnosis using time series analysis of vibrational signals. *Smart Materials & Structures.* 10: 1–6.

N.H.M. Kamrujjaman Serker, Z.S. Wu and S.Z. Li, Damage detection in beam-like structures using noise-polluted distributed static macro strain response, *J. Structure and Infrastructure Engineering,* Accepted.

S.Z. Li and Z.S. Wu, 2007. Development of distributed long-gage fiber optic sensing system for structural health monitoring. *Structural Health Monitoring J.* 16: 133–143.

Safety, Reliability and Risk of Structures, Infrastructures and
Engineering Systems – Furuta, Frangopol & Shinozuka (eds)
© 2010 Taylor & Francis Group, London, ISBN 978-0-415-47557-0

Impact force identification and damage estimation of composite structures

M. Tajima
ESTECH Corporation, Yokohama, Kanagawa, Japan

N. Hu & H. Fukunaga
Tohoku University, Sendai, Miyagi, Japan

ABSTRACT

Detecting impact damages in real-time is an important concern for keeping up structural integrity of
aerospace composite structures since impact damages
may cause extreme degradation of material strength
like compression after impact. The information of
impact force should be a considerably significant indicator in an estimation of structural damages. The
authors proposed (2007) an experimental identification method of impact force acting on CFRP plates
where the relation between impact force history and
corresponding strain responses in CFRP structures
were determined through the impact test using impulse
hammer.

Another issue in a structural health monitoring
based on the identification of impact force is damage
estimation. In designing composite structures, damage
tolerance is one of the considerable issues. A numerical
approach using a finite element model with adaptive cohesive elements (Ref. 2) can predict both the
onset and propagation of delamination with considering matrix damages by avoiding numerical instability,
which can clarify the complicated fracture mechanism
of composite laminates.

In this paper, an identification method of impact
force location and history is developed based on the
experimental transfer matrix and it is verified by
performing the impact test of a stiffened composite panel and then in the drop-weight impact test of
laminated plates. The former test is performed within

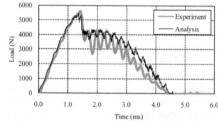

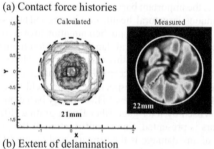

(a) Contact force histories

(b) Extent of delamination

Figure 2. Results of contact force and damage prediction.

the nondestructive energy level. On the other hand, in
the drop-weight impact test damages are observed by
the ultrasonic C-scan inspection. The identification for
the destructive impact force in the drop-weight impact
test is performed using the transfer matrix of intact
structure (Fig. 1). Subsequently, we apply the numerical method using adaptive cohesive elements to the
drop-weight impact problem for 32-plied composite
laminated plates. The validity of the present numerical
method is verified through the comparisons measured
and calculated damages (Fig. 2).

REFERENCES

Tajima, M., Hu, N. & Fukunaga, H. 2007. Experimental
Impact Force Identification of CFRP Stiffened Panels.
*Proceedings of the Sixteenth International Conference on
Composite Materials*: CD-ROM.
Zemba, Y., Hu, N., Hara, E. & Fukunaga, H. 2008. Damage
Propagation Prediction of Quasi-isotropic CFRP Plates
Under Low-velocity Impacts. *Journal of the Japan Society
for Aeronautical and Space Sciences* Vol. 56: 220–227.

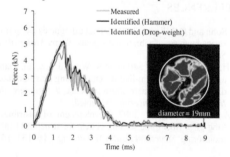

Figure 1. Destructive impact force history identification
result.

Mini-Symposia (MS05) Modelling Seismic Action

Safety, Reliability and Risk of Structures, Infrastructures and
Engineering Systems – Furuta, Frangopol & Shinozuka (eds)
© 2010 Taylor & Francis Group, London, ISBN 978-0-415-47557-0

Absolute accelerogram and record selection criteria for nonlinear seismic analysis of structures

K. Ashkinadze

Bantrel Co., Edmonton, Alberta, Canada

ABSTRACT

Nonlinear time-history analysis using real and synthetic earthquake records (accelerograms) as input action is the most physically consistent and versatile tool for seismic analysis of structures not readily designable by the conventional equivalent static (spectral) method. Applications include combined structures with non-trivial and composite mechanisms of nonlinear response; existing structures undergoing review due to increased seismic requirements; systems of active vibration isolation; and many others.

The principal problem with all models of seismic action is the absence of a criterion for their selection from the pool of available records. There is no assurance that favourable results of a given analysis do not owe to selection of an "accommodating" accelerogram and that another record of the same intensity will not overturn them.

This problem is addressed scarcely in literature. The engineers are referred to building codes, with their methodologies based on equivalent static method, and required to assure that the results of the time-history analysis closely agree with these largely oversimplified methods. This defeats the purpose of the refined nonlinear time-history analysis.

It is desirable to verify reliability of time-history analyses by a method of their own kind. If they can be demonstrated to provide the safety margin required by the code, it would allow their wider use in design practice, benefiting from their versatility, physical consistency and proper consideration of various modes of nonlinear response.

The present paper proposes a new approach to this problem by introducing two new concepts: absolute accelerogram and reduced absolute accelerogram. The response of the structure to selected earthquake records is compared with its response to a specially constructed "absolute" accelerogram, which is defined as the most unfavourable loading among all possible programs of loading not exceeding the specified peak ground acceleration (PGA). It is an objectively

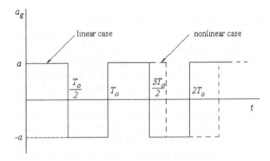

Figure 1. Profile of the absolute accelerogram.

existing, unique time history for any given structure. The absolute accelerogram depends only on deformation and strength properties of the structure itself, which can be reliably evaluated, and does not depend on the largely uncertain seismic situation on the site. The general profile of the absolute accelerogram is shown on Figure 1.

Compared to real seismic actions, the action of the absolute accelerogram on structures is more conservative. To address this problem, a reduced absolute accelerogram is introduced, which has its magnitudes of reactions lowered by a prescribed number of times compared to the absolute accelerogram. The ratio of reduction is chosen in conformance with code-stipulated spectral curves, which assures that the reliability of this method meets the requirements of the code. The proposed method of construction of the reduced absolute accelerogram is by frequency modulation of the input action from resonance state.

Comparison of the rate of structural reaction increase under the action of the reduced absolute accelerogram and selected real or synthesized earthquake records indicates whether the chosen recording is representative for this structure, and allows determining the actually attained margin of safety. Analysis examples are considered that illustrate this concept.

Safety, Reliability and Risk of Structures, Infrastructures and Engineering Systems – Furuta, Frangopol & Shinozuka (eds)
© 2010 Taylor & Francis Group, London, ISBN 978-0-415-47557-0

Critical seismic load inputs for MDOF inelastic structures

Abbas Moustafa & Kazuo Takahashi

Department of Civil Engineering, School of Engineering, Nagasaki University, Nagasaki, Japan

ABSTRACT

The assessment of the seismic performance of buildings under strong ground motions is an important problem in earthquake engineering. The objective of the structural engineer is to design buildings that are safe against possible future earthquakes and are economic at the same time. To achieve this goal, the following criteria should be fulfilled: (1) robust definition of earthquake ground motions for the site, (2) accurate mathematical model for the material behavior, and (3) reliable structural damage descriptors that describe accurately possible structural damage under seismic loads. The accurate definition of design earthquake loads is the first step towards achieving this goal. Earthquake loads are specified in terms of the site response spectra, the time history of the ground acceleration or using the random vibration theory. The inadequate performance of buildings during recent earthquakes, however, has motivated researchers to revise existing methods and to develop new methods for seismic-resistant design of structures.

This paper develops a framework for specifying design earthquake loads for seismic design of multi-degree-of-freedom inelastic buildings using the method of critical excitations. This method relies on the high uncertainty associated with the occurrence of earthquakes and their characteristics (e.g., time, location, magnitude, duration, frequency content, amplitude, etc.), and, also on the safety requirements of important and lifeline structures (e.g., nuclear plants, storage tanks, industrial installations, etc.). An overview on modeling critical earthquakes for nonlinear structures can be found in (Abbas 2006, Takewaki 2006).

In this paper, the earthquake acceleration is expressed as a Fourier series modulated by an envelope function. The coefficients of the series representation are calculated such that the structure inelastic response is maximized subjected to predefined constraints. The constraints contain bounds on the total energy of the earthquake signal, peak values of ground acceleration, velocity and displacement, and upper and lower limits on the Fourier spectra of the ground acceleration. The structure force-displacement relation is taken to be governed by a hysteretic bilinear law. The resulting inverse dynamic problem is tackled using nonlinear optimization techniques through sequential quadratic optimization method. Issues related to various forms of energy dissipated by the inelastic structure are explored. The formulation is demonstrated for a two-story inelastic braced frame structure.

It is shown that critical earthquake loads for the inelastic structure differ from those for the elastic structure. The time variation of the structure deformation differs from that of the elastic system. Unlike the elastic system, the inelastic system dissipates energy through yielding and damping. The study, also, examines the modeling of damping using nonlinear hysteretic model. The effect of the damping ratio, and, also in the strain-hardening ratio on the derived critical acceleration and associated optimal inelastic deformation are also studied.

REFERENCES

Abbas, A.M., 2006. Critical seismic load inputs for simple inelastic structures. *Journal of Sound and Vibration*, 296, 949–967.

Takewaki, I. 2006. Probabilistic critical excitation method for earthquake energy input rate. *Journal of Engineering mechanics*, 132(9) 990–1000.

Safety, Reliability and Risk of Structures, Infrastructures and Engineering Systems – Furuta, Frangopol & Shinozuka (eds)
© 2010 Taylor & Francis Group, London, ISBN 978-0-415-47557-0

Effects of uncertainties in structural properties on the uniform hazard spectra used in the performance based seismic design of structures

M. Niño & A. G. Ayala

Institute of Engineering, UNAM, Mexico, Mexico

ABSTRACT

Over the last few years, non linear dynamic analysis has been frequently the topic of re-search by several groups aiming to the development of simplified methods to predict structural performance under extreme seismic action in an explicit way. However, the seismic demands used in these simplified methods are not consistently defined in as much as they are prescribed by spectral ordinates based on rate of exceedance of a given seismic intensity, dis-regarding any connection with structural performance. This situation is evident in current per-formance based seismic design methodologies which consider as design objectives, pairs of performance levels and seismic design levels defined this way (SEAOC, 1995). Researchers such as Bazzuro and Cornell (1994a and 1994b) and Shome (1999) have tried to correct this drawback by considering indexes related to structural performance in the definition of seismic design levels. However, the procedures investigated by these authors do not consider the va-riability of structural properties (mechanical and dynamical) and their impact on seismic performance.

As real structural properties, in general, differ from those assumed in evaluation and design, this paper presents a formulation to define, in a probabilistic way, seismic demands which consider the effects of this variability on the seismic performance of reinforced concrete structures. It is shown that the structural properties whose dispersion produces a significant effect on the definition of the seismic demands are the vibration period, the lateral strength and the post-yielding to initial stiffness ratio. The proposed procedure considers, for this sake, the statistics of the response of a single degree of freedom system used as reference obtained through Monte Carlo simulation using a set of synthetic accelerograms obtained for a particular site in the lake bed zone of Mexico City through a novel empirical Green function method with two corner frequencies to represent the source. To quantify the relative influence of each of these uncertain variables on the seismic demands, their effect is evaluated by considering each of them separately and all together. The results obtained are compared with the current design spectrum in the Mexico City building code for a life safety performance level characterized by a ductility of four as performance index.

It is concluded that the proposed procedure represents a significant improvement of those proposed in documents such as Vision 2000, and a useful tool for research and further developments on the performance based seismic evaluation and design of existing and new structures.

REFERENCES

Bazzuro and Cornell A. 1994a, Seismic hazard analysis for non-linear structures. I: Methodology, ASCE Journal of Structural Engineering, vol. 120, no. 11, pp. 3320–3344.
Bazzuro and Cornell A., 1994b, Seismic hazard analysis for non-linear structures II: Applications, ASCE Journal of Structural Engineering, vol. 120, no. 11, pp. 3345–3365.
SEAOC 1995, Vision2000: A Framework for Performance Based Design, Structural Engineers Association of California, vol. 1, Sacramento, Cal., USA

Safety, Reliability and Risk of Structures, Infrastructures and Engineering Systems – Furuta, Frangopol & Shinozuka (eds)
© 2010 Taylor & Francis Group, London, ISBN 978-0-415-47557-0

Development of advanced methods for seismic probabilistic safety assessments

G. Thuma
Gesellschaft für Anlagen- und Reaktorsicherheit (GRS) mbH, Köln, Germany

M. Türschmann
Gesellschaft für Anlagen- und Reaktorsicherheit (GRS) mbH, Berlin, Germany

M. Krauss
Bundesamt für Strahlenschutz (BfS), Salzgitter, Germany

ABSTRACT

Development of Advanced Methods for Seismic Probabilistic Safety Assessments

The German Guideline on Safety Reviews (to be carried out on a regular basis) for nuclear power plants requires a seismic probabilistic safety assessment (SPSA) to be performed, if the intensity of the design earthquake at the site is greater than VI (MSK)(BMU 2005, FAK 2005). The corresponding technical reference document on PSA methods recommends an approach based on safety margin factors and provides general guidance on this subject. The SPSA procedure consists of three major steps: (1) the seismic hazard analysis, (2) the determination of failure probabilities of structures, systems and components (SSC), and (3) the development of seismically induced event trees with subsequent calculation of core damage frequencies. Up to now, the technical reference document has not yet reached the same maturity for seismic events as for plant internal events. This is especially true for the screening procedure (part of the second step) and the third step of the SPSA procedure.

For it is hardly possible to analyze all SSC of a nuclear power plant with respect to their seismic vulnerability, the quantity of SSC to be analyzed in detail has to be limited by a suitable screening procedure. This screening procedure should, on the one hand, reduce the number of SCC to be analyzed. Nevertheless, on the other hand, it has to ensure, that all relevant SSC are considered.

The screening procedure for the SPSA under development in Germany is principally based on location-oriented screening methods as applied within the PSA of plant internal fires. To identify those SSC, which might become important during or immediately after an earthquake, one starts with a comprehensive list of all SSC. These SSC are classified according to various criteria (location, type, dependencies on other SCC etc.). Based on this classification plant areas are selected for a walkdown. To ensure that all relevant aspects of the installation conditions are gathered a checklist is used for this plant walkdown. The analysis of the walkdown insights allows for the final determination of the SSC to be analyzed in detail.

Although there is a wide variety of possible event sequences in case of an earthquake, SPSA experience shows that it is reasonable to focus on a limited set of dominant sequences. These sequences can be deduced from the event trees of the PSA for internal events, but the special site-specific characteristics of earthquakes like the possible range of seismic input motions, potential secondary effects (fires, etc.) and seismic interdependencies, have to be included in the modeling of the event sequences.

For all SSC selected in the screening procedure seismically induced failures have to be considered in addition to the usual stochastic failures in the process of calculating unavailabilities. The probability of a seismically induced failure of a SSC depends on the peak ground acceleration, i.e. failure curves are to be used instead of point values (including uncertainties) to calculate the core damage frequency in case of an earthquake.

REFERENCES

BMU (Bundesministerium für Umwelt, Naturschutz und Reaktorsicherheit) 2005. Sicherheitsüberprüfung für Kern-kraftwerke gemäß §19a des Atomgesetzes – Leitfaden Pro-babilistische Sicherheitsanalyse, 31. Januar 2005, Bekanntmachung vom 30. August 2005, Bundesanzeiger, Jahrgang 57, Nummer 207a, 3.

FAK (Facharbeitskreis probabilistische Sicherheitsanalyse für Kernkraftwerke) 2005. Methoden zur probabilistischen Sicherheitsanalyse für Kernkraftwerke, Stand August 2005, BfS-SCHR-37/05, Wirtschaftsverlag NW / Verlag für neue Wissenschaft GmbH, Salzgitter.

Safety, Reliability and Risk of Structures, Infrastructures and Engineering Systems – Furuta, Frangopol & Shinozuka (eds)
© 2010 Taylor & Francis Group, London, ISBN 978-0-415-47557-0

Considerations on the updating of earthquake resistant design codes

H. Sandi

Academy of Technical Sciences of Romania & Institute of Geodynamics of the Romanian Academy, Bucharest, Romania.

1 INTRODUCTION

Some general features of codes are referred to. The constraints to codes for practice, determined by the need of limitation of sophistication, are mentioned too. The main topics dealt with in the paper are presented.

2 CONSIDERATIONS ON THE SPECIFICATION OF SEISMIC ACTION FOR DESIGN

The framework of dealing with representations of seismic action is referred to: representation related to an individual event and representation related to sequences of occurrence cases, and also the conceptual difference of dealing with past events, versus dealing with future, expected, events.

Alternative usual representations of motion for one event: design spectra, design accelerograms and stochastic, are referred to. The approach advocated is postulation of stochastic models and generation on this basis of design accelerograms. Some basic relations concerning random motions are given.

The need to deal with nD representations of seismic hazard is emphasized. Alternative definitions of spaces of accelerograms are referred to. Basic relations on seismic hazard, comparatively for $1D$ and nD representations, are given.

3 OBJECTIVES, OR STRATEGIES, OF ENGINEERING ANALYSES

The use of design accelerograms in engineering analyses is assumed. The alternative objectives or strategies considered for analysis are:

$S.1$: random computational experiment;

$S.2$: examination of sensitivity of output with respect to the variation of some (global, or macroscopic) input parameters;

$S.3$: analysis of seismic vulnerability;

$S.4$: full analysis of risk of exceedance of various limit states (in dependence of the duration of exposure, or of service).

The lack of consideration of these aspects in current codes is emphasized and a solution for gradual implementation is suggested.

4 IMPLEMENTATION OF THE CAPACITY DESIGN STRATEGY

The need of being consistent in the use in practice of the capacity design philosophy is referred to. The specific aspects in relation to which the consistency of codes should be improved are:

a) the need of 3D application of the capacity design principle;

b) consideration of structure and ground as a whole;

c) consideration of the system of nominally structural and nominally non-structural components as a whole in relation to capacity design.

These aspects are discussed, consistency of approach is advocated and some references to relevant studies are presented.

5 SOME COMMENTS ON CODES FOR PRACTICE

The need to consider in codes of interaction between neighboring structures, of consistency in capacity design and of probabilistic concepts is discussed.

6 FINAL CONSIDERATIONS

The orientation of the paper is summarized. Dealing in codes with approaches perhaps not yet ripe for practice, but of conceptual and educational im- portance, is advocated.

Safety, Reliability and Risk of Structures, Infrastructures and
Engineering Systems – Furuta, Frangopol & Shinozuka (eds)
© 2010 Taylor & Francis Group, London, ISBN 978-0-415-47557-0

A study on a simple estimation method of seismic reliability indices against residual deformation for quay walls

T. Nagao

National Institute for Land and Infrastructure Management, Yokosuka, Japan

ABSTRACT

The technical standards for port and harbor facilities in Japan were revised in 2007 and the reliability-based design method was introduced. The target re-liability indices and partial fac-tors for breakwaters are shown in the new technical standards. However, with quay walls of gravity type and sheet pile type, the determin-istic design metohds are still applied in the new tech-naical standards. Residual deformation during earth-quake is the controlling damage mode for quay walls and residual deformation of quay walls is obtained by a nonlinear earthquake response analysis by the 2-dimensional finite element method, that makes prob-abilistic verification corresponding to deformation difficult by analytical techniques. The Monte Carlo simulation technique is considered an effective method of evaluating damage probabil-ity. However, because evaluation of probability by a Monte Carlo simula-tion requires a large number of trials, performing a Monte Carlo simulation by nonlinear earthquake anal-ysis using 2-dimensional finite element method is considered unrealistic from the viewpoint of applica-tion to practical design work, as this method requires a computation time of about 1 hour or so for each trial.

Adoption of the first-order second moment method is a conceivable solution to this problem. In this method, the earthquake response analysis is per-formed only several times, a simple evaluation of the derivative of the performance function is made based on the results, and a simple evaluation of the reliability index is made.

The present research focuses on residual deforma-tion of quay walls, and attempts an evaluation of the probability that deformation will exceed the allow-able value by the first-order second moment method, considering deviations in the S wave velocity of the ground. It was shown that the first-order second moment method is highly applicable to the problem. In particular, when the residual deformation is small, the intercept of a linear regression equation for the initial natural period of the ground and residual de-formation takes a value close to zero. From the viewpoint of reducing the computational load in practical design work, a method of using only the calculated results for one point rather than three points in estimations of the reliability index was studied. A method which assumes that a linear re-gression equation of the natu-ral period of the ground and the residual deformation passes through the ori-gin shows comparatively good agreement with the 3-point approximation method in the range where the reliability index is 1.0-3.0. How-ever, in the region where deformation is 20cm or less, this method tends to overestimate the reliability index. There-fore, a simple estimation method which gives con-servative results for reliability indices in the range of 1.0-3.0 was proposed.

Organized Session (OS23) New Methods for Non-Gaussian and Pulse Problems in Stochastic Dynamics [Session Organized on Behalf of IASSAR Subcommittee SC2]

Safety, Reliability and Risk of Structures, Infrastructures and Engineering Systems – Furuta, Frangopol & Shinozuka (eds)
© 2010 Taylor & Francis Group, London, ISBN 978-0-415-47557-0

Application of the path integration method to reliability problems of dynamic systems

D.V. Iourtchenko
Department of Mathematical Sciences, SPBSPU, S.-Petersburg, Russia

A. Naess & O. Batsevych
Centre for Ships and Ocean Structures & Department of Mathematical Sciences
Norwegian University of Science and Technology, Trondheim, Norway

ABSTRACT

The paper presents a first passage type reliability analysis of nonlinear stochastic single-degree-of-freedom dynamic systems. The path integration method is used to obtain the reliability function and the first passage time.

Reliability and safety are major concerns in designing and developing modern mechanical systems. A system's reliability may be considered as the probability that no system failure occurs within a given time interval. Often the reliability problem is associated with finding the probability that a system's response stays within a prescribed domain, an outcrossing of which leads to immediate failure. A problem of this type is called the first passage problem [Lin & Cai, 1995]. A number of different methods has been proposed to evaluate the first passage time, starting from pioneering works by [Roberts 1976, 1978]. Recently, a new tail-equivalent linearization method has been developed in [Fujimura and Kiureghian, 2007], which may be used for reliability estimates for single as well as multiple-degree-of-freedom (MDOF) systems for stationary inputs.

Special attention should be paid to the reliability of systems, which appears as a result of some design or optimization procedures. Indeed, the purpose of these procedures is to satisfy certain criteria, often not related to the system's reliability. In fact, their implementation may lead to a deterioration of the system's reliability. For instance, consider a stochastic optimal-control problem which aims to reduce the mean response energy of a single-degree-of-freedom (SDOF) undamped, linear oscillator, subjected to a zero-mean external Gaussian white noise, by means of a bounded in magnitude control force. It has been demonstrated in that an optimal control law for a steady-state response is represented by a dry friction law. On the other hand, it has been shown by asymptotic analysis in that a stochastic system with dry friction is less reliable than that of a system with linear damping.

This paper is devoted to a reliability investigation of dynamic systems by application of the numerical path integration (PI) method. The application of the PI method for reliability problems is a relatively novel approach, especially for strongly nonlinear systems. The authors have used the PI method earlier for these systems to estimate the stationary response probability density function of the state space variables [Iourtchenko et. al. 2006]. Later an investigation revealed certain difficulties in dealing with strongly nonlinear systems. Thus, certain development of the PI method is required to accommodate the considered problems [Iourtchenko et. al. 2008]. The PI code is validated by comparing some results to results of Monte Carlo simulations as well as results, obtained for an equivalent linear system. Moreover, a reliability of a first order system subjected to multiplicative and additive noises is considered. For this system we are interested in investigation of an influence of higher order moments' instability on the system's first passage time.

REFERENCES

Lin, Y.K. and Cai, G.Q. (1995). Probabilistic Structural Dynamics. McGraw-Hill., N.Y.

Roberts, J.B. (1976). First passage time for the envelope of a randomly excited linear oscillator. Journal of Sound and Vibraion. Vol.46, No.1, pp. 1–14.

Roberts, J.B. (1978). First passage time for oscillators with non-linear restoring forces. Journal of Sound and Vibraion. Vol.56, No.1, pp. 71–86.

Fujimura K., Kiureghian A. (2007). Tail-equivalent linearization method for nonlinear random vibration. Probabilistic Engineering Mechanics. Vol.22, pp. 63–76.

Iourtchenko D.V., Mo E., Naess A. (2006) Response probability density functions of strongly nonlinear systems by the path integration method. Int. J. of Non-Linear Mechanics. Vol.41, No.5, pp. 693–705.

Iourtchenko D.V., Mo E., Naess A. (2008) Reliability of Strongly Nonlinear Single Degree of Freedom Dynamic Systems by the Path Integration Method. Journal of Applied Mechanics. Vol.75.

Safety, Reliability and Risk of Structures, Infrastructures and Engineering Systems – Furuta, Frangopol & Shinozuka (eds)
© *2010 Taylor & Francis Group, London, ISBN 978-0-415-47557-0*

Simulation of strongly non-Gaussian multi-variate random fields

P. Bocchini
University of Bologna, Bologna, Italy

G. Deodatis
Columbia University, New York, NY, USA

ABSTRACT

The physical quantities involved in the Monte Carlo Simulation of problems in mechanics are often described by a number of correlated variables. For instance, the ultimate strength and ultimate elongation of a cable wire and the various mechanical properties of soils are correlated to a certain degree. Other times, it is useful to substitute an n-dimensional random field (or wave, meaning that one dimension is in the time domain, and the others are in the space domain) by a set of correlated (n - 1)-dimensional fields/processes/waves. For instance, the wind speed acting on a structure is a random wave that is usually modeled as a set of correlated random processes, each of them corresponding to a particular point. In both cases, an algorithm for the simulation of multi-variate (and in general non-Gaussian) random processes/fields is required.

Many authors have proposed methodologies for the simulation of Gaussian vector fields and processes, while only a few have considered also the case of non-Gaussian multi-variate fields/processes. This paper presents an extension of the basic ideas involved in the algorithm proposed by Bocchini and Deodatis (2008) for simulation of uni-variate, strongly non-Gaussian random fields/processes into the case of simulation of homogeneous, one-dimensional, multi-variate (1D-mV) strongly non-Gaussian random fields.

The algorithm belongs to a class of simulation algorithms based on the spectral representation method (Shinozuka and Jan, 1972) and on the translation field theory (Grigoriu, 1995). Therefore, it identifies "underlying Gaussian coherences" or "underlying Gaussian cross-spectra" – as well as "underlying Gaussian auto-spectra" – so that the resulting non-Gaussian vector field, after the translation non-linear mapping, matches the prescribed marginal probability distribution of each component, as well as the target non-Gaussian Cross-Spectral Density Matrix (CSDM). This task is performed by means of an iterative "trial and error" scheme. The proposed approach avoids drawbacks of previous spectral-representation-based versions of the algorithm that had some difficulties matching the prescribed cross-correlation functions, especially in cases of high coherence. The coupling effect among the various components of the vector process determined by the multi-variate Spectral Representation Method is negligible for the purposes of identification of the underlying CSDM. Thus, its elements are analyzed independently, with considerable savings of computational time. Even if the underlying Gaussian cross-spectra are assumed to be real in this work, the algorithm is perfectly capable of considering also the more general case of (complex) target non-Gaussian cross-spectra.

In the numerical examples that are provided, the proposed algorithm has shown high accuracy and high efficiency, especially when a very large number of samples is required (as is the case in typical applications of Monte Carlo Simulation).

REFERENCES

Bocchini, P. and Deodatis, G, 2008. Critical review and latest developments of a class of simulation algorithms for strongly non-Gaussian random fields. Probabilistic Engineering Mechanic, 23(4), 393–407.

Grigoriu, M., 1995. Applied non-Gaussian processes. Prentice Hall.

Shinozuka, M. and Jan, C.M., 1972. Digital simulation of random processes and its application. Journal of sound and vibration, 25(1), 111–128.

Safety, Reliability and Risk of Structures, Infrastructures and Engineering Systems – Furuta, Frangopol & Shinozuka (eds)
© 2010 Taylor & Francis Group, London, ISBN 978-0-415-47557-0

On the derivation of the Fokker-Plank equations by using of fractional calculus

Giulio Cottone & Mario Di Paola
Università degli Studi di Palermo, Palermo, Italy

Francesco Marino
Università degli Studi Mediterranea di Reggio Calabria, Reggio Calabria, Italy

1 INTRODUCTION

Non Linear systems driven by normal or, more generally non-normal white noise processes arise in many problems of engineering interest, such as random vibration theory, earthquake engineering, reliability analysis, to cite just few. Such systems are usually handled by the celebrated Itô stochastic calculus, which mainly consent to study the evolution of the statistics of the response to the system. For instance, the probability density (PDF) of the response might be recovered by means of the so called Fokker-Planck (FP) equation and the characteristic function (CF) evolution is ruled by the Spectral Fokker-Planck (SFP) equation. In case of normal and Poisson white noise excitation there is a spread agreement in literature on the FP and SFP equations, the same cannot be said for stable Lévy excitation. Jespersen et al, 1999 and Metzler & Klafter, 2000 obtained, by means of the so called continuous time random walk (CTRW), the equations ruling the evolution of the PDF. In such equations the Riesz fractional derivative of the PDF appears. Samorodnitsky & Grigoriu, 2003 studied the same system obtaining different behavior for the tails of the stationary PDF.

In this paper a different procedure for deriving the FP equation, based on the fractional calculus is proposed, for Gaussian, Poissonian and stable white noise excitation, getting the Metzler et al' results. The mathematical tool used is the fractional calculus. In recent papers (Cottone & Di Paola, 2008a) and (Cottone et al, 2008b), it has been shown that every PDF or every

CF of a random variable X can be represented by a series whose coefficients are the Riemann-Liouville fractional derivatives of the CF evaluated in 0 of order $\gamma \in C$. Moreover it has been shown that these fractional derivatives are the complex moments of the type $E\left[(\mp iX)^{-\gamma}\right]$. Then in this paper, such complex moment representation of the CF is used in order to derive the SPFK equation of non-linear systems. By means of the same arguments the Kolmogorov-Feller and the Einstein-Smoluchowsky equations are also derived.

REFERENCES

Cottone, G & Di Paola, M. 2008a. On the Use of Fractional Calculus for the Probabilistic Characterization of Random Variables. *Probabilistic Engineering Mechanics*, doi: 10.1016/j.probenmech.2008.08.002.

Cottone, G., Di Paola, M., Pirrotta A. 2008b. Path integral solution by fractional calculus, *Journal of Physics: Conference Series* 96 012007 doi: 10.1088/1742-6596/96/1/012007.

Jespersen, S., Metzler, R., Fogedby, H.G. 1999. Lévy Flights in External Force Fields: Langevin and Fractional Fokker-Planck Equation and their Solutions, *Physical Review*, 59, (3).

Metzler, R., Klafter, J. 2000. The random walk's guide to anomalous diffusion: a fractional dynamics approach, *Physics Reports*, 339, 1–77.

Samorodnitsky, G., Grigoriu, M. 2003. Tails of solutions of certain nonlinear stochastic differential equations driven by heavy tailed Lévy motions, *Stochastic Processes and their Applications*, 105, 69–97.

Safety, Reliability and Risk of Structures, Infrastructures and
Engineering Systems – Furuta, Frangopol & Shinozuka (eds)
© 2010 Taylor & Francis Group, London, ISBN 978-0-415-47557-0

Moment equations and modified closure approximations for a non-linear oscillator under renewal impulse process excitations

M. Tellier
Arup SA, Johannesburg, South Africa

R. Iwankiewicz
Hamburg University of Technology, Hamburg, Germany

ABSTRACT

Linear models are usually accurate to reproduce the dynamic behavior of structures under small amplitude vibrations. The use of non-linear models, however, becomes fundamental in modeling natural impact loads such as a strong ground motion acceleration due to earthquake (Lin, 1967), loading caused by wind, the motion of vehicles on rough ground. The structural response to natural hazard loads as those listed above, may exhibit strongly non-linear characteristics. The simplest model of such excitations is a Poisson train of impulses (Roberts and Spanos, 1990).

The assumption inherent in the Poisson law, that the probability of an event remains constant, is seldom true when actual trains of impulses are considered. Resorting to Erlang renewal driven processes in modeling the impulsive loading phenomena, allows accounting for more realistic distribution of the interarrival times.

The equation governing a non-linear, non-hysteretic oscillator under a random train of impulses can be written as

$$\ddot{X} + f\left(X,\dot{X}\right) = \sum_{i,R=1}^{R_\nu(t)} P_{i,R}\delta\left(t - t_{i,R}\right) \tag{1}$$

Where f(.) is a function of instantaneous values of displacement and velocity and the stochastic excitation is a random train of impulses whose arrival times are driven by an Erlang renewal process $R_\nu(t)$.

The original train of impulses may be replaced by a Poisson driven one with the aid of auxiliary variables (Nielsen, Iwankiewicz and Skjaerbaek, 1995, Tellier and Iwankiewicz, 2005).

The original non-Markov problem is converted into a diffusive one.

With the aid of the generalized Ito's differential rule, the equations governing the evolutions of moments of the augmented state vector can be derived.

If the non-linearities are polynomial functions of displacement and velocity, the equations for moments form an infinite hierarchy. The unknown moments can be evaluated only approximately.

If the non-linearities are other than polynomial, the expectation of the non-linear transformations of the state variables cannot be directly expressed in terms of moments.

A novel closure scheme is here developed that takes into account the specific physical properties of the impulsive load process (Iwankiewicz, Nielsen and Christensen, 1990; Iwankiewicz and Nielsen, 1999). The joint probability density of the augmented state vector is expressed as sum of contributions conditioned on the 'on' and 'off' states of the auxiliary variables.

A comparison between analytical and simulation results shows that the closure scheme devised works well for highly non-Gaussian impulsive excitation processes.

REFERENCES

Iwankiewicz R., S.R.K. Nielsen and P. Thoft-Christensen, 1990. Dynamic response of non-linear systems to Poisson-distributed pulse-trains: Markov approach, Structural Safety, vol. 8, pp.223–238.
Iwankiewicz R. and S.R.K. Nielsen, 1999. Vibration Theory Vol. 4, Advanced methods in stochastic dynamics of non-linear systems, Aalborg University Press, Denmark.
Lin Y.K. 1967. Probabilistic Theory of Structural Dynamics. New York: McGraw-Hill.
Nielsen S.R.K., R. Iwankiewicz and P.S. Skjaerbaek, Moment equations for non-linear systems under renewal-driven random impulses with gamma-distributed interarrival times, IUTAM Symposium on Advances in Non-linear Mechanics, Trondheim, Norway, July 1995, EDS. A. Naess and S. Krenk, Kluwer Academic Publishing, 331–340.
Roberts J.B. and P.D. Spanos, 1990. Random Vibration and Statistical Linearization, Wiley, New York.
Tellier M. and R. Iwankiewicz. 2005. Response of linear dynamic systems to non-Erlang renewal impulses: Stochastic equation approach. Probabilistic Engng Mech 20 (2005): 281–295.

Mini-Symposia (MS09) Life-Cycle Reliability and Optimization of Deteriorating Structures

Safety, Reliability and Risk of Structures, Infrastructures and
Engineering Systems – Furuta, Frangopol & Shinozuka (eds)
© 2010 Taylor & Francis Group, London, ISBN 978-0-415-47557-0

Corrosion damage prediction updating based on visual survey information

Q. Suo & M.G. Stewart

Centre for Infrastructure Performance and Reliability, The University of Newcastle, Callaghan, NSW, Australia

ABSTACT

As concrete cover, concrete strength, surface chloride concentration and other material, environmental and dimensional properties, which influence the corrosion process, are not the same across the whole structure, random field theory can be used to model these basic variables over time and space. Existing models based on stochastic and random field modelling are used for predicting the likelihood and extent of corrosion-induced cracking. These models are modified to include updating of predictions based on visual inspection data, as routine visual inspection can identify the changes that have occurred since previous inspections. The variability of predictions can be minimized and confidence of results would be improved if incorporating new information into the prediction. Bayesian Theorem provides a rational method for incorporating existing information or judgments into predictions of future outcomes.

If n inspections are conducted at times $t_1, t_2, \ldots, t_n$, where $x_i\%$ of surface with crack width $w(t_i)$ exceeding w_i mm is observed at time t_i ($i = 1, 2, \ldots, n$) then the inspection scenario H can be expressed as

$$\left[d_{crack}(t_1) = x_1\% \cap w(t_1) = w_1\right] \cap \left[d_{crack}(t_2) = x_2\%) \cap w(t_2) = w_2\right] \cap$$
$$\cdots \cap \left[d_{crack}(t_n = x_n\%) \cap w(t_n) = w_n\right] \quad (1)$$

where $d_{crack}(t_i)$ is the cracking proportion of a concrete surface, $w(t_i)$ is the crack width at the time t_i and w_i is the threshold value of crack width.

Actual U.S. inspection results for corrosion damage to RC bridge decks (2D random field) exposed to de-icing salts environment are used to illustrate the capabilities of the method, see Table 1. A 70 m long and 3.6 m wide right-hand bridge deck being divided into 250 elements is used as the illustrative example.

Monte-Carlo simulation analysis is used to infer the prior and updated mean and standard deviation of surface cracking damage. The prior and updated mean of $d_{crack}(t)$ is shown is Figure 1.

The results show that the occurrence of surface cracking damage changes future corrosion damage predictions with more accuracy and less variability. The future corrosion damage trends are mainly determined by the existing damage on the deck which can be

Table 1. Inspection survey results (Williamson et al. 2008).

0.5 w/c bridge decks				
Structure no.	t_1 (year)	% damaged at t_1	t_2 (year)	% damaged at t_2
S1 1-1804	34	3.2	37	5.2
S2 1-6101	34	0.0	37	0.0
S3 9-2801	33	1.6	36	3.2
S4 9-6042	34	1.2	37	1.2

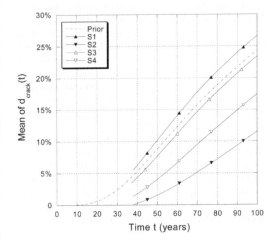

Figure 1. Prior and posterior mean of cracking proportion.

revealed during one or more the visual inspections. The developed models can provide help in planning maintenance and management strategies for RC structures which are subject to corrosion.

Safety, Reliability and Risk of Structures, Infrastructures and Engineering Systems – Furuta, Frangopol & Shinozuka (eds)
© 2010 Taylor & Francis Group, London, ISBN 978-0-415-47557-0

Reliability-based optimal design of prestressed concrete girders under chloride pitting corrosion

V.S. Nguyen & J.S. Kong

School of Civil, Environmental and Architecture Engineering, Korea University, South Korea.

ABSTRACT

Prestress concrete (PC) box girder bridge is widely applied for infrastructure system in many countries. During service time, the performance of structure with considering the environmental effect is necessary to evaluate and analysis in the design process. For solving reliability based design optimization (RBDO) problem in bridge engineering, many of researchers developed the computer programs to perform the optimization. However, these methods are complicated and not convenient for linking two computer programs between general-purpose optimization software ADS and reliability analysis program. This paper presents a method to solve RBDO of prestress concrete box girder bridges using a computer program which integrates the Matlab optimization toolbox and a reliability analysis subroutine. Moreover, the effect of environmental agents is considered in terms of modeling the pitting corrosion phenomena which occur on the post-tensioned tendon during the structural lifetime.

An simply supported PC box girder bridge is considered with span length $L=50$ m, subjected to self-weight dead load and three lane live load HL93 follows AASHTO LRFD code. The results of design variables are shown in Table 1 which X_1 to X_3 are the tendon, shear and torsion reinforcement area, X_4 to X_9 are the geometry dimensions of cross section. The graph of

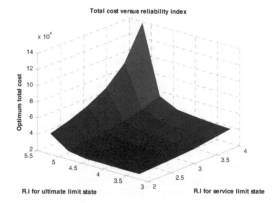

Figure 1. Optimum total cost versus reliability index for ultimate and service limit state.

association effect of reliability index for ultimate limit state and service limit state on the optimum total cost is shown in Figure 1. It can be seen that target reliability indices in the range β_{ser}: 2–3.2 and β_{ult}: 4.4–4.8 may lead to the optimal cost for prestressed concrete structures. These results are in good agreement with the range of target reliability indices suggested in the Euro code and the JCSS code. Moreover, the proposed method to solve RBDO of PC box girder bridge by using optimization toolbox and code development is simple, flexible, practical and appropriate with the bridge engineering design and other field.

Table 1. Results of PC box girder bridge RBDO.

Design variables	Unit	Value	R. index	Value
X_1	cm^2	392.78	β_1	3.1
X_2	cm^2/m	27.76	β_2	2.5
X_3	cm^2/m	5.22	β_3	3.4
X_4	m	2.98	β_4	3.8
X_5	m	6.63	β_5	4.5
X_6	m	3.48	β_6	4.4
X_7	m	0.22	β_7	6.3
X_8	m	0.41	β_8	4.4
X_9	m	0.22	β_9	5.4
Total Cost:	–	32047	β_{10}	4.4

REFERENCES

Ali S. AL-Harthy & Dan M. Frangopol. 1994. Reliability assessment of prestressed concrete beams. *Journal of structure engineering ASCE* 120(1).

American Association of State Highway and Transport Officials, AASHTO LRFD. *2007 specifications for highway-bridges*. Washington, DC.

Samer Barakat et al. 2003. Single objective reliability-based optimization of pre-stressed concrete beams. *Journal of Computers and Structures* 81:2501–2512.

Safety, Reliability and Risk of Structures, Infrastructures and
Engineering Systems – Furuta, Frangopol & Shinozuka (eds)
© 2010 Taylor & Francis Group, London, ISBN 978-0-415-47557-0

Effects of uncertainty on performance of pavement rehabilitation using the dowel bar retrofit technique

Yogini S. Deshpande, Yue Li, Jacob E. Hiller & Yuejun Yin
Michigan Technological University, Houghton, MI, USA

ABSTRACT

Dowel bar retrofitting (DBR) is a pavement rehabilitation technique commonly adopted to increase the service life of pavements in the United States America. This method involves either preventative or corrective maintenance of structural cracks or joints in jointed concrete pavement structures. By saw cutting slots and removing old concrete from these slots, dowel bars can be retrofitted into existing pavements. This allows for the reduction of deflections as wheel loads approach these discontinuities in the pavement slab, thereby reducing faulting between slabs and increasing ride quality for the driver. This rehabilitation method also significantly reduces the bending stresses in each slab, thereby increasing the fatigue life of a concrete pavement.

One major flaw of DBR is related to construction practices of this method. This involves placement of dowels and the concrete fill of the dowel slots, but also of the concrete material selection and mixing process. While properly placed DBR sites have performed admirably over the years (Pierce 1994), poor material and construction practices tend to either immediately reduce load transfer or lend themselves to a breakdown of the concrete slot fill and bond between this material and the older concrete slab (Deshpande 2006). When this happens, the dowels are not allowed to bear against the other slab, thereby reducing their ability to transfer load.

Using finite element analysis software for concrete pavements (ISLAB2000), the material and construction practices of DBR were modeled as a dowel-concrete interaction (DCI) factor, which was first proposed by Friberg (1940). This method models the dowels on a set of springs so that a stiff spring models a well-embedded dowel bar, while a less-stiff set of springs would indicate dowel looseness or poor bond between the concrete slot mix and the old concrete slab. A wide range of input parameters were selected to develop a series of curves to relate the load transfer efficiency determined from deflections (LTE_δ) to this DCI factor for several geometries, material properties, and concrete slab support conditions. From this analysis, a relationship between LTE_δ, DCI, and the stress-based load transfer (LTE_σ) were developed. The use of LTE_σ allows a pavement designer or analyst the ability to determine what effect the load transfer mechanism has on reducing bending stress in the slab to increase the fatigue life.

Analysis of three DBR sites in Michigan (Hiller and Buch 2004) was performed using field test results based on load transfer efficiency of the rehabilitated pavement joint. Best-fit probability distributions and an ANOVA analysis was conducted on these three sites to determine if the materials and construction practices of these three sites were significantly different in terms of the calculated DCI values, which nullifies differences in slab geometries and support conditions between these sites. Correlation coefficients were also calculated between the measured LTE_δ and DCI values to find if these sites were significantly different. Through these analyses, it was observed that the performance efficiency of the DBR technique largely depended upon environmental factors and construction practices adopted at each specific site.

REFERENCES

Deshpande, Y. S. (2006). "Evaluation of Commercial Rapid-Setting Materials and Development of Rapid-Setting Self-Compacting Concrete for Dowel Bar Retrofit Applications" Ph.D., Purdue University.

Friberg, B.F. (1940). Design of Dowels in Transverse Joints of Concrete Pavements, Transactions, ASCE 105, pp. 1076–1116.

Hiller, J., and Buch, N. (2004). "Assessment of Retrofit Dowel Benefits in Cracked Portland Cement Concrete Pavements" Journal of Performance of Constructed Facilities 18(1), 29–35.

Pierce, L. M. (1994). "Portland Cement Concrete Pavement Rehabilitation in Washington State: Case Study." Transportation Research Record (1449), 189–198.

Safety, Reliability and Risk of Structures, Infrastructures and
Engineering Systems – Furuta, Frangopol & Shinozuka (eds)
© 2010 Taylor & Francis Group, London, ISBN 978-0-415-47557-0

Fracture mechanics and reliability based inspection planning for ship structures

Nian-Zhong Chen & Ge Wang

American Bureau of Shipping, Houston, TX, USA

ABSTRACT

Fracture mechanics based reliability assessment and inspection planning for ship structures have been continually receiving much attention since 1980s. However, in general little concern from ship industries has been shown on framing fracture mechanics based rules or guidance for establishing a standard reliability assessment procedure and the corresponding inspection plan for ship structures. This paper tried to present an approach towards practical application for establishing a fracture mechanics and reliability based inspection plan for ship structures.

The developed approach is an integration of fracture mechanics, structural reliability analysis, Bayesian approach, rules and standards from industries and it in general serves the purpose of the practical applications on establishment of a reasonable inspection plan for ship structures. In this approach, the long-term stress range acting on a flaw is assumed to be fitted by a two-parameter Weibull distribution and the corresponding parameters are determined from the rules of American Bureau of Shipping (ABS). The remaining life of structure is predicted by Paris Law based on the British Standard 7910 and a through-thickness failure criterion. The corresponding limit state functions are established with an appropriate stochastic modeling considering initial size of flaw, modeling uncertainty, thickness of plate, material parameters, stress concentration factor, stress range and probability of detection (POD). A first order reliability method coupled with a finite difference method is used to predict the time-variant reliability of the limit state functions. The updated reliability after inspection is computed by the Bayesian approach taking into account the curve of POD. An inspection plan is then set up on the basis of the comparison between the calculated reliability and target reliability index. An example is conducted to demonstrate the capacities of the approach and the code developed.

Fracture mechanics based limit state functions are normally not in a closed form. Monte-Carlo simulation or importance sampling techniques were usually adopted in most of previous studies. In this paper, a new solution with use of a finite difference method incorporated into a first order reliability method for reliability estimation of such implicit limit state functions was developed. This solution not only significantly speeds the computation but also provides sufficient numerical accuracy.

REFERENCES

Chen, N.Z. & Guedes Soares, C. 2007a. Reliability assessment of post-buckling compressive strength of laminated composite plates and stiffened panels under axial compression. *International Journal of Solids and Structures* 44: 7167–7182.

Chen, N.Z. & Guedes Soares, C. 2007b. Reliability assessment for ultimate longitudinal strength of ship hulls in composite materials. *Probabilistic Engineering Mechanics* 22: 330–342.

Moan, T., Ayala-Uraga, E. 2008. Reliability-based assessment of deteriorating ship structures operating in multiple sea loading climates. *Reliability Engineering and System Safety* 93: 433–446.

Wirsching, P.H., Torng, T.Y., Geyer, J.F., Stahl, B. 1990. Fatigue reliability and maintainability of marine structures. *Marine Structures* 3: 265–284.

Safety, Reliability and Risk of Structures, Infrastructures and
Engineering Systems – Furuta, Frangopol & Shinozuka (eds)
© 2010 Taylor & Francis Group, London, ISBN 978-0-415-47557-0

Finite-element modeling via a cohesive-zone approach of reinforced-concrete beams strengthened with fiber-reinforced polymers

G. Alfano

School of Engineering and Design, Brunel University, Uxbridge, UK

F. De Cicco, A. Prota, G. Manfredi & E. Cosenza,

Department of Structural Engineering, University of Naples 'Federico II', Naples, Italy

ABSTRACT

Two-dimensional nonlinear finite-element analysis of reinforced concrete (RC) beams retrofitted with fiber reinforced polymers (FRP) have been carried out up to failure in order to correctly model the interaction between FRP and concrete substrate, which in general can influence the overall performance of a standard RC element.

An isotropic elasto-plastic model for concrete has been used to account for the biaxial-stress state with accuracy. In order to model the bond-slip interaction between the steel reinforcing bars and concrete, and therefore the tension-stiffening effect, use has been made of a cohesive-zone model approach, through a linear-elastic-damage model; the same model has been adopted to account for the possible debonding of the FRP sheet. The investigation of the effects of possible debonding of FRP has been enriched by an experimental campaign on a series of RC beams both with and without external FRP reinforcement.

The experimental data have been compared with numerical results in order to validate the modeling methodology, to numerically evaluate some parameters that cannot be experimentally measured (e.g., the interface-stress profile and the whole strain and stress field in the concrete), and to achieve a deeper understanding of the complex failure modes of the externally reinforced beams.

The effectiveness of the proposed methodology of analysis is confirmed by the ability of the model to numerically reproduce the experimental tests with good accuracy.

Table 1. Geometrical properties of the beams.

Beams	L_1	L_2	B	H	A's	As	FRP sheet
	cm	cm	cm	cm	cm²	cm²	
A1-A2	100	60	15	30	157.1	304.7	no
A3-A4	100	60	15	30	157.1	304.7	no
A5-A6	80	70	15	25	157.1	304.7	no
A7-A8	80	70	15	25	157.1	304.7	no
B1-B2	100	60	15	30	157.1	304.7	yes
B3-B4	100	60	15	30	157.1	304.7	yes
B5-B6	80	70	15	25	157.1	304.7	yes
B7-B8	80	70	15	25	157.1	304.7	yes

REFERENCES

J. Lubliner, J. Oliver, S. Oller and E. Oñate; "A plastic damage model for concrete". *International Journal of Solids and Structures*, 25, 229–326, (1989).

Y. Mi, M.A. Crisfield, G.A.O. Davies and H.B. Hellweg; "Progressive delamination using interface elements". *Journal of Composite Materials,* 32(14): 1246–1272, 1998.

G. Alfano and M.A. Crisfield; "Finite element interface models for the delamination analysis of laminated composites: mechanical and computational issues". *International Journal for Numerical Methods in Engineering*, 50(7): 1701–1736, 2001.

Camanho, P.P. and Davila, C.G. 2002. "Mixed-mode decohesion finite elements for the simulation of delamination in composite material". *NASA/TM-2002-211737*: 1–37.

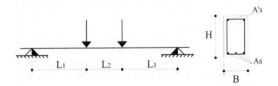

Figure 1. Sketch of tested RC beams.

Safety, Reliability and Risk of Structures, Infrastructures and
Engineering Systems – Furuta, Frangopol & Shinozuka (eds)
© 2010 Taylor & Francis Group, London, ISBN 978-0-415-47557-0

Life cycle performance-based optimal cost allocation methodology for railroad bridge networks

L. Kim
Korea Railroad Research Institute, Uiwang, Korea

H.-N. Cho
Hanyang University, Ansan, Korea

C. Cho
UNICONS Co. Ltd, Seoul, Korea

ABSTRACT

Unlike conventional road bridges, railroad bridges are located in series on a railroad line, which are unable to reduce the operating lanes or construct detour lanes or lines when emergency situations such as natural disasters, train accidents occur during operations. In other words, it may be stated that a problem of a railroad bridge on a route can affect the entire route.

So far, most researches on LCC effective maintenance of bridges have been mainly associated with LCC analysis of a specific bridge to improve maintenance scenario at project level. However, due to the characteristics of the railroad bridges as stated above, a network level maintenance strategies are required prior to commence the project level maintenance for the life cycle effective maintenance and optimal life cycle performance (LCP) of bridges in the network.

In this study, the reliability index is used to represent the performance of a bridge element or system and the effect of repair and rehabilitation. Since a railroad network involves more than hundreds of bridges, Monte Carlo Simulation (MCS) is introduced to evaluate life cycle maintenance costs and performance of lots of elements of each bridge in a line or network at project level.

In order to maintain railroad bridges functionally and effectively at network level, the importance of bridges with a ratio of passengers and freight passing through is defined. It gives priority of the budget allocation to more important bridges. And deterioration weighting factor is considered to allocate higher repair and rehabilitation cost reasonably to more deteriorated elements of bridges on railroad line.

Finally, for optimal cost allocation at network level, Linear programming (LP) is adopted. Linear objective function can be derived to maximize the system performance of network by summing up the performance increase of critical elements of each bridge. It may be constructed by multiplying the factors mentioned above with the improved performance ($\Delta\beta_{ij}$) which represents the effect of repair and rehabilitation and then summing up each result.

The budget limit can be equated with the sum of budget allocation (C_{ij}) for each repair and rehabilitation activities of all the bridges in a network. Also, to ensure bridge performance results in inequality constraints that satisfy required target system reliability of each bridge. Accordingly, a linear programming(LP) problem can be formulated to obtain an optimal or a near optimal solution for improving the performance ($\Delta\beta_{ij}$) of each bridge with optimal repair and rehabilitation activities at network level. LP gives solution easily and effectively as long as it can be expressed as a linear equation form. Also, it can be used very effectively to establish optimal maintenance plan or strategy at network level with numerous bridges such as railroad system. And then, optimal maintenance scenario at network level can be derived through harmonious interaction between project and network level.

In order to verify the construction of optimized maintenance scenarios and the application of budget allocation, the proposed methods are applied to a virtual railroad line, and it is found to be applicable to the real problems for optimal railroad bridge maintenance and management.

REFERENCES

Ang, A. H-S. and Tang, W.H. (1984). *Probability Concepts in engineering Planning and Design*, Vol. I, II, John Wiley & Sons, New York.

Furuta, H., Nakatsu, K. and Frangopol, D. M. (2006). *Optimal cost allocation for improving the seismic performance of road networks*. Proceedings of the 3rd International Conference on Bridge Maintenance, Safety, and Management, 143–144.

Korea Railroad (2005), *Annals railroad statistics*.

Mini-Symposia (MS14) Computational Methods in Stochastic Mechanics
[Session Organized on Behalf of IASSAR Subcommittee SC1]

Safety, Reliability and Risk of Structures, Infrastructures and
Engineering Systems – Furuta, Frangopol & Shinozuka (eds)
© 2010 Taylor & Francis Group, London, ISBN 978-0-415-47557-0

A stochastic finite element method for non-Gaussian random fields

C. Proppe

Institut für Technische Mechanik, Universität Karlsruhe, Karlsruhe, Germany

ABSTRACT

Reliability analysis for problems involving random
parameter fields very often necessitates the solution
of stochastic elliptic boundary value problems. In the
context of the stochastic finite element (FE) method,
approximate solutions of the stochastic boundary
value problem can be viewed as local stochastic
response surfaces that depend on three quantities: the
discretization level of the input random field, discretization parameter for the physical domain and
discretization parameter for the stochastic domain.
Thus, a stochastic response surface can be viewed as
a member of a three parameter family of metamodels.

For non-Gaussian random parameter fields, discretization by polynomial chaos expansions has been
frequently employed. However, recent findings by
Field & Grigoriu (2004) suggest, that polynomial
chaos expansions might be inaccurate for the approximation of higher order moments and the probability
distribution of the non-Gaussian random field, which
might preclude its use for reliability studies.

Therefore, for non-Gaussian fields that can be represented as nonlinear transformations of Gaussian
fields, the truncated Karhunen-Loève expansion of
the non-Gaussian field is introduced in the stochastic
FE formulation. The relationship between the non-
Gaussian random variables of the Karhunen-Loève
expansion and a set of Gaussian random variables is
explicitly computed. The eigenfunctions of the covariance kernel of the non-Gaussian field are evaluated
numerically.

Although sparse grid interpolation could be also
directly applied to the Gaussian random variables of
the Karhunen-Loève representation for the underlying
Gaussian random field, the application of the interpolation to the non-Gaussian random variables is a
more direct alternative and the computation of the non-
Gaussian random variables from the Gaussian random
variables is relatively inexpensive.

Introducing the approximations into the variational formulation of the boundary value problem
and employing local approximations in the stochastic
domain yields together with a collocation scheme an
algebraic problem for the interpolation coefficients of
the approximate solution at the nodes of the FE mesh.
In order to avoid the curse of dimensionality, simulation techniques and stochastic response surfaces
have to be combined: stochastic response surfaces will
direct the simulations and simulations will adapt the
three global parameters of the metamodel.

The developed procedure integrates seamlessly into
non intrusive algorithms, allowing to couple reliability
estimation procedures with existing FE solvers.

REFERENCES

Field Jr., R. V. & Grigoriu, M. 2004. On the accuracy
of the polynomial chaos approximation. *Probabilistic
Engineering Mechanics* 19 (1–2): 65–80.

Safety, Reliability and Risk of Structures, Infrastructures and
Engineering Systems – Furuta, Frangopol & Shinozuka (eds)
© 2010 Taylor & Francis Group, London, ISBN 978-0-415-47557-0

Anisotropic parcimonious polynomial chaos expansions based on the sparsity-of-effects principle

Géraud Blatman

EDF, R&D Division, Site des Renardières, Moret-sur-Loing, France
LaMI, Institut Français de Mécanique Avancée et Université Blaise Pascal,
Campus des Cézeaux, Aubière, France

Bruno Sudret

LaMI, Institut Français de Mécanique Avancée et Université Blaise Pascal,
Campus des Cézeaux, Aubière Cedex, France
Phimeca Engineering S.A., Paris, France

ABSTRACT

Polynomial chaos (PC) expansions allow one to represent explicitly the random response of a mechanical system whose input parameters are modelled by random variables. The PC coefficients may be efficiently computed using a non intrusive regression scheme Berveiller et al., 2006. However, the required number of model evaluations (*i.e.* the computational cost) increases with the PC size, which itself dramatically increases with the number of input variables when the common truncation scheme of the PC expansion is applied (*i.e.* retain all the multivariate polynomials of total degree not greater than a prescribed p).

To circumvent this problem, a truncation strategy based on the use of q-norms with $0 < q < 1$ is proposed. It is motivated by the so-called *sparsity-of-effects principle* (Montgomery, 2004), which states that most models are principally governed by main effects and low-order interactions. The related truncated PC expansions contain a low number of likely important terms compared to the full representation. An anisotropic version is devised in order to further reduce the metamodel complexity. It is aimed at selecting those input random variables with large *sensitivity indices* such as the *Sobol' indices* Sobol', 1993; Sudret, 2008.

Using these truncation strategies, an adaptive algorithm is proposed in order to retain progressively a small number of significant PC coefficients, leading to a *sparse* PC representation. Beside the adaptivity in terms of PC basis, the experimental design is systematically complemented such that the various regression problems are well-posed (Blatman and Sudret 2008. This may be achieved using *sequential sampling* techniques, *e.g. quasi-random numbers* (Niederreiter, 1992).

The method is illustrated by the reliability analysis of a frame structure sketched in Figure 1, featuring 21 *correlated* input random variables. It is shown that the proposed methodology leads to a reduction of the

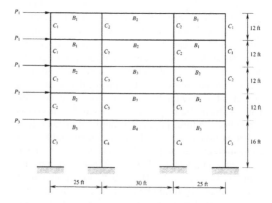

Figure 1. Frame structure

number of finite element runs by a factor 8 compared to traditional full PC expansions.

REFERENCES

Berveiller, M., B. Sudret, and M. Lemaire (2006). Stochastic finite elements: a non intrusive approach by regression. Eur. J. Comput. Mech. 15(1-3), 81–92.

Blatman, G. and B. Sudret (2008). Use of sparse polynomial chaos expansions in adaptive stochastic finite element analysis. Prob. Eng. Mech.. submitted for publication.

Montgomery, D. (2004). Design and analysis of experiments.

Niederreiter, H. (1992). Random number generation and quasi-Monte Carlo methods. Society for Industrial and Applied Mathematics, Philadelphia, PA, USA.

Sobol', I. (1993). Sensitivity estimates for nonlinear mathematical models. Math. Modeling & Comp. Exp. 1, 407–414.

Soize, C. and R. Ghanem (2004). Physical systems with random uncertainties: chaos representations with arbitrary probability measure. SIAM J. Sci. Comput. 26(2), 395–410.

Sudret, B. (2008). Global sensitivity analysis using polynomial chaos expansions. Reliab. Eng. Sys. Safety 93, 964–979.

Safety, Reliability and Risk of Structures, Infrastructures and
Engineering Systems – Furuta, Frangopol & Shinozuka (eds)
© 2010 Taylor & Francis Group, London, ISBN 978-0-415-47557-0

Advanced computational method for reliability analysis of concrete-faced rockfill dam

Qing-Xi Wu & Ming-Zhu Yang
Department of Engineering Mechanics., Hohai University, Nanjing, China

Kui-Zhi Zhao
Department of Geotechnical Engineering, Nanjing Hydraulic Research Institute, Nanjing, China

ABSTRACT

Concrete-faced Rockfill Dam (CFRD) has the advantages of less engineering workload, short construction period, low cost, convenient construction and safe operation. It becomes one of the most chosen dam types, and has the trend of large-scale popularization. The rockfill materials have compli-cated physical and mechanical properties. The parameters (such as elastic modulus, friction angles, etc.) have strong dispersed characteristics, and the variability is large in different zones. The load on the faced rock-fill dam (such as water pressure, seepage pressure, deadweight, etc.) also has some uncertainties. Factors mentioned above give rise to the uncertainties of stress and deformation of CFRD. Therefore, analyzing reliability of CFRD and measuring structural reliability using reliability index is correct and reasonable.

Because of the complexities of the structure of CFRD and the nature of rockfill, it is very difficult to carry through the reliability analysis on this kind of structures. Presently, the satisfactory results cannot be obtained using the conventional methods in struc-tural reliability analysis for large-scale, nonlin-ear structures. Advanced computational method was adopted to calculate the reliability of CFRD and the corresponding calculation formula were derived. The method has the following characteristics: (1) Original 3D nonlinear finite element code can be used directly; (2) Nonlinear response surface function was adopted to substitute the actual complicated limit state function, and the reliability computation can be done briefly, efficiently with better precision. According to a real-life engineering example, the calculation and analyses of anti-crack reliability of the face slab were conducted and better results were obtained.

REFERENCES

Chen Huiyuan. 1985. Analysis of contact element with friction. Journal of Hydraulic Engineering, (4):44–49. (in Chinese)

Duncan. J.M,et al. 1980. Strength Stress-strain & Bulk Modulus Parameters for Finite Element Analysis of Stresses & Movements in Soil Masses. Report No.UCB/GT/80-01/University of California, Berkerly.

Fu Z, Feng J. 1993. Concrete Faced Rockfill Dam. Wuhan: Huazhong University of Technology Press. (in Chinese)

Goodman R.E.. 1968. A model for the mechanics of jointed rock. J. of Soil Mech. & Found. Div., ASCE, 94(3):637–659.

Gu Ganchen, Huang Jinming. 1991. Constitutive model and stress-strain analysis of rockfill in concrete faced fockfill dam.Journal of Hydroelectric Engineering, (1): 12–24. (in Chinese)

Wong F.S. 1985. Slope reliability and response surface method. J. of Geotech. Engrg., ASCE.,111(1)32–53.

Wu Qingxi, Zhuo Jiashou. 2001. A Sequential Response Surface Method with Various f and Its Application to Structural Reliability Analysis. J.of Hohai Univ., 29(2) 75–78. (in Chinese)

Wu Qingxi, Yu Xiaozheng. 2004. Research on the method of reliability analysis of concrete faced rockfill dam. Chinese Journal of Geotechnical Engineering, 26(4): 468–472. (in Chinese)

Ying Wei Liu, Fred Moses. 1994. A sequential response surface method and its application in the reliability analysis of aircraft structural systems. J. of Stru. safety,16:39–46.

Zhu Baili, Shen Zhujiang. 1990. Computation for Mechanics. Shanghai: Shanghai Science and Technology Press. (in Chinese)

Safety, Reliability and Risk of Structures, Infrastructures and Engineering Systems – Furuta, Frangopol & Shinozuka (eds)
© 2010 Taylor & Francis Group, London, ISBN 978-0-415-47557-0

High resolution micrograph synthesis: A parametric texture model and a particle filter

Ramakrishna Tipireddy, Roger Ghanem & Sonjoy Das
University of Southern California, Los Angeles, CA, USA

Somnath Ghosh & Daniel Paquet
The Ohio State University, Columbus, OH, USA

ABSTRACT

This paper presents a method for synthesizing high resolution micrographs from low resolution ones using a parametric texture model and a particle filter. Parameters of the model are computed as a set of joint statistics of the coefficients of a complex wavelet transform. As a first step, the parameters of the model are obtained by decomposing a small number of sample images and computing the joint statistics of the decomposed subbands. Figure 1 shows the low resolution micrograph. High resolution microscopic images are available at locations A and B. High resolution micrograph synthesis is demonstrated by simulating a high resolution micrograph at location C using high resolution microscopic images at locations A and B and the low resolution image at location C.

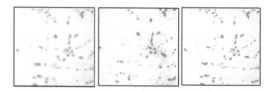

Figure 2. a). Low resolution, magnified digital image at location C. b). Synthesized image at location c). Experimental high resolution image at location C.

The synthesis algorithm generates a random micrograph satisfying the parameters of the texture model by recursively updating the parameters of the input image. Density based Monte Carlo filter is used at each step to update the generated micrograph, using a coarse scale image at that location as an observation. The process is continued until a pre-selected convergence criterion is met. Synthesized images are compared with experimental image. Figure 2 shows low resolution image, synthesized high resolution image and experimental high resolution image.

REFERENCES

Portilla J. and E. P. Simoncelli. A parametric texture model based on joint statistics of complex wavelet coefficients. *International Journal of Computer Vision*, 40(1):49–70, 2000.

Somnath Ghosh, Valiveti D M, Stephen J Harris and James Boileau. A domain partitioning based pre-processor for multi-scale modelling of cast aluminium alloys. *Modelling and simulation in materials science and engineering*, 14 (2006) 1363–1396.

Tanizaki, H 1996, Nonlinear filters: estimation and applications, 2nd edition, Springer Verlag, Berlin.

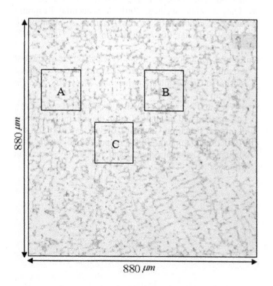

Figure 1. Low resolution, low magnification digital image of cast alluminium alloy W319, for which high resolution images are available at points A and B.

Safety, Reliability and Risk of Structures, Infrastructures and Engineering Systems – Furuta, Frangopol & Shinozuka (eds)
© 2010 Taylor & Francis Group, London, ISBN 978-0-415-47557-0

Multiscale modeling for stochastic forest dynamic

M. Comboul & R. Ghanem
University of Southern California, Los Angeles, CA, USA

ABSTRACT

Our understanding of the effect of varying distur-bance regimes and their interactions on forest succes-sional trajectories is limited. However, we know that neighborhood effects, such as seed dispersal, light availability, and nutrient feedbacks to the soil, all play a role in tree dynamics. These coupled characteristics are defining properties of a complex adaptive system; i.e. the ability to self-organize via local nonlinear interactions and to adjust to external perturbations. Computer simulations are extremely useful for the study of evolving complex systems especially when the accessible observations are scarce. Yet, consider-ing each individual tree interactions can be computa-tionally prohibitive so we need a trade off between detailed mechanistic description and reduced com-plexity. Inspired by the multiscale methods often em-ployed in molecular dynamics simulations, the design of a coarse-grained forest model approximated from a detailed description of the tree interactions produces an effective approach. Our fine scale forest model is described in terms of a Markov Marked Point Process (MPP) as expressed in ?. This model presents several advantages; not only does it offer expandable mecha-nistic properties

through mark additions that feedback into the local interactions, but it also has well-defined statistical properties that can be derived from the en-ergy of the system. As for the coarsening step, we adopted the coarse-graining technique developed in ? and adapted it to the theory of MPPs. Unlike other methods that pre-serve only the slow fluctuations, this process derives a noise model (thus preserving the mi-croscopic fluctu-ations due to the particle interactions) while rescaling the model by increasing the size of the simulation element. The complexity reduction due to the grid coarsening, significantly reduces the required simu-lation time, thereby permitting long-term predic-tions over large forested regions.

REFERENCES

Katsoulakis M., Majda J., Vlachos D., 2003. Coarse-grained stochastic processes and Monte Carlo simu-lations in lattice systems. Journal of Computational Physics, 186(1):250–278.

Särkkä A., Renshaw E., 2006. The analysis of marked point patterns evolving through space and time. Com-putational Statistics & Data Analysis, 51(3):1698–1718.

Organized Session (OS04) Risk Evaluation on Geotechnical and Geo-Environmental Problems

Safety, Reliability and Risk of Structures, Infrastructures and
Engineering Systems – Furuta, Frangopol & Shinozuka (eds)
© 2010 Taylor & Francis Group, London, ISBN 978-0-415-47557-0

Reliability evaluation of a pile foundation system in strength and serviceability limit states

J. Huh & S. Ha
Department of Ocean Civil Engr., Chonnam National University, Yeosu, Korea

A. Haldar
Department of Civil Engr. And Engr. Mechanics, University of Arizona, Tucson, AZ, USA

K. Kwak & J. Lee
Earth Structure & Foundation Engr. Research Div., Korea Inst. of Construction Tech., Koyang, Korea

ABSTRACT

A new efficient and accurate hybrid reliability method is proposed to estimate risk of an axially loaded pile considering realistic pile-soil interaction effect. The soil around the pile is represented by a series of springs. The springs can be characterized by either linear elastic modulus of soil or nonlinear t-z and q-z curves of soil. The system is expected to represent a realistic and efficient load-transfer mechanism. The dual system is then deterministically analyzed using the finite difference method (FDM). Uncertainties associated with load conditions, material and sectional properties of the pile and soil properties are explicitly considered. Uncertainty in soil properties in a layer can be incorporated into the algorithm by considering the stiffness of linear soil springs as random variables when the soil is considered to behave linearly. For the nonlinear soil behavior, the uncertainty in the soil properties is considered by treating f_s and q_p of nonlinear load-transfer curves as random variables, as conceptually shown in Figure 1. q_p and f_s are the unit tip resistance and the unit friction resistance strengths (kN/cm^2) of the soil, respectively.

Since the behavior of such a system is extremely complicated, for the reliability evaluation, a hybrid approach is used by integrating the response surface method (RSM), the FDM, the first-order reliability method (FORM), and an iterative linear interpolation scheme. Since the efficiency of the response surface-based algorithm depends on the intelligent identification of the failure region, the proposed method is integrated with the FORM to locate the failure region. The reliability of the pile-soil system is estimated for both serviceability and strength limit states. Three limit states to be specifically considered are: (1) insufficient soil resistance, (2) excessive settlement of the pile/soil system, and (3) compressive strength failure of the pile.

A concrete pile is considered in the numerical example and the soil profile is considered to be layered. Applicability, accuracy, and efficiency of the proposed reliability algorithm are demonstrated with the help of a realistic example. The results as summarized in Table 1. The accuracy of the method is established by comparing the results with Monte Carlo simulation (MCS). The method is observed to be very efficient since it requires very few deterministic analyses compared to MCS. For the example considered in this study, the serviceability limit state is found to be the most critical and the uncertainty in the soil parameters is very sensitive to the reliability estimation. Accurate estimation of soil properties is very important for the prediction of the behavior of an axially loaded pile-soil system. A considerable amount of resources should be allocated to predict the soil parameters and the serviceability limit states should not be overlooked in practical design of foundations.

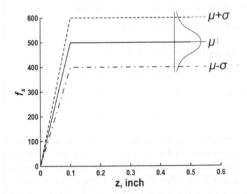

Figure 1. Uncertainty in a f_s-z (t-z) load transfer curve.

Table 1. Reliability analysis results for vertical settlement.

Limit state		Vertical settlement	
MCS	P_f	0.003902 ($\beta \approx 2.660$)	
	NOS	500,000	
No. of R.V.		6 Random variables	
Proposed method	Scheme	1	2
	β	2.633	2.646
	$\approx P_f$	0.004232	0.003989
	Error	−8.5 %	−2.2%
	TNSP	54	103

Safety, Reliability and Risk of Structures, Infrastructures and
Engineering Systems – Furuta, Frangopol & Shinozuka (eds)
© 2010 Taylor & Francis Group, London, ISBN 978-0-415-47557-0

Updating uncertainties in shear strengths with multivariate test data

Jianye Ching
National Taiwan University, Taipei, Taiwan

Yi-Chu Chen
National Taiwan University of Science and Technology, Taipei, Taiwan

Kok-Kwang Phoon
National University of Singapore, Singapore

ABSTRACT

The text in this paper is for visual purpose only. No rights can be taken from this.

Quantifications of uncertainties in soil shear strengths, including undrained shear strength of clay, are essential for geotechnical reliability-based design. Although it is simple to reduce the uncertainties by correlation when the information is one dimensional, it seems challenging to reduce the uncertainties by using multivariate information through correlations. This research proposes a systematic way of achieving multivariate correlations on undrained shear strengths. A set of simplified equations are obtained through Bayesian analysis for the purpose of reducing uncertainties: the inputs to the equations are the results of in-situ or laboratory tests, and the outputs are the updated mean values and coefficients of variation (c.o.v.) of the undrained shear strengths. The analysis results show that the uncertainties in undrained shear strengths can be effectively reduced by incorporating multivariate information.

The undrained shear strength (S_u) considered in this paper is the undrained shear strength determined by CIUC (isotropically consolidated undrained compression) tests. For the model of S_u of clayey soils, the adopted test indices are limited to the following: (a) overconsolidation ratio (OCR); (b) energy-ratio corrected SPT-N value (N_{60}); (c) adjusted CPT reading $q_T'' = q_T' - s_{v0}$, where s_{v0} is the total vertical stress, and q_T is the CPT reading corrected with respect to the pore pressure behind the cone. The updated mean and variance of the logarithm of the undrained shear strengths conditioning on various combination of multivariate test data are listed in the following:

Conditioning on OCR:

$$E\left[\ln\left(S_u\right)\,|\,OCR\right]=0.640\cdot\ln\left(OCR\right)+\ln\left(\sigma_{v0}'\right)-0.874$$
$$Var\left[\ln\left(S_u\right)\,|\,OCR\right]=0.237^2 \qquad (1.)$$

Conditioning on $N60$:

$$E\left[\ln\left(S_u\right)\,|\,N_{60}\right]=0.602\cdot\ln\left(N_{60}\right)+0.243\cdot\ln\left(\sigma_{v0}'\right)+2.363$$
$$Var\left[\ln\left(S_u\right)\,|\,N_{60}\right]=0.277^2 \qquad (2.)$$

Conditioning on qT'':

$$E\left[\ln\left(S_u\right)\,|\,q_T''\right]=0.976\cdot\ln\left(q_T''\right)-2.408$$
$$Var\left[\ln\left(S_u\right)\,|\,q_T''\right]=0.336^2 \qquad (3.)$$

Conditioning on OCR, $N60$:

$$E\left(\ln\left(S_u\right)\,|\,N_{60},OCR\right)=0.373\cdot\ln\left(OCR\right)$$
$$+0.256\cdot\ln\left(N_{60}\right)+0.685\cdot\ln\left(\sigma_{v0}'\right)+0.476 \quad (4.)$$
$$Var\left(\ln\left(S_u\right)\,|\,N_{60},OCR\right)=0.180^2$$

Conditioning on OCR, qT'':

$$E\left(\ln\left(S_u\right)\,|\,q_T'',OCR\right)=0.431\cdot\ln\left(OCR\right)+0.326\cdot\ln\left(q_T''\right)$$
$$+0.674\cdot\ln\left(\sigma_{v0}'\right)-1.418 \qquad (5.)$$
$$Var\left(\ln\left(S_u\right)\,|\,q_T'',OCR\right)=0.194^2$$

Conditioning on $N60$ and qT'':

$$E\left(\ln\left(S_u\right)\,|\,N_{60},q_T''\right)=0.362\cdot\ln\left(N_{60}\right)$$
$$+0.399\cdot\ln\left(q_T''\right)+0.146\cdot\ln\left(\sigma_{v0}'\right)+0.409 \qquad (6.)$$
$$Var\left(\ln\left(S_u\right)\,|\,N_{60},q_T''\right)=0.215^2$$

Conditioning on OCR, $N60$, qT'':

$$E\left(\ln\left(S_u\right)\,|\,N_{60},q_T'',OCR\right)=0.291\cdot\ln\left(OCR\right)$$
$$+0.200\cdot\ln\left(N_{60}\right)+0.220\cdot\ln\left(q_T''\right)+0.534\cdot\ln\left(\sigma_{v0}'\right)-0.187 \quad (7.)$$
$$Var\left(\ln\left(S_u\right)\,|\,N_{60},q_T'',OCR\right)=0.159^2$$

The above results provide estimates for the first two moments of $\ln(S_u)$. Let us denote the estimated mean value and standard deviation of $\ln(S_u)$ by m and s, respectively, then the mean value and c.o.v. of S_u are $\exp(m + s^2/2)$ and $[\exp(s^2) - 1]^0.5$, respectively, by assuming lognormality.

Safety, Reliability and Risk of Structures, Infrastructures and
Engineering Systems – Furuta, Frangopol & Shinozuka (eds)
© 2010 Taylor & Francis Group, London, ISBN 978-0-415-47557-0

Efficient evaluation of slope reliability using importance sampling

Jianye Ching
National Taiwan University, Taipei, Taiwan

Kok-Kwang Phoon
National University of Singapore, Singapore

Yu-Gang Hu
National Taiwan University of Science and Technology, Taipei, Taiwan

ABSTRACT

Evaluating the reliability of a slope is a challenging
task because the possible slip surface is not known
beforehand. Approximate methods via the first-order
reliability method (FORM) provide efficient ways of
evaluating failure probability of the "most probable"
failure surface. The tradeoff is that the failure probability estimates may be biased towards the unconservative
side. The Monte Carlo simulation (MCS) is a viable
unbiased way of estimating the failure probability of a
slope, but MCS is inefficient for problems with small
failure probabilities. This study proposes a novel way
based on the importance sampling technique of estimating slope reliability that is unbiased and yet is much
more efficient than MCS. In particular, the critical
issue of the specification of the importance sampling
probability density function will be addressed in detail.

Several numerical examples are investigated to
verify the proposed novel approach, including the following slope in two clayey soil layers underlain by a
hard soil layer (shown in Figure 1). Furthermore, the
simplified Bishop method of slices is taken as the slope
stability method for this example.

Figure 2 shows the actual failure region F determined by the exhaustive analysis. It is clear that there
are two failure modes from the geometry of the actually failure region. Unfortunately, the FORM method
is only able to identify one mode. More interestingly,
the distances to the origin of the limit state functions
for the two modes are similar, indicating the failure
probabilities of the two failure modes are comparable.

It is interesting to see that both failure modes are captured by the triangles. Those samples are plotted in
Figure 2. It is clear that a large portion of the samples
are failure samples. The MCS method is also taken to
estimate the failure probability. These results are listed
in Table 1 together with the FORM results. It is clear
that the FORM method significantly underestimates
the failure probability although it is computationally
cheap. Both the IS and MCS methods provide unbiased estimates for the failure probability, but the IS
method is obviously more efficient. Furthermore, the
main novelty in this study is to utilize the OMS to
determine suitable locations of the importance sampling probability density function (IS PDF) so that the
IS PDF is much closer to the failure region.

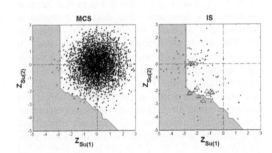

Figure 2. The MCS (N = 10000) and IS (N = 100) samples
in the standard Gaussian space (Example 2).

Table 1. The analysis results from various methods for
Example.

Method	MCS	IS	FORM
Number of sample N	10000	100	
Computer runtime (minutes)	2137	12	1
Estimated failure probability	0.0052	0.0038	0.0016
Estimator c.o.v. (%)	13.8%	20.9%	
Required N to achieve c.o.v. = 20%	4761	109	

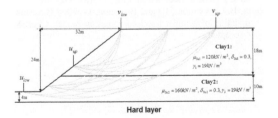

Figure 1. The slope considered in the second example. The
grey lines are the admissible representative slip surfaces.

Safety, Reliability and Risk of Structures, Infrastructures and
Engineering Systems – Furuta, Frangopol & Shinozuka (eds)
© 2010 Taylor & Francis Group, London, ISBN 978-0-415-47557-0

Reliability analysis of differential settlements

Byoung Chan Jung, Paolo Gardoni & Giovanna Biscontin
Zachry Department of Civil Engineering, Texas A&M University, College Station, TX, USA

ABSTRACT

A Bayesian methodology is used to develop an unbiased model that accurately predicts the differential settlements and account for all the prevailing uncertainties. The model is constructed by updating a probabilistic compression model developed by Jung et al. (2008) using new field data. The model developed by Jung et al. was based on a model form developed by Biscontin et al. (2007), which reflects the engineering understanding of the phenomena. As an application, the set of unknown model parameters is calibrated using data from a well-documented case history in Venice, Italy. The database includes soil properties and the field settlements obtained from a full-scale test embankment constructed near Venice. A Bayesian approach is used for the model assessment accounting for both a prior knowledge based on the first set of data and additional data that became available at a later time. The developed model is unbiased and accounts for the inherent uncertainties. The developed probabilistic model is used to assess the conditional probability (fragility) of exceeding a specified differential settlement threshold for a given vertical pressure. Predictive fragilities are developed with special attention given to the treatment and quantification of aleatory and epistemic uncertainties. Sensitivity and importance measures are carried out to identify to which parameter(s) and random variable(s) are key for the reliability of the differential settlement. The developed estimates provide a sound basis for decision about the need to design for differential settlements and the type of intervention most suitable.

The probabilistic model for soil compression is updated to improve the accuracy of the existing model by removing a potential bias incorporating the prevailing uncertainties. The proposed probabilistic model is then used in a formulation to assess the computation of the differential settlement for foundations. The differential settlement predictions based on the probabilistic model compare well with monitored data. Fragility estimates are developed in this study along with sensitivity and importance measures.

Fragility estimates for the differential settlement applied to Treporti Test Site (TTS) in Italy show that the probability of exceedance decrease as the specified threshold settlement increases at a given the vertical pressure.

The sensitivity measures indicate that mean of the slope of K_0-LCC regime, model parameter controlling the curvature, and the reference void ratio for the K_0-LCC of the granular phase have larger effects on fragility while mean of the reference void ratio for the K_0-LCC of the clay-water phase and the model standard deviation have smaller effects over the vertical pressure increase and the specified threshold settlement.

Importance measures are also computed for the random variables in the limit state function of the fragility estimates. The similar trends can be made for the importance measures. In addition, these measures indicate that the slope of K0-LCC regime is the most important random variable then the reference void ratio for the K0-LCC of the granular phase becomes the most important random variable.

REFERENCES

Biscontin, G., Cola, S., Pestana, J. M., and Simonini, P. 2007. Unified compression model for Venice lagoon natural silts. J. Geotec. Geoenviron. Eng., 133(8): 932–942.

Jung, B.-C., Gardoni, P., and Biscontin, G. 2008 Bayesian updating of a unified soil compression model. Georisk (Accepted).

Safety, Reliability and Risk of Structures, Infrastructures and
Engineering Systems – Furuta, Frangopol & Shinozuka (eds)
© 2010 Taylor & Francis Group, London, ISBN 978-0-415-47557-0

Reliability analysis of near field behavior of HLW repository

A. Kobayashi & K. Yamamoto
Kyoto University, Kyoto, Japan

M. Chijimatsu
HAZAMA Corporation, Tokyo, Japan

T. Fujita
Japan Atomic Energy Agency, Tokai, Japan

ABSTRACT

To predict the far future behavior at near field of
high level radioactive waste repository, the numerical
simulation is very helpful. There are, however, many
uncertainties for results of the numerical simulation.
In this paper, the effect of uncertainty of boundary conditions and fundamental mechanical properties i.e.,
elastic modulus and Poisson's ratio, on the mechanical
behavior of near field of the repository was examined.
The method used to examine the error propagation was
the first order second moment method. The mechanical models were the elastic and the damage expansion
models. The reliability of the maximum principal
stress, maximum shear stress at crown of tunnel and the
minimum principal stress at spring line was examined
for one million years.
    As a result, the followings were obtained:

1) For the maximum principal stress at crown, the elastic modulus and Poisson's ratio had a large effect on
   the variance in the elastic and the damage expansion
   models.
2) For the minimum principal stress at spring line, the
   elastic modulus had an effect on the variance in
   both models. In the damage expansion model, the
   Poisson's ratio had a large influence.
3) For the maximum shear stress at crown, the boundary condition had a larger effect than the elastic
   properties in both models.
4) The reliability index of maximum stress gradually
   decreased, that of maximum principal stress gradually increased and that of minimum principal stress
   was small through a long period. This tendency was
   similar for both models as shown in Figures 1 and 2.
5) For damage variable and permeability at crown,
   the horizontal stress on the boundary had a large
   influence.

   While the uncertainty of the prediction is dependent on the mechanical model, it was found from
above results that estimation of boundary conditions
was relatively important to examine the stability and
permeability change around the repository tunnel as
well as the examination of uncertainty of mechanical

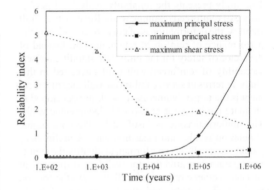

Figure 1. Reliability index in elastic case.

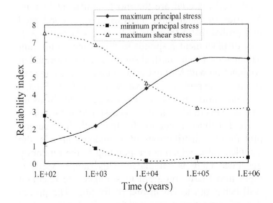

Figure 2. Reliability index in damage model.

properties. Besides, since the reliability of the minimum principal stress was expected to be low, careful
consideration is needed for the prediction of tension
failure and expansion of crack aperture at the spring
line.

REFERENCE

Yamamoto, K., Kobayashi, A. and Aoyama, S.: Numerical
    analysis with damage mechanics for degraded rocks, Inter-
    national Conference on Computational & Experimental
    En-gineering and Sciences, pp.1973–1978, 2005.

Safety, Reliability and Risk of Structures, Infrastructures and Engineering Systems – Furuta, Frangopol & Shinozuka (eds)
© 2010 Taylor & Francis Group, London, ISBN 978-0-415-47557-0

Probabilistic analysis of performance of geogrid reinforced retaining wall

C.N. Liu, J.R. Chen & C.H. Wei

Civil Engineering Department, National Chi Nan University, Nantou, Taiwan

ABSTRACT

This study presents the reliability analysis of performance of reinforced retaining walls. Relative to overall stability, the deformation is of concern considering the performance of reinforced retaining wall. Instead of the factor of safety for the limit state stability, the serviceability of reinforced wall is of interested in this study. In terms of serviceability, the wall face displacement is the target. Numerical models are developed to simulate the behavior of two well documented walls. The numerical code is justified by the comparable simulation results and measured data. A preliminary parametric analysis is performed to identify the significant parameters affecting the wall face displacement of a geosynthetic reinforced wall. It shows the friction angle, bulk density, and elastic modulus of backfill soil, along with elastic modulus of geosynthetic reinforced material are the most significant factors. A series of parametric analysis is then conducted in a probabilistic framework to assess how the wall face displacement responds to these four significant parameters. The numerical simulation is performed in conjunction with Monte Carlo simulation technique to complete the probabilistic parametric analysis. First, a random number between 0 and 1 is generated by using Matlab. This random number is assigned as the cumulative distribution function (CDF). With the assumed probabilistic distributions of a random variable, the sampled value which corresponding to the randomly generated CDF is calculated. This sample value is input into FLAC program to simulate the resultant wall behaviors to finish one realization. The procedure of realization is repeated for multiple times while the analysis results of each realization are collected and input into Matlab for building up the probabilistic distributions of resultant wall behaviors. The statistical characteristics of the randomvariables are listed in Table 1. These random variables are assumed to be normal distributed. 1000 FLAC realizations are conducted in the Monte Carlo simulation. The results of 1000 realizations of maximum facing displacement induced by loadings for wall #1 are shown in Figure 1. The simulation results show that the maximum facing displacement increases with an increase of the applied loadings, while it diverges as the loading becomes

Table 1. Statistical characteristics of random variables used in this study.

Random variable	Mean value	Cov
Soil friction angle	40°	10%
Elastic modulus number of soil	1150	20%
Soil density	1700 kg/m³	5%
Reinforcement stiffness	24 Mpa for wall #1	15%
	12 Mpa for wall #2	

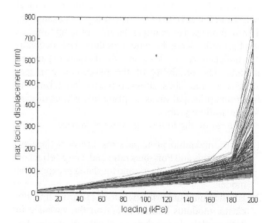

Figure 1. 1000 simulated maximum facing displacement induced by loadings for model wall #1.

large. The relationship between mean value of maximum facing displacement (y, [mm]) and the loading (Q, [kPa]) applied on wall top can be fitted as

$$y = 0.0007Q^2 + 0.386Q + 8.0169 \qquad (1)$$

The coefficient of variation in y is between 12% (for Q = 0) and 16% (for Q = 160 kPa). It increases with an increase of the loading. This relationship is useful to the probabilistic assessment of performance of geosynthetic reinforced retaining wall. It is noted that the relationship presented in Figure 1 and in Equation (1) is specific to the model wall #1 in this study.

General Session (Earthquake Engineering)

Safety, Reliability and Risk of Structures, Infrastructures and
Engineering Systems – Furuta, Frangopol & Shinozuka (eds)
© 2010 Taylor & Francis Group, London, ISBN 978-0-415-47557-0

Seismic vulnerability of a historical masonry building located in Akita prefecture, Japan

C.H. Cuadra
Akita Prefectural University, Yurihonjo, Japan

K. Tokeshi
Politecnico di Torino, Italy

ABSTRACT

Many brick masonry buildings were built during the Meiji period in Japan (by the end of the 19th century), becoming a common technology of construction. During the great Kanto earthquake, which occurred in the year 1923, many brick masonry buildings collapsed dramatically, mainly due to the lack of appropriate reinforcement. Due to the poor seismic performance, this type of construction was abandoned and was replaced by reinforced concrete technology. Therefore, only few masonry constructions remain at present, and have became historical constructions.

Some of these buildings are located at the north east part of Japan, specifically in Akita prefecture, and most of them have been declared local cultural heritages. One of them is located at Ani village, which was constructed in 1879 to serve as residence for a German engineer that worked in a local mining company. In recent years, many regions, cities or towns in Japan have been very active in showing their own particularities; especially in historical and local culture aspects and some masonry buildings are being included in their local activity programs. However, it is necessary to consider the restoration and conservation of these buildings.

For this purpose, the assessment of seismic behavior and its vulnerability as well, are needed. The study of these types of buildings in Japan is essential due to that there are not specific regulations for masonry buildings and that only few researches have been performed in this field.

Under this context, the seismic vulnerability of the target building is assessed in this research. As an initial step to evaluate the seismic vulnerability of these historical buildings, the dynamic characteristics of the target building have been undertaken. For this purpose, microtremor measurements in the building were carried out in this research. For the analytical modeling, mechanical parameters of masonry brick units were estimated by a series of laboratory test on some stock brick units available near the building under study.

Figure 1. Target building (Ani Ijinkan).

The calibration of the mechanical parameters of masonry brick walls in the analytical modeling was performed based on the results obtained from microtremor measurements. The dynamic characteristics obtained from the analytical modeling of target structure shows intricate modes of vibration, which reflect the multiple predominant peaks observed in the transfer functions of microtremor measurements. The multiple predominant frequencies obtained for a certain frequency range are consequence of the effect of brick walls. The procedure has permitted the identification of the probable mode of failure of the concerned structure.

REFERENCES

Cuadra, C., Sato, Y., Tokeshi, J., Kanno, H., Ogawa, J., Karkee, M. B. & Rojas. J. 2005. Evaluation of the dynamic characteristics of typical Inca heritage structures in Machupicchu. *STREMAH IX*, WITPRESS pp. 237–244.

Kanai, J., Tokeshi, K., Cuadra, C., & Karkee. M.B. 2006. Vibration characteristics of buildings using microtremor measurements. *First European Conference on Earthquake Engineering and Seismology* Paper Number: 708.

Safety, Reliability and Risk of Structures, Infrastructures and
Engineering Systems – Furuta, Frangopol & Shinozuka (eds)
© *2010 Taylor & Francis Group, London, ISBN 978-0-415-47557-0*

Seismic resisting mechanism and formulation of traditional timber joints in Japan

H. Tanahashi & Y. Suzuki

Global Innovation Research Organization, Ritsumeikan University, Kusatsu, Japan

ABSTRACT

There are many traditional timber structures, such as temples, shrines and town houses in Japan. Their rational seismic evaluation is the most important in order to mitigate probable damages caused by great earthquakes in the near future. However, the evaluation method has not been established so far because their structural systems and resisting mechanism are much different from modern structures. Also, their seismic resistances are supposed to be insufficient in many cases. Therefore, establishing the evaluation method and seismic reinforcement are urgently needed.

The major structural elements of traditional timber structures are moment resisting frames with semi-rigid joints, mud walls and column rocking restoring forces of thick columns. Among them, moment resistance of the frame with semi-rigid joints is the most important. The structural mechanism of semi-rigid joints is rotational embedment and friction at the contact interfaces inside the joint.

In order to evaluate the semi-rigid joints, we established the formulation of the embedment mechanism using Pasternak model (abbreviated to PM). The PM is an improved and refined model of a continuum compared with the conventional Winkler model. It consists of a shear layer on the Winkler model mechanically, which originated from Pasternak. It can express the surface displacement distributions and strain profiles subjected to partially compressive loads adequately. We, therefore, applied the PM to the elasto-plastic embedment behavior of wood joints, considering the orthotropic properties and the strain softening/hardening.

The densification, or strain hardening more than 50% strains, takes place in the local area beneath the contact interfaces of the joints perpendicular to the grain, and expands to the bottom gradually as the load increases. As a result, the joints show very high ductility.

Based on such facts, we made clear the elasto-plastic embedment mechanism of the semi-rigid joints and formulated their behaviors, introducing stiffness functions which consist of two factors; the increasing factor which means the ratio of the stiffness to that

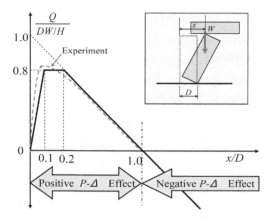

Positive and negative *P-Δ* effects of thick column

Figure 1. Flowchart for setting I, II and III

of the net contact surface and the decreasing factor governed by elasto-plastic characteristics.

The full scale experiments of the frames with semi-rigid joints were carried out and the results are compared with the proposed formulations and discussed. The proposed formulations are shown to be in good agreement with the results if the parameters are assumed appropriately.

Based on their results, we try to evaluate the elasto-plastic restoring force characteristics of the traditional timber frames with semi-rigid joints taking both the positive and negative *P-*Δ effects into account consistently.

Moreover, we discuss the possibilities of collapse failure considering the variability of the restoring force characteristics which are strongly affected by annual ring angles of cross sections of *Nuki* members.

As a result, it is recommended that the restoring forces characteristics should be at least so strong as positive gradient in order to prevent a frame from collapse failure.

Safety, Reliability and Risk of Structures, Infrastructures and
Engineering Systems – Furuta, Frangopol & Shinozuka (eds)
© 2010 Taylor & Francis Group, London, ISBN 978-0-415-47557-0

Reliability-based capacity design of RC bridge system

M. Akiyama & M. Suzuki
Tohoku University, Sendai, Japan

H. Matsuzaki
Tokyo Institute of Technology, Tokyo, Japan

T.H. Dang
Nippon Koei Co., Ltd., Tokyo, Japan

ABSTRACT

Capacity design is a design method that induces a plastic hinge to a member for simple rehabilitation, ensuring its adequate ductility and a capacity hierarchy with other members. By allowing one part of the structure to be damaged and keeping the other parts under elastic behavior, a structure can absorb the seismic energy as well as facilitating rehabilitation after severe earthquakes.

In the seismic design of reinforced concrete (RC) bridge pier and pile foundation system, there exist uncertainties in estimating not only ground motions but also the behavior of RC bridge system. To carry out the capacity design considering these uncertainties, it is necessary to use reliability theory to satisfy the following three conditions. First is the hierarchy between the flexural and the shear capacity that will ensure the ductile failure mode of the members. Second is the condition of the capacity hierarchy between members that will exactly induce the plastic hinge to the appropriate member. Last is the condition that will ensure that the probability of failure is under the allowable value. In this study, a design method to satisfy these conditions has been proposed, and a feasibility study has been carried out.

The following conclusions have been reached in this study.

1) A design method with three types of partial safety factors (γ_I, γ_{II}, γ_{III}) has been proposed. Three factors are determined based on reliability analysis as shown in Figure 1. These are to ensure that (i) all the members under the desirable failure mode are achieved by γ_I, (ii) the damage due to earthquake is induced to only the victim member by γ_{II}, and (iii) the probability of failure under an allowable value is achieved by γ_{III}. Using the proposed method, the capacity design, which allowed the designed structures to absorb the seismic energy under severe ground motions, as well as to facilitate rehabilitation after the earthquake, can be achieved.

2) An application of the proposed method to the design of bridge pier and pile foundation systems has

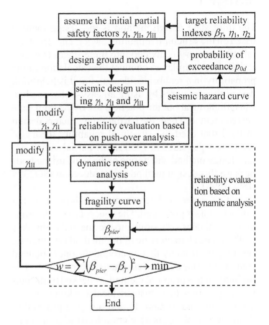

Figure 1. Flowchart for setting I, II and III.

proven that by using reliability evaluation based on push-over and dynamic response analysis, the partial safety factors γ_I, γ_{II}, and γ_{III} can be determined so that the concept of capacity design is realized. Also, the relationships between the partial safety factors and the target reliability indexes were presented.

REFERENCES

Japan Road Association. 2002. Specifications for Highway Bridges, Part V Seismic Design, Tokyo: Maruzen.
National Research Institute for Earth Science and Disaster Pre-vention. J-SHIS (Japan Seismic Hazard Information Sta-tion). http://www.j-shis.bosai.go.jp/

Safety, Reliability and Risk of Structures, Infrastructures and
Engineering Systems – Furuta, Frangopol & Shinozuka (eds)
© 2010 Taylor & Francis Group, London, ISBN 978-0-415-47557-0

A proposal on the determination of partial safety factors for steel piers using seismic hazard maps

Y. Kajita & H. Otsuka
Kyushu University, Fukuoka, Japan

ABSTRACT

The limit state design method is the design method based on the structural reliability theory. The structural reliability theory is divided into three stages (Level I, II, III). It is complex to conduct the calculation of the probability in a real bridge design. So, it is helpful to design the structures by using an easy-to-use method when thinking about the practical use of the limit state design method. Then, the partial safety factors design method that used the structural reliability theory of level I was proposed in 1980's. The partial safety factors design method will be one of the standard methods because this design method is introduced in ISO2394 (ISO, 1998).

In this paper, a determination method of partial safety factors is presented. The concept of the determination method is very similar to the one of the Shiraki's method. The characteristic of our method is to prepare several partial safety factors around the ones obtained from AFOSM method in advance (Robert, 1999). The usability of this proposal method is checked by designing a steel pier to withstand local buckling against a severe ground motion. In addition, the probability distribution of the peak acceleration on the ground surface was calculated by using the previous data of the hypocenter and the attenuation relationship (Hongjun, 1999). In this study, only the earthquake force is treated as an amount of the probability. So, the combination of various loads (for instance, dead load, live load, and earthquake force, etc.) is not referred.

Fig. 1 shows the comparison of the probability density function of the reliability index between AFOSM method and the proposed method. In the case of AFOSM method, the resistant factor and the load factor are 0.97 and 3.56, respectively. On the other hand, in the case of the proposed method, the optimum resistant factor and load factor are 0.40 and 3.30 respectively. It is found from Fig. 1 that the mean value of reliability index in the case of the proposed method is larger than the target reliability index ($\beta_T = 2.2$). So, if our proposed method is used, the number of the bridges which satisfied safety increases.

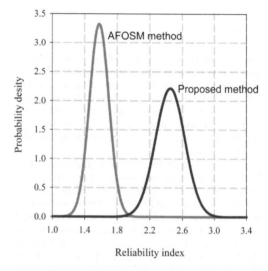

Figure 1. PDF of reliability index (comparison of AFOSM and proposed method).

REFERENCES

ISO: ISO2394 – General principles on reliability for structures –, 1998

Shiraki W, Matsushima M, Dogaki M and Inomo H: A calculation of partial safety factors for reliability based design of steel piers of highway bridges, Proceedings of JCOSSAR 2003, Japan, 2003. pp.961–968

Hongjun SI and Saburoh MIDORIKAWA: Attenuation Relationships of Peak Ground Acceleration and Velocity Considering Effects of Fault Type and Site Condition, 12th World Conference on Earthquake Engineering, 532, (1999)

Robert E. Melchers: Structural reliability analysis and prediction, JohnWiley and Sons Ltd;, 1999

Safety, Reliability and Risk of Structures, Infrastructures and Engineering Systems – Furuta, Frangopol & Shinozuka (eds)
© *2010 Taylor & Francis Group, London, ISBN 978-0-415-47557-0*

Effects of uncertainty on accumulated damage estimation due to aftershocks

Y. Kimura, K. Kawno & N. Kubo
Kagoshima University, Kagoshima, Japan

Y. Nakamura
Nihonkouei, Tukuba, Japan

An available evaluation on aftershocks plays important roles on the performance based design method to seismic motions. In this present study, it is examined about the uncertainty effects on the damage index which is evaluated with combining the ductility factor with the hysteretic energy for the severe damage level of the structure for the SSI system. While there are a little bit influences of the damage index due to after shocks, it is shown that the uncertainty on the maximum seismic acceleration plays important roles on the damage index evaluation. The strength demand spectrum based on the ductility factor is one of the most useful methods which can be treated with nonlinear effects on the structure subjected to seismic forces (Iemura et al (1998)). From the design of view, it is suggested that it is very essential to develop an efficient method on the performance based design method to seismic motions.

It is known that the large scale earthquake is brought about considerable effects on the response due to aftershocks. Since the damage of structure may be closely related to the input seismic energy, it is essential to evaluate accumulated energy on the structure due to aftershocks. Therefore, in order to perform the reliable performance based design, it is important for the nonlinear response situation to clarify the damage evaluations both the main shock and the aftershock. Taking into accounts an increase of the damage of structure by the aftershock, it is important to clarify the accumulated damage on the performance based design due to earthquake. The appropriate estimation of the deterioration due to aftershock is carried out by accumulated damages depending on the nonlinear response situation due to aftershocks.

In this present study, effects on the damage index due to aftershocks are examined. It is indicated that the damage index which is estimated with combining the ductility factor with the hysteretic energy is closely related with accumulated damage estimation of the structure due to aftershocks. The damage index combined the maximum displacement with the hysteretic energy is useful to evaluate the damage situation for the SSI system. The estimation of the damage index is considerable affected by the ground condition and the dynamic characteristics of input seismic motion. It is suggested that an increase of input intensity ratio for aftershock to main shock leads to increase of accumulated damage of the structure. Therefore, it is shown that the fragility on the assigned damage index plays important roles on the available estimation on damage situations due to aftershocks.

REFERENCE

1 Iemura,H., Igarashi,A. & Takahashi,Y., 1998. Ductility and Strength demand for near field earthquake ground motion: Comparative study on the Hyogo-ken Nanbu and the Northridge earthquakes, *Structural safety and Probability* pp.1705–1708

Safety, Reliability and Risk of Structures, Infrastructures and Engineering Systems – Furuta, Frangopol & Shinozuka (eds)
© 2010 Taylor & Francis Group, London, ISBN 978-0-415-47557-0

Use of digital aerial camera images to detect damage to an expressway following an earthquake

Yoshihisa Maruyama & Fumio Yamazaki
Department of Urban Environment Systems, Chiba University, Chiba, Japan

ABSTRACT

Remotely sensed data obtained from satellites and airborne platforms are useful in providing an understanding of the distribution of damage due to natural disasters (Yamazaki 2001). Expressways play an important role in providing access for restoration work in damaged areas, and it is necessary to keep the duration of functional loss to a minimum.

In this study, digital aerial camera images are used to detect expressway damage caused during the 2004 Mid-Niigata earthquake. Firstly, a conventional pixel-based classification was conducted for both the analog and digital image for comparison purposes. Secondly, the procedure proposed by this study was performed in order to detect expressway damages. The results of image processing using analog (Maruyama *et al.* 2006) and digital aerial photographs were compared. A new methodology for detecting damages using digital aerial photographs is presented. The accuracy of the damage estimation and the effectiveness of our method are discussed with comparison to actual damage data.

According to the comparison between the result obtained from analog aerial photo and that from digital aerial image, the median filter was not used in the image processing of the digital aerial image, instead,

(a) Result of image processing

(b) Result of visual inspection

Figure 2. Comparison of the results of (a) image processing of the digital aerial image and (b) visual damage inspection. Damage to the expressway is shown in red.

the errors adjacent to the centerlines and in individual pixels were removed following the two rules set by this study. Following the flowchart in Fig. 1, the damages of expressway were estimated automatically (Fig. 2).

Figure 2 compares the result of image processing with that of visual damage inspection. Almost all damaged sections of the expressway were properly detected through the series of image processing outlined in this study.

REFERENCES

Maruyama, Y., Yamazaki, F., Yogai, H. & Tsuchiya, Y. 2006. Interpretation of expressway damages in the 2004 mid Nii-gata earthquake based on aerial photographs. *Proceedings of the First European Conference on Earthquake Engineer-ing and Seismology*, CD-ROM, 8p, Paper No. 738.

Yamazaki, F. 2001. Applications of remote sensing and GIS for damage assessment. *Proceedings of the 8th International Conference on Structural Safety and Reliability*, CD-ROM, 12p.

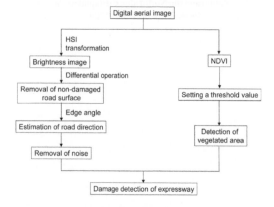

Figure 1. Flowchart of proposed method for detection of ex-pressway damage using digital aerial images.

General Session (Structural Systems &
System Reliability)

Safety, Reliability and Risk of Structures, Infrastructures and
Engineering Systems – Furuta, Frangopol & Shinozuka (eds)
© 2010 Taylor & Francis Group, London, ISBN 978-0-415-47557-0

Modeling uncertainty in seismic response analysis of steel frames buildings with perimeter moment and interior gravity frames

A. Reyes-Salazar & E. Bojórquez
Facultad de Ingeniería, Universidad Autónoma de Sinaloa, Culiacán, Sinaloa, México

D. López-López
Departamento de Ingeniería Civil, Instituto Tecnológico de Sonora, Obregón, Sonora, México

D. De Leon-Escobedo
Facultad de Ingeniería, Universidad Autónoma del Estado de México, Toluca, Estado de México, México

A. Haldar
Department of Civil Engineering and Engineering Mechanics, University of Arizona, Tucson, AZ, USA

ABSTRACT

Modeling uncertainty in the nonlinear seismic response analysis of a structure could be as important as other sources of uncertainty in the load and resistance-related parameters. The issue is comprehensively studied with the help of an arrangement commonly used for steel structures in seismically active regions. For steel buildings, the use of moment resisting steel frames (MRSF) has been popular because they provide maximum flexibility for space utilization and because of their high ductility capacity. However, the characteristics of the basic structural system consisting of MRSF have significantly changed over the years. From the mid 60s to the mid 70s, most connections in steel buildings were assumed to be fully restrained (FR). In the recent past, the use of FR connections were reduced considerably because they were expensive and to eliminate weak-axis connections. FR connections are used only on two frame lines in each direction, usually at the perimeter, and gravity frames are used at the interior (IGF). An important issue that deserves our attention is that perimeter MRSF are usually designed as plane frames to resist the total lateral seismic loading, ignoring the presence of IGF. Due to the action of the rigid floor diaphragm, the IGF, however, will undergo the same lateral deformation as the MRSF. The contribution of the columns in IGF to the lateral resistance of the building could be significant, particularly for buildings with relatively few FRC. Moreover, modeling the buildings as plane frames may not represent the actual behavior of the structure since the participation of some elements are not considered and the contribution of some vibration modes are ignored. Another simplification made in the design of steel buildings with perimeter MRSF and IGF is related to the stiffness of the beam-to-column connection. Conventional analysis and design of steel frames is based on the assumption that beam-to-column connections are either FR or perfectly pinned (PP). Despite these classifications, almost all steel connections used in real frames are essentially semi-rigid (SR) with different rigidities. There is some evidence that some connections, even though considered pinned, can transmit up to 30% of the plastic moment capacity of the beams they are connecting to. The above discussions clearly indicate that there are several sources of uncertainty in the seismic response analysis. Effects of modeling uncertainty are emphasized in the paper. The issues related to structural idealization of steel buildings with perimeter MRSF and IGF presented in this paper are: a) the accuracy of modeling the three-dimensional buildings as plane frames for seismic analysis, b) the effect of the stiffness of the connections of the IGF on the structural response, and c) the level of contribution of the IGF to the lateral resistance. Some steel structures that satisfied all the current seismic requirements proposed in the SAC project are used for this purpose. To quantify the uncertainties, these model structures are excited by twenty recorded earthquake time histories in time domain. The numerical study indicates that modeling the building as planes frames may result in larger interstory shears and displacements and resultant stresses indicating that the design may be conservative. The design is more conservative in terms of resultant stresses. The contribution of IGF to the lateral structural resistance could be significant and the uncertainty in the responses cannot be ignored. The level of uncertainty increases and becomes very high when the stiffness of IGF is considered more appropriately. Based on the results of this study, it can be concluded that the three-dimensional model should be used in seismic analysis, that the IGF should be considered as part of the lateral resistance system, and that the stiffness of the connections should be included in the design of the IGF. Otherwise, the capacity of gravity frames will be overestimated while that of the MRSF will be underestimated.

Safety, Reliability and Risk of Structures, Infrastructures and
Engineering Systems – Furuta, Frangopol & Shinozuka (eds)
© 2010 Taylor & Francis Group, London, ISBN 978-0-415-47557-0

Soil-structure interaction stochastic modelling

D. Novák, L. Miča, B. Teplý & M. Vořechovský
Brno University of Technology, Brno, Czech Republic

J. Buček, R. Rusina & I. Němec
FEM consulting, Ltd, Brno, Czech Republic

ABSTRACT

Soil-structure interaction modelling belongs to the
category of difficult and complex treatment of structural behaviour where the efficiency of computational
model is influenced by many factors. Dominating
factors are firstly the level of computational model
sophistication itself, second uncertainties related to
input data (mainly soil properties). Objective modelling is possible using advanced models of subsoil;
such modelling needs the finite element method and
usually remains at the deterministic level. The efficient surface model and program SOILIN (Bucek et al.
1992) were practically verified in building industry by
many projects and are currently implemented in software systems NEXIS 32 (2001), SCIA ENGINEER
(2008) and RFEM (2006).

The paper describes achievements in utilization of
efficient methods for statistical, sensitivity and reliability analyses of soil-structure interaction. The simulation of random variables was performed by Latin
hypercube sampling using FReET software (Novák
at al. 2008). The aim is a complex probabilistic treatment of SOILIN software: statistical, sensitivity and
reliability (failure probability calculation) analyses.

The utilization of software tools is demonstrated
using example of soil-structure interaction: concrete

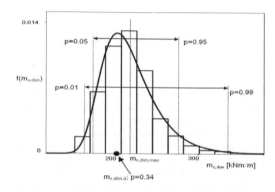

Figure 2. Histogram of the maximum plate design moment.

foundation slab for steel petrol tank (Fig. 1). Variability of input data was restricted to geo-mechanical
variables. Virtual statistical simulation was performed
using 2000 samples. Each simulation consisted of 50
random variables (parameters for each drill and layer).
Resulted histograms with the most suitable PDFs and
percentiles 0.05, 0.95 and 0.01 and 0.99 are shown in
Figure 2 and comparison with deterministic value is
provided, too.

REFERENCES

Bucek, J., Kolár, V. & Obruca, J. 1992. *SOILIN – Settle-
ment and interaction parameters of layered subsoil. User's
Guide*, Brno: FEM consulting, Ltd.
NEXIS 32 (Esa-Prima Win) 2001. *Program system for
the analysis of structure composed of 1D and 2D ele-
ments*. SCIA Group nv, Ind.1007, B-3540 Herk-de-Stad,
Belgium.
Novák, D., Vorechovský, M., & Rusina, M. 2008. *"FREET
version 1.5 – program documentation", User's and Theory
Guides*, Brno/Cervenka Consulting, Prague.
RFEM 2006. *Finite Elements for Plates, Shear Walls, Shells,
Solids and Frameworks*. Tiefenbach: Ingenieur-Software
Dlubal GmbH.
Scia Engineer 2008. *Software System for Analysis, Design
and Drawings of Steel, Concrete, Timber, Aluminium and
Plastic Structures*. Herk-de-Stad: SCIA Group nv.

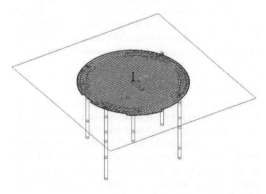

Figure 1. Model of foundation plate and geological drill
holes.

Safety, Reliability and Risk of Structures, Infrastructures and Engineering Systems – Furuta, Frangopol & Shinozuka (eds)
© 2010 Taylor & Francis Group, London, ISBN 978-0-415-47557-0

On-line method of reliability analysis of pipelines with growing defects

L.V. Poluyan, A.V. Bushinskaya, M.G. Malyukova & S.A. Timashev
Science and Engineering Center "Reliability and Safety of Large Systems and Machines",
Ural Branch Russian Academy of Sciences, Ekaterinburg, Russia

ABSTRACT

The results of a holistic study dedicated to developing an Internet-based reliability and remaining life assessment for on-shore main pipelines with growing defects are presented. The reliability analysis is carried out using five types of failure pressure models. Two types of limit state functions are used. The limit state functions (LSF) are comprised of nine parameters that are considered as random variables. These RVs describe the pipe and defect geometry, rate of axial and longitudinal corrosion, strength properties of the pipe wall material, operating pressure. The main reliability analysis is conducted using the Gram-Charlier-Edgeworth (GCE) method of expansion of the limit state function, which can be considered as a generalization of the FOSM and the Taylor series expansion methods, as it is capable of using arbitrary probability density functions (PDF) and takes into account the third and the fourth central moments of the non-Gaussian PDFs. Five types of PDFs are used in the analysis. The Monte Carlo simulation method was used for verifying the results obtained by the G-C-E method. The reliability of each cross-section of the pipe containing a defect is calculated, as well as the reliability of a pipeline segment or the pipeline as a whole. Sensitivity of pipeline reliability to the above nine parameters and their statistical characteristics (PDFs), which describe pipeline and defect geometry, corrosion rates, pipe wall material properties and operation mode is investigated. The expected number and volume of repairs are calculated using some values for the ultimate permissible pipeline failure probability. Comparative study on the computer time needed for calculation of reliability by the G-C-E and the M-C methods is conducted in relation to their applicability for remote Internet usage. A brief description of the Internet-based software that implements the described above technology is given.

Safety, Reliability and Risk of Structures, Infrastructures and Engineering Systems – Furuta, Frangopol & Shinozuka (eds)
© 2010 Taylor & Francis Group, London, ISBN 978-0-415-47557-0

Reliability based inspection and maintenance of pipelines with Markov type corrosion defects growth

L.V. Poluyan, A.V. Bushinskaya, M.G. Malyukova & S.A. Timashev
Science and Engineering Center "Reliability and Safety of Large Systems and Machines", Ural Branch Russian Academy of Sciences, Ekaterinburg, Russia

ABSTRACT

The paper describes a Markov model of corrosion growth of pipe wall defects and its implementation for assessing the conditional probability of pipeline failure and optimizing pipeline repair and maintenance. This pure growth Markov model is of the continuous time, discrete states type. This model is used in conjunction with the geometrical limit state function (LSF) to assess the conditional probability of failure of pressurized pipelines when the main concern is loss of containment. It is shown how to build an empirical Markov model for the length, depth and width of defects, using field data gathered by In-line inspection (ILI) or direct assessment (DA) or by using a combination of a differential equation (DE) that describes defect parameter growth with the Monte Carlo simulation method. As a result of implementation of this approach the probability for the defect parameters being in a given state (analog of a histogram) and the transition intensities (from state to state) are easily derived for any given moment of time. This approach automatically gives an assessment of the probability of failure of a pipeline segment, as it is derived using the data from a specific pipeline length. This model also allows accounting for the pipeline failure pressure LSF. On the basis of this model an algorithm is constructed for optimizing the time of the next inspection/repair. This methodology is implemented to a specific operating pipeline which was several times inspected by a MFL inspection tool. The expected number and volume of repairs depend on the value of the ultimate permissible pipeline failure probability. Sensitivity of pipeline conditional failure rate and optimal repair time to actual growth rate is investigated. A brief description of the software that implements the described above technology is given.

Safety, Reliability and Risk of Structures, Infrastructures and
Engineering Systems – Furuta, Frangopol & Shinozuka (eds)
© 2010 Taylor & Francis Group, London, ISBN 978-0-415-47557-0

Robustness evaluation of timber structures

P.H. Kirkegaard & J.D. Sørensen
Department of Civil Engineering, Aalborg University, Denmark

ABSTRACT

Robustness of structutral systems has obtained a
renewed interest due to a much more frequent use
of advanced types of structures with limited redundancy and serious consequences in case of failure.
The interst has also been facilitated due to recently
severe structural failures such as that at Ronan Point
in 1968 and the World Trade Centre towers in 2001.
In order to minimise the likelihood of such dispro-
portionated structural failures many modern building
codes consider the need for robustness in structures
and provides strategies and methods to obtain robustness. The requirement for robustness is specified in
most buildings codes in a way like the general requirements in the two Eurocodes EN 1990 Eurocode 0:
Basis of Structural Design and EN 1991-1-7 Eurocode
1: Part 1–7 Accidental Actions. However, no specific
criteria are given which can be used to quantify the
level of robustness of a structure which could have a
benefit for design and analysis of structures.

The present paper considers robustness evaluation
of timber structures. A Norwegian sports arena with a
structural system of glulam frames is considered. The
robustness evaluation is based on the framework for
robustness analysis introduced in the Danish Code of
Practice for the Safety of Structures and a probabilistic
modelling of the timber material proposed in the Probabilistic Model Code (PMC) of the Joint Committee
on Structural Safety (JCSS). Due to the framework in
the Danish Code the timber structure has to be evaluated with respect to the following criteria where at
least one shall be fulfilled: a) demonstrating that those
parts of the structure essential for the safety only have
little sensitivity with respect to unintentional loads and
defects, or b) demonstrating a load case with 'removal
of a limited part of the structure' in order to document
that an extensive failure of the structure will not occur
if a limited part of the structure fails, or c) demonstrating sufficient safety of key elements, such that
the entire structure with one or more key elements has
the same reliability as a structure where robustness
is documented by b). By using First-Order Reliability
Methods (FORM) the structural reliability of the structure is estimated at element as well as system level. The
requirement to the safety of the structure is expressed
in terms of an accepted minimum reliability index, i.e.
a target reliability index, proposed by the Joint Committee on Structural Safety (JCSS). In order to simulate
'removal of a limited part of the structure' different
assumed damage scenarios are considered. The results
show that the requirements for robustness of the Norwegian structure are highly related on the modelling
of the snow load used on the structure when 'removal
of a limited part of the structure' is considered.

Safety, Reliability and Risk of Structures, Infrastructures and
Engineering Systems – Furuta, Frangopol & Shinozuka (eds)
© 2010 Taylor & Francis Group, London, ISBN 978-0-415-47557-0

Statistical correlation of steel members for system reliability analysis

H. Idota & L. Guan
Nagoya Institute of Technology, Nagoya, Japan

K. Yamazaki
Takenaka Corporation, Inzai, Japan

ABSTRACT

Statistical correlation of random variables is one of
the most important factors for system reliability analysis, especially because the joint probability function
is significantly affected by the correlation coefficients of random variables. For the reliability of frame
structures, it is necessary to determine the statistical correlation matrix of member strength to evaluate
the ultimate strength of the frames. Although many
studies have examined the effects of statistical correlation on system reliability (e.g., Melchers 1983),
few have collected statistical data on the correlations
for actual building structures. So, the correlation coefficients have been forced to have some assumptions
under engineering judgments.

The purpose of this paper is to propose information
of the statistical correlation on steel member strength
by surveying actual steel buildings. First, to identify individual lots of steel materials, lot-maps for the
actual buildings are presented by tracking the inspection certificates of all members throughout construction. The lot-map is a layout plan of the structural members manufactured from steel material with the same
rolling process. The sample of lot-map is shown in Figure. 1. Next, the correlation coefficients of the yield
strength are analyzed based on the tensile test results
for specimens extracted from the same lots. In addition, the statistical proper-ties of the ultimate strength
of steel moment-resisting frames are examined based
on the results of the Monte Carlo simulations using the
lot-maps. Fi-nally, the paper proposes practical scenarios that can evaluate the effects of statistical correlation
in steel members without using lot-maps.

The results obtained in this paper are presented as
follows:

1) The lot-maps for a steel building structure are
   presented by tracking the inspection certificates
   throughout construction.
2) The statistical properties of the ultimate strength of
   steel moment-resisting frames are examined based
   on the results of Monte-Carlo simulations using
   lot-maps.

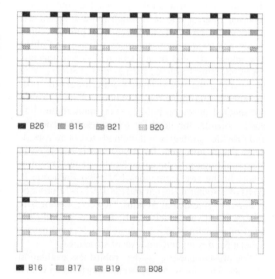

Figure 1. Lot-maps of beam members.

3) Two practical scenarios are proposed to estimate the statistical correlations between member
   strength for calculating the statistics of the ultimate
   strength of frame structures.

1 REFERENCES

Melchers, R. E. 1983. Reliability of parallel structural
systems, Journal of Engineering Mechanics, ASCE,
Vol.109, No.11

Organized Session (OS11) Health Monitoring, Structural Monitoring & Control

*Safety, Reliability and Risk of Structures, Infrastructures and
Engineering Systems – Furuta, Frangopol & Shinozuka (eds)
© 2010 Taylor & Francis Group, London, ISBN 978-0-415-47557-0*

Experimental and numerical studies on evaluation of natural vibration characteristics of 73 year-olds Asahi bridge

M. Komuro

Muroran Institute of Technology, Muroran, Japan

H. Nishi & T. Sato

Civil Engineering Research Institute for Cold Region, Sapporo, Japan

N. Kishi

Muroran Institute of Technology, Muroran, Japan

ABSTRACT

In order to accumulate the basic data for natural vibration characteristics of Asahi bridge (see, Photo.1) which was constructed in 1932, forced vibration test and microtremor measurement were conducted. Here, the natural vibration frequencies and modes of the bridge obtained from the test results were compared with the numerical results obtained from 3D-FE analysis for the dimensions at the beginning in common use to investigate damage levels of the bridge.

Photo 1. Bird's eye view of Asahi Bridge.

Table 1. Comparison of natural vibration frequencies between experimental and numerical results.

Vibration mode		Natural vibration frequency (Hz)		
		FV	MM	FEM
(a) Flexural vibration mode				
Sym.	1	2.63	2.68	2.60
	2	2.95	2.95	2.99
	3	4.80	4.75	4.70
Anti-sym.	1	2.07	2.12	2.09
	2	3.17	–	3.20
(b) Torsional vibration mode				
Sym.	1	3.63	3.61	4.39
	2	3.73	3.73	4.52
Anti-sym.	1	3.14	–	3.79
	2	4.24	4.22	5.02

FV: Forced vibration, MM: Microtremor measurement

Table 1 lists the comparison of the natural vibration frequencies among three results obtained from forced vibration test, microtremor measurement, and numerical analysis. From this table, the maximum error of natural vibration frequency between both tests was less than 3 %. Therefore, natural vibration frequencies of the bridge are properly evaluated by conducting both tests. And the maximum error between numerical results and forced vibration results is less than 2 % for the flexural vibration mode.

Figure 1 shows the comparison among the experimental and numerical results for typical vibration modes specified in this study. It is confirmed that mode shapes in the side span of the symmetrical 2nd flexural vibration (see, Fig. 1b) are a little different among them. However, another mode shapes are almost same to each other.

From this study, following results were obtained: 1) natural vibration frequencies and modes of the bridge can be better evaluated by conducting forced vibration test and microtremor measurement; and 2) since natural vibration frequencies and modes obtained from the experimental results are similar to those obtained from the numerical analysis, damage of the bridge may be negligible practically, even though 73 years has passed since the beginning in common use.

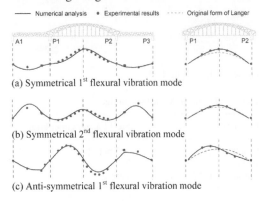

Figure 1. Comparison of the mode shapes between numerical results and experimental results.

Safety, Reliability and Risk of Structures, Infrastructures and Engineering Systems – Furuta, Frangopol & Shinozuka (eds)
© 2010 Taylor & Francis Group, London, ISBN 978-0-415-47557-0

Estimation of bridge vibration characteristics for verification of environmental effects to structural frequencies

B.A. Jawaid, T. Okumatsu, T. Okabayashi, M.R. Ali & T. Shimozuma
Nagasaki University, Nagasaki, Japan

ABSTRACT

Variety approaches have been established for structural health diagnosis with emerged software and hardware based measurement technologies. Among them vibration monitoring is a tool for long-term structural health monitoring with not only structural maintenance but also crisis management. Along with these methods, ambient vibration has been hired for input data representing structural characteristics properly with advantages of measurement simplicity, cost effectiveness, and its suitability. However, there are still leaving much to do such as effects of environmental factors to subject structures. In this paper, we discuss about 1) structural identification methods and its accuracy, 2) developed monitoring system and its application to existing structures, and 3) evaluation of environmental effects against dynamic characteristics of subject structure. We focus on environmental effects among the above, and state the tendency of environmental effects from the measurement data and

identified structural eigen-frequencies. As for site monitoring, oversea-bridge, located in Nagasaki prefecture, is selected as subject structure. Figure 1 shows the bridge front view and the detail of the bridge is shown in Table 1.

We have been conducted experiments and the bridge's ambient vibration has been recorded with accelerometers. Five accelerometers were installed on the bridge to obtain vertical acceleration simultaneously. Two accelerometers were installed at the middle of bridge span for recording the torsional vibration data. Measured data of the three different environmental cases evaluated by each method (ERA, ERA/DC, OMS and BCF). Although each one showed almost same accuracy but ERA/DC method was found better for its higher speed of calculation. Due to this and the space limitations for the paper we have only included the graphical results of (frequency, modal damping and mode shapes) for ERA/ DC method and accuracy evaluation results for all the method in this paper.

Realization theories of structural identification methods were hired to estimate the bridge dynamic characteristics. For obtaining structural dynamic characteristics, coefficient matrices were computed from ambient vibration measurement data and eigenvalues of the corresponding data were analyzed.

By the measurement and analysis, it was revealed that there are subtle changes in estimated frequencies accuracy and mean values. We found that strong windy condition had better accuracy in frequency estimation, where moving vehicle effects frequency estimation accuracy and induces higher vibration modes.

Figure 1. Kabashima bridge.

Table 1. Details of Kabashima bridge

Bridge type	Steel Langer girder bridge
Length	227m(152m)
Width	7.5m
Construction date	1986

REFERENCES

Nagayama, T., Abe, M., Fujino, Y., and Ikeda, K., 2005. Structural identification of a nonproportionally damped system and its application to a full-scale suspension bridge. Journal of Structural Engineering, 131:1536–1545.

Teughels, A., and De Roeck, G., 2004. Structural damage identification of the highway bridge Z24 by FE model updating, Journal of Sound and Vibration. 278:589-610

Wenzel, H. and Pichler, D. 2005. Ambient vibration monitoring. Willey, West Sussex.

Safety, Reliability and Risk of Structures, Infrastructures and Engineering Systems – Furuta, Frangopol & Shinozuka (eds)
© 2010 Taylor & Francis Group, London, ISBN 978-0-415-47557-0

A statistical relative measurement index for damage locating and robustness-to-noise assessment

A.P. Adewuyi & Z.S. Wu
Department of Urban & Civil Engineering, Ibaraki University, Hitachi, Japan

ABSTRACT

Progressive deterioration of civil infrastructure begins once they are built and subjected to continuous normal loading, occasional overloading, adverse environmental conditions and/or extreme natural events such as earthquakes. During the past few decades, a significant amount of research has been conducted in the area of nondestructive damage detection via changes in modal responses of a structure. Extensive review on advances in vibration-based damage identification (VBDI) is reported by Doebling *et al.* (1996). The extension of VBDI techniques to civil infrastructure is currently increasingly receiving wider attention among researchers, infrastructure owners and management agencies.

Accurate measurement and close monitoring of vibration characteristics are very critical for early detection of damage in structures. However, despite the numerous achievements made so far, the search for more reliable strategies for both damage localization and quantification in large-scale civil structures is still in progress due to inevitable variability in measurement data as a result of measurement error and environmental factors. So, the ability to correctly distinguish changes in the modal properties caused by damage from those arising from measurement noise and other environmental conditions is an essential issue requesting serious attention for successful civil structural health monitoring (SHM). Obtaining comprehensive information through innovative sensing system without losing structural integrity in a similar manner to the operation of human nervous system throughout the whole body is essential for successful implementation of SHM programs (Li and Wu, 2007).

Therefore, the process of vibration-based SHM can be fundamentally considered as that of statistical pattern recognition. Summary of various statistical analysis approaches for structural identification based on dynamic measurements can be found in Sohn *et al.* (2003).

This paper presents a statistical non-baseline VBDI algorithm meant to assess the stability of the measurement data, detect and locate damage in civil structures, where variability in dynamic measurements due to electrical noise and environmental influence is often inevitable. The statistical relative measurement index (SRMI) method exploits the regression analysis of peak values of the magnitudes of frequency response function (FRF) of target sensors relative to the reference wherein the statistical features are employed for data reliability assessment and damage localization. Moreover, the ability to effectively manage and make sense of enormous amounts of data collected under continuous monitoring process for an effective diagnostic and/or prognostic system is an added advantage. Through experimental modal analysis of a beam with different structural conditions under succession of random excitations of arbitrary magnitudes at different locations, the robustness of measurements from accelerometers, traditional foil strain gauges and long gauge fiber Bragg grating (FBG) strain sensors to noise is comparatively evaluated. The ability to present the results in an easy-to-interpret graphical format makes SRMI suitable for civil SHM. The ascendancy of long-gauge FBG sensors over the traditional sensors is established for effective damage localization. The limitation of SRMI technique with acceleration response data is also discussed.

REFERENCES

Doebling, S.W., Farrar, C.R., Prime, M.B. and Shevita, D.W. 1996. Damage identification and health monitoring of structural and mechanical systems from changes in their vibration characteristics: A literature review. LANL Report.

Li, S. and Wu, Z.S., 2007. Development of Distributed Long-gage Fiber Optic Sensing System for Structural Health Monitoring. Structural Health Monitoring, 6(2): 133–143

Sohn, H., Farrar, C.R., Hemez, F.M., Shunk, D. D., Stinemates, D. W. and Nadler, B. R. 2003. A Review of Structural Health Monitoring Literature: 1996–2001. *Los Alamos National Laboratory Report*, LA-13976-MS.

*Safety, Reliability and Risk of Structures, Infrastructures and
Engineering Systems – Furuta, Frangopol & Shinozuka (eds)*
© 2010 Taylor & Francis Group, London, ISBN 978-0-415-47557-0

Acceleration-based real-time displacement monitoring method for civil structures

N. Nakata

Johns Hopkins University, Baltimore, MD, USA

ABSTRACT

Dynamic displacement is an important response
parameter for assessment of structures. However, measurement of structural displacements using displacement sensors is difficult in fields due to a requirement
of stationary reference frames. This study presents a
method for monitoring real-time dynamic structural
displacements using acceleration measurements. The
proposed method is based on an experimental model
of acceleration-displacement relationships in the frequency domain that incorporates dynamic characteristics of accelerometers. The model is obtained in a form
of a rational polynomial approximation function, and
can be easily implemented in digital signal processors.
The proposed method is investigated in comparison with a numerical double integration method, and
experimental verifications are conducted for a story-
drift of a building structure under dynamic loadings.
Experimental results show that if signal-to-noise ratios
in the acceleration measurements are relatively high,
the proposed acceleration-based displacement monitoring method provides accurate real-time dynamic
displacements of structures.

Safety, Reliability and Risk of Structures, Infrastructures and Engineering Systems – Furuta, Frangopol & Shinozuka (eds)
© 2010 Taylor & Francis Group, London, ISBN 978-0-415-47557-0

Comparison between coupled local minimizers method and differential evolution algorithm in dynamic damage detections

L. Vincenzi & M. Savoia
DISTART – Structural Engineering, University of Bologna, Bologna, Italy

G. De Roeck
Department of Civil Engineering, Division of Structural Mechanics, Katholieke Universiteit Leuven, Belgium

ABSTRACT

In several areas of civil and mechanical engineering, the maintenance and retrofitting of existing structures requires the diagnostic identification of damages. To this purpose, dynamic testing techniques have received great attention in the engineering community in last three decades. The changes in modal parameters (frequencies and mode shapes) are used to identify damages or local flexibilities of the structure [Teughels et al, 2002].

In a damage detection problem, an updating procedure of the Finite Element model of the structure (model updating) is required, where the unknown model properties are modified in order to obtain numerical predictions as close as possible to a set of measured data, e.g., the modal properties obtained by the dynamic tests. Therefore, an optimization problem must be solved where the objective function to be minimized is defined by the distance between the modal parameters obtained from the experimental tests and those given by a numerical model of the structure. Nevertheless, the objective function is often non-smooth or even discontinuous and may contain multiple local minima, so that very efficient optimization methods are required.

In the present paper, global optimization techniques are used to perform the model updating of a damaged structure. In particular, the Coupled Local Minimizers (*CLM*) method and the Differential Evolution (*DE*) Algorithm are used to detect position and depth of a localized damage, and their numerical performances are compared. The CLM method is a very efficient global gradient-based method originally proposed by Suykens [2001]. The method adopts a number of search points and couples multiple local optimization runs in order to create interaction and information exchange between them. The DE method is a parallel direct search method where N different vectors collecting the parameters of the system are used in the minimization process [Storn and Price, 1997]. The vector population is chosen randomly at the beginning and by adding weighted differences between vectors obtained from the previous population in the subsequent steps.

In the present paper, the performances of the CLM method and the DE algorithm for the damage assessment of a cracked beam through FE model updating are compared. The presence of cracks in a structural member introduces local flexibilities, so modifying its dynamic behaviour. The changes of dynamic characteristics (frequencies and mode shapes) can be measured and subsequently used for damage detection.

The comparison is performed with reference to two benchmark problems: two simply-supported beams under flexural vibrations, with one crack and two cracks to be detected (two and four identification parameters, respectively). The challenge is to find, by FE model updating, the crack location and width (the latter being related to the flexural spring stiffness used to model the crack). The effectiveness of the two algorithms to find the set of optimization parameters has been compared by performing a statistical analysis of the optimization results.

Very good results are obtained by both algorithms. Better performances in term of speed rate and precision are obtained by CLM when number of identified parameters is limited. On the other hand, DE shows good efficiency when the number of unknown parameters increases.

REFERENCES

Storn, R. & Price, K. 1997. Differential Evolution – a simple and efficient heuristic for global optimization over continuous spaces. *Journal of Global Optimization* 11(4): 341–359.

Suykens, J.A.K., Vandewalle, J. & De Moor, B. 2001. Intelligence and cooperative search by coupled local minimizers, *International Journal of Bifurcation and Chaos* 11(8): 2133–2144.

Teughels, A., Maeck, J. & De Roeck, G. 2002. Damage assessment by FE model updating using damage functions. *Computers and Structures* 80(25): 1869–1879.

Safety, Reliability and Risk of Structures, Infrastructures and
Engineering Systems – Furuta, Frangopol & Shinozuka (eds)
© 2010 Taylor & Francis Group, London, ISBN 978-0-415-47557-0

Optimal decision making for structural health monitoring under uncertainty

S. Kim & D.M. Frangopol

Lehigh University, Bethlehem, PA, USA

ABSTRACT

The lifetime performance of structures is affected by deterioration processes under uncertainty. Due to various uncertainties, civil structure managers can not assess and/or predict the structural performance accurately. Structural health monitoring (SHM) can be used as an effective tool to reduce the uncertainty related to assessment and/or prediction of structural performance. Through acquisition of additional relevant information from SHM, updating the monitoring strategy will lead to lower life-cycle cost (Frangopol and Messervey 2007).

In this paper, an approach for cost-effective monitoring planning is proposed by using decision analysis. The maximum expected monetary value associated with total monitoring cost is used as a decision criterion (see Figure 1). The decision tree is formulated by including availability of monitoring data for prediction

and the expected monitoring cost. Based on this decision tree, the decision analysis can be performed. The theory of statistics of extremes is applied to formulate the availability of monitoring data. The availability of monitoring data for prediction is defined as the probability that monitoring data will be usable over the prediction duration. In order to quantify the potential loss statistically, a loss function is introduced. The proposed approach is applied to the monitored data of an existing bridge.

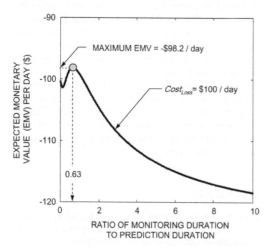

Figure 1. Expected monetary value per day versus ratio of monitoring duration to prediction duration for $Cost_{Loss} = \$100/day$

REFERENCES

Ang, A.H.-S. & Tang, W.H. 1984. *Probability Concepts in Engineering Planning and Design*. Vol. II, John Wiley & Sons.

Cooil, B. & Rust, R. T. 1994. Reliability and expected loss: A unifying principle. *Psychometrika*, 59(2): 203–216.

Farhey, D. N. 2005. Bridge instrumentation and monitoring for structural diagnostics. *Structural Health Monitoring*, 4(4): 301–318.

Frangopol, D. M. & Messervey T. B. 2007. Risk assessment for bridge decision making, *Proceedings of the Fourth Civil Engineering Conference in the Asian Region, CECAR 4, Taipei, Taiwan, June 25–28, 2008* (invited paper); in ASCE Tutorial & Workshop on Quantitative Risk Assessment, Taipei, Taiwan, June 25-28: 37–42.

Frangopol, D.M., Strauss, A. & Kim, S. 2008. Use of monitoring extreme data for the performance prediction of structures: General approach. *Engineering Structures, Elsevier*, 30(12): 3644–3653.

Gumbel, E.J. 1958. *Statistics of Extremes*. Columbia Univ. Press.

Kim, S. & Frangopol, D. M. 2008. Optimal planning of structural performance monitoring under uncertainty (submitted for publication).

Mahmoud, H.N., Connor, R.J. & Bowman, C.A. 2005. *Results of the Fatigue Evaluation and Field Monitoring of the I-39 Northbound Bridge over the Wisconsin River*. ATLSS Report No. 05-04, Lehigh University, Bethlehem, PA.

Rosenkrantz. W. A. 1997, *Introduction to Probability and Statistics for Scientists and Engineers*, The McGraw-Hill Companies, Inc.

Mini-Symposia (MS05) Modelling Seismic Action

Safety, Reliability and Risk of Structures, Infrastructures and
Engineering Systems – Furuta, Frangopol & Shinozuka (eds)
© 2010 Taylor & Francis Group, London, ISBN 978-0-415-47557-0

Earthquakes occurrence time prediction and renewal processes

E. Garavaglia
Department of Structural Engineering, Politecnico di Milano, Milano, Italy

ABSTRACT

During the last decades most countries have developed individual programs of seismic hazard assessment (SHA), aimed at establishing or updating the national seismic codes. Essentially they can be divided into Poisson and non-Poisson models. Poisson models are widely used in engineering seismic hazard analysis, but they may not be appropriate for large earthquakes in the fault-specific case, because any Poisson process is inherently memory less. But some times the Poisson hypothesis cannot be rejected.

The aim of this paper is the modeling of earthquake processes characterized by short and long interoccurrence times between events of the same class of magnitude in a way which is more consistent with the underlying physics. The interoccurrence times process will be modeled as renewal process with a step of memory which keeps track of the last event occurring prior or at time t. The renewal process can be considered the simplest of the Markovian processes, it is sufficiently able to capture the main behavior of the earthquakes: the renewal, and maintains simplicity in analysis also respect the multi states Markovian processes.

By historical catalogue is often evident as interoccurrence times between earthquakes of large magnitude and associated to the same homogeny seismogenetic zone show different behaviors (i.e. a family of shorter interoccurrence times and a family of longer ones), the modeling of it with a mixture distribution able to capture both the behaviors seems to be a good compromise. In this paper a mixture of Exponential and Weibull distributions is proposed and discussed. The estimation of the four parameters present in the mixture is a difficult matter that can compromise the whole modeling. In the paper the choice of distribution used and the parameters optimization are discussed and commented.

The degree of uncertainty, epistemic and statistic, contained in the models is an open question and debated since a lot of years. Without pretension to enter in this debate, the elementary model here proposed could be a first step to incorporate events both quasi-periodic and near each to other.

The credibility of the method proposed will be here investigated fallows the approach proposed by Grandori in (Grandori *et al.*, 1998 and successive) where two different models are put in competition and an indicator, called *credibility*, has been adopted to compare the competing models and identify the winner. Here the mixture model will be put in competition with the Poisson model, and the results obtained are suitable.

The application of the procedure proposed is made on two Italian seismic regions with a degree of homogeneity in earthquake generation: Friuli and Umbria-Marche. In these zones the interoccurrence times behavior changes if changes the magnitude's threshold for which the prediction will be done.

The procedure proposed, applied on the Italian context has been able to capture both Poissonian and non-Poissonian behaviors in the interoccurrence sets analyzed.

The robustness of the method proposed is approached here in a preliminary form, applying the *credibility* procedure also "a posteriori".

REFERENCES

CPTI Working Group 2004. Catalogo Parametrico dei Terremoti Italiani–Versione 2004 (CPTI04) INGV, Bologna on http://emidius.mi.ingv.it/CPTI04/ .

Garavaglia, E., Guagenti, E. & Petrini, L. 2007. The earthquake predictability in mixture renewal models, *Proc. of Società Italiana di Statistica SIS Intermediate Conference 2007 Risk and Prediction*, Venezia, June 6 – 8, 2007, Invited Section, Editor Società Italiana di Statistica, Ed. CLEUP, Venezia, **I**, 361–372.

Grandori G., Guagenti E., & Tagliani A. 1998 A proposal for comparing the reliabilities of alternative seismic hazard models, *Journal of Seismology*, **2**, 27–35.

Safety, Reliability and Risk of Structures, Infrastructures and Engineering Systems – Furuta, Frangopol & Shinozuka (eds)
© *2010 Taylor & Francis Group, London, ISBN 978-0-415-47557-0*

Comparison of seismic hazard map with maximum seismic intensity map based on historical document

S. Sakamoto
Taisei Corporation, Yokohama, Japan

R. Nakamura
Tokyo Electric Power Services Corporation, Tokyo, Japan

K. Shimazaki
The University of Tokyo, Tokyo, Japan

S. Midorikawa
Tokyo Institute of Technology, Yokohama, Japan

ABSTRACT

Two kinds of maps, which describe the seismic potency in Japan, are compared with each other. One is the maximum seismic intensity map based on historical documents. The map is constructed with the intensity distributions of actual past earthquakes, but distributions were not observed systematically. The effective observation period of the map, which is important for understanding the hazard, would be shorter than the actual one. Another one is the hazard map based on the probabilistic hazard analysis. Many hazard maps have been constructed and presented, while they were not verified as a whole map. In this paper, for evaluating the effectiveness of the maximum seismic intensity map and for verifying the hazard map, the comparison was conducted by evaluating the likelihood function of the hazard map for the maximum seismic intensity map.

The hazard maps, which describe the distribution of exceedance probabilities of PGV, are constructed with the method of Headquarters for Earthquake Research Promotion of Japan (2005).

The maximum seismic intensity map is constructed for before 1885 and for after 1885, respectively, because systematical seismic observation started after 1885 in Japan. The map for before 1885 was constructed based on Usami (1999), and the map for after 1885 was constructed with Japan Meteorological Agency (JMA) records. Both maps describe seismic intensities with JMA seismic intensity scales, such as V, VI and VII.

Although systematical observation started after 1885, JMA seismic intensity scales were still judged by human sense until 1995. After 1995, the JMA seismic intensity was defined and associated to JMA seismic intensity scales, and instrumental seismic intensity recordings started. So, it is not sure that JMA scales observed before 1995 correspond well with the JMA definition. For verifying the correspondence, likelihood functions for the maximum intensity map for after 1885 were evaluated. As a result, boundary for

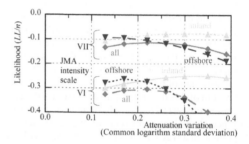

Figure 1. Effect of attenuation variation on the likelihood ac-cording to earthquake type.

scale VII corresponds well with the JMA definition, while boundaries for scales VI and V are overestimated. These boundaries were adopted for estimating the effective observation period of the map for before 1885. The results show that the effective period is about 150~300 years, while its actual period is about 1300 years. It was confirmed that the effective observation period of the map is much shorter than the actual one. Moreover, for examining the difference of the effect of ground motion attenuation variation between earthquake types, the map construction and likelihood evaluations were conducted for each earthquake type: inland earthquakes and offshore earthquakes. Figure 1 shows that likelihood functions for inland earthquakes increase with the attenuation variation while those for offshore earthquakes decrease. It was detected that, for constructing the hazard map, some consideration should be given on the difference of the attenuation variation between earthquake types.

REFERENCES

Headquarters for Earthquake Research Promotion, Japan. 2005. *National Seismic Hazard Maps for Japan.* http://www.jishin.go.jp/main/chousa/06mar_yosoku-e/index-e.htm.

Usami, T., Watanabe, K., Yashiro, K. & Nakamura, R. 1999. Feature of the distribution of regions with intensity III~VII based on the historical earthquake data and active fault (in Japanese). *Historical Earthquake* 15: 35-42.

Safety, Reliability and Risk of Structures, Infrastructures and
Engineering Systems – Furuta, Frangopol & Shinozuka (eds)
© 2010 Taylor & Francis Group, London, ISBN 978-0-415-47557-0

New efficient simulation technique for fully non-stationary stochastic earthquake ground motion model

M. Barbato
*Department of Civil & Environmental Engineering, Louisiana State University at Baton Rouge,
Baton Rouge, LA, USA*

J.P. Conte
*Department of Structural Engineering, University of California at San Diego, La Jolla,
CA, USA*

ABSTRACT

Stochastic processes are commonly used in many engineering fields to model loadings such as earthquake ground motion, wind turbulence, and ocean wave excitation. A large amount of research has been devoted to the development of analytical models and numerical simulation techniques for stochastic processes adequate to represent loadings driven by nature variability. In earthquake engineering, the non-stationarity in time and in frequency content has been recognized as an essential ingredient to realistically capture the loading effects on the structural response (Yeh and Wen 1990). Therefore, great attention has been given to non-stationary earthquake ground motion models, with the aim of accurate and computationally efficient representation of these loadings (Der Kiureghian and Crempien 1989, Conte et al. 1992).

At the same time, Monte Carlo Simulation (MCS) has been widely used for directly solving (often together with variance reduction techniques) and/or checking approximate solutions of random vibration and system stochasticity problems (Shinozuka and Wen 1972). MCS is a general and robust method for solving probabilistic problems for which a deterministic solution is known, but usually is computationally very expensive. A crucial step in MCS is the generation of sample functions of the stochastic processes or random fields involved in the problem. Accurate and efficient sample generation is of paramount importance for a successful application of the MCS technique. A general method for sample generation of stochastic processes or random fields is the Spectral Representation Method (SRM) (Shinozuka and Jan 1972).

In this paper, a new efficient simulation methodology is presented for a fully non-stationary earthquake ground motion stochastic process which has found wide application in earthquake engineering studies (Conte and Peng 1997). This new methodology, based on a "physical" interpretation of the considered stochastic process, is compared with the more general SRM in terms of accuracy and efficiency. The comparison between the newly proposed simulation method and the SRM is made based on an earthquake ground motion model obtained fitting the SEGMM parameters to a historical earthquake. The relative CPU time required for simulating the earthquake ground motion is used as measure of computational efficiency. Accuracy is evaluated by comparing the simulation estimates with the closed-form solutions of statistics of the SEGMM.

From the presented comparison, the newly proposed simulation method for the specific SEGMM considered is found to be more accurate, efficient, flexible and easy to use than the SRM. While the SRM is a general method for stochastic process simulation, the proposed method is specifically developed for and exploits the properties of the considered fully non-stationary SEGMM.

REFERENCES

Conte JP, Pister KS, Mahin SA. 1992. Nonstationary ARMA modeling of seismic motions. *Journal of Soil Dynamics and Earthquake Engineering* 11(7):411–426.

Conte JP, Peng BF. 1997. Fully nonstationary analytical earthquake ground-motion model. *Journal of Engineering Mechanics* (ASCE) 123(1):15–24.

Der Kiureghian A, Crempien J. 1989. An evolutionary model for earthquake ground motion. *Structural Safety* 6:235–246.

Shinozuka M, Jan CM. 1972. Digital simulation of random processes and its application. *Journal of Sound and Vibration* 25(1):111–128.

Shinozuka M, Wen YK. 1972. Monte Carlo solution of nonlinear vibrations. *Journal of American Institute of Aeronautics and Astronautics* 10(1):37–40.

Yeh CH, Wen YK. 1990. Modeling of nonstationary ground motion and analysis of inelastic structural response. *Structural Safety* 8:281–298.

*Safety, Reliability and Risk of Structures, Infrastructures and
Engineering Systems – Furuta, Frangopol & Shinozuka (eds)
© 2010 Taylor & Francis Group, London, ISBN 978-0-415-47557-0*

A procedure for probabilistic seismic hazard analysis which allows to account for Poisson or non-Poissonian models

S. Silvestri, T. Trombetti & G. Gasparini
Department DISTART, University of Bologna, Bologna, Italy

ABSTRACT

In the Performance-Based Seismic Design frame-work (Moehle & Deierlein 2004), the development of an appropriate Probabilistic Seismic Hazard Analysis which allows to statistically characterize the seismic input at the site of an engineering project becomes of crucial importance. The objective of the hazard analysis is to compute, for a given site over a given observation time t, the probability $P[IM \geq \overline{im}]$ of exceeding any particular value ($\overline{im}$) of a specified Intensity Measure (IM).

This paper presents an approach (assumptions + procedure) for Probabilistic Seismic Hazard Analysis, which is alternative to the widely known Cornell's approach (1968) and is developed on the basis of the following concept: the seismic hazard can be represented in terms of an Intensity Measure, which, through specific attenuation laws, is expressed, as a function g of magnitude MS and site-epicentre distance R. The magnitude of the event, which may occur over a given observation time t, and the distance being random variables characterised respectively by probability density functions $f_{MS}(ms)$ and $f_R(r)$. From basic probability theory (Ang & Tang 2007), $f_{IM}(im)$, over a given observation time t, can then be obtained once $IM = g(MS, R)$, $f_{MS}(ms)$ and $f_R(r)$ are known. $IM = g(MS, R)$ is clearly the attenuation law (or ground motion prediction law) at hand. $f_{MS}(ms)$ can be freely chosen according to the selected occurrence model (which may be either the widely-used Poisson process or any other more general Non-Poissonian model). $f_R(r)$ can be easily derived according to the type/geometry of the seismic source zones (line sources or area sources of various shapes).

The procedure has been fully developed with reference to more than one and generically shaped seismic zones.

The peculiarity of this procedure, which leads to the determination of probability functions of a selected ground motion parameter due to the seismic action at a specific site over a given observation time, resides in that (i) the occurrence of seismic events may be schematized either with the widely-used Poisson process or with more general Non-Poissonian models, and (ii) it allows a clear separation between different (aleatory and epistemic) sources of randomness (Abrahmason 2006).

REFERENCES

Abrahamson, N.A. 2006. Seismic Hazard Assessment: Problems with Current Practice and Future Developments, *Proceedings of the First European Conference on Earthquake Engineering and Seismology, 1st ECEES* (a joint event of the 13th ECEE & 30th General Assembly of the ESC), Geneva, Switzerland, 3-8 September 2006, Paper No. 4002 (Keynote Lecture K2).

Ang A., Tang W., 2007. *Probability Concepts in Engineering. Emphasis on Applications to Civil and Environmental Enginering*, 2nd edition, John Wiley & Sons Inc., USA.

Cornell, C.A., 1968. Engineering seismic risk analysis, *Bulletin of the Seismological Society of America* 58: 1583–1605.

Moehle, J., Deierlein, G. G., 2004. A framework methodology for Performance Based Earthquake Engineering, Proceedings of the "13th World Conference on Earthquake Engineering", 13WCEE, Vancouver, B.C., Canada, August 1–6, 2004, Paper No. 679.

Safety, Reliability and Risk of Structures, Infrastructures and Engineering Systems – Furuta, Frangopol & Shinozuka (eds)
© *2010 Taylor & Francis Group, London, ISBN 978-0-415-47557-0*

The role of epsilon for the identification of groups of earthquake inputs of given hazard

T. Trombetti, S. Silvestri & G. Gasparini
Department DISTART, University of Bologna, Bologna, Italy

ABSTRACT

The probabilistic identification of the reference input is fundamental for any sound engineering design. This is particularly true in the case of seismic engineering due to the intrinsic difficulties in the identification of the earthquake input, as well as due to the different performances expected for the structure under such inputs.

Within a Performance Based Seismic Design framework (SEAOC Vision 2000, 1995), the choice of the reference design seismic input (commonly referred to as "earthquake bins") is deeply rooted upon their probabilistic identification. Typically, the earthquake bins are identified by means of earthquake "intensity measures" (*IMs*). IMs consisting of a scalar or vector-valued combination of selected ground motion parameters (*GMPs*) associated to a given probability. In recent years, many research works (Giovenale et al. 2004, Trombetti et al. 2007) have focused on the identification of the optimal IM for earthquake bin creation.

This paper focuses on the identification of groups of uniform hazard (acceleration) time-histories for Performance Based Seismic Design applications. In detail, on the basis of a peculiar Probabilistic Seismic Hazard Analysis, the characteristics that a group of earthquake inputs must possess in order to be associated to a given exceedance probability are obtained. The proposed procedure takes advantage of the information carried by the "epsilon" parameter, and is rooted on a separate treatment of the aleatory variability and the epistemic uncertainty considered in the hazard analysis. The analytical developments allow to identify a condition for the spectral ordinates ("spectral cloud", Fig. 1) of the acceleration time histories, which is valid for a number of structural periods at once, and to quantify (in terms of coefficient of variation of the spectral ordinates) the randomness associated to the epistemic uncertainty (error in the spectral acceleration prediction law).

The statistical characterisation of the spectral cloud (which can be assimilated to a lognormal random process) as here proposed allows to: (1) treat separately and independently the epistemic uncertainty due to the error of the attenuation model from all other time- and

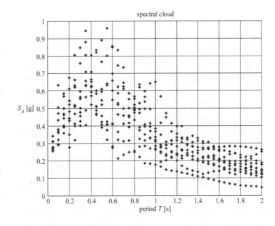

Figure 1. Spectral cloud.

space-related sources of aleatory variability; (2) identify earthquake inputs which retain their significance independently from the period range considered; (3) obtain groups of design earthquake inputs which can be used for different structures and for structures with substantial variations in vibration periods; and (4) link the identification of the seismic hazard strictly to the site, without involving the structure.

REFERENCES

Giovenale, P., Cornell, C.A. & Esteva, L. 2004. Comparing the adequacy of alternative ground motions intensity measures for the estimation of structural responses. *Earthquake Engineering and Structural Dynamics* 33: 951–979.

SEAOC, Vision 2000 Committee 1995. *Performance-based seismic engineering for buildings*. Report prepared by Structural Engineers Association of California, Sacramento, CA.

Trombetti, T., Silvestri, S., Malavolta, D. & Gasparini, G. 2007. A methodology for determination of efficient earthquake bins for Performance Based Seismic Design. *Proceedings of the Tenth International Conference on Applications of Statistics and Probability in Civil Engineering (ICASP10)*, Tokyo, Japan, July 31 – August 3, 2007.

Safety, Reliability and Risk of Structures, Infrastructures and
Engineering Systems – Furuta, Frangopol & Shinozuka (eds)
© 2010 Taylor & Francis Group, London, ISBN 978-0-415-47557-0

An innovative concept on designing low-rise beachfront building capable of reducing earthquake and tsunami damage impacts

I.-K. Chang & N. Liu

College of Architecture, University of Oklahoma, Norman, OK, USA

ABSTRACT

In spite of advances that have been made in search for new and stronger materials, and effective construction methods during the recent decades, failures of conventional low-rise residential and commercial buildings, loss of lives, and widespread economy impacts caused by earthquake, tsunami or flood, tornado or hurricane and wildfire remain a great threat to the mankind.

This paper presents an innovative building design concept that integrates the architecture and the structure in harmony, and applies a method, named "ASID-3DS" that uses "Defusing, and/or Decoupling and/or Dynamic Solutions" for earthquake, tsunami (flood), hurricane (tornado), and wildfire protection designs. The application of the "ASID-3DS method is strategically dependent upon the form, the function, the

site and the loads of the building. A hypothetical low-rise beachfront building is conceptually designed. A sequence of relevant architectural and structural integration designs are presented for the illustration of the ASID-3DS application for the earthquake/tsunami dual protection designs.

The ASID-3DS is presented to draw the attentions of the practicing architects and engineers, as well as, architectural and engineering students. It is believed that any conceptual architecture design conceived with the ASID-3DS integration will surely strengthen the chances of saving lives and minimizing economic loss brought by earthquakes and tsunamis.

Organized Session (OS23) New Methods for Non-Gaussian and Pulse Problems in Stochastic Dynamics [Session Organized on Behalf of IASSAR Subcommittee SC2]

Safety, Reliability and Risk of Structures, Infrastructures and
Engineering Systems – Furuta, Frangopol & Shinozuka (eds)
© 2010 Taylor & Francis Group, London, ISBN 978-0-415-47557-0

An approximate technique for the probability density function of the response of a linear oscillator to Erlang renewal random impulse processes

M. Vasta

University of Chieti-Pescara, Pescara, Italy

ABSTRACT

The analysis of the response of linear and non linear dynamical system to random impulsive process is an important topic in mechanical and civil engineering. It is well known that if a dynamical system is subjected to a Poisson impulse process, the state vector of the system is a non-diffusive, Poisson-driven, Markov process. Its joint probability density function satisfies an integro-differential generalized Fokker-Planck-Kolmogorov equation, also known as Kolmogorov-Feller equation. If however, the driving counting process is other than Poisson, for example a renewal process, then as the increments are not independent, the state vector is not a Markov process. For some classes of impulse processes the problem may be converted into a Markov one, as for the Erlang renewal impulse processes. Conversion of the original non-Markov problem into a Markov one is then done at the expense of introducing auxiliary state variables, which are discrete-valued stochastic processes governed by stochastic differential equations driven by a Poisson process.

In this paper an approximate solution of a general integro-differential forward Chapman-Kolmogorov equation ruling the joint probability density function of the state variables and of the Markov state jump process is considered. The equations ruling the response probability density function are a coupled set of one partial integro-differential equation and $k - 1$ partial differential equations. The equations are solved using an approximate technique devised by considering the evolution of the state variables during small time intervals. The governing integro-differential equations are first converted into partial differential equations and next the method of characteristics is used to find the explicit solution to these equations. Transient marginal probability density functions of the response of a linear oscillator are evaluated from numerically determined joint probability density distribution functions. The validity of the approximate technique is verified by comparing the Markov state probabilities obtained from the joint probability density-distribution functions with those obtained directly from the governing equations for Markov state probabilities.

REFERENCES

Iwankievicz, R. and Vasta, M., 2006. Approximate method for probability density of the response of a linear oscillator to a non-Poisson impulse process. III European on Computational Mechanics, Solids, Structures and Coupled problems in engineering, Lisbon, Portugal.

Iwankievicz, R. and Vasta, M., 2006. Approximate equations for probability density of the response of a linear dynamic system under Erlang renewal impulse process. CISM 2006, pp. 357–365, Rhodes, Greece.

Safety, Reliability and Risk of Structures, Infrastructures and
Engineering Systems – Furuta, Frangopol & Shinozuka (eds)
© 2010 Taylor & Francis Group, London, ISBN 978-0-415-47557-0

On the class of impulse processes generated by two Erlang renewal processes

R. Iwankiewicz & M. Tellier
Hamburg University of Technology, Hamburg, Germany
Arup SA, Johannesburg, South Africa

ABSTRACT

The considered class of random impulse processes is obtained by selecting impulses from the train of impulses driven by an Erlang renewal process. This is done by compounding the random magnitudes of the impulses in the Erlang renewal train with the values of an additional, purely jump, zero-one valued stochastic process which is represented by the following replacement

$$\sum_{i,R=1}^{R(t)} P_{i,R}\delta(t-t_{i,R}) = \sum_{i=1}^{R_\nu(t)} Z(t_i)P_i\delta(t-t_i), \quad (1)$$

where the arrival times t_i are driven by an Erlang renewal process $R_\nu(t)$ with parameters ν, k and $Z(t_i)$ is a value at t_{i-} of the zero-one stochastic variable $Z(t)$ governed by (Iwankiewicz 2003; Tellier & Iwankiewicz 2005; Tellier 2007).

$$dZ(t) = (1-Z)dR_\mu(t) - ZdR_\nu(t), \quad (2)$$

where $R_\mu(t)$ is another Erlang renewal process, with parameters μ, l, independent of $R_\nu(t)$.

The replacement (1) implies the equivalence, with probability 1, of the increments: $dR(t) = Z(t)dR_\nu(t)$.

If $R_\nu(t)$ is an Erlang process, then $dR_\nu(t) = \rho(N_\nu(t))$ $dN_\nu(t)$, where $\rho_\nu(t) = \rho(N_\nu(t))$ is a jump, zero-one valued, stochastic process expressed in terms of some Poisson-driven auxiliary stochastic processes. For the Erlang process with $k = 2$ the function $\rho(N_\nu(t))$ is governed directly by the stochastic equation

$$d\rho_\nu(t) = (1 - 2\rho_\nu(t))dN_\nu(t) \quad (3)$$

In order to characterize the generated impulse process, the general expressions for the probability density functions of the first and of the second waiting time

are derived from the basic principles. based on the fact that the generating processes are Erlang. The mean arrival rate $h(t)$ of the impulses driven by an underlying counting process $R(t)$ is defined as

$$h(t)dt = E[dR(t)] = E[Z(t)dR_\nu(t)] = E[Z(t)\rho_\nu(t)dN_\nu(t)] = E[Z(t)\rho_\nu(t)]\nu dt \quad (4)$$

It is evaluated from the equations (2) and (3) with the aid of the Itô's generalized differential rule. Under the assumption that the underlying process is a renewal process, the mean arrival rate of the impulses is regarded as the renewal density of the hypothetical renewal process and the corresponding probability density function of the inter-arrival times is obtained. For the Erlang processes with integer parameters $k = 1, 2$ and $l = 1, 2$ it is verified that the probability density function of the second waiting time is a convolution of the probability density function of the first waiting time and of the hypothetical probability density function of the inter-arrival times, hence the underlying counting process is a renewal process indeed.

REFERENCES

Iwankiewicz, R. (2003) Response of dynamic systems to random impulses driven by non-Erlang renewal processes. In *Proc. of International Conference on Applied Mechanics and Materials (ICAMM 2003), 21–23 January 2003, Durban, South Africa. Eds. S. Adali, E.V. Morozov and V.E. Verijenko, ISBN1-86840-501-X.*

Tellier, M. and Iwankiewicz, R. (2005). Response of linear dynamic systems to non-Erlang renewal impulses: stochastic equations approach. *Probabilistic Engineering Mechanics* 20(4).

Tellier, M. (2007). *Response of dynamic systems to a class of renewal impulse process excitations: non-diffusive Markov approach (PhD thesis).* University of the Witwatersrand, Johannesburg South Africa.

Safety, Reliability and Risk of Structures, Infrastructures and
Engineering Systems – Furuta, Frangopol & Shinozuka (eds)
© 2010 Taylor & Francis Group, London, ISBN 978-0-415-47557-0

Non-linear systems under Levy white noise handled by path integration method

M. Di Paola
University of Palermo, Palermo, Italy

R. Iwankiewicz
Hamburg University of Technology, Hamburg, Germany

A. Pirrotta
University of Palermo, Palermo, Italy

ABSTRACT

Some real phenomena observed in physics, seismology, electrical engineering, economics and in some other research fields show evident non-Gaussianity either in heavy tails distribution or in the impulsive nature of the recorded samples. According to this new need, the Itô stochastic differential calculus has been extended to Poissonian white noises too (Snyder 1975), providing the equation governing the evolution of the probability density, known as Kolmogorov-Feller equation (Iwankiewicz & Nielsen 1999, Pirrotta 2007). The need for non-Gaussian models, to describe the fluctuations exhibited by non-Gaussian phenomena, has also raised the interest in the so-called α-stable Lévy processes (Grigoriu 2000, Di Paola et al.2007). Linear and non-linear systems driven by external α-stable Lévy white noise processes (formal time derivative of the Lévy motion processes labeled as $L_\alpha(t)$ have been treated, in the past, either in terms of PDF Einstein-Smoluchowsky (ES) equation or in terms of CF (Chechkin & Gonchar 2000, Chechkin et al.2002). However closed-form expressions for the probability density function of dynamical systems driven by all these white noises have been obtained only for very restricted classes. This is the reason for the interest towards numerical methods.

Among the numerical methods for evaluating the PDF the so-called Path Integral Solution (PIS) is an effective tool for evaluating the response in terms of probability density at each time instant, for evaluating moments of various order, energy response PDF, first passage time for strongly non-linear systems. The PIS method mainly consists in evaluating the PDF at a given time instant while a PDF at an earlier close time instant (short time steps) is obtained by solving a convolution integral.

The starting point is the Chapman-Kolmogorov equation, in which the kernel is the transition probability density function. In this way the PDF of the response at the time $(t + \tau)$ may be evaluated when the PDF at an earlier close time instant (t) is already known. The crucial point is to define the kernel according to the system under investigation. In the case of normal white noise, if τ is small, even if the system is non-linear, the transition PDF is almost Gaussian (short-time Gaussian approximation). It follows that the kernel of the integral form is Gaussian and this simplifies the analysis (Barone et al 2008). Path integration method was also applied to dynamics systems driven by Poisson impulse processes (Koyluoglu, Nielsen & Iwankiewicz 1995, Di Paola & Santoro 2008) and by non-Poisson, renewal impulse processes (Iwankiewicz & Nielsen 1996).

The aim of this paper is to investigate the consistency of the Path Integration (PI) method proposed earlier by Naess & Johansen (1991, 1993) for non-linear systems driven by α-stable white noise. It is shown that at the limit for $\tau \to 0$ the Einstein-Smoluchowsky (ES) equation is fully restored. Once the consistency of the PI is demonstrated for the half oscillator, then the extension of the ES equation for MDOF system is found starting from the PI method.

Safety, Reliability and Risk of Structures, Infrastructures and Engineering Systems – Furuta, Frangopol & Shinozuka (eds)
© 2010 Taylor & Francis Group, London, ISBN 978-0-415-47557-0

Accuracy of the narrow-band approximation of stationary wide-band Gaussian processes for extreme value and fatigue analysis

Z. Gao

Centre for Ships and Ocean Structures, Norwegian University of Science and Technology, Norway

T. Moan

Centre for Ships and Ocean Structures and Department of Marine Technology, Norwegian University of Science and Technology, Norway

ABSTRACT

Accurate estimates of extreme values of wide-band Gaussian processes depend on the tail distribution, while fatigue damage prediction relies on cycle counting of the effective stress ranges. It is well-known that the narrow-band approximation gives conservative results for both analyses.

The purpose of this paper is to illustrate the accuracy (overestimation) and the applicability of this approximation based on comparison with the time-domain results. Extreme value statistics and mean fatigue damage estimated by the rainflow method (Matsuishi & Endo, 1968) are used for comparison. The Vanmarcke's bandwidth parameter is used to characterize a wide-band spectrum. In total, the 4200 different spectra analyzed by Gao and Moan (2008) are considered.

Figures 1 and 2 show the ratios of the narrow-band approximation to the time-domain result as a function of bandwidth parameter for three-hour extreme values and fatigue damage, respectively. For fatigue analysis, the slope parameter m of the SN curve is taken as 3.

In general, the overestimation of the narrow-band method and its variation increase with an increasing bandwidth parameter. For extreme value analysis, the narrow-band method produces very accurate results, which agrees well with the conclusion made by Moe and Niedzwecki (2005). However, it becomes more conservative for extremely large bandwidth parameters due to the limited number of peaks in the duration of consideration for these processes. For fatigue analysis, the narrow-band approximation is practically acceptable when the bandwidth parameter is less than 0.5, while fatigue damage might be significantly overestimated for the bandwidth parameters larger than 0.8.

However, a spectrum with a very large bandwidth parameter might not be practically relevant. The maximum obtained bandwidth parameter of linear wave-induced responses considered herein is about 0.59 and the narrow-band method gives only 5% and 20% overestimation for extreme values and fatigue damage, respectively. It indicates that the narrow-band method is a simple and good approximation for practical use.

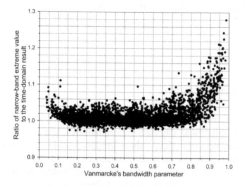

Figure 1. Ratio of the narrow-band approximation of 3-hour extreme values to the time-domain result.

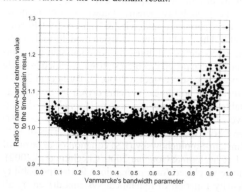

Figure 2. Ratio of the narrow-band approximation of fatigue damage ($m = 3$) to the rainflow result (where RFC denotes the rainflow result)

REFERENCES

Gao, Z. & Moan, T. 2008. Frequency-domain fatigue analysis of wide-band stationary Gaussian processes using a trimodal spectral formulation. *International Journal of Fatigue*: Vol. 30, pp. 1944–1955.

Matsuishi, M. & Endo, T. 1968. Fatigue of metals subjected to varying stress. *Japan Society of Mechanical Engineers*. Fukuoka, Japan.

Moe, G. & Niedzwecki, J.M. 2005. Frequency of maxima of non-narrow banded stochastic processes. *Applied Ocean Research*: Vol. 27, pp. 265–272.

Mini-Symposia (MS09) Life-Cycle Reliability and Optimization of Deteriorating Structures

Safety, Reliability and Risk of Structures, Infrastructures and Engineering Systems – Furuta, Frangopol & Shinozuka (eds)
© 2010 Taylor & Francis Group, London, ISBN 978-0-415-47557-0

Attempt for expansion of bridge lives by Osaka Municipal Government

E. Watanabe
Kyoto University, Kyoto, Japan

T. Yokota & Y. Komatsu
Public Works Bureau, Osaka Municipal Government, Osaka, Japan

H. Furuta
Department of Informatics, Kansai University, Takatsuki, Japan

ABSTRACT

The City of Osaka is known as a Capital of Water. As dated on April 1, 2006, Osaka Municipal Government is taking care of totally 763 bridges with the total bridge length of 47.7 km and total deck area of 72.3 ha. Since Osaka City has been developed near the estuary of rivers such as River Yodo and River Yamato, the number of streams and moats is great and consequently citizens' life and the development of the city depend greatly on the bridges. From this reason and the total number of bridges, Osaka City has been called "City of Eight Hundred and Eight Bridges" and is endowed with many historic bridges (City of Osaka Public Eng.).

The City of Osaka is managing various bridges such as large-scaled, roadway, viaduct over railway and small but important living bridges. These bridges are supporting the social activities of citizens and some of them are landmarks of the city. The majority of the Osaka City bridges were built during the stages of the first city planning and the opening of the World Exposition of Osaka, about 70 and 35 years ago, respectively. Particularly, in the coming 30 years about 100 bridges will be over the age of 100 years and the demand for the renewal or reconstruction of these bridges becomes stronger and thus the city has been concerned about the soaring cost for the maintenance or the renewal.

In view of such budget cut approximately to 40% of the maximum in the past due to economic aggravation of the city, it is becoming difficult to maintain bridges in the traditional method of replacing old bridges with new ones. With these as the background, the maintenance of bridges and the management of road network should avoid concentration of renewal of bridges within a short period of time. Undoubtedly, the preventive maintenance is much desirable than the essential maintenance corresponding to the more serious deteriorating state and costly repair work.

For preventive maintenance, the deterioration process, the future condition state of bridges must be

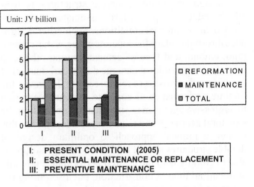

Figure 1. Expected annual reduction of bridge expenditure by preventive maintenance in JY billion.

precisely predicted. The Osaka Municipal Office has started to operate the Bridge Management System (OBMS) since 2005. OBMS executes the optimum planning for the preventive bridge maintenance through the periodical inspection, efficient data acquisition and the priority rule taking into the historic and civic importance.

According to a trial calculation, as shown in Figure 1, the total budget in coming 30 years can be reduced from JY 210 billion to JY 110 billion (average annual budget from JY 7 billion to 3.7 billion) if the preventive maintenance is executed on the aforementioned 100 old bridges but with the renewal of only 15 bridges as compared with the renewal of every one of 100 old bridges (Yokota et. al 2008).

REFERENCES

City of Osaka Public Engineering Works Foundation. http://www.osaka-udce.or.jp/bridge/chap1/index1.htm
Yokota, T., Komatsu, Y. and Nagahashi, S. 2008. Replacement plan of old bridges by Osaka Municipal Government. Bridge and Foundation Engineering, 42(8): 36–38 (in Japanese).

Safety, Reliability and Risk of Structures, Infrastructures and
Engineering Systems – Furuta, Frangopol & Shinozuka (eds)
© *2010 Taylor & Francis Group, London, ISBN 978-0-415-47557-0*

Lifetime probabilistic optimization of steel reinforcement in concrete structures

Fabio Biondini

Department of Structural Engineering, Politecnico di Milano, Milan, Italy

Dan M. Frangopol

Department of Civil and Environmental Engineering, Center for Advanced Technology for Large Structural Systems, Imbt Labs, Lehigh University, Bethlehem, PA, USA

ABSTRACT

The design of durable concrete structures is usually based on simple criteria associated with prescribed environmental conditions. Such criteria introduce threshold values for concrete cover, water-cement ratio, amount and type of cement, among others, to limit the effects of local damage due to carbonation of concrete and corrosion of reinforcement. However, the lifetime performance of concrete structures exposed to aggressive agents may significantly depend on the structural geometry and reinforcement layout. For this reason, a rational approach to optimum design of durable structures should lead to find design solutions able to comply with the desired performance not only at the initial time of construction, but also during the expected lifetime by considering the effects induced by unavoidable sources of damage under uncertainty.

Based on these premises, life-cycle approaches to structural optimization of deteriorating systems have been proposed to highlight the fundamental role played by the time-variant performance in the optimal maintenance planning and selection of the optimum structural design (see for example Frangopol *et al.* 1997, 2002; Kong and Frangopol 2003; Azzarello et al. 2007; Frangopol and Liu 2007; Biondini and Zani 2008). In this paper the life-cycle approach to structural optimization is applied in a probabilistic context to the minimum reinforcement design of concrete cross-sections subjected to diffusive attacks from environmental aggressive agents, like sulfate and chloride, which may lead to deterioration of concrete and corrosion of reinforcement. The time evolution of the structural performance is evaluated by using a general procedure proposed in previous works for concrete structures in aggressive environments (Biondini *et al.* 2004, 2006).

The lifetime probabilistic optimization is formulated to minimize the amount of steel reinforcement under a time-variant design constraint on the lifetime structural reliability. The problem is solved by combining a discrete gradient-based optimization method with a Monte Carlo simulation. The role played by a lifetime approach to structural optimization is shown by comparing the optimal solutions obtained with a classical time-invariant formulation, which considers the initial undamaged state only, and the proposed lifetime formulation, where the time evolution of the structural performance is taken into account. The obtained results show that in a lifetime oriented design the minimum feasible area of reinforcement is not associated with the maximum depth of the steel bars over the concrete cross-section, as it is expected in a classical time-invariant approach, but the location of each single bar with respect to the location of the aggressive agent can play a crucial role in the definition of the optimum structural design.

REFERENCES

Azzarello, L., Biondini, F., and Marchiondelli, A., (2007). Lifetime Optimization of Reinforced Concrete Structures in Aggressive Environments, *Life-Cycle Cost and Performance of Civil Infrastructure Systems*, H.N. Cho, D.M. Frangopol, A.H.-S. Ang (Eds.), Taylor & Francis, 93–102.

Biondini, F., Bontempi, F., Frangopol, D.M., and Malerba, P.G., (2004). Cellular Automata Approach to Durability Analysis of Concrete Structures in Aggressive Environments. *Journal of Structural Engineering*, ASCE, **130**(11), 1724–1737.

Biondini, F., Bontempi, F., Frangopol, D.M., and Malerba, P.G., (2006). Probabilistic Service Life Assessment and Maintenance Planning of Concrete Structures, *Journal of Structural Engineering*, ASCE, **132**(5), 810–825.

Biondini, F., and Zani, G., (2008). Life-Cycle Multi-Objective Optimization of Deteriorating Structures, *Life-Cycle Civil Engineering*, F. Biondini, D.M. Frangopol (Eds.), CRC Press, Taylor & Francis Group.

Frangopol, D.M., Lin, K-Y., and Estes, A.C. (1997). Reliability of reinforced concrete girders under corrosion attack, *Journal of Structural Engineering*, ASCE, **123**(3), 286–297.

Frangopol, D.M., and Liu, M. (2007). Maintenance and management of civil infrastructure based on condition, safety, optimization, and life-cycle cost, *Structure and Infrastructure Engineering*, Taylor & Francis, **3**(1), 29–41.

Frangopol, D.M., Miyake, M., Kong, J.S., Gharaibeh, E.S., and Estes A.C. (2002). "Reliability– and cost–oriented optimal bridge maintenance planning," Chapter 10 in *Recent Advances in Optimal Structural Design*, S. Burns, ed., ASCE, Reston, Virginia, 257–270.

Safety, Reliability and Risk of Structures, Infrastructures and Engineering Systems – Furuta, Frangopol & Shinozuka (eds)
© 2010 Taylor & Francis Group, London, ISBN 978-0-415-47557-0

The role of design and maintenance on the life-cycle reliability of truss structures

Fabio Biondini
Department of Structural Engineering, Politecnico di Milano, Milan, Italy

Dan M. Frangopol
Department of Civil and Environmental Engineering, Center for Advanced Technology for Large Structural Systems, Imbt Labs, Lehigh University, Bethlehem, PA, USA

Elsa Garavaglia
Department of Structural Engineering, Politecnico di Milano, Milan, Italy

ABSTRACT

The structural reliability of deteriorating systems changes over time and, generally, proper maintenance interventions are required to reach suitable levels of life-cycle performance (Frangopol et al. 1997). However, proper design strategies could also contribute to achieve the desired life-cycle reliability without maintenance, or with reduced maintenance activities. In any case, since the structural performance is affected by several sources of uncertainty, the assessment of the life-cycle reliability must be based on a suitable damage modeling and on a probabilistic analysis able to model the main features of the time-variant deterioration process (Biondini *et al.* 2004, 2006).

To investigate the role of the design and maintenance on the life-cycle reliability of aging structures, a probabilistic procedure based on a semi-Markov modeling of the deterioration process is adopted (Howard 1971). This procedure is used to evaluate the life-cycle reliability by considering the effects of selective maintenance, characterized by repair interventions applied only to elements heavily deteriorated and/or characterized by a low safety margin. In fact, for complex structural systems it is crucial to decide which elements require maintenance and to investigate scenarios where repairing involves the deteriorated members only. A selective maintenance requires suitable life-cycle performance indicators to effectively select the members that actually need to be repaired. To this aim, the following condition index μ is defined (Biondini *et al.* 2008):

$$\mu = (1 - \delta) \cdot \frac{R - S}{R} \qquad (1)$$

where δ is a damage index varying in the range $[0,1]$, R is the strength of the considered element, and S is the corresponding loading demand. The index $\mu = \mu(t)$ can be evaluated for each element at each time instant t of the structural lifetime. In this way, based on a proper threshold μ_{min}, the condition $\mu(t) < \mu_{min}$ can be used to identify members heavily deteriorated and/or characterized by a low safety margin. Since μ is a random variable, the above mentioned condition has to be applied in probabilistic terms.

The proposed procedure is applied to compare the life-cycle reliability and life-cycle cost effectiveness of statically determinate and indeterminate truss systems by considering the effects of selective maintenance. The results prove that, when properly dimensioned, the life-cycle cost of statically indeterminate systems can be lower than the cost of statically determinate systems. On the contrary, the benefit of statically indeterminate over statically determinate systems can be reversed if a selective maintenance is applied. However, the positive effect of the selective maintenance can be reduced when the maintenance cost become important. Finally, the crucial role of a proper calibration of the damage model is pointed out. To this aim, the proposed procedure could be integrated with the results of inspection and monitoring activities.

REFERENCES

Biondini, F., Bontempi, F., Frangopol, D.M., and Malerba, P.G., 2004. Cellular Automata Approach to Durability Analysis of Concrete Structures in Aggressive Environments. *Journal of Structural Engineering*, ASCE, **130**(11), 1724–1737.

Biondini, F., Bontempi, F., Frangopol, D.M., and Malerba, P.G., 2006. Probabilistic service life assessment and maintenance planning of concrete structures, *Journal of Structural Engineering*, ASCE **132**(5), 810–825.

Biondini F., Frangopol D.M., and Garavaglia E., 2008. Life-Cycle Reliability Analysis and Selective Maintenance of Deteriorating Structures. *First International Symposium on Life-Cycle Civil Engineering* (IALCCE'08), Varenna, Lake Como, Italy, June 10-14, 2008, F. Biondini & D.M. Frangopol (Eds.), CRC Press, Taylor & Francis Group, London, UK, 483–488.

Frangopol, D.M., Lin, K-Y., and Estes, A.C. 1997. Life-Cycle Cost Design of Deteriorating Structures, *Journal of Structural Engineering*, ASCE, **123**(10), 1390–1401.

Howard, R.A., 1971, *Dynamic Probabilistic Systems*, John Wiley and Sons, New York, NJ, USA.

Mini-Symposia (MS14) Computational Methods in Stochastic Mechanics
[Session Organized on Behalf of IASSAR Subcommittee SC1]

Safety, Reliability and Risk of Structures, Infrastructures and
Engineering Systems – Furuta, Frangopol & Shinozuka (eds)
© 2010 Taylor & Francis Group, London, ISBN 978-0-415-47557-0

Seismic motion incoherency effects on Soil-Structure Interaction (SSI) response of nuclear power plant buildings

Dan M. Ghiocel

Ghiocel Predictive Technologies, Inc., NY, USA

ABSTRACT

The paper discusses key aspects of stochastic modeling of seismic motion incoherency and its implementation in the context of current engineering practice. The paper briefly describes the theoretical basis of stochastic and deterministic incoherent SSI approaches. It should be noted that these incoherent SSI approaches were validated by recent EPRI studies (Short, Hardy, Mertz and Johnson, 2007) and accepted by US NRC for the application to seismic analysis of the new nuclear power plant structures within the United States. The paper illustrates the effects of stochastic ground motion incoherency on seismic SSI responses for a typical nuclear reactor building with no mass eccentricity and a nuclear complex building with significant mass eccentricities founded on a rock site.

Different plane-wave coherency models are considered: i) the Luco-Wong coherency model and, ii) the Abrahamson incoherency models.

The paper illustrates the effects of stochastic ground motion incoherency on seismic SSI responses of a typical axisymmetric nuclear reactor building and nuclear complex building with significant mass eccentricities. To incorporate the stochastic motion incoherency effects on SSI response we used both stochastic and deterministic approaches.

The SSI results are compared for coherent and incoherent seismic inputs to illustrate the effects of motion incoherency on SSI response. The SSI coupling responses of structures are primarily examined to illustrate the additional rocking and torsional motion effects due to the motion incoherency.

The SSI results are considered in terms of both the acceleration response spectra and the structural forces/stresses.

The illustrated SSI examples show that incoherency effects are significant in the high-frequency ranges. As a result of this, the incoherency effects affect less significantly the low-frequency, overall seismic structural responses, namely the structural shear and moments, but much more significantly the high-frequency vibration modes and the in-structure response spectra (ISRS) as shown in Figure 1.

The qualitative effects of motion incoherency effects are: i) for horizontal components a reduction in excitation translation along the input direction concomitantly with an increase in torsional excitation and a slight reduction in foundation rocking excitation, and, ii) for vertical component a reduction in excitation translation in vertical direction concomitantly with an increase of rocking excitation.

REFERENCES

Short, S.A., G.S. Hardy, K.L. Merz, and J.J. Johnson. 2007. Validation of CLASSI and SASSI to Treat Seismic Wave Incoherence in SSI Analysis of Nuclear Power Plant Structures, EPRI Palo Alto, CA, TR-1015110, November.

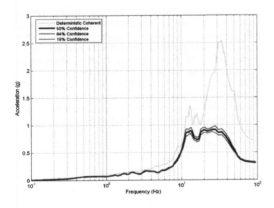

Figure 1. Incoherent vs. Coherent ISRS (5% Damping Ratio).

Safety, Reliability and Risk of Structures, Infrastructures and Engineering Systems – Furuta, Frangopol & Shinozuka (eds)
© 2010 Taylor & Francis Group, London, ISBN 978-0-415-47557-0

On the response variability of beams with large stochastic variations of system parameters

M. Miranda & G. Deodatis

Department of Civil Engineering and Engineering Mechanics, Columbia University, NY, USA

ABSTRACT

A common problem in stochastic structural mechanics is the lack of empirical information available to validate the probabilistic models that are assumed for the uncertain system parameters. Consequently, extensive sensitivity studies with respect to these parameters are typically required. This may lead to prohibitive computational costs and, possibly, to a loss of insight on their relative effect.

In order to address the critical issues mentioned above, the concept of Variability Response Function (VRF) has been proposed in the past as a means of systematically capturing the effect of the second-moment characteristics of the uncertain parameters—specifically, the effect of the spectral density function—on the response variability of structural systems (Shinozuka 1987; Papadopoulos and Deodatis 2006). The existence of the VRF has been mathematically established only for cases of statically determinate structural systems, and for indeterminate systems with small parameter variations by means of first-order Taylor approximations. In both cases, the VRF is a deterministic function that is independent of the probability distribution and of the spectral characteristics of the uncertain parameters. The existence of the VRF for general structural systems (i.e. statically indeterminate systems with large parameter variations) has been recently conjectured on the basis of numerical evidence, but was found to be dependent to a certain degree on parameter variability (Papadopoulos and Deodatis 2006).

In this paper, a Monte Carlo-based methodology is proposed as a generalization of the VRF concept that is applicable to both determinate and indeterminate structures with large stochastic variation in their parameters, without the need of a conjecture (Miranda and Deodatis 2008). Generalized VRFs (GVRFs) are computed numerically using carefully chosen cases of non-Gaussian random fields modeling the uncertain system parameters, selected so as to have a wide range of different probability distributions and spectral characteristics. The GVRFs result from the solution of a system of linear equations combining information on the spectral characteristics and probability distributions of the set of chosen non-Gaussian random fields (matrix of coefficients), and Monte Carlo simulations also based on the characteristics of these fields (right-hand-side vector). Numerical evidence indicates that the GVRFs established with the proposed methodology—although surprisingly close to the classical VRFs—are neither unique nor independent of the probabilistic description of the uncertain system parameters. The deviations from the classical case are found to be slightly sensitive to the non-Gaussian probability distribution, but minimally dependent on the spectral characteristics of the parameters. Taking advantage of this finding, a Fast Monte Carlo methodology is also presented, which approximates very well the GVRFs, while significantly reducing the computational expense involved in their calculation. A numerical example is presented comparing the two proposed methodologies.

REFERENCES

Miranda, M. and G. Deodatis (2008). Generalized variability response functions for statically indeterminate beams with large stochastic variations of bending stiffness. In Proceedings of the Inaugural International Conference of the Engineering Mechanics Institute (EM08), Minneapolis, Minn., May 2008. ASCE. (in CD-ROM).

Papadopoulos, V. and G. Deodatis (2006). Response variability of stochastic frame structures using evolutionary spectra theory. Computer Methods in Applied Mechanics and Engineering 195(9–12), 1050–1074.

Shinozuka, M. (1987). Structural response variability. Journal of Engineering Mechanics 113(6), 825–842.

Safety, Reliability and Risk of Structures, Infrastructures and
Engineering Systems – Furuta, Frangopol & Shinozuka (eds)
© 2010 Taylor & Francis Group, London, ISBN 978-0-415-47557-0

Identification of random geometry for stochastic finite element analysis

G. Stefanou & A. Nouy

*GeM – Research Institute in Civil Engineering and Mechanics, UMR CNRS 6183, University of Nantes,
Centrale Nantes, Nantes, France*

ABSTRACT

In structural analysis, the incorporation of uncertainties inherent in the model seems today essential if one seeks to obtain "reliable" numerical predictions. This necessity led in particular to a rapid development of many ad hoc numerical methods such as stochastic finite elements (Ghanem & Spanos 1991), which provide highly accurate numerical predictions. Uncertainties in material properties or loading are quite well mastered within the context of these techniques. Recently, some numerical strategies have also been proposed in order to deal with geometrical uncertainties (Xiu & Tartakovsky 2006, Nouy et al. 2008). Within a stochastic analysis, these methods only constitute the numerical modeling step. Upstream, the probabilistic model has to be supplied with relevant data, which generally requires experimental campaigns to identify randomness of the model. The identification of a probabilistic model is a very critical point. Indeed, it first requires many samples, constituting a significant set of random outcomes, but also robust identification techniques of random variables or fields. Experimental campaigns or in site measurements are often very expensive, a fact that drastically limits the number of available samples and thus the quality of the probabilistic identification.

In this paper, we propose an efficient methodology in order to identify randomness in the geometry of structures. This method starts from a collection of images, representing different outcomes of the random shape to identify. The basic point is to represent the random geometry in an implicit manner with the level-set technique (Sethian 1999). This technique consists in representing the boundary of a shape with a level-set function, which is the signed distance function to the boundary. This technique is well known and mastered in the context of shape recovery from images. In our context, the random geometry can be characterized by a random level-set function, an outcome of which represents an outcome of the random boundary of the shape to be recovered. The problem of random geometry identification is then equivalent to the identification of a random level-set function, which is a random field. Quite general techniques have been proposed for the identification of random fields (e.g. Desceliers et al. 2006). These techniques are based on a representation of the random field on a polynomial chaos basis. Here, we follow this idea and try to identify a polynomial chaos representation of the random level-set. In particular, we propose efficient numerical strategies in order to identify the coefficients of the decomposition.

The main advantage of the overall methodology is to provide a non-intrusive identification procedure of random geometry in a form suitable for numerical simulation within the eXtended Stochastic Finite Element Method (Nouy et al. 2008). A numerical example will illustrate the overall methodology, from image samples to numerical simulation.

REFERENCES

Ghanem, R. & Spanos, P. D. 1991. *Stochastic finite elements: A spectral approach*. Springer-Verlag, Berlin, 2nd edition: Dover Publications, NY, 2003.

Xiu, D. & Tartakovsky, D.M. 2006. Numerical methods for differential equations in random domains. *SIAM J. on Scientific Computing* 28(3): 1167–1185.

Nouy, A., Clément, A., Schoefs, F. & Moës, N. 2008. An eXtended Stochastic Finite Element Method for solving stochastic partial differential equations on random domains. *Computer Methods in Applied Mechanics and Engineering* 197: 4663–4682.

Sethian, J. 1999. *Level-set methods and fast marching methods: Evolving interfaces in computational geometry, fluid mechanics, computer vision and materials science*. Cambridge University Press, Cambridge, UK.

Desceliers, C., Ghanem, R. & Soize, C. 2006. Maximum likelihood estimation of stochastic chaos representations from experimental data. *Int. J. for Numerical Methods in Engineering* 66(6): 978–1001.

Organized Session (OS04) Risk Evaluation on Geotechnical and Geo-Environmental Problems

Safety, Reliability and Risk of Structures, Infrastructures and Engineering Systems – Furuta, Frangopol & Shinozuka (eds)
© 2010 Taylor & Francis Group, London, ISBN 978-0-415-47557-0

Risk evaluation and reliability-based design of earth-fill dams for overflow due to heavy rains

S. Nishimura & S. Mori
Graduate school of Environmental Science, Okayama University, Japan

ABSTRACT

There are many earth-fill dams for farm ponds in Japan, particularly in the Setouchi, which is the area surrounding the Inland Sea of Japan. Some of them are getting old and decrepit, and have been weakened. Every year, a number of them is damaged by heavy rains and earthquakes, and in a few worst cases, the dams are completely ruined. To mitigate such disasters, improvement works are conducted on the most decrepit earth-fill dams. Since there is a recent demand for low-cost improvements, the development of a design method for optimum improvement works at a low cost is the final objective of this research. A reliability-based design method is introduced here in response to this demand.

In this research, the risk of earth-fills for heavy rains is evaluated. The rainfall intensity is dealt with as a probabilistic parameter, and the statistical model of the annual maximum rainfall intensity is determined. Based on the statistical model, the probability of overflow is calculated. If the water of the reservoir overflows on the earth-fill, it is assumed for the embankment to be ruined here. Based on the estimated probability of failure, the risk of the downstream area of the earth-fill dams is evaluated. Furthermore, the case that the capacity of the water reservoir is increased and the spillway is improved to mitigate the risk of the overflow, is considered, and the expected total cost including the improvement work cost, is estimated. The effect of the improvement is judged based on the total cost. The detail is as the following.

The probability of overflow is defined by following equation.

$$P_f = \mathrm{Prob}\left[h_d < h_p\right] \tag{1}$$

in which h_d is the design overflow head on the spillway bed, and h_p is the maximum overflow head.

Then, the expected total cost is given by the following equation.

$$C_T = C_0 + C_f \cdot E[n] \tag{2}$$

$$E[n] = \begin{cases} \sum_{k=1}^{t}\left[P_{fC}\left(1-P_{fC}\right)^{k-1}\left\{1+(t-k)P_{fI}\right\}\right] & \text{(Current)} \\ t \cdot P_{fI} & \text{(Improved)} \end{cases} \tag{3}$$

in which C_T is expected total cost, n is the frequency of the overflows within the lifetime span t (years), P_{fC} is the probability of overflow corresponding to the current state of the embankment, P_{fI} is the probability of overflow corresponding to the improved state of the embankment, C_0 is improvement cost, and C_f is the failure cost due to flooding.

In this research, the improvement of the spillway is considered, and the improvement make the discharging ability of spillway increased drastically. In Equation (??), the improvement cost C_0 is zero for the current state of the embankment, and it is assumed that when the embankments are broken due to overflow, they are restored to the same level of the improved state of the embankment. The cost C_D is the difference of the costs between the current and the improved states of the embankment as given by the following equation.

$$C_D = C_{TC} - C_{TI} \tag{4}$$

in which C_{TC} and C_{TI} present the expected total costs of the current and improved states. The value of C_D indicates the effect of the improvement work for the spillway.

In an actual example, the difference of the expected total costs between current and improved states, C_D, is 28 million JPY. Therefore, the improvement of the spillway is judged to be effective for the overflow failure. If the rainfall intensity becomes 125% of the current rainfall, the value of C_D becomes 28 times of the current rainfall case, namely, the improvement becomes extremely effective.

*Safety, Reliability and Risk of Structures, Infrastructures and
Engineering Systems – Furuta, Frangopol & Shinozuka (eds)
© 2010 Taylor & Francis Group, London, ISBN 978-0-415-47557-0*

Reliability analysis of unsaturated soil slopes using subset simulation

A. Santoso, K.K. Phoon & S.T. Quek

Department of Civil Engineering, National University of Singapore, Singapore

ABSTRACT

It is important to consider the change in soil matric
suction due to rainfall infiltration as it may eventually lead to slope failure. The change in soil matric
suction is governed by the soil water characteristic
curve (SWCC). Uncertainty of SWCC due to inherent soil variability, measurement errors or modeling
assumptions has been recognized. When this uncertainty is considered, the conventional factor of safety
is no longer sufficient to assess slope stability. A more
realistic indicator is the probability of slope failure,
which can be estimated using reliability analysis. Due
to the physical complexity of unsaturated slope stability problems, simple reliability methods may not
be applicable. A recently proposed general simulation
technique, namely subset simulation, is adopted in this
study. The variability of SWCC is propagated to the
hydraulic conductivity curve using a model suggested
by van Genuchten (1980) and Mualem (1976). The
change in matric suction due to rainfall infiltration
is computed using a finite element transient seepage
analysis. The impact of the variability of SWCC on the
factor of safety and the probability of failure is studied using infinite slope examples. It will be shown that
subset simulation can estimate the probability of slope
failure more efficiently than standard Monte Carlo
simulation.

Safety, Reliability and Risk of Structures, Infrastructures and Engineering Systems – Furuta, Frangopol & Shinozuka (eds)
© 2010 Taylor & Francis Group, London, ISBN 978-0-415-47557-0

A new reliability-based design approach on limit-state factor

Jianye Ching
National Taiwan University, Taipei, Taiwan

ABSTRACT

Recently, reliability-based design approaches emerge as a more reasonable and rigorous way of handling uncertainties. In particular, design approaches based on load and resistance factors (LRFD methods) are increasingly popular. However, for geotechnical problems, it is sometimes not trivial to discriminate loads from resistances. The implementation of LRFD methods seems problematic for these cases. In this paper, a completely different view is taken for reliability-based design. It is shown that sophisticated reliability-based design can be achieved with a single factor, called the limit-state factor. Moreover, the new approach does not require discrimination between loads and resistances but only requires the knowledge of the limit state function. Even more attractively, simple Monte Carlo simulations are sufficient to calibrate the limit-state factor. Several numerical examples are used to verify the proposed approach, including the following example considers a shallow foundation (shown in Figure 1). The results show that the proposed approach is quite promising.

The only limit state considered here is the ultimate bearing capacity:

$$R(Z,\theta) = q \cos \Psi / q_u \qquad (1)$$

The estimated η^* vs. P_F^* relationship is shown in Figure 2. The same examining approach adopted in Example 1 is taken to verify this relationship. Figure 3 show the comparison of the allowable reliability-based design regions S_R and the limit-state design regions S_L.

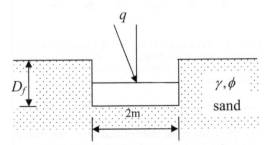

Figure 1. The cross section of the shallow foundation.

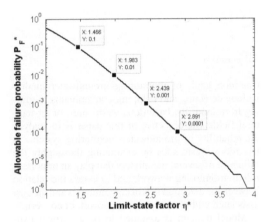

Figure 2. The estimated η^* vs. P_F^* relationship for the limit state.

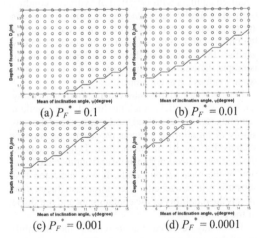

(a) $P_F^* = 0.1$ (b) $P_F^* = 0.01$ (c) $P_F^* = 0.001$ (d) $P_F^* = 0.0001$

Figure 3. The examining results for the limit state.

It can be seen that the two sets match each other very well. This indicates that the resulting limit-state design is indeed very close to the reliability-based design.

Safety, Reliability and Risk of Structures, Infrastructures and
Engineering Systems – Furuta, Frangopol & Shinozuka (eds)
© 2010 Taylor & Francis Group, London, ISBN 978-0-415-47557-0

Reliability assessment of monitoring well networks in stratified groundwater flow systems

K. Inoue, M. Tabata & T. Tanaka
Graduate School of Agricultural Science, Kobe University, Japan

G.J.M. Uffink
Faculty of Civil Engineering & Geosciences, Delft University of Technology, The Netherlands

ABSTRACT

Common goals of monitoring groundwater quality include detecting and mapping contaminants migrating from landfills, hazardous waste sites, or agricultural fields The objective of this paper is to analyze the reliability of groundwater monitoring systems at hazardous waste sites by examining thoroughly the influence of several parameters that play an important role in monitoring network and to assess the relation between the detection loss of contaminants and the cost associated with the monitoring network of concern.

Model domain is assumed to be a vertical two-dimensional aquifer with candidate observation points within a monitoring well. In this study, perfectly stratified aquifers with variation of hydraulic conductivity of parabolic, linear and step function formations are considered where the arithmetic mean value of hydraulic conductivity K is considered to be 20 m/day whereas the variance of the natural logarithm of the isotropic hydraulic conductivity $\ln(K)$ is assigned several values between 0 and 2.

Results obtained from numerical experiments using random walk particle tracking linked with random sampling technique showed the dependence of the reliability of monitoring systems on several parameters such as dispersivity of the medium, heterogeneity of the medium, detection threshold and number and location of the observation points. It was indicated that subsurface heterogeneity is an important factor that affects the reliability of the contaminant and the shape of the plume.

This study revealed that increasing the probability of detection requires that either more monitoring wells be used or that the wells be located relatively close to the source as shown in Figure 1. Moreover, locating a monitoring network closer to the contaminant source is suited to early detection of a potential contaminant release and to reduce the risk of detection loss. On the other hand, in addition to the risk, it was demonstrated that evaluation of cost is essential for decision-making to install monitoring system. As shown in Figure 2, risk and cost substantially varies

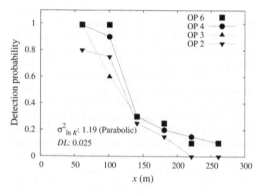

Figure 1. Relation between the location of well and detection probability under the parabolic spatial variation of hydraulic conductivity.

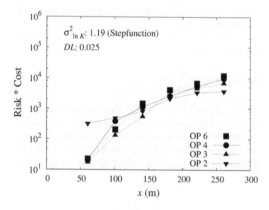

Figure 2. Results of risk-cost as a function of monitoring system location for different variances, monitoring points, and detection limits under the porous formations of step function type.

according to the number and location of observation points. Hence, risk and cost assessment plays an important role in the stage of aquifer management or initial detection of groundwater contamination.

Safety, Reliability and Risk of Structures, Infrastructures and Engineering Systems – Furuta, Frangopol & Shinozuka (eds)
© 2010 Taylor & Francis Group, London, ISBN 978-0-415-47557-0

Effect of uncertainty in soil properties on seismic settlement of earth structure

A. Wakai
Gunma University, Kiryu, Japan

S. Nishimura
Okayama University, Okayama, Japan

S. Tani
National Institute for Rural Engineering, Tsukuba, Japan

ABSTRACT

In the seismic design of embankment, it is usually assumed that the soil in the embankment is homogeneous. However, such a simple homogeneous analysis with uniform material may often provide us an inappropriate prediction, which is much different from the actual phenomena. The predicted result is strongly affected by the statistical uncertainty of each material parameter and their spatial variation in the embankment. It indicates that the mean values always could not be adopted as the representative of the heterogeneous soil properties in the embankment.

In this study, the residual settlement of a heterogeneous embankment, induced by a strong earthquake, is analyzed in the parametric studies with the dynamic elasto-plastic finite element analysis. The Monte Carlo simulation, based on a huge number of trials for randomly varying soil parameters, is performed to investigate the sensitivity of the uncertainty in material properties in the embankment. Such a simple probabilistic approach is more reliable for a complicated problem with strong material nonlinearity, because a conventional finite element code can be used in each trial calculation.

Based on the conclusions obtained from the Monte Carlo simulation performed in this study, we may develop a more rational seismic design procedure based on the reliability in the future. Its conceptual figure is shown as bellows.

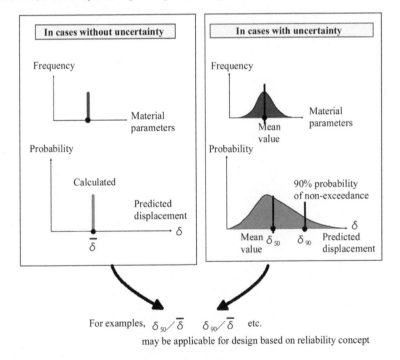

Figure 1. Conceptual figure of application for design based on reliability concept.

General Session (Earthquake Engineering)

Safety, Reliability and Risk of Structures, Infrastructures and
Engineering Systems – Furuta, Frangopol & Shinozuka (eds)
© 2010 Taylor & Francis Group, London, ISBN 978-0-415-47557-0

Optimal multi-objective seismic design of a highway bridge by selective use of nonlinear static and dynamic analyses

D. Vamvatsikos

Department of Civil & Environmental Engineering, University of Cyprus, Nicosia, Cyprus

ABSTRACT

A methodology is introduced for the optimal multi-criteria performance-based seismic design of highway bridges by selectively employing nonlinear static and dynamic analysis to emulate full Incremental Dynamic Analysis (IDA). IDA is a novel analysis method that can thoroughly estimate the seismic demands and limit-state capacities of structures under seismic loads by subjecting a structure to a suite of ground motion records that are suitably scaled to several levels of intensity (Vamvatsikos & Cornell 2002). By combining the results with probabilistic seismic hazard analysis the mean annual frequencies of exceeding each limit-state are estimated. However, IDA is computer-intensive and it easily becomes prohibitively expensive for optimization purposes.

Simplified methodologies have recently emerged that are able to accurately approximate the results of IDA at a fraction of its cost (Vamvatsikos & Cornell 2005). Such methods can be employed in an optimization framework to rapidly evaluate the potential of candidate designs to lie close to the opti-mum. Thus, we can discard most candidate designs and form an elite set suitable for full-scale IDA evaluation. IDA-level accuracy is retained but the computational load is significantly decreased. Moreover, the user may tune the optimization towards either speed or accuracy by decreasing or increasing, respectively, the size of the elite set.

This unique analysis capability is applied on a typical two-span, single-column bent high-way bridge using the NSGA-II algorithm for multi-criteria optimization (Deb et al. 2000) that can accurately evaluate alternate designs to select the ones that satisfy multiple criteria in a Pareto sense. The criteria selected are the material cost and the mean annual frequency of exceeding the Global Instability limit-state. Additionally, design constraints are placed on the return periods of the Immediately Operational, Operational, Life-Safe and Collapse Prevention limit-states (Porter 2002). The intent is to satisfy some standard operational and safety requirements while minimizing simultaneously the bridge cost and the frequency of

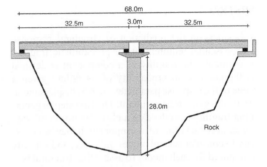

Figure 1. The bridge to be designed.

collapse, allowing the engineer to choose the best tradeoff. The final result is a sophisticated tool for optimal design that achieves a compromise between the speed of nonlinear static methods and the accuracy of nonlinear dynamic methods to generate Pareto-optimal designs for structures under seismic loading.

REFERENCES

Deb K., Pratap A., Agarwal S. & Meyarivan T. 2000. A fast and elitist multi-objective genetic algorithm: NSGA-II. *KanGAL Report No. 200001.* Indian Institute of Technology, Kanpur, India.

Porter K. 2002. Draft Bridge Decision Variables. *Draft Report on I-880 PEER Testbed.* Pacific Earthquake Engineering Research Center.

Vamvatsikos D. & Cornell C.A. 2002. Incremental Dynamic Analysis. *Earthquake Engineering and Structural Dynamics,* 31(3): 491–514.

Vamvatsikos D. & Cornell C.A. 2005. Direct estimation of the seismic demand and capacity of MDOF systems through Incremental Dynamic Analysis of an SDOF Approximation. *ASCE Journal of Structural Engineering,* 131(4): 589–599.

Safety, Reliability and Risk of Structures, Infrastructures and Engineering Systems – Furuta, Frangopol & Shinozuka (eds)
© 2010 Taylor & Francis Group, London, ISBN 978-0-415-47557-0

The seismic fragility assessment of the feed water tanks plant using robust prediction concept of structural response

P. Varpasuo

Fortum Nuclear Services Ltd, Espoo, Finland

ABSTRACT

Robustness in the prediction of structural response is an essential requirement for probabilistic fragility assessment. The fragility of a component is defined as the conditional probability of its failure given a value of the response parameter, such as displacement, strain, stress or stress resultant. The first step in generating fragility is to develop a clear definition of what constitutes the failure of a component. It may be necessary to consider several modes of failure, and fragility is required for each mode. To assess the structural performance, it is important that robust predictions are made that treat all the uncertainties, from modeling applicable loads to modeling the structural behavior.

For large tanks such as the feed water tank, a failure of the support system or a plastic collapse of the pressure boundary is considered to be the dominant failure mode. Structures can be considered to fail functionally when the inelastic deformations under seismic loads are estimated to be sufficient to potentially interfere with the operability of equipment attached to the structure or fractured sufficiently for equipment attachments to fail. The event and fault trees should appropriately reflect these failure conditions.

The fragility of large components is directly developed from the seismic response analysis results. The component fragility for a particular failure mode is expressed in terms of the ground-acceleration capacity A. The fragility is therefore the probability at which the random variable A is less than or equal to a specified value, a. The ground-acceleration capacity is, in turn, modeled as

Equation 1 $A = \underline{A}\, \varepsilon_R \varepsilon_U$

In Eq. (1) A is the median ground-acceleration capacity, ε_R is variable representing the inherent randomness about A, and ε_U is a random variable representing the uncertainty in the median value. It is assumed that both ε_R and ε_U are log-normally distributed with logarithmic standard deviations β_R and β_U, respectively.

REFERENCES

Säteilyturvakeskus Asiakirja No. A213/73, Säteilyturvakeskuksen lausunto Loviisan Ydinvoimalaitoksen käyttöä koskevasta lupahakemuksesta, Liite 1, Loviisan Ydinvoimalaitoksen Turvallisuusarvio, 3.7.2008.

ESTIMATION OF SEISMIC HAZARD IN LOVIISA NUCLEAR POWER PLANT SITE, Fortum Nuclear Services Ltd., 12 December 2007, FNS-169181, LO1-T84252-1.

Bender B. , Perkins David M. 1987. SEISRISK III: A Computer Program for Seismic Hazard Estimation, U. S. Geological Survey Bulletin 1772, United States Government Printing Office; Washington.

ABAQUS version 6-7.1, Users documentation, Simulia Inc., Dassault Systèmes, 2004, 2007.

YVL 2.6 , Seismic design requirements for nuclear power plants, Finnish Center for Radiation Protection, (STUK), December 2001.

NRC Perspective on Structural Mechanics Simulation Software, Gordon S. Bjorkman, SMiRT20 Preconference Topical Workshop, Identifying Structural Issues in Advanced Reactors, Hotel Haikko Manor, Porvoo, Finland Nov. 5 – 6, 2008.

Safety, Reliability and Risk of Structures, Infrastructures and
Engineering Systems – Furuta, Frangopol & Shinozuka (eds)
© 2010 Taylor & Francis Group, London, ISBN 978-0-415-47557-0

Modeling of phase spectrum of earthquake motion using the concept of group delay time

Tadanobu Sato
Section of Disaster Management and Social Service, Kobegakuin University, Kobe, Japan

Tomohiro Kawanishi & Yoshitaka Murono
Railway Technical Research Institute, Earthquake & Structural Engineering, Kunitachi, Tokyo, Japan

Wataru Isaka & Hitoshi Furuta
Department of Informatic, Kansai University, Takatsuki, Osaka, Japan

ABSTRACT

To simulate realistic earthquake motions, we need not only amplitude spectrum but also phase spectrum. To investigate the non-stationary characteristics of earthquake motion, its phase characteristics should be modeled properly. The concept of group delay time is used to develop a simulation model of phase spectrum of earthquake motion. The mean and variance of group delay time are assumed to be expressed as functions of the earthquake magnitude, traveling path and local effects. We have already derived the regression coefficients of the mean and standard deviation of the group delay time of earthquake motion (Sato et. al. 2000, Sato et. al.2003), but the equations used in the former studies were similar to the attenuation equations of the peak acceleration, velocity and displacement of earthquake motion. These regression equations do not compatible with the physical process of wave propagation phenomenon.

In this paper we proposed new regression equation of group delay time based on the physical consideration. When the earthquake motion is expressed by a convolution of the source time function, the time function of traveling path and the time function of local effect, the phase spectrum of earthquake motion is given as the sum of phase spectrum of these three functions. The group delay time of earthquake motion is also expressed as the sum of group delay time of three functions because the group delay time is the first order derivative of phase spectrum with respect to the frequency. The functions used for the regression analyses here are a linear combination of hypocenter distance, parameter expressing local effect and exponential function of earthquake magnitude.

To calculate mean and variance of group delay time we deconvolute the time history of earthquake motion into several band passed time histories using Meyer (1967) wavelet. Assuming ergodic characteristic of group delay time process in the each band frequency domain we calculate mean and variance of group delay time from band passed time history. Almost ten frequency band of Meyer's wavelet is selected to cover the frequency range of earthquake engineering interest. Using these data we determine the parameter of regression equation of mean and variance of group delay time.

The stochastic characteristic of group delay time is also investigated using observed earthquake motion. We provide a reasonably acceptable background for the assumption of Gaussian process of group delay in the frequency domain to simulate a sample group delay time.

Finally, we demonstrate the efficiency of our model by simulating sample earthquake motions compatible with a response spectrum (Annnaka et.al., 1982).

REFERENCES

Sato, T., Murono, Y. & Nishimura, A. 2002, Phase spectrum modeling to simulate design earthquake motion, *Journal of Natural Disaster Science*, 24(2), 91–100.
Sato, T., Murono, Y. & Murakami, M. 2003, Modeling of phase spectra for simulation of near-fault design earthquake, *Proceedings of the Advancing Mitigation Technologies and Disaster Response for Lifeline Systems*, No.25, 769–778.
Meyer Y. (1967), *Wavelets and operator*, Cambridge University Press
Annaka, T., Yamazaki, F. and Katahira, F.1997. Proposal of estimation equation of maximum acceleration and response spectrum using the records observed by the 87 type strong motion seismograph, *Proc. 24th Earthquake Engineering Symposium, 197.7,* 161–164, JSCE (in Japanese).

Safety, Reliability and Risk of Structures, Infrastructures and Engineering Systems – Furuta, Frangopol & Shinozuka (eds)
© 2010 Taylor & Francis Group, London, ISBN 978-0-415-47557-0

Method to simulate phase spectrum of earthquake motion by using a stochastic differential equation

Tadanobu Sato
Kobegakuin University , Kobe, Japan

Cong Zhang & Lingyi Y. Lu
Southeast University, Nanjing, P.R. China

ABSTRACT

This paper presents a new method to simulate a phase spectrum of earthquake motion by using the concept of group delay time and stochastic differential equation. It is known that the phase spectrum is an important characteristic of earthquake motion. Although amplitude and phase spectra are essential to simulate an earthquake motion there have been very few research efforts conducted on the modeling of phase spectrum comparing with the modeling of amplitude spectrum.

In the first part, we introduce a method to calculate group delay time, which is a function of frequency, from observed earthquake motions. Using dataset of observed earthquake motions we develop a database of group delay time to be used for regression analysis of parameters controlling a stochastic differential equation defined in the second part of this paper.

In the second part we assume that the stochastic characteristic of group delay time is expressed by a stochastic differential equation of which mean and square processes are defined by ordinary differential equations. The coefficients of stochastic differential equation can be expressed by the coefficients of its mean and square process. Because these coefficients are function of frequency we assume ergodic characteristic of group delay time in frequency domain. Then an algorithm is developed to identify the coefficient functions that controlling the stochastic differential equation. For simplicity, the coefficient functions are assumed to be expressed by power series of frequency. Two identification schemes such as least square method and Levenberg-Marquardt method are used to identify coefficients of the power series. After storing these power series coefficients into parameter dataset we developed regression equations for these coefficients as function of earthquake magnitude, epicenter distance and local effects.

In the third part, the Milstein approximation scheme is used to solve the stochastic differential equation of group delay time. After we check the stochastic characteristics of simulated sample group delay times comparing with those of observed earthquake motions we demonstrate efficiency of developed method to simulate group delay time of earthquake motion. A sample phase spectrum is obtained by integrating a sample group delay time with respect to the frequency. The efficiency of the simulation model of group delay time is also demonstrated by simulating design earthquake motion compatible with design response spectra.

REFERENCES

Brigham, E.O. 1979. *The Fast Fourier Transform*. Shanghai: Shanghai Scientific and Technical Publishers.
GB50011-2001, 2001. *Code for seismic design of buildings in the People's Republic of China*. Beijing: China Construction Industry Press.
Kawashima, K. & Aizawa, K. 1986. Modification of earthquake response spectra with respect to damping, *J. Struct. Mech. Earthquake Eng.* JSCE, No. 344.
Kwashima, K. & Aizawa, K. 1986. Modification of earthquake response spectra with respect to damping. *J.Struct. Mech. Earthquke Eng.* JSCE, No.344/1-1:351–355.
Mikosch, Thomas 1998. *Elementary stochastic calculus with finance in view*. Singapore: World Scientific Publishing Co. Pte. Ltd.
SATO, Tadanobu, MURONO, Yoshitaka & NISHIMURA, Akihiko 2002. Phase spectrum modeling to simulate design earthquake motion, *J. Natural Disaster Science*, Vol. 24, No.2.

Safety, Reliability and Risk of Structures, Infrastructures and Engineering Systems – Furuta, Frangopol & Shinozuka (eds)
© 2010 Taylor & Francis Group, London, ISBN 978-0-415-47557-0

Physical theory model to simulate cyclic behavior of braces

M. Dicleli & E.E. Calik
Department of Engineering Sciences, Middle East Technical University, Ankara, Turkey

ABSTRACT

During an earthquake, the seismic energy is mainly dissipated within the braces in steel braced frames. Thus, the overall performance of such frames depends largely on the hysteretic behavior of the braces. The hysteretic behavior of steel braces involves complex physical phenomena such as; yielding in tension, progressive lengthening called growth effect, inelastic buckling in compression as well as deterioration of the buckling capacity due to Bauschinger effect and the residual kink within the brace. It is important to have an accurate analytical model of this complex hysteretic behavior to correctly predict the seismic response of steel braced frames particularly for performance-based design purposes.

Many experimental and analytical studies have been conducted on the hysteretic behavior of steel braces The experimental studies provided a wide range of data which have been used by many researchers to develop analytical models to simulate the hysteretic behavior of braces under severe cyclic load reversals. The developed analytical models can be categorized in three groups; finite element, phenomenological and physical theory models. Finite element models are computationally quite expensive. In contrast, phenomenological models, which are based on simplified hysteretic rules that mimic the experimental cyclic axial force-deformation relationship of braces, are computationally more efficient than finite element models. Nonetheless, their range of application is limited to braces having experimental data. On the other hand, physical theory models combine the advantages of both finite element and phenomenological models. Therefore, these models are more suitable for the analytical formulation of axial force-deformation hysteretic behavior of steel braces. However, most of the existing physical theory models are either limited by several specific empirical coefficients available only for certain brace types for the simulation of the growth and Bauschinger effects or fail to accurately simulate such effects for various brace types. Thus a simple physical theory model that addresses all the limitations stated above is required.

In this study, a simple, yet an efficient physical theory model that can be used to simulate the inelastic

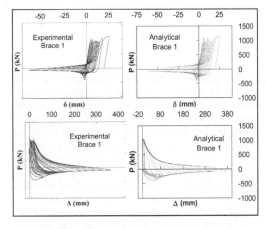

Figure 1. Comparison of experimental and analytical axial force versus axial and transverse direction hysteresis loops of W-Section braces ($KL/r = 120$).

cyclic axial force-axial deformation and axial force–transverse deformation relationships of steel braces is developed. The model consists of a brace idealized as a pin ended member with a plastic hinge located at its mid-length. Input parameters of the model are based only on the properties of the brace. and is broadly applicable to steel braces with various section types and slenderness ratios. It is observed that the analytically obtained axial force versus axial displacement as well as axial force versus transverse displacement hysteresis loops compare reasonably well with the experimental ones as demonstrated in Figure 1 below.

*Safety, Reliability and Risk of Structures, Infrastructures and
Engineering Systems – Furuta, Frangopol & Shinozuka (eds)
© 2010 Taylor & Francis Group, London, ISBN 978-0-415-47557-0*

Seismic performance of eccentrically braced frame with vertical link

M. Dicleli

Department of Engineering Sciences, Middle East Technical University, Ankara, Turkey

A. Mehta

Department of Civil Engineering and Construction, Bradley University, Peoria, IL, USA

ABSTRACT

Steel structures are designed for ductility, where the earthquake energy is dissipated in plastic hinges. Under this design philosophy, when a conventional steel structure such as moment resisting frame (MRF) or chevron braced frame (CBF) is subjected to a strong earthquake, yielding of the plasticized zones and the resulting large permanent displacements may induce nearly irreparable damage to the structure. Thus, it may be required to rebuild the essential structural members of the structure following a strong earthquake.

MRF and CBF are commonly used in the construction of steel buildings. The MRF has large ductility capacity compared to other frame types. However, it requires large member sizes to keep the lateral drifts within code-mandated limits. Even then, the flexibility of the MRF may result in large drift-induced nonstructural and structural damage under seismic loading. Consequently, costly post-earthquake rehabilitation of the structure may be required. On the other hand, CBF possesses high elastic stiffness to prevent large drifts. Material saving could also be achieved as the frame members are subjected to less bending effect due to the presence of the braces. However, its ability to dissipate energy solely depends on the unstable hysteretic behavior of the braces due to buckling effects producing loss of lateral stiffness and strength of the frame. This in turn, results in soft-story formations, instability and hence damage to the frame members.

In light of the above discussion, it is clear that in addition to the advantages of each frame type, there are numerous disadvantages. Furthermore, both frames are prone to substantial structural damage during a strong earthquake. Consequently, a novel frame system is needed to ensure satisfactory post-earthquake performance of steel buildings under service loads with minimal rehabilitation cost.

In this research, the seismic performance of an efficient energy dissipating eccentrically braced steel frame (EEDBF) in relation to that of MRF and CBF is studied. The frame is intended to combine the advantages of MRF and CBF and eliminate most of the disadvantages pertinent to these frames. Seismic analyses of the three frames are conducted to assess the performance of the EEDBF compared to that of MRF and CBF.

The analyses of the frames revealed that MRF generally yield large inter-story drifts compared to the other frames. On the other hand CBF displays a good response for low to moderate intensity earthquakes and for lower number of stories. Nonetheless, for high intensity earthquakes and for larger number of stories, a sudden deterioration in the strength and stiffness of the frame is observed. Furthermore, the behavior of the CBF is found to be highly dependent on the brace contribution to the overall strength of the frame, the slenderness of the braces and the frequency characteristics of the ground motion. On the other hand, EEDBF displays a more stable behavior over a wide range of structural and ground motion properties. It also exhibits a more even distribution of earthquake input energy over the height of the frame and generally yields lower drifts as compared to the other frames studied. Moreover, the behavior of the EEDBF is independent of the brace properties. Thus, EEDBF combines the advantages of both CBF and MRF and therefore, displays an overall more desirable behavior as compared to the other frames. Damage analyses of the frames also revealed that the EEDBF generally exhibits less damage and larger reserve lateral deformation capacity compared to the other frames. Thus, in the event of an earthquake, it is anticipated that the yielding of the SE in the EEDBF will prevent buckling of the braces and minimize damage to the structural components of the frame. The damaged SE can be easily replaced for a relatively small cost. This may result in a minimal post earthquake rehabilitation cost compared to the other frames.

*General Session (Structural Systems &
System Reliability & Lifeline Risk Assessment)*

Safety, Reliability and Risk of Structures, Infrastructures and
Engineering Systems – Furuta, Frangopol & Shinozuka (eds)
© *2010 Taylor & Francis Group, London, ISBN 978-0-415-47557-0*

Application of optimal system reliability concepts to risk prediction of rock slopes

M.P. Enright & A. Ghosh
Southwest Research Institute, San Antonio, TX, USA

ABSTRACT

The need for reliability assessment of rock slopes is becoming increasingly recognized in the geotechnical community, particularly where significant losses are possible in terms of safety and economic risk. As design codes and standards are gradually upgraded to include quantitative reliability requirements, the demand for probabilistic risk quantification methodologies continues to increase. As part of this effort, more accurate reliability estimation methods are needed to bridge the gap between complex deterministic models and sophisticated uncertainty quantification methods to provide a realistic assessment of these potentially high-consequence engineering projects.

A system reliability approach can be used to assess the reliability of rock slopes in which a sliding rock mass is modeled as a number of interconnected blocks. The computation time associated with reliability prediction of general systems may require significant computations, particularly if correlation among members and postfailure material behavior are considered. Although the time required for deterministic computations has been dramatically reduced by the availability of highly efficient computers, it is still a significant issue for simulation-based risk assessment of systems with relatively small failure probabilities.

A method has recently been developed for sampling-based system reliability predictions that identifies the number of Monte Carlo samples needed for individual members and failure modes. It is an extension of a technique developed previously for mutually exclusive failure modes. It consists of optimal allocation of Monte Carlo samples to individual members based on initial estimates of the associated failure probabilities. The optimal solution is determined by minimizing the variance of the system. This method has the potential to substantially reduce the total number of simulations required for a specified accuracy level.

In this paper, the optimal methodology is applied to the efficient risk prediction of rock slopes modeled as structural systems. An algorithm is presented for allocating samples to block failures based on estimates of the individual block failure probabilities. The results

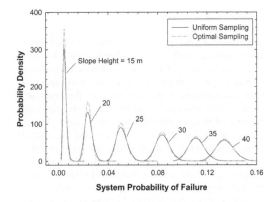

Figure 1. System probability of failure probability densities associated with uniform and optimal sampling for representative rock slope.

can be applied to efficient risk prediction of general rock slopes.

REFERENCES

Enright, M.P., Millwater, H.R., and Huyse, L., "Adaptive optimal sampling methodology for reliability prediction of series systems," AIAA Journal, Vol. 44, No. 3, 2006, pp. 523–528.

Goodman, R.E., "Introduction to rock mechanics," second edition, John Wiley & Sons, NY, 1989.

Hendawi, S. and Frangopol, D.M., "System reliability and redundancy in structural design and evaluation," Structural Safety, Vol. 16, No. 1, 1994, pp. 47–71.

Hoek, E. and Bray, J.W., "Rock Slope Engineering," revised third edition, The Institute of Mining and Metallurgy, 1981.

Jimenez-Rodriguez, R., Sitar, N., and Chacon, J., "System reliability approach to rock slope stability," International Journal of Rock Mechanics & Mining Sciences, Vol.43, 2006, pp. 847–859.

*Safety, Reliability and Risk of Structures, Infrastructures and
Engineering Systems – Furuta, Frangopol & Shinozuka (eds)*
© 2010 Taylor & Francis Group, London, ISBN 978-0-415-47557-0

The use of variance decomposition in dimension reduction for stochastic structural systems

Sanjay Arwade & Mohammadreza Moradi
University of Massachusetts, Amherst, MA, USA

ABSTRACT

In considering the response of structural systems to environmental loads it is possible to treat the structure as filter with random parameters that transforms a load into a response. The load may itself also contain uncertainties. One challenge in the treatment of uncertain structural systems is the typically large number of random parameters in the system. A technique for rapidly identifying a subset of the random system parameters that are particularly important in determining system response would aid in efficient evaluuation of structural reliability. Existing method for indentifying such parameter subsets include sensitivity analysis and principal components analysis. Here the SobolÕ decomposition is implemented as a method for identifying the important subset of system random parameters. The Sobol decomposition expresses a function of many random variables as the summation of a series of functions of all permutations of the set of random variables, and, most importantly, decomposes the total system variance in partial variances quantifying the contribution of each individual random variable and all permutations of the random variables. By setting a threshold value for the fraction of the total variance that should be captured one can use the SobolÕ partial variances to identify a subset of the random variables that drive a specified fraction of the total response uncertainty. This approach to using the Sobol decomposition leads to the possibility of reducing the dimension of the space of random parameters in which the system resides. Using the Sobol functions an approximation to the original function can be obtained. Some issues that arise in using the Sobol decomposition in this way are the rate of convergence of the reduced order representation and accurate estimation of the partial variances. In the absense of an analytic form for the response the partial variances must be estimated by Monte Carlo Simulation in a high dimensional parameter space. This above approach to dimension reduction and approximation of systems with many random parameters is applied to the problem of the collapse load of structural frames with random material strength and stiffness. It is shown that convergence of the estimates of the partial variances is fast, and that an approximation to the true response function can be obtained with the inclusion of only a relatively small subset of the system parameters.

Safety, Reliability and Risk of Structures, Infrastructures and Engineering Systems – Furuta, Frangopol & Shinozuka (eds)
© *2010 Taylor & Francis Group, London, ISBN 978-0-415-47557-0*

Efficient reliability analysis of general systems by sequential compounding using Dunnett-Sobel correlation model

W.-H. Kang & J. Song
University of Illinois, Urbana, IL, USA

ABSTRACT

For risk management of an engineering system, it is essential to evaluate the likelihood of its failure efficiently and accurately. Suppose the probability of a system event is given by the following multivariate normal integral:

$$P(E_{sys}) = \int_{\Omega} \varphi_n(\mathbf{z}; \mathbf{R}) \, d\mathbf{z} \qquad (1)$$

where E_{sys} is the system event; Ω denotes the domain of the system event defined in the space of multivariate standard normal random variables $\mathbf{Z}$; $\sigma_n(.)$ is the joint probability density function of Z; and $\mathbf{R}$ is the correlation coefficient matrix. Evaluation of the multifold integration in Equation 1 is challenging especially when the definition of the system event is complex; the system has a large number of components; and/or the component events have significant statistical dependence.

To overcome these challenges, this paper proposes a new multi-scale system reliability analysis method, which is applicable to any general systems including series, parallel, cut-set and link-set systems. The method sequentially compounds the components in a sub-parallel or sub-series system into a single equivalent component. During the compounding process, the reliability index of the compound event and the correlation coefficient between the new compound event and the other events are updated by use of the Dunnett-Sobel correlation model (Dunnett & Sobel 1955). A heuristic procedure is also proposed to overcome the constraint of the Dunnett-Sobel model, which will be validated through extensive parametric studies in the future.

The accuracy of the proposed method is demonstrated through various numerical examples in comparison with other existing methods. For example, Figure 1 shows the probability of a parallel system with ten components having equal reliability index but all different correlations. The proposed method shows a good agreement with the Monte Carlo simulation with 10^8 samples compared with the product of conditional

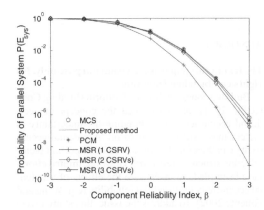

Figure 1. Comparison of the proposed method, the MSR method and the PCM method for parallel systems consisting of 10 components with unequal correlation coefficients.

marginal method (Pandey 1998) and the matrix-based system reliability method (Kang & Song 2008).

The proposed method enables efficient computation of the multivariate normal integral in Equation 1 because it does not involve sampling or multi-fold numerical integrations. Another important merit of the proposed approach is that one can find the correlation coefficients between sub-systems accurately, e.g. correlation between cut sets or link sets, which may provide important insights on the statistical correlation between failure modes.

REFERENCES

Dunnett, C.W. & Sobel, M. 1955. Approximations to the probability integral and certain percentage points of a multivariate analogue of Student's t-distribution. *Biometrika* 42: 258–260.

Kang, W.-H. & Song, J. 2008. Evaluation of multinormal integral and sensitivity by matrix-based system reliability method. *Proc. 10th AIAA Nondeterministic Approaches Conference, Schaumburg, IL, 7–10 April 2008.*

Pandey, M.D. 1998. An effective approximation to evaluate multinormal integrals. *Structural Safety* 20: 51–67.

Safety, Reliability and Risk of Structures, Infrastructures and Engineering Systems – Furuta, Frangopol & Shinozuka (eds)
© 2010 Taylor & Francis Group, London, ISBN 978-0-415-47557-0

Accidental eccentricity effect on structural reliability

H.Y. Chang
Department of Civil and Environmental Engineering, National University of Kaohsiung, Kaohsiung, Taiwan

K.C. Lin
National Center for Research on Earthquake Engineering, Taipei, Taiwan

H.Y. Yeh & J.Y. Chen
Department of Civil and Environmental Engineering, National University of Kaohsiung, Kaohsiung, Taiwan

ABSTRACT

The text in this paper is for visual purpose only. No rights can be taken from this.

Building codes, such as the Uniform Building Code (ICBO 1991), has introduced the concept of accidental eccentricity, for estimating the uncertainty in system mass, stiffness and so on. Accidental eccentricity can therefore be considered an important factor affecting structural reliability. However, limited efforts have been made to investigate the effects of accidental eccentricity on structural reliability (e.g. Wang and Foliente 2002). To address the issue, this study investigated the effects of accidental torsion in seismic reliability assessment. The analyzed structures are a 6-story and a 20-story regular steel office buildings (Wei 2006). The eccentricity in a floor plan was simulated by shifting the mass form the centroid by 5% of the dimension normal to earthquake shaking. The eccentricity along building heights was replicated by Latin hypercube sampling. That allowed considering a more realistic distribution of live load mass and the accidental torsion. The effect of ignoring accidental torsion and the use of code accidental eccentricity were also assessed.

The fragilities for immediate occupancy and life safety were evaluated using 0.7% and 2.5% interstory drift limits (θ) (FEMA 2000). Two limit-state probabilities are considered of great importance. One is the 10% failure probability for exceeding the limit state of immediate occupancy. The other is the 50% failure probability for exceeding the limit state of life safety. These two limit-state PGAs were assessed and compared in detail. Analysis results show that accidental torsion may play a minor role in seismic reliability assessment. However, the fragilities for low-rise buildings and at small drift limits may be underestimated if ignoring the accidental torsion effects. On the other hand, the code recommendation of 5% mass eccentricity at all floors may provide conservative estimates. More-over, the trend will increase in high-rise buildings at small drift limits.

Table 1. Limit-state PGAs for 6- and 20-story buildings.

6-story building		NONE	EC	REC
$\theta \geqq 0.7\%$	PGA (g)	0.171	0.144	0.157
$P_f = 10\%$	Normalization	1.089	0.966	1.000
$\theta \geqq 2.5\%$	PGA (g)	0.625	0.585	0.620
$P_f = 50\%$	Normalization	1.008	0.944	1.000
20-story building		NONE	EC	REC
$\theta \geqq 0.7\%$	PGA (g)	0.310	0.243	0.298
$P_f = 10\%$	Normalization	1.040	0.815	1.000
$\theta \geqq 2.5\%$	PGA (g)	1.343	1.393	1.429
$P_f = 50\%$	Normalization	0.940	0.975	1.000

Note: there are three eccentricity cases. NONE: without eccentricity; EC: with 5% mass eccentricity at all floors (i.e. code recommendation for accidental torsion); REC: with 5% mass eccentricity at arbitrary floors (i.e. the accidental torsion induced by an unfavorable live load mass distribution).

REFERENCES

Curadelli, R. O. and Rieba J. D. (2004). "Reliability based assessment of the effectiveness of metallic dampers in buildings under seismic excitations." Engineering Structures, 26, 1931–1938.

Federal Emergency Management Agency (2000). FEMA-356: Prestandard and commen-tary for the seismic rehabilitation of buildings. Washington, D. C.

Lin, B. Z. and Tsai, K. C. (2006). Platform of inelastic structural analysis for 3D systems – PISA3D R2.0.2 users' manual. National Center for Research on Earthquake Engineering.

Uniform Building Code (UBC). (1991). International Conferences of Building Officials (ICBO), Whitteier, California.

Wang, C.H. and Foliente G. C. (2002). "Seismic Reliability of Low-Rise Nonsymmetric Woodframe Buildings "Journal of Structural Engineering, ASCE, 132(5), 733–744.

Wei, C.Y. (2006). "A study of local-buckling BRB and cost performance of BRBF." Thesis for the M.S. degree at National Taiwan University.

Safety, Reliability and Risk of Structures, Infrastructures and
Engineering Systems – Furuta, Frangopol & Shinozuka (eds)
© 2010 Taylor & Francis Group, London, ISBN 978-0-415-47557-0

Seismic reliability assessment of water supply systems

M.B. Javanbarg
Graduate School of Engineering, Kyoto University, Kyoto, Japan

S. Takada
Department of Civil Engineering, Kobe University, Kobe, Japan

ABSTRACT

Reliability assessment of water networks is a complex process because many issues have to be taken
into consideration. These include possible variations
in demands, reliability of individual components and
their locations, fire flow requirements and their locations to mention but a few. Further complications arise
from the fact that it is difficult to consider useful
performance measures for the network under heavy
seismic damage with different damage states.

Markov et al. (1994) developed a special algorithm
for the hydraulic analysis of the seismically damaged
network and calculated serviceability measures for
the auxiliary water supply system in San Francisco.
Hwang et al. (1998) performed a hydraulic simulation
analysis to assess the serviceability of the water supply system in the city of Memphis. The serviceability
of a system was determined from the connectivity and
flow analysis of a seismically damaged network which
was established from the Monte Carlo simulation.
Shi (2006) and Wang (2006) developed a hydraulic
network model for earthquake simulation of water
network operated by the Los Angeles Department of
Water and Power. The model accounted for flows and
pressures in a heavily damaged system and provided a
method for simulating pipeline leakage and breakage.

We have developed a representative hydraulic
model which is able to evaluate the performance of
water supply systems considering hydraulic analysis
and modeling both breakage and leakage as well as
accounting for various leakage scenarios in pipeline
systems (Javanbarg & Takada 2007). In the method,
two reliability measure were considered presenting
seismic performance of each node within the system;
availability index which is the ratio of the output available water pressure in damaged network to the required
pressure at each demand node within the undamaged
network, and serviceability index which is the ratio of
the output available water flow demand in damaged
network to the required water flow demand at each
demand node within the undamaged network. In this
study we have applied the proposed model to seismic

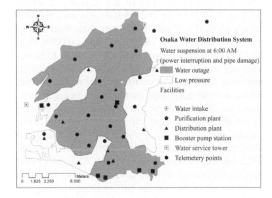

Figure 1. Simulated water suspension pattern by both power
interruption and pipeline failure impacts at 6:00 AM.

performance assessment of Osaka City water network
(Fig. 1).

REFERENCES

Hwang, H.M.H, Lin, H. & Shinozuka, M. 1998. Seismic Per-
formance Assessment of Water Delivery Systems. *Journal
of Infrastructure Systems*, ASCE, 4(3): 118–125.
Javanbarg, M.B. & Takada, S. 2007. Redundancy model for
water supply systems under earthquake environments. In
*Proc. of the 5th International Conf. on Seismology and
Earthquake Engineering*, Tehran, Iran, May.
Markov, I. J., Grigoriu, M. & O'Rourke, T. D. 1994. *An
Evaluation of Seismic Serviceability of Water Supply Net-
works with Application to San Francisco Auxiliary Water
Supply System*. Tech. Report, NCEER-94-0001, Multidis-
ciplinary Center foe earthquake Engineering Research,
Buffalo, New York.
Shi, P. 2006. Seismic Response Modeling of Water Supply
Systems. *PhD Dissertation*, Cornell University, Ithaca,
NY.
Wang, Y. 2006. Seismic Performance Evaluation of Water
Supply Systems. *PhD Dissertation*, Cornell University,
Ithaca, NY.

Safety, Reliability and Risk of Structures, Infrastructures and
Engineering Systems – Furuta, Frangopol & Shinozuka (eds)
© 2010 Taylor & Francis Group, London, ISBN 978-0-415-47557-0

Reliability analysis of infrastructure and lifeline networks using OBDD

M.B. Javanbarg, C. Scawthorn, J. Kiyono & Y. Ono
Graduate School of Engineering, Kyoto University, Kyoto, Japan

ABSTRACT

The increased complexity of the real-life networks
such as infrastructure and lifeline networks requires
new analytical methods to evaluate their behavior. In
the last decades, Binary Decision Diagrams (BDD) has
provided an extraordinary efficient method to represent and manipulate Boolean functions. Akers (1978)
proposed BDD as a powerful representation of truth
tables with an easy implementation. Since his work,
there has been a tremendous amount of literature
in which BDD has been utilized for different applications. Bryant (1986) applied the BDD concept to
Boolean function manipulation. Rauzy (1993), Coud-
ert & Madre (1994) and Sinnamon & Andrews (1997)
applied BDD to reliability analysis of fault trees. In this
study, we apply the Ordered Binary Decision Diagram
(BDD) to network reliability analysis of infrastructure
networks to achieve the exact terminal-pair reliability.

The effectiveness of this approach has been demonstrated by performing experiment on several lifeline
networks. The Boston major highways network has
been considered as the first case study. Taleb-Agha
(1977) analyzed the seismic reliability of this lifeline system using series systems in parallel (SSP)
network. The same network was also modeled by Selcuk and Yucemen (1999) with the aid of a shortest
path algorithm using breadth first search (BFS) of
graph. Another case study has been considered from
Ching and Hsu (2007) who developed a novel source-
destination (O-D) connectivity reliability method for
lifeline networks. As the last case study, we have
applied OBDD to reliability analysis of the Kobe city
water network (Javanbarg & Takada, 2007). The results
of nodal reliability using OBDD are compared with the
results of both deterministic and probabilistic models
(Javanbarg & Takada 2007) as presented in Figure 1.
It has been found that, OBDD is able to evaluate the
exact terminal-pair reliability for a real-life network
which is defined as probability that the nodes in the
network can communicate to each other, taking into
account the possible failures of network links.

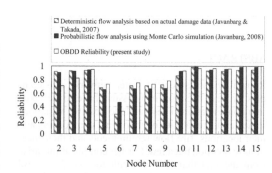

Figure 1. Comparison between the OBDD reliability results
and results of the case study from Javanbarg & Takada (2007).

REFERENCES

Akers, S.B. 1978. Binary Decision Diagrams. *IEEE Trans.
Comput.* C-27(6): 509–516.
Bryant, R. 1986. Graph Based Algorithms for Boolean Func-
tion Manipulation. *IEEE Trans. Comput.*, 35(8): 677–691.
Ching, J. & Hsu, W. 2007. An Efficient Method for Evaluat-
ing Origin-Destination Connectivity Reliability of Real-
World Lifeline Networks. *Computer-Aided Civil and
Infrastructure Engineering*, 22: 584–596.
Coudret, O. & Madre, J.C. 1994. Metaprime: An Interactive
Fault-Tree Analyzer, *IEEE Trans. Rel.*, 43(1): 121–127.
Javanbarg, M.B. & Takada, S. 2007. Redundancy model for
water supply systems under earthquake environments. In
*Proc. of the 5th International Conf. on Seismology and
Earthquake Engineering*, Tehran, Iran, May.
Rauzy. A. 1993. New Algorithms for Fault Tolerant Trees
Analysis. *Reliability Engineering and System Safety*,
50(59): 203–211.
Selcuk, A.S. & Yucemen, M.S. 1999. Reliability of lifeline
networks under seismic hazard. *Reliability Engineering
and System Safety*. 65: 213–227.
Sinnamon, R.M. & Andrew, J.D. 1997. Improved Accuracy in
Quantitative Fault Tree Analysis. *Quality and Reliability
Engineering International*, 13: 285–292.
Taleb-Agha, G. 1977. Seismic risk analysis of lifeline net-
works, *Bull. Seism. Soc. Am.* 67: 1625–1645.

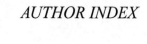

AUTHOR INDEX

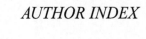

AUTHOR INDEX

Author index

T - #0062 - 071024 - C0 - 254/178/46 [48] - CB - 9780415475570 - Gloss Lamination